Electrocatalysis for Membrane Fuel Cells

Electrocatalysis for Membrane Fuel Cells

Methods, Modeling, and Applications

Edited by Nicolas Alonso-Vante and Vito Di Noto

WILEY-VCH

Editors

Prof. Nicolas Alonso-Vante
University of Poitiers
IC2MP-UMR-CNRS 7285
4 rue Michel Brunet
F-86073 Poitiers Cedex 9
France

Prof. Vito Di Noto
University of Padova
Department of Industrial Engineering
Via Marzolo 9
I-35131 Padova
Italy

Cover Image: Courtesy of Vito Di Noto, Nicolas Alonso Vante and Keti Vezzù

Library of Congress Card No.: applied for

British Library Cataloguing-in-Publication Data
A catalogue record for this book is available from the British Library.

Bibliographic information published by the Deutsche Nationalbibliothek
The Deutsche Nationalbibliothek lists this publication in the Deutsche Nationalbibliografie; detailed bibliographic data are available on the Internet at <http://dnb.d-nb.de>.

Print ISBN: 978-3-527-34837-4
ePDF ISBN: 978-3-527-83055-8
ePub ISBN: 978-3-527-83056-5
oBook ISBN: 978-3-527-83057-2

Typesetting: Straive, Chennai, India
Printing: CPI Group (UK) Ltd, Croydon, CR0 4YY

C9783527348374_040624

Contents

Preface

In electrochemical energy converters such as low-temperature membrane fuel cells, the slow kinetics of oxygen reduction (ORR) represents one of the main reasons for a high overpotential in a fuel cell, e.g. polymer electrolyte membrane (PEM) type. Despite this inherent phenomenon, in these systems, fuel cells constitute (i) a cornerstone in the energy technologies of the present twenty-first century for transportation and stationary applications and (ii) one of the two pillars, together with electrolyzers, of the future Green Hydrogen Economy of the world. These scenarios are comforted by the rapid advances in the development of materials based on noble and non-noble electrocatalytic materials that encompass a bunch of applications operating in a wide pH range (acidic and alkaline). In this context, in order to find a utility, to the knowledge obtained to date, for current and future researchers in this field of activity, the repository of such an avalanche of information is thus a central resource to be transmitted with a global perspective. It is for this reason that the present book aims to consolidate and transmit this knowledge while providing the necessary forum to complement what is published daily in specialized journals. Thus, the contributions of experts working in both academic and industrial research and development will serve as a reference source for the fundamentals and applications of fuel cells, establishing the state-of-the-art and disseminating research advances within a scope corresponding to textbooks for undergraduate and graduate students.

This book, devoted to fuel cell electrocatalysis, will, we hope, further the development and application of this exciting technology on the road to the successful establishment of a clean and sustainable energy economy in the twenty-first century. For the reader's convenience, this book, with a total of 15 chapters, is organized in seven sections, namely Overview, Fundamentals, State of the Art, Physical–Chemical Characterization, Modeling, Protocols, and Systems.

The first chapter discusses how application requirements and system-level considerations create constraints on fuel cell materials and electrocatalysts, with the goal of informing more strategic and impactful research and development efforts. In the second chapter, the discussion is centered on how an atomically designed catalyst surface efficiently produces protons and electrons from hydrogen on the anode and water from oxygen, protons, and electrons on the cathode. In the third chapter, insights are provided on how fundamental electrochemistry can be exploited to guide fuel cell research, whereas the fourth chapter discloses the quantification of the kinetic descriptors that determine the activity and stability of the anode and cathode electrocatalysts, providing analytical methods and electrochemical set-ups as supports. Moreover, in Chapter 5, the author discusses some means for protecting catalytic sites in order to maintain high performance in the light of recent data from the literature. Chapter 6, furthermore, puts into relevance the state-of-the-art of platinum group metal (PGM)-free ORR catalysts. Herein, the authors provide an overview of important parameters that influence the catalysis of ORR with well-defined ORR catalysts. In Chapter 7, recent development in electrocatalysts for the hydrogen oxidation reaction (HOR) is put on the floor, emphasizing the state-of-the-art PGM- and non-PGM-based electrocatalysts for the HOR in alkaline conditions. An important ingredient in the proton exchange membrane fuel cell (PEMFC) system is the polymeric electrolyte. In this context, Chapter 8 describes the features that a membrane must exhibit to be implemented in a fuel cell. This chapter ends with a comprehensive overview of the mechanisms of ion conduction proposed for fuel cell membranes, followed by a brief summary outlining the perspectives of the research in this field. The characteristics of ORR electrocatalyst support (carbon-based and oxide-based) have been analyzed in Chapter 9. Of importance, in all interface research, is the *in operando* technique, and/or probing under real fuel cell operating conditions is offered in Chapter 10 with the use of X-ray absorption spectroscopy (XAS). Theoretical modeling and computation to unravel the local reaction environment are given in Chapter 11. This chapter addresses this complex issue by introducing some basic concepts of electrochemical interfaces, especially the surface charging relation. The authors highlight the electrocatalytic interfaces pertaining to the role of chemisorption-induced surface dipoles that could cause nonmonotonicity in the surface charging behavior. The electrocatalytic materials research protocols for investigating fuel cell reactions are deployed in Chapters 12 and 13. In sum, the correct evaluation of fuel cell reactions, selection of reference electrodes, durability tests of PGM-free materials, and fuel cell testing procedures are put forward in the light of the most advanced literature data research. The last section of the book presents Chapters 14 and 15. These chapters analyze the fundamentals of fuel cell simulation by means of a mono-dimensional analytical model considering multiphase water transport affecting the electrical conductivity properties of the cell membrane, whereas Chapter 15 analyzes the optimization of the operative conditions and the prediction of the system durability that back the design of the PEMFC stack and components of the balance of the plant.

The editors appreciate the contributing authors of this book, who maintained high scientific standards.

N. Alonso-Vante acknowledges financial support from the European Union (ERDF) and "Région Nouvelle Aquitaine."

V. Di Noto thanks the financial support of EIT Raw Materials, project Alpe, and Graphene Flagship, Core 3, of the European Union.

Nicolas Alonso-Vante
University of Poitiers, IC2MP-UMR CNRS 7285
Poitiers, France

Vito Di Noto
University of Padova, Department of Industrial Engineering
Padova, Italy

Part I

Overview of Systems

1

System-level Constraints on Fuel Cell Materials and Electrocatalysts

Elliot Padgett and Dimitrios Papageorgopoulos

Hydrogen and Fuel Cell Technologies Office, Office of Energy Efficiency and Renewable Energy, U.S. Department of Energy, 1000 Independence Ave., SW, Washington, D.C., 20585, USA

1.1 Overview of Fuel Cell Applications and System Designs

Fuel cells are anticipated to play an important role in the future clean energy economy as versatile energy conversion devices across many applications and sectors. Fuel cells have important current and potential applications in three broad areas: (i) transportation powertrains, in vehicles such as cars, buses, trucks, rail locomotives, ships, and aircraft; (ii) stationary power systems, such as distributed power generation, backup power, and combined heat and power (CHP) systems; and (iii) specialty applications such as material handling equipment as well as portable systems for auxiliary power or devices such as personal electronics or mobile communications equipment. While fuel cells for these diverse applications have some common foundations, the systems for each application have different requirements and priorities, which call for different system designs and technologies to meet them. The development of advanced, application-relevant materials and electrocatalysts is essential to overcoming the technical challenges that remain to bring fuel cells into widespread adoption and realization of their potential. This chapter discusses how application requirements and system-level considerations create constraints on fuel cell materials and electrocatalysts, with the goal of informing more strategic and impactful research and development efforts. The primary focus will be on transportation applications and polymer electrolyte membrane (PEM) fuel cells, but other applications and fuel cell types will also be included for context and comparison.

1.1.1 System-level Fuel Cell Metrics

It is useful to begin by covering the typical high-level metrics for fuel cell systems, which provide a basis for comparing different fuel cell types, application requirements, and alternative technologies as well as for benchmarking technological

Electrocatalysis for Membrane Fuel Cells: Methods, Modeling, and Applications, First Edition.
Edited by Nicolas Alonso-Vante and Vito Di Noto.

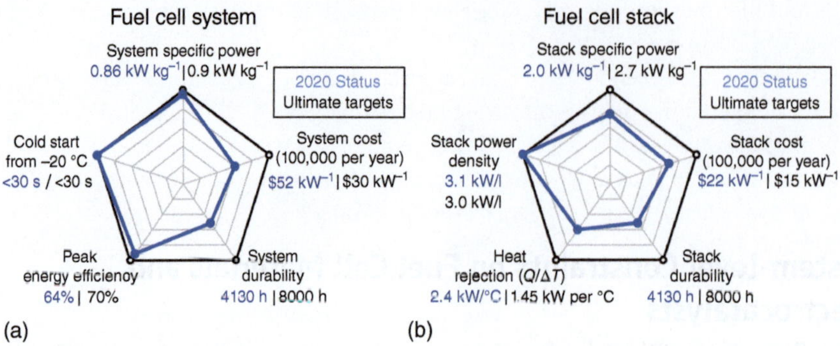

Figure 1.1 Diagrams summarizing the current status of automotive fuel cell systems (a) and stacks (b) relative to DOE targets. Source: Reproduced from U.S. Department of energy [4] / https://www.hydrogen.energy.gov/pdfs/20005-automotive-fuel-cell-targets-status.pdf / Public domain.

progress. These metrics are commonly used as specifications for fuel cell products and targets for fuel cell research, development, and demonstration (RD&D) programs [1–3]. For instance, system and stack-level targets for automotive fuel cells set by the U.S. Department of Energy (DOE), along with respective status estimates, are illustrated in Figure 1.1 [4]. The most used metric categories include cost, durability, efficiency, system size, and flexibility. Each of these, as well as specific metrics, will be described below.

There are several different metrics in common use that describe the size of fuel cell systems, combining the power output and physical mass or volume of the system. Power output may be given as gross power – the total electrical power output of the fuel cell stack – or as net power – the power output of the stack minus the power consumption of the supporting balance of plant (BOP). This distinction must be specified to avoid confusion and may be included in the power units (as kW_{gross} or kW_{net}, for example). To address application-driven system size and weight restrictions, the power output can be given as an absolute total, per unit weight of the system (this is known as the specific power, with units such as $kW\,kg^{-1}$), or per unit volume of the system (this is known as the power density, with units such as $kW\,l^{-1}$).

The energy conversion efficiency of a fuel cell system can be specified in either the electrical power output per fuel input (e.g. $kWh\ kg_{H_2}^{-1}$) or as a percentage of the fuel's lower heating value. Fuel cells generally are more efficient at low power than at high power, and the efficiency is closely tied to the fuel cell performance, as the same mechanisms of voltage loss decrease both. There are therefore different definitions for system efficiency specified at different performance levels, most commonly at the peak efficiency (at low power), peak or rated power, or an average efficiency over a particular duty cycle.

Fuel cell durability or lifetime is commonly specified as the number of hours of operation before a certain level of degradation is reached. While, in practice, the tolerable level of degradation will vary depending on the user's needs, it is also useful to use standardized end of life definitions such as 10% voltage degradation at rated power for benchmarking purposes. It is also important to recognize that degradation rates and lifetimes for fuel cell systems will depend on the duty cycle and stressors of each application.

The cost of a fuel cell system is an important metric but is more challenging to determine than other metrics that are rooted in the physical or engineering parameters of the system. The actual cost of deployed fuel cell systems is of interest in business transactions and to assess current market competitiveness. The projected cost of fuel cell systems using earlier stage, lab-demonstrated technologies, and larger manufacturing scales is also useful for tracking advances in technology and informing research and development (R&D) needs. The cost of fuel cell systems is commonly specified per power output (e.g. $\$\ kW_{net}^{-1}$) although this metric depends on both the system size and the definition of the system boundaries (fuel storage, power electronics, and other components are commonly excluded from the fuel cell system, although system definitions vary).

Flexibility and robustness are umbrella concepts that encompass many different potential metrics for the ability of the fuel cell system to adjust to provide power as it is required. These include the time required to start the fuel cell system, its capability to start and sustain power under cold or hot conditions, its ability to quickly adjust to varying power demands, and the reliability of the system.

It is important to recognize that the various aspects of fuel cell systems that are described by these metrics are interrelated. For instance, an alteration to a fuel cell system that lowers its cost may also impact its power output, efficiency, and durability. Composite metrics that constrain related metrics in a particular way can therefore also be useful. For example, DOE has introduced a "durability-adjusted cost" metric for automotive fuel cells, which describes the cost of an 80-kW_{net} fuel cell system that also meets the requirements for 8000 hour on-road durability [5].

1.1.2 Fuel Cell Subsystems and Balance of Plant (BOP) Components

Fuel cells require supporting BOP equipment to provide high performance and durability, including the supply of air and fuel, cooling, and system monitoring and control. It is important to understand the common subsystems and components used for these purposes. State-of-the-art fuel cell system designs are generally proprietary, but representative model systems have been developed to provide public information. For instance, the DOE has funded the development of a model automotive fuel cell system in a collaboration between Strategic Analysis, Inc. and Argonne National Laboratory and with feedback from the U.S. DRIVE (Driving Research and Innovation in Vehicle efficiency and Energy sustainability) Partnership [6, 7]. This model system is a useful resource for understanding the subsystems and components in transportation fuel cell systems. Similar model systems are being developed for medium- and heavy-duty vehicles [8–10] and have also been developed for stationary and other fuel cell types [11–14].

Example diagrams of fuel cell systems for automotive and heavy-duty vehicle applications are shown in Figure 1.2. These diagrams provide a representative illustration of typical BOP components and subsystems in transportation fuel cell systems.

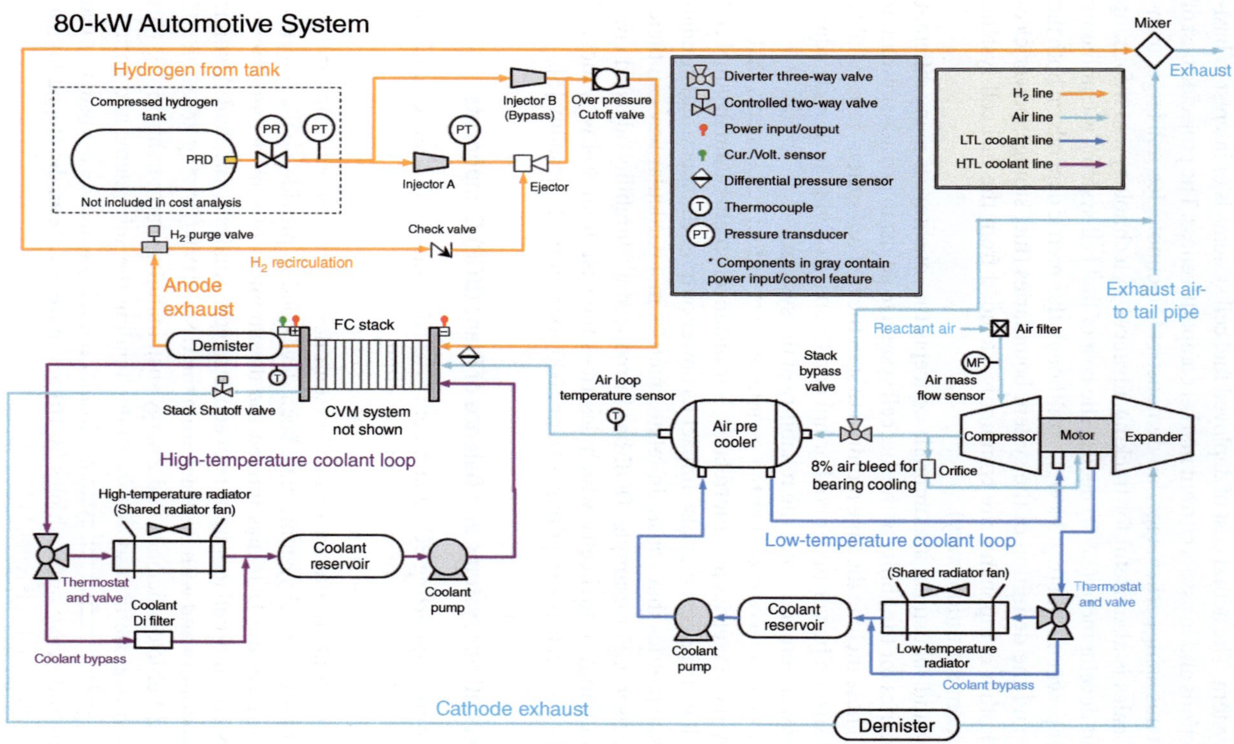

Figure 1.2 Schematics of representative fuel cell system designs for automotive (top) and heavy-duty vehicle (HDV) applications (bottom) illustrating major balance of plant components and subsystems. Source: Reproduced from Ref. [6].

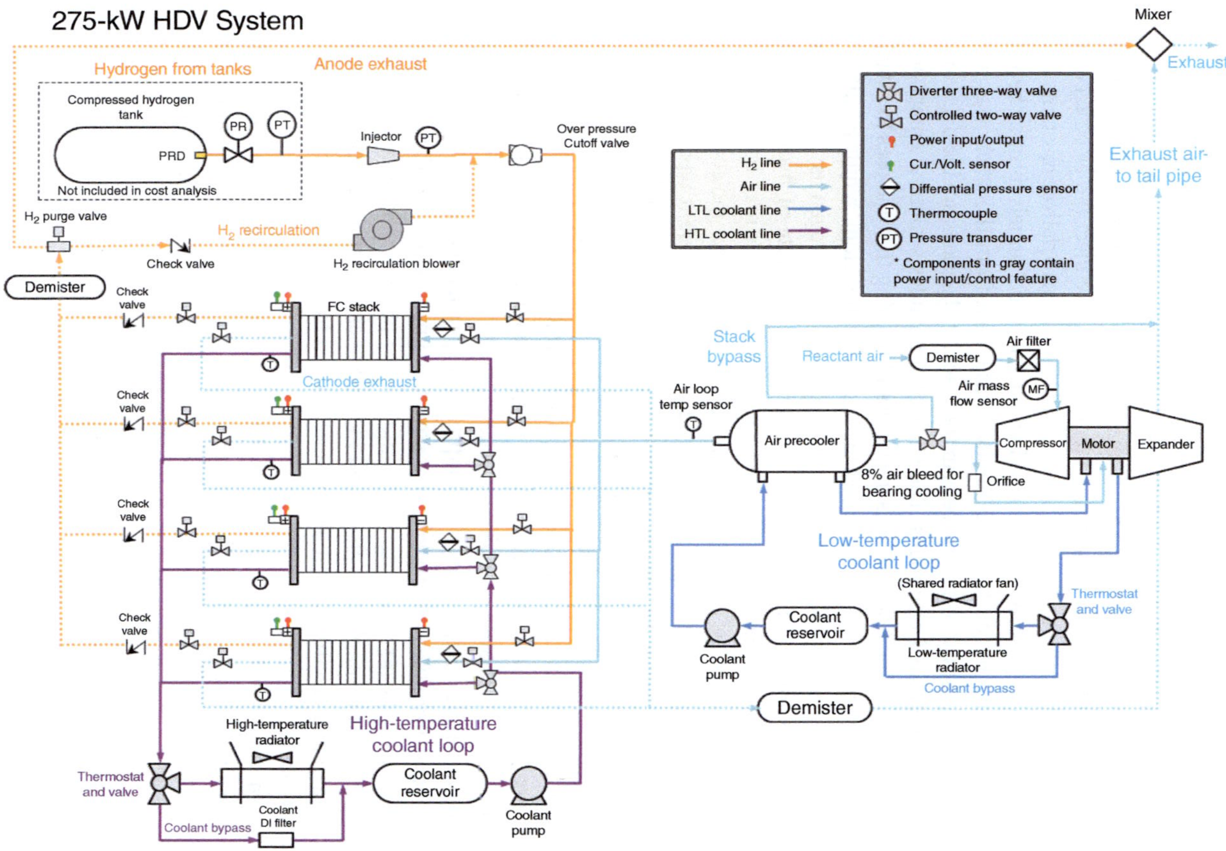

Figure 1.2 *(Continued)*

As the power-producing component, the stack is the heart of the fuel cell system. The stack is a collection of individual galvanic cells, each of which provides <1 V when operating, connected in series to create a power device that provides a higher, more useful voltage. In some applications, multiple stacks may be used together for modular, higher power systems. Each cell contains a membrane electrode assembly (MEA), which is the electrochemically active stack component, with diffusion media (gas diffusion layers and microporous layers) on each electrode encouraging uniform distribution of reactants over the MEA and removal of water. MEAs are connected in the stack by electrically conductive bipolar plates to collect the electric current produced, which include flow channels facing the MEAs to deliver reactants to the electrodes. Bipolar plate assemblies also include coolant channels running between (and separated from) the MEAs to remove waste heat from the stack. Gas manifolds distribute gases between the cells in the stack, and gaskets are included to seal gas within the desired electrodes. The stack also includes structural components, including tie rods that hold the cells together and housing that encloses the stack.

Fuel must be supplied and prepared for the fuel cell system, requiring different BOP components depending on the type of fuel. If hydrogen is the fuel, the preparation required is minimal: the pressure and flow rate of hydrogen to the stack must be controlled, and in some cases the hydrogen may be humidified. Unused hydrogen may also be recirculated. More complex molecules, such as alcohols or hydrocarbons, may also be used as fuel for different types of fuel cell systems. High-temperature fuel cells (e.g. solid oxide) can use complex fuels directly, reforming them internally in the stack. Low-temperature fuel cells may include an external reformer, which produces hydrogen from the fuel and must also remove by-products, such as carbon monoxide, that are harmful to the fuel cell.

Fuel cell systems also require air to be supplied to the fuel cell stack. To enable high-power performance, the air is typically pressurized by a compressor, which may be a simple compressor or a compressor-expander module, which recoups some energy from the pressurized outlet stream to improve overall efficiency. Compression heats the air supply, so precooling may be necessary before the air enters the stack. The air supply may also be humidified to ensure optimal performance of membrane and electrodes, and air must be filtered to remove potentially harmful contaminants.

The thermal management subsystem removes waste heat from the fuel cell system, using coolant to transfer heat from the stack (and other BOP components, such as the compressor, as needed) to the radiator. Thermal management subsystems typically consist of pumps, coolant lines, and radiators, although the radiator is sometimes considered to be external to the fuel cell system. Multiple thermal management subsystems may be used, such as a high-temperature loop for the fuel cell stack and a low-temperature loop for the air processing subsystem.

The fuel cell system also includes components used to monitor and control the system. Numerous sensors are used in fuel cell systems, including stack voltage and current monitors, pressure and temperature sensors at different points in the system, and hydrogen sensors to detect leaks. These sensors provide information to the system controller, which directs the system to deliver requested power, while maintaining safe operation and avoiding conditions that may degrade the fuel cell.

Several other systems and components commonly accompany the fuel cell system, but are not considered a part of it, such as the fuel storage system, power electronics, and hybrid batteries. The boundaries between the fuel cell system and these other systems necessary for applications are often not defined consistently. However, these external systems have minimal impact on the choice of fuel cell materials and so will not be covered in this chapter.

1.1.3 Comparison of Fuel Cell Systems for Different Applications

The design of fuel cell systems and the technologies used vary significantly between different applications. For transportation fuel cell systems, flexibility and fast startup are critical, making PEM fuel cells the preferred technology. For automotive applications, the fuel cell system is typically sized to provide around 100 kW rated power and is usually accompanied by a hybrid energy storage battery to support transient and peak power demands. Larger fuel cell systems (hundreds of kW to several MW) with multiple stacks are used for heavy-duty vehicles, while smaller (up to tens of kW) but similar systems are used for material handling vehicles such as forklifts. Transportation fuel cell systems are typically direct hydrogen fueled, making the fuel supply subsystem relatively simple. The compressed air supply and heat rejection are both very important to enable high power density and specific power. A low cost is important for the fuel cell system to compete with incumbent combustion engine technologies. Durability is also a key concern for transportation fuel cells, as powertrains are required to endure thousands of hours of operation for automobiles and tens of thousands of hours for heavy-duty vehicles. The relative importance of different system metrics varies significantly between different transportation applications as well. For example, automotive fuel cell developers prioritize lowering capital cost and improving high-power performance to enable system size and cost reductions. In contrast, full lifecycle costs are important for commercial, heavy-duty vehicles, making durability and efficiency important priorities. Furthermore, for heavy-duty vehicle applications that carry heavy loads, the fuel cell system needs to be designed to deliver high power for more sustained periods, which can create more harsh conditions for fuel cell materials.

Stationary fuel cell systems vary widely in scale from <1 kW "micro-CHP" residential systems to large multimegawatt systems. For backup power systems, flexibility and responsiveness are critical, so PEM fuel cells are typically used. Because backup power systems operate only a small fraction of the time, capital cost dominates their overall cost. For distributed power and CHP applications, systems are typically operated continuously for very long periods, making durability and efficiency very important. The fuel processing system is also important for stationary fuel cell systems fueled by methane (natural gas or biogas). Stationary fuel cells have minimal constraints on the system size or weight.

For specialty applications such as material handling, fuel cells must provide at least equivalent performance without significant changes in functionality, size, and counterbalance weight compared to the incumbent technology. They must provide short bursts (15–20 seconds) of high power for lifting a heavy load, plus sustained power to drive the equipment. On the other hand, portable fuel cell systems typically have low power requirements. However, they are subject to extreme system size and weight limitations, and often are designed to minimize the required BOP, for example by operating at or near ambient pressure. Because these constraints impact the fuel storage system as well, liquid fuels are of interest for these applications. Cost may or may not be a serious constraint depending on the application; for consumer electronics, low costs are required to compete with Li-ion batteries, which have seen rapidly falling prices in recent years. However, for military or other specialty applications high costs may be acceptable.

1.2 Application-derived Requirements and Constraints

This section covers constraints on fuel cell operation and material choices that are imposed by the system and application requirements. Fuel cell materials must meet all system-level requirements simultaneously, which makes some otherwise promising materials infeasible. The most fundamental requirement of a fuel cell system is to provide the power demanded by the application. This requirement includes two broad categories: (i) maximum power performance, either instantaneous or sustained, and (ii) flexibility to deliver power under a variety of conditions and in response to changing demand. The fuel cell system, components, and materials must also be durable to provide the required performance not only initially but also after extensive use and exposure to potentially damaging conditions. Finally, fuel cell systems must be available at low cost to be competitive with alternative power systems, considering both initial capital and operating costs. It is important to note that performance, durability, and cost are interrelated, which allows for trade-offs between the three, depending on the lifecycle requirements of the application.

1.2.1 Fuel Cell Performance and the Heat Rejection Constraint

Cell-level performance is a fundamental issue underlying the system-level power density, specific power, cost, and efficiency. The fuel cell system must be sized to deliver the power required depending on the nature of the application and the system architecture. For a fuel-cell-dominant hybridization scheme, the fuel cell must deliver the required sustained maximum power, as the relatively small battery can add to the peak power for a limited period of time. For a battery-dominant hybridization scheme, the fuel cell instead must deliver the average power required, with the battery supplying power for peak demand. For example, fuel-cell-dominant automotive fuel cells operate most of the time at low-power conditions, where the system is most efficient, but occasionally require a high rated power (such as for highway merging). This makes rated power important because it drives system size requirements and is directly related to cost.

The voltage loss mechanisms that determine fuel cell performance, illustrated in Figure 1.3a, have been thoroughly described in many other texts on fuel cells and electrochemistry [15, 16], so we will only briefly recap them here. The ideal potential

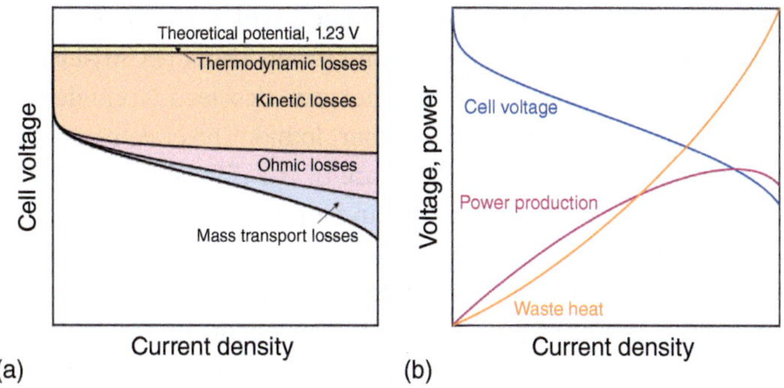

Figure 1.3 Illustrations of (a) the voltage loss mechanisms that contribute to the fuel cell polarization curve and (b) the relationship among the fuel cell voltage, electrical power production, and waste heat that must be rejected.

(for a perfectly reversible process) is determined by the overall thermodynamics of the fuel cell reaction, corresponding to 1.23 V for a hydrogen-oxygen fuel cell at standard ambient temperature and pressure [15]. Nonstandard thermodynamic conditions modify this potential as described by the Nernst equation, but the correction is generally small (on the order of 10 mV) for PEM fuel cells. The actual voltage of an operating fuel cell depends on the current density and is determined by voltage losses from reaction kinetics, Ohmic resistance, and mass transport.

The largest loss for low-temperature fuel cells under typical conditions is due to the slow kinetics of the oxygen reduction reaction (ORR) on the cathode. ORR kinetics for PEM fuel cells are well described by the Tafel approximation [17] and have a roughly logarithmic dependence on the current density, growing rapidly at low current density and then varying slowly at moderate to high current densities. ORR kinetic losses are impacted by the intrinsic activity of a catalyst material, the active surface area in the fuel cell electrode, and interactions with the polymer electrolyte in the electrode (the ionomer), which may coat the active surface. By contrast, the kinetics of the hydrogen oxidation reaction (HOR) at the fuel cell anode are extremely rapid, and HOR kinetic losses for PEM fuel cells are typically negligible, even with extremely low catalyst loadings. For high-temperature fuel cells kinetic challenges are minimal.

Ohmic losses increase linearly with current density in proportion to the overall resistance of the cell. In PEM fuel cells the membrane is typically the primary source of ohmic resistance, with the electrode ionomer also contributing significantly under some conditions. The carbon-based materials commonly used for gas diffusion media and catalyst supports contribute minimal resistance, although contact resistances and less-conductive, corrosion-resistant alternative materials may contribute significant ohmic losses. For high-temperature, solid oxide fuel cells, the ohmic resistance of the ceramic electrolyte typically dominates overall losses.

Mass transport-related losses are negligible at low current density but grow rapidly at high current density. The primary source of transport losses for hydrogen-air fuel cells is oxygen diffusion in the cathode. This includes both bulk oxygen transport through the electrode and local oxygen transport resistance associated with oxygen diffusion to a limited number of catalytically active sites, which is a particularly important and challenging problem for low-platinum group metal (PGM)-loaded electrodes [18]. Inadequate removal of product water can also lead to condensation or "flooding," leading to significant mass transport losses. The effectiveness of mass transport is determined by the porous structure of the diffusion media and electrodes, including the catalyst support structure and ionomer dispersion.

Beyond simple single-cell performance, heat rejection puts an important constraint on performance. As illustrated in Figure 1.3b, as the voltage losses increase at higher current densities, the efficiency of energy conversion in the fuel cell declines. Consequently, the increase in power output slows, eventually reaching a peak at high current density and high voltage loss. This also leads to an accelerating growth in the amount of waste heat produced by the fuel cell.

During steady-state operation, the fuel cell system must remove all waste heat produced by the fuel cells stack. For brief periods, the stack can be allowed to generate excessive heat if it is at a relatively low temperature, so higher power is possible in transient operation than the continuous power rating. The heat Q rejected from radiator can be simply described by Newton's law of cooling:

$$Q = hA\Delta T \tag{1.1}$$

where h is the heat transfer coefficient, A is surface area, and $\Delta T = T_c - T_a$ is the difference between the coolant temperature T_c and the ambient temperature T_a. For a given radiator, h and A are fixed, so $Q/\Delta T$ must stay below a certain value. This makes a particular value of $Q/\Delta T$ a metric to describe radiator capacity, which is limited for vehicle applications. To meet this heat rejection constraint, it is possible to either lower the amount of waste heat produced or raise the operating temperature. This sets a practical limit on the feasible fuel cell operating conditions [19].

A simple formula relates the $Q/\Delta T$ heat rejection metric to rated (continuous) power operating conditions, particularly the cell voltage V_r, which determines the fraction of energy converted to waste heat, and the stack coolant outlet temperature T_c, which determines how effectively that heat can be rejected through the radiator:

$$\frac{Q}{\Delta T} = \frac{P_g(V_i - V_r)/V_r}{T_c - T_a} \tag{1.2}$$

where P_g is the gross power rating of the stack, and V_i is the ideal cell voltage. This relationship is illustrated in Figure 1.4, which shows the dependence of the cell voltage at rated power V_r on T_c and $Q/\Delta T$, using standard assumptions [20] for an automotive fuel cell system (90 kW_{gross} rated power, 40 °C ambient temperature, and 1.23 V ideal cell voltage). In general, lower voltages (and therefore higher current and power densities) can be used with either a higher coolant temperature or higher $Q/\Delta T$ (i.e. radiator size). The DOE has set a target for fuel cell heat rejection of $Q/\Delta T \leq 1.45$ kW per °C to enable use of practically sized automotive radiators for fuel cell vehicles.

Meeting this automotive heat rejection constraint creates a strong motivation for using higher temperature (e.g. 94 °C) and higher pressure (e.g. 2.5 bar) operating conditions at rated power [19]. Raising the temperature allows higher current

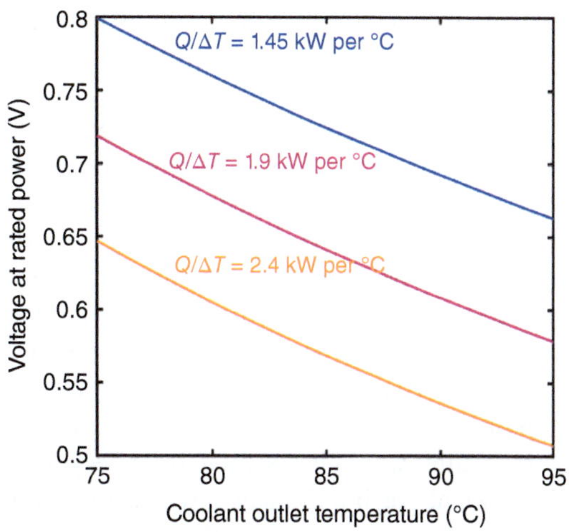

Figure 1.4 Plot of the relationship imposed by the heat rejection constraint among cell voltage at rated power, coolant outlet temperature, and the $Q/\Delta T$ metric.

(and power) density by relaxing the voltage limitation imposed by heat rejection, for example from 0.76 V at 80 °C to 0.67 V at 94 °C. Raising the operating pressure also enables higher power density by improving kinetics and mass transport, although this performance boost must be balanced against higher parasitic power losses from air compression.

The heat rejection constraint also translates into constraints on the set of viable fuel cell materials. For heat-rejection-constrained fuel cell applications such as transportation, high-power performance beyond the heat rejection limit is not useful, and the cell cannot be operated at the "maximum" point of the power curve shown in Figure 1.3b. This reinforces the importance of high-activity catalyst materials; if catalyst's onset potential is too low, the cell voltage will be low at any practical current density, and it will be infeasible to meet the heat rejection constraint. Materials must also be selected to minimize ohmic and mass transport losses to extend high-power performance near the rated-power voltage limit. The heat rejection constraint also creates a motivation to develop materials, especially membranes and ionomers, that provide adequate performance and durability at higher operating temperatures. It may come as a surprise that fuel cell voltages near 0.6 V (roughly 50% efficient) may be impractical for heat rejection reasons, given that this efficiency greatly exceeds that of internal combustion engines. However, the lower operating temperature of PEM fuel cells and lack of significant heat removal in the exhaust stream make heat rejection somewhat more challenging in comparison to combustion engines [10].

1.2.2 Startup, Flexibility, and Robustness

In addition to meeting the maximum power requirements of each application, the fuel cell system must deliver the appropriate level of power when needed under all relevant conditions. This is not a significant constraint for some applications, such as distributed power generation or CHP systems, which may always run at essentially constant power. However, it is a critical requirement for many transportation systems, which must be able to start up quickly and adapt to variations across their duty cycles.

Transportation fuel cells generally must start up from significantly lower ambient temperature than standard operation (around 80 °C), including possibly very low temperatures (e.g. −40 °C) for compatibility with cold climates. The fuel cell system must provide adequate power within a short time from startup, and while some amount of energy goes to heating the stack to operating temperatures, excessive energy consumption during startup must be avoided. As will be discussed below, startup and shutdown can also create conditions, such as fuel starvation, that may degrade the fuel cell materials and therefore must be managed [21, 22]. While system startup is largely a matter of system design and engineering, it also has some important implications for viable fuel cell materials. For instance, higher operating temperatures would be useful to improve performance and heat rejection issues as discussed above, but challenges with startup from low temperatures (e.g. poor ionic conductivity) prevent the use of many intermediate and high-temperature materials for transportation applications.

The requirement for flexible and robust operation also imposes some constraints on fuel cell materials, especially related to water management. Rapid water production at high current densities, especially at relatively low temperatures, can lead to electrode flooding and performance losses. To address this issue, the porosity, thickness, and hydrophobicity of fuel cells must be engineered, and ideally materials for catalysts, catalyst supports, and diffusion media can be designed to enable water rejection without flooding. Conversely, dry conditions can also pose a challenge for performance, as low humidity can lead to performance losses from decreased conductivity of polymer membranes and ionomers as well as decreased catalyst utilization. Dry conditions can be avoided to some extent by humidification of the inlet gas streams, or by design of the stack to enable self-humidification of the cells, but performance loss under dry conditions is an important limitation on the feasible upper temperature for PEM fuel cells, as significant humidification becomes impractical at temperatures approaching the boiling point of water. Ultimately, materials development to improve tolerance to a variety of operating conditions has an important role in enabling the flexibility of fuel cell systems.

1.2.3 Fuel Cell Durability

The conditions inside a fuel cell are intrinsically degrading to materials. The strongly acidic electrolyte and oxidizing potentials lead to both chemical and electrochemical stresses. Furthermore, changes in temperature and humidity within the cell lead to swelling of the membrane and mechanical stress on the MEA. Materials degradation caused by these stresses leads to degraded performance over time and in some cases catastrophic failure of the fuel cell. The catalysts, membrane, and bipolar plates are all at risk for serious degradation, making durability a key concern for their materials. The various degradation mechanisms of fuel cell materials have been discussed extensively in the research literature [23] and remain a topic of ongoing interest. The major degradation mechanisms will only be briefly summarized here.

Metal-based fuel cell catalysts suffer from dissolution, especially at high electrode potentials and during voltage cycling. Noble metals, such as platinum, are generally redeposited somewhere in the cell, leading either to catalyst mass loss if the redeposition occurs in the membrane or to catalyst coarsening if the redeposition occurs in the electrode, since the Ostwald ripening mechanism tends to favor migration from smaller particles to larger ones. Both catalyst mass loss and coarsening lead to loss of the active catalyst surface area. Less noble metals, such as cobalt and nickel, tend to remain ion-exchanged into the membrane or ionomer when dissolved, poisoning its ionic conductivity. Catalyst coarsening can also occur through migration and coalescence of catalyst nanoparticles. The catalyst support can also suffer from degradation, especially for carbon materials when the cathode approaches open-circuit voltage or when the anode experiences high potentials due to cell reversal [21]. These circumstances do not generally occur during normal operation of the fuel cell but may occur during uncontrolled startup or shutdown, or due to fuel starvation [24].

Fuel cell membranes experience oxidative chemical damage caused by free radicals generated at high electrode potentials, which leads to thinning of the membrane and eventually pinhole formation. This degradation mechanism can be dramatically accelerated by contaminant cations such as iron or copper that catalyze Fenton reactions, decomposing hydrogen peroxide generated in the ORR into free radicals. Because Fenton catalyst cations are a risk factor for membrane degradation, dissolution of these metals from other components, such as stainless-steel bipolar plates or catalyst materials, must be carefully avoided. Membrane degradation can also be caused by mechanical stresses, especially due to swelling from humidity cycling in the cell, or from imperfections in the electrodes such as bumps. State-of-the-art fuel cell membranes typically contain a reinforcement layer, making them more resilient to mechanical stresses.

Bipolar plates may degrade through dissolution of their base metals, especially for stainless steel plates, which can lead to membrane degradation as noted above. Bipolar plates can also form passivating surface layers that can increase contact resistance and lead to performance losses. Metallic bipolar plates therefore require coatings, commonly using precious metals, to ensure corrosion resistance and conductivity.

Fuel cell systems are generally designed to mitigate materials degradation by controlling the operating conditions to especially avoid degrading conditions. System mitigation strategies commonly include limiting time spent at high temperatures, restricting the voltage window to avoid high potentials when idling or at low power, and controlling the startup and shutdown procedures to avoid exposing the electrodes to high potentials. Sensors, for example cell voltage monitors and temperature or humidity sensors, are typically part of integrated diagnostics to actively track the current degradation state of the fuel cell system. Furthermore, prognostic algorithms can be implemented and integrated for advanced control and projection of fuel cell remaining lifetime during operation [25]. System mitigation strategies often involve some degree of compromise for fuel cell performance or flexibility. For instance, limiting the operating temperature leads to less-effective heat rejection and lower rated power. Limiting the maximum operating voltage limits the range of output power and maximum possible efficiency. However, extensive monitoring and controls add to the system complexity and cost. As such, it is preferrable to achieve as much durability as possible with material solutions to limit reliance on system mitigation.

Fuel cell applications generally require adequate performance to be maintained over long periods of time. For example, the DOE has set ultimate targets for fuel cell system lifetime at 8000 hours for light-duty vehicles, 30 000 hours for heavy-duty trucks, and 80 000 hours for large (100 kW to 3 MW) stationary power systems [1, 26]. Materials and components must be stress tested in advance to ensure that they are durable enough for these applications. Because it is impractical to run routine testing as long as the application lifetime, accelerated stress tests (ASTs) are necessary for R&D purposes. ASTs, such as those recommended by the DOE [1], generally are designed following one of two broad approaches. The first of these is to attempt to represent the conditions created by real duty cycles for a particular application. This approach can provide a relatively direct demonstration of a material or component's fitness for that application. However, because fuel cell systems generally incorporate mitigation strategies to avoid degrading conditions, a drawback of this approach is that it is difficult to both represent a realistic system including mitigation and also give adequate acceleration for practical testing. The second approach is to target a particular degradation mechanism. This approach enables highly accelerated testing by focusing on deliberately degrading conditions, while also providing scientific clarity by selecting a single mechanism. However, mechanism-targeted ASTs can neglect potentially interacting degradation mechanisms, which may require separate, combined ASTs to capture. Also, it is not straightforward to translate from mechanism-targeted AST results to expected lifetime. Modeling that includes the scientific basis of degradation can have a useful role in projecting lifetime from AST results and exploring mitigation strategies [27].

1.2.4 Cost

While cost constraints are unlike the constraints described above, which have a physical or chemical basis, the need for competitive fuel cell system cost puts important limitations on feasible materials. The important cost factors for fuel cell systems include intrinsic material costs and manufacturing costs, which contribute to up-front capital costs, as well as ongoing operating costs.

Similar to fuel cell system designs, information on manufacturing costs for real fuel cell systems and components is generally proprietary. However, useful guidance can be derived from publicly reported cost estimates based on bottom-up analysis, input from industry experts, or a combination of these. For instance, James et al. [6] have performed extensive cost analysis based on the model automotive fuel cell system discussed above using industry input to inform manufacturing assumptions. This analysis projected the cost of fuel cell systems and components at different manufacturing scales, as illustrated in Figure 1.5, assuming the use of state-of-the-art technology currently demonstrated at laboratory scale that is expected to be employed for future, high-volume manufacturing. This approach is necessarily more speculative than assuming "on-road" technology that is currently commercially available, which lags laboratory-scale technology by several

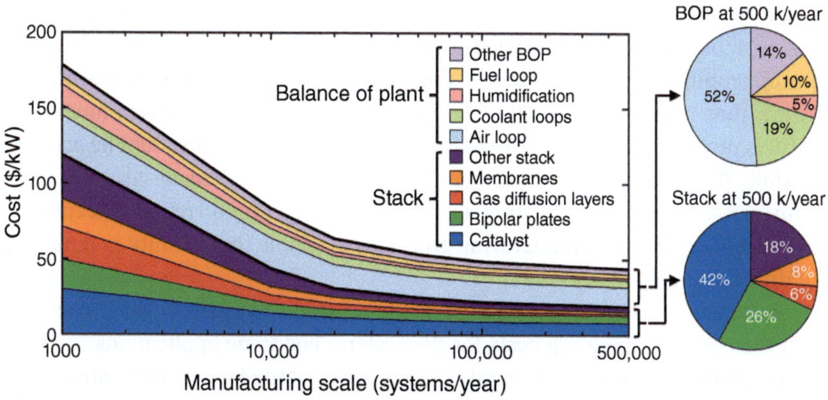

Figure 1.5 Breakdown of fuel cell system cost by major stack and balance of plant (BOP) components and subsystems, projected to different manufacturing scales. Pie charts at right show the fraction of stack and BOP costs due to different components at 500 000 systems per year manufacturing scale. Source: Adapted from Ref. [6].

years, but it is most useful for identifying high-impact research and development opportunities. Similar cost analysis is also available for other fuel cell types, such as stationary fuel cells [11–14, 28].

Manufacturing costs generally decline with manufacturing scale for specialized BOP and fuel cell stack components. Overall, Figure 1.5 shows that increasing manufacturing volume from current levels (thousands of systems per year) to high volume (hundreds of thousands of systems per year) is expected to decrease fuel cell system costs by roughly two-thirds even without any advances in technology. However, not all costs can be decreased simply by manufacturing scale-up. Some components, such as the cooling subsystem components, are already manufactured at high volume and available off-the-shelf and are therefore not expected to see significant cost reductions as fuel cell systems are manufactured at higher scale. Furthermore, fuel cell stack components commonly use several intrinsically expensive materials, which set a limit on the cost reductions that can be achieved through high-volume manufacturing. The most notable examples are platinum-group-metal (PGM) catalysts, which provide the highest performance and durability among currently available catalyst materials, and bipolar plates containing titanium, 316 stainless steel, and PGM coatings for improved corrosion resistance and conductivity. Together, catalysts and bipolar plates account for most of the cost of fuel cell stacks at high manufacturing volume according to the analysis illustrated in Figure 1.5. Consequently, reducing the number of costly materials required for high-performance fuel cell stacks and the development of low-cost alternative materials are key research priorities.

Even without expensive feedstock materials, viable materials and components must be manufacturable without adding significant expense. Thus, they must be designed with scalability in mind. For instance, membranes are expected to be a significant contributor to fuel cell stack costs even at high volume manufacturing because the fluoropolymer chemistry required to synthesize conventional polymer electrolyte membranes is challenging and expensive. In general, materials that involve synthesis or processing that includes many steps, challenging processes, or procedures that cannot be translated from small batches to large-scale production may add prohibitively large costs for fuel cell applications. Furthermore, manufacturing that includes steps that are hazardous or potentially environmentally harmful is problematic for implementation at scale. Therefore, it should be avoided if possible. The cost and environmental impact of fuel cell materials and components can also be mitigated somewhat through recycling or reuse. Selection of materials and design of manufacturing processes to enable recyclability or reuse are therefore of interest.

Beyond up-front cost impacts, fuel cell materials and system designs can also impact the ongoing operating costs for the system. For some applications, such as medium- and heavy-duty commercial vehicles or distributed power generation, lifetime fuel costs can be significantly greater than up-front capital costs [26]. In these applications, the efficiency of the fuel cell system therefore plays an important role in overall lifetime costs, making it advantageous to use more expensive materials or components that improve efficiency, such as higher catalyst loadings, high-activity catalysts, or BOP components with reduced parasitic power requirements.

1.3 Material Pathways to Improved Fuel Cells

Understanding of application-driven constraints provides necessary guidance for efforts to research and develop advanced fuel cell materials and electrocatalysts. All major constraints and requirements must be met simultaneously by a viable fuel cell system. As such, these requirements can rule out the viability of some materials, while making other advancements highly impactful for the technology. Further improvements to oxygen reduction catalysts have a very important role to play in advancing fuel cell technologies.

Catalyst advancements have significant potential to improve fuel cell performance. ORR kinetics generally remain the largest source of voltage loss for PEM fuel cells, and local transport issues with cost-effective, low-PGM loadings make catalyst design a key factor in high-power performance as well. Improving the intrinsic activity of the catalytically active surface is important, but also interactions among the catalyst, support, and ionomer in the operational MEA environment are essential to realizing high performance in real fuel cell applications.

Catalysts are also currently the single largest contributor to fuel cell stack cost, and catalyst improvements are necessary to bring down fuel cell cost to achieve widespread use and high-volume manufacturing. Two general pathways are available for significant cost reductions. The first of these is to improve the performance of Pt-based catalysts in low-PGM fuel cells. Performance improvements both allow the PGM loading to be reduced, and the overall stack size to be reduced, enabling reduction in all other stack costs. The second pathway is the development of low-cost, alternative PGM-free catalysts, such as iron-nitrogen-carbon-based materials [29]. If these catalysts advance to the point of reaching near-parity with low-PGM catalysts on performance and durability, they can replace PGM-based catalysts to enable a large cost reduction and avoidance of materials scarcity issues.

Catalyst durability is a primary factor limiting fuel cell lifetime. The necessary, cost-motivated use of low-PGM loadings leaves little margin of error for catalyst degradation, making good catalyst durability a necessity. Furthermore, increasing interest in heavy-duty applications that demand exceptionally long lifetimes makes fuel cell durability a top R&D priority. These applications also have rigorous demands for performance and efficiency, which make aggressive system mitigation strategies (e.g. strong voltage clipping or temperature limitations) undesirable. Advances in the durability of high-performance catalysts and other fuel cell materials will be essential for fuel cell success in these applications.

Beyond hydrogen-fueled PEM fuel cells, there are several long-term, transformational pathways for fuel cell advancement, including alkaline membrane, intermediate temperature, direct liquid, and reversible fuel cell technologies. These technologies are at an early stage of R&D but have the potential to dramatically improve the competitiveness of fuel cell technologies across applications by expanding the capabilities of fuel cells and lowering their cost and dependence on rare materials such as platinum. Alkaline membrane fuel cells present the opportunity to combine the high power density of PEM fuel cells with alkaline conditions that offer possible compatibility with less-expensive materials. Intermediate temperature fuel cells, which operate in the 200–500 °C temperature range above the operating temperatures of PEM and phosphoric acid fuel cells but below those of solid oxide and molten carbonate fuel cells, offer the potential advantages of high efficiency, low-cost materials, fuel flexibility, and effective heat rejection. Direct liquid fuel cells, which consume fuels such as methanol, ethanol, or dimethyl ether in low-temperature cells, could address significant challenges with hydrogen fuel storage and infrastructure. Reversible fuel cells, which can operate in either fuel cell or electrolysis modes, are a promising technology to enable long-duration energy storage either for grid leveling supporting greater adoption of intermittent renewable energy sources or for remote, off-grid applications. These approaches will require the development of new catalyst systems and approaches to integrate these catalysts into devices with materials and chemical environments dramatically different from hydrogen-fueled PEM fuel cells. While the most probable applications for these long-term fuel cell technologies are still unknown, consideration of the possible relevant systems-level constraints can help guide their strategic advancement toward useful applications.

1.4 Note

The views and opinions of the authors expressed herein do not necessarily state or reflect those of the United States Government or any agency thereof. Neither the United States Government nor any agency thereof, nor any of their employees, makes any warranty, expressed or implied, or assumes any legal liability or responsibility for the accuracy, completeness, or usefulness of any information, apparatus, product, or process disclosed, or represents that its use would not infringe privately owned rights.

Acronyms

AST	accelerated stress test
BOP	balance of plant
CHP	combined heat and power
DOE	U.S. Department of Energy
HDV	heavy-duty vehicle
HOR	hydrogen oxidation reaction
MEA	membrane electrode assembly
ORR	oxygen reduction reaction
PEM	polymer electrolyte membrane
PGM	platinum group metal
R&D	research and development
RD&D	research, development, and demonstration
U.S. DRIVE	Driving Research and Innovation in Vehicle efficiency and Energy sustainability

Symbols

ΔT	difference temperature
A	surface area
H	heat transfer coefficient
P_g	gross power rating of the stack
Q	heat
T_a	ambient temperature
T_c	coolant temperature
V_i	ideal cell voltage
V_r	cell voltage

References

1 U.S. Department of Energy (2016). Hydrogen and fuel cell technologies office multi-year research, development, and demonstration plan, Section 3.4: Fuel cells. https://www.energy.gov/eere/fuelcells/downloads/hydrogen-and-fuel-cell-technologies-office-multi-year-research-development (accessed 16 June 2021).

2 Fuel Cells and Hydrogen Joint Undertaking (2020). State-of-the-art and future targets. https://www.fch.europa.eu/soa-and-targets (accessed 13 December 2020).

3 New Energy and Industrial Technology Development Organization (2018). NEDO's technology development roadmap for fuel cells and hydrogen. https://www.nedo.go.jp/english/news/AA5en_100334.html (accessed 13 December 2020).

4 Padgett, E. and Kleen, G. (2020). Automotive fuel cell targets and status. U.S. Department of Energy Hydrogen and Fuel Cell Technologies Program Record # 20005. https://www.hydrogen.energy.gov/pdfs/20005-automotive-fuel-cell-targets-status.pdf (accessed 16 June 2021).

5 Kleen, G. and Padgett, E. (2021). Durability-adjusted fuel cell system cost. U.S. Department of Energy Hydrogen and Fuel Cell Technologies Program Record #

21001. https://www.hydrogen.energy.gov/pdfs/21001-durability-adjusted-fcs-cost .pdf (accessed 16 June 2021).

6 James, B.D., Huya-Kouadio, J.M., Houchins, C., DeSantis, D.A. (2018). Mass production cost estimation of direct H_2 PEM fuel cell systems for transportation applications: 2018 update. Strategic Analysis Inc. https://www.energy.gov/sites/prod/files/2019/12/f70/fcto-sa-2018-transportation-fuel-cell-cost-analysis.pdf (accessed 16 June 2021).

7 Thompson, S.T., James, B.D., Huya-Kouadio, J.M. et al. (2018). Direct hydrogen fuel cell electric vehicle cost analysis: system and high-volume manufacturing description, validation, and outlook. *Journal of Power Sources* 399: 304–313. https://doi.org/10.1016/j.jpowsour.2018.07.100.

8 James, B.D. (2020). Fuel cell systems analysis. *U.S. DOE Hydrogen and Fuel Cells Program 2020 Annual Merit Review and Peer Evaluation Meeting*, Washington, D.C. (19–21 May 2020). https://www.hydrogen.energy.gov/pdfs/review20/fc163_james_2020_o.pdf (accessed 16 June 2021).

9 Ahluwalia, R.K., Wang, X., Peng, J.-K., and Papadias, D. (2020). Fuel cell system optimization for rail and maritime applications. *U.S. DOE Hydrogen and Fuel Cells Program 2020 Annual Merit Review and Peer Evaluation Meeting,* Washington, D.C. (19–21 May 2020). https://www.hydrogen.energy.gov/pdfs/review20/fc329_ahluwalia_2020_p.pdf (accessed 16 June 2021).

10 Ahluwalia, R.K., Wang, X., and Peng, J.-K. (2020). Fuel cell system modeling and analysis. *U.S. DOE Hydrogen and Fuel Cells Program 2020 Annual Merit Review and Peer Evaluation Meeting*, Washington, D.C. (19–21 May 2020). https://www .hydrogen.energy.gov/pdfs/review20/fc017_ahluwalia_2020_o.pdf (accessed 16 June 2021).

11 Battelle Memorial Institute (2016). Manufacturing cost analysis of 100 and 250 kW fuel cell systems for primary power and combined heat and power applications. https://www.energy.gov/sites/prod/files/2016/07/f33/fcto_battelle_mfg_cost_analysis_pp_chp_fc_systems.pdf (accessed 16 June 2021).

12 Battelle Memorial Institute (2016). Manufacturing cost analysis of PEM fuel cell systems for 5- and 10-kW backup power applications. https://www.energy.gov/sites/prod/files/2016/12/f34/fcto_cost_analysis_pem_fc_5-10kw_backup_power_0 .pdf (accessed 16 June 2021).

13 James, B.D. and DeSantis, D.A. (2015). Manufacturing cost and installed price analysis of stationary fuel cell systems. Strategic Analysis Inc. https://www .sainc.com/assets/site_18/files/publications/sa%202015%20manufacturing%20cost %20and%20installed%20price%20of%20stationary%20fuel%20cell%20systems_rev3 .pdf (accessed 16 June 2021).

14 Wei, M., Lipman, T., Mayyas, A., et al. (2014). A total cost of ownership model for low temperature PEM fuel cells in combined heat and power and backup power applications. Lawrence Berkeley National Laboratory. https://eta-publications.lbl.gov/sites/default/files/wei_-_a_total_cost_of_ownership_model_for_low_temp_pem_fuel_cells_-_final.pdf (accessed 16 June 2021).

15 O'Hayre, R., Cha, S.-W., Colella, W., and Prinz, F.B. (2016). *Fuel Cell Fundamentals*, 3e. Hoboken, NJ: Wiley.

16 Dicks, A.L. and Rand, D.A.J. (2018). *Fuel Cell Systems Explained*, 3e. Hoboken, NJ: Wiley.

17 Gasteiger, H.A., Kocha, S.S., Sompalli, B., and Wagner, F.T. (2005). Activity benchmarks and requirements for Pt, Pt-alloy, and non-Pt oxygen reduction catalysts for PEMFCs. *Applied Catalysis B: Environmental* 56: 9–35. https://doi.org/10.1016/j.apcatb.2004.06.021.

18 Kongkanand, A. and Mathias, M.F. (2016). The priority and challenge of high-power performance of low-platinum proton-exchange membrane fuel cells. *Journal of Physical Chemistry Letters* 7: 1127–1137. https://doi.org/10.1021/acs.jpclett.6b00216.

19 Ahluwalia, R.K., Wang, X., and Steinbach, A.J. (2016). Performance of advanced automotive fuel cell systems with heat rejection constraint. *Journal of Power Sources* 309: 178–191. https://doi.org/10.1016/j.jpowsour.2016.01.060.

20 USDRIVE (2017). Fuel cell technical team roadmap. https://www.energy.gov/sites/prod/files/2017/11/f46/FCTT_Roadmap_Nov_2017_FINAL.pdf (accessed 13 December 2020).

21 Reiser CA, Bregoli L, Patterson TW, Yi JS, Yang JD, Perry ML, Jarvi TD (2005) A reverse-current decay mechanism for fuel cells. *Electrochemical and Solid-State Letters* 8:A273. https://doi.org/10.1149/1.1896466.

22 Perry, M.L., Patterson, T., and Reiser, C. (2019). Systems strategies to mitigate carbon corrosion in fuel cells. *ECS Transactions* 3: 783–795. https://doi.org/10.1149/1.2356198.

23 Borup, R., Meyers, J., Pivovar, B. et al. (2007). Scientific aspects of polymer electrolyte fuel cell durability and degradation. *Chemical Reviews* 107: 3904–3951. https://doi.org/10.1021/cr050182l.

24 Patterson, T.W. and Darling, R.M. (2006). Damage to the cathode catalyst of a PEM fuel cell caused by localized fuel starvation. *Electrochemical and Solid-State Letters* 9: A183. https://doi.org/10.1149/1.2167930.

25 Sutharssan, T., Montalvao, D., Chen, Y.-K. et al. (2017). A review on prognostics and health monitoring of proton exchange membrane fuel cell. *Renewable and Sustainable Energy Reviews* 75: 440–450. https://doi.org/10.1016/j.rser.2016.11.009.

26 Marcinkoski, J. (2019). Hydrogen class 8 long haul truck targets. U.S. Department of Energy Hydrogen and Fuel Cell Technologies Program Record # 19006. https://www.hydrogen.energy.gov/pdfs/19006_hydrogen_class8_long_haul_truck_targets.pdf (accessed 16 June 2021).

27 Ahluwalia, R.K., Wang, X., Peng, J.-K. et al. (2021). Achieving 5,000-h and 8,000-h low-PGM electrode durability on automotive drive cycles. *Journal of the Electrochemical Society* 168: 044518. https://doi.org/10.1149/1945-7111/abf507.

28 Ramsden, T. (2013). An evaluation of the total cost of ownership of fuel cell powered material handling equipment. National Renewable Energy Laboratory. https://www.nrel.gov/docs/fy13osti/56408.pdf (accessed 21 June 2021).

29 Thompson ST, Wilson AR, Zelenay P, Myers DJ, More KL, Neyerlin KC, Papageorgopoulos D (2018) ElectroCat: DOE's approach to PGM-free catalyst and electrode R&D. *Solid State Ionics* 319:68–76. https://doi.org/10.1016/j.ssi.2018.01.030.

2

PEM Fuel Cell Design from the Atom to the Automobile

Andrew Haug and Michael Yandrasits

3M Center, Building 201-02-N-19, St. Paul, MN 55144-1000, USA

2.1 Introduction

Proton-Exchange Membrane (PEM) Fuel Cells continue to be a key component to a renewable, zero-emission energy conversion future. Modern fuel cell systems emerged between the late 1950s and mid-1960s, driven by the nascent United States Space Program. Phosphoric acid, alkaline, and PEM fuel cell technologies competed for the right to fly to the moon. The proton-exchange membrane fuel cell (PEMFC) system using a sulfonated polystyrene ionomer membrane flew on several Gemini missions 1. Ongoing issues led to the choice of alkaline fuel cells (AFCs) for the Apollo missions and subsequent Space Shuttle program [1]. Their combination of reliability, efficiency, and relative light weight led to their choice over batteries and other fuel cell systems; 110 kg Apollo AFC systems produced 1.42 kW operating on H_2 and O_2 at 4 atm pressure, while the 118 kg units of the Space Shuttle produced 7–12 kW at platinum group metal (PGM) loadings totaling 30 mg_{PGM} cm^{-2} [1–3]. As advances in membrane and electrode processing improved, the subsequent power density improvements led to increased PEMFC development work in the late 1980s. Much of the early work on durability mechanisms was translated from decades earlier PAFC system developments. Over decades, failures from mechanical (membrane puncture), chemical (ionomer breakdown), electrochemical (catalyst dissolution), stack-level (fuel maldistribution), and system-related (start-up and shutdown procedures) mechanisms have been identified. Improved solutions are continually being developed. Expensive PGM loadings dropped 30× to roughly 0.9 mg cm^{-2} by 2005, and to as low as 0.125 mg cm^{-2} in 2020 [4]. First-generation vehicles such as the Toyota Mirai utilize ~0.36 mg_{Pt} cm^{-2} [5] or about 30 g of Pt per car [6]. This is nearly a 100× decrease from the 35 mg cm^{-2} PEMFC system first used to provide electrical power to the Gemini spacecraft [7]. Even more dramatically, fuel cell system cost per kW ranged from $40,000–100,000/kW of the 1960s and 1970s [8]. For a production of 1–500 K systems per year, 2002 costs equated to $275/kW [9] to ~$800/kW [10, 11], and 2020 costs dropped further to $40–65/kW [10].

Electrocatalysis for Membrane Fuel Cells: Methods, Modeling, and Applications, First Edition.
Edited by Nicolas Alonso-Vante and Vito Di Noto.

Today, there are PEMFC systems generating watts to megawatts, powering sensors, electric bicycles, forklifts, automobiles, heavy-duty vehicles, drones, trains, and container ships. New applications are being evaluated daily. As hydrogen production transitions from gray to green and blue, and as costs continue to fall, PEMFC and proton-exchange membrane water electrolyzer (PEMWE) systems will further play a critical role in renewable energy generation and storage. These systems will be core components of a growing hydrogen economy that in turn will be a critical part of the overall electron economy. Before 2020 and the COVID epidemic, projected PEMFC market growth has regularly expected to be well over 15% per year [12, 13]. However, the significant renewable energy stimulus investments by Europe [14, 15] and Asia in 2020 will likely accelerate growth over these earlier estimates.

PEMFC is fundamentally governed by the anode Hydrogen Oxidation Reaction (HOR) and the cathode Oxygen Reduction Reaction (ORR):

$$H_2 \rightarrow 2H^+ + 2e^- \quad \text{ANODE}$$

$$O_2 + 4H^+ + 4e^- \rightarrow 2\,H_2O \quad \text{CATHODE}$$

The thermodynamic potential governing the HOR and ORR reactions on platinum is 1.223 V at 25 °C. Thus, a 12.23-mV gain equates to a 1% efficiency improvement. System performance is defined by stack performance minus parasitic losses. Fuel cell stack performance is defined by the overpotential (efficiency) losses of the reactions combined with reactant and product transport resistance losses. The sum of these losses gives PEMFC performance, known as a "polarization curve," as shown in Figure 2.1. The cathodic ORR reaction is roughly a million times slower than the HOR reaction [16, 17] and thus responsible for the majority of the kinetic efficiency losses. To counter these losses, the cathode utilizes the majority of the precious metal catalyst content and subsequently accounts for significant system cost.

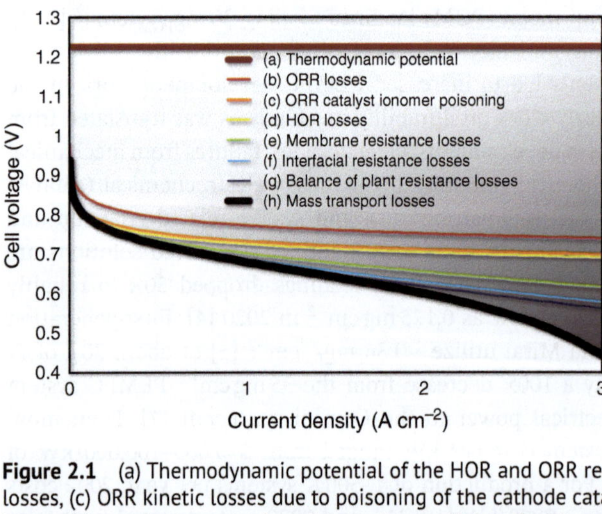

Figure 2.1 (a) Thermodynamic potential of the HOR and ORR reactions, (b) ORR kinetic losses, (c) ORR kinetic losses due to poisoning of the cathode catalyst by ionomer, (d) HOR losses, (e) membrane protonic resistance losses, (f) interfacial protonic resistance losses, (g) balance of plant resistance losses, (h) reactant gas mass transport resistance losses.

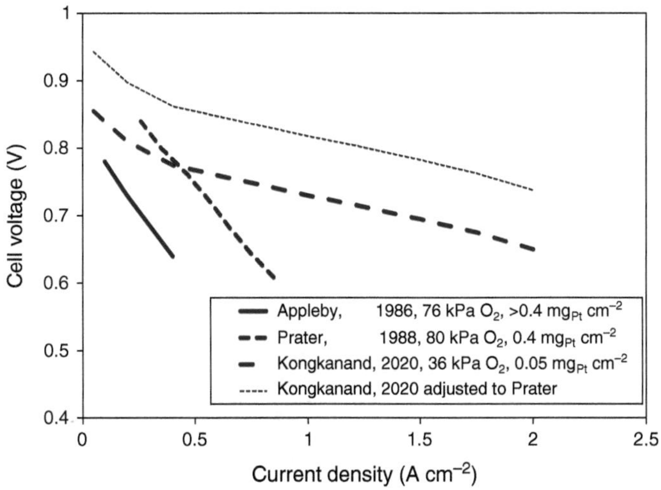

Figure 2.2 Historic performance curves in H_2/air at various temperatures and pressures [7, 18, 19]. The oxygen partial pressure of each curve has been extrapolated. Loading, oxygen concentration, and the Tafel slope were used to adjust Kongkanand performance to Prater conditions and Pt content.

The remaining losses are material transport related. Resistance losses occur from proton transport across the membrane and through the electrodes to catalyst surfaces. Electronic resistance losses occur as electrons transport through electrodes, gas diffusion layers (GDLs), and flowfields. Gas transport resistance occurs as hydrogen and oxygen transport through flowfields, GDLs, electrodes, and ionomer covering the catalyst. Removal of product water as liquid and vapor must be considered at every level. The best system designs can provide higher system power outputs for less than one-fourth of the total precious metal loadings [18] of nonoptimized systems, allowing kinetic and transport efficiency gains to be achieved at lower system cost.

A snapshot of these polarization curve improvements over the past 20+ years is shown in Figure 2.2. Improvements in every system category allowed up to 4× current density improvements at 0.7 V utilizing one-eighth of the catalyst, translating to a $32 \times \mathrm{W}\,g_{PGM}^{-1}$ improvement – all while increasing system durability from hundreds to thousands of hours.

The representative PEMFC system is shown in Figure 2.3. More detailed automotive and backup power system designs can be found in [10, 11]. Key components include:

(1) Fuel (hydrogen) storage tank, typically a carbon fiber composite;
(2) Inlet fuel and air humidifiers for optimal stack operation;
(3) Coolant system consisting of coolant (water), pumps, and heat exchangers;
(4) Pressure valves, compressors, and regulators for higher power density pressurized systems;
(5) Fuel recirculation to maximize fuel efficiency;
(6) Power conditioning components and motor.

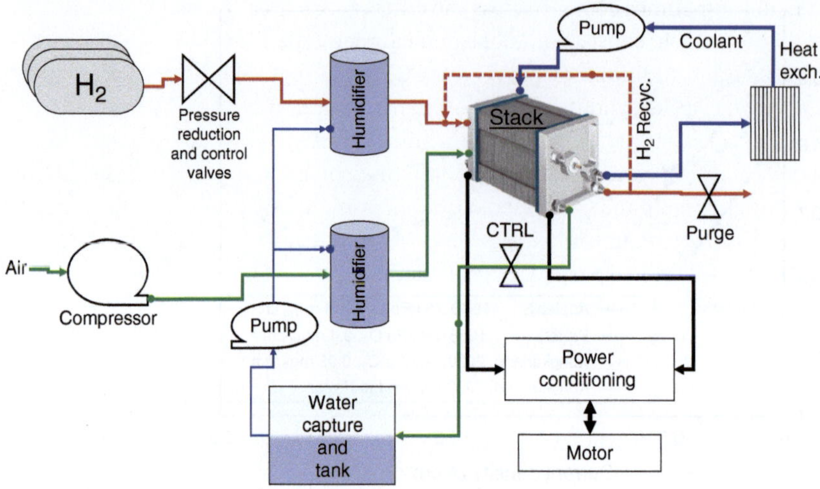

Figure 2.3 Design of a pressurized, water-cooled PEMFC system [20, 21] containing a PEMFC stack, hydrogen tanks, pressure control valves, hydrogen recirculation system, air compressor, inlet gas humidifiers, coolant heat exchanges, coolant pumps, power conditioner, and motor.

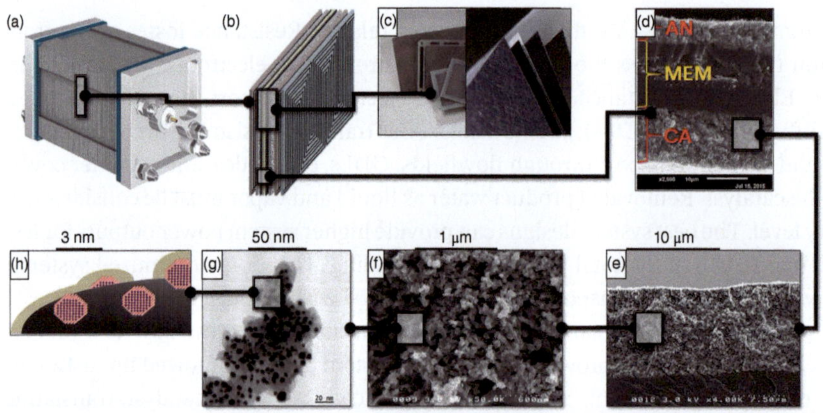

Figure 2.4 PEMFC stack (a) composed of many individual cells (b) of which the core MEA (c) is composed of GDLs, gaskets, and a CCM (d) containing anode and cathode electrodes (e) that are porous catalyst ionomer films (f) composed of metal on carbon catalysts (g) bound by nano-thick ionomer film where the catalyst metal ideally is an intermetallic arrangement of Pt and an alloying material such as cobalt (h).

The PEMFC stack is the core of the system. Shown in Figure 2.4a is a scalable power generation unit consisting from <10 up to >100 repeating cells, allowing the developer to readily scale to the appropriate power output, efficiency, volume, weight, and/or cost targets. These cells, Figure 2.4b, are also scalable in surface area from as little as 10 cm^2 for smaller applications (drones, portable power, etc.) to ≥300 cm^2 for automotive and heavy-duty-vehicle applications. These cells consist of a membrane-electrode assembly (MEA) as well as flowfields (FFs) to supply fuel, oxidant, and sometimes coolant across the entire active area. Typically, the MEA as shown in Figure 2.4c consists of anode and cathode GDLs, a catalyst-coated membrane (CCM), and additional gasketry. The CCM (Figure 2.4d) is composed of anode and cathode electrodes and the proton-exchange membrane (PEM). The anode and cathode electrodes are typically 2–15 μm thick porous layers, Figure 2.4e,f, composed of a catalyst mixed with an ion-conducting polymer. The PEM is typically an 8–20-μm thick membrane composed of ionomer and often an inert perfluorinated mechanical reinforcement, Figure 2.4d. Individual catalyst particles are typically composed of 2–5 nm PGM particles such as platinum, Figure 2.4g, dispersed atop 12–50 nm carbon support particles. A state-of-the-art PGM catalyst is an atomically organized Pt-alloy such as intermetallic PtCo, Figure 2.4h.

Design and quality control at every level and interface is necessary to ensure that millions of systems provide uniform gas, water, proton, and heat flow to and away from every catalyst particle in every cell spanning the entire multicell stack under every operating condition for their multiyear lifetimes.

The intent of this section is to provide an outline of the various materials and interfaces present in the PEMFC system and then demonstrate how these components are integrated to optimize efficiency, maximize durability, and minimize cost.

2.2 The PEMFC Catalyst

Platinum-based catalysts have been shown to be the most efficient for PEMFCs and PAFCs. Early PEMFC catalyst development drew significantly from PAFC that occurred between the late 1960s and the 1980s. During this time, United Technologies, Johnson Matthey, Tanaka, and many others extensively studied platinum ORR and HOR activity, performance, and stability. To enhance efficiency, PAFC developers evaluated platinum dispersed across various high surface area carbons [22–24] as well as alloys of Pt with cobalt, iridium, chromium, nickel, ruthenium, rhodium, and other metals [25, 26]. Indeed, UTC built and sold stationary >100 kW PAFC systems containing PtCoCr catalyst dispersed on carbon in the 1980s and 1990s [27].

With the reemergence of PEMFC technology in the 1980s, catalyst work accelerated. Driven by platinum's high cost ($13 and $73 per gram between 2000 and 2021) and limited supply [28], developments focused on attaining the maximum performance, efficiency (activity), and durability from a minimum amount of catalyst material. To that end, Pt dispersed as nanoparticles over different nanoparticle supports including carbon, conductive ceramics [29, 30], or as nano-thin films over inert organic whiskers [31] is used to maximize active surface area per gram PGM. Currently, carbon blacks are almost exclusively used as the conductive support. Common carbon types include Vulcan XC-72, Ketjen Black, Acetylene Blacks, and graphitized versions of these carbons – several of which are shown in Figure 2.5 (top), all of which have diameters ≤50 nm. Many others have been investigated but are too unstable (Black Pearls 2000), expensive (carbon nanotubes), or contain insufficient surface area for sufficient Pt distribution.

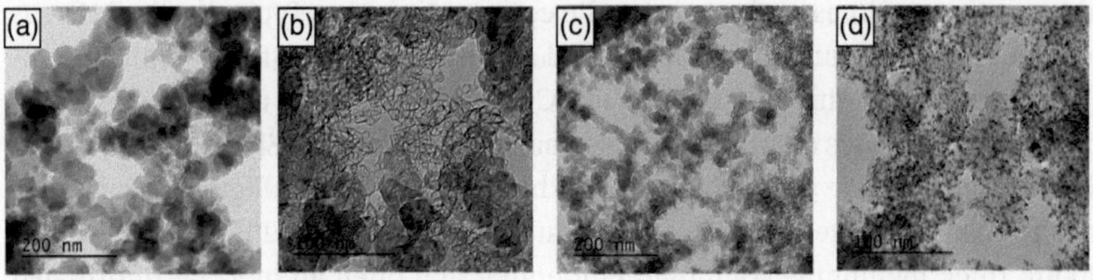

Figure 2.5 (a) Vulcan XC72 carbon, (b) Graphitized Vulcan XC72 carbon, (c) Acetylene Black carbon, and (d) 50 wt% Pt on Vulcan XC72.

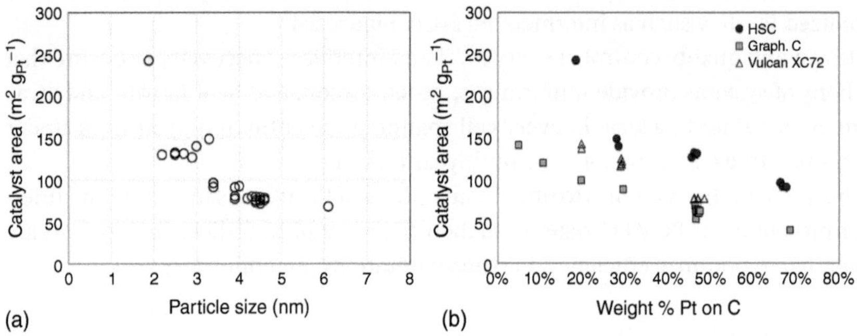

Figure 2.6 Platinum surface area as measured by CO adsorption (a) for HSC catalysts of different mean particle sizes (measured by XRD). Platinum surface area vs. various wt% Pt deposited over carbon support types (b).

Platinum is most commonly dispersed over nano-supports by mixing the support in chloroplatinic acid and continually reducing metal out of solution [32, 33] onto the support surface as nanoparticles until the desired weight percent metal (e.g. 20, 30, 50, and 70 wt%) is achieved. An example of 50 wt% Pt on Vulcan XC-72 (Tanaka 10V50E) is shown in Figure 2.5d. Pt nanoparticle size may vary from less than 2 to over 8 nm, as shown in Figure 2.6a. As the Pt particle size drops, metal specific surface areas can exceed 200 $m^2\ g_{PGM}{}^{-1}$ as measured by CO adsorption. This value translates to roughly 80–100 $m^2\ g_{PGM}{}^{-1}$ for the same catalyst measured by in-cell methods such as cyclic voltammetry of a cathode electrode in an operating PEMFC under saturated conditions. If desired, a heat treatment typically in excess of 600 °C may be applied to the catalyst in a reducing environment such as hydrogen [34]. This causes the metal to migrate and coalesce, increasing mean particle size. Overall, catalysts with in-cell measured surface areas of >40 $m^2\ g_{Pt}{}^{-1}$ are typically desirable. Higher metal surface areas are typically achieved on HSC materials (~800 $m^2\ g^{-1}$) as shown in Figure 2.6b as compared to Vulcan XC-72 (300 $m^2\ g^{-1}$), graphitized carbons (50–200 $m^2\ g^{-1}$), or other LSC materials. The greater carbon surface area may limit Pt particle "clumping" during deposition. Such clumping is most pronounced on faceted graphitized carbons of high metal loading where Pt particles preferentially sit on the facet interfaces [35]. Clumped Pt particles may decay faster as they need not migrate to begin coalescing and losing surface area during operation.

Because of the ORR's sluggishness, a cathode containing 8× more Pt than the anode will suffer 10–20× greater kinetic voltage efficiency losses – up to 400 mV as shown in Figure 2.1 [17]. For the ORR, the Pt surface structure offers the optimal combination of adequate oxygen adsorption binding strength [36] and relatively weak intermediate binding strengths of reaction intermediates and products. These factors work together to limiting catalyst surface coverage (i.e. poisoning) by intermediates, thereby presenting a maximum available surface for subsequent oxygen adsorption. This optimization is graphically represented by a Volcano plot [36] with aforementioned optimizations pointing to a "peak" catalyst activity upon which platinum is closest, followed by palladium. Pt particle surfaces are composed of {100} and {111} facets as well as step edges. It is well known that {111} surfaces are as much as 10× more active than {100} [37, 38]. Particles <2 nm contain predominantly step edges of low activity. Particles >5 nm may contain significant {111} faceting [39] but are relatively low in overall surface area. Pt-particle size effects have been well characterized in PAFC systems, where Pt specific activity (measured in Amps per catalyst particle area) increases by ~3× as particle size increases from 2 to 10 nm [24, 40]. Evidence of a higher concentration active facets is demonstrated in Figure 2.7a for Pt/HSC and PtCo/HSC. Unfortunately, however, as the particle size grows, the specific surface area drops, Figure 2.6. Thus, mass activity (specific activity × catalyst area/mass), which is directly correlatable to operational efficiency, peaks at cell surface areas between 40 and 60 $m^2\ g^{-1}$ correlating to 2–4.5 nm. This is consistent with the optimal particle size of ~2–4 nm that has been espoused for over 30 years [40, 41].

ORR kinetics may be estimated by a simplified version of the Butler–Volmer Equation [42, 43] known as the Tafel equation. PEMFC ORR Tafel slopes are logarithmically measured at typically 60–65 mV per decade [41] of current density. Based on the Tafel slope and assuming no transport losses, doubling the specific catalyst surface area (by doubling Pt loading and/or area) leads to a moderate 19–20 mV efficiency gain.

The pH in PEMFC electrodes ranges between −0.5 and +1. In accordance with the Pourbaix diagram for Pt [44] (a measurement of the Pt oxidation state as a function of pH and potential), a certain degree of dissolution to Pt^{2+} ions can occur as potentials rise above 0.85 V [45, 46]. This dissolution becomes relevant since

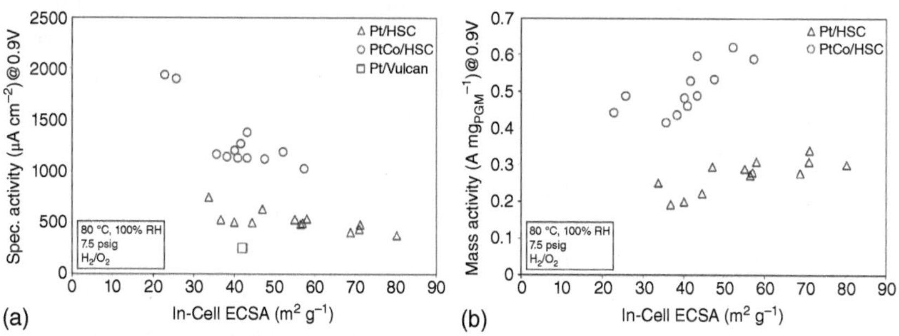

Figure 2.7 Cathode catalyst specific activity (a) and mass activity (b) as a function of electrochemical surface area (ECSA) measured for CCMs containing cathodes of 0.25 $mg_{Pt}\ cm^{-2}$, 3M800 ionomer, and I/C = 1.0.

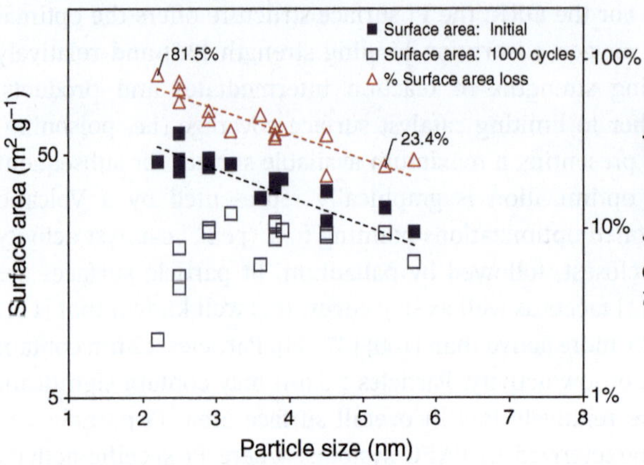

Figure 2.8 Metal ECSA loss rate vs. catalyst particle size: 0.25 mg_{Pt} cm^{-2} PEMFC cathode electrodes; 0.6–1.2 V triangle wave cycles, 80 °C, 100% RH, 100 mV s^{-1}.

cathodic potentials may exceed 0.9 V when the PEMFC system is near open-circuit voltage (for example an idling car). Over thousands of cycles above this potential, smaller Pt particles (<3 nm) are significantly less stable than larger particles as shown in Figure 2.8. A combination of higher surface area available for dissolution and an increased concentration of less-stable step edge sites leads to accelerated rates of surface area (and performance) loss. Support type has been shown to impact dissolution rates, with Pt on LSCs losing surface area faster than Pt on HSCs [18]. As shown by K. Moore [47], the close proximity of Pt particles on LSC enables coalescence, bypassing the additional metal migration (Oswald ripening) portion of the metal decay mechanism. Pt dissolution has also been shown to decay faceted Pt structures into more spherical forms [48], likely erasing the activity gains they offer. Further, a significant amount of Pt may, as Pt^{2+}, migrate away from the PEMFC cathode altogether and into the membrane [49]. For these durability reasons, in addition to the activity benefits stated above, the ideal mean catalyst particle size should be closer to 3–4 nm.

Because HOR kinetics on Pt are much faster than the ORR [16, 17], systems using pure H_2 fuel may use very low Pt-content anodes (0.02–0.05 mg cm^{-2}) and still achieve low overpotential (efficiency) losses. While much less common now than in the 1990s for emissions and efficiency reasons, some PEMFC systems still utilize reformed natural gas (reformate) fuel containing ppm levels of CO. CO at as little as 10 ppm can adsorb over >90% of all anode Pt sites under PEMFC conditions and subsequently reduce overall performance by over 100 mV [50]. Thus, for reformate, the anode electrode employs more CO-tolerant alloys such as PtRu at significantly higher loadings (≥0.20 mg_{PGM} cm^{-2}).

While carbon thermodynamically begins to oxidize above 0.207 V, carbon oxidation becomes significant typically above 1.2 V [51]. Improper cell and system operation may temporarily send anode or cathode potentials above 1.4 V, rapidly oxidizing the carbon support. Oxidation rates of amorphous carbons of HSC materials are roughly 10× faster than graphitized carbon [51], and these scale with surface area.

Further, dispersing Pt atop carbon accelerates the oxidation rate compared to bare carbon as Pt likely catalyzes carbon oxidation [51, 52]. As the carbon surface oxidizes, it becomes more hydrophilic. Carbon will ultimately oxidize to CO and CO_2. This loss of carbon can result in the catalyst cleaving from the support surface, resulting in performance loss. To limit this loss, catalyst carbon support choice must be properly mated with electrode and system operation.

Alloying Pt with additional metals has been shown to further increase ORR reaction efficiency, i.e. specific activity. Cobalt, nickel, iron, iridium, molybdenum, chromium, rhodium, gold, and other elements have been studied in binary and tertiary combinations with Pt [41]. Many Pt-alloys in various nanoparticle configurations have been evaluated to generate a Volcano plot [53] similar to that discussed for single-element catalysts with Pt_3Co, Pt_3Ni, Pt_3Fe, and other alloys showing enhanced activity over Pt 59. Faceted $Pt_3Ni\{111\}$ has demonstrated the maximum activity, achieving 10× activity gain over ordered Pt{111} surfaces.

Non-Pt catalysts such as Pd [54] and PGM-free catalysts [41, 55, 56] have also been studied and developed for over 40 years as a way to lower PEMFC cost. Significant work has been focused on the development of catalysts made from the pyrolysis of carbon, nitrogen, and metals such as Fe [57, 58]. While significant progress has been made, these materials do not yet approach the performance or efficiency of Pt-based catalysts [59].

The alloying process itself is not straightforward, and not all alloys offer the same performance and stability. For example, manufacturing an ideal PtCo catalyst requires several additional steps compared to Pt. First, Pt and an excess of the required Co are deposited on a carbon surface. Alloying then occurs at temperatures typically above 850 °C to ideally create an ordered "intermetallic" alloy of Pt and Co where the majority of Pt and Co atoms then arrange in an ideal lattice. Significant optimization of processing conditions is required to achieve an optimal particle size and minimize particle size distribution. Cobalt near the particle surface is then "washed" in an appropriate acid concentration at a set temperature and over an ideal time to produce a more durable, Pt-rich particle surface. Such intermetallic PtCo particles have demonstrated extremely high activities [60–62].

While Pt-alloys such as PtCo and PtNi have been shown to be more active, they typically decay more significantly in PAFC 28 and PEMFC cathode environments. Ni, Co, Ru, and other transition metals have different degrees of stability as dictated by their Pourbaix diagrams [44]. Unalloyed surface cobalt may begin to dissolve as soon as the catalyst is introduced to the ionomer. Further, as Pt dissolves during PEMFC operation, it will leave Co unalloyed and readily soluble. A Pt_3Co particle may lose >80% of its cobalt during its operating lifetime [27].

Alternately, combining Pt with iridium (typically as IrO_x, where $x \sim 2.0$) has been shown to increase catalyst and electrode stability [63]. IrO_x is thermodynamically much more stable, even at potentials as high as 1.7 V [44].

As the HOR and ORR are surface reactions, significant work has been done to replace the less-utilized core of the catalyst particle with cheaper metals [64–66]. These types of catalysts are known as core–shell catalysts. Core–shell catalysts have been made in which between one and five surface layers of the particle is Pt and

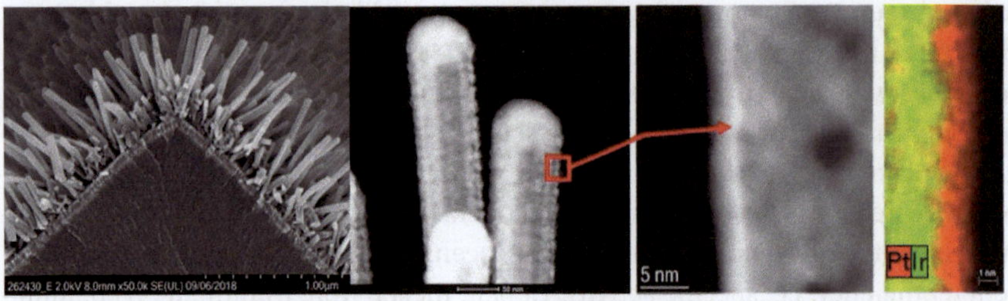

Figure 2.9 NSTF pyrelene red whiskers on Kapton substrate (left), NSTF coated with catalyst (middle and middle right). An example of layered Pt over Ir catalyst film (right).

inner layers are Pt-alloy, Cu, Fe, Pd, or other metals. Exceptionally high in-cell values >100 $m^2\ g_{PGM}^{-1}$ have been reached [66]. Additionally, certain core materials can contract the surface Pt–Pt distance to better optimize surface reactions. Indeed, core–shells such as Cu/Pt{111} place very high on the Volcano plot [53]. However, catalyst cores such as Cu, Co, and others are not inherently stable over long periods in a PEMFC, and may dissolve out, collapsing the particle, and resulting in significant loss of surface area and performance.

The more complex catalytic materials are increasingly difficult to scale to mass production. Ultimately, hundreds of kilograms per month of these nanoparticle metals on nanoparticle carbons must be made at low cost and tight specifications. Ensuring that a Pt/C catalyst is within a product specification likely requires meeting a particle size distribution (measured by XRD), a surface area measurement ($m^2\ g^{-1}$ measured by CO adsorption), and a carbon surface area measurement ($m^2\ g^{-1}$ measured by N_2 BET). A PtCo/C nanoparticle additionally requires specifying Pt:Co atomic ratio and alloying extent. A Pt core–shell material further requires ensuring shell uniformity over the multiple billions of trillions of nanoparticles contained in 10,000 FC vehicles. Such extra specifications add to manufacturing complexity and product cost.

Many alternate catalyst structures have been developed. Pt and Pt-alloys have been deposited on stable supports including conductive ceramics such as titanium oxide, niobium-doped titanium oxide, titanium nitride, and others [29, 67, 68]. Nanostructured thin film (NSTF) catalysts deposited on organic perylene red (PR) have shown unique promise as an alternate system. Here, PR is grown into nanowhiskers, and then catalyst is deposited upon them using physical vapor deposition (PVD) [31]. Pt, Pt-alloys such as PtNi, layered materials such as Pt over Ir, Figure 2.9 (right), have been made with high precision [69]. With this catalyst system, up to 30 $m^2\ g_{PGM}^{-1}$ in-cell, electrochemical area has been achieved with high activity and exceptional durability.

2.3 The Electrode

Electrodes are typically 2–15 μm thick porous layers. They contain a sufficient catalyst loading to achieve efficiency, power, and durability targets. This catalyst is intimately mixed with sufficient ionomer to transport protons to and from all catalyst sites. Electrode porosity must be sufficient to further transport gas to and water from all reactant sites. The goal of PEMFC electrode design is to optimally combine these core materials and properties to maximally utilize and maintain catalyst surface area over a wide range of operating conditions. In this way, electrodes can minimize precious metal loadings and reduce system cost.

Pt/C and Pt-alloy/C electrodes originated in the 1960s and 1970s for PAFC systems. Since PAFCs utilize liquid electrolyte, their electrodes require a specific combination of metal on carbon and polytetrafluoroethylene (PTFE). Sintering these materials at precise conditions creates an appropriate balance of hydrophilicity and -phobicity required to fill roughly 15–35% of pores with phosphoric acid – enabling ionic transport without compromising gas transport. In early PEMFC electrodes, ionomer was frequently coated on top of a similar version of this PAFC electrode, creating poor ionomer distribution and poor catalyst utilization. Thus, to achieve sufficient performance, 4–10 $mg_{Pt}\ cm^{-2}$ [16, 41, 70–72] was used. Then Wilson and Gottesfeld [73] mixed ionomer and catalyst with solvent as an ink. They then coated and dried the ink to form an electrode with drastically improved material distribution and catalyst utilization. This enabled lowering overall MEA loadings by 10× to ~0.8 $mg_{Pt}\ cm^{-2}$. By the early 2000s, Gasteiger demonstrated that anode loadings of 0.05 $mg_{Pt}\ cm^{-2}$ were sufficient when using H_2 fuel. Improved catalysts, ionomers, and ionomer distribution enabled sufficient performance and durability to be achieved with less than 0.2 $mg_{Pt}\ cm^{-2}$ [61].

PEMFC electrodes transport four distinct material phases (electrons, ions, gases, and liquid water) to and from the catalyst particle surface. Furthermore, this transport must be maintained for hundreds to ten-thousands of hours at temperatures ranging from −40 to 100 °C. They are relatively thin to minimize transport resistance and contain a combination of micro- and nanopores. Connected micropores are necessary for transporting reactant gas into and product water out of each electrode. Connected electrically conductive material is required to transport electrons between the GDL and the catalyst reaction sites. Finally, connected ionomer film enables proton transport between the membrane and electrode catalyst sites. Porous electrodes containing these phases are shown in Figures 2.4d–f and 2.10.

Electrodes are typically manufactured by first combining catalyst, ionomer, and solvent to form an ink. This ink is then coated, or sometimes sprayed, atop a substrate

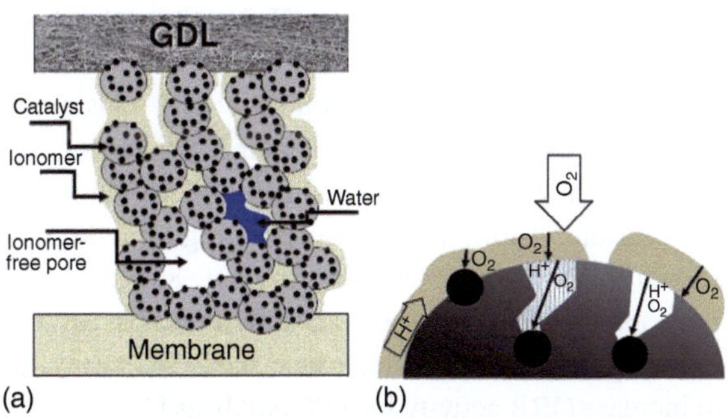

Figure 2.10 PEMFC electrode (a) and composite M/C representation of possible catalyst–ionomer interactions (b).

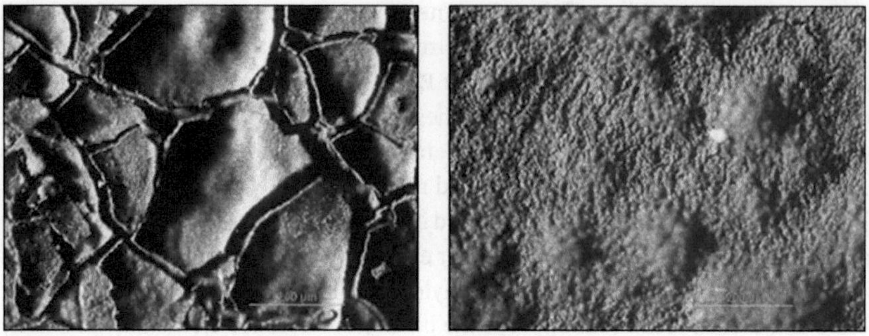

(a) (b)

Figure 2.11 Optical images of electrodes coated on a GDL substrate showing severe "mudcracking" (a) and no cracks (b).

such as a PEM, GDL, or release liner such as a Teflon film. For these electrodes, the amount of catalyst content, known as loading (in units of milligrams per geometric cm^2 or $mg_{Pt}\ cm^{-2}$), may be varied simply by adjusting the thickness of the electrode. Optimal electrode manufacture balances many variables including Pt or Pt-alloy type, metal particle size, wt% metal on carbon, carbon type, catalyst loading, ionomer type, ionomer-to-carbon weight ratio (I/C), solvent system, ink mixing method and duration, ink solids content and viscosity, coating method, and coating substrate. Examples of these variables and their impact will be discussed.

Examples of poor- and high-quality electrode coatings on a GDL are shown in Figure 2.11. Improper choices in solvent, substrate, materials, and processing can lead to a range of electrode problems from "mudcracking" (accelerating decay as discussed in Section 2.6) to yield loss, poor performance, and reduced durability.

Design choices can significantly impact cathode catalytic activity. Overall PEMFC activity and efficiency improvements may be achieved by increasing cathode Pt loading, utilizing increased surface area catalysts, and incorporating more active Pt-alloys. Further, Pt particles may exist largely in nanopores [35] of HSC supports, compared to the carbon surface for LSCs. Surface Pt is likely in contact with ionomer as shown in Figure 2.10b. Interactions between the ionomer and catalyst are well known to reduce Pt activity by 70% to over 90% [38, 74–77], reducing PEMFC performance as represented by Figure 2.1c. The protogenic end groups [78], and to some extent the side chain ether groups [79], of some ionomers coordinate with the Pt surface and reduce ORR activity. As shown in Figure 2.7, Pt on HSC is typically 2–4× more active because the Pt resides largely within the carbon support pores, limiting contact with ionomer [80]. This applies to Pt-alloys as well. As an example, by doubling catalyst loading, converting from LSC to HSC supports, and replacing Pt with intermetallic PtCo, activity gains of 10× (65 mV or ~5% efficiency) are possible. Additionally, incorporating ionic liquids into catalyst pores has been shown to limit ionomer poisoning [81, 82]. Further, some imide-based ionomers have shown gains in the H_2–air activity region (low current densities) [52]. Finally, reducing the ratio of ionomer to carbon (I/C ratio) in the electrode has been shown to increase ORR activity for LSC catalysts [76].

Without proper design, cathode electrodes incur increased transport resistances and eventual limitations as shown in Figure 2.1. Its pores, ionomer film, and carbon are tortuously connected, reducing transport rates vs. intrinsic material values. Commonly a Bruggeman correction $[(1-\varepsilon)/\tau)]^x$ is used [83, 84] in electrodes to correlate transport values with intrinsic material properties, where ε is porosity, τ is tortuosity, and $x = 1$–1.5. More rigorous studies further quantify this value adding limitations such as the percolation threshold [85], a minimum quantity required for transport. Further, gas transport through cathode pores is multicomponent [83] as O_2 diffuses through N_2 and product H_2O vapor.

There are many types of electrode transport losses that can become significant without proper design. The electrodes such as those represented by Figure 2.10a may contain significant amounts of ionomer-free pores (catalyst agglomerates). This is more common in HSC carbons where significant nonpolar carbon area and van der Waals forces lead to such agglomeration [86]. Poor proton transport through large ionomer-free agglomerates may limit catalyst utilization and activity [83, 87, 88]. Insufficient ionomer humidification may further reduce overall proton conductivity through the electrode, increasing transport resistance losses. Further, pores containing sufficient hydrophilic surface area (from metal particles, low EW ionomer, or oxidized carbon) can flood, also represented in Figure 2.10a. This can limit gas transport to reactant sites. Finally, oxygen and proton transport may be limited close to the catalyst particle. Figure 2.10b shows that the metal particle may be covered with ionomer, requiring oxygen to "locally" dissolve and diffuse through to the catalyst surface. This local transport resistance can be measured [89, 90]. As electrode surface area decreases and current densities grow, these local transport resistances become increasingly significant [66]. Catalyst particles within carbon pores are likely to suffer further limited protonic conductivity due to lack of ionomer present. Oxygen diffusion into the pore may compete with product water leaving the pore, increasing transport resistance. Finally, electrodes that are too thin may create excessive capillary pressure as high quantities of liquid product water are created in small volumes [83, 91] while electrodes overly thick (>15 μm) may incur additional losses from longer material transport distances [92].

At every level, transport within the electrode must be optimized. A sufficient I/C ratio is chosen to provide bulk proton transport while minimizing gas transport losses. Ionomer EW is chosen to prevent electrode flooding even at high current densities where pore capillary pressures may be high. Improved materials and processing may allow for decreased ionomer content, increasing gas transport through the electrode and near catalyst surfaces [93]. It has been proposed and shown from GISAXS [93] that at thicknesses approaching electrode ionomer films (<20 nm), classic PFSA ionomer does not phase segregate into micellular domains. For this reason, O_2 transport mechanisms (for example along the backbone) may be disrupted and reduced. Multiple solutions have been proposed, including developing ionomers with decreased density – especially as thin films [94, 95]. Ionomers of increased oxygen permeability are an area of active development [96], and many have demonstrated improved PEMFC performance [97–99].

Heat and pressure factor into transport as well. Protonic and gas transport (even through ionomer) increase with increased temperature [83, 100]. Liquid water may be significantly present at lower temperature (<50 °C) operation and can significantly impede gas transport by filling pores. This can be countered by utilizing more hydrophobic ionomers [52]. Increasing operating pressure from ambient to 2–3 atm can greatly increase oxygen concentrations, leading to enhanced transport and activity – but at increased system cost and complexity.

Electrodes must maintain high activity and transport over thousands of hours. Carbon support oxidation, discussed in Section 2.2, effects electrode operation in several ways:

(1) Carbon oxidation is a surface reaction that results in an increase in the concentration of surface oxides on the carbon [101, 102]. These oxides increase pore hydrophilicity, causing increased pore flooding and transport losses under cooler, wetter operating conditions;
(2) Cleaving catalyst metal from the carbon surface breaks the metal-to-support electrical contact, deactivating the metal particles. Electrode catalyst surface area is then reduced, reducing performance;
(3) Carbon ultimately oxidizes to CO_x resulting in electrode thinning and an increase in the I/C ratio. This oxidation, in combination with #1 and #2, imposes further gas transport limitations.

Again, the addition of IrO_x or alloys such as $IrRuO_x$ in the cathode electrode [103] can inhibit performance decay. These oxygen evolution reaction (OER) catalysts begin to significantly oxidize water as potentials rise above 1.3–1.4 V. The competing OER reaction acts to limit the cathode rise that may occur during start-up/shutdown style events (discussed in Section 2.7) and thus limit carbon oxidation [104–106].

Finally, operating conditions influence decay rates. Both carbon corrosion and metal dissolution rates [107] increase with increasing temperature. Carbon corrosion is very sensitive to temperature, with corrosion rates doubling roughly every 25 °C [51]. Lowering relative humidity has also been shown to limit metal dissolution rates [108] as shown by in-house data in Table 2.1.

Table 2.1 Impact of cell temperature and relative humidity on metal dissolution rates: 30 wt% PtCo/HSC, 3M800 ionomer, I/C = 1.0, 0.25 mg_{Pt} cm^{-2}.

Temperature (°C)	RH (%)	% ECSA loss
40	100	50
60	100	61
80	100	80
80	66	58
80	42	25

Accelerated protocol: 0.6–1.2 V, triangle wave, 100 mV s^{-1}, 2500 cycles, 80 °C, 100% RH, H_2/N_2.

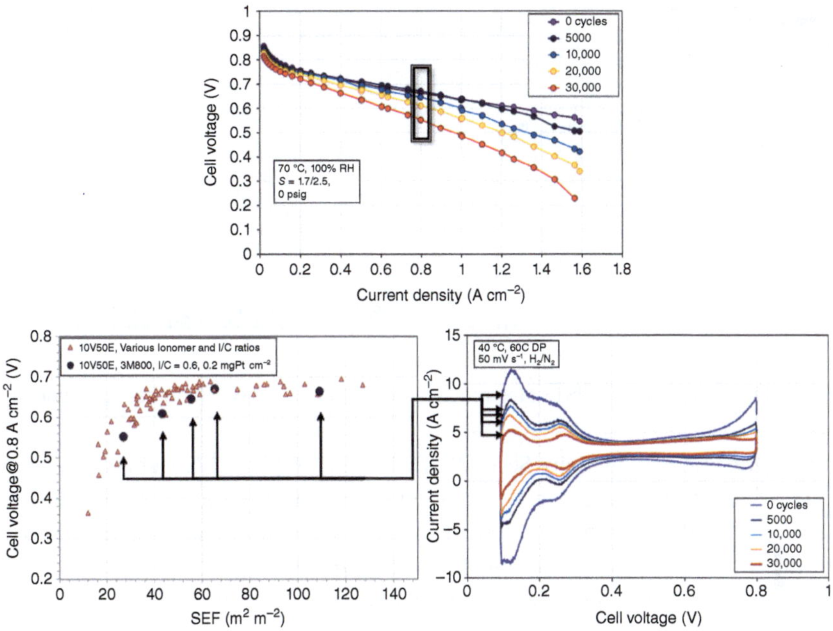

Figure 2.12 Performance impact (left, center) and catalyst surface area loss (center, right) of the DoE metal AST cycling protocol for a cathode electrode containing TKK 10V50E 50 wt% Pt on Vulcan catalyst, 3M800 ionomer, I/C = 0.6, and an initial loading of 0.2 mg_{Pt} cm^{-2} compared to alternate 10V50E cathode electrode formulations and states of decay.

To effectively iterate toward more durable electrode solutions, accelerated decay methods must be used. The US DoE has developed highly accelerated metal and support dissolution tests [109] known as accelerated stress tests or ASTs, allowing catalysts and electrodes to be evaluated in days instead of years. By rapidly cycling through naturally occurring decaying conditions such as 0.95 V (metal decay focused) and 1.5 V (carbon support focused) at elevated temperature, metal and carbon decay can be rapidly achieved and compared. Figure 2.12 (right) shows loss of cathode ECSA during the DoE metal AST. This leads to loss of activity and increased transport losses shown in Figure 2.12 (left). Comparing this cathode electrode to many similar formulations, it is clear that there is a minimum catalyst surface area (ECSA*loading) below which performance drops precipitously (Figure 2.12 (middle)). Here, it is believed that restricted oxygen transport local to the catalyst surface becomes dominant and limiting [110]. As well as limiting the rate catalyst surface area loss, lowering cathode I/C ratio has been shown to maintain high current operation at very low surface areas [52]. The standard NSTF electrode, Figure 2.9, in part by being ionomer free, demonstrates high current operation at exceedingly low surface areas [111].

For Pt-alloy cathodes, as Pt dissolution occurs, the transition metal alloy (cobalt for example) also dissolves. While Pt precipitates as cathode potential lowers, dissolved cobalt may remain an ion. Firstly, Co^{2+} ions complex with protogenic groups

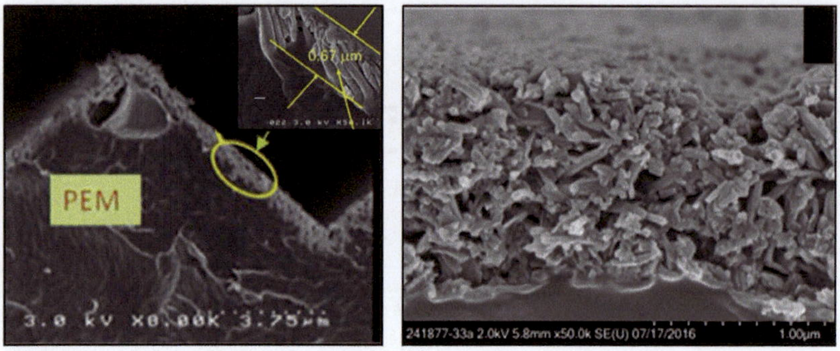

(a) (b)

Figure 2.13 Thin standard NSTF electrode (a) and dispersed NSTF + ionomer electrode (b).

reducing ionomer conductivity. Secondly, Co^{2+} ions complex with water can migrate like H^+. They migrate with complexed water to the cathode electrode reaction sites under load. Unlike H^+, which then reduces, Co^{2+} remains with its complexed water near the catalyst, leading to additional capillary pressure and pore flooding.

Many alternate electrode designs have been investigated. An NSTF PEMFC electrode shown in Figure 2.13a contains no carbon or ionomer and, at ~0.4 μm thick, is 5–30 times thinner than traditional electrodes. Because of this thinness, reactant transport distances are much shorter. Thus, only modest ionic conductivity is needed and is hypothesized to occur across the catalyst metal film [95] covering the "whiskers." Similarly, gas transport resistances and local transport resistance are minimal [112]. The elimination of electrode ionomer significantly reduces the local transport resistance compared to traditional electrodes. As the whiskers are hydrophilic, "flooding" of the cathode electrode with product water can reduce low-temperature operational efficiency [63]. More recently, NSTF powder-based electrodes have remedied this issue, Figure 2.13b, by integrating whiskers with more hydrophobic carbon and ionomer [52]. More recent designs utilizing Pt layered over iridium as the cathode catalyst have shown little or no metal or carbon dissolution-based decay while demonstrating exceptionally high activity [63].

2.4 Membrane

A PEM is a barrier to gas transport and electrical conduction and a transport medium for protons and water. It must operate for thousands of hours at temperatures between −40 and >100 °C, <30% to 100% relative humidity, and at pH <1. Additionally, it must withstand chemical degradation from hydrogen peroxide, hydroxyl radicals, and other reactive species. The ionomer solid polymer electrode (SPE) manufactured as a PEM is what uniquely defines the PEMFC in terms of operating conditions and limitations.

Membrane technology can be separated into (i) the ionomer material; (ii) membrane processing; and (iii) MEA integration (discussed in Section 2.6). Dupont was

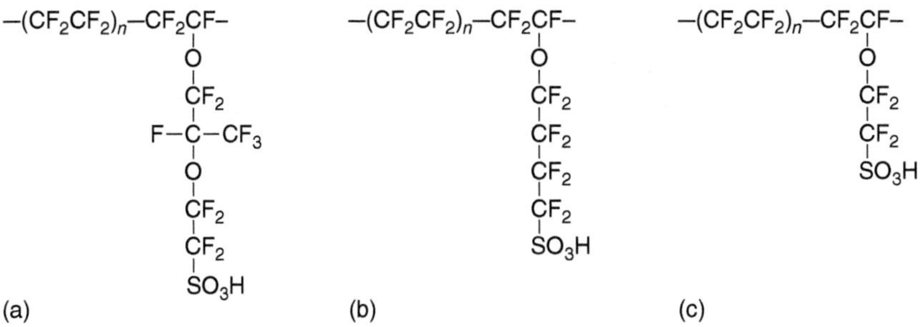

Figure 2.14 Chemical structures of the three main PFSA ionomers used in PEM fuel cells. Long (a), medium (b), and short (c) side chains.

the first to develop membranes in the 1960s under the ionomer tradename Nafion® [113, 114]. Dow [115, 116], Solvay, 3M, Asahi Glass [117], W.L. Gore, and others have developed and commercialized membranes with Nafion, Figure 2.14a, and similar ionomers with shorter side chains (SSCs) of two or four carbons in length, as shown in Figure 2.14b,c.

In the past 20 years, many additional ionomer types have been developed. Bis(sulfonyl) imides were first incorporated by Wainright [118] into the side chain (LSC) ionomer as shown in Figure 2.15a, and three months later this protogenic group with a sulfonic acid was incorporated into the side chain to create the multiacid side chain (MASC) ionomer perfluoro-imide acid (PFIA) [119, 120] as shown in Figure 2.15b, enabling lower equivalent weight ionomers. Two and even three bis(sulfonyl) imides were incorporated into the 3M MASC as perfluoro ionene chain extended (PFICE) variants [121], as shown in Figure 2.15c where Rf′ = 2–3. These achieved high protogenic group concentrations. Proton conducting group concentration in the ionomer is typically characterized by the equivalent weight

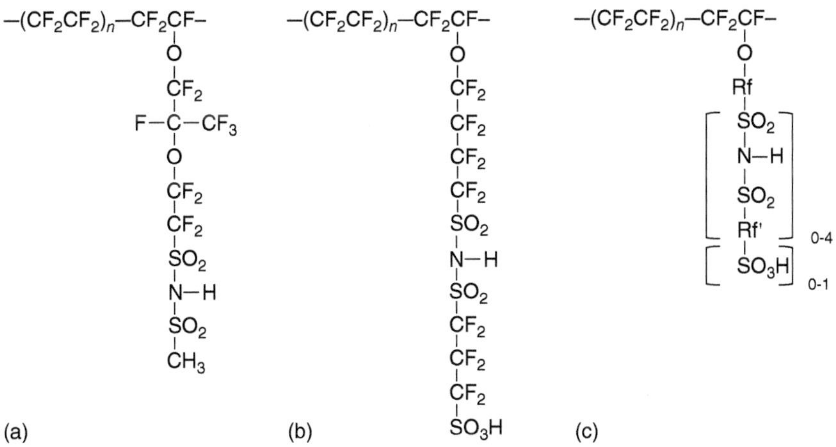

Figure 2.15 Examples of perfluoro bis(sulfonyl)imide ionomer (a), 3M's perfluoro-imide acid (PFIA) (b), and a generalized ionomer structure (c) that may contain several bis(sulfonyl) imide units in the side chain.

(EW) of polymer per mole of acid (g mol^{-1}) and sometimes by ion-exchange capacity (IEC, or 1/EW [122]). In order to increase oxygen permeability, nonprotogenic terminal groups have been incorporated into the ionomer bis(sulfonyl)imide side chain [123, 124], and cyclic structures have been incorporated into the backbone [97, 99]. Ionomers with aromatic backbones such as sulfonated polysulfone (BPSH) [125] have been developed with the goal of decreasing oxygen permeability. Other ionomer variations utilize branched side chains [126].

Ionomers may be cast from solution or extruded to create the PEM. Inherently, the ionic side chains form phase-separated domains to generate channels and micelles (on the length scale of ones to tens of nm) of protogenic groups [127–129]. Protons conduct through these tortuous channels by either the Grotthuss hopping or vehicular mechanisms [130–133], driven by ionic potential between anode and cathode. Increased water content in the ionomer typically increases ion conductivity. It is for this reason the reactant gases are humidified. In addition, lowering the equivalent weight will increase the protonic conductivity of the ionomer [127, 133] as shown in Figure 2.16. Overall membrane resistance may be further decreased by reducing its thickness. Many other factors such as processing, thermal history, film thickness, and others can influence this conductivity as well but are beyond the scope of this discussion.

The first PEMs were extruded to thicknesses on the order of 175 μm. In order to reduce overall proton transport resistance and increase performance, PEMs today are as thin as 10 μm [134]. A standard Nafion 1100EW ionomer sulfonate end group holds ~3 water molecules at 40% RH and about 12 at 100% RH [133, 135] leading to varying amounts of swelling during routine operation as shown in Figure 2.17b. Repeating swell and contraction from RH variations, especially in-plane, may generate mechanical stress [136] that can induce membrane damage. Over time, this repeated stress can lead to fatigue-based mechanical failure [137–139] in the form of cracks and tears – commonly against adjacent materials such as gaskets.

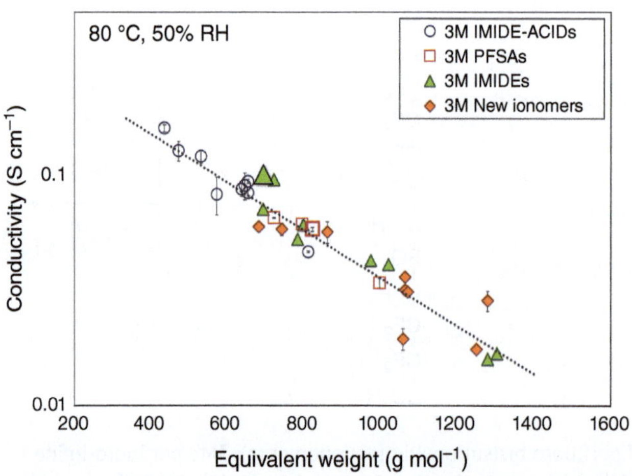

Figure 2.16 Conductivity vs. equivalent weight for 3M ionomers of various types.

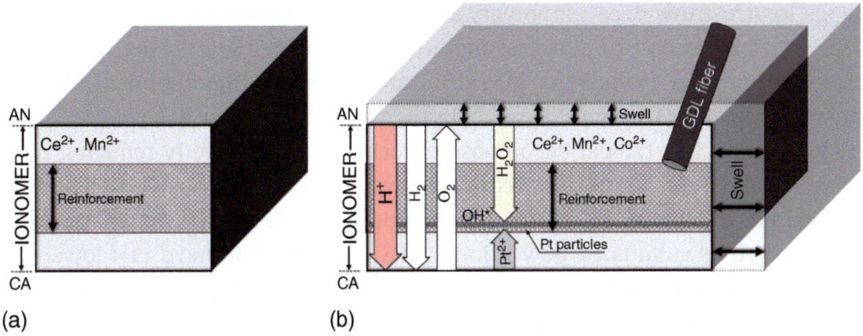

Figure 2.17 Material components of a manufactured PEM before operation (a). Materials within a PEM and mechanisms occurring within a PEM during operation (b).

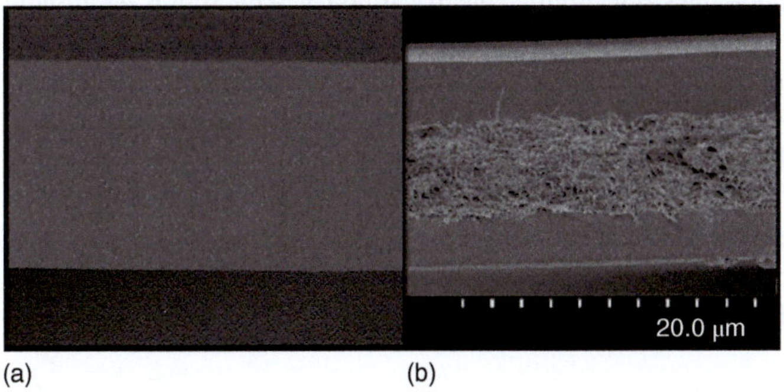

Figure 2.18 Unreinforced membrane (a). 3M membrane reinforced with electrospun fluoropolymer nanofiber (b).

As ionomer EW is decreased, ionomers retain more water resulting in more severe swell [140, 141] likely accelerating this mechanical decay mechanism. Processing, thermal annealing [142, 143], and the addition of mechanical reinforcement layers [107, 144, 145] such as that shown in Figure 2.18 are utilized to minimize in-plane swell [146] and increase membrane lifetime. These reinforcements [107, 145], such as Gore's expanded polytetrafluoroethylene (ePTFE) [147], act to limit membrane swell to the through-plane direction.

Ultimately, there is a lower EW limit for standard PFSAs. They become significantly water soluble at roughly 600EW, and while MASC ionomers demonstrate mechanical improvements in this range [52], this presently remains the lower limit of PEM EW and conductivity.

Further, ionomers can chemically decompose during PEMFC operation – the mechanism of which is described in Section 2.6. This mechanism ultimately generates hydroxy radicals within the membrane [133, 148] that can react with, and ultimately "unzip" the ionomer beginning at the end groups, or by cleaving the side chain most commonly at the $-OCF_2-$ groups. This results in thinning the membrane, compromising mechanical strength [149]. The weakened membrane ultimately fails by "pinhole" formation or tears often from the stress of humidity cycling. To arrest this decay, ionomers are typically modified by fluorinating end groups to reduce chemical attack at these locations [133, 150]. Further, additives such as cerium and manganese [151–155] are commonly incorporated into the membrane directly or by initially placing them in adjacent MEA layers 5. These inorganic ions act as radical decomposition catalysts, significantly reducing concentrations of damaging species. Reducing oxygen crossover has also been shown to reduce peroxide formation and chemical decay rates. Low oxygen-permeable ionomers such as sulfonated polysulfone (BPSH) – shown to have 11× lower O_2 crossover – have extended lifetimes in accelerated open circuit voltage membrane decay tests [156]. Additionally, chemical and mechanical membrane degradation occur simultaneously and together significantly decrease time-to-failure vs. either mechanism alone [137].

In addition to the durability implications of gas crossover, hydrogen fuel crossing over to the cathode does not produce electrical power. This fuel is essentially lost, lowering system fuel efficiency [108, 157]. Crossover increases as the membrane becomes thinner [158]. Ultimately this leads to a minimum membrane thickness of 10–15 μm [5, 159–162]. While hydrocarbon ionomer membranes such as BPSH [163] have shown reduced hydrogen and oxygen crossover, they come with the penalty of weaker protogenic group acidity and higher resistivity as a function of relative humidity vs. fluorinated ionomers [164–166]. Ultimately, like many aspects of PEMFC design, an optimization among conductivity, durability, efficiency, and cost must be considered when undertaking PEM design.

2.5 The GDL

The GDL is situated between the flowfield and catalyst layer as shown in Figure 2.19. It enables electrons, reactant gas, and product water to rapidly transport to and from catalyst layer. Similar to all other PEMFC components, it must be uniform, inexpensive, and durable. Unique to the GDL is the multilayer design necessary to serve these transport functions. Properties such as porosity (and pore size distribution), degree of compressibility, and combination of hydrophobicity and hydrophilicity are required to readily transport liquid water and reactant gas.

The modern PEMFC gas diffusion layer evolved from similar versions used in PAFC systems [167–169]. In the PAFC system, excess phosphoric acid electrolyte

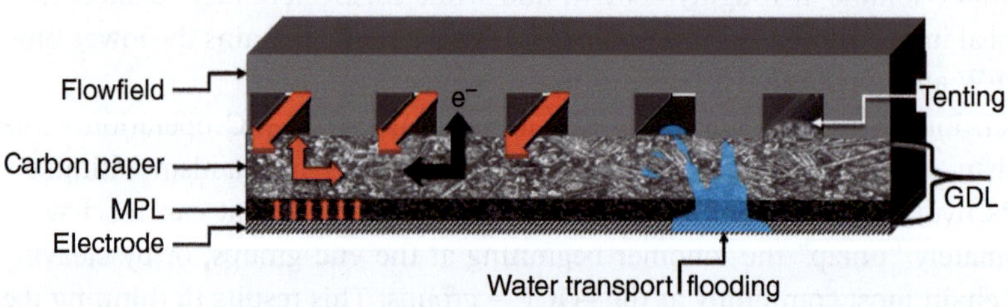

Figure 2.19 The PEMFC GDL is situated between the flowfield and electrode layers and is composed of a carbon paper and microporous layer (MPL).

was stored as a reservoir in the hydrophilic pores of a carbon paper. This reservoir enabled the replenishment of CCM electrolyte as it slowly evaporated from the system over its lifetime. PTFE, fluorinated ethylene propylene (FEP), polyvinylidene difluoride (PVDF), or other fluoropolymers [170] were added to both the paper and the attached ~50–100 μm electrode, and then the combination was sintered to achieve an optimal amount of open hydrophobic porosity. In this way, these PAFC gas diffusion electrodes (GDEs) achieved an optimal combination of liquid, gas, and electronic transport. The current PEMFC GDL is similar to this PAFC GDE in the method of preparation [171, 172].

PEMFC GDLs today are between 100 and 400 μm thick and typically composed of a carbon-paper "backing" layer and a microporous layer (MPL). The MPL purposely penetrates into (also known as overlaps) the backing layer to ensure electrical connectivity. The backing layer constitutes the majority of the GDL thickness while the MPL is typically 25–50 μm thick. While woven carbon cloth has been used for its exceptional flexibility and processibility, industry almost exclusively uses carbon paper (examples of which are shown in Figure 2.20) as the backing layer. These papers currently offer the best combination of thinness, in-plane and through-plane transport, uniformity, and processibility.

The carbon paper often begins as cellulosic paper made with some polyacrylonitrile fiber that is then converted into carbon at process temperature up to 1350 °C in a nitrogen atmosphere [173]. A small quantity of resin solution may also be added. This resin collects at fiber–fiber interfaces and, upon carbonization, likely improves conductivity at these connecting points as well as altering transport properties. GDL papers with different amounts of resin are shown in Figure 2.20 (top). Backing layer porosity is typically >75%, and carbon fibers range from 5 to 10 μm in diameter. A hydrophobic treatment, usually composed of 5–20 wt% [174–176] fluoropolymer, is applied and sintered to create optimal hydrophobicity.

For the MPL, a mixture of carbon particles (often graphitized nanoparticles or graphite flakes) are typically mixed with 10–35% fluoropolymer [170, 176–178] and

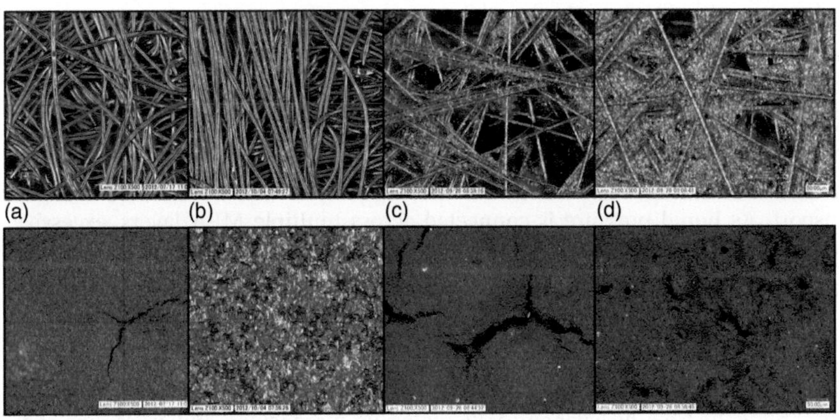

Figure 2.20 Carbon paper (top) and microporous layer (bottom) for (a) Freudenberg H2315, (b) Freudenberg H2415, (c) SGL 25BC, and (d) AvCarb MB30.

solvent to form a slurry. This is then applied atop one side of the paper and dried to form the MPL. Additional sintering may occur for this layer, or the paper and MPL may be sintered together at conditions similar to PAFC paper–electrode systems [178–180]. Examples of various MPLs are shown in Figure 2.20 (bottom), where (b) is unique in that it likely incorporates plate-like carbon. Many other additional materials and treatments have been investigated for this layer. Yang et al. [170] compiled an extensive listing of MPL materials including carbon particle types [181], graphene [182, 183] and carbon nanotubes [184] to alter porosity and tortuosity and impact gas transport. Alternately, SnO_2, TiO_2, and other oxides have been added to promote liquid water transport [185].

Much of the active catalyst layer is located beneath the flowfield (FF) ribs as shown in Figure 2.19. An FF rib may be 0.5–2 mm in width – 50–1000× greater than the electrode thickness. Thus, the thicker GDL must provide the material transport paths beneath these ribs. Proper GDL design involves choosing a minimum backing thickness to provide sufficient volume for reactant gas and product water not only to move through-plane to or from the flowchannel but also in-plane into and out of areas underneath these FF ribs.

Reactant gases and water vapor typically move via multicomponent diffusion frequently represented by the Stephan–Maxwell equations [186]. Porosity and tortuosity impact these transport rates. Thus, for the backing layer, a high degree of connected macropores containing hydrophobic fluoropolymer is ideal for rapid gas in- and through-plane transport. The added hydrophobicity acts to limit liquid water from residing in these macropores [187]. Hydrophobicity in the MPL is similarly critical, but because the mean pore size on the order of ~100× smaller [186] transport tends to be mainly perpendicular to the electrode as shown in Figure 2.19.

Liquid water typically transports through the backing layer via percolation [188, 189] and is affected by pore hydrophobicity and capillary pressure [190]. Achieving some degree of microporosity (<1 μm) that is fluoropolymer-free and less hydrophobic can be beneficial in the backing layer. In these micropores, reducing hydrophobicity enables liquid water to preferentially gather in and percolate through without compromising gas flow in the macropores. This strategy can similarly be applied to the MPL to aid in liquid water transport [191].

Hence, an optimal quantity of, and sintering conditions for, the chosen fluoromaterial is critical for both GDL layers [192, 193]. Too much overall pore hydrophobicity requires a high liquid (capillary) pressure to achieve percolation-based liquid water transport. As liquid pressure is connected across multiple MEA layers, excessive buildup of capillary pressure in the MPL may lead to flooding of the underneath electrode layer [168, 174, 177, 194] which contains less hydrophobic, ionomer-dominated pores. Oppositely, too little hydrophobicity allows easy liquid transport but, at low temperature operation, leads to excessive flooding and gas transport limitations [195].

By varying the backing layer and MPL materials, dimensions, and processing conditions, an ideal combination of liquid and gas transport can be tailored for specific locations in the GDL (anode vs. cathode). Similarly, these choices will also

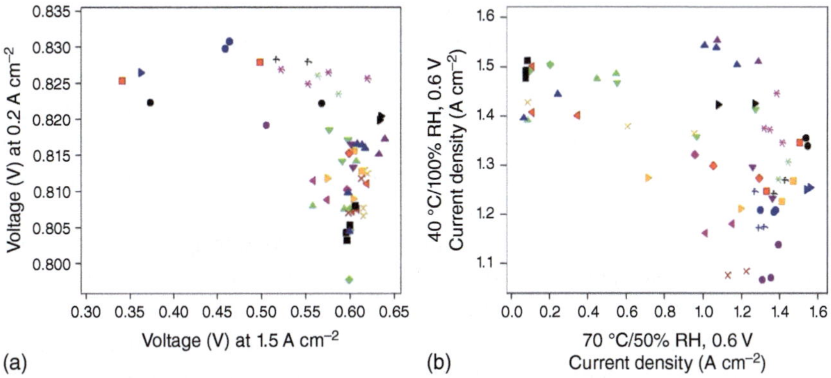

Figure 2.21 Performance vs. various cathode GDL types: 70 °C, 50% RH (a, *y*-axis) and 70 °C, 100% RH (a, *x*-axis) and 40 °C, 100% RH and 70 °C, low RH (b). Pt anode, PtCo/HSC cathode, 3M800 membrane.

have an impact on system operating conditions. Consequently, GDL product types vary significantly in performance in order to meet different operating environments. Figure 2.21b tests GDL product types for cell where the cathode operates under cold/wet (40 °C/100% RH) vs. hot/dry (70 °C/50% RH) conditions. An ideal cathode GDL retains enough water at low current density operation to achieve high efficiency (Figure 2.21a, *y*-axis)), while removing enough product water at high current densities to maintain high gas transport and efficiency (Figure 2.21 (left, *x*-axis)).

Further, anode and cathode GDLs are likely to be designed differently for the following reasons: (i) Hydrogen diffusion is roughly 6× faster than oxygen, (ii) The cathode utilizes air (21% O_2) compared to a typical anode of near pure H_2, and (iii) water production occurs on the cathode side. Because of these factors, much less volume is required to transport high concentrations of fast-moving hydrogen beneath anode flowfield ribs. Alternately, for the portion of the air-breathing cathode where the half the oxygen has already been consumed, and the operating conditions are 70 °C, 100% RH, and 0 psig, the GDL will receive gas from the flowfield containing just 7.2 vol% O_2. Because of the possibility of these low reactant gas concentrations, many cathode GDLs are designed to be thicker, more porous, and less tortuous in order to ensure efficient transport even when high power is requested.

In addition to gas and liquid material transport, the GDL must readily conduct electrons between the catalyst layer and the flowfield, enabling electrons to flow to and from the catalyst beneath the flowchannels or ribs as shown in Figure 2.19. Individual fibers of carbon papers may be more than 100 μm apart – 7–50× the thickness of a catalyst layer. Thus, the MPL improves electrical conduction between catalyst layers and paper fibers.

GDLs may be characterized by many methods. Porosity, pore size distribution, and tortuosity can be determined for both liquid and vapor transport [196]. Methods such as mercury porosimetry [197] can measure overall pore size distribution, while contact porosimetry [194] can distinctly evaluate distributions of hydrophilic and -phobic pores by utilizing partially wetting (water) and fully wetting (octane) liquids.

Further, newer imagery methods such as X-ray CT scanning can now take 3D images of GDLs even while in operation [196, 198] allowing real-time visualization of water generation and transport.

In addition to specifications required for optimal performance, GDLs require additional specific mechanical requirements. Firstly, GDLs must be flexible and robust to withstand the rolling and unrolling during coating, layering, bonding, sealing, and cutting. The backing layer fibers must be oriented in-plane since through-plane facing fibers may puncture the membrane [199, 200] during cell compression. These punctures inevitably lead to small short circuits that compromise performance and result in cell decay. Further, the backing layer must withstand 10–30% compressibility without fiber breakage through the operational lifetime. GDL compression ensures good fuel cell layer electrical contact and reduces resistance losses. Finally, under compression GDLs must exhibit minimal "tenting" [201]. Tenting occurs when a GDL is compressed and then bows into the flowchannel, as shown in Figure 2.19. This can restrict gas flow, leading to added flowfield pressure drop and flow nonuniformity.

There are several limitations to today's GDLs. While GDLs contain specifications on fiber diameter, density, porosity, and pore-size distribution, fiber-based materials are inherently random. This can lead to difficulties in achieving uniformity in thickness, pore size distribution, and other metrics. Further, GDLs are inherently expensive due to the high energy cost required for carbonization and graphitization.

2.6 CCM and MEA

The CCM, as shown in Figures 2.4e and 2.22a, is typically composed of a membrane plus anode and cathode electrodes. The MEA, shown in Figure 2.22a,b, is composed of the CCM plus anode and cathode GDLs and subsequent gasketry.

Critical to CCM and MEA optimization is integration. Choices in anode, cathode, and membrane layers, along with gasketing arrangements, can significantly

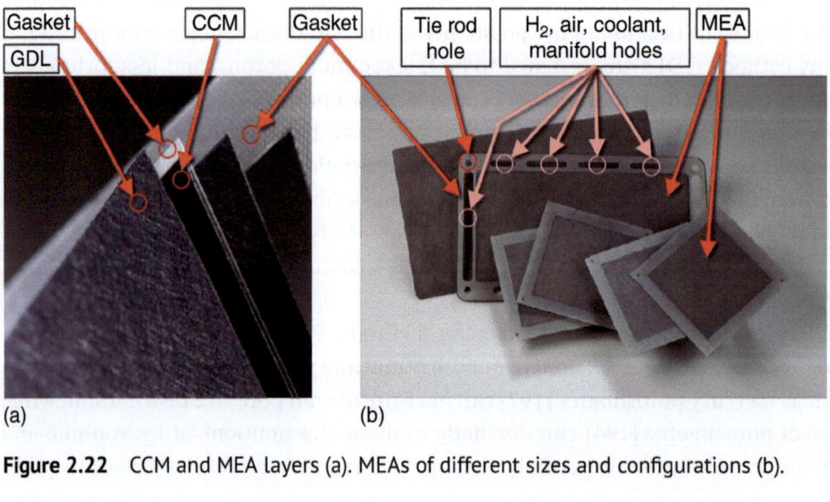

Figure 2.22 CCM and MEA layers (a). MEAs of different sizes and configurations (b).

influence performance, operational robustness, and durability [83, 202–204]. Additionally, between every one of these layers is an interface that must be optimized for performance, durability, and manufacturability. Interfaces may be between layers (through-plane) or where layers end or overlap (in-plane).

High CCM and MEA performance over a wide range of operating conditions (−40 to 100 °C and <25–100%RH) requires optimal water management [205–207]. Water may be produced at the cathode electrode at rates between <0.1 and >3.0 A cm^{-2}. Reactant gases may be dry or fully humified, ambient or pressurized. During start-up, operating temperatures may be <0 °C, resulting in cathode ice formation [208]. For all these situations, and in all layers, water in its various phases must be present but not significantly accumulate. Water must readily transport into, through, and out of the MEA to achieve optimal performance and durability.

For example, excess cathode electrode product water must exit the electrode to prevent flooding. Because the cathode also generates heat, both temperature and humidity gradients create the water concentration gradients enabling gas phase transport [83, 85, 209]. Lower operating temperatures reduce this mechanism by limiting maximum water vapor concentrations. At saturated vapor conditions, excess product water is removed as a liquid and thus is driven by hydraulic pressure. However, there are resistances to liquid transport into and through the adjacent layers (the membrane and MPL). Water moves through the membrane by three competing mechanisms: (i) electroosmotic drag, where water is pulled by protons across the membrane toward the cathode [210, 211], (ii) a liquid hydraulic pressure gradient can drive water in either direction, and (iii) a water vapor concentration gradient between the two sides of the membrane can drive water [83] in either direction across the membrane. Oppositely, liquid water leaves via percolation through the MPL after building sufficient pressure to surpass the percolation threshold (described in Section 2.5).

Within the CCM, material choices further influence the way water transports and its effects on performance. Incorporating lower EW ionomers, with lower water contact angles, into an electrode increases electrode pore hydrophilicity [212, 213]. Subsequently, at lower operating temperatures where liquid water is present, lower EW electrodes may more readily flood, leading to performance loss.

Additionally, the choice of a Pt alloy for the cathode (such as PtCo) vs. Pt further influences water transport in the CCM. As PtCo degrades, Co^{2+} ions are formed that can complex with the protogenic groups of the ionomer and reduce proton conductivity [214]. Water will associate with the Co^{2+} ion [215]. Concentration gradients may attempt to drive Co^{2+} toward the membrane. However, under load, like H^+, Co^{2+} will be electroosmotically driven to the cathode. However, unlike H^+, which is consumed, Co^{2+} and its complexed water remain in the cathode electrode, increasing the risk of flooding, reducing water transport toward the anode, and reducing ionic conductivity. Further, Co^{2+} has been shown to reduce local oxygen transport to cathode catalyst [216]. These effects all act to reduce cathode and cell performance, especially at higher current densities [217]. Cationic peroxide scavengers such as cerium and manganese ions act similarly to dissolved cobalt ions.

As discussed in Section 2.5, CCM and GDL choices must be made in unison to create the optimal MEA. Anode and cathode GDLs may be purposely paired to increase humidification of the CCM, enhance cathode water removal, and optimize reactant gas flow. For example, because hydrogen transport is fast relative to water vapor, a higher tortuosity anode GDL (resin-filled paper and/or graphite containing MPL) may aid in increasing anode humidification without significantly impacting hydrogen transport resistance. Further, anode GDL hydrophilic treatments may enable a capillary-based liquid pressure gradient to develop across the MEA. This gradient can then "pull" liquid water out of the cathode, across the membrane, and toward the anode. These choices may be combined with an "open GDL" on the cathode side composed of a carbon paper backing containing little carbonized resin or fluoropolymer and a low tortuosity MPL to minimize cathode oxygen transport resistance. The integrated effect of these layer choices is an MEA optimized for high performance under a wide range of operating conditions.

The aforementioned CCM and MEA layers must all be properly assembled through processes including multilayer coating, thermal lamination, and other means. Poor interfaces can add further resistance to both protons and water transporting from one layer to another (e.g. membrane to electrode). In the case of Figure 2.23, insufficient CCM lamination temperature resulted in lower performance during lower temperature operation. Poor cathode–membrane–anode interfaces may interfere with water removal through the membrane, increasing cathode electrode capillary pressure and inducing flooding-based cell performance losses. Additionally, significant work has been done optimizing GDL compression to minimize interfacial conductivity losses without compromising performance [218, 219]. Ideal GDL compression must also account for changes over time due to material fatigue and decay [185, 220].

CCM and MEA assembly must be tailored to their materials' characteristics. For example, internal work has measured the alpha transition temperature of Nafion 1000EW at 89 °C, compared to 100 °C for 3M800EW and 106 °C for

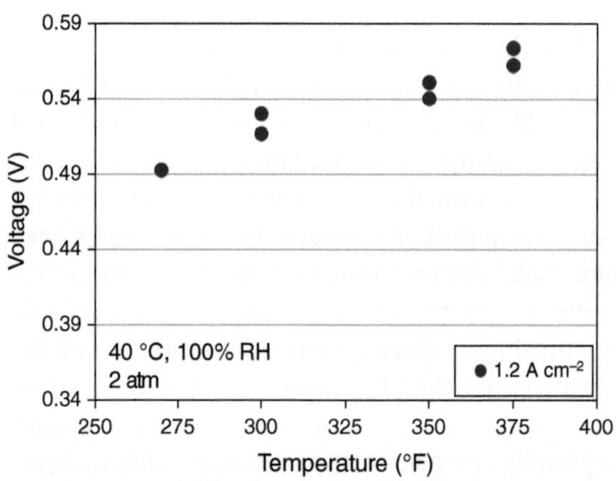

Figure 2.23 Impact of temperature on roll-to-roll lamination temperature on performance. Electrode and membrane contain 3M800EW. Performance: 40 °C, 100% RH, AN = CA = 2 atm_{abs}, stoic [AN/CA] = 2/2.5 H_2/air.

3M1000EW. An ideal membrane/electrode interface is one in which ionomer is continuous between membrane and electrode and not just physically attached. Thus, to achieve a more ideal interface using 3M ionomers, a higher bonding temperature is typically required.

Further, several fuel cell decay mechanisms develop across multiple layers. Membrane chemical degradation is such an example. Oxygen crossing over from the cathode electrode can form H_2O_2 at the low potentials of the anode/membrane interface [221] as shown in Figure 2.17b. A fraction of this H_2O_2 diffuses back into the membrane. There, it interacts with a unique "band" of platinum that can form within the membrane. This platinum has dissolved within the cathode electrode over time, migrated into the membrane, and then precipitated out at a location dictated by potentials created by hydrogen and oxygen crossover [222, 223]. At this platinum band, H_2O_2 likely reacts to form hydroxyl radicals [224–226] as shown in Figure 2.17 which then can break down the ionomer by mechanisms described in Section 2.4.

Additionally, electrode quality can accelerate the membrane decay mechanism. Electrode cracks as little as a few microns thick offer a direct path for cathode oxygen to reach the membrane [227], enhancing crossover rates and anode H_2O_2 formation. Added MEA layers may also inhibit crossover-based membrane decay. A proposed CCM solution to prevent H_2O_2 formation is to incorporate a low-porosity, ionomer-rich electrode sublayer between the catalyst and membrane to preferentially consume reactant gases before they might diffuse into and crossover the membrane [228].

Electrode cracking may lead to mechanical failure as well. Over time and cycling of operating humidity, membranes may "creep" into electrode cracks. These "pinch points" may accelerate mechanical failure of the membrane, leading to "pinholes" and electrical shorts. CCMs containing defect-free electrodes and membrane reinforcements are important in limiting such accelerated decay mechanisms. Additionally, a carbon paper fiber wrongly oriented toward the membrane may, over time and under compression, creep through the membrane causing an electrical short and a pinhole enabling direct gas crossover.

Finally, the typical cycling of humidification during operation leads to repeated shrinking and expanding of the membrane. As stated previously, RH cycling can lead to membrane fatigue and failure, often at "pinch points" where the edges of compressed GDL or gaskets press against the membrane. Appropriate CCM, GDL, and gasketry designs such as those shown in Figure 2.24 can mitigate these "pinch point" stresses by diffusing through-plane pinch. For example, gasketry can mitigate GDL compression-based pinch in a staggered way by utilizing thin (<10 μm) gaskets of differing dimensions as shown in Figure 2.24c.

MEA commercialization requires bonding millions of parts with areas >300 cm^2 and tolerances of <1 mm at six-sigma levels of quality. If just 5% of the product is lost at each MEA manufacture process stage – electrode coating and membrane coating, CCM manufacture, and MEA assembly – the overall cost can increase nearly 15%.

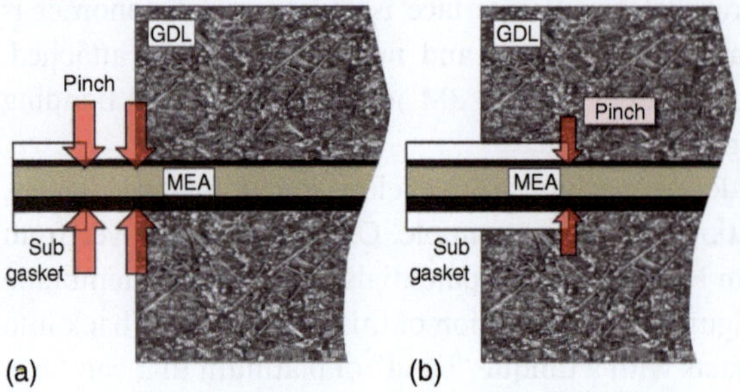

Figure 2.24 Different MEA subgasket designs. The CCM subgasket may not overlap the GDL (a), may uniformly overlap the GDL (b, a), or may overlap the GDL in a staggered way (b, c).

2.7 Flowfield and Single Fuel Cell

It is at the flowfield (FF) and single cell level where the largely 1D of through-plane transport discussed thus far meets the additional 2D of reactant, water, and heat distribution across the MEA active area, creating the 3D nature of PEMFC operation. The active area here is defined as the geometric catalyst-containing area in each cell where electrochemical reactions occur. This area may range from <5 to >300 cm^2.

An FF is an electrically conductive layer designed to uniformly distribute fuel, oxidant, or coolant across the active area of a single cell [229]. It further redistributes heat and water across the MEA to minimize flooding, under-humidification, and overheating across the cell. FFs transmit heat and electrons to the adjacent cells in the stack. In one FF variation, liquid water can be transported between coolant and gaseous FF layers to humidify gas streams and remove excess liquid [230]. FFs ultimately remove excess heat and product water from each cell of the stack. Thus, FFs must be electrically and thermally conductive, chemically and thermally robust, appropriately hydrophobic to prevent flooding of their channels with product water, readily manufacturable, inexpensive, minimal in height, and optimally designed for their system. A single, or unitized, cell consists, Figure 2.25b, of the MEA, anode and cathode FFs, seals and gasketry, and often a coolant FF.

FFs are typically composed of stamped metal with passivated coatings, carbon composite, or even porous carbon composite [229]. The flowfield covers the entire cell active area and connects to gas or coolant manifolds. Gasketry and/or sealing adhesives prevent overboard gas leakage and "unitize" flowfields and the MEA into a single cell. FFs are the most volume- and weight-intensive repeating component of the fuel cell stack. In applications such as drones and automobiles, minimizing space and volumetric power density is important. Thus, these applications use thin FFs that are typically based on stamped metal. Stationary and backup power systems, which have less stringent weight and volume targets, have additional flowfield choices.

Reactant gases and coolant (water) are introduced to each FF from manifolds designed to be within the FF, Figure 2.25 (coolant), or external to it, Figure 2.25

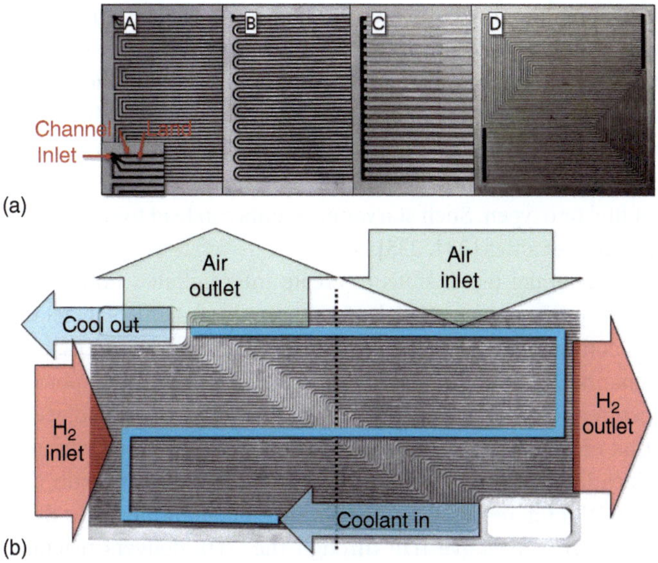

Figure 2.25 (a) A quad-serpentine lab-scale cell (A), a bi-serpentine (B), straight-channel flow (C), and nine-channel, narrow width flowfield (D). A large-scale flowfield with internal manifolding for coolant (b).

(hydrogen). As shown in Figures 2.19 and 2.25, many FF designs are composed of ribs and channels that act to direct flow. These FF designs may be serpentine, straight channel, or a mixture thereof. Channel-less designs include pin-type [229, 231] or quite exotic [5, 232]. Several levels of FF design must be concurrently optimized to cell size, power output, and system operation: (i) individual FF channel/rib/pin design, (ii) individual FF design, and (iii) the interaction of anode, cathode, and coolant FFs together.

A good example of this design process is the cathode FF. Here, air typically enters at stoichiometries of 1.8 to >3.0× that of the requested current density. For rib/channel-based FFs, oxygen is consumed along the flowchannel path, while product heat and water accumulate. This leads to variation in temperature, pressure, humidity, and oxygen concentration across the MEA. An ambient pressure system with no coolant may limit the variation in these properties by increasing air flowrate to a stoichiometry of >5.0, increasing evaporative cooling to limit maximum cell temperatures and to limit cell oxygen concentration variation. Alternately, the overall cathode pressure may be increased to between 200 and 350 kPa_{abs} while keeping the flowrate at 2.0, thereby increasing both inlet and exit oxygen concentrations. High flowrate solutions require a larger flowchannel design shown in Figure 2.25c in order to prevent excessive pressure drop, while the pressurized system may use narrower channels like those shown in Figure 2.25d.

FFs must be designed for an optimal pressure drop. Reactant gases move along the flowchannels, beneath the ribs, and through the GDL by pressure drop. Resistance to gas and coolant flow must be overcome by increasing inlet pressure. Fans, or more energy-intensive compressors, generate the necessary inlet pressures to "push" these fluids through the flowfield. Liquid water may collect in the anode flowfield as reactant hydrogen is consumed and in either flowfield from product water generation. Hydrophobic FF coatings and hydrophobic GDL surfaces lead to droplet formation in and along flowchannel walls. Without significant flow and pressure, droplets may become slugs [233], blocking gas flow. This can lead to nonuniform flow from channel to channel, and ultimately regions of MEA active area become starved of fuel or oxygen. Such starvation events can lead to significant performance and durability penalties [234, 235].

Additionally, designing channel dimensions to create interdigitated flow (IDF) can significantly improve performance. IDF is where some amount of reactant gas preferentially flows convectively beneath flowfield ribs, through the GDL, and into an adjacent flowchannel. Combining narrow flowchannels and thick, porous GDLs (Figure 2.25d) can reduce gas pressure drop beneath an FF rib to an adjacent channel to less than that created as gas moves along a lengthy serpentine channel to the same adjacent location. In this situation, some convective IDF beneath the FF rib and through the GDL preferentially occurs. Orders of magnitude faster than diffusion-based gas transport, convective IDF through the GDL delivers reactant gases to the MPL with minimal transport resistance losses. Similarly, product water can be removed more effectively from beneath FF ribs by this convective flow. IDF also can be mechanically created by "dead-ending" alternating parallel gas flowchannels, forcing flow to pass through the GDL to the adjacent channel to complete the flowfield path.

In comparison to the cathode, the anode FF requires smaller flowchannels as hydrogen enters the flowfield at a high concentration. Designs are coordinated with the system choices to ensure some pressure drop occurs across the flowfield. Reasonable pressure drop ensures uniform flow from cell to cell and also that condensed water does not block individual channel flow.

Coolant flowchannels serve primarily to redistribute heat across the cell and remove excess heat. They can carry away heat accumulated by reactant gas flowfields and then bring that heat to warm the cooler reactant gas inlet regions. Higher power applications generate more heat and thus often require higher coolant flow that must be accommodated by adjusting flowchannel size and path design.

As an example, at FF inlets, dry reactant gases enter the FF and initially pass over the MEA. In this region, they evaporate product water, cooling the MEA. The reactant gas proceeds to carry product heat and water across the cell, building temperature and humidity. Coolant similar carries waste heat through and out of the cell. By sending in hydrogen, air, and coolant along optimal paths as shown in Figure 2.25b (i) the heat capacity of the coolant can be used to deliver heat to the gas inlets to prevent excess evaporative cooling, (ii) the humidified gas exit flow can provide humidification opposite a dry inlet gas inlet to increase membrane humidification, (iii) localized regions of excessive heat (hotspots) can be prevented, and (iv) product water can efficiently be removed without excessive flooding or flow maldistribution. Without proper thermal management (i) temperatures in large cells may exceed even 100 °C, rapidly increasing catalyst and membrane decay, and (ii) channel flooding may occur as a cold inlet coolant passes over a hot, saturated reactant, which may lead to flow maldistribution and loss of power.

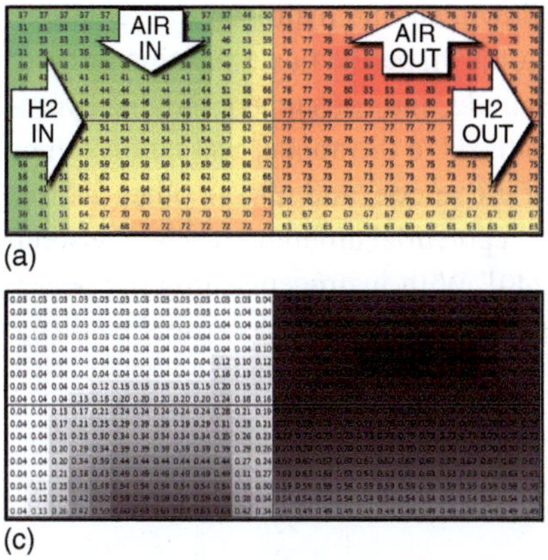

Figure 2.26 Hypothetical operating conditions for an alternate design of the single cell shown in Figure 2.25. The temperature (a) and humidification (b) profiles are used to estimate the fraction of cathode catalyst surface area loss (c) for a 2.5-nm Pt particle Pt/C cycled 1000 times at 80 °C, 100% RH from 0.6 to 1.2 V, and 20 mV s^{-1} in H_2/N_2.

Besides these more extreme cases, variations in pressure, temperature, and humidification across the active area will locally affect rates of metal dissolution, support corrosion, gas crossover, ionomer chemical attack, membrane mechanical stress, and other decay mechanisms. This is best demonstrated by the hypothetical example presented in Figure 2.26. Here, the hydrogen FF is single pass. The air FF is a two-pass system that is remixed in an external manifold before being fed to the second pass. Further, there is no coolant FF present. The cell is operated at a moderate average current density. Dry inlet gases will cause evaporative cooling of product water resulting in MEA conditions as low as 31 °C and 27% RH near the inlet regions. Hydrogen and air in the FF accumulate heat and product water, warming and humidifying regions further down the flowchannel paths. A hotspot near the flow outlets reaches 83 °C and 100% RH for this case. Extrapolating the metal dissolution rates vs. *T* and RH from Table 2.1 to the conditions of this hypothetical cell reveals <5% catalyst surface area loss near anode and cathode inlets compared to >80% surface area loss at the hotspot. Such variation in material decay rates may significantly accelerate overall cell performance loss and reduce time to failure compared to a more optimal single cell design.

Severe decay may arise from the channel flooding described earlier. Water-filled anode FF channels may block fuel from reaching local regions of the anode MEA, creating fuel starvation [236]. Under the load requested by adjacent cells, and without fuel, local anode overpotentials will rise to >1.5 V where the next oxidizable material, carbon, will electrochemically oxidize at rates sufficient to provide the requested current. In this situation, carbon may be consumed in just minutes, severely damaging the anode, and drastically reducing performance. As previously stated, iridium can be incorporated in the anode to mitigate these decay mechanisms [237]. However, this mechanism can best be eliminated by proper cell, stack, and system design to prevent fuel starvation situations from occurring [238].

In addition, when a PEMFC stack is shut down, air and hydrogen can remain in the flowchannels and MEA. Over extended shutdown periods, external air may enter the cathode, cross over the membrane to the anode, and consume remaining hydrogen. Upon start-up, when hydrogen is reintroduced to the anode, a situation can be created where hydrogen is in one portion of the anode, and air the other. At this H_2–air front, and with air on the cathode side, a potential differential arises between the H_2 and air portions of the anode [239, 240]. With hydrogen setting the potential at near 0 V, the air portion of the anode can reach ~0.6 V. Because the voltage

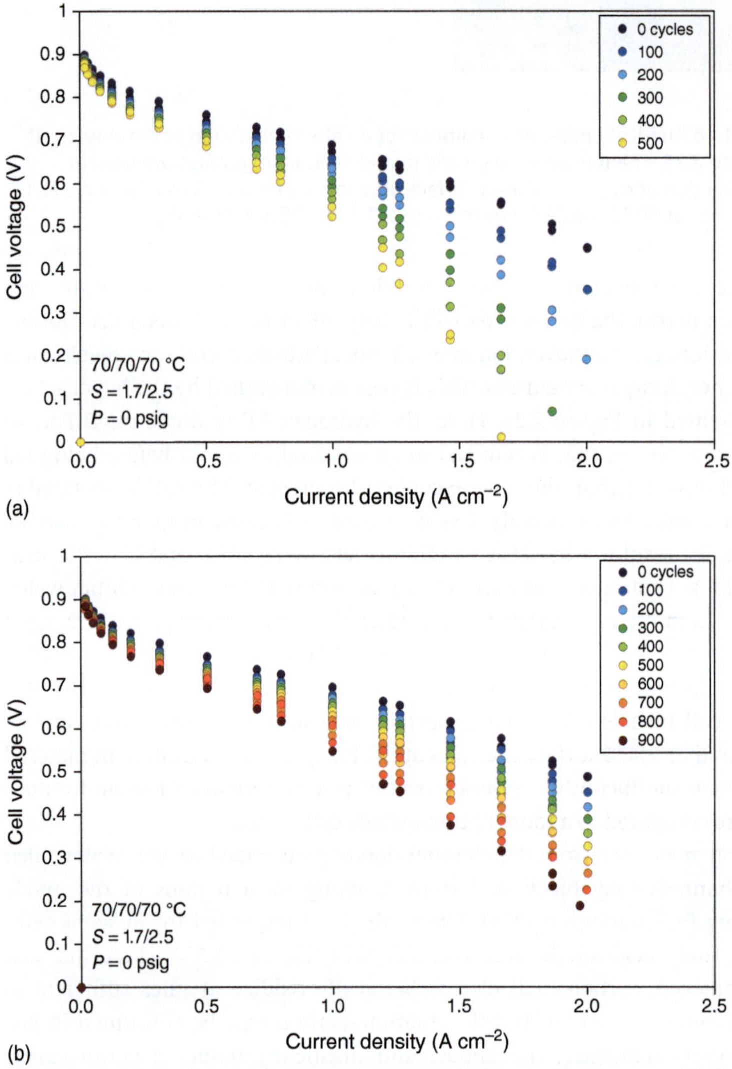

Figure 2.27 Performance after SUSD cycles carried out at 70/70/70 °C and 0 psig for a cathode containing (a) 0.25 mg_{Pt} cm^{-2} of PtCo/HSC and (b) 0.25 mg_{Pt} cm^{-2} of PtCo/HSC with 0.01 mg_{Ir} cm^{-2}; 50 cm^2 active area; 25 sccm air is fed into the anode during shutdown, and 800 sccm H_2 during start-up to simulate H_2–air SUSD fronts.

between H_2 and air across the membrane must be set by the anode-H_2-to-cathode-air typical voltage of ~0.9 V, a cathode potential across from the anode air regions may reach ~1.5 V, causing cathode carbon support corrosion [241]. Over hundreds of such unmitigated "start-ups," cathode decay occurs leading to significant loss of performance over time. Iridium, this time incorporated into the cathode, can again mitigate this decay mechanism by promoting water electrolysis to compete with carbon support corrosion [104] as shown in Figure 2.27.

2.8 Stack and System

Individual cells are typically sized and stacked to achieve system power requirements. For example, stacking fifty 300 cm^2 cells operating at 1.0 A cm^{-2} at 0.67 V/cell provides roughly 10 kW of power at roughly 54% efficiency. Thus, 8.5 kW of waste heat is generated that must be removed from the stack and system. For the same cell design system, a 500-cell stack would produce 100 kW, enough to power an automobile. Stacked cells are aligned and held together using tie rods connected to rigid endplates. Internal manifolds such as those shown in Figures 2.17 and 2.28 will extend the entire height of the stack. External manifolds similarly extend the length of the stack. Uniform flow across all cells is maintained by (i) ensuring that the pressure drop across each manifold is low (typically <10%) compared to that across flowfield and (ii) ensuring uniform pressure drop across all cells. A combination of gasketry like that shown in Figure 2.17, adhesive sealing, and compression prevents leakage out of the perimeter of individual cells and "overboard" into the larger system compartment.

Systems are designed to the size, weight, and volume requirements of the application. Stationary, backup power, and cargo shipping allow for larger volume and weight footprints. Here simpler, unpressurized systems may be utilized – especially if low-power outputs (<1.0 A cm^{-2}) are required. Automotive, heavy-duty vehicle, and drone applications have very tight volume and weight requirements and may utilize thin stamped-metal FF plates. Automotive and heavy-duty trucks with 100–300 kW power requirements may operate at reactant pressures of 2–3 atm to increase power and efficiency, enable uniform cell-operating conditions, and reduce cost [242]. "Motive" applications such as forklifts requiring 5–40 kW power have system requirements that lay in between the automotive and

Figure 2.28 Rendering highlighting internal air and hydrogen manifolds extending through a 16-cell stack.

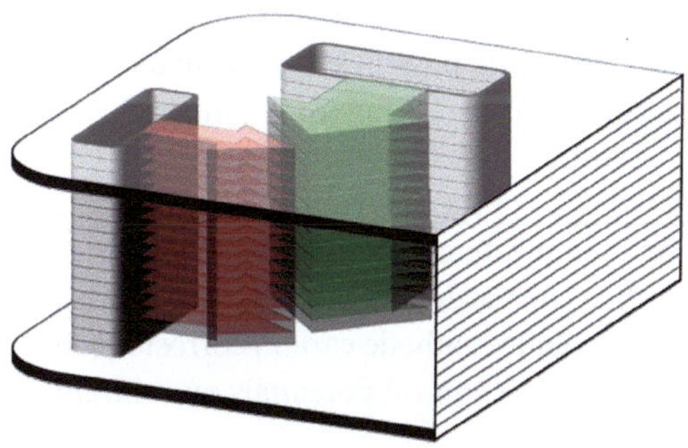

stationary extremes. Stationary, HDV, and shipping applications require 25–100 kh lifetimes while drones and backup power may be significantly less. Finally, some systems such as submarines and stationary units may utilize hydrogen-rich reformed fossil fuels (also known as reformate). Finally, stationary, motive, and automotive applications must design for pollution and debris. SO_2 and CO can poison PEMFC anode catalysts at <10 ppm.

Further, fuel exhausted from the system equates to lost fuel efficiency. Many systems recirculate exhausted fuel back to the anode inlet to reduce lost fuel [242]. Other systems "dead end" the anode by closing the anode exit and essentially operating most of the time at 100% fuel utilization [243]. In both cases, nitrogen from cathode air can cross over the membrane and increase in concentration (to >30%) in the anode. Thus, routine anode purging is designed into these systems – flushing out anode reactants at regular intervals with fresh hydrogen. In this way, only a few percent of hydrogen may be lost to exhaust.

These system requirements factor into cell size, precious metal content and type, number of cells, FF design, and types (coolant plates every cell, every *n*th cell, or none). Ambient pressure stacks operating at lower current densities ($<1\,A\,cm^{-2}$) may specifically be designed to not need a coolant system. Eliminating coolant plates, pumps, exhaust water condensers, and valves greatly simplifies the system. High air flow enables the necessary stack cooling to be achieved solely by product water evaporation. Reduced decay rates from these cooler conditions also may lower catalyst loading and cost. This design choice must be weighed against the thicker cathode flowfields and lower power operation, resulting in higher stack volume per kW. In contrast, an HDV may be required to generate >250 and 100–250 kW of heat in a limited volume. To achieve this power density, the system may operate at high pressure to limit stack size and at ≥90 °C at high power to limit the radiator size required to dissipate the heat [244]. To achieve 25 kh lifetimes while undergoing 1000s of start-up–shutdowns and lakhs of idle and acceleration cycles, HDVs further require very durable catalysts at higher cathode electrode loadings [244]. High power, low system volume applications like automobiles also require elevated pressures and advanced cooling systems to achieve power, efficiency, and cost targets. At 2–3 $A\,cm^{-2}$ under acceleration, a 300-cm^2 active area cell may produce 1 kW of heat per cell. Some stack space may be saved by inserting coolant flowfields every second or third MEA, but at the cost of nonuniform temperatures (and decay) cell to cell.

For the reasons above, cell-to-cell performance within a single stack can vary [245, 246] by over 8% [247]. Additionally, designs can lead to cool reactant temperatures near endplates or to fuel maldistributions reducing performance in particular stack locations [248]. Cell performance variation can grow as stacks age as well [249, 250], likely arising from temperature and humidity variations.

The stack may fail when the first cell fails. Thus, system strategies to both mitigate cell decay mechanisms and prevent accelerated decay of a single cell are critical. For example, to complement material strategies such as cathode iridium addition, methods of starting up and shutting down PEMFC stacks can be employed to prevent or reduce cathode carbon corrosion [251]. For example, voltage "clipping" can be utilized to control potentials and minimize high cathode potentials [107, 235].

Similarly, as carbon corrosion is temperature sensitive, ensuring that hydrogen–air anode fronts occur only under cool conditions can minimize carbon oxidation rates.

Loss in performance of one portion of a single cell can bring down stack performance. For example, if one fraction of a cell cathode is denied oxygen, it will produce no current. As a result, no electrons are propagated from that portion of the anode to an adjacent cathode. Therefore, that section of adjacent cathode will produce no current either, and so on. Similarly, the failure of a single cell can mean the failure of an entire stack by interrupting the propagation of current from cell to cell.

Thus, quality control of stacks and surrounding system components is critical to manufacturing thousands and millions of fuel cell products. Achieving six-sigma quality effectively means 3.4 failures per one million products. Because a single cell failure can cause an entire stack to fail, achieving six sigma for a 500-cell stack means *one cell failure per 170 million cells.* Every square millimeter of every electrode, every atomically oriented catalyst, catalyst particle size, and catalyst particle distribution over the support; every ionomer lot equivalent and molecular weight, and every electrode formulation and loading; every square millimeter of CCM and every GDL fiber orientation; every flowfield channel, every seal material and application, and every single cell build; every stack assembly process and compression; and every system pump, radiator, software, fuel tank, valve, and every sensor *must function to a specified lifetime* for (in the case of automobiles) aggressive drivers, city drivers, ocean side (salt water in air) drivers, cross-country drivers, high-altitude drivers, high-speed drivers, and for drivers who leave their cars parked outside at −40 °C.

Acronyms

AFC	alkaline fuel cell
AST	accelerated stress test
BPSH	sulfonated polysulfone
CCM	catalyst-coated membrane
DoE	Department of Energy
ECSA	electrochemical surface area
EW	equivalent weight
FEP	fluorinated ethylene propylene
FF	flowfield
GDE	gas diffusion electrode
GDL	gas diffusion layer
HDV	heavy-duty vehicle
HOR	hydrogen oxidation reaction
HSC	high surface-area carbon
IDF	interdigitated flow
IEC	ion-exchange capacity
LSC	low surface-area carbon
MASC	multiacid side chain
MEA	membrane-electrode assembly

MPL	microporous layer
NSTF	nanostructured thin film
ORR	oxygen reduction reaction
PAFC	phosphoric acid fuel cell
PEM	proton-exchange membrane
PEMFC	proton-exchange membrane fuel cell
PEMWE	proton-exchange membrane water electrolyzer
PFIA	perfluoro-imide acid
PFICE	perfluoro ionene chain extended
PGM	platinum group metals (Pt, Ir, Pd, Ru, Rh, and Os)
PR	pyrelene red
PTFE	polytetrafluoroethylene
PVDF	polyvinylidene difluoride
RDE	rotating disk electrode
RH	relative humidity
SPE	solid polymer electrolyte
SUSD	start-up/shut down
XRD	X-ray diffraction

References

1 Warshay, M. and Prokopius, P.R. (1989). The fuel cell in space: yesterday, today, tomorrow. NASA Technical Memorandum 102366.

2 Behling, N.H. (2013). History of proton exchange membrane fuel cells and direct methanol fuel cells. In: *Fuel Cells, Current Technology Challenges and Future Research Needs*, 423–600. Elsevier.

3 EG&G Technical Services, Inc. (2004). Alkaline fuel cell. In: *Fuel Cell Handbook*, 7e, 4-1. US Dept. of Energy https://www.netl.doe.gov/sites/default/files/netl-file/FCHandbook7.pdf#page=117&zoom=100,0,0.

4 Pollet, B.G., Kocha, S.S., and Staffell, I. (2019). *Current Opinion in Electrochemistry* 16: 90–95.

5 Borup, R., More, K., and Weber, A. (2018). FC135: FC-PAD: fuel cell performance and durability consortium. DoE AMR review (14 June 2018).

6 Haraeus Precious Metals (2018). PGM market analysis. https://www.heraeus.com/media/media/hpm/doc_hpm/precious_metal_update/en_6/20181031_PGM_Market_Analysis.pdf (accessed January 2021)

7 Appleby, A.J. and Yeager, E.B. (1986). *Energy* 11: 137.

8 Schaeffer, G.J. (1998). Fuel cells for the future: a contribution to technology forecasting from a technology dynamics perspective. PhD thesis. Universiteit Twente.

9 Breakthrough Technologies Institute, Inc. (2012). 2011 fuel cell technologies market report. Breakthrough Technologies Institute, Inc. (July 2012).

10 James, B. (2018). 2018 Cost projections of PEM fuel cell systems for automobiles and medium-duty vehicles. S DoE webinar (25 April 2018).

11 Battelle (2016). Manufacturing cost analysis of PEM fuel cell systems for 5- and 10-kW backup power applications. DOE contract no. DE-EE0005250 (October 2016).

12 Adroit Market Research Report (2019). https://www.adroitmarketresearch.com/industry-reports/fuel-cell-market (accessed January 2021)

13 Grandview Research Market Report (2020). https://www.grandviewresearch.com/industry-analysis/fuel-cell-market (accessed January 2021)

14 Bloomerg (2020). Europe approves biggest green stimulus in history. http://Bloomerg.com (21 July 2020). https://www.bloomberg.com/news/articles/2020-07-21/eu-approves-biggest-green-stimulus-in-history-with-572-billion-plan#:~:text=European%20governments%20approved%20the%20most,to%20renewable%20energy%20and%20agriculture.

15 Rathi, A. (2020). Germany shows the world how to green the stimulus. Washington Post (17 July 2020). https://www.washingtonpost.com/business/on-small-business/germany-shows-the-worldhow-to-green-the-stimulus/2020/06/16/6a8ffde8-afba-11ea-98b5-279a6479a1e4_story.html (accessed January 2021)

16 Gasteiger, H.A. and Yan, S.G. (2004). *Journal of Power Sources* 127: 162–171.

17 Vielstich, W., Lamm, A., and Gasteiger, H.A. (2003). Chapter 46. In: *Handbook of Fuel Cells: Fundamentals, Technology, and Applications*, vol. 3 (ed. W. Vielstich, A. Lamm, and H.A. Gasteiger), 593. Wiley.

18 Kongkanand, A. (2020). Highly-accessible catalysts for durable high-power performance. DoE AMR (30 May 2020).

19 Prater, K. (1990). *Journal of Power Sources* 29: 239.

20 Wang, Y., Diaz, D.F.R., Chen, K.S. et al. (2020). *Materials Today* 32: 178–203.

21 Jahnke, T., Futter, G., Latz, A. et al. (2016). *Journal of Power Sources* 304: 207–223.

22 Connolly, J.F., Flannery, R.J., and Meyers, B.L. (1967). *Journal of the Electrochemical Society* 114: 241–243.

23 Blurton, K.F., Kunz, H.R., and Rutt, D.R. (1978). *Electrochimica Acta* 23: 183–190.

24 Bregoli, L.J. (1978). *Electrochimica Acta* 23: 489–492.

25 Ross, P.N., Kinoshita, K., Scarpellino, A.J., and Stonehart, P. (1975). *Journal of Electroanalytical Chemistry and Interfacial Chemistry* 59: 177–189.

26 Watanabe, M., Tsurumi, K., Mizukami, T. et al. (1994). *Journal of the Electrochemical Society* 141 (10): 2659.

27 Aindow, T.T., Haug, A.T., and Jayne, D. (2011). *Journal of Power Sources* 196: 4506–4514.

28 U.S. Geological Survey (2020). Mineral commodity summaries. https://pubs.usgs.gov/periodicals/mcs2020/mcs2020-platinum.pdf (accessed January 2021)

29 Ramani, V.K. (2018). Corrosion-resistant non-carbon electrocatalyst supports for PEFCs. DoE Annual Merit Review, Project FC145 (14 June 2018).

30 Wang, G., Niangar, E., Huang, K. et al. (2015). *ECS Transactions* 69: 1179.

31 Debe, M.K. (2012). *Nature* 486: 43–51.

32 van Dam, H.E. and van Bekkum, H. (1991). *Journal of Catalysis* 131: 335–349.

33 Wang, D., Liu, Y., Huang, J., and You, T. (2012). *Journal of Colloid and Interface Science* 367: 199–203.

34 Bett, J.A., Kinoshita, K., and Stonehart, P. (1974). *Journal of Catalysis* 35: 307–316.

35 More, K.L., Cullen, D.A., Sneed, B.T., and Reeves, K.S. (2016). New fuel cell materials: characterization and method development. DoE Annual Merit Review Meeting, Project FC020 (8 June 2016).

36 Nørskov, J.K., Rossmeisl, J., Logadottir, A. et al. (2004). *Journal of Physical Chemistry B* 108: 17886–17892.

37 Stamenkovic, V.R., Fowler, B., Mun, B.S. et al. (2007). *Science* 315 (5811): 493–497.

38 Stacey, J., Regmi, Y.N., Leonard, B., and Fan, M. (2017). *Renewable and Sustainable Energy Reviews* 69: 401–414.

39 Tritsaris, G.A. (2011). *Catalysis Letters* 141: 909–913.

40 Kinoshita, K. (1992). *Electrochemical Oxygen Technology*. New York: Wiley.

41 Gasteiger, H.A., Kocha, S.S., Sompalli, B., and Wagner, F.T. (2005). *Applied Catalysis B: Environmental* 56: 9–35.

42 Newman, J. and Thomas-Alyea, K.E. (2004). *Electrochemical Systems*, 3e. New York: Wiley.

43 Weber, A., Balasubramanian, S., and Das, P.K. (2012). Proton exchange membrane fuel cells. In: *Fuel Cell Engineering*, Advanced in Chemical Engineering, vol. 41 (ed. K. Sundmacher), 65. Elsevier.

44 Pourbaix, M. (1974). *Equilibria in Aqueous Solutions*, 2e. Texas, USA: NACE.

45 Darling, R.M. and Meyers, J.P. (2003). *Journal of the Electrochemical Society* 150: A1523.

46 Wang, X.X., Kumar, R., and Myers, D.J. (2006). *Journal of Electrochemical and Solid-State Letters* 9: A225.

47 K. Moore's AMR report, Project 020, 2016 DOE Hydrogen and Fuel Cells Annual Merit Review Meeting, June 8, 2016.

48 Sun, Y., Polani, S., Luo, F. et al. (2021). *Nature Communications* 12: 5984.

49 Ferreira, P.J., Shao-Horn, Y., Morgan, D. et al. (2005). *Journal of the Electrochemical Society* 152 (11): A2256–A2271.

50 Springer, T.E., Rockward, T., Zawodzinski, T.A., and Gottesfeld, S. (2001). *Journal of the Electrochemical Society* 148 (1): A11–A23.

51 Kinoshita, K. (1988). *Carbon: Electrochemical and Physochemical Properties*, 1e. Wiley.

52 Haug, A.T. (2019). DoE Annual Merit Review Meeting, Project FC155 (April).

53 Stephen, I.E.L., Bondarenko, A.S., Grønbjerg, U. et al. (2012). *Energy and Environmental Science* 5: 6744.

54 Fernandez, J.L., Walsh, D.A., and Bard, A.J. (2005). *Journal of the American Chemical Society* 127: 357–365.

55 Faubert, G., Côté, R., Dodelet, J.P., and Lefevre, M. (1999). *Electrochimica Acta* 44: 2589.

56 Ross, P.N. and Wagner, F.T. (1982). Prospects for the development of non-noble metal catalysts for hydrogen-air fuel cells. *LBL-14192*. Lawrence Berkeley Laboratory, Berkeley, CA.

57 Holby, E.F. and Zelenay, P. (2016). *Nano Energy* 29: 54–64.
58 Chen, Z. (2011). *Energy and Environmental Science* 4: 3167–3193.
59 Thompson, S.T., Wilson, A.R., Zelenay, P. et al. (2018). *Solid State Ionics* 319: 68–76.
60 Spendelow, J.S. (2019). Advanced electrocatalysts through crystallographic enhancement. FCTO Webinar (31 July 2019).
61 Kongkanand, A. (2019). DOE EERE FCTO Annual Merit Review, Project FC144.
62 Spendelow, J. (2019). DOE EERE FCTO Annual Merit Review, Project FC161.
63 Steinbach, A.J. (2019). DoE Annual Merit Review Meeting, Project FC143 (April 2019).
64 Zhang, J., Mo, Y., Vukmirovic, M.B. et al. (2004). *Journal of Physical Chemistry B* 108: 10955–10964.
65 Adzic, R.R., Zhang, J., Sasaki, K. et al. (2007). *Topics in Catalysis* 46: 249–262.
66 Kongkanand, A., Subramanian, N.P., Yu, Y. et al. (2016). *ACS Catalysis* 6: 1578–1583.
67 He, C., Sankarasubramanian, S., Matanovic, I. et al. (2019). *Chemistry-Sustainability-Energy-Materials* 12 (15): 3468.
68 Daudt, N.F. (2020). *Renewable and Sustainable Energy* 9: 18.
69 Steinbach, A.J. (2018). DoE Annual Merit Review Meeting, Project FC143 (14 June 2018).
70 Murphy, O.J., Hitchens, G.D., Manko, D.J. et al. (1994). *Journal of Power Sources* 47: 353.
71 Ralph, T.R., Hards, G.A., Keating, J.E. et al. (1997). *Journal of the Electrochemical Society* 144: 3845.
72 Srinivasan, S., Manko, D.J., Koch, H. et al. (1990). *Journal of Power Sources* 29: 367–387.
73 Wilson, M.S. and Gottesfeld, S. (1992). *Journal of Applied Electrochemistry* 22.
74 Pivovar, B. (2014). DOE EERE FCTO Annual Merit Review, Project FC007.
75 Subbaraman, R., Strmcnik, D., Stamenkovic, V., and Markovic, N.M. (2010). *Journal of Physical Chemistry C* 114: 8414–8422.
76 Shinozaki, K., Morimoto, Y., Pivovar, B.S., and Kocha, S.S. (2016). *Journal of Power Sources* 325: 745.
77 Takeshita, T., Kamitaka, Y., Shinozaki, K. et al. (2020). *Journal of Electroanalytical Chemistry* 871: 114250.
78 Gottesfeld, S., Raistrick, I.D., and Srinivasan, S. (1987). *Journal of the Electrochemical Society* 134: 1455–1462.
79 Kodama, K., Motobayashi, K., Shinohara, A. et al. (2018). *ACS Catalysis* 8: 694–700.
80 Yarlagadda, V., Carpenter, M.K., Moylan, T.E. et al. (2018). *ACS Energy Letters* 3: 618–621.
81 Snyder, J., Livi, K., and Erlebacher, J. (2013). *Advanced Functional Materials* 23: 5494–5501.
82 Huang, K., Song, T., Morales-Collazo, O. et al. (2017). *Journal of the Electrochemical Society* 164: F1448–F1459.
83 Weber, A. and Newman, J. (2004). *Chemical Reviews* 104 (10): 4679–4726.

84 Bruggeman, D.A.G. (1935). *Annals of Physics* 24: 636.

85 Weber, A., Darling, R., and Newman, J. (2004). *Journal of the Electrochemical Society* 151 (10): A1715–A1727.

86 Zhao, J., Schaefer, D.W., Shi, D. et al. (2005). *The Journal of Physical Chemistry. B* 109: 23351–23357.

87 Rao, R. and Rengaswamy, R. (2006). *Journal of Power Sources* 158: 110–123.

88 Siegel, N.P., Ellis, M.W., Nelson, D.J., and Von Spakovsky, M.R. (2003). *Journal of Power Sources* 115: 81.

89 Baker, D.R. (2009). *Journal of the Electrochemical Society* 156: B991.

90 Nonoyama, N., Okazaki, S., Weber, A.Z. et al. (2011). *Journal of the Electrochemical Society* 158: B416.

91 Song, D., Wang, Q., Liu, Z. et al. (2004). *Journal of Power Sources* 126: 104–111.

92 Marr, C. and Lir, X. (1999). *Journal of Power Sources* 77: 17–27.

93 Haug, A.T. (2018). DoE Annual Merit Review Meeting, Project FC155 (14 June 2018).

94 Ono, Y., Ohma, A., Shinohara, K., and Fushinobu, K. (2013). *Journal of the Electrochemical Society* 160 (8): F779–F787.

95 Bi, C. and Yang, Y. (2021). *Journal of Physical Chemistry C* 125 (1): 464–480.

96 Yandrasits, M.A., Lindell, M.J., and Hamrock, S.J. (2019). *Current Opinion in Electrochemistry* 18: 90–98.

97 Yamada, K., Hommura, S., and Shimohira, T. (2012). *ECS Transactions* 50 (2): 1495–1501.

98 Kinoshita, S. (2014). *ECS Transactions* 64: 371–375.

99 Rolfi, A., Oldani, C., Merlo, L. et al. (2018). *Journal of Power Sources* 396: 95–101.

100 Broka, K. and Ekdunge, P. (1997). *Journal of Applied Electrochemistry* 27: 117–123.

101 Paik, C.H., Jarvi, T.D., and O'grady, W.E. (2004). *Electrochemical and Solid-State Letters* 7 (4): A82–A84.

102 Roen, L.M., Paik, C.H., and Jarvi, T.D. (2004). *Electrochemical and Solid-State Letters* 7 (1): A19–A22.

103 Min, M., You, E., Lee, S.W., and Pak, C. (2021). *Journal of Power Sources* 488: 1–8.

104 Atanasoski, R., van der Vliet, D., Cullen, D., and Atanasoska, L. (2015). Durable catalysts for fuel cell protection during transient conditions. *DoE DE-EE0000456 Final Report*.

105 Srivastava, R. and Strasser, P. (2009). *ECS Transactions* 25: 565–571.

106 Atanasoski, R.T., Cullen, D.A., Vernstrom, G.D. et al. (2013). *ECS Electrochemistry Letters* 2: 25–28.

107 Borup, R., Meyers, J., Pivovar, B. et al. (2007). *Chemical Reviews* 107: 3904–3951.

108 Mathias, M.F., Makharia, R., Gasteiger, H.A. et al. (2005). *Electrochemical Society Interface* 14: 24–35.

109 DoE accelerated stress test protocols and targets. Office of Energy Efficiency and Renewable Energy. https://www.energy.gov/eere/fuelcells/doetechnical-targets-polymer-electrolyte-membrane-fuel-cell-components (accessed June 2021).

110 Kongkanand, A., Wenbin, G., and Mathias, M.F. (2018). Proton-exchange membrane fuel cells with low-Pt content. In: *Encyclopedia of Sustainability Science and Technology* (ed. R.A. Meyers). Springer https://doi.org/10.1007/978-1-4939-2493-6_1022-1.

111 Kongkanand, A., Gu, W., and Mathias, M.F. (2019). Proton-exchange membrane fuel cells with low-Pt content. In: *Fuel Cells and Hydrogen Production*, Encyclopedia of Sustainability Science and Technology Series (ed. T. Lipman and A. Weber), 323–342. New York, NY: Springer.

112 Kongkanand, A. and Mathias, M. (2016). *Journal of Physical Chemistry Letters* 7: 1127–1137.

113 Heitner-Wirguin, C. (1996). *Journal of Membrane Science* 120: 1–33.

114 Robertson, M.A.F. and Yeager, H.L. (1997). Structure and properties of perfluorinated ionomers. In: *Ionomers: Synthesis, Structure, Properties and Applications* (ed. M.R. Tant, K.A. Mauritz, and G.L. Wilkes), 290–330. Springer.

115 Xu, G. (1993). *Polymer* 25: 397.

116 Mauritz, K.A. and Moore, R.B. (2004). *Chemical Reviews* 104 (10): 4535–4586.

117 Doyle, M. and Rajendran, G. (2003). Perfluorinated membranes, Chapter 30. In: *Handbook of Fuel Cells Fundamentals, Technology and Applications*, vol. 3, Part 3 (ed. W. Vielstich, H.A. Gasteiger, and A. Lamm), 351. Chichester, UK: Wiley.

118 Wainright, J.S., Savinell, R.F., DesMarteau, D.D. et al. (1994). *Proceedings - Electrochemical Society* 94: 265–273.

119 Schaberg, M.S., Abulu, J.E., Haugen, G.M. et al. (2010). *ECS Transactions* 33: 627–633.

120 Bawagan, A.D., Hamrock, S.J., Schaberg, M. et al. (2014). *Vibrational Spectroscopy* 75: 213–217.

121 Yandrasits, M., Lindell, M., Schaberg, M., and Kurkowski, M. (2017). *ECS Interface* 26: 49–53.

122 Moukheiber, E., De Moor, G., Flandin, L., and Bas, C. (2012). *Journal of Membrane Science* 389: 294–304.

123 Shinohara, A. and Kudo, K. (2013). ECS Abstracts#1282, 224th ECS Meeting.

124 Kodama, K. and Shinohara, A. (2014). *Journal of the Electrochemical Society* 161 (5): F649–F652.

125 Chang, Y., Brunello, G.F., Fuller, J. et al. (2013). *Polymer Chemistry* 4: 272.

126 Isamu Kaneko, Atsushi Watakabe, Jyunichi Tayanagi, Susumu Saito. Fluorosulfonyl group-containing compound, method for its production and polymer thereof. US7667083B2. Filed:2009.

127 Gierke, T.D., Munn, G.E., and Wilson, F. (1981). *Journal of Polymer Science: Polymer Physics* 19: 1687.

128 Hsu, W.Y. and Gierke, T.D. (1983). *Journal of Membrane Science* 13: 307.

129 Yandrasits, M., Lindall, M., and Hamrock, S. (2019). *Current Opinion in Electrochemistry* 18: 90–98.

130 Kreuer, K.D., Paddison, S.J., Spohr, E., and Schuster, M. (2004). *Chemical Reviews* 10 (104): 4637–4678.

131 Paddison, S.J. (2003). *Annual Review of Materials Research* 33: 289–319.

132 Hickner, M.A. (2012). *Journal of Polymer Science Part B: Polymer Physics* 50 (1): 9–20.

133 Kosuglu, A. and Weber, A. (2017). *Chemical Reviews* 117: 987–1104.

134 Borup, R. and Weber, A. (2019). FC-PAD: fuel cell performance and durability consortium. *LA-UR-19-22970*.

135 Kusoglu, A. and Weber, A.Z. (2012). *Polymers for Energy Storage and Delivery: Polyelectrolytes for Batteries and Fuel Cells*, vol. 1096. Washington, DC: American Chemical Society.

136 Knights, S.D., Colbow, K.M., St-Pierre, J., and Wilkinson, D.P. (2004). *Journal of Power Sources* 127: 127–134.

137 Li, Y., Dillard, D.A., Case, S.W. et al. (2009). *Journal of Power Sources* 194: 873–879.

138 Mench, M.M., Kumbur, E.C., and Veziroglu, T.N. (2012). *Polymer Electrolyte Fuel Cell Degradation*. Elsevier Ltd.

139 Vengatesan, S., Fowler, M.W., Yuan, X.Z., and Wang, H. (2011). *Journal of Power Sources* 196 (11): 5045–5052.

140 Hamrock, S. (2011). Membranes and MEA's for dry, hot operating conditions. DoE AMR (11 May 2011).

141 Hamrock, S. (2010). Some durability considerations for proton exchange membranes. DoE High Temperature Membrane Working Group.

142 Zawodzinski, T.A., Springer, T.E., Uribe, F., and Gottesfeld, S. (1993). *Solid State Ionics* 60: 199–211.

143 Frey, T. and Linardi, M. (2004). *Electrochimica Acta* 50: 99–105.

144 Kolde, J.A., Bahar, B., Wilson, M.S., et al. (1995). Proceedings of the 1st International Symposium on Proton Conducting Membrane Fuel Cells, Vol. 95, pp. 193–201.

145 Liu, W., Suzuki, T., Mao, H., and Schmiedel, T. (2013). *ECS Transactions* 50 (2): 51–64.

146 Wang, J., Yang, M., Dou, P. et al. (2014). *Industrial and Engineering Chemistry Research* 53 (36): 14175–14182.

147 Cleghorn, S., Kolde, J., and Liu, W. (2003). Catalyst coated composite membranes. In: *Handbook of Fuel Cells – Fundamentals, Technology and Applications*, vol. 3 (ed. W. Vielstich, H. Hasteiger, and A. Lamm). New York: Wiley https://doi.org/10.1002/9780470974001.f303049.

148 Curtin, D.E., Lousenberg, R.D., Henry, T.J. et al. (2004). *Journal of Power Sources* 131: 41–48.

149 Gittleman, C.S., Coms, F.D., and Lai, Y.H. (2012). *In Polymer Electrolyte Fuel Cell Degradation*. Boston: Academic Press.

150 Zhou, C., Guerra, M.A., Qiu, Z.M. et al. (2007). *Macromolecules* 40 (24): 8695–8707.

151 Frey, M.H., Hamrock, S.J., Haugen, G.M., and Pham, P.T. (2009) Fuel cell membrane electrode assembly. US Patent 7,572,534, filed 20 September 2004 and issued 11 August 2009.

152 Pierpont, D.M., Frey, M.H., and Hamrock, S.H. (2012). (3 M Innovative Properties Co). Durable fuel cell membrane electrode assembly with combined additives. US Pat. 8,137,828 March 20.

153 Trogadas, P., Parrondo, J., and Ramani, V. (2008). *Electrochemical and Solid-State Letters* 11: B113–B116.

154 Coms, F., Liu, H., and Owejan, J.E. (2008). *ECS Transactions* 16: 1735–1747.

155 Pierpont, D.M., Hamrock, S.J., and Frey, M.H. (2012). Durable fuel cell membrane electrode assembly with combined additives. US Patent 8,137,828, filed 23 December 2008 and issued 20 March 2012.

156 Sethuraman, V., Weidner, J.W., Haug, A.T., and Protsailo, L. (2008). *Journal of the Electrochemical Society* 155 (2): B119–B124.

157 Baik, K.D., Hong, B.K., and Kim, M.S. (2013). *Renewable Energy* 57: 234–239.

158 Perry, M. and Fuller, T. (2002). *Journal of the Electrochemical Society* 149 (7): S59–S67.

159 Gittleman, C.S., Kongkanand, A., Masten, D., and Gu, W. (2019). *Current Opinion in Electrochemistry* 18: 81–89.

160 Kienitz, B. (2021). *International Journal of Hydrogen Energy* 46: 11176–11182.

161 Gittleman, C. (2009). Automotive perspective on PEM evaluation. DoE High Temperature Membrane Working Group Meeting (18 May 2009).

162 Francia, C., Ijeri, V.S., Specchia, S., and Spinelli, P. (2011). *Journal of Power Sources* 196: 1833–1839.

163 Hickner, M.A., Ghassemi, H., Kim, Y.S. et al. (2004). *Chemical Reviews* 104: 4587–4612.

164 Goldbach, J., Mountz, D., Zhang, T., et al. (2010). Improved, low-cost, durable fuel cell membranes, DoE AMR (8 June 2010).

165 McGrath, J.E. (2006). Advanced materials for proton exchange membranes. DoE AMR.

166 Protsailo, L. and Haug, A.T. (2005). Development of high temperature membranes and improved cathode membranes and improved cathode catalysts for PEM fuel cells. DOE Hydrogen Program Review (23–26 May 2005).

167 Passalacqua, E. (1992). *International Journal of Hydrogen Energy* 17 (10): 771–776.

168 Appleby, A.J. (1986). *Energy* 11 (2): 13–94.

169 Hara, N., Tsurumi, K., and Watanabe, M. (1996). *Journal of Electroanalytical Chemistry* 413: 81–88.

170 Yang, Y., Zhou, X., Li, B., and Zhang, C. (2021). *International Journal of Hydrogen Energy* 46: 4259–4282.

171 Watanabe, M. (1985). *Journal of Electroanalytical Chemistry* 195: 81–93.

172 Maoka, T. (1988). *Electrochimica Acta* 33 (3): 379–383.

173 Mathias, M., Roth, J., Fleming, J., and Lehnert, W. (2003). *Handbook of Fuel Cells*. Wiley.

174 Lin, G.Y. and Nguyen, T.V. (2005). *Journal of the Electrochemical Society* 152: A1942–A1948.

175 Giorgia, L., Antolini, E., Pozio, A., and Passalacqua, E. (1998). *Electrochimica Acta* 43 (24): 3675–3680.

176 Cindrella, L., Kannan, A.M., Lin, J.F. et al. (2009). *Journal of Power Sources* 194: 146–160.

177 Park, S., Lee, J.W., and Popov, B.N. (2008). *Journal of Power Sources* 177: 457–463.

178 Breault, R.D. and Goller, G.J. (1979). Automated Catalyst Processing for Loud Electrode Fabrication for Fuel Cells. US Patent 4,233,181 (11 November 1982).

179 Breault, R. D., Harding, R. P., and Kemp, F. S (1977). US4043933A - Method of fabricating a fuel cell electrode. United Technologies Corporation.

180 Kohlmayr, G.M. and Stonehart, P. (to United Technologies Corp.) (1979). Method of fabricating a fuel cell electrode. US Patent 4,163,811 (B/7/1979).

181 Antolini, E., Passos, R.R., and Ticianelli, E.A. (2002). *Journal of Power Sources* 109 (2): 477–482.

182 Leeuwner, M.J., Patra, A., Wilkinson, D.P., and Gyenge, E.L. (2019). *Journal of Power Sources* 423: 192–202.

183 Latorrata, S., Mariani, M., Stampino, P.G. et al. (2020). *Materials Today: Proceedings* 31: 426–432.

184 Schweiss, R., Steeb, M., Wilde, P.M., and Schubert, T. (2012). *Journal of Power Sources* 220: 79–83.

185 Cindrella, L., Kannan, A.M., Ahmad, R., and Thommes, M. (2009). *International Journal of Hydrogen Energy* 34: 6377.

186 Weber, A. (2010). *Journal of Power Sources* 195: 5292–5304.

187 Lim, C. and Wang, C.Y. (2004). *Electrochimica Acta* 49: 4149–4156.

188 Sinha, P.K. and Wang, C.Y. (2007). *Electrochimica Acta* 52: 7936.

189 Gostick, J.T., Ioannidis, M.A., Fowler, M.W., and Pritzker, M.D. (2009). *Electrochemistry Communications* 11: 576–579.

190 Lee, J., Yip, R., Antonacci, P. et al. (2015). *Journal of the Electrochemical Society* 162: F669.

191 Ahn, M., Cho, Y.H., Cho, Y.H. et al. (2011). *Electrochimica Acta* 56: 2450.

192 Pasaogullari, U., Wang, C.Y., and Chen, K.S. (2005). *Journal of the Electrochemical Society* 152: A1574–A1582.

193 Das, P., Santamaria, A.D., and Weber, A.Z. (2015). *Procedia Engineering* 105: 751–756.

194 Gostick, J.T., Fowler, M.W., Ioannidis, M.A. et al. (2006). *Journal of Power Sources* 156: 375–387.

195 Passalacqua, E., Squadrito, G., Lufrano, F. et al. (2001). *Journal of Applied Electrochemistry* 31 (4): 449–454.

196 Zenyuk, I., Parkinson, D.Y., Connolly, L.G., and Weber, A.Z. (2016). *Journal of Power Sources* 328: 364–376.

197 Lobato, J., Cañizares, P., Rodrigo, M.A. et al. (2008). *Journal of Applied Electrochemistry* 38: 793–802.

198 Zenyuk, I., Parkinson, D.Y., Hwang, G., and Weber, A.Z. (2015). *Electrochemistry Communications* 53: 24–28.

199 Gittleman, C.S., Coms, F.D., and Lai, Y.-L. (2011). Membrane durability: physical and chemical degradation. In: *Modern Topics in Polymer Electrolyte Fuel Cell Degradation* (ed. M.M. Mench, E.C. Kumbur, and T.N. Veziroğlu), 15–88. Amsterdam: Academic Press.

200 Ulsh, M. (2019). *International Journal of Hydrogen Energy* 44: 8533–8547.

201 Owejan, J., Gagliardo, J.J., Sergi, J.M. et al. (2009). *International Journal of Hydrogen Energy* 34: 3436–3444.

202 Um, S. and Wang, C.Y. (2004). *Journal of Power Sources* 125: 40–51.

203 Weber, A.Z. and Newman, J. (2005). *Journal of the Electrochemical Society* 152: A677.

204 Tiss, F., Chouikh, R., and Guizani, A. (2013). *International Journal of Hydrogen Energy* 38: 8532–8541.

205 Springer, T.E., Wilson, M.S., and Gottesfeld, S. (1993). *Journal of the Electrochemical Society* 140 (12): 3513.

206 Wang, Z.H., Wang, C.Y., and Chen, K.S. (2001). *Journal of Power Sources* 94 (1): 40.

207 Dai, W., Wang, H., Yuan, X.Z. et al. (2009). *International Journal of Hydrogen Energy* 34: 9461–9478.

208 Dursch, T.J., Trigub, G.J., Liu, J.F. et al. (2013). *International Journal of Heat and Mass Transfer* 67: 896–901.

209 Nguyen, T.V. and White, R.E. (1993). *Journal of the Electrochemical Society* 140: 2178.

210 Springer, T.E., Zawodzinski, T.A., and Gottesfeld, S. (1991). *Journal of the Electrochemical Society* 138 (8): 2334–2342.

211 Nguyen, T.V. and White, R.E. (1993). *Journal of the Electrochemical Society* 140 (8): 2178–2186.

212 Abbou, S., Tajiri, K., Medici, E. et al. (2018). *ECS Transactions* 86 (13): 163–169.

213 Zawodzinski, T.A., Gottesfeld, S., Shoichet, S., and McCarthy, T.J. (1993). *Journal of Applied Electrochemistry* 23 (1): 86–88.

214 Okada, T., Nakamura, N., Yuasa, M., and Sekine, I. (1997). *Journal of the Electrochemical Society* 144: 2744.

215 Borup, R., Kusoglu, A., Neyerlin, K.C. et al. (2020). *Current Opinon in Electrochemitry* 21: 192–200.

216 Braaten, J., Kongkanand, A., and Litster, S. (2017). *ECS Transactions* 80: 283–290.

217 Cai, Y., Kongkanand, A., Gu, W., and Moylan, T.E. (2015). *ECS Transactions* 69 (17): 1047–1061.

218 Lee, W.K., Ho, C.H., Van Zee, J.W., and Murthy, M. (1999). *Journal of Power Sources* 84: 45–51.

219 Bazylak, A., Sinton, D., Liu, Z.S., and Djilali, N. (2007). *Journal of Power Sources* 163: 784–792.

220 Park, J., Oh, H., Ha, T. et al. (2015). *Applied Energy* 155: 866–880.

221 Paulus, U.A., Schmidt, T.J., Gasteiger, H.A., and Behm, R.J. (2001). *Journal of Electroanalytical Chemistry* 495: 134–145.

222 Gummalla, M., Atrazhev, V.V., Condit, D. et al. (2010). *Journal of The Electrochemical Society* 157 (11): B1542.

223 Burlatsky, S.F., Atrazhev, V., Cipollini, N. et al. (2006). *ECS Transactions* 1: 239.

224 Healy, J., Hayden, C., Xie, T. et al. (2005). *Fuel Cells* 5 (2): 302–308.

225 Danilczuk, M., Lancucki, L., Schlick, S. et al. (2012). *ACS Macro Letters* 1: 280–285.

226 Rodgers, M.P., Bonville, L.J., Kunz, H.R. et al. (2012). *Chemical Reviews* 112: 6075–6103.

227 Mittal, V. and Motupally, S. (2008). Membrane durability in PEM fuel cells. High Temperature Working Group Meeting, DoE AMR, South Windsor, CT (9 June 2008).

228 Burlatsky, S., Cipollini, N. E., Condit, D. A., et al., (2009). Protective and precipitation layers for PEM fuel cell. French Patent WO2009/085327A2, filed 03 January 2008 and issued 09 July 2009.

229 Porstmann, S., Wannemacher, T., and Drossel, W.G. (2020). *Journal of Manufacturing Processes* 60: 366–383.

230 Weber, A. and Darling, R. (2007). *Journal of Power Sources* 168: 191–199.

231 Guo, N., Leu, M.C., and Koylu, U.O. (2013). *International Journal of Hydrogen Energy* 38: 6750–6761.

232 Kim, J., Luo, G., and Wang, C.Y. (2017). *Journal of Power Sources* 365: 419–429.

233 Owejan, J., Trabold, T.A., Jacobson, D.L. et al. (2007). *International Journal of Hydrogen Energy* 32: 4489–4502.

234 Taniguchi, A., Akita, T., Yasuda, K., and Miyazaki, Y. (2004). *Journal of Power Sources* 130: 42.

235 Perry, M.L., Patterson, T., and Reiser, C. (2006). *ECS Transactions* 3: 783.

236 Wu, J., Yuan, X.Z., Martin, J.J. et al. (2008). *Journal of Power Sources* 184: 104–119.

237 Atanasoski, R.E., van der Vliet, D., Cullen, D., and Atanasoska, L. (2011). Durable catalysts for fuel cell protection during transient conditions. DoE Hydrogen Program, FY2010 Annual Progress Report: DOE/GO-102011-3178, pp. 825–829.

238 Daud, W.R.W., Rosli, R.E., Majlan, E.H. et al. (2017). *Renewable Energy* 113: 620–638.

239 Reiser, C.A., Bregoli, L., Patterson, T.W. et al. (2005). *Electrochemical and Solid-State Letters* 8: A273.

240 Meyers, J.P. and Darling, R.M. (2006). *Journal of the Electrochemical Society* 153 (8): A1432–A1442.

241 Makharia, R., Kocha, S., Yu, P. et al. (2006). *ECS Transactions* 1 (8): 3.

242 Wang, G., Yu, Y., Liu, H. et al. (2018). *Fuel Processing Technology* 179: 203–228.

243 Yang, Y., Zhang, X., Guo, L., and Liu, H. (2017). *International Journal of Hydrogen Energy* 42: 24435–24447.

244 James, B.D. (2020). Fuel cell systems analysis. Project ID#163. DoE Annual Merit Review (30 May 2020).

245 Ahn, S.-Y., Shin, S.J., Ha, H.Y. et al. (2002). *Journal of Power Sources* 106: 295–303.

246 Miller, M. and Bazylak, A. (2011). *Journal of Power Sources* 196: 601–613.

247 Zhu, W., Payne, R.U., Cahela, D.R., and Tatarchuk, B.J. (2004). *Journal of Power Sources* 128: 231–238.

248 Pei, P., Yuan, X., Chao, P., and Wang, X. (2010). *International Journal of Hydrogen Energy* 35: 3147–3151.

249 Bose, A., Babburi, P., Kumar, R. et al. (2013). *Journal of Power Sources* 243: 964–972.

250 Wang, G., Huang, F., Yu, Y. et al. (2018). *International Journal of Hydrogen Energy* 43: 4471–4481.

251 Oyarce, A., Zakrisson, E., Ivity, M. et al. (2014). *Journal of Power Sources* 254: 232–240.

Part II

Basics – Fundamentals

3

Electrochemical Fundamentals

Vito Di Noto, Gioele Pagot, Keti Vezzù, Enrico Negro, and Paolo Sgarbossa

University of Padova, Section of Chemistry for Technologies (ChemTech), Department of Industrial Engineering, Via Marzolo 9, I-35131 Padova, Italy

3.1 Principles of Electrochemistry

Electrochemistry is a branch of physical chemistry that studies the chemical processes occurring when electrons are exchanged between different chemical species, i.e. redox reactions. In this case the chemical energy (ΔG_{rxn}) of certain species is converted into electric energy or work (w_{el}) [1–5]. Here, a brief background to understand most of the phenomena modulating the operation of electrochemical devices, with a particular focus on fuel cells, is provided. For more detailed information, the reader can take into consideration the specialized literature cited at the end of this chapter [2, 3, 5]. Figure 3.1 shows a simple galvanic cell separating the electrochemical reaction into two different half-reactions using two physically separated half-cells constituted by (i) two metal electrodes, M and P, each inserted into a half-cell in contact with an M^{x+} and P^{y+} ions in solution, respectively; (ii) an electric conductor, which connects the two electrodes to an external load; and (iii) a porous barrier, or a salt bridge filled with a suitable electrolyte, that separates the half-cells and allows ions to migrate between the electrode compartments, thus maintaining the system's electric neutrality.

3.2 The Role of the First Faraday Law

The electrochemical process occurring in the galvanic cell is modulated by the standard Gibbs free energy (ΔG°) characterizing the following general chemical reaction [2]:

$$\nu' \cdot M^{x+}_{(aq)} + \nu \cdot P^{y+}_{(aq)} \rightarrow \nu' \cdot M^{(x+\nu)+}_{(aq)} + \nu \cdot P^{(y-\nu')+}_{(aq)} \qquad \Delta G^\circ < 0 \tag{3.1}$$

M^{x+} (P^{y+}) and $M^{(x+\nu)+}$($P^{(y-\nu')+}$) are the reduced (oxidized) and oxidized (reduced) species, respectively, and ν and ν' are the stoichiometric coefficients. The above

Electrocatalysis for Membrane Fuel Cells: Methods, Modeling, and Applications, First Edition.
Edited by Nicolas Alonso-Vante and Vito Di Noto.

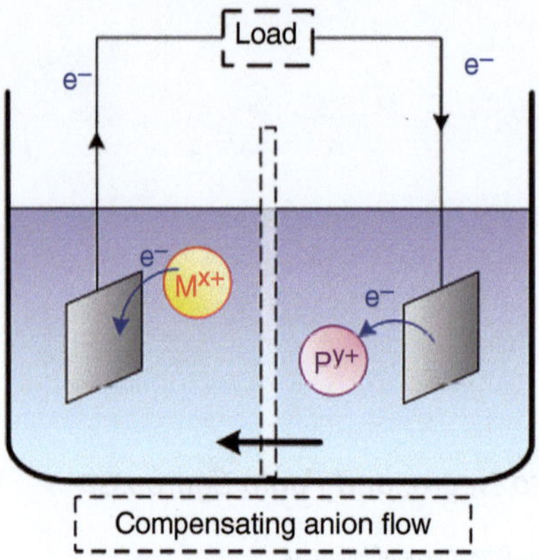

Figure 3.1 Example of a basic electrochemical cell, displaying the interconnection between the two electrodes through the external load resistor and a porous barrier allowing for the spontaneous cell reaction to proceed.

overall reaction can be split into two half-reactions (one of oxidation and the other of reduction) [2]:

$$\begin{cases} M^{x+}_{(aq)} \rightarrow M^{(x+\nu)+}_{(aq)} + \nu \cdot e^- & \text{oxidation} \\ P^{y+}_{(aq)} + \nu' \cdot e^- \rightarrow P^{(y-\nu')+}_{(aq)} & \text{reduction} \end{cases} \tag{3.2}$$

These half-reactions indicate that the amount of substance obtained at each electrode is proportional to the number of moles of exchanged electrons (n_e). The latter depends on the current flow (i) that passes through the cell over a period of time δt:

$$n_e = \frac{i \cdot \delta t}{N_A \cdot e_0} \tag{3.3}$$

N_A is the Avogadro's constant, and e_0 is the electron charge in Coulombs. $N \cdot e_0 = F \approx 96{,}485\ \text{C mol}^{-1}$ is the Faraday's constant (F), which represents the total charge in Coulombs of 1 mol of electrons. If we take into account Eq. (3.1), and

$$\frac{n_M}{\nu'} = \frac{n_P}{\nu} = \frac{n_e}{\nu \cdot \nu'} \tag{3.4}$$

where n_M and n_P are the moles of M and of P, respectively, then for each electrode the moles of reagents that are consumed during the *electrochemical* process are:

$$n_M = \frac{1}{\nu \cdot F} \int_0^t i \cdot dt \tag{3.5}$$

and

$$n_P = \frac{1}{\nu' \cdot F} \int_0^t i \cdot dt \tag{3.6}$$

The maximum theoretical capacity (Q_T) for the anodic and cathodic process depends on the moles of substance transformed electrochemically at the electrodes, which are, respectively,

$$Q_{T,M} = \int_0^t i \cdot dt = \nu \cdot F \cdot n_M \tag{3.7}$$

and

$$Q_{T,P} = \int_0^t i \cdot dt = \nu' \cdot F \cdot n_P \tag{3.8}$$

On the basis of Eq. (3.5) we can derive some important considerations:

(A) when $\nu \cdot n_M$ is equal to $\nu' \cdot n_P$, $Q_{T,M} = Q_{T,P} = Q_T$. In this case, the composition of the cell is perfectly balanced, and the reagents are completely consumed at the same moment in both half-cells;
(B) on the other hand, if $\nu \cdot n_M > \nu' \cdot n_P$ or $\nu \cdot n_M < \nu' \cdot n_P$, then $Q_{T,M} \neq Q_{T,P}$, and the cell is completely unbalanced. It is to be noted that in the previous case the cell is "*positively limited*" or "*cathodically limited*," while in the second case it is "*negatively limited*" or "*anodically limited*." In these conditions, the maximum capacity that can be extracted from an electrochemical cell is limited by only one of the two half-reactions that are occurring in the process. In general, the limiting process is also the rate-determining process of the overall redox reaction.

3.3 Electric Double Layer and the Formation of a Potential Difference at the Interface

In an electrochemical cell, when a metal electrode (i.e. an electron conductor) is immersed into an electrolyte, an interface between the two phases is formed, and two layers of opposite polarity appear [3–5], i.e. the double layer, Figure 3.2. The structure of the solution inside the double layer is more complicated than that of the metal electrode. In particular, in the portion of the solution closer to the electrode two different regions can be identified: (i) an ***inner layer*** (inner Helmholtz plane), which may contain ions with a low degree of solvent coordination and presenting strong chemical bonding interactions with the metal atoms on the surface, together with oriented solvent molecular dipoles and (ii) a ***time-averaged*** diffuse layer (outer Helmholtz plane), where ions are attracted to or repelled from the electrode/electrolyte interface by electrostatic forces, and where they are also affected by thermal collisions. The effective depth of the diffuse layer is a function of the electrolyte concentration, temperature, dielectric constant of the solvent, etc. The formation of an electric double layer at a metal–solution interface induces a particular arrangement of ions in the region near to the electrode surface, which results in a variation in the electric potential as a function of the distance from the interface. The structure of the double layer may affect significantly the rates of electrochemical reactions [6].

In the case of electrochemical cells characterized by significant direct currents, the main mechanism on the basis of the established potential difference at the phase boundaries is the charge transfer (e.g. electron transfer between two phases and ion transfer between a metal and a solution including its ions).

For example, an inert metal, such as platinum, which is immersed into a solution where ions at different oxidation states are dissolved (e.g. Fe^{2+} and Fe^{3+}), induces the activation of charge transfer processes. At the beginning, when the two phases (i.e. the metal and the solution) are separated, there is no potential difference between the metal and the electrolyte. As soon as they come into contact, two opposite processes can be observed: (I) an electron is transferred from the conduction band of the

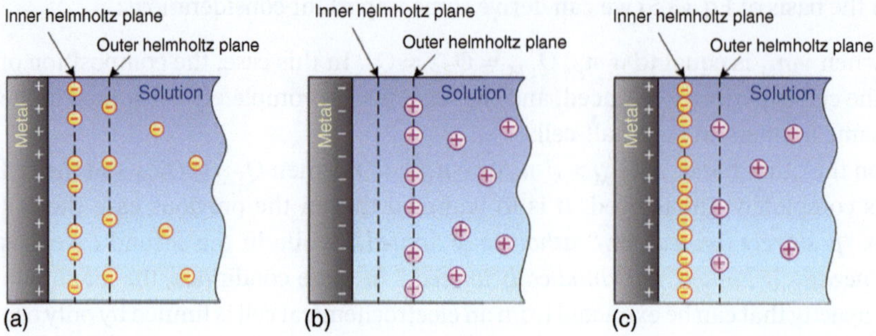

Figure 3.2 Structure of the electric double layer under different conditions of electrode polarization: (a) metal positively charged, anions present at the inner Helmholtz plane (chemically interacting with the metal) and in the diffuse double layer beyond the outer Helmholtz plane; (b) metal negatively charged, inner Helmholtz plane empty, cations in the outer Helmholtz plane; and (c) metal positively charged, strong adsorption of anions in the inner Helmholtz plane, balancing cations in the diffuse layer.

metal to the Fe^{3+} ion, inducing a reduction event and (II) an electron is transferred from the Fe^{2+} ion to the metal, resulting in an oxidation reaction. If the kinetics of two processes (i.e. I and II) are different, the two phases become progressively charged. In particular, if process I has a lower activation energy, the platinum metal turns positively charged, and the solution gets a negative charge. Thus, a potential difference is established between the two phases. This potential difference influences the charge transfer kinetics. Indeed, the positively charged platinum metal inhibits the reduction of Fe^{3+} ions and promotes the oxidation of Fe^{2+} ions. The charge transfer event proceeds, and the potential difference varies till the kinetics of processes I and II assume the same value. At this condition the electrochemical equilibrium is reached, a double layer is formed, and the potential difference (i.e. equilibrium potential difference) becomes stable.

Two concepts deserve to be pointed out at this stage: (a) First, the quantity of the charge transferred between the two phases in order to reach the equilibrium potential difference is typically very small, and the actual change in the composition of the solution is negligible. For example, it can be demonstrated that when Pt electrode is immersed into an Fe^{2+}/Fe^{3+} solution, a reduction from 10^{-9} to 10^{-10} mol cm^{-2} of Fe^{3+} takes place. (b) Second, the charge transfer kinetics are very important, since high rates are necessary to reach the equilibrium.

3.4 The Cell

The cell is an electrochemical system able to separate physically into two compartments the two half-reactions, see Eq. (3.2), characterizing an overall electrochemical process, see Eq. (3.1). The two compartments are set apart by a porous/electrolyte separator, which only allows the migration of suitable ions to balance the charge of the system [1, 2], see Figure 3.3.

If we consider that the overall redox reaction occurring in the cell, shown in Eq. (3.1), is reversible, the chemical reaction can be reversed. Correspondingly,

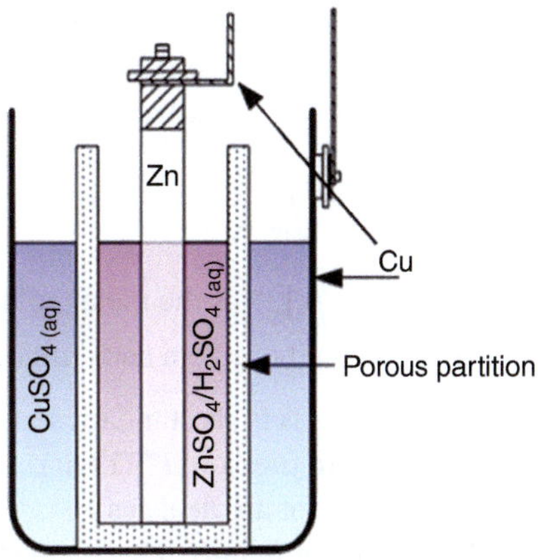

Figure 3.3 Example of a cell: scheme of the Daniell cell.

the current would flow in the opposite direction. This outcome can be obtained by applying to the electrodes an external potential difference higher than that of the cell. As the current drawn from a reversible cell tends to zero, the potential difference measured between the two electrodes is known as the electromotive force (E_{cell}), or *e.m.f.*, of the cell. The *e.m.f.* of any particular cell is a quantitative measure of the tendency of the cell reaction to occur spontaneously, and it is correlated with the free energy expended by this electrochemical process. Indeed, the maximum electric work that can be obtained from the cell for 1 mol of reactant is

$$w_{el,max} = nFE \tag{3.9}$$

By definition, $w_{el,max}$ is equal to the Gibbs free energy of the reaction (ΔG_{rxn}) associated with 1 mol of reactant, thus

$$\Delta G_{rxn} = -nFE \tag{3.10}$$

Galvanic cells can only supply electric work equal to the free energy change associated with the redox reaction occurring during the electrochemical process when the current flow tends to zero. The cell voltage under load is always smaller than the *e.m.f.*, so only part of the thermodynamically available work can be exploited.

3.5 The Spontaneous Processes and the Nernst Equation

The direction to which the reaction moves spontaneously at a constant T imposes a negative ΔG_{rxn}, where

$$(\Delta G_{rxn}) = \Delta(H - TS) = \Delta H_{rxn} - T\Delta S_{rxn} = G(\text{products}) - G(\text{reagents}) \tag{3.11}$$

For an overall electrochemical reaction at T and P constant,

$$(\Delta G_{rxn})_{T,P} = \int_{\text{reactants}}^{\text{products}} dG = \int_{\text{reactants}}^{\text{products}} \sum_i \mu_i dn_i \tag{3.12}$$

where μ_i is the chemical potential of the ith species, and dn_i are the changes in the number of moles of each component. It is important to notice that the chemical potential is evaluated relative to the chemical potential of the standard state μ_i^0:

$$\mu_i = \mu_i^0 + \int_{P^\circ}^{P} \left(\frac{\partial \mu_i}{\partial P}\right)_{T,n_i} \mathrm{d}P = \mu_i^0 + \int_{P^\circ}^{P} V_{m,i}\mathrm{d}P \tag{3.13}$$

where $V_{m,i} = \left(\frac{\partial \mu}{\partial P}\right)_{T,n_i}$ is the molar volume of the ith component on P. The standard state depends on the system and defines the reference chemical potential as follows:

(1) *Gases*: A gas is in its standard state at a given temperature when its pressure is the standard pressure P° (1 bar):
$\mu_i^0 = \mu_i(P^\circ)$ for an ideal gas
(2) *Pure solids or liquids*: The standard state refers to pure solid or liquids at the standard ambient pressure of 1 bar:

$$\mu_i^0 = \mu_i(P^\circ)$$

(3) *Liquids in solution*: The standard state of solutions corresponds to that of pure liquids, regardless of whether the liquid is the solvent or the solute:
$\mu_i^0 = \mu_i(P^\circ)$ liquids in solution
(4) *Solid or gas solutes in solution*: In this case, the standard state is chosen to be the solute in solution at some fixed concentration. For many systems, the standard state corresponds to 1 M (1 mol of solute per liter of solution, 1 M) or 1 m (1 mol solute per kilogram of solvent, 1 m). In this case

$$\mu_i^0 = \lim_{[\text{ith}]\to 1M} \mu_i(\text{sol}) \quad \text{for both solid(s) or gas(es) in solution.}$$

To transform Eq. (3.13) into an equation useful for experiments we have the possibility to evaluate its integral in two different ways:

(a) $V_m = {RT}/{P}$ for ideal gases. In this case, $\mu_i = \mu_i^0 + RT\ln\frac{P}{P^\circ}$ (3.14)

(b) $V_m = constant$ for pure, incompressible liquid or solids. In this case, $\mu_i = \mu_i^0 + V_m(P - P^\circ)$

For nonideal systems:

$$\mu_i = \mu_i^\circ + RT\ln a_i \tag{3.15}$$

where a_i is an empirical factor usually indicated as the "*activity*" of ith species. For ideal gases the activity is a dimensionless figure (P/P°). If the empirical potential is expressed in terms of mole fractions, x_i, the dimensionless activity coefficient γ_i is defined:

$$\gamma_i = \frac{a_i}{x_i} \tag{3.16}$$

When $\gamma_i = 1$, the activity is equal to the mole fraction (ideal solution). However, the compositions of real systems are often expressed as molarities (M_i) or molalities (m_i), and in these cases:

$$a_i = \gamma_i\frac{M_i}{M^0} \text{ and } a_i = \gamma_i\frac{m_i}{m^0} \tag{3.17}$$

where M^0 and m^0 are the standard molarity and molality, respectively.

Now, if we consider Eqs. (3.15) and (3.12), we can obtain

$$\Delta G_{\mathrm{rxn}} = \sum_i \nu_i \mu_i^0 + RT \sum_i \nu_i \ln a_i = \Delta G_{\mathrm{rxn}}^0 + RT \ln \xi \tag{3.18}$$

where $\Delta G_{\mathrm{rxn}}^0 = \sum_i \nu_i \mu_i^0$, ν_i are the stoichiometry coefficients, which are positive for products and negative for reagents. The activity ratio, ξ, is given by

$$\ln \xi = \sum_i \nu_i \ln a_i = \ln \left(\prod_i a_i^{\nu_i} \right) \tag{3.19}$$

When an electrochemical reaction is at equilibrium, $\Delta G_{\mathrm{rxn}} = 0$, and Eq. (3.18) becomes

$$\Delta G_{\mathrm{rxn}}^0 = -RT \ln \xi = -RT \ln K_{\mathrm{eq}} \tag{3.20}$$

where K_{eq} is the equilibrium constant of the reaction.

Combining Eq. (3.19) with Eqs. (3.18) and (3.10) the general equation (3.1) is obtained:

$$E = E^0 - \frac{RT}{nF} \ln \xi = E^0 - \frac{RT}{nF} \ln \prod_i a_i^{\nu_i} \tag{3.21}$$

Here, $E°$ is the standard potential ($\Delta G_{\mathrm{rxn}}^0 = -nFE^0$), and Eq. (3.21) is the well-known Nernst equation.

For a spontaneous reaction it is expected that $E > 0$ and $\Delta G_{\mathrm{rxn}} < 0$.

3.6 Representation of an Electrochemical Cell and the Nernst Equation

An electrochemical cell is written from left to right on the basis of the following assumptions [1, 2]:

i. the electrode with the lower reduction potential (i.e. the negative electrode or anode) is written to the left. This is the half-cell where the oxidation process takes place;
ii. moving from the left to the right follows the electrolyte. This is the component acting as an ion sink, capable of absorbing or expelling ions following the electrode processes;
iii. after the electrolyte follows the cathode. This is the electrode with the higher reduction potential (i.e. the positive electrode or cathode). This is the half-cell where the reduction process takes place;
iv. each interface is indicated by a single vertical line. This represents an electrode–electrolyte phase separation;
v. a double vertical line is drawn when two types of interfaces are present. An example is when two half-cells are separated by means of a salt bridge or a porous pot not in direct contact with the electrode compartment;

vi. for each species its physical state and its activity (i.e. the concentration or partial pressure) should be indicated.

Following these rules, positive charges flow from the left to the right of the cell across the electrolyte, and the *e.m.f.* (cell potential, $E°_{cell}$) is $E^0_{cell} = E^0_C - E^0_A$, where E^0_C and E^0_A are the standard potentials of the half-reactions at the cathode and anode, respectively. The E^0_{cell} has the same voltage of the cathode electrode, when the $E°$ of anode is assumed equal to zero. For example, a proton-exchange membrane fuel cell (PEMFC) fed with hydrogen and oxygen gases can be represented as follows:

$$\underbrace{\mathrm{Pt}|\mathrm{H_2}(p_1\ \mathrm{atm}),\mathrm{H^+}}_{\text{anode}}|\underbrace{\text{ion conducting membrane}}_{\text{membrane}}|\underbrace{\mathrm{H_2O_{(vap)}}(p_2\ \mathrm{atm}),\mathrm{O_2}(p_2\ \mathrm{atm})|\mathrm{Pt}}_{\text{cathode}}$$

or

$$\mathrm{Pt}\,|\,\mathrm{H_2}(p_1\ \mathrm{atm}),\mathrm{H^+}\|\mathrm{H_2O_{(vap)}}(p_2\ \mathrm{atm}),\mathrm{O_2}(p_2\ \mathrm{atm})\,|\,\mathrm{Pt}$$

with || = ion conducting membrane.

For a hydrogen–oxygen fuel cell, the two half-reactions and the overall reaction are

$$\begin{cases} -1\mathrm{x}\ [2\mathrm{H^+} + 2\mathrm{e^-} \leftrightarrows \mathrm{H_2}] & E^0_{\mathrm{H^+/H_2}} = +0.00\ \mathrm{V} \\ \frac{1}{2}\mathrm{O_2} + 2\mathrm{e^-} + 2\mathrm{H^+} \leftrightarrows \mathrm{H_2O} & E^0_{\mathrm{O_2/H_2O}} = +1.229\ \mathrm{V} \end{cases}$$

$$\mathrm{H_2} + \frac{1}{2}\mathrm{O_2} \leftrightarrows \mathrm{H_2O} \tag{3.22}$$

On the basis of the relationship between ΔG_{rxn} and *e.m.f.*, for a hydrogen-oxygen PEMFC, $E^0_{cell} = 1.229$ V.

The application of the Nernst Equation (3.21), rewritten as

$$E = E^0 + \frac{RT}{nF}\ln\frac{\prod_{i=1}^{r} a^{\nu_i}_{\mathrm{reagents}}}{\prod_{i=1}^{p} a^{\nu_i}_{\mathrm{products}}} \tag{3.23}$$

to the half-reactions of Eq. (3.22), in the convention of a reduction process, yields

$$\begin{cases} E_{\mathrm{H^+/H_2}} = E^0_{\mathrm{H^+/H_2}} + \frac{RT}{2F}\ln\frac{[\mathrm{H^+}]^2}{p_{\mathrm{H_2}}} \\ E_{\mathrm{O_2/H_2O}} = E^0_{\mathrm{O_2/H_2O}} + \frac{RT}{2F}\ln\frac{p^{1/2}_{\mathrm{O_2}}[\mathrm{H^+}]^2}{a_{\mathrm{H_2O}}} \end{cases} \tag{3.24}$$

The *e.m.f.* of the overall reaction (E_{cell}) is:

$$E_{\mathrm{cell}} = E_{\mathrm{O_2/H_2O}} - E_{\mathrm{H^+/H_2}} = \left(E^0_{\mathrm{O_2/H_2O}} - E^0_{\mathrm{H^+/H_2}}\right) + \frac{RT}{2F}\ln\frac{p_{\mathrm{H_2}}p^{1/2}_{\mathrm{O_2}}}{a_{\mathrm{H_2O}}} \tag{3.25}$$

the latter corresponds to the application of Eq. (3.23) to the overall reaction, described in Eq. (3.22).

It should be noted that in general (i) $a_{\mathrm{H_2O}} = 1$; and (ii) the performances of the electrodes depend on the partial pressure of reagents and on pH. Thus, Eqs. (3.24) can be rewritten as

$$\begin{cases} E_{\mathrm{H^+/H_2}} = E^0_{\mathrm{H^+/H_2}} - \frac{RT}{0.4343\,F}\mathrm{pH} - \frac{RT}{2F}\ln p_{\mathrm{H_2}} \\ E_{\mathrm{O_2/H_2O}} = E^0_{\mathrm{O_2/H_2O}} - \frac{RT}{0.4343\,F}\mathrm{pH} + \frac{RT}{4F}\ln p_{\mathrm{O_2}} \end{cases} \tag{3.26}$$

highlighting that the best performance of a cell, at a given reagent partial pressure, is obtained when the pH at the anode (A) is coinciding with that of the cathode (C) electrode.

3.7 The Electrochemical Series

To better understand the importance of Eq. (3.21), let us assume $E^0_{H^+/H_2}$ for the anodic half-reaction [2, 4]:

$$2H^+_{(aq)} + 2e^- \rightleftarrows H_{2(g)} \qquad E^0_{H^+/H_2} = 0.0\ V$$

By combining the above H^+/H_2 standard reaction with all the redox couples M/M^{n+} in a cell with the following configuration

$$Pt\,|\,H_2(1\ atm), H^+(1\ M)||M^{n+}(1\ M)\,|\,M$$

it is possible to determine the standard electrode potentials of all the half-reactions reported in Table 3.1.

From Table 3.1, it is easy to conclude that the standard potentials of the half-reactions taking place in a fuel cell mounting an acidic electrolyte and fed with hydrogen and oxygen gases:

$$\begin{cases} 2H^+ + 2e^- \rightarrow H_2 & E^0 = +0.00\ V \\ \frac{1}{2}O_2 + 2e^- + 2H^+ \rightarrow H_2O & E^0 = +1.229\ V \end{cases} \tag{3.27}$$

allow to find the direction of the H_2/O_2 overall reaction, highlighting that:

(1) $E^0_{O_2/H_2O} > E^0_{H^+/H_2}$ and $\Delta G^0_{O_2/H_2O} < \Delta G^0_{H^+/H_2}$. This demonstrates that the reduction of oxygen is more spontaneous than that of H^+. Therefore, it is expected that in a fuel cell fed with H_2 and O_2, H_2 is oxidized, and O_2 is reduced. The overall reaction is then

$$\frac{1}{2}O_2 + H_2 \rightarrow H_2O, \text{where } E^0_{cell} = E^0_c - E^0_A = E^0_{\frac{O_2}{H_2O}} - E^0_{\frac{H^+}{H_2}} = +1.229\ V \tag{3.28}$$

(2) to obtain the correct stoichiometry in Eqs. (3.27), the half-reaction

$$O_2 + 4e^- + 4H^+ \rightarrow 2H_2O$$

is multiplied by ½. It is to be noticed that in the evaluation of E^0_{cell}, $E^0_{O_2/H_2O}$ must not be multiplied by the coefficient ½.

The study of the oxidation process of methanol can be carried out in a similar way, using the following half-reactions:

$$\begin{cases} CO_2 + 6H^+ + 6e^- \rightarrow CH_3OH + H_2O & E^0_{CO_2/CH_3OH} = -0.03\ V \\ O_2 + 4e^- + 4H^+ \rightarrow 2H_2O & E^0_{O_2/H_2O} = +1.229\ V \end{cases} \tag{3.29}$$

Also in this case, the reduction of oxygen is the most spontaneous process. The overall reaction, which can be exploited in a direct methanol fuel cell (DMFC), is

Table 3.1 Standard potentials of reduction in aqueous solution at 25 °C.

Half-reaction	E^0 (V)	Half-reaction	E^0 (V)
$Li^+ + e^- \rightleftarrows Li$	−3.040	$S_4O_6^{2-} + 2e^- \rightleftarrows 2S_2O_3^{2-}$	0.081
$K^+ + e^- \rightleftarrows K$	−2.931	$S + 2H_2O^+ + 2e^- \rightleftarrows H_2S + 2H_2O$	0.142
$Ca^{2+} + 2e^- \rightleftarrows Ca$	−2.868	$Sn^{4+} + 2e^- \rightleftarrows Sn^{2+}$	0.151
$Na^+ + e^- \rightleftarrows Na$	−2.7109	$Cu^{2+} + e^- \rightleftarrows Cu^+$	0.153
$Mg^{2+} + 2e^- \rightleftarrows Mg$	−2.372	$Hg_2Cl_2 + 2e^- \rightleftarrows 2Hg + 2Cl^-$	0.2681
$H_3O^+ + e^- \rightleftarrows H_2O + H$	−2.100	$Cu^{2+} + 2e^- \rightleftarrows Cu$	0.342
$Al^{3+} + 3e^- \rightleftarrows Al$	−1.662	$O_2 + 2H_2O + 4e^- \rightleftarrows 4OH^-$	0.401
$Ti^{2+} + 2e^- \rightleftarrows Ti$	−1.630	$Cu^+ + e^- \rightleftarrows Cu$	0.521
$ZnO_2^{2-} + 2H_2O + 2e^- \rightleftarrows Zn + 4OH^-$	−1.215	$I_2 + 2e^- \rightleftarrows 2I^-$	0.535
$Mn^{2+} + 2e^- \rightleftarrows Mn$	−1.185	$O_2 + 2H_3O^+ + 2e^- \rightleftarrows H_2O_2 + 2H_2O$	0.682
$Cr^{2+} + 2e^- \rightleftarrows Cr$	−0.913	$Fe^{3+} + e^- \rightleftarrows Fe^{2+}$	0.771
$2H_2O + 2e^- \rightleftarrows H_2 + 2OH^-$	−0.8277	$Hg_2^{2+} + 2e^- \rightleftarrows Hg$	0.7973
$Te + 2H_3O^+ + 2e^- \rightleftarrows H_2Te + 2H_2O$	−0.783	$Ag^+ + e^- \rightleftarrows Ag$	0.7996
$Zn^{2+} + 2e^- \rightleftarrows Zn$	−0.7618	$2NO_3^- + 4H_3O^+ + 2e^- \rightleftarrows N_2O_4 + 6H_2O$	0.803
$Cr^{3+} + 3e^- \rightleftarrows Cr$	−0.744	$NO_3^- + 3H_3O^+ + 2e^- \rightleftarrows HNO_2 + 4H_2O$	0.934
$As + 3H_3O^+ + 3e^- \rightleftarrows AsH_3 + 3H_2O$	−0.608	$NO_3^- + 4H_3O^+ + 3e^- \rightleftarrows NO + 6H_2O$	0.957
$H_3PO_2 + H_3O^+ + e^- \rightleftarrows P + 3H_2O$	−0.508	$Br_2 + 2e^- \rightleftarrows 2Br^-$	1.065
$Fe^{2+} + 2e^- \rightleftarrows Fe$	−0.447	$Pt^{2+} + 2e^- \rightleftarrows Pt$	1.118

$Cr^{3+} + e^- \rightleftarrows Cr^{2+}$	−0.407	$MnO_2 + 4H_3O^+ + 2e^- \rightleftarrows Mn^{2+} + 6H_2O$	1.224
$Cd^{2+} + 2e^- \rightleftarrows Cd$	−0.403	$O_2 + 4H_3O^+ + 4e^- \rightleftarrows 6H_2O$	1.229
$Se + 2H_3O^+ + 2e^- \rightleftarrows H_2Se + 2H_2O$	−0.400	$Cr_2O_7^{2-} + 14H_3O^+ + 6e^- \rightleftarrows 2Cr^{3+} + 21H_2O$	1.232
$Tl^+ + e^- \rightleftarrows Tl$	−0.336	$Cl_2 + 2e^- \rightleftarrows 2Cl^-$	1.358
$Co^{2+} + 2e^- \rightleftarrows Co$	−0.277	$ClO_3^- + 6H_3O^+ + 6e^- \rightleftarrows Cl^- + 9H_2O$	1.451
$Ni^{2+} + 2e^- \rightleftarrows Ni$	−0.257	$PbO_2 + 4H_3O^+ + 2e^- \rightleftarrows Pb^{2+} + 6H_2O$	1.455
$Mo^{3+} + 3e^- \rightleftarrows Mo$	−0.200	$MnO_4^- + 8H_3O^+ + 5e^- \rightleftarrows Mn^{2+} + 12H_2O$	1.507
$Sn^{2+} + 2e^- \rightleftarrows Sn$	−0.1374	$HClO + H_3O^+ + e^- \rightleftarrows \frac{1}{2}Cl_2 + 2H_2O$	1.611
$Pb^{2+} + 2e^- \rightleftarrows Pb$	−0.1263	$H_2O_2 + 2H_3O^+ + 2e^- \rightleftarrows \frac{1}{2}Cl_2 + 2H_2O$	1.776
$2H_3O^+ + 2e^- \rightleftarrows 2H_2O + H_2$	0	$Co^{3+} + e^- \rightleftarrows Co^{2+}$	1.83
$NO_3^- + H_2O + 2e^- \rightleftarrows NO_2^- + 2OH^-$	0.010	$F_2 + 2e^- \rightleftarrows 2F^-$	2.866

obtained by multiplying the half-reaction involving CO_2 by −3/2 and then adding the oxygen reduction. The result is the following:

$$CH_3OH + \frac{3}{2}O_2 \rightarrow CO_2 + 2H_2O$$

The net cell potential is $E^0_{cell} = E^0_c - E^0_A = E^0_{\frac{O_2}{H_2O}} - E^0_{\frac{CO_2}{CH_3OH}} = +1.199$ V. It should also be noted that in this case E^0_{cell} is determined without multiplying $E^0_{O_2/H_2O}$ by −3/2.

In summary, the Nernst equation is an important tool to link the cell potential to the nature of the electroactive species (i.e. E^0) and to their activities. In practice, E_{cell} diminishes as the reaction proceeds, and the reagents are converted into products.

3.8 Dependence of the E_{cell} on Temperature and Pressure

The dependence of the cell potential (E_{cell}) on temperature and pressure is deduced from $\Delta G = -nFE_{cell}$ and the Gibbs free energy equation [1, 2, 7]:

$$dG = -SdT + Vdp \tag{3.30}$$

where

$$\left(\frac{\partial \Delta G}{\partial T}\right)_P = -\Delta S \text{ or } \left(\frac{\partial E_{cell}}{\partial T}\right)_P = \frac{\Delta S}{nF} \tag{3.31}$$

and

$$\left(\frac{\partial \Delta G}{\partial P}\right)_T = \Delta V \text{ or } \left(\frac{\partial E_{cell}}{\partial P}\right)_T = -\frac{\Delta V}{nF} \tag{3.32}$$

At constant pressure, the potential gradient of the cell can be either positive or negative on the basis of the sign of the entropy variation during the reaction. It is observed that ΔS, and thus the temperature coefficient of the E_{cell}, does not change significantly on T when no gases are produced during the reaction.

The temperature coefficient of Eq. (3.31) describes the heat generated or absorbed by a cell under operation conditions. Indeed, a galvanic cell operating at a small current in order to guarantee a reversible equilibrium of processes in the cell absorbs or yields a finite amount of heat for the reaction conversion. The heat generation or absorption by the cell is known as the Peltier effect. This is the result of the difference between the enthalpy change (ΔH) and the free energy change (ΔG). For a reversible reaction, the released energy, $-\Delta H$, does not coincide with the amount of energy converted into work (i.e. the electric work) or to the heat equivalent to the work, ΔG.

If more energy is released by the system into the external environment ($\Delta H < \Delta G$), the excess $\Delta G - \Delta H$ is (i) released by the cell isothermally or (ii) heating the cell. On the contrary, if more energy is converted into work when released by the reaction, the cell cools downs, and external heat should be supplied to the cell to keep its temperature constant. Taking all together, the heat generation of a reversible cell is $\Delta G - \Delta H$, while the heat consumption is $Q_{hc} < 0$.

If we consider that $\Delta H = \Delta G + T\Delta S$, the amount of heat generated is $Q_{hg} = \Delta G - \Delta H = -T\Delta S$, which by Eq. (3.31) allows to obtain

$$Q_{hg} = T\left(\frac{\partial \Delta G}{\partial T}\right)_P = -nFT\left(\frac{\partial \Delta E}{\partial T}\right)_P = -T\Delta S \tag{3.33}$$

and $\Delta H = \Delta G - T\left(\frac{\partial \Delta G}{\partial T}\right)_P$

By integrating Eqs. (3.31) and (3.32), when no gas is produced for a reaction and either T or P is constant:

$$(I)\ P = \text{constant;}\ \int_{298}^{T} \mathrm{d}E_p = \int_{298}^{T} \frac{\Delta S}{nF}\mathrm{d}T; \qquad E_T = E_{298} + \frac{(T-298)\Delta S}{nF} \tag{3.34}$$

$$(II)\ T = \text{constant;}\ \int_{P_0}^{P} \mathrm{d}E_p = -\int_{P_0}^{P} \frac{\Delta V}{nF}\mathrm{d}p; \qquad E_P = E_{P_0} - \frac{1}{nF}\int_{P_0}^{P} \frac{\Delta V}{nF}\mathrm{d}p$$

In II, for a fuel cell operating at a low pressure ($P < 10$ atm), $\Delta V = nRT/P$, yielding

$$E_P = E_{P_0} + \frac{RT}{F}\int_{P_0}^{P} \frac{\mathrm{d}p}{p} = E_{P_0} + \frac{RT}{F}\ln\left(\frac{P}{P_0}\right) \tag{3.35}$$

Results indicate that E_P increases on P of reagents. In summary, Eqs. (3.34) and (3.35) show clearly that E_T changes linearly on T, and E_P changes logarithmically on P.

3.9 Thermodynamic Efficiencies

In the previous section it was discussed that for a reversible reaction ΔH and ΔG correspond to the overall released enthalpy and the energy converted into work by a cell [8], respectively. On this basis, the thermodynamic or ideal efficiency of energy conversion is [8]:

$$\eta_{\text{thermo}} = \frac{\Delta G}{\Delta H} = 1 - T\left(\frac{\Delta S}{\Delta H}\right) \tag{3.36}$$

In typical spontaneous reactions, $\Delta H < 0$, and (i) for $\Delta S < 0$, we obtain efficiencies lower than 1, while (ii) for $\Delta S > 0$, the thermodynamic efficiency is expected to be above 100%. In this latter condition, it is expected that the heat absorbed by the system from the environment is converted into work. Now, if we consider that $\Delta H = -nFE_H^0$, i.e. the overall ΔH is transformed into work, and $\Delta G = -nFE_G^0$, the actual energy converted into work, Eq. (3.36), is

$$\eta_{\text{thermo}} = \frac{E_G^0}{E_H^0} \tag{3.37}$$

which, for a number of reactions used in fuel cells, yields the efficiency values summarized in Table 3.2.

It should be noticed that in general the reactions summarized in Table 3.2 present $\Delta S < 0$ ($T\Delta S < 0$ is the amount of heat released to the environment). The oxidation of

Table 3.2 Thermodynamic data of fuel cell reactions under standard condition at 25 °C. η_{thermo} is the ideal efficiency defined by Eqs. (3.36) and (3.37).

Fuel cell	Reaction	n	$-\Delta H^0$ (kJ mol^{-1})	$-\Delta G^0$ (kJ mol^{-1})	E^0_H (V)	E^0_G (V)	η_{thermo} (%)
Hydrogen	$H_2 + \frac{1}{2}O_2 \rightarrow H_2O$	2	286.0	237.3	1.481	1.229	83.0
CO	$CO + \frac{1}{2}O_2 \rightarrow CO_2$	2	283.1	257.2	1.173	1.066	90.9
Formic acid	$HCOOH + \frac{1}{2}O_2 \rightarrow CO_2 + H_2O_{(l)}$	2	270.3	285.5	1.402	1.48	105.6
Formaldehyde	$CH_2O_{(g)} + O_2 \rightarrow CO_2 + H_2O_{(l)}$	4	561.3	522.0	1.452	1.350	93.0
Methanol	$CH_3OH + \frac{3}{2}O_2 \rightarrow CO_2 + 2H_2O_{(l)}$	6	726.6	702.5	1.255	1.214	96.7
Methane	$CH_4 + 2O_2 \rightarrow CO_2 + 2H_2O_{(l)}$	8	890.8	818.4	1.153	1.060	91.9
Ammonia	$NH_3 + \frac{3}{4}O_2 \rightarrow \frac{1}{2}N_2 + \frac{3}{2}H_2O_{(l)}$	3	382.8	338.2	1.324	1.170	88.4
Hydrazine	$N_2H_4 + O_2 \rightarrow N_2 + 2H_2O_{(l)}$	4	622.4	602.4	1.612	1.560	96.8
Zinc	$Zn + \frac{1}{2}O_2 \rightarrow ZnO$	2	348.1	318.3	1.805	1.650	91.4

formic acid with an ideal efficiency of more than 100% is an exception. Indeed, in this case, ΔG is larger than ΔH, and the conversion of chemical energy into electricity occurs by absorption of heat from the environment.

The overall efficiency of a fuel cell during operation can be determined considering the actual cell potential, $E(i)$, at a fixed current i, and all the possible dissipative phenomena influencing its operation:

$$E(i) = E^0 - iR_m - \sum_{j=1}^{n} \eta_j(i) \tag{3.38}$$

where R_m is the membrane resistance, and $\eta_j(i)$ are the losses occurring at the electrodes associated with the charge transfer diffusion and kinetic events. The overall efficiency, Eq. (3.38), of an operating fuel cell can thus be rewritten as

$$\eta_{cell} = \frac{E(i)}{E^0_H} = \eta_{thermo} - \frac{\left[iR_m + \sum_{j=1}^{n} \eta_j(i)\right]}{E^0_H} \tag{3.39}$$

Now, for an H_2/O_2 fuel cell under practical operating conditions the current densities are usually measured at 0.8 or 0.6 V. In both of these two cases, using the data of Table 3.2, $\eta_{cell}\% = {}^{0.8}/_{1.48} \cdot 100 = 54.1\%$ and $\eta_{cell}\% = {}^{0.6}/_{1.48} \cdot 100 = 40.54\%$ are obtained. η_{cell} values at different temperatures are obtained using the $E(i)$ values determined by Eqs. (3.34) and (3.36). Taking all together the overall effective efficiency of a cell, η_{eff}, can be expressed as

$$\eta_{eff} = \eta_{thermo} \cdot \eta_{cell} \cdot \eta_{rxn} = 50 \div 55\% \tag{3.40}$$

Here, η_{rxn} is the reaction yield given by $\eta_{rxn} = \frac{m_{rxn}(j)}{m_{rxn,T}}$, where $m_{rxn}(j)$ is the mass of reactant consumption, and $m_{rxn,T}$ is the total mass provided to a fuel cell for its operation.

Actually, for an operating fuel cell, η_{rxn} is the fuel utilization efficiency given by

$$\eta_{rxn} = \frac{i/_{nF}}{\rho_{fuel}} \tag{3.41}$$

where i is the current generated by the fuel cell, and ρ_{fuel} is the rate of the fuel supplied to the cell in mol/s. Eq. (3.41) states that (i) when a large excess of fuel is supplied to a cell its overall efficiency is compromised and (ii) it is necessary to adjust the fuel rate in order to provide the cell only a slight excess of fuel. This is usually carried out taking as reference the current density actually delivered by the cell.

The stoichiometric ratio, β, is a useful parameter to describe the composition of reactants feeding a fuel cell (i.e. a fuel cell operating with 1.5 times more fuel than is needed to use 100% of the fuel has $\beta = 1.5$). Thus, by definition $\beta = \frac{\rho_{fuel}}{i/nF}$, and the fuel utilization efficiency is

$$\eta_{rxn} = \frac{1}{\beta} \tag{3.42}$$

In summary, η_{eff} is finally obtained substituting Eqs. (3.37), (3.39), and (3.42) into Eq. (3.40):

$$\eta_{eff} = \eta_{thermo} \cdot \eta_{cell} \cdot \eta_{rxn} = \left(\frac{\Delta G}{\Delta H}\right) \cdot \frac{E(i)}{E_H^0} \cdot \frac{i}{(nF \cdot \rho_{fuel})} \tag{3.43}$$

which, for a cell operating at $E = 0.8$ V under a fixed stoichiometric condition ($\eta_{rxn} = 0.95$), becomes

$$\eta_{eff} = \left(\frac{\Delta G}{\Delta H}\right) \cdot \frac{E(i)}{E_H^0} \cdot \frac{1}{\beta} = 50 \div 55\% \tag{3.44}$$

Finally, the efficiency of a whole device must include η_{system}, which describes all the losses resulting from the other components of the devices ($\eta_{supplies}$):

$$\eta_{system} = \eta_{eff} \cdot \eta_{supplies} = 45 \div 50\% \tag{3.45}$$

3.10 Case Study – The Impact of Thermodynamics on the Corrosion of Low-T FC Electrodes

The electrodes of low-T fuel cells, including proton-exchange membrane fuel cells (PEMFCs) and anion-exchange membrane fuel cells (AEMFCs), typically operate at extreme values of pH [9]. Specifically, the electrodes of PEMFCs and AEMFCs run in a strongly acidic and a strongly alkaline environment, respectively. The abundant $H^+_{(aq)}$ and $OH^-_{(aq)}$ species that are found in these conditions are involved in the electrochemical processes taking place at the electrodes (e.g. the hydrogen oxidation reaction, HOR, at the anode and the oxygen reduction reaction, ORR, at the cathode). Thus, pH plays a crucial role to set the equilibrium potential of each electrode and ultimately determines the overall thermodynamic potential of the FC [10, 11]. However, it must be highlighted that the potential of a certain electrode is also capable to trigger a number of other electrochemical phenomena involving the materials with which the electrode is made, with a particular reference to the electrocatalyst

(EC). State-of-the-art ECs for both the HOR and ORR and for both PEMFCs and AEMFCs typically consist of Pt nanoparticles supported on carbon nanoparticles exhibiting a large surface area and are indicated herein as "*Pt/C ECs*" [12]. In the case of AEMFCs, HOR ECs usually also include Ru alloyed with Pt and are labeled herein as "*PtRu/C ECs*" [13].

The case of PEMFCs, as depicted in Figure 3.4a, is now considered. The cathode EC is meant to promote the ORR, whose equilibrium potential is equal to 1.23 V *vs.* SHE. This value is larger than the equilibrium potentials of the processes associated with the dissolution of Pt and with the formation of PtO_x species (e.g. PtO and PtO_2), which are poor ORR ECs in acidic conditions [14]. Hence, in principle, an EC bearing Pt active sites would be unable to promote effectively the ORR. However, in practice the cathode of a PEMFC operates in the potential range between 0.5 and 0.8 V, mainly due to the large kinetic overpotential of the ORR [10]. In these conditions, the electrode potential is low enough to prevent the dissolution of Pt species and the formation of platinum oxides; platinum is present at the PEMFC cathode as Pt(0), which is thermodynamically stable and is a good ORR electrocatalyst [11, 15]. The equilibrium potential of the anode is 0 V *vs.* SHE. Also in these conditions Pt is present as a thermodynamically stable Pt(0), which is an excellent EC for the HOR [9, 16]. In conclusion, in an operating PEMFC Pt is present as Pt(0) both at the anode and at the cathode and is expected to be stable. Pt dissolution/formation of PtO_x species are actually important phenomena as concerns EC degradation, since they are on the basis of the aggregation of the Pt nanoparticles and the resulting loss of electrochemically active surface area (ECSA). However, Pt dissolution/formation of PtO_x species are mostly triggered by situations where the electrode potential becomes very large (e.g. in transients during PEMFC start/stop and at open-circuit potential, OCP) and not by "*normal*" operation [17]. The cathode potential of a PEMFC is significantly larger than the equilibrium potential of the CO_2/C redox couple, see Figure 3.4a. Thus, in operating conditions the carbonaceous EC support is prone to oxidation, yielding $CO_{x(s)}$ species (i) characterized by a poor electrical conductivity and (ii) favoring the detachment of the Pt nanoparticles bearing the active sites [18]. These are two additional important mechanisms involved in the degradation of the cathodic EC, which are further exacerbated by the highly oxidizing environment due to the O_2 reactant at the cathode. State-of-the-art ORR ECs are able to operate at all the cathodes of PEMFCs due to the significant "*kinetic*" tolerance of the carbonaceous supports to oxidation. Still, major efforts are dedicated by modern research to raise the tolerance to oxidation of the supports for ORR ECs, with the purpose to extend the durability of the corresponding PEMFC [19]. On the other hand, the anode potential of PEMFCs (*ca.* 0 V *vs.* SHE, also considering the small kinetic overpotentials associated with the HOR promoted by an EC bearing Pt active sites) falls in the stability potential window of graphitic carbon, see Figure 3.4a [11]. Hence, in principle, the carbonaceous support of conventional state-of-the-art Pt/C ECs used at the anode of PEMFCs is not prone to oxidative corrosion during normal PEMFC operation.

The electrochemical phenomena involved in AEMFC operation and EC degradation are quite similar to those described above for PEMFCs, see Figure 3.4b. Some

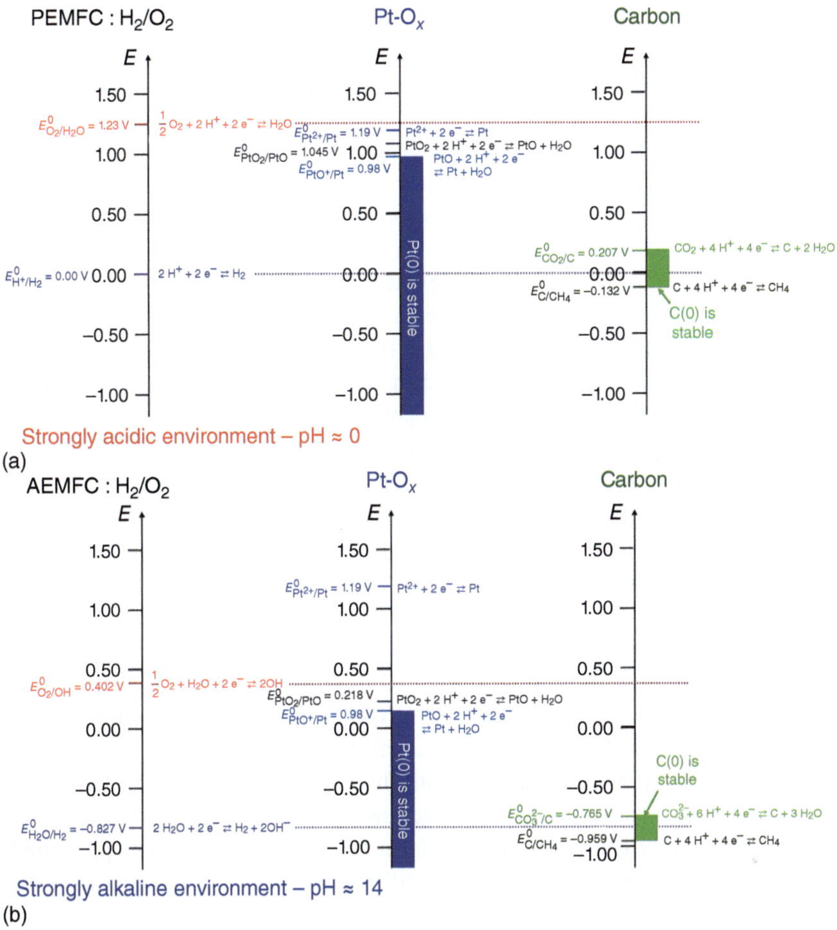

Figure 3.4 Impact of pH on the equilibrium potentials of the main electrochemical processes involved in the operation and electrode degradation reactions of PEMFCs (a) and AEMFCs (b).

differences are present due to the involvement of $OH^-_{(aq)}$ and species that are stable in the alkaline environment (e.g. CO_3^{2-} anions instead of the $CO_{2(g)}$ species found in the acidic environment). The pH affects significantly the equilibrium potentials of most processes involved in FC operation. For instance, the equilibrium potential of the ORR is equal to 1.23 and 0.40 V *vs.* SHE in acidic and alkaline conditions, respectively, see Figure 3.4b. However, it is shown that pH tends to shift the equilibrium potentials of all the processes involved in AEMFC operation and EC degradation by similar figures [11]. Consequently, most of the considerations reported above about the electrochemical phenomena prompting EC degradation retain their validity. In the case of platinum, it is shown that the equilibrium potential associated with Pt dissolution (that is not affected by pH) is much higher than the equilibrium electrode potential arising from the ORR. Thus, platinum dissolution is not expected to take place during normal AEMFC operation. On the other hand, the equilibrium potential associated with the formation of PtO_x species is still lower than that of the cathodic electrode. However, also in the case of AEMFCs the cathodic electrode operates at a significantly lower potential than that determined by thermodynamics due to the ORR kinetic overpotentials on the Pt active sites of the Pt/C EC. In these conditions Pt is found as thermodynamically stable Pt(0) able to operate as a good EC for the ORR. Also in the case of AEMFCs: (i) the formation of PtO_x species and consequent agglomeration of Pt active sites are mostly triggered by potential swings during start–stop cycles and OCP conditions; and (ii) the platinum at the anode is found as a thermodynamically stable Pt(0) able to promote very effectively the HOR without triggering thermodynamically driven EC degradation phenomena [11]. In the specific case of AEMFCs, the Ru found at the anode is found as thermodynamically stable Ru(0) [11]. Similar to a PEMFC, in an AEMFC: (i) the cathode potential is significantly larger than the equilibrium potential of the CO_3^{2-}/C redox couple; and (ii) the anode potential falls in the stability potential window of graphitic carbon, see Figure 3.4b. Thus, also in the case of AEMFCs, normal cell operation triggers the thermodynamically driven corrosion of the carbon support of the ORR EC as exacerbated by the oxidative environment while it is not expected to prompt significant corrosion phenomena on the anodic EC support.

In conclusion, during normal operation of both PEMFCs and AEMFCs, thermodynamics indicates that EC degradation is prompted by the oxidation of the carbonaceous support of the cathodic Pt/C EC. Potential swings during start–stop cycles and OCP conditions would also favor the degradation of the cathodic EC by means of Pt dissolution (for PEMFCs only)/formation of PtO_x species (for both PEMFCs and AEMFCs) that promote the agglomeration of Pt active sites. The potential at which the anodic Pt/C or PtRu/C EC operates in PEMFCs and AEMFCs, respectively, is not expected to trigger degradation phenomena prompted by simple thermodynamics in both the Pt/PtRu active sites and the carbonaceous support.

3.11 Reaction Kinetics and Fuel Cells

3.11.1 Correlation Between Current and Reaction Kinetics

Three different types of charge migration processes take place in an operating fuel cell: (i) the flow of electrons (e.g. in electrode materials, terminal connectors, and load resistance); (ii) the flow of ions in the electrolyte (e.g. aqueous solutions, solid electrolyte, and membranes); and (iii) the charge transfer reactions at the electrolyte/electrode interface [1, 2].

At the *steady state*, charge neutrality is present in each macroscopic section of the cell, or, in other words, in each cross section of the circuit the total flow of charge must be the same. Thus, the rate of the electron flow in the external circuit is equal to the rate of charge transfer at each electrode/electrolyte interface. For a general electrode process:

$$O_1 + ne^- \rightarrow R_1$$

$n{\cdot}F$ coulombs are required to reduce 1 mol of O_1. Thus, the rate of reduction of O_1 at the electrode is given by

$$\frac{dn_{O_1}}{dt} = \frac{1}{nF} \cdot \frac{dQ}{dt} = \frac{1}{nF} \cdot i \tag{3.46}$$

where n_{O_1} are the moles of O_1, Q is the total charge, and i is the total current. Eq. (3.46) represents the rate of reduction of O_1 [mol s^{-1}]. Similarly, during the oxidation process the following reaction occurs:

$$R_2 \rightarrow O_2 + n'e^-$$

The rate of oxidation of R_2 species is equal to

$$\frac{dn_{R_2}}{dt} = \frac{1}{n'F} \cdot \frac{dQ}{dt} = \frac{1}{n'F} \cdot i \tag{3.47}$$

If the kinetics of a process is not fast enough to maintain the rate of the other processes occurring in the electrochemical cell, it defines the *current limiting process*. Thermodynamics can predict the direction of a reaction and the *e.m.f.* characterizing the investigated process. Nevertheless, it cannot give any indications on the kinetics and rate of reactions.

3.11.2 The Concept of Exchange Current

The exchange current (i_0) parameter plays a crucial role in the evaluation of the activity of a particular electrode/electrolyte interface [1, 2]. To clarify this concept, we consider what happens when a zinc electrode is immersed in a solution containing $ZnSO_4$. In this case, the following two charge transfer processes occur simultaneously at the electrode/electrolyte interface:

$$\begin{gathered} Zn^{2+}_{(aq)} \xrightarrow{2e^-} Zn_{(s)} \\ Zn_{(s)} \xrightarrow{-2e^-} Zn^{2+}_{(aq)} \end{gathered} \tag{3.48}$$

The first reaction (a reduction) directly correlates with the cathodic current (i.e. i_c), while the second reaction corresponds to the anodic current of the process (i.e. i_a). At equilibrium, i.e. when no net current is flowing from the electrode ($i = i_c - i_a = 0$), the rate associated with the cathodic process is equal to that of the anodic process; thus, $i_c = i_a = i_0$. In this condition, the parameter i_0 is defined as the "*exchange current*". When a positive potential is applied to the Zn electrode, it becomes easier to oxidize, and its reduction is more difficult. In terms of currents, this means that $i_a > i_c$; thus, $|i_{Zn}| = i_c - i_a \neq 0$.

3.12 Charge Transfer Theory Based on Distribution of Energy States

A very general theory describing the driving force of the charge transfer processes occurring at the electrode/electrolyte interface is based on the description of the overlap occurring between the electronic states of the electrode and the electronic states of reactants present in solution [20–25]. These phenomena are responsible of the current i and the exchange current i_0 and thus of the kinetics characterizing the electrode reactions. To discuss this topic in general, here we would like to (i) analyze the electronic structure of the electrode and (ii) elucidate the electronic configuration of electroactive species present in the electrolyte. The driving force for an electron transfer process is in the migration of one electron from an occupied state to an empty state of another species at the same energy. If the process is a reduction, the occupied state is on the electrode, and the empty state is expected on the electroactive component present in solution (O). For the oxidation process the opposite process takes place. These concepts will be clear if we analyze Figure 3.5 in detail.

Figure 3.5 shows that the number of the electronic state from $\mathcal{E}$ to $\mathcal{E} + d\mathcal{E}$ energy range is

$$d\mathcal{N}(\mathcal{E}) = A \cdot \mathfrak{D}(\mathcal{E})d\mathcal{E} \tag{3.49}$$

where A is the area of the electrode in contact with the electrolyte, and $\mathfrak{D}(\mathcal{E})$ is the density of states (the number of electronic states per unit energy per area, [area·energy]$^{-1}$, in cm^{-2} eV^{-1}) [26]. For a material at the absolute zero temperature,

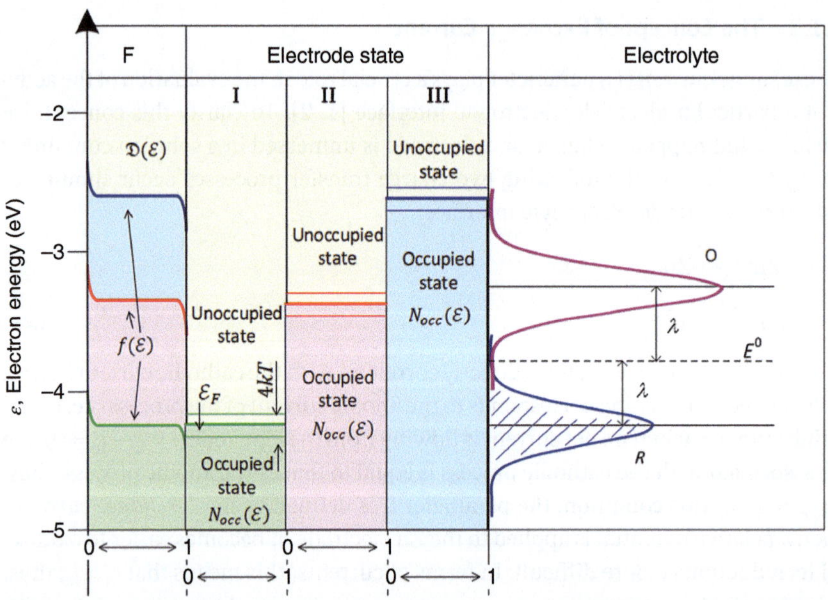

Figure 3.5 Energy map of electronic states on a metal electrode and on electroactive species O and R present in the electrolyte phase. The ordinate shows the absolute electron energy, $\mathcal{E}$. $\mathcal{E}_F$ is the energy of the Fermi level. $f(\mathcal{E})$ is the Fermi–Dirac distribution function. The Gaussian profiles in the electrolyte side correspond to the probability functions $\mathcal{P}_O(\lambda, \mathcal{E})$ and $\mathcal{P}_R(\lambda, \mathcal{E})$ of O and R species, respectively. The electron energy corresponding to the standard potential of redox species is $E° \approx -3.8$ eV with a $\lambda = 0.3$ eV. In the three electrode states (I, II, and III), the occupied states are shown by colored shading ($\mathcal{N}_{occ}(\mathcal{E}) = f(\mathcal{E}) \cdot \mathfrak{D}(\mathcal{E})$). F shows in yellow shading the density of state, i.e. the number of states for unit of energy, $\mathfrak{D}(\mathcal{E})$.

the energy of the highest filled state corresponds to the energy of Fermi level. This coincides with the chemical potential of the electron in this state. As temperature rises, the thermal energy stimulates the transition of some of the electrons above $\mathcal{E}_F$, creating vacancies in states below this latter energy. The thermal filling of the states centered at around $\mathcal{E}_F$ is described by the Fermi–Dirac distribution function [26]:

$$f(\mathcal{E}) = \frac{1}{1 + e^{(\mathcal{E}-\mathcal{E}_F)/kT}} \tag{3.50}$$

where $f(\mathcal{E})$ is the probability that at a given temperature a state of energy $\mathcal{E}$ is occupied by an electron, and k is the Boltzmann constant. At $\mathcal{E} < \mathcal{E}_F, f(\mathcal{E}) \approx 1$, while at $\mathcal{E} \gg \mathcal{E}_F, f(\mathcal{E}) \approx 0$, see Figure 3.5F. Thus, the sigmoidal profile $f(\mathcal{E})$ of Figure 3.5F delimits the occupied density of states below $f(\mathcal{E})$ from the unoccupied density of state which are above $f(\mathcal{E})$ and which are marked with a yellow background, Figure 3.5F. At $\varepsilon = \varepsilon_F, f(\mathcal{E}) = 1/2$. The states with energies comprised in the range $\varepsilon_{F\_}\ 2kT \leq \varepsilon \leq \varepsilon_F + 2kT$ present $f(\mathcal{E})$ values ranging from 0 to 1. The number of electrons from $\mathcal{E}$ and $\mathcal{E} + d\mathcal{E}$ is given by

$$\mathrm{d}N_e = 2A \cdot N_{\mathrm{occ}}(\mathcal{E})\mathrm{d}\mathcal{E} \tag{3.51}$$

where $N_{\mathrm{occ}}(\mathcal{E}) = f(\mathcal{E})\mathfrak{D}(\mathcal{E})$ is the density of the state function occupied by electrons with units of [area·energy]$^{-1}$, i.e. cm^{-2} eV^{-1}. Concurrently, the density of unoccupied states is

$$N_{\mathrm{unocc}}(\mathcal{E}) = [1 - f(\mathcal{E})]\mathfrak{D}(\mathcal{E}) \tag{3.52}$$

As shown in Figure 3.5, the Fermi level changes as the potential (E) of the electrode is changed, see Figure 3.5, I, II, and III. Higher energies, $\mathcal{E}$, of electrons in the electrode correspond to more negative $\Delta\mathcal{E}$ ($\Delta\mathcal{E} = -\nu FE$). The opposite behavior is observed as $\mathcal{E}$ is lowered.

These electrode modulations take place, thanks to the charging phenomena of the metal electrode. Therefore, all states are concurrently shifted depending on the external potential. Consequently, $\mathfrak{D}(\mathcal{E})$ is mostly independent of ε_F, and the distribution of states around the Fermi level at all potentials is the same. Thus, the profile of $f(\mathcal{E})$, which depends on ε_F, shifts up and down on ε_F, see Figure 3.5F.

In the electrolyte, R and O species correspond to two species with states filled with one electron or empty, respectively. It should be highlighted that R and O species are in the electrolyte, and they are able to exchange electrons with the electrode only if their electronic functions are able to interact with that of the electrode. This is possible only if R and O species are adsorbed on the surface of electrode. This allows us to express the density of states of R and O species in terms of their concentration, rather than their total number. At any moment, the number of electrons provided by R species to the electrode can be distributed over an energy range described by a concentration density function $\mathfrak{D}_R(\lambda, \mathcal{E})$ with the units of [volume·energy]$^{-1}$ (cm^{-3} eV^{-1}). In the energy range $E - (E + \mathrm{d}E)$, $\mathfrak{D}_R(\lambda, \mathcal{E})\mathrm{d}\mathcal{E}$ is the infinitesimal number concentration of R species. This function is proportional to the overall surface concentration of R, $C_R(0, t)$:

$$\mathfrak{D}_R(\lambda, \mathcal{E})d\mathcal{E} = N_A C_R(0, t)\mathcal{P}_R(\lambda, \mathcal{E})d\mathcal{E} \tag{3.53}$$

where N_A is the Avogadro's number, and $\mathcal{P}_R(\lambda, \mathcal{E})$ is the probability density function in [energy]$^{-1}$ units. As expected, normalization of $\mathcal{P}_R(\lambda, \mathcal{E})$ yields

$$\int_{-\infty}^{+\infty} \mathcal{P}_R(\lambda, \mathcal{E})d\mathcal{E} = 1 \tag{3.54}$$

and the integral of $\mathfrak{D}_R(\lambda, \mathcal{E})$ over all $\mathcal{E}$ gives the total number concentration of states,

$$\int_{-\infty}^{+\infty} \mathfrak{D}_R(\lambda, \mathcal{E})d\mathcal{E} = \int_{-\infty}^{+\infty} N_A C_R(0, t)\mathcal{P}_R(\lambda, \mathcal{E})d\mathcal{E} = N_A C_R(0, t) \tag{3.55}$$

Following the same procedure it is easy to obtain the infinitesimal population of O:

$$\mathfrak{D}_O(\lambda, \mathcal{E})d\mathcal{E} = N_A C_O(0, t)\mathcal{P}_O(\lambda, \mathcal{E})d\mathcal{E} \tag{3.56}$$

which, after normalization of $\mathcal{P}_O(\lambda, \mathcal{E})$ as described above for R and integration, yields $N_A C_O(0, t)$, the total number concentrations of O states.

The rate of reduction of O states (r_O) from occupied electron states on the electrode in the energy range $\mathcal{E} - \mathcal{E} + d\mathcal{E}$ is a fraction of the total rate of reduction. This fraction is indicated as the "local rate" r_O at $\mathcal{E}$. At this infinitesimal energy range electrons from occupied states on the electrode in the time interval Δt can make the transition on states on species O in the same energy range. Therefore, the rate of reduction is given by the number of occupied electron states of the electrode involved in the electron-exchange counts in the time, Δt. For $\Delta t \to 0$, the rate of reduction is the instantaneous rate, which does not appreciably change: (i) the number of unoccupied states in the electrolyte and (ii) the energy of the unoccupied levels of O species in terms of their internal vibrational and rotational motions (i.e. Δt is at or below the time scale of vibrational modes). On this basis, $r_R(\mathcal{E})$, the "*local rate*" of reduction is

$$r_R(\mathcal{E}) = \frac{\Omega_R(\mathcal{E})AN_{occ}(\mathcal{E})d\mathcal{E}}{\Delta t} \tag{3.57}$$

where $AN_{occ}(\mathcal{E})d\mathcal{E}$ is the number of electrons exchanged, and $\Omega_R(\mathcal{E})$ is the probability of these electrons to carry out the transition to an unoccupied state on O:

$$\Omega_R(\mathcal{E}) = \eta_R(\mathcal{E}) \cdot \mathfrak{D}_O(\lambda, \mathcal{E}) \tag{3.58}$$

$\eta_R(\mathcal{E})$ is a proportionality function with units of [volume·energy] (i.e. cm^3 eV), and $\mathfrak{D}_O(\lambda, \mathcal{E})$ is the density of state of O. By integrating all the infinitesimal energy ranges the total rate of reduction, r_R, is

$$r_R = \nu \int_{-\infty}^{+\infty} \eta_R(\mathcal{E}) \cdot \mathfrak{D}_O(\lambda, \mathcal{E})AN_{occ}(\mathcal{E})d\mathcal{E} \tag{3.59}$$

Here, $\nu = 1/\Delta t$ is the electron transfer frequency. By substituting Eq. (3.58) into Eq. (3.59), we obtain

$$r_R = k_f^R[AN_A C_O(0, t)] \tag{3.60}$$

where $k_f^R = \nu \int_{-\infty}^{+\infty} \eta_R(\mathcal{E}) \cdot \mathfrak{D}_O(\lambda, \mathcal{E})f(\mathcal{E})\mathfrak{D}(\mathcal{E})d\mathcal{E}$ is the forward kinetic constant of reaction for R.

k_f^R is expressed in electron or chemical species per second. If we divide Eq. (3.60) by AN_A, we can obtain the conventional rate of reduction, in mol cm^{-2} s^{-1}:

$$r_R = k_f^R \cdot C_O(0, t) \tag{3.61}$$

Following the same procedure, it is easy to derive the rate constant for the oxidation of R:

$$r_{\mathrm{O}} = k_{\mathrm{b}}^{\mathrm{O}} \cdot C_{\mathrm{R}}(0, t) \tag{3.62}$$

with $k_{\mathrm{b}}^{\mathrm{O}} = \frac{\nu}{AN_A} \int_{-\infty}^{+\infty} \eta_{\mathrm{O}}(\mathcal{E}) \cdot \mathfrak{D}_{\mathrm{R}}(\lambda, \mathcal{E})[1 - f(\mathcal{E})]\mathfrak{D}(\mathcal{E}) d\mathcal{E}$.

In Figure 3.5, the distribution of states of R species does not overlap the electronic function of unoccupied density of states of the electrode. Therefore, the integral of Eq. (3.62) is practically zero, and $k_{\mathrm{b}}^{\mathrm{O}}$ is negligible.

Thus, the electrode is in a reducing condition with respect to the O/R redox couple. If the electrode potential is moved toward a more positive value, the Fermi level is shifted downward to a more negative value, see Figure 3.5 I, where the R states start overlapping with the unoccupied electron states of the electrode. In these conditions the integral of Eq. (3.62) is different from zero, and $k_{\mathrm{b}}^{\mathrm{O}}$ becomes significant in the redox process.

The formalism here reported is very general and, by Eqs. (3.59)–(3.62), demonstrates that the description of the electron function of the electrode or the surface of a possible electrocatalyst deposited onto an electrode support plays a crucial role in modulating the kinetic properties of the interface electrode/electrolyte.

To determine the probability densities, $\mathcal{P}_{\mathrm{O}}(\lambda, \mathcal{E})$ and $\mathcal{P}_{\mathrm{R}}(\lambda, \mathcal{E})$, of the O and R electroactive species present in the electrolyte we can use the Marcus theory. We start assuming that [1, 2, 4, 27–29]:

(a) the electron transfer events occur entirely from the Fermi level;
(b) the redox reaction takes place on a multidimensional surface defining the standard free energy of the system in terms of the nuclear coordinates of reactant, product, and solvent species.

Changes of nuclear coordinates of O and R and solvent take place owing to their vibrational and rotational motion together with reorientation, local fluctuation relaxations, and long-range migration of solvent molecules. A quantitative analysis of these phenomena on the reaction coordinate, q, can be performed assuming that:

(a) the reactant species are at a fixed distance from each other;
(b) the standard free energies, G_{O}^0 and G_{R}^0, of O and R species, respectively, depend quadratically on the reaction coordinate, q, as follows [27]:

$$G_{\mathrm{O}}^0 = \frac{k}{2}(q - q_{\mathrm{O}})^2$$
$$G_{\mathrm{R}}^0 = \frac{k}{2}(q - q_{\mathrm{R}})^2 + \Delta G^{\circ} \tag{3.63}$$

with q_0 and q_{R} the equilibrium atomic coordinates of O and R, and k the force constant of their bond lengths. It should be noticed that for an electrode reaction, ΔG^0 is given by $-nF(E - E^{\circ})$.

A typical free energy plot representing the assumptions summarized by Eq. (3.63) is shown in Figure 3.6.

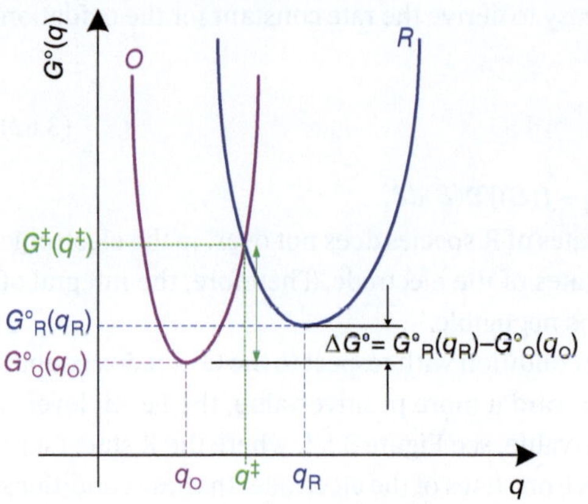

Figure 3.6 G^0 on q for an electron transfer reaction. For a heterogeneous reaction the O profile is the superposition of the energies for species O and for an electron on the electrode at the Fermi level at the potential E. $\Delta G^0 = F(E - E^0)$ is the standard free energy change for the reaction. The changes in q could represent changes in bond lengths within the electroactive species or within the host or surrounding solvent shells.

The transition state corresponds to the coordinate $q^\ddagger$, where O and R profiles intersect each other giving rise to an electron transfer process with free energies:

$$G^\circ_\mathrm{O}(q^\ddagger) = \frac{k}{2}(q^\ddagger - q_\mathrm{O})^2$$

$$G^\circ_\mathrm{R}(q^\ddagger) = \frac{k}{2}(q^\ddagger - q_\mathrm{R})^2 + \Delta G^\circ \tag{3.64}$$

If we consider that (i) at $q^\ddagger$, $G^\circ_\mathrm{O}(q^\ddagger) = G^\circ_\mathrm{R}(q^\ddagger)$ and (ii) at q_0, $G^\circ_\mathrm{O}(q_\mathrm{O}) = 0$, Eqs. (3.63) and (3.64) yield

$$q^\ddagger = \frac{q_\mathrm{R} + q_\mathrm{O}}{2} + \frac{\Delta G^\circ}{k(q_\mathrm{R} - q_\mathrm{O})} \tag{3.65}$$

$$\Delta G^\ddagger_\mathrm{f} = G^\circ_\mathrm{O}(q^\ddagger) - G^\circ_\mathrm{O}(q_\mathrm{O}) = G^\circ_\mathrm{O}(q^\ddagger) \tag{3.66}$$

By substituting $q^\ddagger$, Eq. (3.65) into Eq. (3.64) and this latter into Eq. (3.66), we can obtain [1]

$$\Delta G^\ddagger_\mathrm{f} = \frac{\sigma}{4}\left[1 + \frac{\Delta G^0}{\sigma}\right]^2 \tag{3.67}$$

where $\sigma = \frac{k}{2}(q_\mathrm{R} - q_\mathrm{O})^2$.

For an electrode reaction, Eq. (3.67) can be rewritten as

$$\Delta G^\ddagger_\mathrm{f} = \frac{\sigma}{4}\left[1 - \frac{F(E - E^0)}{\sigma}\right]^2 \tag{3.68}$$

It should be noticed that Eq. (3.67) is very general if we include into ΔG^0 all the other energy contributions, here not considered, such as transport phenomena, ion pairing interactions, and the electrostatic work to transport a charged reactant to a suitable position near the electrode. σ is a crucial parameter (it is the molecular reorganization parameter that represents the energy necessary to transform a specific nuclear configuration of the reactant and of the solvent into another one associated with the product state). To better describe the system, the parameter σ can be split into two contributions, σ_i and σ_S:

$$\sigma = \sigma_i + \sigma_S \tag{3.69}$$

σ_i is to the reorganization of O species, and σ_S the reorganization of the solvent molecules. In the harmonic approximation σ_i is the superposition of the normal modes of all the reactant contributions:

$$\sigma_i = \sum_j \frac{1}{2} k_j (q_{\mathrm{O}j} - q_{\mathrm{R}j})^2 \tag{3.70}$$

Here, the k_jth and the q_jth are the force constant and the displacements of normal coordinates, respectively.

If we rewrite Eq. (3.68) in terms of the electron energy

$$\Delta G_{\mathrm{f}}^{\ddagger} = \frac{\sigma}{4}\left[1 - \frac{\mathcal{E} - E^{\circ}}{\sigma}\right]^2 \tag{3.71}$$

where E° is the standard potential of the O/R redox couple, it results in $\Delta G_{\mathrm{f}}^{\ddagger} = 0$ at $\mathcal{E} = E^0 + \sigma$. Therefore, at the Fermi level the maximum *"local rate"* of reduction is obtained when $E_{\mathrm{F}} = E^0 + \sigma$. If the Fermi level assumes any other energy value $\mathcal{E}$, from Eqs. (3.56) and (3.71), and $k_{\mathrm{f}} = \nu_n k_{\mathrm{el}} \cdot e^{-\Delta G_{\mathrm{f}}^{\ddagger}/k_B T}$ (with ν_n the nuclear frequency factor, and k_{el} the electronic transition coefficient), and considering that $\eta_R(E^0) = \eta_R(E^0 + \sigma)$, it results in

$$\frac{r(E_F = \mathcal{E})}{r(E_F = E^0 + \sigma)} = e^{-\frac{\sigma}{4k_B T}\left(1 - \frac{\mathcal{E} - E^0}{\sigma}\right)^2} = \frac{\mathfrak{D}_{\mathrm{O}}(\lambda, \mathcal{E})}{\mathfrak{D}_{\mathrm{O}}(\lambda, E^0 + \sigma)} \tag{3.72}$$

This is a Gaussian distribution function of energy density of states having a mean value at $\mathcal{E} = E^0 + \sigma$ and a standard deviation of $\Sigma = (2\sigma kT)^{1/2}$, see Figure 3.5. In addition, from Eq. (3.56), we can obtain

$$\frac{\mathfrak{D}_{\mathrm{O}}(\lambda, \mathcal{E})}{\mathfrak{D}_{\mathrm{O}}(\lambda, E^0 + \sigma)} = \frac{\mathcal{P}_{\mathrm{O}}(\sigma, \mathcal{E})}{\mathcal{P}_{\mathrm{O}}(\sigma, E^0 + \sigma)} \tag{3.73}$$

and

$$\mathcal{P}_{\mathrm{O}}(\sigma, \mathcal{E}) = \mathcal{P}_{\mathrm{O}}(\sigma, E^0 + \sigma) \cdot e^{-\frac{(\mathcal{E} - E^0 - \sigma)^2}{4\sigma k_B T}}. \tag{3.74}$$

The normalization of Eq. (3.74) carried out in accordance with $\mathcal{P}_{\mathrm{O}}(\sigma, \mathcal{E})$ function, Eq. (3.54), allows to determine $\mathcal{P}_{\mathrm{O}}(\sigma, E^0 + \sigma) = (4\pi\sigma k_B T)^{-1/2}$, thus yielding

$$\mathcal{P}_{\mathrm{O}}(\sigma, \mathcal{E}) = (4\pi\sigma k_B T)^{-1/2} \cdot e^{-\frac{(\mathcal{E} - E^0 - \sigma)^2}{4\sigma k_B T}} \tag{3.75}$$

Following the same derivation, it is easy to obtain

$$\mathcal{P}_{\mathrm{R}}(\sigma, \mathcal{E}) = (4\pi\sigma k_B T)^{-1/2} \cdot e^{-\frac{(\mathcal{E} - E^0 + \sigma)^2}{4\sigma k_B T}} \tag{3.76}$$

Equations (3.75) and (3.76) show that the energy distribution of states of R and O species present in the electrolyte exhibits a Gaussian function with a similar shape which is centered at $E^0 + \sigma$ and $E^0 - \sigma$ energy, respectively, see Figure 3.5. As expected, the Gerischer–Marcus model described above [20–25]:

(a) at equilibrium converges to the Nernst equation;

$$\frac{k_{\mathrm{b}}}{k_{\mathrm{f}}} = e^{f(E - E^0)} = e^{-\frac{\mathcal{E} - E^0}{k_B T}}$$

(b) in general provides equations for k_b and k_f that are based on component functions able to describe real electrode/electrolyte systems, Eqs. (3.61) and (3.62);
(c) is able, including the Marcus theory [1, 2, 4, 27–29] without work terms, to define the distribution functions $\mathcal{P}_O(\sigma, \mathcal{E})$ and $\mathcal{P}_R(\sigma, \mathcal{E})$ of electroactive species present in the electrolyte, Eqs. (3.75) and (3.76);
(d) demonstrates that the reorganization energy, σ, has an important effect on the expected current–potential response curves.

3.12.1 The Butler–Volmer Equation

As demonstrated above, see Section 3.11, an efficient electron transfer process between two different phases occurs when the acceptor and the donor electroactive species in the electrolyte exhibit an energy density distribution exactly fitting with that of the Fermi level in the metal electrode [1–5], see Figure 3.5.

For a general electrode process:

$$O_{(aq)} + ne^- \rightleftharpoons R \tag{3.77}$$

Combining Eqs. (3.46) and (3.47) with Eqs. (3.61) and (3.62) allows to write the rates of forward (reduction) and backward (oxidation) reactions as

$$i_f = k_f nFAC_O(0,t) \text{ and } i_b = k_b nFAC_R(0,t) \tag{3.78}$$

where i_f and i_b are the current values of cathodic and anodic processes; k_f and k_b are the cathodic and anodic heterogeneous rate constants described above, see Section 3.11 in the explanation of Eqs. (3.61) and (3.62); A is the area of the electrode/electrolyte interface; and $C_O(0,\ t)$ and $C_R(0,\ t)$ are the surface concentrations of O and R electroactive species, respectively. After some mathematical rearrangements, Eqs. (3.61) and (3.62) of k_f and k_b, respectively, can be transformed into

$$k_f = A_{0,f} \cdot e^{-\Delta G_f^{\ddagger}/RT}$$
$$k_b = A_{0,b} \cdot e^{-\Delta G_b^{\ddagger}/RT} \tag{3.79}$$

where $A_{0,f}$ and $A_{0,b}$ are the heterogeneous rate constants of the redox processes, and $\Delta G_f^{\ddagger}$ and $\Delta G_b^{\ddagger}$ are the cathodic and anodic activation energies for the processes of Eq. (3.77).

When E is changed by ΔE, the relative energy of the electron in the electrode changes by $-F(E - E°)$. Thus, it is expected that the O curve moves up or down by this amount.

In Figure 3.7, it is shown that changing E by ΔE the curve of O moves downward by a fraction of the total energy, yielding a profile with a new value of activation energy ($\Delta G_f^{\ddagger}$).

A careful analysis of Figure 3.7 shows that after this downward shift the intersection point of O and R curves is modified, and concurrently the new anodic barrier is now

$$\Delta G_b^{\ddagger} = \Delta G_{0,b}^{\ddagger} - (1-\alpha)F(E - E^0) \tag{3.80}$$

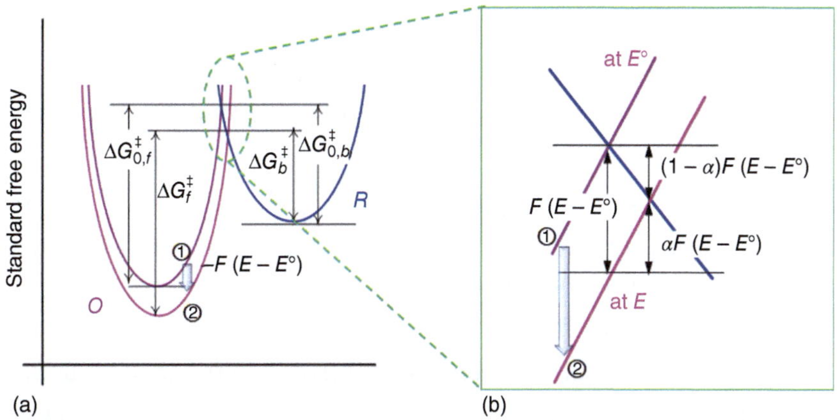

Figure 3.7 Effect of potential modulation on the standard free energy of activation of oxidation and reduction reactions. ①x02780; is the original O curve. ②x02781; is the curve shifted downward after changing E by ΔE [30].

while that of the cathodic barrier is

$$\Delta G_{\mathrm{f}}^{\ddagger} = \Delta G_{0,\mathrm{f}}^{\ddagger} + \alpha F(E - E^{0}) \tag{3.81}$$

where α is the transfer coefficient that ranges from 0 to 1.

Inserting the activation energies of Eqs. (3.80) and (3.81) into (3.79), we obtain

$$\begin{aligned} k_{\mathrm{f}} &= A_{0,\mathrm{f}} \cdot e^{-\Delta G_{\mathrm{f}}^{\ddagger}/RT} . e^{-\frac{\alpha F(E-E^{0})}{RT}} \\ k_{\mathrm{b}} &= A_{0,\mathrm{b}} \cdot e^{-\Delta G_{\mathrm{b}}^{\ddagger}/RT} . e^{+\frac{(1-\alpha)F(E-E^{0})}{RT}} \end{aligned} \tag{3.82}$$

The first two terms of these equations are independent of the potential and correspond to the rate constants $k_{\mathrm{f},0}$ and $k_{\mathrm{b},0}$ at $E = E^{0}$. Thus, Eq. (3.82) can be simply rewritten as

$$\begin{aligned} k_{\mathrm{f}} &= k_{\mathrm{f},0} \cdot e^{-\frac{\alpha F(E-E^{0})}{RT}} \\ k_{\mathrm{b}} &= k_{\mathrm{b},0} \cdot e^{+\frac{(1-\alpha)F(E-E^{0})}{RT}} \end{aligned} \tag{3.83}$$

Now, on the basis of Eq. (3.78), the forward and backward rates are

$$r_{\mathrm{f}} = \frac{i_{\mathrm{f}}}{nFA} = k_{\mathrm{f}} C_{\mathrm{O}}(0, t) \text{ and } r_{\mathrm{b}} = \frac{i_{\mathrm{b}}}{nFA} = k_{\mathrm{b}} C_{\mathrm{R}}(0, t) \tag{3.84}$$

These, when the interface electrode/electrolyte is at equilibrium, and $C_{\mathrm{O}}^{*} = C_{\mathrm{R}}^{*} = C_{\mathrm{O}}(0, t) = C_{\mathrm{R}}(0, t)$ (C_{O}^{*} and C_{R}^{*} are the bulk concentrations of R and O species, respectively), yield

$$r_{\mathrm{f}} = r_{\mathrm{b}} \text{ and } k_{\mathrm{f}} C_{\mathrm{f},O}(0, t) = k_{\mathrm{b}} C_{\mathrm{b},R}(0, t) \tag{3.85}$$

At equilibrium, $E = E°$ and $k_{\mathrm{f}} = k_{\mathrm{b}}$. In this case, E^{0} is the potential where the forward and backward rates have the same value, i.e. $k^{0} = k_{\mathrm{f}} = k_{\mathrm{b}}$ and $k°$ is the standard rate constant. Taking into account the latter considerations, k_{f} and k_{b} can now be easily expressed as

$$k_f = k° \cdot e^{-\alpha f(E-E^0)}$$
$$k_b = k° \cdot e^{(1-\alpha)f(E-E^0)} \tag{3.86}$$

where $f = \frac{F}{RT}$. Using Eq. (3.84) is now easily obtained an expression for the net reaction (r):

$$r = r_f - r_b = k_f C_O(0,t) - k_b C_R(0,t) = \frac{i}{nFA} = \frac{i_f - i_b}{nFA} \tag{3.87}$$

By substituting Eqs. (3.86) into Eq. (3.87) the complete current potential equation known as Butler–Volmer equation is achieved:

$$i = nFAk° \cdot [C_O(0,t)e^{-\alpha f(E-E°)} - C_R(0,t)e^{-(1-\alpha)f(E-E°)}] \tag{3.88}$$

At equilibrium, the net current is zero, and $C_O(0,t) = C_R(0,t) = C_O^* = C_R^*$; thus,

$$i = 0 = i_f - i_b, i_f = i_b = i_0 \tag{3.89}$$

and

$$e^{f(E_{eq}-E^0)} = \frac{C_O^*}{C_R^*} \text{ with } E_{eq} = E^0 + \frac{RT}{nF} \ln \frac{C_O^*}{C_R^*} \tag{3.90}$$

It should be noticed that in this condition (i.e. net current equal zero) there is at the electrode still a balanced faradic activity which, on the basis of the second Eq. (3.89), corresponds to i_0, the exchange current. This parameter can be derived using Eqs. (3.88)–(3.90), for i_f at $E = E_{eq}$:

$$i_0 = nFAk° C_O^* \cdot e^{-\alpha f(E_{eq}-E^0)} \tag{3.91}$$

The substitution into Eq. (3.91) of Eq. (3.90) yields

$$i_0 = nFAk° C_O^{*(1-\alpha)} C_R^{*\alpha} \tag{3.92}$$

which states that the exchange current is an important parameter proportional to $k°$. In the particular case where $C_O^* = C_R^* = C$, this becomes more evident:

$$i_0 = nFAk° C \tag{3.93}$$

Actually, in the experiments it is preferred to use the exchange current normalized to the electroactive area, i.e. the exchange current density, $J_0 = i_0/A$. Dividing Eq. (3.88) by Eq. (3.92) and using the Nernst equation (3.90), we can easily obtain

$$i = i_0 \left[\frac{C_O(0,t)}{C_O^*} e^{-\alpha f\eta} - \frac{C_R(0,t)}{C_R^*} e^{(1-\alpha)f\eta} \right] \tag{3.94}$$

where $\eta = E - E_{eq}$ is the overpotential with respect to E_{eq}. To analyze the profiles of Eq. (3.94), let us assume that $\frac{C_O(0,t)}{C_O^*} = \frac{C_R(0,t)}{C_R^*} = 1$. In this approximation, the profiles of the curves described by Eq. (3.94) depend on k^0 and α parameters, see Figure 3.8.

When $\alpha = 0.5$, the current density increases symmetrically on η, Figure 3.8a. On the other hand, for $k° =$ constant, the symmetry of the curve depends on α, Figure 3.8b.

An interesting feature of the profiles presented in Figure 3.8 is the coincidence of all curves near the origin, which can be used to determine the value of i_0. In practice, the exchange current is a measure of any system's ability to deliver a net current without a significant energy loss due to the presence of any activation process. It can be calculated as the derivative of the current with respect to η, when η is close to zero:

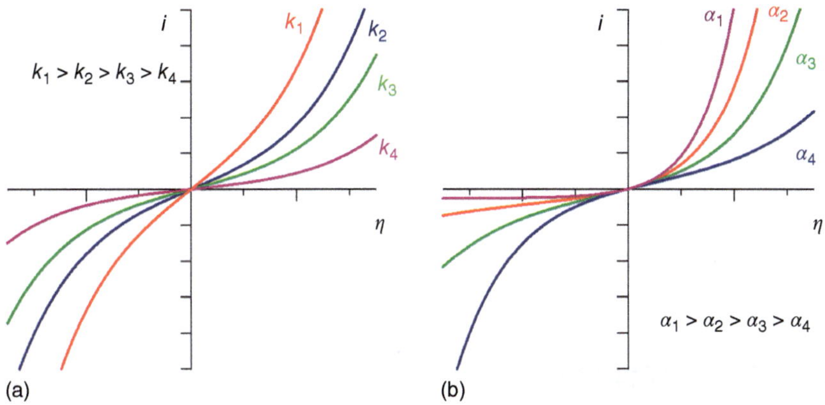

Figure 3.8 Influence of $k°$ on the shape of the current–potential curves, $\alpha = 0.5$ (a). Influence of α on the shape of the current–potential profiles, $k° =$ constant (b).

$$\frac{\mathrm{d}i}{\mathrm{d}\eta} = i_0 \left\{ \frac{-\alpha nF}{RT} \exp\left(\frac{-\alpha nF}{RT}\eta\right) - \frac{(1-\alpha)nF}{RT} \exp\left(\frac{(1-\alpha)nF}{RT}\eta\right) \right\} \tag{3.95}$$

$$\left(\frac{\mathrm{d}i}{\mathrm{d}\eta}\right)_{\eta=0} = i_0 \left\{ \frac{-\alpha nF}{RT} - \frac{(1-\alpha)nF}{RT} \right\} = -\frac{i_0 nF}{RT} \tag{3.96}$$

Experimentally, i_0 can be determined by taking the tangent to the current/potential curve at the origin. In practice, when η is very small, the general expression for the current η, Eq. (3.94), can be approximated to the relationship

$$i = -i_0 \frac{nF\eta}{RT} \tag{3.97}$$

since $\exp(-x) \approx 1 - x$. Another crucial parameter is the *charge transfer resistance* or *polarization resistance*, R_p, which is defined as the inverse of the current/potential gradient at $\eta = 0$.

$$R_\mathrm{p} = \frac{1}{\left(\frac{\mathrm{d}i}{\mathrm{d}\eta}\right)_{\eta=0}} \tag{3.98}$$

and

$$R_\mathrm{p} = \frac{RT}{nFi_0} \tag{3.99}$$

For $k° \gg 1 \rightarrow R_\mathrm{p} \cong 0$.

The symmetry factor, α, can be obtained by means of appropriate electrochemical experiments carried out at the equilibrium and using Eq. (3.92) rewritten as

$$\log\left({}^{i_0}/{}_{C_\mathrm{O}^*}\right) = \log(k° nFA) - \alpha \log\left(C_\mathrm{O}^*/C_\mathrm{R}^*\right) \tag{3.100}$$

Thus, by plotting $\log(i_0/C_\mathrm{O}^*)$ *vs.* $\log(C_\mathrm{O}^*/C_\mathrm{R}^*)$, the slope of the curve is exactly α. Finally, after the determination of α and i_0, the heterogeneous rate constant, k_0, of a general redox process can be measured by

$$k° = \frac{i_0}{nFAC_O^{*1-\alpha} C_\mathrm{R}^{*\alpha}} \tag{3.101}$$

When the reaction is carried out at the standard potential of the redox couple, Eq. (3.101) becomes

$$k^\circ = \frac{i_0}{nFAC_O^*} = \frac{i_0}{nFAC_R^*} \tag{3.102}$$

and, if $C_O^* = C_R^* = 1$,

$$k^\circ = \frac{i_0}{nFA} \tag{3.103}$$

3.12.2 The Tafel Equation

From the definitions of i_f and i_b, see Eq. (3.84), we can evaluate the ratio

$$\frac{i_f}{i_b} = \exp\left(-\frac{nF\eta}{RT}\right) \tag{3.104}$$

which taking the logarithm becomes

$$\log_{10}\left(\frac{i_f}{i_b}\right) = -\frac{n}{0.059}\eta \text{ at } 25^\circ\text{C} \tag{3.105}$$

Thus, for an overpotential of −118 mV and a one-electron transfer process, $i_f/i_b = 100/1$. Thus, for overpotentials of ±200 mV and above, only either the cathodic or the anodic contribution to the total net current is relevant. In these conditions, for a large cathodic overpotential we can write

$$i = i_f - i_b \approx i_f = i_0 \exp\left(-\alpha\frac{nF\eta}{RT}\right) \rightarrow \ln i = \ln i_0 - \alpha\frac{nF\eta}{RT} \tag{3.106}$$

Using Eq. (3.106) and expressing the overpotential *vs.* ln *i*, we have

$$\eta = (2.303\,{}^{RT}/_{\alpha nF})\log i_0 - (2.303\,{}^{RT}/_{\alpha nF})\log i \tag{3.107}$$

which is the traditional form of the well-known Tafel equation [31]. This is a useful tool for the evaluation of the kinetic parameters of an electrochemical process. Tafel plots, see Figure 3.9, report the log *i* as a function of η and allow to determine the exchange current, i_0, by extrapolating the linear portions of the curves at $\eta \rightarrow 0$, Figure 3.10. From the same experiment, it is easy to measure α by analyzing the gradient of the curve profiles.

In Figure 3.9, the possible negative deviations from linearity at very large overpotentials are associated with limitations imposed by the mass transfer phenomena.

The Tafel relationships are not observed for processes with fast electrode kinetics. On the contrary, when electrode kinetics are sluggish, and significant activation overpotentials are required, good Tafel relationships are detected. This aspect highlights that the Tafel analysis is an important tool to study processes with totally irreversible kinetics such as the oxygen reduction reaction (ORR) exploited in the operation of most fuel cells.

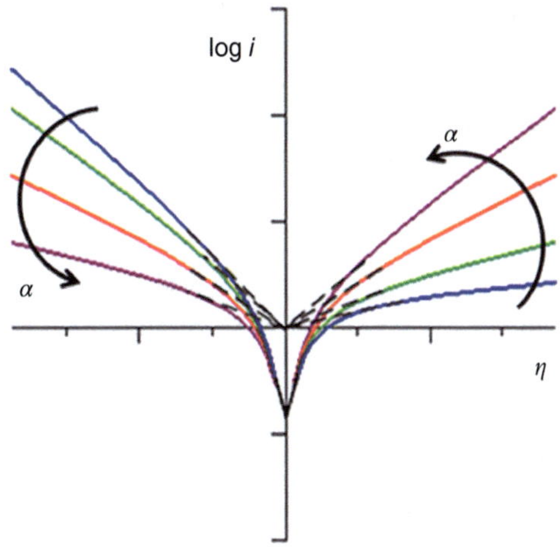

Figure 3.9 Tafel plot: log i vs. η for a general case. $\eta = E - E_{eq}$ is the overpotential.

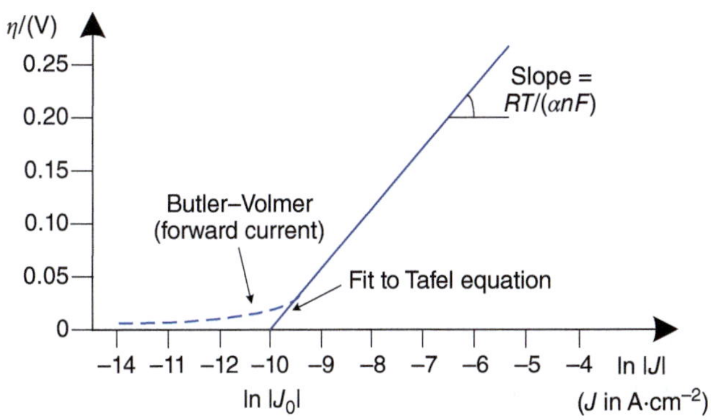

Figure 3.10 η vs. ln$|J|$ of a hypothetical electrochemical reaction. At high η, a linear fit of data with Tafel equation, Eq. (3.107), allows to determine J_0 and α. The Tafel approximation deviates from linearity at low η.

3.12.3 Interplay Between Exchange Current and Electrocatalyst Activity

To improve the performance of an electrocatalysts, it is crucial to increase the value of i_0. From Eq. (3.91), which is described in terms of the forward reaction [10], we have:

$$J_0 = nFC_R^* \cdot k_f^{\circ} e^{-\frac{\Delta G_f^{\ddagger}}{RT}} \tag{3.108}$$

where J_0 is the exchange current density of a reactant, R, at equilibrium.

A careful analysis of Eq. (3.108) shows that the strategies to raise the performance of an electrocatalyst are:

1. to raise the reactant concentration, C_R^*;
2. to reduce the forward activation barrier, $\Delta G_f^{\ddagger}$;
3. to increase the temperature, T;
4. to increase the density of electrochemically active sites,

Table 3.3 Exchange current densities for Hydrogen Oxidation Reaction (HOR) on various metal active sites.

Active sites	Electrolyte	J_0 (A cm^{-2})
Pt	Acid	10^{-3}
Pt	Alkaline	10^{-4}
Pd	Acid	10^{-4}
Rh	Alkaline	10^{-4}
Ir	Acid	10^{-4}
Ni	Alkaline	10^{-4}
Ni	Acid	10^{-5}
Ag	Acid	10^{-5}
W	Acid	10^{-5}
Au	Acid	10^{-6}
Fe	Acid	10^{-6}
Mo	Acid	10^{-7}
Ta	Acid	10^{-7}
Sn	Acid	10^{-8}
Al	Acid	10^{-10}
Cd	Acid	10^{-12}
Hg	Acid	10^{-12}

$T = 300$ K, 1 atm H_2.
The values are normalized to the surface area of metal electrodes [16, 32].

to improve the electronic and bifunctional effects affecting the sites, thus increasing k_f°.

In fuel cells we are producing a large amount of net current. This condition is similar to an irreversible reaction process typically described by a Tafel equation, see Eqs. (3.106) and (3.107), in which the forward reaction is dominating.

Figure 3.10 shows an example of Tafel plot. At high η, we observe the expected linear behavior, while at low η it deviates from Tafel behavior. J_0 and α are determined using the intercept and the slope of the linear fit of Figure 3.10, respectively. Results for the Hydrogen Oxidation Reaction (HOR) and the Oxygen Reduction Reactions (ORR) of different types of electrocatalysts are summarized in Tables 3.3 and 3.4, respectively.

Pt active sites result among the most active for both the HOR and the ORR, regardless of pH. However, even on the best Pt ECs the ORR activity is at least six orders of magnitude lower than that of the HOR. In addition, with respect to pure Pt, the ORR exchange current density of Pt nanoparticles supported on carbonaceous species in "*Pt/C*" ECs is increased, see Table 3.3. A further improvement is registered by alloying such carbon-supported Pt nanoparticles with suitable first-row transition metals (e.g. Fe, Co, and Ni). As a result of these latter trends in

Table 3.4 Exchange current densities for oxygen reduction reaction (ORR) on various metal active sites.

Active sites	Electrolyte	J_0 (A cm^{-2})
Metal active surface in acid electrolyte		
Pt	Acid	10^{-9}
Pd	Acid	10^{-10}
Ir	Acid	10^{-11}
Rh	Acid	10^{-11}
Au	Acid	10^{-11}
Pt alloys in PEMFC		
Pt/C	Nafion	$3 \cdot 10^{-9}$
PtMn/C	Nafion	$6 \cdot 10^{-9}$
PtCr/C	Nafion	$9 \cdot 10^{-9}$
PtFe/C	Nafion	$7 \cdot 10^{-9}$
PtCo/C	Nafion	$6 \cdot 10^{-9}$
PtNi/C	Nafion	$5 \cdot 10^{-9}$

Values are normalized to the *real surface area* of the metal. The exchange current density for the ORR is orders of magnitude smaller than that of the HOR, although the same group of metals shows the highest activity for both reactions. Pt alloys show a slightly higher performance over pure Pt in a PEMFC environment [15].

the values of exchange current densities, it is revealed that even for the low-T FCs mounting the best state-of-the-art ECs: (i) the ORR overpotential is much larger than the HOR overpotential (*ca.* 300 mV *vs. ca.* 30 mV); (ii) the Pt loading at the FC anode is significantly lower than that at the cathode (*ca.* 0.05 mg cm^{-2} or less *vs. ca.* 0.3–0.4 mg cm^{-2}); and (iii) it is very difficult to reduce the ORR overpotentials below the values afforded by Pt/C ECs [33]. These figures explain why modern research: (i) with respect to HOR ECs is much more strongly focused on the development of ORR ECs beyond the state of the art and allowing for a minimization of the loading of Pt and (ii) despite decades of effort failed to result in really breakthrough ORR ECs. Indeed, the best developmental Pt-based ECs hardly yield an ORR overpotential lower than *ca.* 50 mV in comparison with state-of-the-art Pt/C ECs, and even in that case such improvements are very often registered only in *ex situ* half-cell experiments and not in a full-fledged FC running in operating conditions [34].

3.13 Conclusions

A detailed understanding of fundamental electrochemistry is extremely useful in the study of electrochemical energy conversion devices, with a particular reference to FCs. First of all, it allows to reach a comprehensive understanding of (i) the physical structure of an FC and (ii) the specific roles played by its various components, elucidating the various phenomena on the basis of device operation. Equilibrium electrochemistry, culminating with the Nernst equation and its applications, yields a clear picture of the fundamental processes exploited by the FC for its functioning, and how such processes are affected by the environmental conditions especially in terms of T, pressure of the gaseous reactants, and pH. As a corollary, it becomes possible to gauge in detail the efficiency of the FC as a whole, and how the latter can be maximized by a suitable choice of reactants and operating conditions. Equilibrium electrochemistry also allows for the identification of the most relevant parasitic phenomena associated with FC operation, with a particular reference to those involved in the degradation of the electrode configurations. This critical information can then be exploited to develop specific strategies to mitigate such degradation phenomena and extend FC durability. The study of charge transfer by taking into account the distribution of energy states is a crucial tool to delve into the basic features associated with redox processes, and how the latter are influenced by the fundamental electronic characteristics of the involved electroactive species. Electrochemical kinetics as modeled by the simple, yet clear and effective Butler–Volmer equation and its various approximations (e.g. the Tafel equation) allow to quantify some of the most relevant intrinsic features of the electrochemical processes involved in FC operation (e.g. the exchange current density J_0, the transfer coefficient α, and the activation barrier $\left(\Delta G_i^{\ddagger}\right)$) and correlate them with the experimental overpotentials. This information highlights the main targets to be pursued by the research and also suggests some possible strategies to tackle the shortcomings of modern FCs. Taken all together, basics of electrochemistry are a crucial tool for understanding fundamental phenomena controlling the operations of materials in the FC and in development of new FCs exhibiting a performance and durability beyond the state of the art.

Acronyms

AEMFCs	anion-exchange membrane fuel cells
HOR	hydrogen oxidation reaction
ORR	oxygen reduction reaction
PEMFCs	proton-exchange membrane fuel cells

Symbols

0	standard conditions
‡	transition state
$\mathfrak{D}(\mathcal{E})$	density of states, [area·energy]$^{-1}$, in cm^{-2} eV^{-1}
$\mathfrak{D}_O(\lambda, \mathcal{E})$	density of state of O species
$\mathfrak{D}_R(\lambda, \mathcal{E})$	density of state of R species
$\mathcal{E}$	absolute electron energy
$\mathcal{E}_F$	energy of Fermi level
$\mathcal{N}_{occ}(\mathcal{E})$	occupied state at energy level, $\mathcal{E}$, [area·energy]$^{-1}$, in cm^{-2}·eV^{-1}

$\mathcal{N}_{\mathrm{unocc}}(\mathcal{E})$	unoccupied state at energy level, $\mathcal{E}$, [area·energy]$^{-1}$, in $\mathrm{cm}^{-2}\cdot\mathrm{eV}^{-1}$
$\mathcal{P}_O$	probability function of O species
$\mathcal{P}_{\mathrm{O}}(\sigma, \mathcal{E})$	energy distribution of states of O species present in the electrolyte
$\mathcal{P}_{\mathrm{R}}$	probability function of R species
$\mathcal{P}_{\mathrm{R}}(\sigma, \mathcal{E})$	energy distribution of states of R species present in the electrolyte
α	transfer coefficient
β	stochiometric ratio ($\beta = \frac{\rho_{\mathrm{fuel}}}{i/nF}$)
γ_{i}	activity coefficient of the ith component
δt	period of time
ΔG_{rxn}	free Gibbs energy of a reaction
$\Delta G_{\mathrm{b}}^{\ddagger}$	anodic activation energies
$\Delta G_{\mathrm{f}}^{\ddagger}$	cathodic activation energies
ΔS_{rxn}	change of entropy during a reaction
ΔV	change in volume
$\eta_{\mathrm{O}}(\mathcal{E})$	overpotential of O species
$\eta_{\mathrm{R}}(\mathcal{E})$	overpotential of R species
η_{cell}	efficiency of a cell
η_{eff}	overall effective efficiency of a cell
η_j	jth losses
η_{rxn}	reaction yield
η_{supplies}	efficiency due to the losses of the other components of the device
η_{system}	efficiency of whole device
η_{thermo}	thermodynamic or ideal efficiency of energy conversion
μ^0	chemical potential of the standard state
μ_i	chemical potential of the ith species
μ_i^0	chemical potential of the standard state of the ith component
ν	electron transfer frequency
ν and ν'	stochiometric coefficients
ν_{n}	nuclear frequency factor
ξ	activity ratio
ρ_{fuel}	rate of the fuel supplied to the cell in mol/s
σ	parameter ($\sigma = \frac{k}{2}(q_{\mathrm{R}} - q_0)^2$)
σ_S	parameter associated with the reorganization of the solvent
σ_i	parameter associated with the reorganization of O species
$\Omega_{\mathrm{R}}(\mathcal{E})$	probability of an electron to move to an unoccupied state on O
a_{i}	activity of the ith component
a	activity
$A_{0,\mathrm{b}}$	heterogeneous rate constant of the anodic process
$A_{0,\mathrm{f}}$	heterogeneous rate constant of the cathodic process
$C_{\mathrm{O}}(0, t)$	overall surface concentration of O species
C_{O}^{*}	concentration of O species at the equilibrium

$C_R(0, t)$	overall surface concentration of R species
C_R^*	concentration of R species at the equilibrium
dn_i	change in the number of moles of the ith component
dT	change in temperature
dp	change in pressure
e_0	electron charge in coulombs.
E	potential
E^0	standard potential
E_A^0	standard potential of anode
E_G^0	standard potential from standard free energy, $G°$
E_H^0	standard potential from standard enthalpy, $H°$
E_c^0	standard potential of cathode
E_{cell}^0	standard potential of a cell or standard *e.m.f.*
E_{cell} or *e.m.f.*	electromotive force
E_{eq}	potential at the equilibrium
E_p	potential at p value of pressure
E_T	potential at T value of temperature
$f(\mathcal{E})$	Fermi–Dirac distribution function
$f = \frac{F}{RT}$	factor
F	Faraday's constant (96,490 C mol^{-1})
$G°_O$	standard free energies of O species
$G°_R$	standard free energies of R species
H	enthalpy
i	current
i_0	exchange current
i_b	current value of anodic process
i_f	current value of cathodic process
J	current density
J_0	exchange current density
k	force constant of bond lengths for O or R species
k_B	Boltzmann constant ($1.380649 \cdot 10^{-23}$ m^2 kg s^{-2} K^{-1})
k_b	backward kinetic constant of reaction
k_{el}	electronic transition coefficient
k_f	forward kinetic constant of reaction
k_f^O	forward kinetic constant of reaction for O
k_f^R	forward kinetic constant of reaction for R
K_{eq}	equilibrium constant
m^0	molalities at standard condition
m_i	molality of the ith component
m_{rxn}	mass of reagent consumption
$m_{rxn,T}$	total mass provided to a fuel cell for its operation
$M^{(x+v)+}$ ($P^{(y-v')+}$)	reduced (oxidized) species
M^{x+} (P^{y+})	oxidized (reduced) species
M^0	molarities at standard condition
M_i	molarities of the ith component

n_e	moles of electrons
n_M	moles of M
n_P	moles of P
N_A	Avogadro's constant
N_e	number of electrons
O_i	oxidant state of the *i*th reagent
p_i	partial pressure of the *i*th component
P	pressure
q	atomic coordinate
$q^{\ddagger}$	atomic coordinate of transition state
q_O	equilibrium atomic coordinate of O
q_R	equilibrium atomic coordinate of R
Q_{hc}	heat consumption
Q_{hg}	heat generated by a reaction
Q_T	maximum theoretical capacity
$Q_{T,i}$	maximum theoretical capacity of the *i*th species
r_O	rate of reduction of O state
r_R	rate of reduction of R state
R	universal constant of gas (8.31446 J mol^{-1} K^{-1} or 0.082057 l atm mol^{-1} K^{-1})
R_i	reduced state of the *i*th reagent
R_m	membrane resistance
R_p	charge transfer resistance or polarization resistance
S	entropy
T	temperature in K
vap	vapor state
w_{el}	electric energy, or work
$w_{el,max}$	maximum electric work
x	mole fraction
x_i	mole fraction of the *i*th component

References

1 Bard, A.J., Faulkner, L.R., and White, H.S. (2001). *Electrochemical Methods: Fundamentals and Applications*, 2e. New York: Wiley.

2 Atkins, P.W. (1986). *Physical Chemistry*, 3e. New York: W.M. Freeman and Company.

3 Bockris, J.O.′.M. and Amulya, K.R. (2002). *Modern Electrochemistry 2B:Electrodics in Chemistry, Engineering, Biology and Environmental Science*, 2e. New York, NY: Springer.

4 Bockris, J.O.′.M., Amulya, K.R., and Gamboa-Aldeco, M. (2002). *Modern Electrochemistry 2A: Fundamentals of Electrodics*. New York, NY: Springer.

5 Bockris, J.O.′.M. and Amulya, K.R. (2002). *Modern Electrochemistry, Ionics*, 2e, vol. 1. New York, NY: Springer.

6 Karlberg, G.S., Rossmeisl, J., and Nørskov, J.K. (2007). Estimations of electric field effects on the oxygen reduction reaction based on the density functional theory. *Physical Chemistry Chemical Physics* 9 (37): 5158–5161.

7 EG&G Technical Services, I.S.A.I.C (2004). *Fuel Cell Handbook*. Morgantown, West Virginia: U.S. Department of Energy, Office of Fossil Energy, National Energy Technology Laboratory.

8 Vielstich, W., Lamm, A., and Gasteiger, H.A. (ed.) (2003). *Handbook of Fuel Cells: Fundamentals, Technology, Applications*, vol. 1. New York: Wiley.

9 Durst, J., Siebel, A., Simon, C. et al. (2014). New insights into the electrochemical hydrogen oxidation and evolution reaction mechanism. *Energy and Environmental Science* 7 (7): 2255–2260.

10 Ryan O'Hayre, S.-W.C., Colella, W., and Prinz, F.B. (2016). *Fuel Cell Fundamentals*, 3e. New York: Wiley.

11 Pourbaix, M. (1974). *Atlas of Electrochemical Equilibria in Aqueous Solutions*. Houston, TX, USA: National Association of Corrosion Engineers.

12 Shao, M., Chang, Q., Dodelet, J.P. et al. (2016). Recent advances in electrocatalysts for oxygen reduction reaction. *Chemical Reviews* 116 (6): 3594–3657.

13 Hamo, E.R., Singh, R.K., Douglin, J.C. et al. (2021). Carbide-supported PtRu catalysts for hydrogen oxidation reaction in alkaline electrolyte. *ACS Catalysis* 11 (2): 932–947.

14 Zhang, L., Li, H., and Zhang, J. (2014). Kinetics of oxygen reduction reaction on three different Pt surfaces of Pt/C catalyst analyzed by rotating ring-disk electrode in acidic solution. *Journal of Power Sources* 255: 242–250.

15 Damjanovic, A. and Brusic, V. (1967). Electrode kinetics of oxygen reduction on oxide-free platinum electrodes. *Electrochimica Acta* 12 (6): 615–628.

16 Vielstich, V., Lamm, A., and Gasteiger, H.A. (2003). *Handbook of Fuel Cells – Fundamentals, Technology and Applications*, vol. 2. Chichester: Wiley.

17 Durst, J., Lamibrac, A., Charlot, F. et al. (2013). Degradation heterogeneities induced by repetitive start/stop events in proton exchange membrane fuel cell: inlet vs. outlet and channel vs. land. *Applied Catalysis B: Environmental* 138-139: 416–426.

18 Castanheira, L., Silva, W.O., Lima, F.H. et al. (2015). Carbon corrosion in proton-exchange membrane fuel cells: effect of the carbon structure, the degradation protocol, and the gas atmosphere. *ACS Catalysis* 5 (4): 2184–2194.

19 Sui, S., Wang, X., Zhou, X. et al. (2017). A comprehensive review of Pt electrocatalysts for the oxygen reduction reaction: nanostructure, activity, mechanism and carbon support in PEM fuel cells. *Journal of Materials Chemistry A* 5 (5): 1808–1825.

20 Levich, V.G. (1966). *Advances in Electrochemistry and Electrochemical Engineering* 4: 249.

21 Hush, N.S. (1971). *Reactions of Molecules at Electrodes, in Reactions of Molecules at Electrodes*. London: Wiley.

22 Wolfgang Schmickler, E.S. (2010). *Interfacial Electrochemistry*. Heidelberg: Springer Berlin.

23 Miller, C.J. (1995). *Physical Electrochemistry, Chapter 2*. New York: I. Rubinstein, M. Dekker.

24 Gerisher, H. (1970). *Physical Chemistry: An Advanced Treatise*, vol. 9A (ed. D.H.H. Eyring and W. Jost). New York: Academic.

25 Gerisher, H. (1961). *Advances in Electrochemistry and Electrochemical Engineering* 1: 139.

26 Kittel, C. (2004). *Introduction to Solid State Physics*, 8e. New York: Wiley.

27 Marcus, R.A. and Siddarth, P. (1992). Theory of electron transfer reactions and comparison with experiments. In: *Photoprocesses in Transition Metal Complexes, Biosystems and Other Molecules. Experiment and Theory*, 49–88. Berlin: Springer Dordrecht.

28 Trube, H. (1970). *Electron Transfer Reactions of Complex Ions in Solutions*. New York: Academics.

29 Erberson, L. (1987). *Electron Transfer Reactions in Organic Chemistry*. Berlin: Springer-Verlag.

30 O'Hayre, R., Cha, S.W., Colella, W., and Prinz, F.B. (2016). Chapter 3 – Fuel cell reaction kinetics. In: *Fuel Cell Fundamentals*, 3e, 77–116. Hoboken, New Jersey: Wiley.

31 Tafel, J. (1905). *Über die Polarisation bei kathodischer Wasserstoffentwicklung. Zeitschrift für Physikalische Chemie* 50: 641.

32 Berger, C. (1968). *Handbook of Fuel Cell Technology*. Englewood Cliffs, N.J.: Prentice-Hall.

33 Gasteiger, H.A., Kocha, S.S., Sompalli, B. et al. (2005). Activity benchmarks and requirements for Pt, Pt-alloy, and non-Pt oxygen reduction catalysts for PEMFCs. *Applied Catalysis B: Environmental* 56 (1-2 SPEC. ISS): 9–35.

34 Zhang, X., Li, H., Yang, J. et al. (2021). Recent advances in Pt-based electrocatalysts for PEMFCs. *RSC Advances* 11 (22): 13316 13328.

4

Quantifying the Kinetic Parameters of Fuel Cell Reactions

Viktoriia A. Saveleva[1,2], Juan Herranz[1], and Thomas J. Schmidt[1,3]

[1] *Paul Scherrer Institute, Electrochemistry Laboratory, Forschungsstrasse 111, 5232 Villigen, Switzerland*
[2] *The European Synchrotron, ESRF, 71 Avenue des Martyrs, CS 40220, 38043 Grenoble Cedex 9, France*
[3] *ETH Zürich, Institute for Physical Molecular Sciences, Vladimir-Prelog-Weg 2, 8093 Zürich, Switzerland*

4.1 Introduction

The H_2-oxidation reaction and O_2-reduction reaction (HOR, ORR) taking place in polymer electrolyte fuel cell (PEFC) anodes and cathodes have been the subject of continued research since the inception of electrocatalysis, both due to their fundamental importance and in sight of their decisive role in the development of this energy conversion technology. The continued development of Pt-based catalysts with enhanced dispersions and intrinsic catalytic activities, combined with improvements in the ionic conductivities, chemical and mechanical resistances of perfluorosulfonic acid (PFSA) membranes and ionomers, has led to PEFCs capable of reaching power densities $\geq 1\,W\,cm^{-2}$ at $\geq 0.65\,V$. While these advancements pave the way for the application of these devices in the automotive sector [1], further decreases in the amount (i.e. loading) of these expensive catalysts (ideally $<0.1\,mg_{Pt}\,cm^{-2}$) are required in order to foster the extended commercialization of PEFCs [2]. This passes by further enhancing the ORR activity of Pt-based catalysts (as well as improving the mass transport properties of their catalyst layers) [2, 3], which is customarily done by alloying Pt with other transition metals (e.g. Co and Ni) as to tune the O-bonding properties of the resulting catalyst surface [4]. This loading reduction may also affect the amount of Pt implemented in PEFC anodes, likely bringing it to values $\ll 50\,\mu g_{Pt}\,cm^{-2}$ at which HOR kinetics may limit the device's performance. Complementarily, ORR catalysts of the so-called Me/N/C kind, in which a transition metal (Me) such as Fe or Co is atomically dispersed within the pores of a high surface area carbon forming N-coordinated sites, are attracting increasing attention as possible, inexpensive replacements to these Pt-based materials, but their activity and stability still need to be improved to render them device implementable [5].

In order to fulfill these demanding requirements, a better understanding of the reaction mechanisms and kinetic descriptors determining the activity and stability of this variety of catalysts is needed. Complementarily, the establishment of reactivity

Electrocatalysis for Membrane Fuel Cells: Methods, Modeling, and Applications, First Edition.
Edited by Nicolas Alonso-Vante and Vito Di Noto.

metrics allowing for an unambiguous comparison among materials is of paramount importance. To this end, it is worth considering that the kinetics of both reactions (i.e. HOR and ORR) are typically regarded as complying with the Butler–Volmer equation, according to the classic formalism [6]:

$$i_{\text{kin}}(T, C_{\text{i}}) = i_0(T, C_{\text{i}}) \cdot \mathcal{R} \cdot \left[e^{\frac{\alpha_a \cdot F}{R \cdot T} \cdot \eta_{\text{kin}}} - e^{-\frac{\alpha_c \cdot F}{R \cdot T} \cdot \eta_{\text{kin}}} \right] \tag{4.1}$$

where $i_{\text{kin}}(T, C_{\text{i}})$ corresponds to the current that one would expect under exclusive kinetic control at a given temperature (T) and reactant concentration (C_{i}), $i_0(T, C_{\text{i}})$ corresponds to the catalyst's intrinsic exchange current density at the same T and C_{i} conditions, and $\mathcal{R}$ refers to the roughness factor, viz. the ratio between the surface area of the catalyst exposed to the reaction environment and the electrode's geometric area. As for α_{a} and α_{c}, these represent the reactions of anodic and cathodic transfer coefficients, respectively, while the kinetic overpotential term (η_{kin}) corresponds to the difference between the potential at which the kinetically controlled reaction is proceeding (E) and its reversible potential (E_{rev}; the reader is referred to Chapter 3 for more detailed information on the derivation of this thermodynamic quantity) and also represents purely kinetic overpotential. Finally, F and R stand for the Faraday and gas constants, respectively.

Furthermore, the dependency of the exchange current density on the reaction temperature and reactant concentration is expressed by the following Equation [7]:

$$i_0(T, C_{\text{i}}) = i_0^* \left(T^*, C_i^*\right) \cdot \left(\frac{C_{\text{i}}}{C_i^*}\right)^{\gamma} \cdot e^{-\frac{E_{\text{act}}}{R \cdot T} \cdot \left(1 - \frac{T}{T^*}\right)} \tag{4.2}$$

where T^* and C_{i}^* correspond to reference temperature and reactant concentration conditions, i_0^* is the corresponding exchange current density at these reference conditions, γ is the reaction order with regard to the reactant concentration, and E_{act} is the reaction's activation energy.

Additionally, in sluggish reactions that onset far away from E_{rev} (see the dashed line in Figure 4.1a), the anodic vs. cathodic exponential terms in Eq. (4.1) have a negligible effect on the reverse reaction branches. As a result, for a kinetically hindered process, like the ORR, the anodic term in the Butler–Volmer equation can be neglected, resulting in a Tafel-like equation:

$$i_{\text{kin}}(T, C_{\text{i}}) = -i_0(T, C_{\text{i}}) \cdot \mathcal{R} \cdot e^{-\frac{\alpha_c \cdot F}{R \cdot T} \cdot \eta_{\text{kin}}} = -i_0(T, C_{\text{i}}) \cdot \mathcal{R} \cdot 10^{-\frac{\alpha_c \cdot F}{2.3 \cdot R \cdot T} \cdot \eta_{\text{kin}}} \tag{4.3}$$

which can in terms be rearranged into the following, well-established relation between the overpotential and the logarithm of the current:

$$\begin{aligned} \log_{10}|i_{\text{kin}}(T, C_{\text{i}})| &= \log_{10}|\, i_0(T, C_{\text{i}}) \cdot \mathcal{R}| - \frac{\alpha_c \cdot F}{2.3 \cdot R \cdot T} \cdot \eta_{\text{kin}} \\ &= \log_{10}|\, i_0(T, C_{\text{i}}) \cdot \mathcal{R}| - \frac{1}{b_c} \cdot \eta_{\text{kin}} \end{aligned} \tag{4.4}$$

where the b_{c} term is often referred to as the Tafel slope. Thus, as illustrated in Figure 4.1b, it is a common practice to infer the values of α_{a} and α_{c} from the slopes in the linear parts of the so-called Tafel plots featuring the logarithm of the

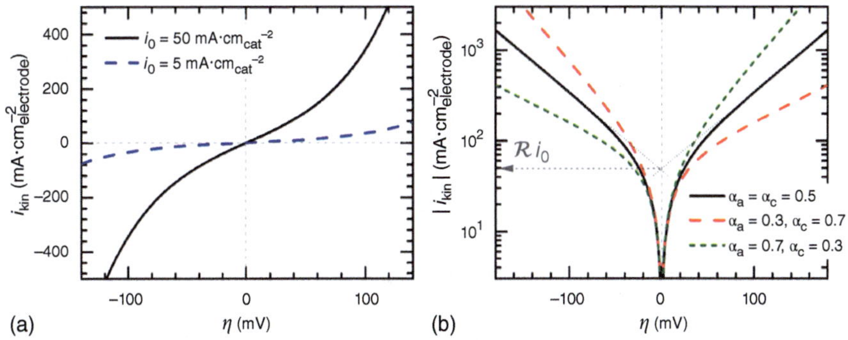

Figure 4.1 Relations between the overpotential and the kinetically controlled current (so-called polarization curves) computed using the Butler–Volmer equation (see Eq. (4.1)). Panel "a" displays the effect of a 10-fold variation in the exchange current density (i_0) for a hypothetical reaction occurring at 298 K, with anodic and cathodic transfer coefficients of 0.5 and an idealized roughness factor of 1 $\mathrm{cm}^2_{\mathrm{cat}}\ \mathrm{cm}^{-2}_{\mathrm{electrode}}$. Panel "b" displays the corresponding Tafel plot (i.e. with the absolute values of the kinetic currents in a logarithmic scale) for the polarization curve with $i_0 = 50\ \mathrm{mA\ cm}^{-2}_{\mathrm{electrode}}$, along with the asymmetry in the anodic and cathodic branches caused by increasing or decreasing the values of their transfer coefficients. One must be aware that the anodic and cathode transfer coefficients do not need to sum up to 1, as these parameters are defined experimentally. Note how $\mathcal{R}i_0$ corresponds to the crossing of the linearly extrapolated anodic and cathodic branches at zero overpotential (i.e. at E_{rev}), as indicated by Eq. (4.4).

current vs. the (over)potential. Moreover, extending these linear regions to zero overpotential yields the corresponding value of $\mathcal{R}i_0$ (cf. Figure 4.1b). Notably, this extrapolation must be performed with care when applied to sluggish reactions with current onsets far from E_{rev}, since it can entail large imprecisions in the resulting $\mathcal{R}i_0$ values [7]. In such cases, it is customary to assess the reaction kinetics based on i_{kin} values at an (over)potential that can be directly interpolated within the Tafel regime (i.e. in which the measured current is under kinetic control, as it is often done for the ORR at a potential of 0.9 V vs. reversible hydrogen electrode [RHE]).

Finally, the so-determined $\mathcal{R}i_0$ (or $\mathcal{R}i_{\mathrm{kin}}$, if interpolating within the Tafel slope) can be used to isolate the corresponding exchange current density (viz. i_{kin}) if knowing the value of the roughness factor. For the nanoparticle-based catalysts currently used in commercial PEFCs, the latter corresponds to the product of the catalyst loading per electrode area (L_{cat}, e.g. in $\mathrm{mg_{Pt}\ cm^{-2}_{electrode}}$) and its electrochemical surface area (ECSA), which corresponds to the catalyst-mass-normalized surface of these nanoparticles in contact with the electrolyte (correspondingly, in $\mathrm{cm^{-2}_{Pt}\ mg^{-1}_{Pt}}$), i.e.

$$\mathcal{R} = L_{\mathrm{cat}} \cdot \mathrm{ECSA} \tag{4.5}$$

Notably, this ECSA metric is not applicable to the Me/N/C catalysts discussed above, since one cannot define an electrolyte-exposed area for their atomically dispersed Me–N$_x$ sites. Instead, for these kinds of materials it is more convenient to refer to their site density (SD, in sites $\mathrm{cm^{-3}_{MeNC}}$), which is directly related to these active centers' intrinsic activity. The latter is in turn determined based on their turnover frequency (TOF). The latter corresponds to the number of electrons exchanged per site and second at a given potential and operative condition (i.e. with units of e^- $site^{-1}$ s^{-1}), and thus for these Me/N/C catalysts the preexponential terms of the Butler–Volmer equation can be reformulated as follows [8, 9]:

$$i_0(T, C_i) \cdot \mathcal{R} = \frac{L_{MeNC} \cdot SD \cdot TOF_0(T, C_i) \cdot e^-}{\rho_{electrode}} \tag{4.6}$$

where L_{MeNC} is the catalyst loading (in mg_{MeNC} $cm^{-2}_{electrode}$), TOF_0 would correspond to the turnover frequency at E_{rev} (i.e. analogous to i_0), e^- is the charge of an electron ($1.6 \cdot 10^{-19}$ C), and $\rho_{electrode}$ is the density of the catalyst processed as a porous electrode (in mg_{MeNC} $cm^{-3}_{electrode}$).

While the determination of HOR and ORR kinetic parameters would ideally be carried out directly in the PEFC environment of envisaged application, in practice this entails several sources of potential errors [7, 10–13]. In the first place, hydrogen crossover through the PFSA membrane results in a parasitic current [7] in the order of $\approx$1–5 mA cm^{-2}_{geom} that needs to be accounted for upon assessing ORR kinetics through separate measurements at the same temperature and pressure [14]. Additionally, while the two-electrode configuration intrinsic to PEFC measurements of the MEA is well fitted to ORR kinetic studies, given the ultrahigh exchange current density of the HOR on Pt (*vide infra*), the investigation of HOR/HER kinetics in an H_2:O_2-fed PEFC would require the use of a reference electrode [15, 16]. Finally, gradients in the concentrations of reactants and products, as well as in the distribution of water across the PEFC's flow field, generally from the inlet to the outlet [17] and also under channel and land [18, 19], lead to inhomogeneities in the current distribution across the electrodes.

The combination of these shortcomings with the time- and resource-demanding procedures required for the preparation of PEFC electrodes and hardware often translate in the use of other electrochemical techniques to assess these PEFC reactions' kinetic parameters. Thus, in this chapter we first look into the procedures to estimate these materials' ECSA/SD and then provide an overview of the variety of setups that have been used for quantifying HOR and ORR kinetics.

4.2 Electrochemical Active Surface Area (ECSA) Determination

One of the critical steps in electrocatalysis is the quantification of the number of active sites or (for nanoparticle-based catalysts) active surface area on which the electrochemical reaction takes place. The latter can sometimes be estimated based on the *ex situ* characterization of the catalyst, as it is often done for the carbon-supported metal nanoparticle catalysts customarily used in PEFC electrodes, and for which the active area can be defined based on the nanoparticles' dispersion and metal loading. For the former, the nanoparticles' size distribution can be assessed by measuring the diameter of multiple (ideally $\gg$100) nanoparticles through transmission electron microscopy [20–22], from which one can determine the number- and surface-averaged diameters ($\overline{d_N}$, $\overline{d_S}$) [23, 24]:

$$\overline{d_N} = \frac{\sum_{i=1}^{n} l_i d_i}{\sum_{i=1}^{n} l_i} \tag{4.7}$$

$$\overline{d_S} = \frac{\sum_{i=1}^{n} l_i d_i^3}{\sum_{i=1}^{n} l_i d_i^2}, \tag{4.8}$$

where l_i stands for the number of particles having a diameter d_i.

Either of these diameters can be subsequently converted to a total surface area (SA) value knowing the shape of the nanoparticles and their density (ρ). Thus, in case of spherical geometry, the SA can be defined as

$$SA = \frac{6}{\rho \overline{d}} \tag{4.9}$$

Notably, the best agreement between the SA values derived from Eq. (4.9) and the *in situ* determined ECSAs relevant for device implementation (*vide infra*) is obtained when the average diameter used in the above calculation corresponds to $\overline{d_s}$. However, this approach is prone to large imprecisions when applied to materials with hard-to-define nanoparticle geometries (e.g. unsupported nanochains) [25, 26] and/or densities (such as core-shell particles) and is not extensible to Me/N/C catalysts, since their active sites consist of hard-to-resolve, atomically dispersed metal atoms coordinated by nitrogen functionalities (see Chapters 13 and 6) [27]. Most importantly, the SA of a catalyst involved in a kinetic process once processed as a fuel cell electrode strongly depends on the distribution of triple phase boundaries within it, and thus this key property should ideally be determined directly in the PEFC setup. Thus, in the following sections we discuss these *in situ* approaches for determining this ECSA, which is generally assessed using methods based on surface faradaic reactions such as underpotential deposition (UPD) processes or the adsorption of prove molecules (e.g. CO stripping) or on non-faradic double-layer capacitances.

4.2.1 ECSA Determination Using Underpotential Deposition

UPD is the deposition of atoms (e.g. hydrogen or metals such as copper or mercury) onto an electrode surface in (sub)monolayer quantities at potentials more positive than the equilibrium potential for the reduction (and bulk deposition) of the metal [20, 28–30]. Assuming that a full coverage of the surface can be attained prior to this bulk deposition process, the charge associated with this electrochemical reduction (and the subsequent oxidation of the adsorbed monolayer) can be normalized with regard to a surface area-specific charge to yield an ECSA value [8]. In what follows, we focus on the specific use of hydrogen- or Cu-UPD for the quantification of the ECSA of Pt-based catalysts, since these are the two most usual UPD methods featured in the PEFC and electrocatalysis literature.

4.2.1.1 Hydrogen Underpotential Deposition (H_{UPD})

The underpotential deposition of H on Pt electrodes is a well-known feature of this noble metal that proceeds according to the equation:

$$Pt + H^+ + e^- \rightleftarrows Pt - H_{UPD} \tag{4.10}$$

This is well illustrated in Figure 4.2, displaying a cyclic voltammogram (CV) recorded on a Pt-based electrode immersed in an aqueous medium; there, the current features corresponding to this H_{UPD} process (and sensitive to the crystalline facets in the Pt-nanoparticles' surface) [32] can be observed at potentials ≤0.4 V vs. RHE and down to ≈0.05 V vs. RHE (i.e. slightly above the H_2-evolution onset) at

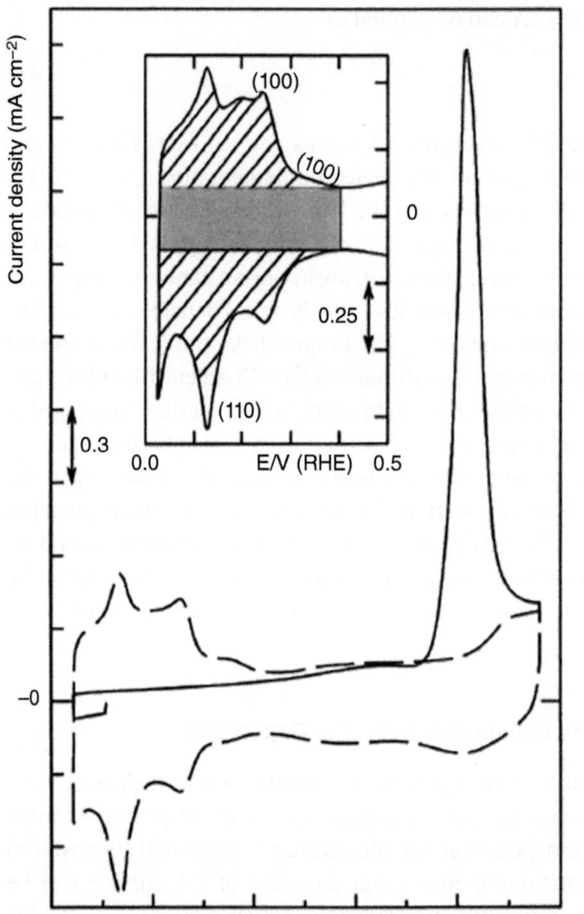

Figure 4.2 Cyclic voltammogram (dashed line) and CO-stripping potential scan (solid line) on a Pt/C electrode recorded in 0.5 M H_2SO_4 with a Pt loading of 28 μg cm^{-2}. The inset features a magnification of the hydrogen adsorption/desorption process in which the gray area corresponds to double-layer charging, and the dash-patterned one to the hydrogen adsorption and desorption charges. The measurements were performed at 25 °C with a scan rate of 20 mV s^{-1}. Source: Reproduced from Ref. [31] with permission of IOP Publishing.

which a quasi-complete coverage of the Pt-surface with H_{UPD} can be assumed. As a result, these charges of hydrogen adsorption and desorption are related to the number of surface platinum atoms and the catalyst's ECSA. The latter is then estimated by normalizing the adsorption/desorption charge (obtained by integrating the current in the H_{UPD} potential region discussed above using a horizontal baseline, as illustrated in the inset of Figure 4.2) by the value of the charge associated with the formation of a hydrogen monolayer on a polycrystalline platinum surface. The latter is generally accepted to correspond to $210\,\mu C\,cm^{-2}$ but can strongly depend on the surface's individual crystal facets. In this regard, the CV in Figure 4.2 corresponds to the voltammetric response of carbon-supported Pt nanoparticles that can be regarded as consisting of Pt(111), Pt(100), and Pt(110) facets, whereby the two last orientations are responsible for the peaks at ≈0.25 and ≈0.15 V vs. RHE in Figure 4.2 (see inset).

In spite of the simplicity of this method, one should be careful when applying it to the study of Pt nanoparticles of sizes <3.5 nm [33] for which it can lead to a significant underestimation of the ECSA with regard to the values estimated with the CO-stripping technique discussed below (i.e. approx. twofold lower ECSAs estimated by H_{UPD} for ≈1.5 nm nanoparticles). A similar trend has been observed for Pt-based alloy nanoparticles, for which the ECSA determination by H_{UPD} could be as low as ≈70% vs. CO stripping [34–37].

4.2.1.2 Copper Underpotential Deposition (Cu_{UPD})

While the H_{UPD} method works well for pure Pt systems, the determination of the ECSA of multicomponent, Pt-based materials (i.e. alloys) is less straightforward and depends on the electrochemical properties of the other metal, for which the formation of surface oxides at low potentials (i.e. <0.4 V vs. RHE) may overlap with the H_{UPD} peaks. In this case, other methods such as the CO-stripping method discussed below or the underpotential deposition of copper [38, 39] reviewed here can be used. Copper atoms possess similar atomic radii as Pt (0.128 vs. 0.138 nm for Cu vs. Pt, respectively), so one can assume the adsorption of a single Cu atom to each surface Pt-atom through a two-electron process [40]:

$$Pt + Cu^{2+} + 2e^- \rightleftarrows Pt - Cu_{UPD} \tag{4.11}$$

In this case, the charge associated with the monolayer formation of copper should correspond to $420\,\mu C\,cm^{-2}$ (i.e. double the $210\,\mu C\,cm^{-2}$ considered for the one-electron H_{UPD} process, *vide supra*) [41]. Chiefly, rotating ring disc electrode (RRDE) measurements on Pt(111) and Pt(100) single crystals have confirmed that [42, 43], in the presence of halide anions (e.g. Cl^- and Br^-), the total amount of copper underpotentially deposited on these facets corresponds to a close to complete monolayer coverage, thus validating the use of this normalization charge. An example of the cyclic voltammograms recorded on a polycrystalline platinum electrode with or without Cu^{2+} ions present in the solution is shown in Figure 4.3, where peaks II_C and I_C represent the UPD process and subsequent growth of a bulk copper layer on the electrode surface, correspondingly, followed by their removal upon inversion of the scan direction, which leads to peaks I_A and II_A. Similarly,

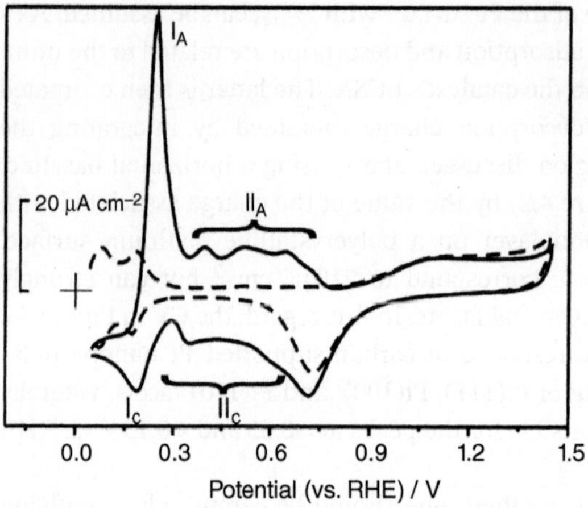

Figure 4.3 Cyclic voltammograms recorded at a scan rate of $10\,mV\,s^{-1}$ on a polycrystalline Pt electrode in a solution of $0.1\,mol\,dm^{-3}$ H_2SO_4 (dashed line) and with an additional $0.5 \times 10^{-3}\,mol\,dm^{-3}$ of $CuSO_4$ (solid line). Features I_C/I_A correspond to bulk copper deposition and stripping, respectively, while II_C/II_A are caused by the underpotential adsorption and desorption of Cu on Pt. Source: Reproduced from Ref. [40] with permission of American Chemical Society.

to the facet sensitivity of the H_{UPD} process discussed above, the presence of four distinct peaks in the Cu_{UPD} region is associated with variations in the adsorption energies of copper on different Pt-sites. Most importantly, the need to perform these measurements using a solution containing Cu^{2+} implies that this Cu_{UPD} method can be easily applied in the rotating disc electrode (RDE) setup discussed below but is not suited for the PEFC configuration, in which the presence of such transition metal can poison the PFSA membrane and ionomer [44].

4.2.2 ECSA Quantification Based on the Adsorption of Probe Molecules

4.2.2.1 CO Stripping

The electrochemical oxidation of a preadsorbed CO monolayer (i.e. the so-called CO-stripping method) is a widely studied ECSA quantification method that has been shown to be more reliable than H_{UPD} measurement not only on Pt-based catalysts but on materials based on others metals such as Ir, Rh, or Pd. [33–37, 45] The adsorption of a CO monolayer on Pt is triggered by bubbling gaseous CO in the electrolyte while applying a given potential that determines/defines the CO coverage [46], with 0.1 V vs. RHE being regarded as a value that leads to full CO coverage of a polycrystalline Pt-surface. Prior to the stripping step, and while maintaining this potential hold, the electrolyte is purged with an inert gas to remove any dissolved CO. Once this is completed, the potential is scanned up to an inversion potential typically $\geq$1 V vs. RHE, yielding the CO-stripping voltammogram displayed in Figure 4.2. Interestingly, the latter (which was recorded on a high surface area, carbon-supported Pt electrode) features a CO-stripping peak at $\approx$0.81 V vs. RHE, which is higher than the typical values obtained for polycrystalline Pt (≈0.72 V vs. RHE) [47] and (similar to H_{UPD}) [31] attests this technique's facet sensitivity. Interestingly, the onset of CO oxidation coincides with that of (hydr)oxide formation on the Pt-surface (dashed curve recorded in CO-free electrolyte in Figure 4.2) according to [47]

$$H_2O \rightarrow OH_{ads} + H^+ + e^- \tag{4.12}$$

This behavior can be rationalized by the involvement of the adsorbed hydroxide in the CO oxidation reaction following the equation [48]:

$$CO_{ads} + OH_{ads} \rightarrow CO_2 + H^+ + e^- \tag{4.13}$$

Following the same procedure as in H_{UPD} measurements, the CO-stripping current is integrated using a CV recorded in the same, CO-free and inert-saturated electrolyte (at the same scan rate) as a baseline. The resulting charge is then used to estimate the corresponding ECSA using a specific charge of 420 μC cm^{-2} assumed for the formation of a CO monolayer on Pt as the normalization basis.

4.2.2.2 NO_2^-/NO Sorption

The evaluation of the ECSA of the active site density in Me/N/C materials is not a trivial task and currently relies on the strong interaction between their atomically dispersed, nitrogen-coordinated (so-called Me–N_x) sites and the nitrite anion, which form a stably poisoned catalyst adduct that can be used as a site density probe [49, 50]. Analogously to the CO-stripping measurements on Pt-based catalysts discussed above, the catalytic activity of an NO-poisoned Me/N/C-catalyst can be recovered through a reductive stripping of NO to ammonia (in a five-electron process) at potentials ≤0.2 V vs. RHE, thus allowing the quantification of the number of the active sites (see scheme in Figure 4.4) [50, 51]. However, this method possesses two main disadvantages, namely (i) it requires mild acidic conditions (pH ≈ 5) that are not representative of the PEFC environment (in which the operative pH is ≈0), and (ii) it is unclear whether the sites probed by nitrite are the same as those involved in the O_2-reduction process.

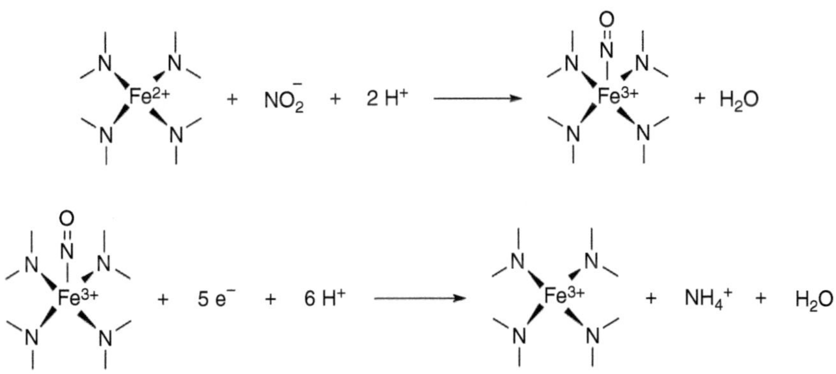

Figure 4.4 Nitrite adsorption and stripping steps on Fe–N_x sites. Source: Reproduced from Ref. [50]/ Springer Nature / CC By 4.0.

4.2.3 Double-layer Capacitance Measurements and Other Methods

As an alternative to the nitrite-sorption measurements discussed above, the ECSA of Pt-free, Me/N/C catalysts can be assessed to some extent by evaluating their double-layer capacitance, as it is widely done for the analysis of metal oxide-based electrocatalysts. Unfortunately, unlike in the UPD and CO-stripping methods discussed above, there is no unambiguous normalization charge relating this double-layer capacitance (which may not only be determined by the catalysts' surface area but also by the presence of redox-active surface species) and the catalysts' SD. Moreover, this kind of measurement is exclusively sensitive not only to the Me–N_x sites on which the reaction is believed to take place but also to the response of the carbon matrix in which these are embedded. It is possibly because of these reasons that we are unaware of any studies displaying a direct correlation between the double-layer capacitance (or derived ECSA/SD) of these kinds of materials and their ORR activity. Nevertheless, the analysis of the double-layer capacity can still be very useful, e.g. in degradation studies of these materials in which it can provide insight into the corrosion of their carbon matrix [52–54].

4.2.4 ECSA Measurements in a PEFC: Which Method to Choose?

For Pt-based catalysts, determining the ECSA through the underpotential hydrogen adsorption/desorption reaction using cyclic voltammetry is a very convenient method not only in RDE experiments but also in a PEFC setup containing a membrane-electrode assembly (MEA). In the latter PEFC experiments, this H_{UPD} method can be applied by flowing N_2 at the electrode in which the ECSA wants to be evaluated and H_2 at the opposite side (as to render the latter a pseudo-reference, dynamic hydrogen electrode) [55]. To illustrate this, a comparison of the CVs recorded on the same catalyst in RDE and different PEFC setups is shown in Figure 4.5 [56]. In these MEA configurations, one can clearly notice a positive shift of the HER-onset potential with increasing nitrogen flow, caused by the concomitant, a positive shift of the HOR/HER Nernst potential, and resulting in a narrower CV integration window that leads to underestimated ECSA values. On the contrary, the presence of liquid (in RDE or a water-purged PEFC experiment) shifts this hydrogen evolution onset to lower potentials, cf. Figure 4.5. Thus, to minimize this effect, the nitrogen flow should be restricted to 1 sccm cm^{-2} of active area [56]. Hydrogen crossover through the PFSA membrane also leads to the fluctuations in hydrogen concentration in the cathode compartment of PEFC affecting the HER onset and ECSA, consequently.

These inaccuracies become particularly relevant for electrodes with an ultralow Pt loading ($<10\,\mu g_{Pt}\,cm^{-2}_{geom}$) and a concomitantly small capacitive current. In order to avoid these, one should (i) use scan rates $\geq 100\,mV\,s^{-1}$ and/or (ii) use diluted hydrogen in the anode compartment to reduce hydrogen crossover through the membrane. Note that the application of fast scan rates is justified by the small charge transfer resistance of the H_{UPD} process on Pt in acid environments [57, 58], which implies that it is possible to measure CVs at scan rates up to $1 - 10\,V\,s^{-1}$

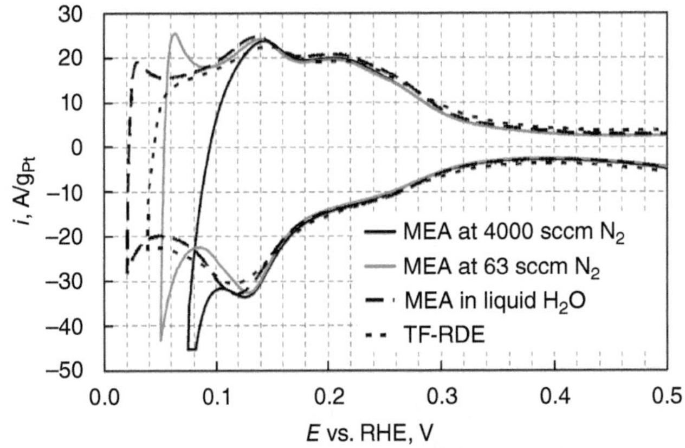

Figure 4.5 Pt-mass-normalized cyclic voltammograms recorded on Pt-based electrodes in 0.1 M $HClO_4$ (dashed line) and in a PEFC with two configurations: nitrogen-purged cases (black and gray solid lines) at different nitrogen concentrations and liquid water-purged case (dotted black line). All CVs were recorded at a scan rate of 20 mV s^{-1} and 25 °C. The catalyst loadings were 12.8 μg_{Pt} cm^{-2}_{geom} in the RDE setup, 0.37 mg_{Pt} cm^{-2}_{geom} in the N_2-purged PEFC, and 0.43 mg_{Pt} cm^{-2}_{geom} in the liquid water-purged PEFC. Source: Reproduced from Ref. [56] with permission of IOP Publishing.

without destabilizing the H_{UPD} adsorption isotherm. For metals such as Rh or Pd, characterized by two orders of magnitudes slower H_{UPD} rates [58, 59] than Pt, fast scan rates could lead to lower charge densities associated with hydrogen deposition. On top of this, metal contamination during PEFC operation can lead to the poisoning of the PFSA-based ionomer that acts as the electrolyte in the CLs, reducing catalyst utilization and the accuracy of the ECSA analysis using this UPD method [60]. Alternatively, as discussed above, CO stripping is more reliable than H_{UPD} and can shed additional light into effects such as the electronic properties of the catalyst, the location of the Pt nanoparticles on the carbon support [61], and the surface structure of Pt-based alloy [35, 62, 63]. However, such CO-stripping measurements in PEFC configuration remain scarce due to the additional infrastructure required to perform such measurements in a safe way.

4.3 H_2-Oxidation and Electrochemical Setups for the Quantification of Kinetic Parameters

Hydrogen is customarily used as the fuel in PEFCs, where it is electrochemically oxidized at the anode according to the following Equation:

$$H_2 \rightarrow 2H^+ + 2e^- \tag{4.14}$$

Due to its ultrafast kinetics on Pt in acid [64], the HOR requires low amounts of Pt catalyst at the anode (down to ≈0.05 mg_{Pt} cm^{-2}_{geom}) before becoming a significant contributor to the efficiency losses in PEFC performance. Concomitantly, these extremely fast kinetics in acid electrolytes imply that the investigation of the HOR reaction kinetics and the corresponding mechanism required for modeling and understanding of electrode degradation mechanisms (e.g. upon H_2 starvation and cell reversal) are very challenging, as further described below. Despite these difficulties, it is generally accepted that the HOR and complementary, H_2-evolution reaction (HER) proceed through a combination of two out of three following microscopic steps [65]:

$$\text{Tafel step: } H_2 \rightleftarrows 2H_{ads} \tag{4.15}$$

$$\text{Heyrovsky step: } H_2 \rightleftarrows 2H_{ads} + e^- + H^+ \tag{4.16}$$

$$\text{Volmer step: } H_{ads} \rightleftarrows H^+ + e^- \tag{4.17}$$

whereby the first step, so-called Tafel step, does not imply an electron transfer process, and the resulting surface intermediate (adsorbed hydrogen, H_{ads}) does not depend on the solution pH [66]. Interestingly, the possible assimilation of this H_{ads} intermediate to underpotentially deposited hydrogen has been extensively debated in the literature. Due to the fact that H_{UPD} and H_{ads} are considered as energetically indistinguishable, the former can be the surface species involved in the H_2-oxidation/evolution reaction [67, 68].

As discussed in this chapter's introduction, in order to unveil the exact HOR/HER mechanism, one needs to quantify the reactions' exchange current density, activation energy, and anodic and cathodic transfer coefficients. Tafel slopes can be used to identify the reaction's rate-determining step (RDS), but great care must be taken upon evaluation of these slopes, since they can be affected by changes in the surface coverage of adsorbed hydrogen within the considered potential window [69–71]. The latest experimental evidence suggests that the HOR/HER on Pt-based electrodes proceeds through the dissociative adsorption of hydrogen (Tafel reaction, Eq. (4.15)) combined with a one-electron transfer (Volmer reaction, Eq. (4.17)) RDS in a mechanism that is not affected by the surface structure of Pt nor the electrolyte's pH [72]. Notably, this assignment of the HOR to the Tafel [67] step remains controversial, with other recent studies suggesting that this should correspond to the Volmer step [73]. Since a more detailed discussion of these mechanistic aspects is provided in Chapters 3 and 5 of this book (as well as some recent reviews [74, 75]), in what follows we focus instead on discussing the different electrochemical setups used for the quantification of this reaction's kinetic parameters, which in many cases have also been extensively applied to the study of the ORR.

4.3.1 Rotating Disc Electrodes (RDEs)

An RDE is composed of a cylinder of a conductive material (e.g. the metal surface of choice or a conductive substrate such as glassy carbon on which the catalyst is immobilized) embedded at the center of a cylindrical, insulating holder (often made of polytetrafluoroethylene [PTFE]) in such a way that both cylinders' circular ends remain coplanar, Figure 4.6a. In this configuration, the controlled rotation

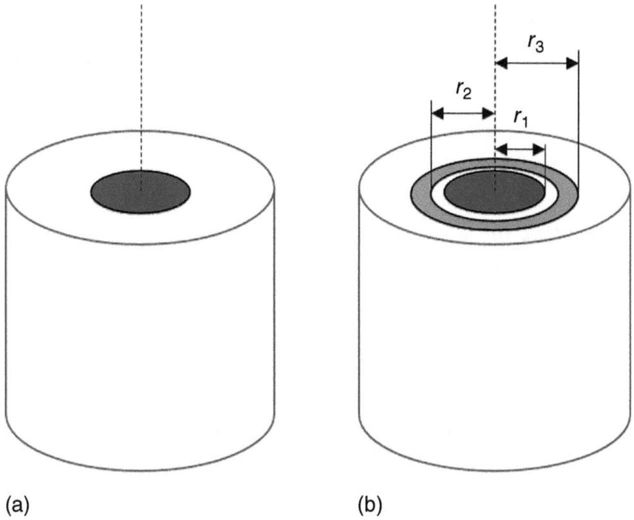

Figure 4.6 Schemes of RDE (a) and RRDE (b) setups.

of the electrode along the disc's axis causes a constant flux of electrolyte from its bulk and toward the RDE's surface, along with its evacuation sideways of the electrode. This key feature leads to a laminar flow of electrolyte toward the RDE surface that results in a well-defined boundary layer along which the reactant's concentration decreases linearly between its bulk value and that at the electrode's surface. Specifically, the thickness of this boundary layer in RDE experiments is a function of the electrode's angular rotation rate (ω, in rad s^{-1}), the electrolyte's kinematic viscosity (ν, in cm^2 s^{-1}), and the diffusion coefficient of the reactant in the same electrolyte (D, in cm^2 s^{-1}):

$$\delta_{BL} = 1.61 \cdot D^{1/3}\nu^{1/6}\omega^{-1/2} \tag{4.18}$$

Correspondingly, the maximum current density determined by the mass transport properties of the RDE (or limiting current, i_{lim}) can be defined by the following equation originally developed by Levich [76]:

$$i_{lim} = nFDc_{bulk}/\delta_{BL} = 0.62 \cdot nFD^{2/3}\nu^{-1/6}\omega^{1/2}c_{bulk} \tag{4.19}$$

where n is the number of electrons involved in the electrochemical reaction, and c_{bulk} is the concentration of the reactant in the bulk of electrolyte. Thus, for the typical RDE rotation rates of 10^2–10^4 rpm and electrolyte diffusion coefficients and viscosities customary for aqueous electrolytes ($D \approx 1 \cdot 10^{-5}$ cm^2 s^{-1} and $\nu \approx 1 \cdot 10^{-2}$ cm^2 s^{-1}), the diffusion-layer thickness is in the order of ≈5–50 μm.

Subsequently, this i_{lim} and the measured current density can be used to compute the so-called kinetic current density, i_{kin}, which corresponds to the current that one would measure if the reactant concentration on the electrode surface were to be equal to that in the electrolyte bulk (i.e. in the absence of mass transport limitations). For a reaction with a reaction order of 1 with regard to the reactant concentration [47], this i_{kin} can be derived using the well-known Koutecky–Levich equation:

$$\frac{1}{i} = \frac{1}{i_{lim}} + \frac{1}{i_{kin}} \tag{4.20}$$

On this basis, in the past two decades RDE voltammetry has become the preponderant technique of choice for assessing the kinetic parameters of the electrochemical reactions (i.e. HOR and ORR) taking place in low-temperature PEFC. Specifically, this kinetic analysis was greatly facilitated by the development of the thin-film (TF-)RDE technique [31, 77] in which the catalyst of choice is processed into an ink and deposited on an inert, conductive substrate (typically, glassy carbon) with the loading required to yield a catalyst layer (CL) of thickness ≪1 μm. This minimized thickness translates in negligible diffusion gradients through the CL, allowing the direct use of Eq. (4.20) to quantify i_{kin}.

Chiefly, when (TF-)RDE is used to study the HOR kinetics on Pt in acidic electrolytes, the so-measured, ohmic-drop-corrected polarization curves directly match the Nernstian diffusion overpotential [78, 79]:

$$\eta_{diff} = \frac{RT}{2F} \ln\left(1 - \frac{i}{i_{lim}}\right) \tag{4.21}$$

This implies that the HOR/HER currents on Pt electrodes in acidic electrolytes are completely under mass transport control, leading to underestimated i_0 values and false mechanistic interpretations [67, 80–82]. On the contrary, the HOR/HER on Pt in alkaline electrolytes are two orders of magnitude slower than in acid [72], allowing for the quantification of the kinetic parameters using RDE voltammetry [79]. Most importantly, this implies that the reliability of i_0 values derived from RDE measurements is limited to those reactions in which this kinetic parameter lays below the limiting current densities inherent to this technique (i.e. ≤1 mA cm^{-2}). This mass transport limitation is primarily attributed to the limited solubility of H_2 (and other PEFC-relevant gases, such as O_2) in aqueous electrolytes (≤1 mmol l^{-1} at room temperature), according to the Levich equation (Eq. (4.19)) [83], and ultimately the accurate evaluation of HOR/HER i_0-values on Pt electrodes in acidic electrolytes must be done using other mass-transport-free methods that we discuss in the following sections [79].

4.3.2 Hydrogen Pump (PEFC) Approach

In an electrochemical hydrogen pump (sometimes referred to as "hydrogen compressor"), the measurements are performed in a PEFC setup in which the HOR occurs at the anode, and the HER at the cathode [64]. To this end, Pt electrocatalysts of different loadings are used on both sides of the MEA, whereby the low-loaded side (with ≤0.003 mg_{Pt} cm_{geom}^{-2}) serves as the working electrode, and the high-loaded side of the MEA (with ≥0.1 mg_{Pt} cm_{geom}^{-2}) as a counter/reference electrode for which negligible kinetic losses can be assumed. The i_0-values derived from the use of this hydrogen pump approach are ≈200 mA cm_{Pt}^{-2} at 313 K [72] and ≈600 mA cm_{Pt}^{-2} at

353 K [64] and thus ≈10- to ≈100-fold larger than what has been reported based on RDE measurements [64]. Apart from Pt-based electrodes, this H_2-pump method has been successfully applied to other electrocatalytic materials such as Pd and Ir [72, 80] for which the exchange current densities from RDE experiments had typically been underestimated due to the uncorrected mass transport effects.

4.3.3 Ultramicroelectrode Approach

Another way to overcome the mass transport issues of the RDE setup is to decrease the size of the electrode down to the (sub-)micron scale [80]. In this case, the mass transport coefficient m for a microelectrode with radius r can be estimated as [76]

$$m = \frac{HD}{r} \tag{4.22}$$

whereby the coefficient H depends on the shape of the exposed microelectrode (equal to 1 for a hemispherical microelectrode, and to $4/\pi$ for a disc-shaped microelectrode), and D is the reactant's diffusion coefficient ($\approx 3.7 \cdot 10^{-5}$ cm^2 s^{-1} for H_2 in a sulfuric acid solution). To illustrate the advantages of this approach vs. RDEs, the mass transport coefficient in the latter setup can be determined as follows [84]:

$$m = 0.62 D^{2/3} \nu^{-1/6} \omega^{1/2} \tag{4.23}$$

where ν is the kinetic viscosity of the electrolyte ($\approx 1.1 \cdot 10^{-2}$ cm^2 s^{-1} for 0.1 M H_2SO_4), and ω is the rotation rate. Assuming a value of 1000 rpm for the latter parameter, the same mass transport coefficient could be achieved for a relatively large UME with a radius of 25 μm. For a more realistic, hemispherical microelectrode with a radius of 1 μm, mass transport rates >0.37 cm s^{-1} that would require unfeasible RDE rotation rates >10 000 rpm can be achieved.

Taking advantage of these features, the first use of a Pt microelectrode with a diameter of about 100 μm was performed by Bagotzky et al. in 1973 [66]. A smooth microelectrode was placed as to form a thin electrolyte film above the main electrolyte reservoir, Figure 4.7. Due to the reduced electrolyte film thickness, hydrogen diffusion toward the electrode surface is not hindered, and the transfer of impurities from the solution to the catalyst surface is very slow, partially eliminating the possibility of catalyst deactivation due to the adsorption of impurities.

Microelectrodes have also been replaced by electrodes made of individual Pt particles supported on small carbon substrates. In this configuration, Chen et al. [80] observed the presence of two limiting current plateaus similar to the ones first described by Bagotzky et al. [66] that were assigned to a mass transport limitation preceding the electron transfer or to a preceding chemical transformation coupled to a rapid electrochemical reaction [80].

4.3.4 Scanning Electrochemical Microscopy (SECM) Approach

Scanning electrochemical microscopy (SECM) is a versatile technique that can be used to obtain topographic images of surfaces and to study the electronic conductivity and chemical activity of substrates [85, 86]. In SECM, an ultramicroelectrode

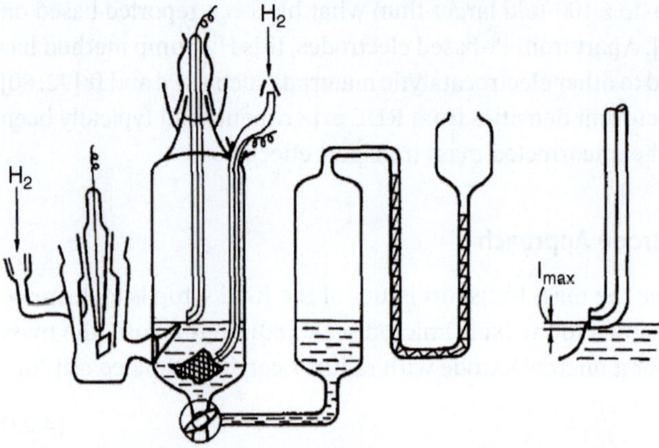

Figure 4.7 Drawing of the cell (left) and detailed magnification of the platinum microelectrode (right) used for HOR kinetics studies. Source: Reproduced from Ref. [66] with permission of Elsevier.

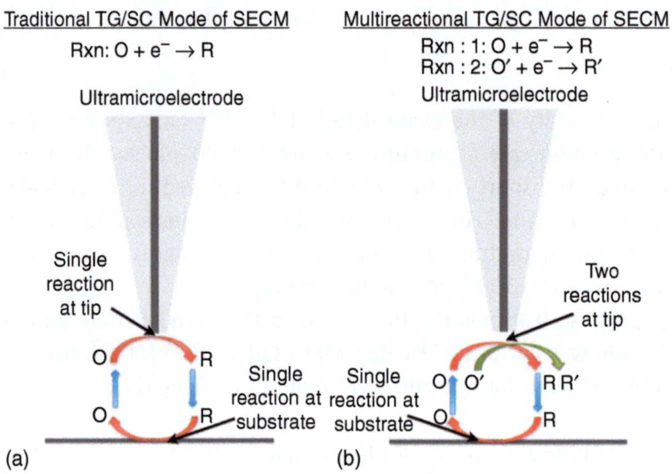

Figure 4.8 Scheme of an SECM operated in TG/SC mode, with a single vs. multiple reactions taking place at the tip electrode being displayed in panels (a) vs. (b), respectively, and a single reaction involving one of these products happening at the substrate in both cases. Source: Reproduced from Ref. [87] with permission of American Chemical Society.

(UME) with a radius of a few nanometers to 25 μm (i.e. the tip) is scanned toward the surface of a biased substrate (see Figure 4.8 for a scheme of the SECM working principle). Similar to the UME approach, the minimized electrode dimensions allow steady-state kinetic measurements at high mass transport rates and the concomitant investigation of fast reactions.

There are two general modes of SECM operation: the so-called generation/collection (G/C) and feedback modes [88]. In the G/C mode, the tip is located at distances ≤10 tip radii from the substrate, and both tip and substrate currents are monitored. In the subcase of tip generation/substrate collection (TG/SC), the tip is used to generate a reactant that is detected at the substrate electrode. For HOR/HER studies, this implies that the tip potential is held at a value (e.g. −0.7 V vs. RHE) corresponding to the diffusion-limited production of hydrogen, so that as the tip is brought within a few tip radii of the substrate (≈1 μm) this H_2 diffuses to the substrate and is oxidized there [89]. Notably, since this substrate electrode is considerably larger than the tip, 100% collection efficiency (defined as the ratio of substrate and tip currents) is systematically achieved. It has been shown that kinetic measurements performed in this SECM configuration are free of artifacts caused by ohmic-drop and DL charging that are associated with the other electrochemical techniques traditionally used to study HOR/HER kinetics. The derivation of the kinetic parameters is based on the quantitative theory developed for different modes of SECM operation, fitting experimental tip currents versus interelectrode distances [90]. As a result, exchange current densities on polycrystalline Pt electrodes extracted from these measurements in 0.01 M $HClO_4$ are in the range of ≈60 mA cm^{-2}, with Tafel slopes of 117 mV $decade^{-1}$ corresponding to a cathodic transfer coefficient of ≈0.5 [89]. Moreover, SCEM setups can also be used for the discrimination of partial currents in simultaneous reactions. As an example of this, Figure 4.8b features two reduction reactions simultaneously taking place at the tip, whereby only one of the resulting products can be oxidized on the substrate [88]. For more details about the multiple possibilities offered by the combinations of tip vs. substrate modes, the readers are referred to recent comprehensive reviews on the topic [88, 90, 91].

4.3.5 Floating Electrode Method

The floating electrode (FE) approach combines the high mass transport rates characteristic of gas-diffusion electrodes in fuel cells with a flat, uniform catalyst deposition process. The first extensive experimental studies of FEs were conducted in the context of phosphoric acid fuel cells in the 1960s and 1980s [92–95] and were supported by the corresponding theoretically models [96, 97]. Recently, the FE method has also been adapted to PEFC studies [98, 99], and an example of such a setup is displayed in Figure 4.9 [100]. The FE consists of a 12-μm thick, hydrophobic polycarbonate membrane with an optimized pore geometry (pore diameter of 400 nm, tortuosity of one) and coated with a 100-nm layer of gold as a conductive catalyst substrate. The catalyst is deposited on top of this Au-substrate in the form of a highly uniform ≤600-nm-thick layer deposited close to the substrate pores using vacuum filtration [100]. Essentially, the presence of the hydrophobic layer allows diffusion of the gas toward the catalyst while preventing the wetting of the porous channels with liquid electrolyte, leading to a ≈ threefold increase in mass transport of reactant gases to this CL [101]. Alternatively, Chen et al. [99] sputtered a ≈ 3 nm layer of Pt (leading to a roughness factor of 1.45) onto the surface of a substrate to achieve an ultrathin catalyst for their temperature-dependence HOR/HER kinetic studies.

The absence of limiting currents in experiments using FEs allows resolving the distinct potential/current features of the HOR/HER [101–104] and the ORR [105, 106].

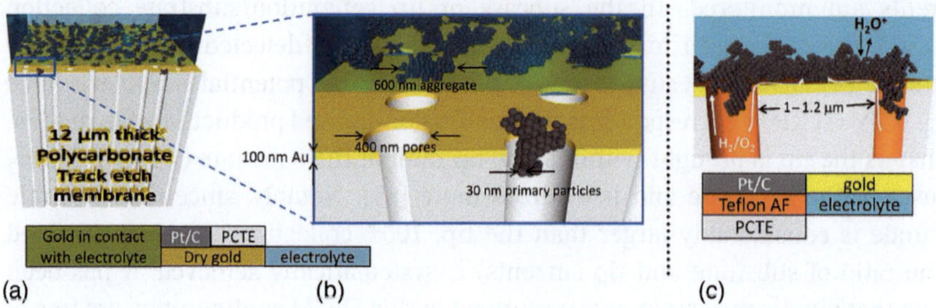

Figure 4.9 Scheme of the microstructure of the floating electrode (a) with expansion of the blue region (b) showing the porous polycarbonate membrane covered with a thin Au-layer, and the particles of catalyst deposited in this substrate's pores. Panel (c) shows a schematic of the FE's cross section. Source: Reproduced from Ref. [100] with permission of American Chemical Society.

In spite of these advantages, there are many factors affecting the FE performance that should be carefully considered, such as the accurate determination of the ECSA and loading determination in such ultrathin catalyst layers, the effect of break-in procedure, or the careful and reproducible preparation of the FE and particularly of the hydrophobic layer.

4.3.6 Methods Summary

To summarize this discussion of the different approaches currently used for determining the kinetic parameters of the HOR/HER in acidic media, Figure 4.10 displays an Arrhenius plot of the i_0 values measured on polycrystalline Pt and

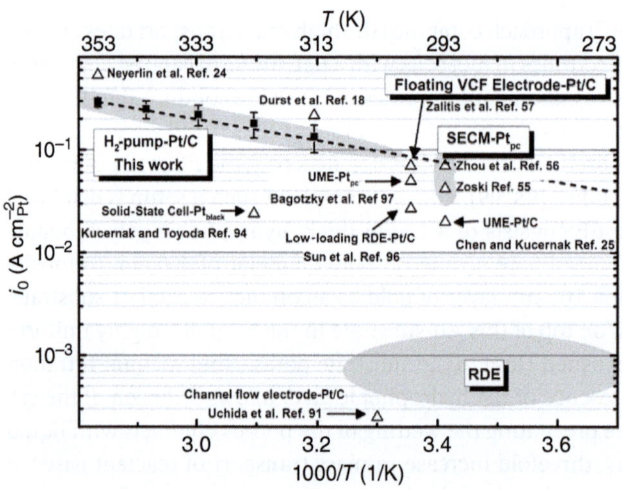

Figure 4.10 Arrhenius plots of the HOR/HER exchange current densities (i_0) obtained on various Pt electrodes using different electrochemical setups. Note that the references quoted in the figure do not correspond to the equivalent numbers in this chapter and can be found in the original work of Durst et al. [78]. Source: Reproduced from Ref. [78]/IOP Publishing/CC BY 4.0.

various carbon-supported Pt-nanoparticle catalysts using these techniques [78]. One can clearly see there that the mass transport limitations inherent to the RDE measurements lead to two orders of magnitude lower exchange current densities with regard to the other methods described above. Moreover, the expected linear relation between these i_0 values and the corresponding inverses of the temperatures at which the measurements were performed is only fulfilled by the self-consistent H_2-pump measurements by Durst and collaborators and a couple of studies implementing FE or SECM approaches [85, 101]. Based on this linear data in Figure 4.10, an activation energy of $16 \pm 2\,kJ\,mol^{-1}$ can be estimated for the HOR/HER on Pt in acidic electrolytes.

4.4 ORR Kinetics

The ORR follows a complex reaction scheme, whereby the reduction of oxygen can proceed either via a four-electron step directly to water or in a two-electron step to form hydrogen peroxide (see scheme in Figure 4.11). Additionally, the latter H_2O_2 can be further reduced to H_2O or chemically decompose (i.e. disproportionate) into H_2O and O_2. It is generally accepted that the ORR follows a direct four-electron mechanism on polycrystalline Pt electrodes [108–110], although a significant amount of H_2O_2 can be produced on Pt(111) and Pt(100) facets at potentials at which these surfaces start being covered with H_{UPD} (i.e. <0.4 V vs. RHE) [32, 111]. On the other hand, the Me/N/C catalysts discussed above do tend to produce significant amounts of H_2O_2 [112], and the subsequent decomposition of this ORR by-product into aggressive hydroperoxyl radicals is believed to be at the origin of a part of the instability issues affecting these materials [113, 114].

4.4.1 ORR Mechanism Studies with RRDE Setups

Rotating ring-disc electrodes are excellently suited for studying the ORR pathway. Specifically, the coaxial Pt- or Au-ring electrode acts as a detector of the hydrogen peroxide formed at the disc electrode during the oxygen reduction by polarizing it at a potential at which the electrooxidation of this H_2O_2 is under diffusion control

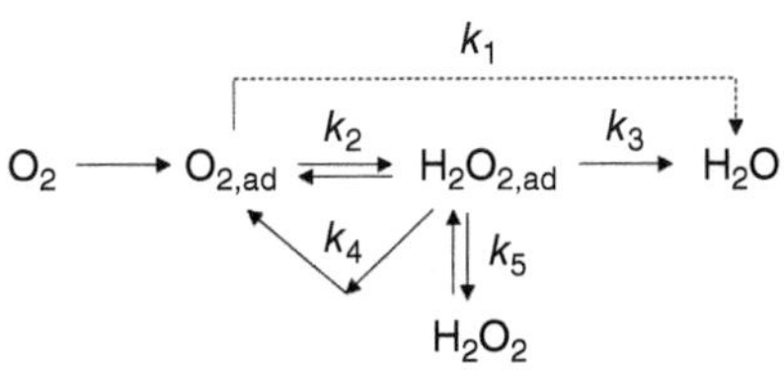

Figure 4.11 Scheme depicting the possible pathways of the oxygen reduction reaction, whereby the O_2 molecule is first adsorbed on the catalyst surface (forming $O_{2,ad}$), to then get reduced either to H_2O or H_2O_2 still adsorbed on the active site ($H_2O_{2,ad}$). In the latter case, the peroxide can be desorbed to the electrolyte, further electroreduced to H_2O, or disproportionate into O_2 and H_2O. Source: Reproduced from Ref. [107] with permission of John Wiley & Sons.

(typically, 1.2–1.5 V vs. RHE) [77]. Similar to the diffusion limited disc current, see Eq. (4.19), the corresponding ring current is defined as

$$I_R = 0.62nF\pi\left(r_3^3 - r_2^3\right)^{2/3} D^{2/3} \nu^{-1/6} \omega^{1/2} c_{\text{bulk}} \tag{4.24}$$

where r_2 and r_3 are the internal and external radii of the ring electrode, respectively, see Figure 4.6b. A key feature of any RRDE is its collection efficiency (N), which corresponds to the fraction of the species formed at the disc that reaches the ring and participates in the reaction taking place there, according to the equation:

$$N = -\frac{I_R}{I_D} \tag{4.25}$$

Theoretically, this N depends exclusively on the dimensions of the ring and disc electrodes [84], but imprecisions in the components' radii and planarity translate in the need to experimentally determine its value. This is typically done taking advantage of a fast redox reaction, like Fe^{2+}/Fe^{3+}; thus, using $K_3[Fe(CN)_6]$ dissolved in an alkaline, aqueous solution as the Fe^{3+} source, Fe^{3+} is reduced to Fe^{2+} at the disc, and the fraction of the latter ferrous cation reaching the ring (i.e. N) is oxidized back to Fe^{3+} [77].

If applicable, the dependency of the collection efficiency on the electrode's rotation rate can be used to evaluate the lifetime of intermediates formed at the disc electrode [115]. Moreover, the thin-film RDE approach discussed above can also be successfully extended to RRDE [116], and thus the percentage fraction of the hydrogen peroxide produced during the ORR can be then estimated using the formalism [77]:

$$x_{H_2O_2}[\%] = 100 \cdot \frac{2 \cdot I_R/N}{I_D + I_R/N} \tag{4.26}$$

An example of the currents recorded at the ring and disc electrodes upon studying the reduction of O_2 on a carbon-supported Pt catalyst immobilized at the disc is shown in Figure 4.12. As illustrated by the ring currents, the ORR on this material proceeds exclusively via a four-electron mechanism in the potential range ≥ 0.6 V relevant for most PEFC applications. Complementarily, an increase in hydrogen peroxide formation is observed at potentials <0.6 V vs. RHE and specifically in the H_{UPD} region, caused by the (partial) blocking of the Pt-surface by adsorbed hydrogen [77] (*vide supra*).

4.4.2 ORR Pathway on Me/N/C ORR Catalysts

As shown in Chapter 13, the metal in Me/N/C ORR catalysts is typically distributed into two groups of species: atomically dispersed Me–N_x moieties and Me-based particles such as carbon-encapsulated, reduced metal, oxides, and/or carbides. Such a complex and diverse composition makes the study of these materials very challenging, and their low active site concentration translates in the need for thick catalyst layers that may introduce artifacts upon R(R)DE studies (see below). When applied to Fe-based Me/N/Cs, such RRDE works have shown that the ORR on Fe/N/C catalysts exclusively containing Fe–N_x sites follows a direct four-electron mechanism

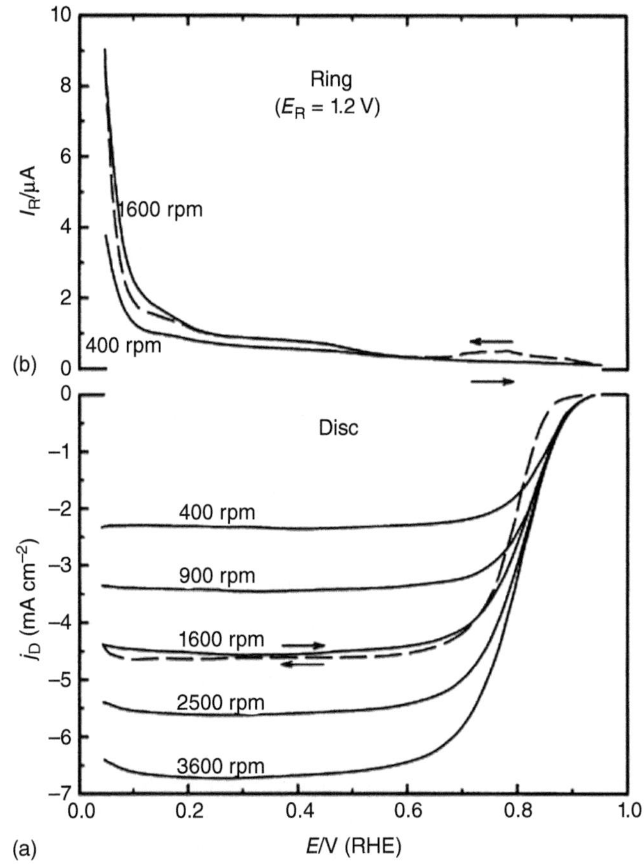

Figure 4.12 Oxygen reduction current density recorded on Pt/C catalyst in O_2-saturated 0.5 M H_2SO_4 at 60 °C in an RRDE setup, using a disc catalyst loading of 14 μg_{Pt} cm^{-2} and at a scan rate of 5 mV s^{-1}. The corresponding ring currents were recorded simultaneously at 400 and 1600 rpm, with the ring potential held at 1.2 V vs. RHE. Only positive-going potential sweeps are shown except for the measurement performed at 1600 rpm, whereby the dashed lines represent the negative going scan. Source: Reproduced from Ref. [77] with permission of Elsevier.

[117–119], whereas Fe-free analogues prepared under the same synthesis conditions cannot convert oxygen directly to water [119] but instead feature high H_2O_2 yields [119]. Moreover, H_2O_2-electroreduction studies showed that carbon-encapsulated Fe-based particles [118] and Fe–N_x-sites with a pyridinic N-coordination environment are active for this peroxide reduction reaction (PRR) [119]. While the catalyst synthesis route defines the catalyst composition [113, 120–124], both direct and indirect ORR pathways may coexist in the obtained materials on Fe–N_x sites and Fe-based particles, respectively [119, 125]. Moreover, similar results have also been obtained on Co-based materials of the same catalyst family (i.e. Co/N/C) [126, 127].

On top of this, some groups have reported a catalyst loading effect on the hydrogen peroxide yield in RRDE measurements [118, 128]. It was then shown that at low disc catalyst loadings of 20–40 μg cm^{-2}, the hydrogen peroxide molecules formed by the Me/N/C catalyst have a higher probability of diffusing away from the catalyst layer and into the electrolyte, resulting in a large ring current and concomitantly high H_2O_2 yield [128]. In the case of thick electrodes resulting from catalyst disc loadings ≥800 μg cm^{-2}, though, hydrogen peroxide may be further reduced on the PRR-active catalytic sites present in the catalyst or chemically decompose into H_2O and O_2 upon diffusion across the catalyst layer. As a result, the majority of the hydrogen peroxide formed cannot reach the electrolyte and be further detected at the RRDE's ring, leading to an apparently low H_2O_2 yield [129, 130].

4.4.3 ORR Kinetics: Methods

The half-cell electrochemical methods developed for the kinetics studies of the ORR can be divided into two groups, namely (i) those facilitating large limiting currents and (ii) those with finite, yet well-defined limiting currents [131]. The methods in the first group have limiting currents that are large compared to the raw currents in the vicinity of the reaction's onset, so that the latter can be directly regarded as kinetic currents. Techniques such as the floating electrode methods [101] discussed in the HOR section as well as gas-phase/wall-jet electrode [17] can be included in this class. As for the methods in the second group, these operate on the basis of relatively small but well-defined limiting currents and thus include the R(R)DE [131, 132] measurements discussed above or channel flow dual electrodes [133] in which oxygen diffuses through a rather thick (>10 μm) boundary layer that results in a well-defined limiting current transformable into a kinetic current applying the Koutecky–Levich equation.

The R(R)DE method remains the preferred approach for the analysis of ORR kinetics. Comparatively, the derivation of such kinetic information in PEFC experiments implies further complications resulting from H_2-crossover and ohmic-drop as well as mass transport losses that can be lessened using thin membranes and low catalyst loadings. Thus, in what follows we further discuss the complementarity of the ORR-kinetic parameters extracted from such RDE vs. PEFC studies on Pt-based vs. Me/N/C catalysts.

4.4.3.1 Pt-based Electrodes

The comparison among mass-normalized ORR-current densities derived from RDE vs. PEFC measurements is typically done at a potential of 0.9 V vs. RHE at which the influence of mass transport on the PEFC performance is generally negligible [77]. Nevertheless, such PEFC data still need to be carefully corrected for ohmic losses (typically derived from electrochemical impedance measurements) and H_2-crossover through the PFSA membrane, and the currents measured in the RDE (also ohmic-drop corrected) need to be transformed into the corresponding i_{kin} values using Eq. (4.20). On top of this, one should account for two effects related to the differences in the experimental setups. First, PEFC measurements are customarily performed at temperatures of ≈80 °C, whereas RDE is usually carried out at room temperature and up to a maximum of 60 °C. As discussed in the introduction, one can expect these temperature differences to lead to a shift of the catalyst's i_0 (as well as of E_{rev}) [134]. From an experimental point of view, this temperature effect on the mass activity of Pt/C catalysts has been investigated using RDE [77, 134], microelectrodes [111, 135], and flow electrodes [133], and it was shown that the ORR activity on Pt in a non-anion-adsorbing electrolyte (0.1 M $HClO_4$) in RDE measurements at 60 °C is equivalent to that in a PEMFC at 80 °C [8].

4.4.3.2 Pt-free Catalysts: RDE vs. PEFC Kinetic Studies

Unlike Pt-based catalysts in which the ORR activities in RDE vs PEFC measurements are typically comparable, certain Me/N/C materials possess relatively high ORR activities in the PEFC but an approx. one- to two-order magnitude poorer performance in RDE experiments [136–139]. A first reason for this could again be the temperature difference between PEFC and RDE studies, but considering the activation energy of ≈10 kJ mol^{-1} derived from the Arrhenius plots of these catalysts, only a twofold activity difference would be expected from 25 to 80 °C (assuming a fixed O_2-concentration) [136]. Alternatively, this discrepancy could also stem from differences in the TOF of the ORR-active Fe–N_x sites in liquid vs. solid (i.e. PFSA-based) electrolytes. This was explained by invoking an effect of the adsorption of sulfate/sulfonate ions on protonated NH^+-groups neighboring to these Fe–N_x sites on their TOF. Specifically, while the protonation itself would result in a approx. Fivefold TOF increase [140], the subsequent adsorption of the HSO_4^- in H_2SO_4-based electrolyte or of perfluoro sulfonate groups in the PEFC cathode ionomer would invalidate this effect and decrease again these sites' TOF. Notably, the higher mobility of the (bi)sulfate anions in the liquid electrolyte would lead to a quasi-instantaneous activity decay of the catalysts in RDE experiments, whereas in the PEFC this effect would be delayed by the slowed mobility of the sulfonic groups in the PFSA ionomer.

4.5 Concluding Remarks

In summary, the quantification of the ECSA of Pt-based catalysts is well established and preponderantly relies on the recording of cyclic voltammograms and evaluation of their H_{UPD} charges, which can nevertheless entail certain imprecisions solvable through the CO-stripping approach. On the other hand, the equivalent quantification of the site density in Me/N/C catalysts remains a subject of ongoing research, with the one available method available for this task (nitrate reduction) being unamenable to the PEFC environment.

As for the quantification of the PEFC reactions' kinetic parameters, multiple setups have been developed to curtail the difficulties posed by the ultrafast rates of the HOR on Pt-surfaces in acidic media, which hinder the use of RDEs for this purpose. Among those methods (i.e. UMEs and FEs), H_2-pump setups seem to systematically lead to the most reliable results (based on the Arrhenius plot in Figure 4.10) but require a demanding preparation of MEAs and use of PEFC hardware that render the possibility of using other, simpler methods appealing. Finally, R(R)DEs have become the go-to technique for a fast and reproducible assessment of ORR kinetics, but users should keep in mind the possible sources of artifacts associated with them (e.g. excessively thick catalyst layers), and that it does not fully mimic the more complex processes taking place in a real PEFC electrode.

Acronyms

CL	catalyst layer
CV	cyclic voltammogram
DL	double layer
ECSA	electrochemical surface area
FE	floating electrode
G/C	generation/collection
HER	hydrogen evolution reaction
HOR	hydrogen oxidation reaction
MEA	membrane-electrode assembly
ORR	oxygen reduction reaction
RDS	rate-determining step
RDE	rotating disc electrode
RHE	reversible hydrogen electrode
RRDE	rotating ring disc electrode
SA	surface area
SD	site density
SECM	scanning electrochemical microscopy
TOF	turnover frequency
TF	thin film
TG/SC	tip generation/substrate collection
PEFC	polymer electrolyte fuel cell
PRR	peroxide reduction reaction
PTFE	polytetrafluoroethylene
PFSA	perfluorosulfonic acid
UPD	underpotential deposition
UME	ultramicroelectrode

Symbols

b_c	Tafel slope
C_i	reactant concentration
C_i^*	reference reactant concentration
c_{bulk}	bulk concentration
D	diffusion coefficient
d_i	diameter
$\overline{d_N}$	number-averaged diameter

$\overline{d_S}$ surface-averaged diameter
E potential
E_{rev} reversible potential
E_{act} reaction activation energy
e^- electron charge
F Faraday constant
H shape-dependent coefficient
I_R ring current
I_D disc current
i_{kin} kinetically controlled current
i_0 intrinsic exchange current density
$i_0{}^*$ exchange current density at reference conditions
i_{lim} limiting current
l_i number of particles
L_{cat} catalyst loading per electrode area
m mass transport coefficient
N collection efficiency
n number of electrons involved in the electrochemical reaction
R gas constant
$\mathcal{R}$ relative roughness factor
r radius
T temperature
T^* reference temperature
α_a anodic transfer coefficient
α_c cathodic transfer coefficient
γ reaction order
δ_{BL} boundary layer thickness
η_{kin} kinetic overpotential
η_{diff} Nernstian diffusion overpotential
ν kinematic viscosity
$\rho_{electrode}$ electrode density
x percentage fraction
ω rotation rate

References

1 Kojima, K. and Fukazawa, K. (2015). Current status and future outlook of fuel cell vehicle development in Toyota. *ECS Transactions* 69: 213–219. https://doi.org/10.1149/06917.0213ecst.

2 Kongkanand, A. and Mathias, M.F. (2016). The priority and challenge of high-power performance of low-platinum proton-exchange membrane fuel cells. *Journal of Physical Chemistry Letters* 7: 1127–1137. https://doi.org/10.1021/acs.jpclett.6b00216.

3 Yarlagadda, V., Carpenter, M.K., Moylan, T.E. et al. (2018). Boosting fuel cell performance with accessible carbon mesopores. *ACS Energy Letters* 3: 618–621. https://doi.org/10.1021/acsenergylett.8b00186.

4 Greeley, J., Stephens, I.E.L., Bondarenko, A.S. et al. (2009). Alloys of platinum and early transition metals as oxygen reduction electrocatalysts. *Nature Chemistry* 1: 552–556. https://doi.org/10.1038/nchem.367.

5 Osmieri, L., Park, J., Cullen, D.A. et al. (2021). Status and challenges for the application of platinum group metal-free catalysts in proton-exchange membrane fuel cells. *Current Opinion in Electrochemistry* 25: 100627–100639. https://doi.org/10.1016/j.coelec.2020.08.009.

6 Guidelli, R., Compton, R.G., Feliu, J.M. et al. (2014). Defining the transfer coefficient in electrochemistry: an assessment (IUPAC technical report). *Pure and Applied Chemistry* 86: 245–258. https://doi.org/10.1515/pac-2014-5026.

7 Neyerlin, K.C., Gu, W., Jorne, J., and Gasteiger, H.A. (2006). Determination of catalyst unique parameters for the oxygen reduction reaction in a PEMFC. *Journal of the Electrochemical Society* 153: A1955–A1963. https://doi.org/10.1149/1.2266294.

8 Gasteiger, H.A., Kocha, S.S., Sompalli, B., and Wagner, F.T. (2005). Activity benchmarks and requirements for Pt, Pt-alloy, and non-Pt oxygen reduction catalysts for PEMFCs. *Applied Catalysis B: Environmental* 56: 9–35. https://doi.org/10.1016/j.apcatb.2004.06.021.

9 Kramm, U.I., Herranz, J., Larouche, N. et al. (2012). Structure of the catalytic sites in Fe/N/C-catalysts for O_2-reduction in PEM fuel cells. *Physical Chemistry Chemical Physics* 14: 11673–11688. https://doi.org/10.1039/c2cp41957b.

10 Ticianelli, E.A., Derouin, C.R., Redondo, A., and Srinivasan, S. (1988). Methods to advance technology of proton exchange membrane fuel cells. *Journal of the Electrochemical Society* 135: 2209–2214.

11 Mukerjee, S. and Srinivasan, S. (1993). Enhanced electrocatalysis of oxygen reduction on platinum alloys in proton exchange membrane fuel cells. *Journal of Electroanalytical Chemistry* 357: 201–224.

12 Wesselmark, M., Wickman, B., Lagergren, C., and Lindbergh, G. (2010). Hydrogen oxidation reaction on thin platinum electrodes in the polymer electrolyte fuel cell. *Electrochemistry Communications* 12: 1585–1588. https://doi.org/10.1016/j.elecom.2010.08.037.

13 Subramanian, N.P., Greszler, T.A., Zhang, J. et al. (2011). Pt-oxide coverage-dependent oxygen reduction reaction (ORR) kinetics. *ECS Transactions* 41: 985–1007.

14 Kocha, S.S., Yang, J.D., and Yi, J.S. (2006). Characterization of gas crossover and its implications in PEM fuel cells. *AIChE* 52: 1916–1925. https://doi.org/10.1002/aic.

15 Liu, Z., Wainright, J.S., Huang, W., and Savinell, R.F. (2004). Positioning the reference electrode in proton exchange membrane fuel cells: calculations of primary and secondary current distribution. *Electrochimica Acta* 49: 923–935. https://doi.org/10.1016/j.electacta.2003.10.004.

16 Ohs, J.H., Sauter, U., Maass, S., and Stolten, D. (2012). The effect of the reference electrode position on the measurement of half cell polarization in proton-exchange membrane fuel cells. *Journal of the Electrochemical Society* 159: F181–F186. https://doi.org/10.1149/2.006207jes.

17 Kucernak, A.R. and Toyoda, E. (2008). Studying the oxygen reduction and hydrogen oxidation reactions under realistic fuel cell conditions. *Electrochemistry Communications* 10: 1728–1731. https://doi.org/10.1016/j.elecom.2008.09.001.

18 Zhang, J., Kramer, D., Shimoi, R. et al. (2006). In situ diagnostic of two-phase flow phenomena in polymer electrolyte fuel cells by neutron imaging: Part B. Material variations. *Electrochimica Acta* 51: 2715–2727. https://doi.org/10.1016/j.electacta.2005.08.010.

19 Reum, M., Freunberger, S.A., Wokaun, A., and Büchi, F.N. (2009). Measuring the current distribution with sub-millimeter resolution in PEFCs: II. Impact of operating parameters. *Journal of the Electrochemical Society* 156: B301–B310. https://doi.org/10.1149/1.3043422.

20 Trasatti, S. and Petrii, O.A. (1992). Real surface area measurements in electrochemistry. *Journal of Electroanalytical Chemistry* 327: 353–376. https://doi.org/10.1016/0926-860x(96)80148-7.

21 Ferreira, P.J., la O', G.J., Shao-Horn, Y. et al. (2005). Instability of Pt/C electrocatalysts in proton exchange membrane fuel cells. *Journal of the Electrochemical Society* 152: A2256–A2271. https://doi.org/10.1149/1.2050347.

22 Guilminot, E., Corcella, A., Charlot, F. et al. (2007). Detection of Pt^{z+} ions and Pt nanoparticles inside the membrane of a used PEMFC. *Journal of the Electrochemical Society* 154: B96–B102. https://doi.org/10.1149/1.2388863.

23 Lööf, P., Stenbom, B., Nordén, H., and Kasemo, B. (1993). Rapid sintering in NO of nanometre-sized Pt particles on γ-Al_2O_3 observed by CO temperature-programmed desorption and transmission electron microscopy. *Journal of Catalysis* 144: 60–76. https://doi.org/10.1006/jcat.1993.1314.

24 Thomas, J.M. and Thomas, W.J. (2015). *Principles and Practice of Heterogeneous Catalysis*, 2e. Wiley-VCH.

25 Henning, S., Kühn, L., Herranz, J. et al. (2016). Pt-Ni aerogels as unsupported electrocatalysts for the oxygen reduction reaction. *Journal of the Electrochemical Society* 163: F998–F1003. https://doi.org/10.1149/2.0251609jes.

26 Henning, S., Kühn, L., Herranz, J. et al. (2017). Effect of acid washing on the oxygen reduction reaction activity of Pt–Cu aerogel catalysts. *Electrochimica Acta* 233: 210–217. https://doi.org/10.1016/j.electacta.2017.03.019.

27 Chung, H.T., Cullen, D.A., Higgins, D. et al. (2017). Direct atomic-level insight into the active sites of a high-performance PGM-free ORR catalyst. *Science* 357: 479–484. https://doi.org/10.1126/science.aan2255.

28 Nart, F.C. and Vielstich, W. (2010). Noramlization of porous active surfaces. In: *Handbook of Fuel Cells*, 1–14. https://doi.org/10.1002/9780470974001.f203023.

29 Oviedo, O.A., Reinaudi, L., Garcia, S., and Leiva, E.P.M. (2016). *Underpotential Deposition: From Fundamentals and Theory to Applications at the Nanoscale.* Springer.

30 Herrero, E., Buller, L.J., and Abruña, H.D. (2001). Underpotential deposition at single crystal surfaces of Au, Pt, Ag and other materials. *Chemical Reviews* 101: 1897–1930. https://doi.org/10.1021/cr9600363.

31 Schmidt, T.J., Gasteiger, H.A., Stäb, G.D. et al. (1998). Characterization of high-surface-area electrocatalysts using a rotating disk electrode configuration. *Journal of the Electrochemical Society* 145: 2354–2361. https://doi.org/10.1149/1.1838642.

32 Markovic, N.M., Gasteiger, H.A., and Ross, P.N. (1995). Oxygen reduction on platinum low-index single-crystal surfaces in sulfuric acid solution: rotating ring-Pt(hkl) disk studies. *The Journal of Physical Chemistry* 99: 3411–3415. https://doi.org/10.1021/j100011a001.

33 Maillard, F., Pronkin, S.N., and Savinova, E.R. (2009). Influence of size on the electrocatalytic activities of supported metal nanoparticles in fuel cell-related reactions. In: *Handbook of Fuel Cells: Advances in Electrocatalysis Materials, Diagnostics and Durability*, vol. 5 and 6 (ed. W. Vielstich, H.A. Gasteiger, and H. Yokokawa), 1090. Wiley.

34 Stamenkovic, V.R. (2015). Annual merit review, DOE: hydrogen and fuel cells and vehicle technologies, Washington, DC (8–12 June 2015).

35 Van Der Vliet, D.F., Wang, C., Li, D. et al. (2012). Unique electrochemical adsorption properties of Pt-skin surfaces. *Angewandte Chemie, International Edition* 51: 3139–3142. https://doi.org/10.1002/anie.201107668.

36 Shao, M., Odell, J.H., Il Choi, S., and Xia, Y. (2013). Electrochemical surface area measurements of platinum- and palladium-based nanoparticles. *Electrochemistry Communications* 31: 46–48. https://doi.org/10.1016/j.elecom.2013.03.011.

37 Rudi, S., Cui, C., Gan, L., and Strasser, P. (2014). Comparative study of the electrocatalytically active surface areas (ECSAs) of Pt alloy nanoparticles evaluated by H_{upd} and CO-stripping voltammetry. *Electrocatalysis* 5: 408–418. https://doi.org/10.1007/s12678-014-0205-2.

38 Scortichini, C.L. and Reilley, C.N. (1983). Surface characterization of Pt electrodes using underpotential deposition of H and Cu. *Journal of Electroanalytical Chemistry and Interfacial Electrochemistry* 152: 255–260.

39 Quiroz, M.A., Meas, Y., Lamy-Pitara, E., and Barbier, J. (1983). Characterization of a ruthenium electrode by underpotential deposition of copper. *Journal of Electroanalytical Chemistry and Interfacial Electrochemistry* 157: 165–174. https://doi.org/10.1016/S0022-0728(83)80387-8.

40 Green, C.L. and Kucernak, A. (2002). Determination of the platinum and ruthenium surface areas in platinum-ruthenium alloy electrocatalysts by underpotential deposition of copper. I. Unsupported catalysts. *The Journal of Physical Chemistry B* 106: 1036–1047. https://doi.org/10.1021/jp0131931.

41 Schlaup, C. and Horch, S. (2013). Study of underpotential deposited Cu layers on Pt(111) and their stability against CO and CO_2 in perchloric acid. *Physical Chemistry Chemical Physics* 15: 19659–19664. https://doi.org/10.1039/c3cp52649f.

42 Markovic, N. and Ross, P.N. (1993). Effect of anions on the underpotential deposition of Cu on Pt(111) and Pt(100) surfaces. *Langmuir* 9: 580–590. https://doi.org/10.1021/la00026a037.

43 Marković, N.M., Grgur, B.N., Lucas, C.A., and Ross, P.N. (1998). UPD of Cu on Pt(100): effects of anions on adsorption isotherms and interface structures. *Electrochimica Acta* 44: 1009–1017. https://doi.org/10.1016/S0013-4686(98)00205-9.

44 Andolfatto, F., Durand, R., Michas, A. et al. (1994). Solid polymer electrolyte water electrolysis: electrocatalysis and long-term stability. *International Journal of Hydrogen Energy* 19: 421–427. https://doi.org/10.1016/0360-3199(94)90018-3.

45 Maillard, F., Eikerling, M., Cherstiouk, O.V. et al. (2004). Size effects on reactivity of Pt nanoparticles in CO monolayer oxidation: the role of surface mobility. *Faraday Discussions* 125: 357–377. https://doi.org/10.1039/b303911k.

46 Cuesta, A., Couto, A., Rincón, A. et al. (2006). Potential dependence of the saturation CO coverage of Pt electrodes: the origin of the pre-peak in CO-stripping voltammograms. Part 3: Pt(poly). *Journal of Electroanalytical Chemistry* 586: 184–195. https://doi.org/10.1016/j.jelechem.2005.10.006.

47 Rheinländer, P.J., Herranz, J., Durst, J., and Gasteiger, H.A. (2014). Kinetics of the hydrogen oxidation/evolution reaction on polycrystalline platinum in alkaline electrolyte reaction order with respect to hydrogen pressure. *Journal of the Electrochemical Society* 161: F1448–F1457. https://doi.org/10.1149/2.0501414jes.

48 Marković, N.M., Schmidt, T.J., Grgur, B.N. et al. (1999). Effect of temperature on surface processes at the Pt(111)–liquid interface: hydrogen adsorption, oxide formation, and CO oxidation. *The Journal of Physical Chemistry B* 103: 8568–8577. https://doi.org/10.1021/jp991826u.

49 Malko, D., Kucernak, A., and Lopes, T. (2016). Performance of Fe–N/C oxygen reduction electrocatalysts toward NO_2^-, NO, and NH_2OH electroreduction: from fundamental insights into the active center to a new method for environmental nitrite destruction. *Journal of the American Chemical Society* 138: 16056–16068. https://doi.org/10.1021/jacs.6b09622.

50 Malko, D., Kucernak, A., and Lopes, T. (2016). *In situ* electrochemical quantification of active sites in Fe/N/C non-precious metal catalysts. *Nature Communications* 7: 13285–13294. https://doi.org/10.1038/ncomms13285.

51 Primbs, M., Sun, Y., Roy, A. et al. (2020). Establishing reactivity descriptors for platinum group metal (PGM)-free Fe-N-C catalysts for PEM fuel cells. *Energy & Environmental Science* 13: 2480–2500. https://doi.org/10.1039/d0ee01013h.

52 Oh, H.S., Oh, J.G., Haam, S. et al. (2008). On-line mass spectrometry study of carbon corrosion in polymer electrolyte membrane fuel cells. *Electrochemistry Communications* 10: 1048–1051. https://doi.org/10.1016/j.elecom.2008.05.006.

53 Zana, A., Speder, J., Reeler, N.E.A. et al. (2013). Investigating the corrosion of high surface area carbons during start/stop fuel cell conditions: a Raman study. *Electrochimica Acta* 114: 455–461. https://doi.org/10.1016/j.electacta.2013.10.097.

54 Choi, C.H., Baldizzone, C., Grote, J.P. et al. (2015). Stability of Fe-N-C catalysts in acidic medium studied by operando spectroscopy. *Angewandte Chemie, International Edition* 54: 12753–12757. https://doi.org/10.1002/anie.201504903.

55 Gasteiger, H.A., Gu, W.B., Makharia, R. et al. (2003). Fuel cell technology and applications. In: *Handbook of Fuel Cells: Fundamentals, Technology, Applications*, vol. 3, 593.

56 Carter, R.N., Kocha, S.S., Wagner, F. et al. (2007). Artifacts in measuring electrode catalyst area of fuel cells through cyclic voltammetry. *ECS Transactions* 11: 403–410. https://doi.org/10.1149/1.2780954.

57 Sibert, E., Faure, R., and Durand, R. (2001). High frequency impedance measurements on Pt(111) in sulphuric and perchloric acids. *Journal of Electroanalytical Chemistry* 515: 71–81. https://doi.org/10.1016/S0022-0728(01)00639-8.

58 Łosiewicz, B., Jurczakowski, R., and Lasia, A. (2012). Kinetics of hydrogen underpotential deposition at polycrystalline platinum in acidic solutions. *Electrochimica Acta* 80: 292–301. https://doi.org/10.1016/j.electacta.2012.07.019.

59 Duncan, H. and Lasia, A. (2007). Mechanism of hydrogen adsorption/absorption at thin Pd layers on Au(111). *Electrochimica Acta* 52: 6195–6205. https://doi.org/10.1016/j.electacta.2007.03.068.

60 Garrick, T.R., Moylan, T.E., Carpenter, M.K., and Kongkanand, A. (2017). Electrochemically active surface area measurement of aged Pt alloy catalysts in PEM fuel cells by CO stripping. *Journal of the Electrochemical Society* 164: F55–F59. https://doi.org/10.1149/2.0381702jes.

61 Shinozaki, K., Yamada, H., and Morimoto, Y. (2011). Relative humidity dependence of Pt utilization in polymer electrolyte fuel cell electrodes: effects of electrode thickness, ionomer-to-carbon ratio, ionomer equivalent weight, and carbon support. *Journal of the Electrochemical Society* 158: B467–B478.

62 Becknell, N., Kang, Y., Chen, C. et al. (2015). Atomic structure of Pt_3Ni nanoframe electrocatalysts by *in situ* X-ray absorption spectroscopy. *Journal of the American Chemical Society* 137: 15817–15824. https://doi.org/10.1021/jacs.5b09639.

63 Chattot, R., Martens, I., Scohy, M. et al. (2020). Disclosing Pt-bimetallic alloy nanoparticle surface lattice distortion with electrochemical probes. *ACS Energy Letters* 5: 162–169. https://doi.org/10.1021/acsenergylett.9b02287.

64 Neyerlin, K.C., Gu, W., Jorne, J., and Gasteiger, H.A. (2007). Study of the exchange current density for the hydrogen oxidation and evolution reactions. *Journal of the Electrochemical Society* 154: B631–B635. https://doi.org/10.1149/1.2733987.

65 Krischer, K. and Savinova, E.R. (2008). Fundamentals of electrocatalysis. In: *Handbook of Heterogeneous Catalysis* (ed. G. Ertl, H. Knözinger, F. Schüth, and J. Weitkamp), 1873. https://doi.org/10.1002/9783527610044.hetcat0101.

66 Bagotzky, V.S. and Osetrova, N.V. (1973). Investigations of hydrogen ionization on platinum with the help of micro-electrodes. *Journal of Electroanalytical*

Chemistry and Interfacial Electrochemistry 43: 233–249. https://doi.org/10.1016/S0022-0728(73)80494-2.

67 Conway, B.E. and Tilak, B.V. (2002). Interfacial processes involving electrocatalytic evolution and oxidation of H_2, and the role of chemisorbed H. *Electrochimica Acta* 47: 3571–3594. https://doi.org/10.1016/S0013-4686(02)00329-8.

68 Schmidt, T.J., Ross, P.N., and Markovic, N.M. (2002). Temperature dependent surface electrochemistry on Pt single crystals in alkaline electrolytes: Part 2. The hydrogen evolution/oxidation reaction. *Journal of Electroanalytical Chemistry* 524–525: 252–260. https://doi.org/10.1039/b208322a.

69 Lasia, A. and Rami, A. (1990). Kinetics of hydrogen evolution on nickel electrodes. *Journal of Electroanalytical Chemistry and Interfacial Electrochemistry* 294: 123–141. https://doi.org/10.1016/0022-0728(90)87140-F.

70 Herraiz-Cardona, I., Ortega, E., Vázquez-Gómez, L., and Pérez-Herranz, V. (2012). Double-template fabrication of three-dimensional porous nickel electrodes for hydrogen evolution reaction. *International Journal of Hydrogen Energy* 37: 2147–2156. https://doi.org/10.1016/j.ijhydene.2011.09.155.

71 Herraiz-Cardona, I., González-Buch, C., Valero-Vidal, C. et al. (2013). Co-modification of Ni-based type Raney electrodeposits for hydrogen evolution reaction in alkaline media. *Journal of Power Sources* 240: 698–704. https://doi.org/10.1016/j.jpowsour.2013.05.041.

72 Durst, J., Siebel, A., Simon, C. et al. (2014). New insights into the electrochemical hydrogen oxidation and evolution reaction mechanism. *Energy and Environmental Science* 7: 2255–2260. https://doi.org/10.1039/c4ee00440j.

73 Bhowmik, T., Kundu, M.K., and Barman, S. (2016). Palladium nanoparticle-graphitic carbon nitride porous synergistic catalyst for hydrogen evolution/oxidation reactions over a broad range of pH and correlation of its catalytic activity with measured hydrogen binding energy. *ACS Catalysis* 6: 1929–1941. https://doi.org/10.1021/acscatal.5b02485.

74 Herranz, J., Durst, J., Fabbri, E. et al. (2016). Interfacial effects on the catalysis of the hydrogen evolution, oxygen evolution and CO_2-reduction reactions for (Co-) electrolyzer development. *Nano Energy* 29: 4–28. https://doi.org/10.1016/j.nanoen.2016.01.027.

75 Davydova, E.S., Mukerjee, S., Jaouen, F., and Dekel, D.R. (2018). Electrocatalysts for hydrogen oxidation reaction in alkaline electrolytes. *ACS Catalysis* 8: 6665–6690. https://doi.org/10.1021/acscatal.8b00689.

76 Gileadi, E. (2011). *Physical Electrochemistry*. Wiley-VCH.

77 Paulus, U.A., Schmidt, T.J., Gasteiger, H.A., and Behm, R.J. (2001). Oxygen reduction on a high-surface area Pt/Vulcan carbon catalyst: a thin-film rotating ring-disk electrode study. *Journal of Electroanalytical Chemistry* 495: 134–145. https://doi.org/10.1016/S0022-0728(00)00407-1.

78 Durst, J., Simon, C., Hasché, F., and Gasteiger, H.A. (2015). Hydrogen oxidation and evolution reaction kinetics on carbon supported Pt, Ir, Rh, and Pd Electrocatalysts in acidic media. *Journal of the Electrochemical Society* 162: F190–F203. https://doi.org/10.1149/2.0981501jes.

79 Sheng, W., Gasteiger, H.A., and Shao-Horn, Y. (2010). Hydrogen oxidation and evolution reaction kinetics on platinum: acid vs alkaline electrolytes. *Journal of the Electrochemical Society* 157: B1529–B1536. https://doi.org/10.1149/1.3483106.

80 Chen, S. and Kucernak, A. (2004). Electrocatalysis under conditions of high mass transport: investigation of hydrogen oxidation on single submicron Pt particles supported on carbon. *The Journal of Physical Chemistry B* 108: 13984–13994. https://doi.org/10.1021/jp048641u.

81 Breiter, M., Vielstich, W., Lamm, A., and Gasteiger, H.A. (2003). *Handbook of Fuel Cells: Fundamentals, Technology and Applications*, vol. 2. New York: Wiley.

82 Lasia, A., Vielstich, W., Gasteiger, H.A. et al. (2010). *Handbook of Fuel Cells: Fundamentals, Technology and Applications*. United States: Wiley.

83 Zheng, J., Yan, Y., and Xu, B. (2015). Correcting the hydrogen diffusion limitation in rotating disk electrode measurements of hydrogen evolution reaction kinetics. *Journal of the Electrochemical Society* 162: F1470–F1481. https://doi.org/10.1149/2.0501514jes.

84 Bard, A.J. and Faulkner, L.R. (2001). *Electrochemical Methods: Fundamentals and Applications*, 2e. New Jersey: Wiley https://doi.org/10.1016/B978-0-12-381373-2.00056-9.

85 Zhou, J., Zu, Y., and Bard, A.J. (2000). Scanning electrochemical microscopy: Part 39. The proton/hydrogen mediator system and its application to the study of the electrocatalysis of hydrogen oxidation. *Journal of Electroanalytical Chemistry* 491: 22–29. https://doi.org/10.1016/S0022-0728(00)00100-5.

86 Bard, A.J., Fan, F.R.F., Kwak, J., and Lev, O. (1989). Scanning electrochemical microscopy. Introduction and principles. *Analytical Chemistry* 61: 132–138. https://doi.org/10.1021/ac00177a011.

87 Leonard, K.C. and Bard, A.J. (2013). The study of multireactional electrochemical interfaces via a tip generation/substrate collection mode of scanning electrochemical microscopy: the hydrogen evolution reaction for Mn in acidic solution. *Journal of the American Chemical Society* 135: 15890–15896. https://doi.org/10.1021/ja407395m.

88 Zoski, C.G. (2016). Review—advances in scanning electrochemical microscopy (SECM). *Journal of the Electrochemical Society* 163: H3088–H3100. https://doi.org/10.1149/2.0141604jes.

89 Zoski, C.G. (2003). Scanning electrochemical microscopy: investigation of hydrogen oxidation at polycrystalline noble metal electrodes. *The Journal of Physical Chemistry B* 107: 6401–6405. https://doi.org/10.1021/jp027436g.

90 Bard, A.J. and Mirkin, M.V. (2012). *Scanning Electrochemical Microscopy*, 2e, 670. CRC Press.

91 Bard, A.J. (2007). The rise of voltammetry: from polarography to the scanning electrochemical microscope. *Journal of Chemical Education* 84: 644–651. https://doi.org/10.1021/ed084p644.

92 Austin, L.G. and Almaula, S. (1967). An experimental study of the mode of operation of porous gas- diffusion electrodes with hydrogen fuel. *Journal of the Electrochemical Society* 114: 927–933.

93 Giner, J., Parry, J.M., Smith, S., and Turchan, M. (1969). Methods for characterizing the structure and electrochemical behavior of Teflon-bonded Pt electrodes methods for characterizing the structure and electrochemical behavior of Teflon-bonded Pt electrodes. *Journal of the Electrochemical Society* 116: 1692–1696.

94 Kunz, H.R. and Gruver, G.A. (1975). The catalytic activity of platinum supported on carbon for electrochemical oxygen reduction in phosphoric acid. *Journal of the Electrochemical Society* 122: 1279–1287.

95 Maoka, T. (1988). Electrochemical reduction of oxygen on small platinum particles supported on carbon in concentrated phosphoric acid—I. effects of platinum content in the catalyst layer and operating temperature of the electrode. *Electrochimica Acta* 33: 371–377.

96 Bjornbom, P. (1987). Modelling of a double-layered PTFE-bonded oxygen electrode. *Electrochimica Acta* 32: 115–119.

97 Giner, J. and Hunter, C. (1969). The mechanism of operation of the Teflon-bonded gas diffusion electrode: a mathematical model. *Journal of the Electrochemical Society* 116: 1124–1130.

98 Antolini, E., Giorgi, L., Pozio, A., and Passalacqua, E. (1999). Influence of Nafion loading in the catalyst layer of gas-diffusion electrodes for PEFC. *Journal of Power Sources* 77: 136–142.

99 Chen, Y., Li, M., Liao, L. et al. (2009). A thermostatic cell with gas diffusion electrode for oxygen reduction reaction under fuel cell relevant conditions. *Electrochemistry Communications* 11: 1434–1436. https://doi.org/10.1016/j.elecom.2009.05.023.

100 Lin, X., Zalitis, C.M., Sharman, J., and Kucernak, A. (2020). Electrocatalyst performance at the gas/electrolyte Interface under high-mass-transport conditions: optimization of the "floating electrode" method. *ACS Applied Materials & Interfaces* 12: 47467–47481. https://doi.org/10.1021/acsami.0c12718.

101 Zalitis, C.M., Kramer, D., and Kucernak, A.R. (2013). Electrocatalytic performance of fuel cell reactions at low catalyst loading and high mass transport. *Physical Chemistry Chemical Physics* 15: 4329–4340. https://doi.org/10.1039/c3cp44431g.

102 Kucernak, A.R. and Zalitis, C. (2016). General models for the electrochemical hydrogen oxidation and hydrogen evolution reactions: theoretical derivation and experimental results under near mass-transport free conditions. *Journal of Physical Chemistry C* 120: 10721–10745. https://doi.org/10.1021/acs.jpcc.6b00011.

103 Zalitis, C.M., Kucernak, A.R., Sharman, J., and Wright, E. (2017). Design principles for platinum nanoparticles catalysing electrochemical hydrogen evolution and oxidation reactions: edges are much more active than facets. *Journal of Materials Chemistry A* 5: 23328–23338. https://doi.org/10.1039/c7ta05543a.

104 Zalitis, C.M., Sharman, J., Wright, E., and Kucernak, A.R. (2015). Properties of the hydrogen oxidation reaction on Pt/C catalysts at optimised high mass transport conditions and its relevance to the anode reaction in PEFCs and cathode reactions in electrolysers. *Electrochimica Acta* 176: 763–776. https://doi.org/10.1016/j.electacta.2015.06.146.

105 Markiewicz, M., Zalitis, C., and Kucernak, A. (2015). Performance measurements and modelling of the ORR on fuel cell electrocatalysts – the modified double trap model. *Electrochimica Acta* 179: 126–136. https://doi.org/10.1016/j.electacta.2015.04.066.

106 Jackson, C., Smith, G.T., Markiewicz, M. et al. (2018). Support induced charge transfer effects on electrochemical characteristics of Pt nanoparticle electrocatalysts. *Journal of Electroanalytical Chemistry* 819: 163–170. https://doi.org/10.1016/j.jelechem.2017.10.010.

107 Marković, N.M., Schmidt, T.J., Stamenković, V., and Ross, P.N. (2001). Oxygen reduction reaction on Pt and Pt bimetallic surfaces: a selective review. *Fuel Cells* 1: 105–116. https://doi.org/10.1002/1615-6854(200107)1:2<105::aid-fuce105>3.3.co;2-0.

108 Kinoshita, K. (1992). *Electrochemical Oxygen Technology*, 448. Wiley.

109 Yeager, E. (1984). Electrocatalysts for O_2 reduction. *Electrochimica Acta* 29: 1527–1537. https://doi.org/10.1016/0013-4686(84)85006-9.

110 Sui, S., Wang, X., Zhou, X. et al. (2017). A comprehensive review of Pt electrocatalysts for the oxygen reduction reaction: nanostructure, activity, mechanism and carbon support in PEM fuel cells. *Journal of Materials Chemistry A* 5: 1808–1825. https://doi.org/10.1039/C6TA08580F.

111 Grgur, B.N., Marković, N.M., and Ross, P.N. (1997). Temperature-dependent oxygen electrochemistry on platinum low-index single crystal surfaces in acid solutions. *Canandian Journal of Chemistry* 75: 1465–1471. https://doi.org/10.1139/v97-176.

112 Sun, Y., Silvioli, L., Sahraie, N.R. et al. (2019). Activity-selectivity trends in the electrochemical production of hydrogen peroxide over single-site metal-nitrogen-carbon catalysts. *Journal of the American Chemical Society* 141: 12372–12381. https://doi.org/10.1021/jacs.9b05576.

113 Goellner, V., Armel, V., Zitolo, A. et al. (2015). Degradation by hydrogen peroxide of metal-nitrogen-carbon catalysts for oxygen reduction. *Journal of the Electrochemical Society* 162: H403–H414. https://doi.org/10.1149/2.1091506jes.

114 Choi, C.H., Lim, H.K., Chon, G. et al. (2018). The Achilles' heel of iron-based catalysts during oxygen reduction in an acidic medium. *Energy & Environmental Science* 11: 3176–3182. https://doi.org/10.1039/c8ee01855c.

115 Herranz, J., Garsuch, A., and Gasteiger, H.A. (2012). Using rotating ring disc electrode voltammetry to quantify the superoxide radical stability of aprotic Li-air battery electrolytes. *Journal of Physical Chemistry C* 116: 19084–19094. https://doi.org/10.1021/jp304277z.

116 Schmidt, T.J. and Gasteiger, H.A. (2010). Rotating thin-film method for supported catalysts. In: *Handbook of Fuel Cells*, 1–18. https://doi.org/10.1002/9780470974001.f203024.

117 Jaouen, F. and Dodelet, J.P. (2009). O_2 reduction mechanism on non-noble metal catalysts for PEM fuel cells. Part I: experimental rates of O_2 electroreduction, H_2O_2 electroreduction, and H_2O_2 disproportionation. *Journal of Physical Chemistry C* 113: 15422–15432. https://doi.org/10.1021/jp900837e.

118 Choi, C.H., Choi, W.S., Kasian, O. et al. (2017). Unraveling the nature of sites active toward hydrogen peroxide reduction in Fe-N-C catalysts. *Angewandte Chemie, International Edition* 56: 8809–8812. https://doi.org/10.1002/anie.201704356.

119 Artyushkova, K., Serov, A., Rojas-Carbonell, S., and Atanassov, P. (2015). Chemistry of multitudinous active sites for oxygen reduction reaction in transition metal-nitrogen-carbon electrocatalysts. *Journal of Physical Chemistry C* 119: 25917–25928. https://doi.org/10.1021/acs.jpcc.5b07653.

120 Rojas-Carbonell, S., Artyushkova, K., Serov, A. et al. (2018). Effect of pH on the activity of platinum group metal-free catalysts in oxygen reduction reaction. *ACS Catalysis* 8: 3041–3053. https://doi.org/10.1021/acscatal.7b03991.

121 Jaouen, F., Charreteur, F., and Dodelet, J.P. (2006). Fe-based catalysts for oxygen reduction in PEMFCs. *Journal of the Electrochemical Society* 153: A689–A698. https://doi.org/10.1149/1.2168418.

122 Ramaswamy, N. and Mukerjee, S. (2012). Fundamental mechanistic understanding of electrocatalysis of oxygen reduction on Pt and non-Pt surfaces: acid versus alkaline media. *Advances in Physical Chemistry* 2012: 1–17. https://doi.org/10.1155/2012/491604.

123 Artyushkova, K., Rojas-Carbonell, S., Santoro, C. et al. (2019). Correlations between synthesis and performance of Fe-based PGM-free catalysts in acidic and alkaline media: evolution of surface chemistry and morphology. *ACS Applied Energy Materials* 2: 5406–5418. https://doi.org/10.1021/acsaem.9b00331.

124 Jaouen, F., Proietti, E., Lefèvre, M. et al. (2011). Recent advances in non-precious metal catalysis for oxygen-reduction reaction in polymer electrolyte fuel cells. *Energy & Environmental Science* 4: 114–130. https://doi.org/10.1039/c0ee00011f.

125 Jia, Q., Ramaswamy, N., Tylus, U. et al. (2016). Spectroscopic insights into the nature of active sites in iron–nitrogen–carbon electrocatalysts for oxygen reduction in acid. *Nano Energy* 29: 65–82. https://doi.org/10.1016/j.nanoen.2016.03.025.

126 Olson, T.S., Pylypenko, S., Fulghum, J.E., and Atanassov, P. (2010). Bifunctional oxygen reduction reaction mechanism on non-platinum catalysts derived from pyrolyzed porphyrins. *Journal of the Electrochemical Society* 157: B54–B63. https://doi.org/10.1149/1.3248003.

127 Ziegelbauer, J.M., Olson, T.S., Pylypenko, S. et al. (2008). Direct spectroscopic observation of the structural origin of peroxide generation from co-based pyrolyzed porphyrins for ORR applications. *Journal of Physical Chemistry C* 112: 8839–8849. https://doi.org/10.1021/jp8001564.

128 Bonakdarpour, A., Lefevre, M., Yang, R. et al. (2008). Impact of loading in RRDE experiments on Fe–N–C catalysts: two- or four-electron oxygen reduction? *Electrochemical and Solid-State Letters* 11: B105–B108. https://doi.org/10.1149/1.2904768.

129 Koslowski, U.I., Abs-Wurmbach, I., Fiechter, S., and Bogdanoff, P. (2008). Nature of the catalytic centers of porphyrin-based electrocatalysts for the ORR: a correlation of kinetic current density with the site density of Fe-N_4 centers. *Journal of Physical Chemistry C* 112: 15356–15366. https://doi.org/10.1021/jp802456e.

130 Médard, C., Lefèvre, M., Dodelet, J.P. et al. (2006). Oxygen reduction by Fe-based catalysts in PEM fuel cell conditions: activity and selectivity of the catalysts obtained with two Fe precursors and various carbon supports. *Electrochimica Acta* 51: 3202–3213. https://doi.org/10.1016/j.electacta.2005.09.012.

131 Shinozaki, K., Zack, J.W., Richards, R.M. et al. (2015). Oxygen reduction reaction measurements on platinum electrocatalysts utilizing rotating disk electrode technique. *Journal of the Electrochemical Society* 162: F1144–F1158. https://doi.org/10.1149/2.1071509jes.

132 Shinozaki, K., Zack, J.W., Pylypenko, S. et al. (2015). Benchmarking the oxygen reduction reaction activity of Pt-based catalysts using standardized rotating disk electrode methods. *International Journal of Hydrogen Energy* 40: 16820–16830. https://doi.org/10.1016/j.ijhydene.2015.08.024.

133 Wakabayashi, N., Takeichi, M., Itagaki, M. et al. (2005). Temperature-dependence of oxygen reduction activity at a platinum electrode in an acidic electrolyte solution investigated with a channel flow double electrode. *Journal of Electroanalytical Chemistry* 574: 339–346. https://doi.org/10.1016/j.jelechem.2004.08.013.

134 Parthasarathy, A., Srinivasan, S., Appleby, A.J., and Martin, C.R. (1992). Temperature dependence of the electrode kinetics of oxygen reduction at the platinum/Nafion interface: a microelectrode investigation. *Journal of the Electrochemical Society* 139: 2530–2537. https://doi.org/10.1149/1.2221258.

135 Beattie, P.D., Basura, V.I., and Holdcroft, S. (1999). Temperature and pressure dependence of O_2 reduction at Pt|Nafion® 117 and Pt|BAM® 407 interfaces. *Journal of Electroanalytical Chemistry* 468: 180–192. https://doi.org/10.1016/S0022-0728(99)00164-3.

136 Jaouen, F., Goellner, V., Lefèvre, M. et al. (2013). Oxygen reduction activities compared in rotating-disk electrode and proton exchange membrane fuel cells for highly active Fe–N–C catalysts. *Electrochimica Acta* 87: 619–628. https://doi.org/10.1016/j.electacta.2012.09.057.

137 Ebner, K., Herranz, J., Saveleva, V.A. et al. (2019). Fe-based O_2-reduction catalysts synthesized using Na_2CO_3 as a pore-inducing agent. *ACS Applied Energy Materials* 2: 1469–1479. https://doi.org/10.1021/acsaem.8b02036.

138 Lefèvre, M., Proietti, E., Jaouen, F., and Dodelet, J.P. (2009). Iron-based catalysts with improved oxygen reduction activity in polymer electrolyte fuel cells. *Science* 324: 71–74. https://doi.org/10.1126/science.1170051.

139 Proietti, E., Jaouen, F., Lefèvre, M. et al. (2011). Iron-based cathode catalyst with enhanced power density in polymer electrolyte membrane fuel cells. *Nature Communications* 2: 416–425. https://doi.org/10.1038/ncomms1427.

140 Herranz, J., Jaouen, F., Lefèvre, M. et al. (2011). Unveiling N-protonation and anion-binding effects on Fe/N/C catalysts for O_2 reduction in proton-exchange-membrane fuel cells. *Journal of Physical Chemistry C* 115: 16087–16097. https://doi.org/10.1021/jp2042526.

5

Adverse and Beneficial Functions of Surface Layers Formed on Fuel Cell Electrocatalysts

Shimshon Gottesfeld

[1]*Fuel Cell Consulting, Ltd., 3404 Rosendale Rd., Niskayuna, NY 12309, USA*
[2]*Fuel Cell Consulting, Ltd., 6 Haogen St., Herzliya 4655103, Israel*

5.1 Introduction

Advantages of higher specific energy and energy density impact the choice of electric power sources for larger mobile platforms, which require larger quantities of energy stored onboard. The focus in fuel cell applications for transport has consequently shifted recently to larger vehicles, covering buses and trains, as well as ships and airplanes. The development of fuel cell power sources for heavy trucks and buses requires, however, addressing a new set of specs for the fuel cell power system [1]. The on-the-road milage of buses and heavy trucks over their lifetime is 10 times larger than that of a passenger vehicle, and consequently, the long-term stability demands from the fuel cell stack are more stringent. The MEA for such stacks will have to exhibit minimized performance loss over a 10^6-mile-long road of a heavy vehicle, calling in turn for the development of supported electrocatalysts of improved stability as a major fuel cell research and development (R&D) activity. This chapter offers a discussion of this central target of fuel cell technology development, looking at possible means of protection against various modes of performance loss observed frequently for fuel cell electrodes, including surface deactivation, nanocatalyst detachment which is typically followed by catalyst agglomeration, and metal catalyst electrodissolution.

In this discussion, the role of catalyst surface layers is of particular interest. As an example, native surface oxides may protect a metal catalyst against electrodissolution but at the same time deactivate the catalyst metal surface, requiring in turn a significant overpotential for it to be removed at a significant penalty in fuel cell conversion efficiency [2]. Adverse and beneficial functions of catalyst surface layers may come intertwined almost as a rule, and therefore, the idea of adding a surface overlayer to protect electrocatalysts against various modes of loss and deactivation can be described as bold. Along these lines, several recent reports have described "catalyst capping" as a potential approach for reduction of the rate of catalyst particles detachment and agglomeration while maintaining the activity of the noncapped

Electrocatalysis for Membrane Fuel Cells: Methods, Modeling, and Applications, First Edition.
Edited by Nicolas Alonso-Vante and Vito Di Noto.

catalyst [3–5]. Reference [3] describes ORR electrocatalyst capping by mesoporous silica, and Refs. [4, 5] describe, respectively, hydrogen oxidation and hydrogen reduction electrocatalyst capping by carbon. In fact, related albeit somewhat different technical solutions based on coverage of the catalyst by an inert overlayer were proposed much earlier, aiming to improve selectivity when two parallel processes tend to take place at the same electrode surface, for example, coating the anode catalyst by an overlayer in a water electrolyzer fed by chloride-containing water to lower the rate of chlorine evolution vs. that of oxygen evolution [6]. Such increase in the ratio of oxygen evolution reaction (OER)/chlorine evolution reaction (ClER) can be achieved even by an inert mass transport barrier which would limit substantially the rate of chloride mass transport while the rates of water transport (in acid) or of OH^- ions (in alkaline media) across the same barrier remain high, thanks to the much higher concentrations of these anode reactants in the anode feed.

Use of catalyst capping in pursuit of improved long-term integrity is intrinsically challenging because the cap inevitably adds a mass transport barrier that could significantly impede reactant access to the catalyst surface, and anchoring of such an overlayer on the catalyst surface is likely to cause blocking of part of the catalyst surface. However, early ventures into this field provided some support for possible application of capped catalysts, albeit clearly requiring further work to reveal the full value of capping. Importantly, the real value of this concept can be judged only by comparative evaluation, where this concept is weighed against alternative means of catalyst protection under given electrochemical conditions. Such comparative evaluation is offered here, for example, for protection against Pt/C catalyst deterioration by capping vs. an alternative mode of protection by replacement of the carbon support with a ceramic support, and for use of optimized fuel cell operation routines that minimize exposure to high potentials to lower the rate of Pt electrodissolution, as alternative to capping of the fuel cell cathode catalyst.

With recent reports of beneficial effects of specific types of catalyst caps coming out, it is worthwhile to examine here the adverse and beneficial functions of the variety of catalyst surface layers that play key roles in fuel cell electrode processes. Starting with the chemisorbed oxide layers that block and thereby deactivate catalyst surface sites, the adsorbed O or OH species formed at open circuit on a Pt or Pt alloy surface are the reason for a substantial fraction of the overall loss in the energy conversion process in low-temperature H_2/O_2 fuel cells. The degree of site blocking by these oxide surface species is lowered under current by electroreduction of the surface oxide, driven by increase in the cathode overpotential. The threshold cathode potential for onset of this reductive electroactivation process determines the "onset potential" for the faradaic oxygen reduction reaction (ORR) process at Pt [2]. The overall voltage loss of a polymer electrolyte membrane fuel cell (PEMFC) associated with a certain faradaic ORR current demand, therefore, has a substantial component, of ~300 mV, defined by the onset of generation of active Pt metal sites. The ORR-blocking oxide species chemisorbed over Pt surface sites under open circuit conditions is hence a catalyst surface layer with the following descriptors:

"native" – as it forms by intrinsic ingredients of the cell,

"adverse" – as significant voltage loss is incurred in uncovering metal sites,

"also beneficial" – in protecting against electrodissolution, and
"removable *in situ*" – electro-reduceable *in situ* (albeit at significant loss).

The presence of such a chemisorbed oxide layer on Pt at open circuit is the consequence of a relatively high binding energy of oxygen to Pt metal and (even) to Pt alloys, which nevertheless strikes the best combination among metal catalysts of relative ease of surface O or OH reduction required for Pt surface activation and of formation of adsorbed OOH at the Pt metal site in the first step of the ORR in acid.

Different descriptors apply to ionomer thin films that form over the surface of catalyst agglomerates and create a significant mass transport barrier at low catalyst loadings (see Section 5.5.2). In the latter case such set of descriptors will be:

"native,"
"adverse," and
"not removable *in situ*."

Finally, when it comes to catalyst caps introduced over the active surface area to improve longevity by slowing catalyst particle agglomeration and/or dissolution, the set of descriptors for the effective catalyst cap would be:

"not native" (added as extra ingredient to the catalyst or catalyst layer),
"potentially beneficial" (in maintaining catalyst integrity and performance stability).

This chapter begins with a discussion of catalyst capping as a newly proposed approach to protection of electrocatalysts, following earlier introduction for protection against agglomeration of catalysts in high-temperature chemical reactors.

5.2 Catalyst Capping in Heterogeneous Catalysis and in Electrocatalysis

The introduction of the Pt@mSiO_2 catalyst for CO oxidation [7] was one of the first uses in heterogeneous catalysis of a core–shell nanocatalyst where the core is catalytically active, and the shell is as an inert ultrathin protective coating (or "cap"). The porous silica shell formed over the Pt nanoparticles was shown in that work to significantly retard the agglomeration of Pt nanoparticles, which otherwise would take place at high rate at the high temperatures characteristic of a CO chemical oxidation reactor. The mesoporous inorganic silica shells prepared by the steps shown in Figure 5.1 allowed the use of this capped catalyst up to temperatures of 750 °C while preventing the thermally activated Pt agglomeration process and keeping good access of the gaseous reactants to the Pt catalyst surface through the mesoporous silica coat.

Clear proof of the operational benefit of catalyst capping by silica was provided as shown in Figure 5.2. The latter figure shows the CO oxidation activity of Pt@mSiO_2 as function of the reactor temperature and, for comparison, the CO oxidation activity of TTAB-capped Pt nanoparticles.

In the low-temperature regime, the activity of the Pt@mSiO_2 catalyst is seen to be identical to that of Pt nanoparticles capped by the highly open thin layer of TTAB

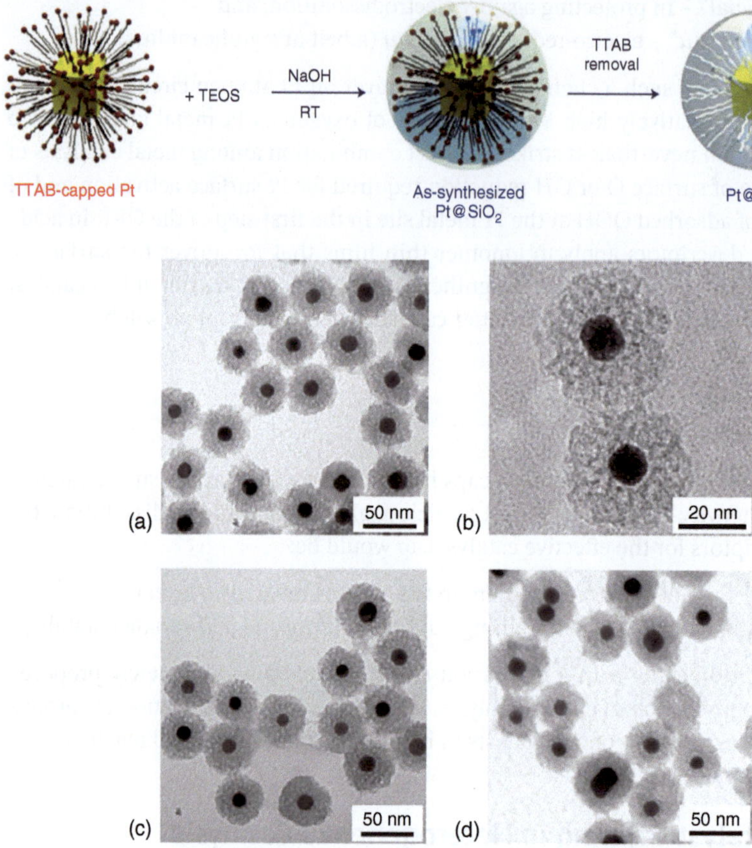

Figure 5.1 Top: The synthesis of composite Pt@mSiO_2 nanoparticles; Bottom: TEM images of the silica-capped Pt particles demonstrating protection of the as-prepared capped Pt particles (a, b) against sintering on exposure to temperatures of 500 °C (c) and 750 °C (d). Source: Reproduced from Ref. [7] with permission of Springer Nature.

(see Figure 5.1a), proving that the mesopores in the mesoporous silica (mSiO_2) caps secure high rates of reactant access to the Pt core. As the temperature is raised further to reach the onset of "ignition" of the CO oxidation process, the activity of the Pt@mSiO_2 catalyst continues to increase with further increase in the temperature, whereas the activity of Pt catalyst particles lacking the protective oxide cap is lost as a result of extensive sintering [7]. This type of cap, developed for Pt nanocatalysts used in a thermal oxidation process, can be described as a monofunctional protective cap, defending against a single dominant mode of catalyst activity loss – in this case of particle sintering. The (single) function of this protective cap is to slow down catalyst sintering by creating physical barriers around the Pt cores, while securing effective reactant access and product removal to/from the Pt catalyst core through pores in the silica cap introduced by the mode of preparation described in Figure 5.1.

It seems that this and similar developments of capped catalysts for use in high T catalytic reactors served as inspiration for the soon-to-follow attempts to implement

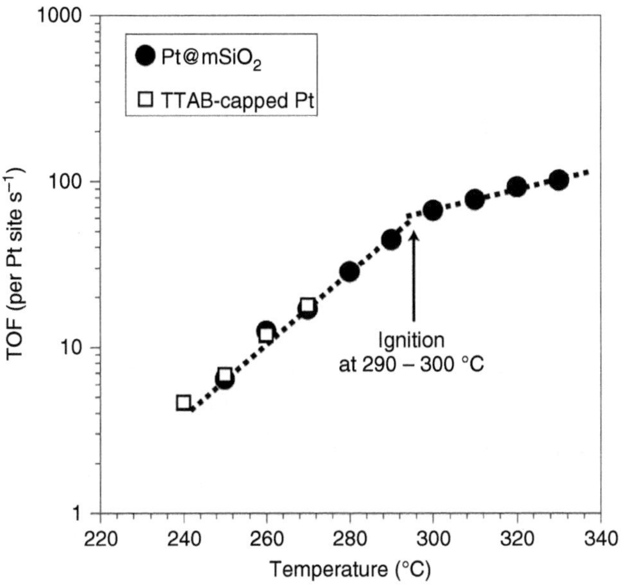

Figure 5.2 The activity in CO thermal oxidation of Pt nanoparticles capped by porous silica ($mSiO_2$) vs. the activity in the same process of Pt nanoparticles capped by a highly open thin layer of TTAB (see Figure 5.1a). Source: Reproduced from Ref. [7] with permission of Springer Nature.

a similar approach for the protection of *electrocatalysts*. In fact, some of the protective caps described for electrocatalysts in low T cells were similarly made of ultrathin layers of microporous silica or another microporous oxide. In fuel cells, catalyst agglomeration can take place by processes that do not require significant thermal activation; for example, particle growth can occur by dissolution–reprecipitation ("Ostwald ripening"). Also, catalyst activity loss in low T fuel cells may result from loss of catalyst mass by electrodissolution of the metal catalyst and/or by corrosion of the catalyst support followed by particle detachment. Additionally, surface contamination may play a significant role in catalyst deactivation. A protective cap designed for electrocatalysts in a low T fuel cell should therefore provide defense against potential damage by various modes of catalyst deactivation and loss, when in contact with the relevant electrochemical environment and under the relevant operation conditions.

A schematic presentation offered by Esposito [8] and duplicated in Figure 5.3 describes various modes of electrocatalyst activity loss and the different functions that may be required of a protective cap in a low T electrochemical cell.

Unlike the case of a heterogeneous catalyst in a chemical reactor, the electrocatalyst metal particle must be in contact with an electronically conducting material. Consequently, the metal electrocatalyst (M in Figure 5.3) capped by a silica overlayer will have to be in direct electronic contact with a carbon or electronically conducting oxide support ("support" in Figure 5.3), to be secured by exposure of the metal surface adjacent the e-conducting support (see Figure 5.3). This demand of direct

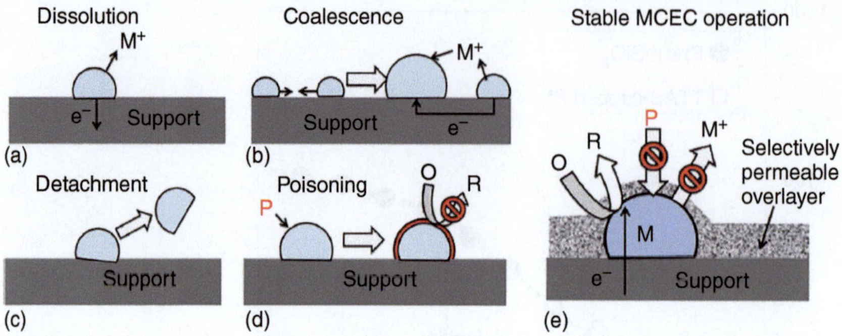

Figure 5.3 Common mechanisms of electrocatalyst nanoparticle degradation including (a) electrochemical oxidative dissolution, (b) agglomeration (sintering) of smaller particles into larger ones, (c) physical detachment from the support, and (d) poisoning by adsorption of poison P onto an active electrocatalyst surface, all significantly reducing a desired electroreduction process: $O + e^- \rightarrow R$; (e) a schematic presentation of a metal nanoparticle M covered by a selectively permeable cap which would protect the catalyst particle from all modes of degradation (a–d). Source: Reproduced from Ref. [8] with permission of the American Chemical Society.

contact between a partly exposed metal surface and the support material applies to capping by $mSiO_2$, whereas a more recent approach to electrocatalyst capping is based on ultrathin carbon [4, 5], rather than ceramic cap overlayers, and consequently, electronic contact to the substrate is made by the carbon cap, rather than exposed surface of the metal nanocatalyst.

The important modes of electrode deactivation in an electrochemical cell using an acid electrolyte are metal catalyst loss by anodic dissolution, according to

$$M \rightarrow M_{aq}^{+n} + ne^- \tag{5.1}$$

and carbon support disintegration driven by a corrosion cell, according to, for example:

ORR at an adjacent metal catalyst site

$$O_2 + 4e^- + 4H^+ \rightarrow 2H_2O \tag{5.2a}$$

carbon oxidation at a carbon site

$$C + 2H_2O \rightarrow CO_2 + 4e^- + 4H^+ \tag{5.2b}$$

and the resulting overall cell process; $C + O_2 \rightarrow CO_2$

One key question regarding a catalyst cap is:

* *Can a protective cap slow down the process* (*Eq.* 5.1) *of anodic dissolution*?

Takenaka et al. [3] suggested that a cap covering a Pt electrocatalyst particle could keep, next to the catalyst surface, a certain concentration of Pt ions generated initially in the electrodissolution process, thereby limiting substantially the steady-state rate of electrodissolution. A zero-order analysis is offered next for the cap properties required to achieve such effective confinement of metal ions next to the metal catalyst under steady-state cell current.

The kinetically controlled rate of the electrodissolution process (Eq. (5.1)) will be typically very high in acid electrolytes, even under a small anodic overpotential. When a fuel cell ORR current demand, J, defines a cathode potential of $E_{cath}(J)$, the target concentration of Pt^{+2} ions needed to be kept next to the Pt metal surface to prevent electrodissolution must maintain an effective overpotential smaller than zero for process (Eq. (5.1)) at the cathode potential $E_{cath}(J)$, to totally prevent it. This means that the following condition should hold:

$$E_{cath}(J) < E_{eq\left(\frac{Pt+2}{Pt}\right)} \tag{5.3a}$$

where

$$E_{eq\left(\frac{Pt+2}{Pt}\right)} = E^0_{\left(\frac{Pt+2}{Pt}\right)} + \frac{2.303RT}{2F}\log[Pt^{+2}]$$

and, given that $E^0_{\left(\frac{Pt+2}{Pt}\right)} = 1.18$ V vs. reversible hydrogen electrode (RHE), the condition described by Eq. (5.3a) can be rewritten as

$$E_{cath}(J) < 1.18 + \frac{2.303RT}{2F}\log[Pt^{+2}] \tag{5.3b}$$

Consequently, the local concentration ensuring zero dissolution current is given by

$$\log[Pt^{+2}] > \frac{2F}{2.303RT}[E_{cath(J)} - 1.18\text{ V}] \tag{5.3c}$$

This suggests that when the cathode potential of a PEM fuel cell under current is, for example, 0.85 V, the minimum concentration of Pt^{+2} that needs to be maintained next to the metal catalyst in order to prevent electrodissolution is given by $\log[Pt^{+2}] = -(0.33/0.030)$, i.e. $[Pt^{+2}] = 10^{-11}$ M. Such ultralow concentration of the ion is likely to be present next to the catalyst surface without the benefit of a protective cap, and this is expected to be true even when considering that E^0 (Pt^{+2}/Pt) is somewhat negatively shifted when the metal is in fine particle form [9].

When E_{cath} is close to the open circuit potential of an air cathode in acidic systems (~1.0 V), the local concentration required next to the nanocatalyst surface to stop further electrodissolution is given by $\log[Pt^{+2}] > -(0.18/0.030)$ or $[Pt^{+2}] > 10^{-6}$ M, and with a 10^{-6} M solution of Pt^{+2} being kept next to the metal Pt catalyst by confinement under/in a porous cap, the estimated steady-state rate of diffusion of the Pt ions outward through a 10-nm-thick cap is ~0.1 mA cm^{-2} of cathode surface area, assuming an effective diffusion coefficient of Pt ions through the cap of 10^{-7} cm^2 s^{-1}. This rate of diffusion-controlled loss is indeed much lower than the rate of Pt electrodissolution expected at 1.0 V under kinetic control. However, with a total loading of Pt of 0.1 mg cm^{-2} of cathode area, even this strongly reduced rate of cathode catalyst loss can last for only several minutes before all the Pt metal catalysts in the cathode have electro-dissolved. Even before considering the expected adverse effects of such a cap on oxygen access, these zero-order calculations explain why catalyst capping may not be the most straightforward means of preventing Pt loss by electrodissolution.

Reported results, discussed further below in Section 5.4, describe reduced rate of agglomeration of Pt ORR catalysts following capping with SiO_2, arguing that such capping provides meaningful protection against electrodissolution as well [3]. It should be noted, however, that the loss by electrodissolution expected for the capped Pt catalysts at constant potential may be quite faster than the loss by electrodissolution observed during potential multicycling: with the Pt ions generated being kept confined next to the catalyst surface, they may be readily redeposited during the negative going half-cycle. For this reason, the test for effective protection against electrodissolution should be done at constant potential rather than by AST with potential multicycling.

A possible alternative way to prevent significant loss of Pt by electrodissolution is to minimize exposure of the Pt cathode catalyst to potentials higher than 0.85 V [10]. This means, in turn, consideration of possible replacement of the air in the fuel cell cathode by nitrogen when the fuel cell circuit is opened.

Moving on to the other common mechanism of catalyst deactivation in low T fuel cells, the relevant question here is:

* *Can a protective cap prevent the detachment of metal catalyst nanoparticles as a result of carbon substrate corrosion which is typically followed by nanoparticle agglomeration?*

While electrodissolution is a central mode of performance loss of platinum-group metal (PGM) ORR catalysts under high cathode potentials, the loss of PGM catalyst performance under lower potentials takes place mostly by catalyst particle detachment and agglomeration. The expected role of a cap is then to prevent the detachment and/or migration of catalyst nanoparticles in a fuel cell electrode under current by "anchoring" the particles onto the substrate. Along these lines, Chatenet and coworkers [4] demonstrated an effect of carbon capping over a carbon-supported Pd catalyst in reducing the rate of catalyst agglomeration during accelerated stability test (AST) performed in an alkaline environment. Alkaline electrolytes accelerate the corrosion of carbon supports in contact with an active ORR catalyst, and consequently, effective protection of a Pd/C catalyst against corrosion in an alkaline medium is of value.

Finally, in addition to the loss of catalyst mass by electrodissolution and loss of catalyst surface area by agglomeration, surface passivation is another significant mode of catalyst performance loss which plays a particularly important role in limiting the activity of non-PGM catalysts in hydroxide-exchange membrane fuel cell (HEMFC) anodes. The possible mitigation of surface passivation of non-PGM catalysts is the subject of the next section.

5.3 Passivation of PGM/TM and Non-PGM HOR Catalysts and Its Possible Prevention

A recent overview provides a wide perspective of hydrogen oxidation reaction (HOR) electrocatalysis in alkaline media, with the larger part devoted to PGM catalysts and the other part to nickel and nickel alloy catalysts [11]. The discussion in this section highlights specific aspects to do with surface layers and the possible protection against *non*-PGM catalyst deactivation and provides some updates based on very recent reports in the field. Effective protection against performance loss of catalysts in HEMFC anodes is an important goal to pursue: an active and stable non-PGM anode catalyst will make the HEMFC a *truly PGM-free polymer electrolyte fuel cell* [2]. Optimized electrolyte alkalinity (pH ~13) reduces significantly the rate of electrodissolution of Ni and of Co; however, this advantage is gained at a penalty of surface deactivation as metal sites are blocked by anodic formation of chemisorbed hydroxide and/or oxide species. Among transition metals, nickel and nickel alloy anode catalysts have been investigated to largest degree, although unalloyed Ni, whether unsupported or carbon supported, exhibits by itself very low HOR activity. Ni does not chemisorb H atoms reversibly at low η_{an} as required to achieve high rates of HOR at low anode overpotentials, apparently so because of excessive Ni–H binding energy [12, 13]. This has called to date for the addition of a second metal component to form a binary HOR anode catalyst, where the second component chemisorbs H atoms reversibly at low $\eta_{an,}$ (e.g. Pd or Ru), and the OH groups electrosorbed at low η_{an} on the Ni domains of such a binary catalyst surface provide the oxidative function required in the final (Volmer) step of the HOR. An example of such a PGM/Ni alloy catalyst is the RuNi catalyst described recently in Ref [13], see Figure 5.4. Obviously, reaching acceptable anode performance by addition of a PGM at 0.2 mg cm^{-2} [13] does not get the HEMFC close to being "PGM-free," and the HEMFC with the RuNi anode catalyst reached performance similar to that of a HEMFC with a PtRu-catalyzed anode of same PGM loading only at $\eta_{an} < 0.1$ V (Figure 5.4).

As mentioned above, zero HOR activity is observed in Figure 5.4 for a carbon-supported (unalloyed) Ni electrode, whereas an Ru/C anode exhibits moderate HOR activity around 0 V vs. RHE with the HOR current increasing relatively slowly with increase in η_{an} and then falling with further increase above ~0.12 V.

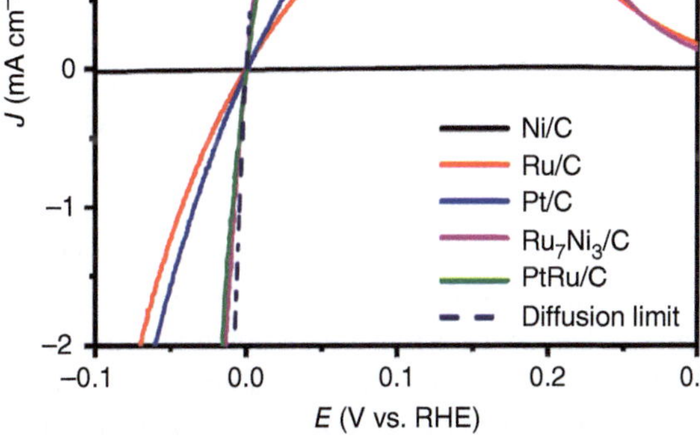

Figure 5.4 HOR polarization curves of Ru/C, Ni/C, Pt/C, PtRu/C, and Ru_7Ni_3/C recorded in 0.1 M KOH at 1 mV s^{-1} at an RDE rotating at 1600 RPM. Source: Reproduced from Ref. [13]. Springer Nature/CC BY 4.0.

The relatively slow rise in the current and the fall above 0.12 V are the result of buildup of surface Ru hydroxide over the Ru metal with the increase in η_{an}, with the onset of such surface oxidation taking place near 0 V vs. RHE. In clear contrast to the negligible, or limited HOR activity of Ni/C and Ru/C, respectively, the binary catalyst Ru_7Ni_3/C exhibits significant HOR activity, as seen in the same figure. The steeper rise in the current near 0 V when using the RuNi anode catalyst is likely enabled by the bifunctional activity of these binary alloys, where OH groups formed on Ni surface sites by electrooxidation of OH^- anions facilitate the oxidation of H atoms formed on adjacent Ru sites by dissociation of H_2. The HOR anode reaction at the bifunctional RuNi catalyst in the prepassivation potential region can be schematically described by the steps:

$$\mathrm{Ru} + \frac{1}{2}\mathrm{H_2} \rightleftarrows \mathrm{RuH_{ads}} \tag{5.4a}$$

$$\mathrm{Ni} + \mathrm{OH^-} \rightleftarrows \mathrm{NiOH_{ads}} + \mathrm{e^-} \tag{5.4b}$$

Followed by

$$\mathrm{RuH_{ads}} + \mathrm{NiOH_{ads}} \rightleftarrows \mathrm{H_2O} + \mathrm{Ru} + \mathrm{Ni} \tag{5.4c}$$

with the overall expected anode process:

$$\frac{1}{2}\mathrm{H_2} + \mathrm{OH^-} \rightleftarrows \mathrm{H_2O} + \mathrm{e^-}$$

Above ~0.10 V, electrooxidative passivation is observed for the binary alloy, with likely contributions from the formation of both Ni and Ru surface hydroxides and/or oxides.

Figure 5.5 describes the performance of a HEMFC using the Ru_7Ni_3/C as anode catalyst at loading of 0.2 mg_{Ru} cm^{-2}, side by side with the performance of the HEMFC when using other anode catalysts. The performance of the fuel cell with the RuNi alloy as anode catalyst looks relatively impressive, with max power higher than that achieved using PtRu in the anode at the same PGM loading, and with voltage stability demonstrated at 0.5 A cm^{-2} (Figure 5.5). This moderate current demand corresponds, however, to a low anode potential, in the alloy prepassivation region, thereby avoiding buildup of hydroxide and/or oxide on the RuNi catalyst surface. Prevention of oxidative deactivation of such RuNi catalysts is still to be demonstrated under currents higher than 0.5 A cm^{-2}, which are expected in actual operation of a HEMFC stack.

Possible protection from deactivation of non-PGM anode catalysts, including unalloyed Ni and alloys of Ni with other transition metals (TMs), is discussed next. One such non-PGM anode catalyst of significant HOR activity in 0.1 KOH around 0 V vs. RHE was reported back in 2012 – see Figure 5.6 [14].

Pursuit of this CoMoNi alloy as anode catalyst for HOR in alkaline electrolyte [14] is well in line with the target of a truly PGM-free HEMFC. It is, therefore, worthwhile to clarify what makes this particular non-PGM catalyst unusually HOR active, albeit in a narrow anode potential range of 0–0.12 V vs. RHE, and what is the reason for the loss of activity above 0.12 V. Sheng et al. [14] reported based on density functional theory (DFT) calculations that NiMo is expected to exhibit a

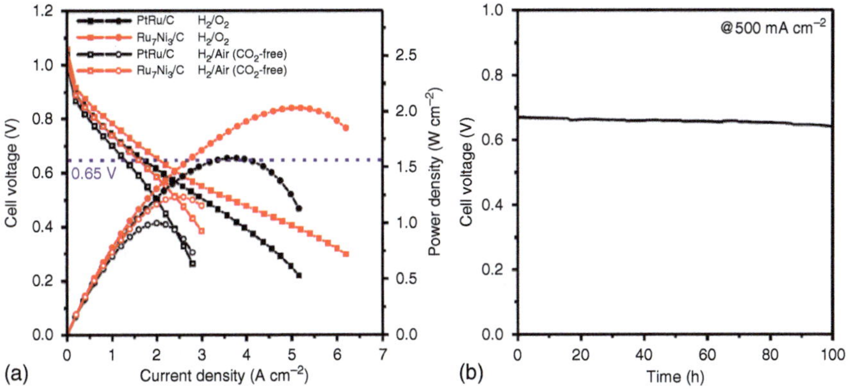

Figure 5.5 HEMFC performance with Ru_7Ni_3/C as anode catalyst: (a) voltage–current and power–current plots with this and with other HEMFC anode catalysts, (b) HEMFC voltage change at constant current with Ru_7Ni_3/C as the anode catalyst. Source: Reproduced from Ref. [13]. Springer Nature/CC BY 4.0.

Figure 5.6 Anodic H_2 oxidation in 0.1 M KOH using ternary alloys of Co, Mo, and Ni as anode catalysts. Source: Reproduced from Ref. [14] with permission of the Royal Society of Chemistry.

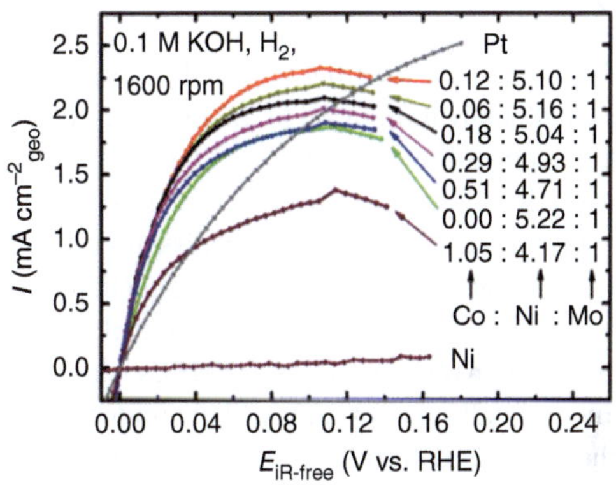

significantly lower H binding energy than either Ni or Mo, as required to ensure faster exchange of chemisorbed H atoms near 0 V. Ni and Co sites in the CoMoNi alloy surface are expected, according to their similar Pourbaix diagrams, to be partly covered by both H and OH species in the narrow potential range of high HOR activity exhibited by the ternary catalyst. The Pourbaix diagram and the surface Pourbaix diagram for Ni are provided in Figure 5.7.

OH groups on Ni (and Co) sites likely facilitate the oxidation of adsorbed H atoms at low η_{an} in the final step (Volmer) of the HOR; however, these OH surface groups are also the precursors of $Ni(OH)_2$ and $Co(OH)_2$ formation on the metal surface at pH 13, starting to form at ~50 mV vs. RHE, Figure 5.7. Consequently, the combination of these three transition metals in a single alloy catalyst achieves optimization of the H binding energy at $\eta_{an} < 0.12$ V but also supports a process of surface passivation with an onset at a similar, low value of η_{an}.

The probability of severe surface passivation of a TM *anode* catalyst is high: unlike the electrodissolution of ORR catalysts, which is slowed down with increase in η_{cath},

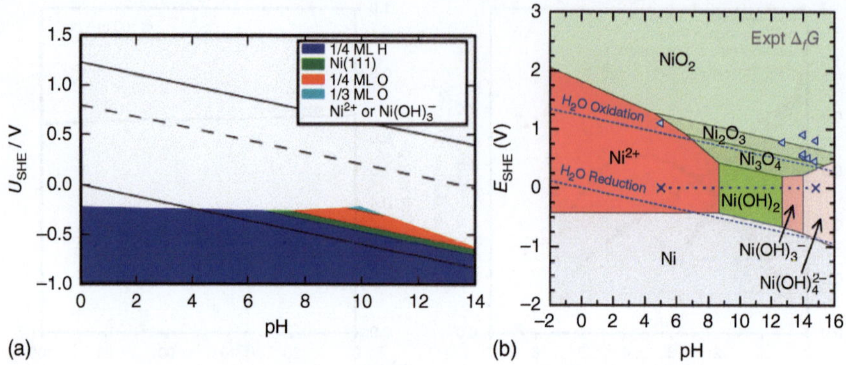

Figure 5.7 (a) The surface Pourbaix diagram for Nickel. Source: Reproduced with permission from Hansen et al. [15]. Copyright© ("2008") the Royal Society of Chemistry. (b) The phase Pourbaix diagram for Ni ([I] = 10^{-6} mol l^{-1}, 298.15 K, 1 bar). Source: Reproduced from Ref. [16] with permission of the American Chemical Society.

the oxidative process that causes deactivation of HOR catalysts is accelerated by the overpotential η_{an} which drives the fuel cell anode process. Consequently, if/when the anode overpotential is raised to maintain a certain current demand following partial passivation of a TM anode catalyst, such raising of η_{an} will further increase the degree of metal surface oxidation bringing the cell current further down. This scenario is indeed manifested in Figures 5.5 and 5.6.

At pH >13, Ni metal electrodissolution starts at potentials very close to 0 V vs. RHE, forming hydroxy or oxy-*anions* (Figure 5.7). Loss of Ni from an anode catalyst by electrodissolution will hence take place in aqueous base solutions of concentration higher than 0.1 M. In HEMFCs, significant electrodissolution of an Ni or Co anode catalyst is not highly likely under ordinary fuel cell operation conditions, as the local pH within the ionomer does not typically exceed 13.

The above description of modes of performance loss for anode catalysts in alkaline fuel cells and HEMFCs leads to consideration of the following possible ways to slow down such anode deactivation processes:

(a) Non-PGM transition metal alloys
  (i) Operation at low anode overpotentials – thereby keeping the rate of anode surface passivation low enough for long periods. This could be achieved, in principle, by operating at low anode current densities, using high loadings of the anode catalyst. With the much lower cost per unit mass of a non-PGM anode catalyst, use of higher anode loadings of several mg cm^{-2} (geo.) should not be prohibitive provided the thicker catalyst layer to be expected does not present very high transport barriers.
  (ii) Hydrogen transport enhancement – A higher intrinsic rate of HOR per unit surface area of the anode catalyst would help to generate a demand of HOR current at a lower anode overpotential and, thereby, drop the rate of Ni catalyst passivation at that demand current. A prime requirement for such improved selectivity of HOR vs. catalyst surface passivation is to secure

effective H_2 gas access to the highest fraction of the overall catalyst surface area. This would require, in turn, securing electrolyte-free micropores in the anode catalyst layer to be introduced using polytetrafluoroethylene (PTFE) as component of the catalyst layer. Another possible approach is the use of electrospun fibers, confining the catalyst and ionomer to the fibers while leaving the interfiber microgaps as gas conduits [17].

(iii) Capping of catalyst particle – A catalyst cap could add a thin layer of selective permeation properties in front of the surface of the metal catalyst to ensure high rate of permeation of hydrogen gas while the liquid water product of the HOR process is collecting at the catalyst surface. This may call for the use of mixed hydrophobic and -philic components of the microcap structure, as described by Takenaka et al. [3].

(b) PGM/Ni binary alloys:
Enhanced activity at low anode potentials is apparently achieved by use of binary PGM/Ni alloys (or mixtures) as the anode catalysts. However, the drawback of better performing PGM/TM catalysts is the catalyst cost and accessibility, particularly with the PGM component of the alloy used to reach the performance seen in Figure 5.5 as high as 0.2 mg_{Ru} cm^{-2}. Achieving similar HEMFC performance with a PGM/Ni anode catalyst of much lower PGM loading may secure some advantage vs. purely PGM catalysts of similar loading.

5.4 Literature Reports on Fuel Cell Catalyst Protection by Capping

5.4.1 Protection of ORR Pt catalysts Against Agglomeration by an Ultrathin Overlayer of Mesoporous SiO_2 or Me−SiO_2

Takenaka et al. [3] reported the use of mesoporous SiO_2 for the protection of Pt ORR catalysts against performance loss following a selected routine of multicycling between 0.6 and 1.0 V. The silica cover layer was prepared along the lines described in [7] for the protection of supported Pt catalysts in high-temperature chemical reactors. In the latter work on ORR Pt catalysts [3] commercial Pt/carbon black (Pt/CB) catalysts, uncovered and covered with mesoporous silica, were subjected to an AST regime of potential multicycling in 0.1 M $HClO_4$. Transmission electron microscopy (TEM) images reported in the latter work revealed substantial agglomeration of the uncovered Pt/CB catalyst under the AST regime employed, with the agglomeration largely prevented by the nanometer thick $mSiO_2$ cover layers. This work thereby provided proof that Pt agglomeration, which likely starts by catalyst nanoparticle detachment from the CB substrate, as a result of carbon corrosion, can be restricted by the anchoring effect of an SiO_2 overlayer. The authors further claimed that a protective layer of Me-SiO_2 prepared from *methyl*-tetraethoxysilane exhibited a better protective function than porous SiO_2 layers prepared from tetraethoxysilane. They argued that the more hydrophobic nature of the Me-SiO_2 ultrathin overlayer is beneficial in enabling effective side-by-side oxygen and water transport through

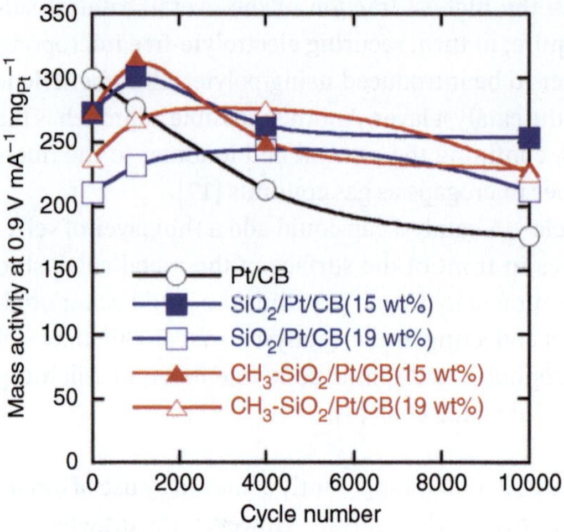

Figure 5.8 Comparison of the changes in mass activity measured at 0.9 V for Pt ORR catalysts supported on carbon black, and for such catalysts further coated by protective ultrathin layer of porous silica. Source: Reproduced from Ref. [3] with permission of Elsevier.

the silica layer to/from the ORR catalyst surface. Rendering selective transport characteristic to such porous protective layers could be an important added value, allowing the layer to function not only in prevention of catalyst agglomeration but also adding transport selectivity that could support electrode process selectivity, for example increasing the rate of the HOR vs. the rate of anode surface oxidation in an Ni-containing anode catalyst.

As seen in Figure 5.8, after 10,000 cycles, the loss of mass activity is seen to be significantly smaller for the capped catalysts vs. the uncapped catalyst. Also noted is the need of overlayer thickness optimization to avoid excessive transport loss. The initial rise in the performance of capped catalysts with potential cycling indicates some microfracturing or microdetachment of the silica cap as an apparent condition for initial activation.

5.4.2 Protection by Carbon Caps Against Catalyst Detachment and Catalyst Passivation Under Ambient Conditions

Several recent reports by Chatenet and coworkers (see as example Ref. [4]) describe the application of catalyst caps for protection against instability in alkaline fuel cells. This search for proper protection of Pt/C and Pd/C catalysts in alkaline electrolytes followed findings by Chatenet and coworkers [18] that PGM nanoparticle catalysts are particularly prone in alkaline electrolytes to detachment from carbon supports because of the high rate of corrosion of such supports in alkaline electrolytes.

It is suggested in Ref. [4] that more robust PGM HOR catalysts can be prepared by adding a carbon cap around the metal nanoparticle, specifically in the case of a palladium metal catalyst. Carbon-coated Pd catalysts were synthesized from the mixtures of a carbonaceous complex of Pd (II) with Vulcan XC-72 carbon, and then the dry powder containing the Pd precursor and the carbon was pyrolyzed under Ar by ramping the temperature from room temperature to 450 °C and holding at that temperature for 30 minutes. The so-formed carbon-capped Pd/C nanoparticles are

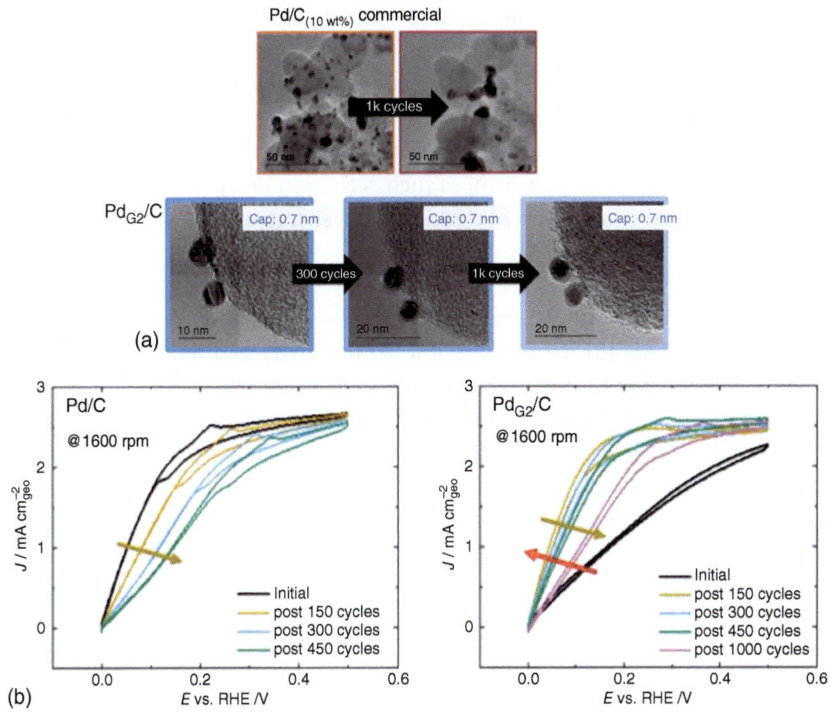

Figure 5.9 (a) Substantial change in the morphology of uncapped Pd (top) following 1k AST cycles (top) prevented by carbon capping (bottom). (b) Changes in the HOR activity with cycling of a Pd/C electrode for an uncapped (left) and a carbon-capped electrode (right). Source: Reproduced from Ref. [4] with permission of IOP Publishing.

reported to exhibit improved stability following an accelerated stress test routine of 1000 repeated steps, each of 3 s at 0.1 V and 3 s at 1.23 V vs. RHE, see Figure 5.9. The strong morphology changes seen in Figure 5.9a following such AST routine and the prevention of these changes by catalyst capping mirror the high-temperature scenario described in Figure 5.1, with the nonconducting silica cap used in the high T chemical reactor replaced by an electronically conducting carbon cap for a catalyst in a polymer electrolyte fuel cell. Once such an ultrathin (0.7 nm) carbon cap can be made sufficiently permeable by initial cycling across a limited potential range while avoiding detachment, Figure 5.9b (right), it may ultimately prove to provide effective capping for an HOR fuel cell catalyst.

It is argued in Ref. [4] that the AST regime used, of stepping between 0.1 and 1.23 V, is appropriate for testing the stability of an anode Pd/C catalyst in an H_2/air fuel cell, because an anode catalyst is inevitably exposed to potentials above 1 V vs. RHE when it is hydrogen starved locally and/or when exposed to both hydrogen and oxygen along the anode active surface area on turning the cell on from rest. This presumption is not fully defensible, as system-level solutions are available to prevent such anode catalyst overoxidation scenarios. Furthermore, taking full advantage of such solutions seems imperative in this case, because of the vulnerability of the carbon cap itself to corrosion.

As seen in Figure 5.9b, the rate of anode performance loss was about halved following capping. Being an early attempt to introduce and test capping of electrocatalysts, further advancement in the perfection of this technique must follow to achieve a more substantial cut in the rate of catalyst detachment and agglomeration. Another concern is the need of initial activation of the carbon-capped catalyst by potential muticycling, see Figure 5.9b, as such activation likely involves microfracturing of the cap and when repeated could cause over-fracturing and complete loss of the carbon cap.

Protection of Ni electrocatalysts in alkaline electrolyte by carbon capping has been described recently in the context of hydrogen evolution (HER) catalysts for water electrolyzers [5]. Ni exhibits low HER activity and tends to passivate as a result of the buildup of surface hydride under cathodic overpotentials (see top part of Figure 5.7). However, unlike the oxidative surface passivation which has prevented to date the use of Ni, or PGM-free Ni alloy in an alkaline fuel cell, Ni has been used for a long time in cathodes of commercial, liquid alkaline *electrolyzers*. Ref. [5] describes capped, binary Ni/Ni-oxide HER catalysts prepared recently by thermal treatment of carbon-supported Ni nanoparticles chelated with cupferron. The thermal treatment of the chelated Ni nanoparticles results in an embedded nickel metal atom with a clear graphitic coating, reported in Ref. [5] to engender improved stability vs. both oxidative passivation and subsurface hydride formation. Figure 5.10 provides a comparison between polarization characteristics of HER cathodes using three different Ni catalysts: NFK – carbon-capped nickel nanoparticles mixed with Ketjen black and Ketjen black-supported Ni and NiCr. The HER performance of Pt/C is added as reference. The results in Figure 5.10 show that the carbon-capped Ni catalyst (NFK) exhibits an anode overpotential lower by 100 mV at 30 mA cm^{-2} vs. NiCr/KB and lower by 120 mV vs. Ni/KB although at

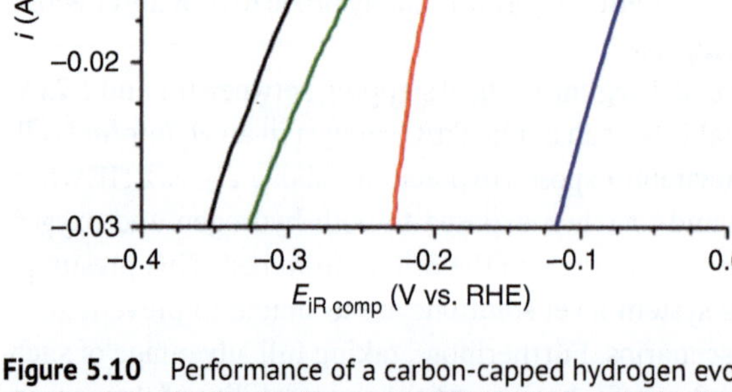

Figure 5.10 Performance of a carbon-capped hydrogen evolution (HER) Ni catalyst shown side by side with the performance of Pt/C, Ni/C, and NiCr/C HER catalysts. Source: Reproduced from Ref. [5]. IOP Publishing/CC by 4.0.

the same current it is, however, 100 mV larger vs. that of Pt/C. The main difference in the state of nickel in the capped catalyst vs. the uncapped Ni/KB catalyst under ambient conditions, as determined by X-ray photoelectron spectroscopy (XPS), is the large fraction of Ni^0 in the capped catalyst and $Ni(OH)_2$ in the uncapped Ni/KB. The authors suggested that the protection of the metallic state in the capped catalyst from oxidation, as confirmed under ambient conditions, is also likely the explanation for the higher activity of the capped catalyst *in situ*, i.e. in a cathode of an alkaline electrolyzer under HER current. This hypothesis is open to question because the $Ni(OH)_2$ found in the Ni/KB catalyst under ambient conditions is likely to be reduced when serving *in situ* as an active hydrogen evolving cathode catalyst. The paper also describes critical effects of the precise protocol of preconditioning by potential cycling on the HER performance of the capped catalyst. It seems that the HER performance and performance stability reported in the paper are the combined result of the mode/nature of catalyst capping and the specific preconditioning protocol applied. Of particular significance in this regard is the possible role of the potential multicycling routine in microfracturing of the carbon cap as key for the activation of the capped catalyst.

In summarizing these recent examples of attempts to protect catalysts by capping, the following observations are offered:

- The approach of protection of fuel cell catalysts by capping is still to show clear value based on demonstrated advances in performance and/or performance stability vs. best-in-class uncapped catalysts.
  - Some protection by capping against detachment and agglomeration has been demonstrated for Pt ORR catalysts capped by ultrathin overlayers of silica and Pd HOR catalysts capped by ultrathin overlayers of carbon.
    - However, the merit of this mode of protection against detachment and agglomeration needs to be carefully weighed in each case against the alternative of replacing carbon by an oxide or carbide support – see discussion below.
  - The recent work on carbon protection of an Ni HER catalyst provides some evidence for catalyst protection against (*ex situ*) passivation. For reduction to practice, optimized preconditioning protocols must be established for such capped Ni HER catalysts, including the frequency of such preconditioning required to keep the catalyst active. Also, confirmation is required of cap overall structural integrity under repeated activation pretreatments by some selected potential multicycling routine.
    - Additionally, the merit of cap-protected Ni HER catalysts of the type described in [5] needs to be evaluated for applicability to the Ni-based HER catalysts of highest activity which have been demonstrated recently in industrial scale water electrolyzers. The latter, reportedly highly stable catalysts, contain, in addition to Ni, other transition metals at various levels and are reported to catalyze generation of H_2 at $1\,A\,cm^{-2}$ at very low overpotential of $<100\,mV$. Further extending the long-term stability of such advanced HER Ni catalysts by carbon capping will be the proper test for the value of carbon capping.

- Stabilization by capping of a non-PGM *HOR* catalyst, which could open the door for a PGM-free HEMFC, is still to be demonstrated. The main mode of deactivation of such catalysts is oxidative passivation, and possible ways other than capping to tackle this severe barrier are discussed in Section 5.5.

5.5 Other Means for Improving the Performance Stability of Supported Electrocatalysts

5.5.1 Replacement of Carbon Supports by Ceramic Supports

The first part of this section discusses an alternative approach for the protection of supported catalysts against detachment by corrosion: the replacement of the carbon support by a ceramic support which is significantly less prone to corrosion under fuel cell operation conditions. The choice between capping of a carbon-supported catalyst and, instead, supporting the uncapped catalyst on an oxide or carbide substrate requires proper comparative evaluation.

Alternative support materials for fuel cell catalysts have been gaining attention by offering the essential properties of high surface area, high electrical conductivity and thermal stability, and, above all, their corrosion resistance. Strong metal–support interactions identified in oxide-supported Pt catalysts [11] suggest possible alloy formation between Pt and the support in the cases of Ti and W oxides, explaining the strong bonds of the Pt catalyst to some oxide supports. ITO has also been demonstrated as effective Pt catalyst support in the ORR. While exhibiting superior immunity to corrosion under potentials experienced ordinarily by an ORR cathode, the stability of these oxide supports is, however, typically challenged under more negative potentials, as a result of electroreduction of the oxide. Consequently, an ongoing research activity has been devoted to further improvement of the properties of ceramic supports for Pt catalysts, including carbides and nitrides, among others. The following example of recent advancements in this area describes the synthesis and characterization of a TiNC ceramic of optimized composition serving as Pt catalyst support in acid and in a PEMFC MEA [19].

The TEM image of this TiCN-supported catalyst is provided in Figure 5.11a, showing support particle sizes like those in Pt/C catalysts. Figure 5.11b shows side-by-side ORR current potential plots recorded in acid solution with Pt/TiCN ORR and Pt/C ORR catalysts of Pt loading of 36 μg cm^{-2}. The current–potential plots recorded for the Pt/TiCN catalyst show significant improvements in performance and performance stability vs. the Pt/C catalyst. The most striking is the much better stability exhibited by Pt/TCN following AST using multicycling between 0.6 and 1.3 V. While the Pt/C catalyst loses its activity practically completely after this AST routine, the Pt/TiCN catalyst is only slightly affected. The other property is the higher ORR activity in 0.5 M H_2SO_4 exhibited by Pt/TiCN vs. that measured at Pt/C. The reason offered in the paper for the higher beginning of life (BOL) activity of Pt/TiCN is catalyst-support interactions, which lower the Pt–O_{ads} binding energy. This interpretation is supported by XPS measurements in the Pt 4f region [19].

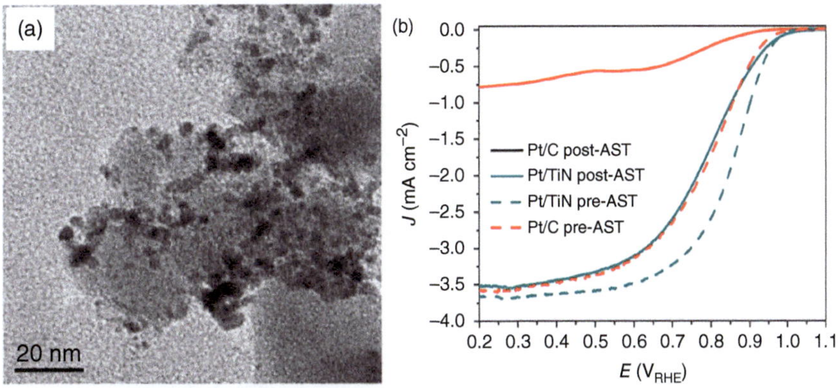

Figure 5.11 (a) TEM image of a TiCN-supported Pt catalyst. (b) ORR current–potential plots in 0.5 M H_2SO_4 recorded with Pt/TiCN and Pt/C ORR catalysts of Pt loading of 36 μg cm^{-2}. Source: Reproduced from Ref. [19] with permission of Elsevier.

Another example of successful prevention of electrode deactivation by replacing a carbon catalyst support with a ceramic support was demonstrated recently in alkaline environment for a HEMFC Pt cathode catalyst [20]. When carbon-supported Pt catalysts were used for the HEMFC air cathode, voltage loss of rate increasing continuously with time could not be prevented. In contact with Pt, carbon corrosion is known to be severe in the ORR potential range, particularly so in aqueous alkaline media, and as these results demonstrated, vulnerability of the carbon also remains high in contact with an OH^- ion-conducting ionomer. In contrast, on replacement of the carbon catalyst support in the cathode by a ceramic support, the severe fall of the voltage over the first 200 hours of constant current operation was completely eliminated [20].

5.5.2 Protection of Pt Catalysts by Enclosure in Mesopores

Placing PEMFC catalyst nanoparticles inside the mesopores of HSC catalyst supports as protection against catalyst surface blocking by ionomer nanoshells.

The significant increase in the voltage loss measured at higher cell currents on reduction of the PEMFC catalyst loading under 0.2 mg_{Pt} cm^{-2} [21] presented at the time a significant challenge for the efforts to lower further the cost of polymer electrolyte fuel cell stacks and the risk of relying on a metal of limited global availability. Replacement of partially graphitized carbon, e.g. Vulcan XC-72, with high surface area mesoporous carbon as support for Pt-based catalyst particles in the polymer electrolyte fuel cell (PEFC) cathode was consequently the first approach demonstrated to improve significantly the high current performance of air cathodes of ultralow loadings of Pt.

Figure 5.12 gives an example of this technical advancement, and explanation is provided in the captions. The ability to maintain cathode performance at Pt loadings as low as ~0.1 mg cm^{-2} by use of a mesoporous, high surface area carbon (HSC) support is explained by the unique microarchitecture of the supported catalyst

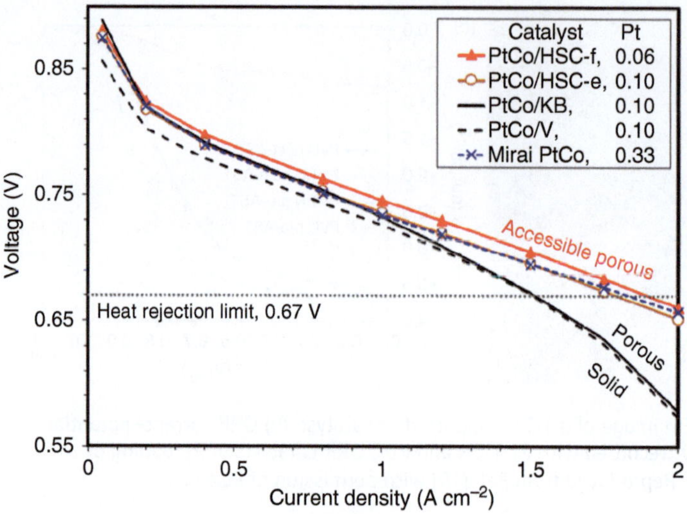

Figure 5.12 Polarization curves recorded for H_2/air fuel cells, with cathodes catalyzed by ultralow loadings of PtCo supported on different types of carbon. Use of mesoporous HSCs as the catalyst support lowers the overall oxygen transport resistance by limiting the degree of catalyst coverage by recast ionomer. The limiting current seen for ultralow cathode loadings is consequently lifted, allowing to equate the performance of a cathode of 0.1 mg_{Pt} cm^{-2} to that of the cathode of 0.3 mg_{Pt} cm^{-2} used earlier in a Toyota Mirai PEFC stack. Source: Reproduced from Ref. [22] with permission of the American Chemical Society.

and catalyst layer, achieved using the mesoporous HSC catalyst supports – a breakaway from the microstructure considered for a long time as prerequisite for highly performing PEFC catalyst layers, i.e. an ensemble of ionomer-coated catalyst agglomerates. The latter, "conventional" microstructure, ensures direct contact between the catalyst particle and the ion-conducting medium; however, the ionomer shell around the catalyst agglomerates adds an oxygen transport barrier, shown to become prohibitively large when the loading drops under 0.2 mg_{Pt} cm^{-2}. Evidence has been provided that such a microshell of ionomer in direct contact with the catalyst agglomerate can also act as blocker of active sites, lowering the rate of the electrocatalytic process at the ionomer-blocked surface sites. HSCs, and, particularly, mesoporous HSCs, can secure a *different microenvironment* around the catalyst particle by "hosting" the Pt particle in a mesopore within the carbon particle.

A schematic presentation of different possible distributions of Pt and ionomer in an HSC agglomerate is offered in Figure 5.13. The figure suggests that the Pt particles on the outer wall of a mesopore will be readily covered by a shell of ionomer and will therefore be less accessible by oxygen and less active in the ORR reaction. In contrast, Pt particles inside a mesopore are likely free of an ionomer shell and can maintain high interfacial activity, although their overall ORR activity will depend on effective transport of reactants and ions into and out of the mesopore (described in Figure 5.12 as "accessible pore").

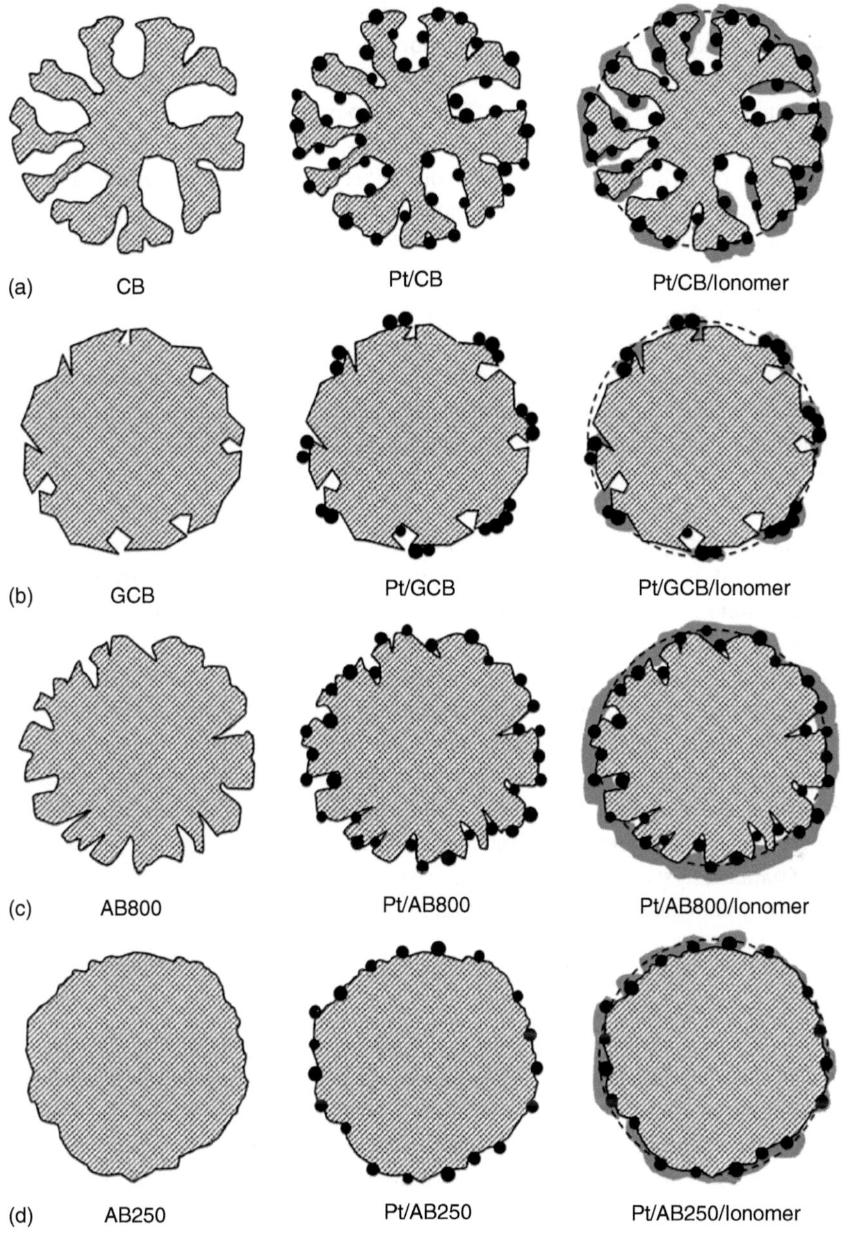

Figure 5.13 A schematic presentation of different possible distributions of Pt and ionomer in a catalyst agglomerate consisting of HSC-supported Pt nanoparticles and recast ionomer. Pt particles that form within the mesopore can remain free of an ionomeric surface coat, thanks to the limited access of the ionomer into the mesopore when the catalyst and ionomer are intermixed to form the catalyst "ink." Source: Reproduced from Ref. [23]. Elsevier/CC BY 4.0.

The high BOL mass activity of Pt achieved with mesoporous carbon supports implies that a catalyst particle located inside the mesopore and, therefore, having no direct contact with the ionomer is still somehow accessed by protons required to maintain ORR activity at this particle. One possible form of proton access is proton (or hydroxide) migration along pore walls, and considering the micropore wall is made of carbon, such migration could likely take place by a process akin to electroosmotic drag of aqueous ions located in the outer, diffuse component of the double layer. Further results reported recently show a ×10 larger mass activity of Pt supported on HSCs vs. that of Pt supported on graphitized carbon as well as a different effect of activation cycles on the observed mass activity of HSC-supported vs. Vulcan-supported Pt. A possible reason for the observed increase in the mass activity with number of activation cycles in the case of Pt/HSC is generation by the activation cycles of an aqueous ionomer dispersion which is possibly required for effective transport of protons to Pt particles hosted in mesopores.

The case of Pt particles, which are deposited within microvoids of a mesoporous carbon support and are thereby protected from direct contact with the ionomer when the supported catalyst is mixed with the ionomer, is an example of a *carbon envelope* around the catalyst particle that acts as a protective mass transport barrier. While the rate of ionomer permeation into the mesopore is drastically reduced, access by both oxygen and protons is maintained high when the proper choice of HSC is made. This nanoarchitecture of carbon-supported catalysts is different than that of "capped catalysts" prepared by coating a catalyst single atom, or nano particle, by an ultra-thin, well-adherent layer of carbon or graphite. The catalyst in a carbon mesopore does not undergo thermal treatment during or after deposition of the metal atom(s) and consequently secures free access to a significant fraction of the catalyst surface. Whether this difference in nanoarchitecture is the full explanation or not, this form of nanocatalyst confinement in mesopores has been to date the most successful demonstration of catalyst protection in an operating fuel cell electrode.

5.6 Conclusions

The demand to maintain high PEFC performance over very long durations requires consideration of novel means for catalyst protection against electrodissolution, detachment, agglomeration, and passivation. Along these lines, some recent reports have described "catalyst capping" as a potential approach for reducing the rate of catalyst activity loss over extended cell operation times. Attempts at such capping have been made for Pt/C ORR cathode catalysts using ultrathin surface overlayers of $mSiO_2$ and Pd/C HOR catalysts using ultrathin surface overlayers of carbon. The reported results indicate that such capping reduces the rate of catalyst agglomeration likely, thanks to mechanical anchoring of the Pt catalyst particle onto the carbon substrate. The results reported to date are still insufficient to evaluate the possible practical value of such catalyst capping in securing long-term performance stability. Furthermore, such evaluation must be comparative, requiring consideration of alternative means for extending the stability of a fuel cell catalyst. One such alternative is the stabilization of supported catalysts by replacement of the carbon with a ceramic support. The "Big Prize" for effective catalyst protection, which is still to be awarded, should be given for substantial improvement in the performance and performance stability of a non-PGM anode catalyst in an alkaline environment. The main mode of deactivation of TM anode catalysts in alkaline media is metal surface passivation, and curbing this surface process in a HEMFC anode under current is a tall challenge. Furthermore, this passive surface layer is the reason for the vastly superior stability of TM catalysts in alkaline environments, being therefore a good example for the intertwined adverse and beneficial functions of catalyst surface layers.

Abbreviations

AST	accelerated stability test
BOL	beginning of life
ClE	chlorine evolution reaction
HEMFC	hydroxide conducting membrane fuel cell
HER	hydrogen evolution reaction
HOR	hydrogen oxidation reaction
HSC	high surface area carbon
$mSiO_2$	mesoporous silica
ORR	oxygen reduction reaction
PEFC	polymer electrolyte fuel cell
PGM	platinum group metal
TM	transition metal

References

1 Cullen, D.A., Neyerlin, K.C., Ahluwalia, R.K. et al. (2021). New roads and challenges for fuel cells in heavy-duty transportation. *Nature Energy* 6 (5): 462–474. https://doi.org/10.1038/s41560-021-00775-z.

2 Gottesfeld, S. (2022). Editors' choice—review—polymer electrolyte fuel cell science and technology: highlighting a general mechanistic pattern and a general rate expression for electrocatalytic processes. *Journal of the Electrochemical Society* 169 (12): 124518. https://doi.org/10.1149/1945-7111/acada3.

3 Takenaka, S., Goto, M., Masuda, Y. et al. (2018). Improvement in the durability of carbon black-supported Pt cathode catalysts by silica-coating for use in PEFCs. *International Journal of Hydrogen Energy* 43 (15): 7473–7482. https://doi.org/10.1016/j.ijhydene.2018.02.159.

4 Doan, H., De Moraes, R.S. and Chatenet, M. (2022). (Invited) Hydrogen oxidation reaction (HOR) on carbon-capped Pd/C in alkaline media. ECS Meeting Abstracts MA2022-01 (49):2049. doi:10.1149/MA2022-01492049mtgabs.

5 Doan, H., Kendrick, I., Blanchard, R. et al. (2021). Functionalized embedded monometallic nickel catalysts for enhanced hydrogen evolution: performance and

stability. *Journal of the Electrochemical Society* 168 (8): 084501. https://doi.org/10.1149/1945-7111/ac11a1.

6 Tong, W., Forster, M., Dionigi, F. et al. (2020). Electrolysis of low-grade and saline surface water. *Nature Energy* 5 (5): 367–377. https://doi.org/10.1038/s41560-020-0550-8.

7 Joo, S.H., Park, J.Y., Tsung, C.-K. et al. (2009). Thermally stable Pt/mesoporous silica core–shell nanocatalysts for high-temperature reactions. *Nature Materials* 8 (2): 126–131. https://doi.org/10.1038/nmat2329.

8 Esposito, D.V. (2018). Membrane-coated electrocatalysts—an alternative approach to achieving stable and tunable electrocatalysis. *ACS Catalysis* 8 (1): 457–465. https://doi.org/10.1021/acscatal.7b03374.

9 Myers, D.J., Wang, X., Smith, M.C., and More, K.L. (2018). Potentiostatic and potential cycling dissolution of polycrystalline platinum and platinum nano-particle fuel cell catalysts. *Journal of the Electrochemical Society* 165 (6): F3178. https://doi.org/10.1149/2.0211806jes.

10 Harzer, G.S., Schwämmlein, J.N., Damjanović, A.M. et al. (2018). Cathode loading impact on voltage cycling induced PEMFC degradation: a voltage loss analysis. *Journal of the Electrochemical Society* 165 (6): F3118. https://doi.org/10.1149/2.0161806jes.

11 Campos-Roldán, C.A. and Alonso-Vante, N. (2019). The hydrogen oxidation reaction in alkaline medium: an overview. *Electrochemical Energy Reviews* 2 (2): 312–331. https://doi.org/10.1007/s41918-019-00034-6.

12 Sheng, W., Myint, M., Chen, J.G., and Yan, Y. (2013). Correlating the hydrogen evolution reaction activity in alkaline electrolytes with the hydrogen binding energy on monometallic surfaces. *Energy & Environmental Science* 6 (5): 1509–1512. https://doi.org/10.1039/C3EE00045A.

13 Xue, Y., Shi, L., Liu, X. et al. (2020). A highly-active, stable and low-cost platinum-free anode catalyst based on RuNi for hydroxide exchange membrane fuel cells. *Nature Communications* 11 (1): 5651. https://doi.org/10.1038/s41467-020-19413-5.

14 Sheng, W., Bivens, A.P., Myint, M. et al. (2014). Non-precious metal electrocatalysts with high activity for hydrogen oxidation reaction in alkaline electrolytes. *Energy & Environmental Science* 7: 1719–1724. https://doi.org/10.1039/c3ee43899f.

15 Hansen, H.A., Rossmeisl, J., and Norskov, J.K. (2008). Surface Pourbaix diagrams and oxygen reduction activity of Pt, Ag and Ni(111) surfaces studied by DFT. *Physical Chemistry Chemical Physics* 10 (25): 3722–3730. https://doi.org/10.1039/B803956A.

16 Huang, L.F., Hutchison, M.J., Santucci, R.J. Jr., et al. (2017). Improved electrochemical phase diagrams from theory and experiment: the Ni–water system and its complex compounds. *Journal of Physical Chemistry C* 121 (18): 9782–9789. https://doi.org/10.1021/acs.jpcc.7b02771.

17 Brodt, M., Wycisk, R., Dale, N., and Pintauro, P. (2016). Power output and durability of electrospun fuel cell Fiber cathodes with PVDF and Nafion/PVDF

binders. *Journal of the Electrochemical Society* 163 (5): F401. https://doi.org/10.1149/2.0711605jes.

18 Lafforgue, C., Maillard, F., Martin, V. et al. (2019). Degradation of carbon-supported platinum-group-metal electrocatalysts in alkaline media studied by in situ Fourier transform infrared spectroscopy and identical-location transmission electron microscopy. *ACS Catalysis* 9 (6): 5613–5622. https://doi.org/10.1021/acscatal.9b00439.

19 Lori, O., Kozhushner, A., Honig, H.C., and Elbaz, L. (2023). Enhanced oxygen reduction and fuel cell performance and durability of ultra-low loading Pt-supported high surface area titanium nitro-carbide. *Journal of Power Sources* 559: 232620. https://doi.org/10.1016/j.jpowsour.2022.232620.

20 Weiss, C.M., Setzler, B.P., and Yan, Y. (2022). Study of cathode gas diffusion architecture for improved oxygen transport in hydroxide exchange membrane fuel cells. *Journal of the Electrochemical Society* 169 (5): 054505. https://doi.org/10.1149/1945-7111/ac6446.

21 Kusoglu, A. and Weber, A.Z. (2017). New insights into perfluorinated sulfonic-acid ionomers. *Chemical Reviews* 117 (3): 987–1104. https://doi.org/10.1021/acs.chemrev.6b00159.

22 Kongkanand, A. and Mathias, M.F. (2016). The priority and challenge of high-power performance of low-platinum proton-exchange membrane fuel cells. *The Journal of Physical Chemistry Letters* 7 (7): 1127–1137. https://doi.org/10.1021/acs.jpclett.6b00216.

23 Park, Y.-C., Tokiwa, H., Kakinuma, K. et al. (2016). Effects of carbon supports on Pt distribution, ionomer coverage and cathode performance for polymer electrolyte fuel cells. *Journal of Power Sources* 315: 179–191. https://doi.org/10.1016/j.jpowsour.2016.02.091.

Part III

State of the Art

6

Design of PGM-free ORR Catalysts: From Molecular to the State of the Art

Naomi Levy and Lior Elbaz

Bar-Ilan University, Chemistry Department, 1 Max and Anna Webb, Ramat-Gan, 5290002, Israel

6.1 Introduction

Platinum group metal-free (PGM-free) molecular catalysts (MCs) have been of great interest in the past decade in the attempt to replace the PGM catalysts in fuel cells, to reduce the price of this technology. The pioneering work by R. Jasinski suggested the use of bioinspired macromolecules such as metallophthalocyanines as catalysts for oxygen reduction reaction (ORR), in order to replace the scarce Pt at the fuel cells' cathodes [1]. After this work, numerous publications came out on other similar complexes such as metalloporphyrins and metallocorroles. In this chapter, we give a current perspective on the effect of different parameters in these PGM-free MC macrocycles on the electrocatalysis of ORR, after several decades of research in the field.

The parameters that we address here can be divided into four categories: (i) the influence of molecular changes within the complex, (ii) the influence of neighboring molecules, (iii) the physical and/or chemical interactions between the catalyst and its support material, and (iv) the effect of pyrolysis of MCs on their catalytic performance.

To understand the different effects on the ORR, it is important first to understand the different catalytic reaction mechanisms with transition metal complexes. We divide them into two main categories: the *two-electron mechanism* and the *four-electron mechanism*. Scheme 6.1 shows a simplified description of these mechanisms in acidic (red) and alkaline (blue) environments. In the case of MCs, it has been established that the first step of the reaction in acidic environment is always the first electron reduction of the MC's metal center, and only then the catalytic reaction begins with the formation of an oxygen adduct on the metal center. In alkaline solutions the initial stage of the complex begins with the formation of hydroxyl anion coordination to the metal center as an axial ligand [2]. This ligand will easily leave the complex once the first electron is transferred to the metal center, followed by the reduction reaction. There has also been extensive work done on that specific initial stage, at different pH levels, emphasizing the importance of the MCs' environment and the impact of the electrolyte on the system [3].

Electrocatalysis for Membrane Fuel Cells: Methods, Modeling, and Applications, First Edition.
Edited by Nicolas Alonso-Vante and Vito Di Noto.

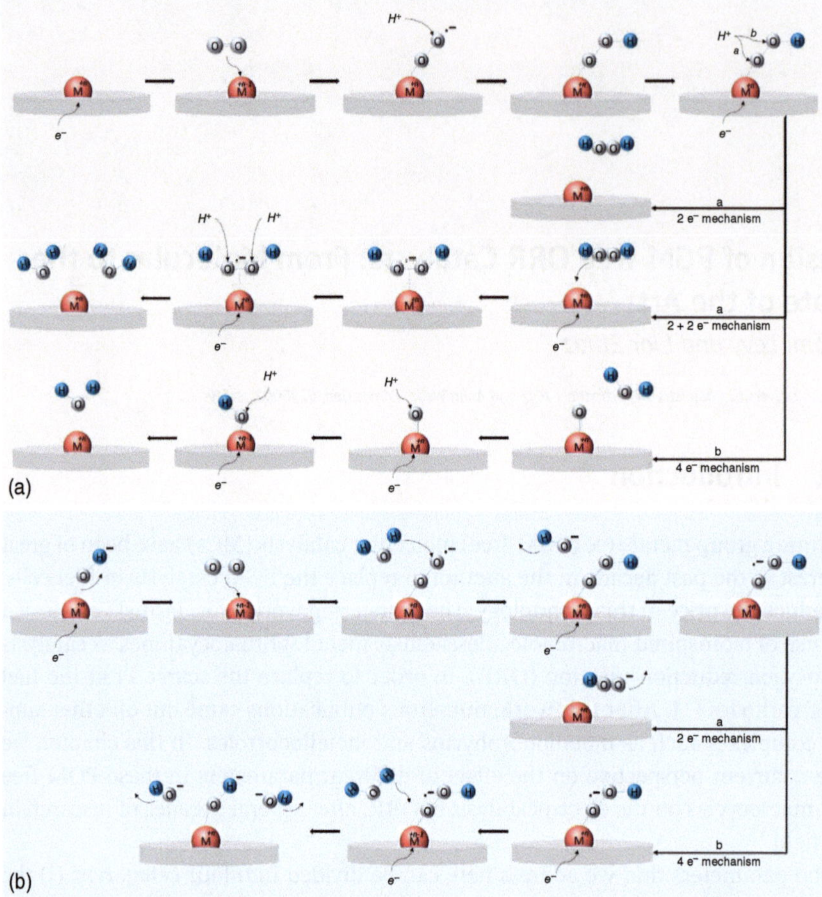

Scheme 6.1 Suggested ORR mechanisms of MCs in (A) acidic solution and (B) alkaline solution.

Although, as mentioned above, the pH of the solution plays an important role in the reaction mechanism [2–4], it is also important to point out that the data presented in this chapter deals with either extremely acidic or extremely basic conditions that are relevant to polymer electrolyte fuel cells (PEFCs) and alkaline exchange membrane fuel cells (AEMFCs), respectively. Other work in neutral and close to neutral pH environment which can be used in certain fuel cells is not covered here.

In both extreme cases, the stage at which the mechanism of 2, consecutive 2 + 2, or $4e^-$ transfer, is determined is the protonation of the peroxo intermediate. For example, in the case of low pH, after an oxygen molecule bonds to the metal center of the catalyst, and the first proton attaches to the external oxygen (the oxygen farthest from the metal center), the second proton will now have two sites where it can attach to: (i) the oxygen closest to the metal center – the "internal oxygen atom" or (ii) the second oxygen that already has a proton – the "external oxygen atom."

Whether it attaches to one or the other determines the overall mechanism of the ORR, since the hydrolysis of the peroxide species to generate H_2O_2 (in the case of acidic environment) competes with O—O bond cleavage. If that proton attaches to the internal oxygen atom, the production of hydrogen peroxide would be favored and that would lead to a system where the mechanism is of a $2e^-$ transfer. On the other hand, if the proton attaches to the external oxygen atom, the O—O bond will weaken and that would bring upon the desired $4e^-$ transfer mechanism to produce water [5]. In this chapter, we present several cases where the system is designed such that one mechanism dominates the other.

6.2 The Influence of Molecular Changes Within the Complex

The basic structure of most macrocyclic MCs for ORR is a conjugated macrocycle constructed of four pyrrole rings connected by either an amine or methane bridge, Figure 6.1, and a metal center, usually a first-row transition metal. Figure 6.2 shows the most common MCs: porphyrin, phthalocyanine, and corrole (Figure 6.2a–c, respectively).

6.2.1 The Role of the Metal Center

In recent years, the role of transition metals embedded in heat-treated PGM-free catalysts has been under debate. There were some studies that suggest that ORR can be catalyzed efficiently with metal-free catalysts [6]. It has been proven since

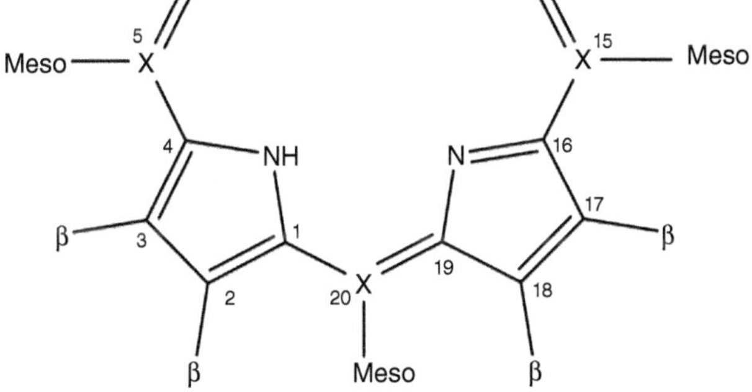

Figure 6.1 A general structure of MCs where X can be either a carbon or a nitrogen atom, carbon positions 2,3,7,8,12,13,17, and 18 are β positions, and positions 5,10,15, and 20 are the Meso positions.

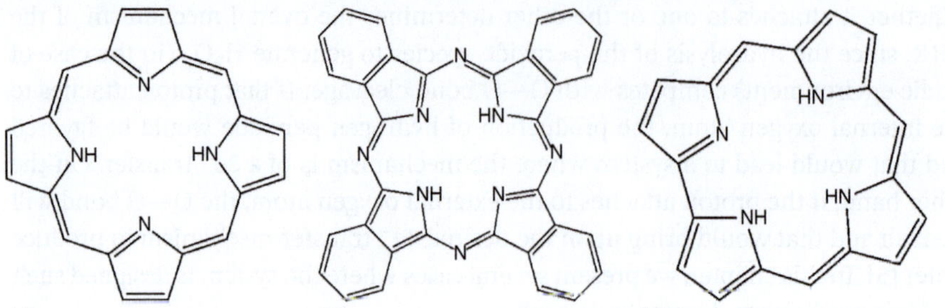

Figure 6.2 The molecular structure of porphyrin (a), phthalocyanine (b), and corrole (c). Molecular catalysts (MC).

that this fact is not completely true for acidic condition, but possible in alkaline conditions due to the presence of quinone-like moieties on carbonaceous surfaces, while certain transition metals can significantly enhance the performance also in alkaline conditions [7]. In many cases, the catalysts in such studies had traces of metals from their synthesis process, which could not be neglected and played a role in the electrocatalysis of ORR with these materials [8]. Other studies with heat-treated PGM-free catalyst try to focus on the role of the metal center, and some say it is not the catalytic center and serve only as an enabler for the reaction. This ongoing debate hinders the progress in the field to some extent. One way to solve it is to extrapolate conclusions from studies of well-defined molecular catalysts. In the case of MCs, the role of the metal center was studied for more than half a century, and it was proven to be the catalytic center for ORR with macrocyclic compounds such as phthalocyanines, porphyrins, and corroles [9]. The effect of the metal center in MCs has been thoroughly studied, mostly with first-row transition metals such as Mn, Fe, Co, Cu, and Ni, which can also be found in biological systems [10]. As seen in Scheme 6.1, the ORR mechanism with MCs in acidic solutions is such that the metal in the complex is first reduced, only then the oxygen molecule binds to it, making the reduction of the metal center the thermodynamic limiting step, which sets the onset potential of the reaction as can be seen, for example, in Figures 6.3 and 6.4 [11, 12]. Both cases

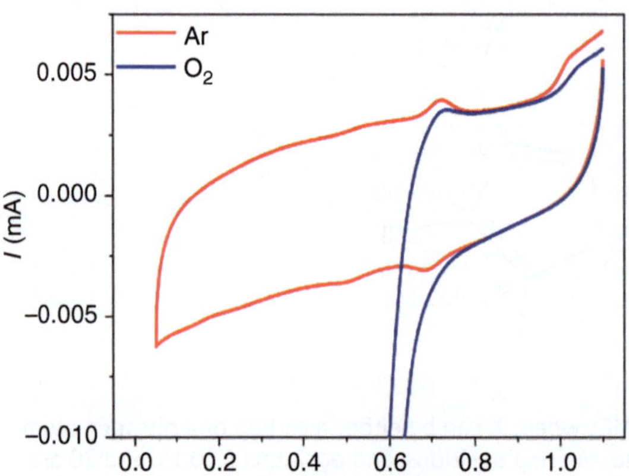

Figure 6.3 Cyclic voltammetry of Co-corrole deposited on a glassy carbon electrode in Ar and O_2 atmosphere in a 0.5 M H_2SO_4 solution. Source: Reproduced from Ref. [11] with permission of the American Chemical Society.

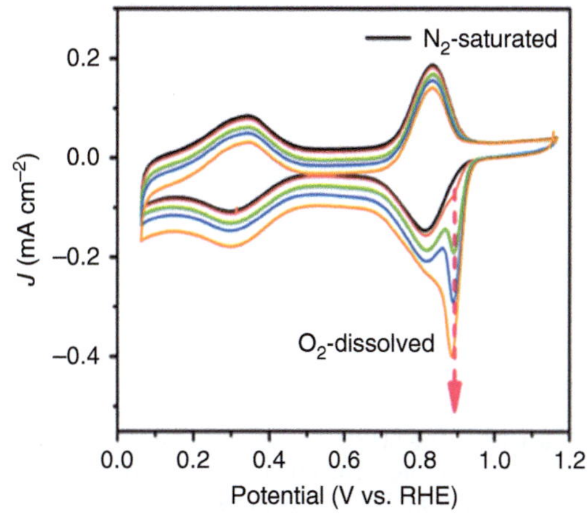

Figure 6.4 Cyclic voltammetry of Fe-phthalocyanine on an acetylene black surface in an N_2 and O_2 atmosphere in a 0.1 M KOH solution. Source: Reproduced from Ref. [12] / Springer Nature / CC BY 4.0.

show the reversible redox couple associated with the metal center's redox reaction under deaerated environment (around 0.7 and 0.85 V vs. RHE for Figures 6.3 and 6.4, respectively), and the increase in cathodic currents in oxygen-saturated environment attributed to ORR. The potential at which ORR takes place is the same as the reduction potential of the metal center, hence the dependence of the reaction overpotential on the redox potential of the metal center in MCs. In addition to controlling the ORR overpotential, the metal center and the electron density around it may also affect the reaction's turnover frequency as will be discussed later. This ORR mechanism is different than that with PGM catalysts such as Pt, where the binding of oxygen, which forms an oxide layer on the metal, is the limiting step [13]. It is therefore necessary to understand the molecular orbital arrangement and energy levels of the lowest unoccupied molecular orbital (LUMO), into which this first electron transfers. Calculations indicated that the LUMO energy (eV) in both porphyrin and corrole complexes decreases upon increase in atomic number from Mn to Fe, Co, Ni, and Cu along the first-row transition metals [14]. Theoretically, if this had been the only relevant parameter, the best performing complex would have been the copper complex (out of the five metals discussed above). This is in fact not the case. When looking into the computational studies of these complexes, it appears that the LUMO is dominantly on the metal center only in the case of Mn, Fe, and Co. In the case of Ni and Cu, the LUMO is mainly on the surrounding aromatic rings. The shapes of the LUMOs hence indicate that the first electron reduction of the MC will occur on the metal center for Mn, Fe, and Co, but will occur on the ligand for Cu and Ni, and in their case, only the second electron will go to the metal center and enable the formation of an oxygen adduct and eventually reduce oxygen. In the case study of metallocorroles, this prediction was confirmed by looking at the singly occupied highest molecular orbitals (SOHMOs) of the 1e$^-$ reduced corroles. Figure 6.5 shows that the molecular orbital is mainly located on the metal center for Mn, Fe, and Co, but on the ligand in the case of Ni and Cu [15, 16]. These studies also clearly show that the Mn complex is not as catalytically active as Fe and Co. The reason for this lies in the molecular orbital energy levels, after the first electron reduction. The "new"

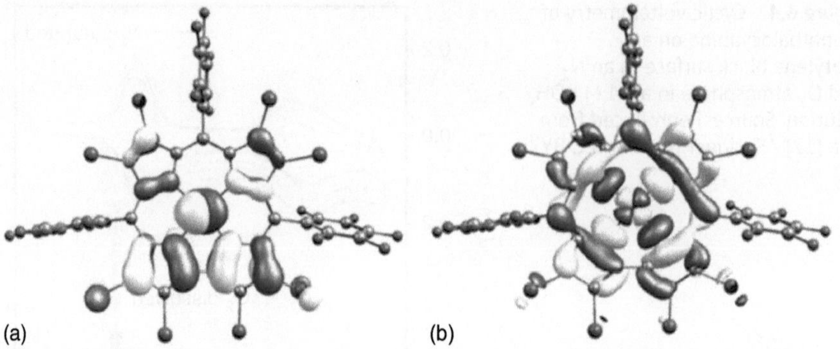

Figure 6.5 The SOHMO isovalue surfaces for (a) Co-corrole with (which is similar to that of Mn and Fe corroles) and (b) Cu-corrole (which is similar to that of Ni corrole). Source: Reproduced from Ref. [15] with permission of John Wiley & Sons.

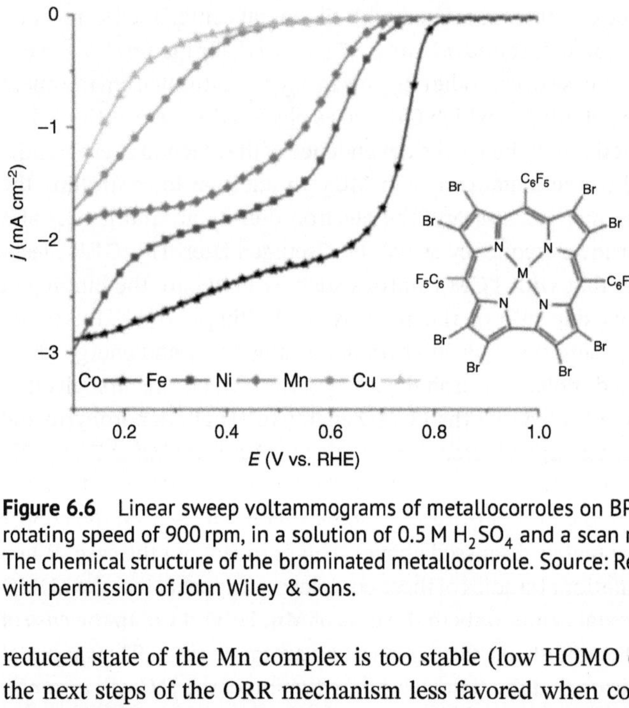

Figure 6.6 Linear sweep voltammograms of metallocorroles on BP2000 carbon with a rotating speed of 900 rpm, in a solution of 0.5 M H_2SO_4 and a scan rate of 5 mV s^{-1}. Inset: The chemical structure of the brominated metallocorrole. Source: Reproduced from Ref. [15] with permission of John Wiley & Sons.

reduced state of the Mn complex is too stable (low HOMO energy level), making the next steps of the ORR mechanism less favored when compared to Fe and Co [15, 17–22].

These findings were confirmed experimentally, as can be seen in Figure 6.6, for the rotating ring electrode (RDE) measurements of metallocorroles with different metal centers, showing that the cobalt and iron clearly display much higher ORR performance compared to the other tested metal centers. Similar conclusions were reached in theoretical study done by Orellana et al. [17], who studied the ORR catalytic activity of four transition metalloporphyrins (Mn, Fe, Co, and Ni) adsorbed on carbon nano tubes (CNTs). This work followed the work of Shi and Zhang [23] that showed that the catalytic activity of transition metal macrocycles toward ORR

Table 6.1 Bond distance (*d*), first ionization energy (IE), and magnetic moment (*m*) of the transition metal porphyrins.

Molecule	d_{M-N} (Å)	d_{N-C} (Å)	IE (eV)	*m* (μB)
MnP	2.017	1.398	5.39	3
FeP	1.99	1.397	5.98	2
CoP	1.986	1.394	6.72	1
NiP	1.968	1.395	6.76	0

Source: Reproduced from Ref. [17] with permission of the American Physical Society.

is directly related to the ionization energy/potential and the O_2 binding energy to the metal center. As seen in the case of corroles, the ionization energy of first-row transition-metalloporphyrins increases in the order of Mn < Fe < Co < Ni (Table 6.1). These energies slightly change when the same porphyrins are attached to CNTs, but the trend stays the same.

In order to be able to fully predict the ORR catalytic activity of macrocyclic complexes, in addition to the calculations of the reduction potentials, the binding energies of the dioxygen molecule to the metal center were studied. One way to calculate this interaction is to calculate the bond length between the two oxygen atoms. The longer the dioxygen bond, the lower the stretching frequency, and the lower the energy that would be needed to break that bond. Based on these calculated relative stretching frequency, the O=O bond weakens in the following order: MnP > FeP > CoP > NiP [17]. Figure 6.7 shows the spin-density plots for O_2 adsorbed on the metal center of metalloporphyrins adsorbed on CNTs, having similar properties to those presented in the case of corroles. According to this, the Mn complex shows the most likely initial conditions toward ORR, exhibiting a longer and weaker O—O bond, followed by the Fe and Co complexes. The Ni complex, on the other hand, shows the weakest interaction with the oxygen molecule and the highest O_2 stretching frequency. It is hence not surprising that most of the work done with MCs mainly deal with Co and Fe complexes. As hinted earlier, the vast knowledge that was accumulated on well-defined MCs can be used to guide the design of other classes of PGM-free catalysts such as M–N–Cs.

6.2.2 Addition of Substituents to MCs

In this section, we discuss how different substituents on the backbone of MCs influence their ORR activity. As discussed in Section (6.2.1), the electron density around the metal center can affect the redox potential of the MC, and as a consequence, it will affect the ORR overpotential. The lower the LUMO and the higher the electron density around the metal center, the lower the ORR overpotential with MCs. The tunable structure of some macrocyclic complexes allows to add electron-withdrawing/donating substituents on the peripheral ring of the ligand, which could further lower the LUMO energy of the metal center. There are two

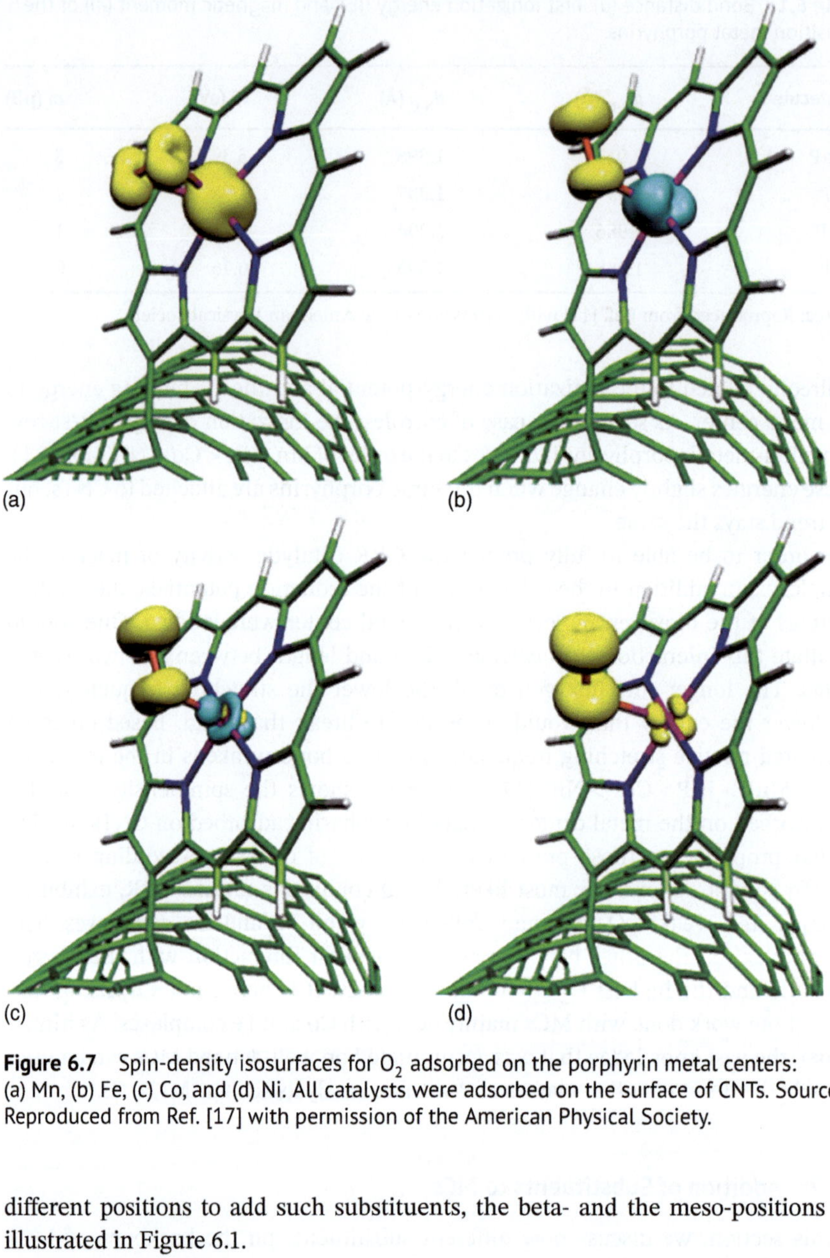

Figure 6.7 Spin-density isosurfaces for O_2 adsorbed on the porphyrin metal centers: (a) Mn, (b) Fe, (c) Co, and (d) Ni. All catalysts were adsorbed on the surface of CNTs. Source: Reproduced from Ref. [17] with permission of the American Physical Society.

different positions to add such substituents, the beta- and the meso-positions as illustrated in Figure 6.1.

6.2.2.1 Beta-substituents

Beta substituents are the closest electron-donating/withdrawing moieties on the MC ring to the metal center. Their character can directly affect the catalytic performance of system. Elbaz et al. [24] studied a series of complexes with beta substituents with increasing electron-withdrawing tendency. In this work, they synthesized Co-corroles with a series of four beta-substituents: H, F, Cl, and Br. Such halide

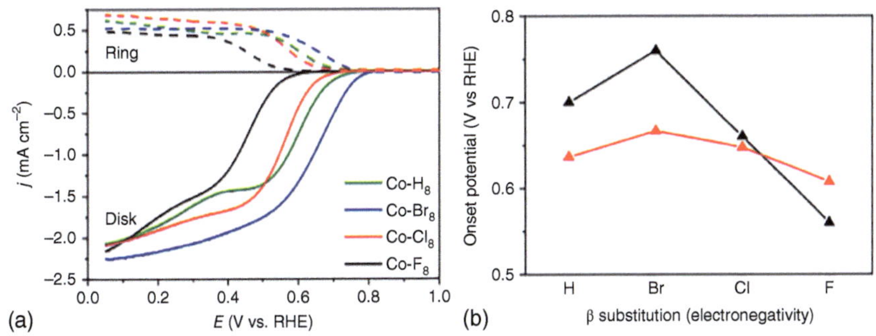

Figure 6.8 (a) RRDE measurements of Β-halogenated Co-corroles in 0.1 M KOH at a rotation speed of 900 rpm and a scan rate of 5 mV s^{-1}. (b) Comparison of the Co-corroles onset potentials from RRDE in 0.1 M KOH (black) and 0.5 M H_2SO_4 (red). Source: Reproduced from Ref. [24] with permission of The Royal Society of Chemistry.

substituents have competing electron push–pull effects: electron-withdrawing effect from the complex to the halide and π-donating effect from it to the complex [25, 26]. Substituents having high electron affinity like halides could on one hand "pull" the electrons from the complex, lower the energy of the complexes LUMO, and hence lower the ORR overpotential. On the other hand, these same substituents will have the tendency to "push" electrons toward the aromatic molecule via its π-system, having the opposite effect on the ORR overpotential. The ORR overpotential of the corroles in this study decreased in the following order: Br > Cl ~ H > F, Figure 6.8a, with Br being the best in terms of overpotential although it is not the most electronegative of the four different substituents. This implies that the effect of the halides could be a combination of both σ-withdrawing and π-donating effects, and that optimization of the two is necessary to reach low overpotentials.

The trend, which is clear from the results presented in Figure 6.8, could only be detected when the MCs were deposited on glassy-carbon electrodes and was similar for acidic and alkaline solutions. Once adsorbed on a high surface area porous carbon such as BP2000, this trend disappeared, and the ORR catalytic performance of all β-substituted corroles appeared to be almost the same. It is concluded therefore that such electron-donating/withdrawing effects are relatively weak and are easily dominated by other effects resulting from the interaction between MCs and the support material supports. These other effects will be discussed later.

Tasca et al. [27, 28] studied the effect of electron-withdrawing substituents in phthalocyanines and used a ciano group to take this case to an extreme in terms of electronegativity. In their work, the MCs were adsorbed on CNTs and tested both in acidic and in alkaline conditions. As with the corroles, the ciano substituents showed a similar effect on the ORR to the halogenated phthalocyanines. Another series of β-substituted MCs that was tested was based on the different substituents found in biological heme complexes [29]. Figure 6.9 shows an example of two biological β-substituted porphyrins: heme A and heme B, found in hemoproteins. Dey et al. [29] compared similar synthetic porphyrins and showed the influence of the presence of two ester groups on β-positions of a porphyrin, compared

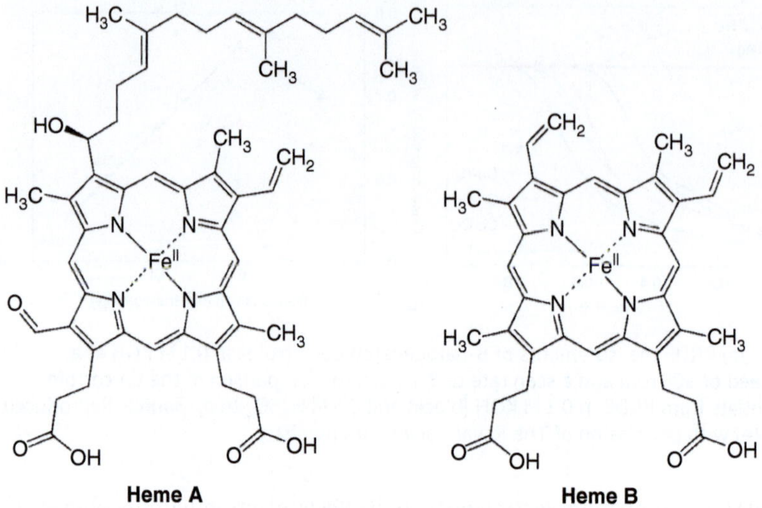

Figure 6.9 Chemical structure of heme A and heme B molecules.

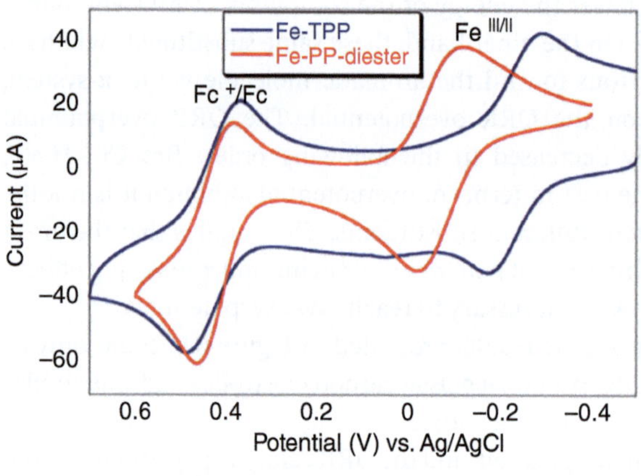

Figure 6.10 CV of the Fe-TPP and Fe-PP-diester complexes in MeOH solvent (containing an excess amount of imidazole) having 100 mM TBAP as the supporting electrolyte, glassy carbon as the working electrode, and Fc^+/Fc as the internal standard. Source: Reproduced from Ref. [29] with permission of The Royal Society of Chemistry.

to the same complex without these two substituents (Figure 6.10). The cyclic voltammogram of the iron porphyrin having the diester groups (Fe-PP-diester) shows a quasi-reversible wave with $E_{1/2}$ of −0.49 V vs. Fc^+/Fc. The reduction potential of unsubstituted iron porphyrin (Fe-TPP) under the same conditions is −0.66 V vs. Fc^+/Fc. Therefore, the Fe-PP-diester shows 170 mV more positive shift relative to the Fe-TPP, indicating that the former is easily reduced due to the electron-withdrawing groups on its ring.

6.2.3 Meso-substituents

in macrocyclic MCs was also studied using computational and electrochemical methods. As in the case of β-substituted macrocyclic MCs, the focus of these studies was mostly on the influence of the substituents on the molecular orbital energy levels after the attachment of dioxygen molecule. A study of a series of meso-substituted corroles showed that those having meso-substituents with a strong electron-withdrawing character were most readily reduced [11]. In some cases, the studied MCs did not show the same trend as the calculated ones, and, as in the case of β-substituted MCs, once adsorbed on a high surface area, porous carbon substrates, these differences between the substituents disappeared [11, 30, 31].

One specific exception was observed with a group of meso-substituted MCs categorized as "Hangman" complexes. These MCs are usually substituted in one meso-position with a rather long molecule that has a specific edge group. This group can be of great importance for the ORR mechanism and much less to reducing the reaction overpotential, since it can strongly affect the selectivity of the ORR catalyst to the 2, 2 + 2, or $4e^-$ mechanism. The hangman group will usually include a proton transfer group such as a carboxy group to enable the protonation on the second oxygen of the peroxo ligand (M–O–O*–H), leading to the favored $4e^-$ reduction mechanism to water in low pH [32, 33].

6.2.4 Axial Ligands

The coordination number of the most common transition metal centers used in MCs, Fe(III) and Co(III), is six. This means that they have two axial positions that can be used to bind other ligands or an oxygen molecule or both. The out-of-plane fifth and sixth coordination positions on the metal center can affect ORR electrocatalysis in several ways: It can affect the energy level of the catalyst's LUMO (similar to substituents discussed before); it can interfere or induce the bonding of the oxygen to the metal and hence change the catalytic activity (for example some ligands such as NO and CO are regarded as poisons since they bind strongly on both axial positions and block them from any interaction with oxygen molecules, hence making the catalysts inactive); it can impact the planarity of the complex, deforming its electronic structure and changing its electrochemical response, and it can act as a molecular wire and facilitate the electron transfer from the electrode to the metal center, while acting as an anchor of the macrocycle to the surface, which may also affect the durability of the MC among other parameters [34, 35]. Since PGM-free durability in general, and that of MCs in particular, is one of the impediments of using them in commercial acidic fuel cells, it is important to note that adding a fifth ligand to the metal center, especially one that is attached to the surface such as quinones [13, 36, 37], can stabilize the metal in the complex. This stabilization is very important in cases where protons attack the pyrrole ligands in the MC, which can result in metal leaching and complete loss of activity.

In the synthesis of PGM-free MCs, the presence of axial ligands usually comes from the use of these molecules as stabilizers or counterions to the metal center. In many cases, molecules such as pyridine and triphenyl phosphine and various counter ions such as chloride are used to maintain the electroneutrality of the complex, but it is commonly assumed that these ligands leave the catalyst once placed in an aqueous solution, and replaced by ions from the electrolyte.

Biological systems were also intensively investigated to understand the role of axial ligands in the electrocatalysis of ORR. A number of different axial ligands can be found in heme complexes in nature: a neutral imidazole found in histidine amino acids, an anionic thiolate in cysteine amino acids, and an anionic phenolate in tyrosine amino acids [37–40]. The influence of these three families of ligands was investigated with Fe(III)-porphyrins. The anionic ligands, thiolate, and phenolate tend to negatively shift the redox potential of the metal center by 500 mV compared to the imidazole, when measured in polar organic solutions. This can be attributed to the changes in electron density on the metal, caused by the larger "push effect" of the anionic ligands, whereas the neutral imidazole induces a much smaller effect [41]. These findings are important, but since they were conducted at moderate pH conditions, they are less applicable for fuel cell applications.

Some of the main studies of axial ligands at extreme pH conditions were conducted by Zagal et al. [42, 43] where the catalytic surface was modified by covalently bonding the catalyst to a gold surface using different axial ligand. At first, they studied the effect of the ligand by comparing the ORR catalytic performance of iron phthalocyanine (FePc) adsorbed on a gold surface and of that covalently attached to the gold surface via self-assembled monolayers (SAMs) of 4-aminothiophenol (ATP) or 1-(4-mercaptophenyl)-2,6-diphenyl-4-(4-pyridyl) pyridinium tetrafluoroborate (MDPP). In Figure 6.10, they show that the coordinated FePc displays better ORR catalytic activity in terms of overpotential than the nonligated one. The source of this enhancement is attributed to an electron-withdrawing effect of the axial ligand, which lowers the electron density on the Fe center. A similar system, where the FePc was attached to CNTs via a pyridine (neutral) axial ligand, also showed a decrease in the electron density on the Fe site and an improvement in the activity when compared to the same MC without the axial ligand [44].

Following this work, two types of pyridinium molecules as electron-withdrawing ligands were tested with two phthalocyanines: one without electron-pulling substituents (FePc) and the other with chlorine substituents (16(Cl)FePc), Figure 6.11. For a so-called neutral complex such as FePc, any electron-withdrawing axial group is expected to lower its electron density, hence lower its LUMO and its ORR overpotential. For this reason, the "up" pyridinium ligand, having the stronger pulling effect, shows a better activity than the "down" pyridinium ligand, Figure 6.12, right. On the other hand, a complex that is rich in electron-withdrawing substituents such as 16(Cl)FePc is at its optimal state when it comes to catalytic activity, and any additional electron-withdrawing ligand will decrease the ORR catalytic activity, Figure 6.13.

Another important phenomenon relating to the addition of these ligands is their influence on the ORR mechanism. They tend to increase the selectivity toward the $4e^-$ pathway. This was shown in several cases, including the ones presented above, by calculating the number of electrons for the ORR process using the Randles–Ševčik equation for irreversible processes controlled by mass transport values [42]. These calculations show the high tendency toward the $4e^-$ transfer in all complexes with a fifth ligand.

In addition to changing the catalytic activity of the MCs, as can be seen in the examples above, the use of a fifth ligand is also for anchoring the complex to the surface. Many systems have been investigated where ligands are covalently bonded to the support on one side and coordinated to the metal center of the MC complex on

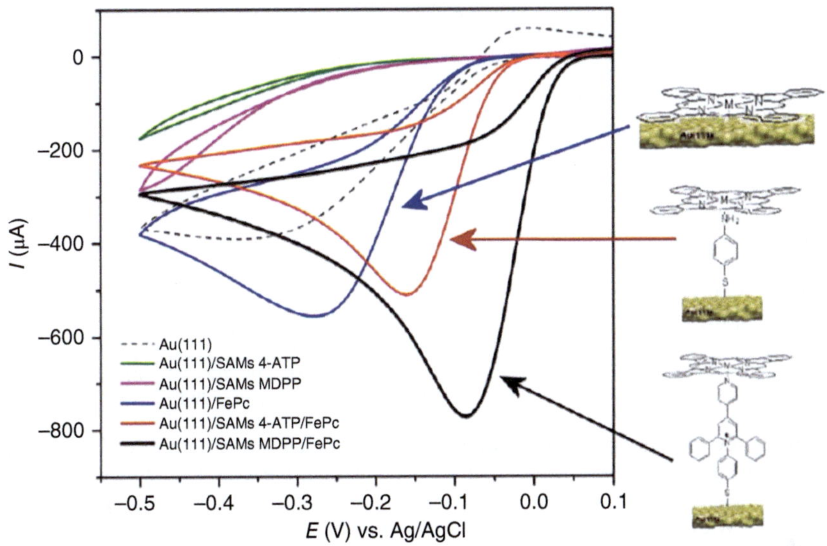

Figure 6.11 Structure of the two phthalocyanins (FePc and 16(Cl)FePc), pyridinium ligands (up and down), and the molecular systems composed of their various combinations. Source: Reproduced from Ref. [42] with permission of the American Chemical Society.

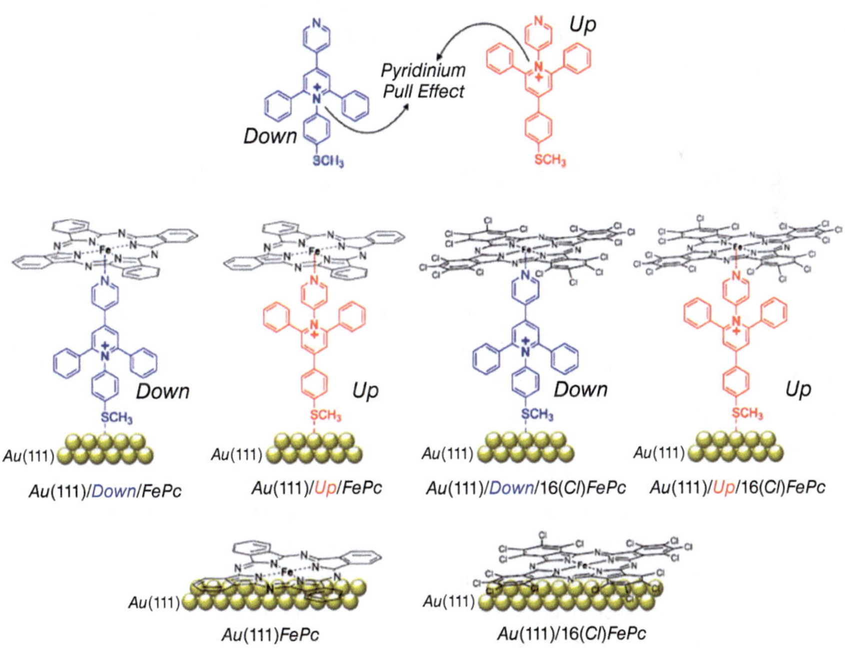

Figure 6.12 Electroreduction of O_2 on a clean Au(111) surface and modified with FePc, with and without 4-ATP and MDPP SAMs. Measurements conducted in O_2 saturated 0.1 M NaOH, $dE/dt = 0.05\ V\ s^{-1}$. Source: Reproduced from Ref. [43] with permission of the American Chemical Society.

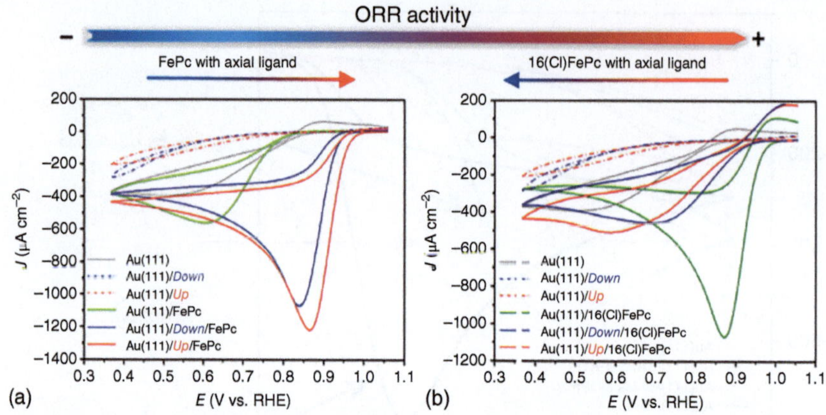

Figure 6.13 Cyclic voltammetry for ORR on Au(111) modified with different configurations of pyridinium ligands, with and without (a) FePc and (b) 16(Cl)FePc. O_2 saturated 0.1 M NaOH; $dE/dt = 0.05\ V\,s^{-1}$. Gray curve represents Au(111) in (a) and (b). Source: Reproduced from Ref. [42] with permission of the American Chemical Society.

the other [17, 35, 42, 43, 45–49]. In some cases, the conductive surface such as carbons, at which the catalyst is adsorbed on, will have surface moieties that can attach to the MC as a fifth axial ligand [50]. One common example is quinone-like surface moieties, which are very common on carbon supports. These surface moieties are also known to affect the ORR potential and the reaction selectivity, similar to the examples given before [13, 36, 37, 51].

6.3 Cooperative Effects Between Neighboring MCs

When focusing on a single catalytic site, it appears that the turn-over frequencies (TOF) of PGM catalysts are much higher than those of PGM-free catalysts [52, 53]. This difference in TOFs is at least one order of magnitude. In order to make PGM-free ORR catalysts a viable alternative to PGMs, it is needed to make sure that the PGM-free-based systems will produce the same power density. Since TOF is an intrinsic property, the best way to compensate for the relatively low TOF of PGM-free catalysts is by increasing the active site density [53]. This will essentially increase the mass activity of the system but will not change the performance per active site. This is a reasonable solution, given the fact that the catalytic sites in MCs are atomically dispersed in the catalytic matrix, and that their loading can be increased significantly compared to traditional loading in fuel cell cathodes with a marginal effect on the system cost, if any, since they are composed of earth-abundant materials. Having said that, it is important to notice that in practice, increasing the number of active sites of MCs brings upon several effects that do improve the catalytic activity per site. These usually refer to the effect where two neighboring sites have on each other's ORR kinetics and reaction mechanism. Whether they are dimers, polymers, or any other molecular frameworks, the density of active catalytic sites and the proximity of two or more catalytic sites can dramatically influence their ORR activity and selectivity.

The first effect relates to a synergetic relationship between two neighboring catalytic sites. In the case of catalysis following the 2 + 2 electron transfer reaction mechanism, one site will reduce oxygen to peroxide, and its neighboring site will "catch" the peroxide and further reduce it to water. Aside from a possible neighboring synergetic effect, the overall design of the system, as in the case study of "packman" catalysts, enables the higher kinetics and higher selectivity of each catalytic site and of the system as a whole.

6.3.1 Bimetallic Cofacial Complexes – "Packman" Complexes

In nature, there are two classes of enzymes that are known to efficiently reduce oxygen utilizing multi-metal-centered catalytic sites: cytochrome *c* oxidase (C*c*O) and multicopper oxidase (MCO) [54–60]. The active sites of C*c*O enzymes each contain an Fe-bound heme *a*3 and a proximal Cu center. MCO enzymes also feature redox-active components in their active sites, each of which include four Cu centers. Of the two O_2-reducing proteins, MCOs operate closer to the thermodynamic potential for ORR than CcOs, which require an overpotential of ~400 mV [61–68].

These biological systems have been the key inspiration for the design of many ORR MCs. Mimicking them and reaching the selectivity and relatively low overpotentials of their protein systems can be very challenging from a synthetic standpoint, and in extrapolating their performance at close to neutral pH environment to either very high or very low pH. One attempt to do so is the use of bimetallic catalysts, where two metal centers might work synergistically to reduce oxygen at lower overpotentials and achieve $4e^-$ selectivity. Early work done by Collman and Chang et al. [69, 70] led to the development of Packman porphyrins, which are cofacial bisporphyrins covalently bridged and tied together. These systems are considered rigid in relation to the distance between the two metal centers, allowing the perfect binding of oxygen within the Packman "mouth." Later, the optimization of these systems was extensively studied with different bridge molecules and different metal centers such as iron, cobalt, copper, and combinations thereof [71–76].

Over the years, the effect of the multi-metal-centered catalytic sites, outside of the biological environment and in aqueous systems at extreme pH conditions, needed for fuel cell applications, has been investigated. The contribution of the copper in CcO, for example, is thought to be an electron source in the ORR in such proteins [77]. In electrochemical systems, where these MCs are adsorbed or covalently connected on the surface of an electrode, the contribution of the copper or any second metal in the system is somewhat different. It is calculated to be more of a proton donor, leading to the addition of the second proton which increases the selectivity for the $4e^-$ reduction pathway.

6.3.2 MC Polymers

As mentioned above, it is of great importance to increase the catalyst's site density, especially in systems of PGM-free ORR catalysts. Increasing the surface area and allowing a higher quantity of catalyst to adsorb on the surface is only one way of doing so. The second, and more efficient, way to increase the active site density is using the catalyst as a monomeric unit and polymerizing it to form a catalytic framework. The polymerization can be either chemical or electrochemical, to form polymer films, aerogels, or metal–organic frameworks (MOFs). In this section, we discuss these different types.

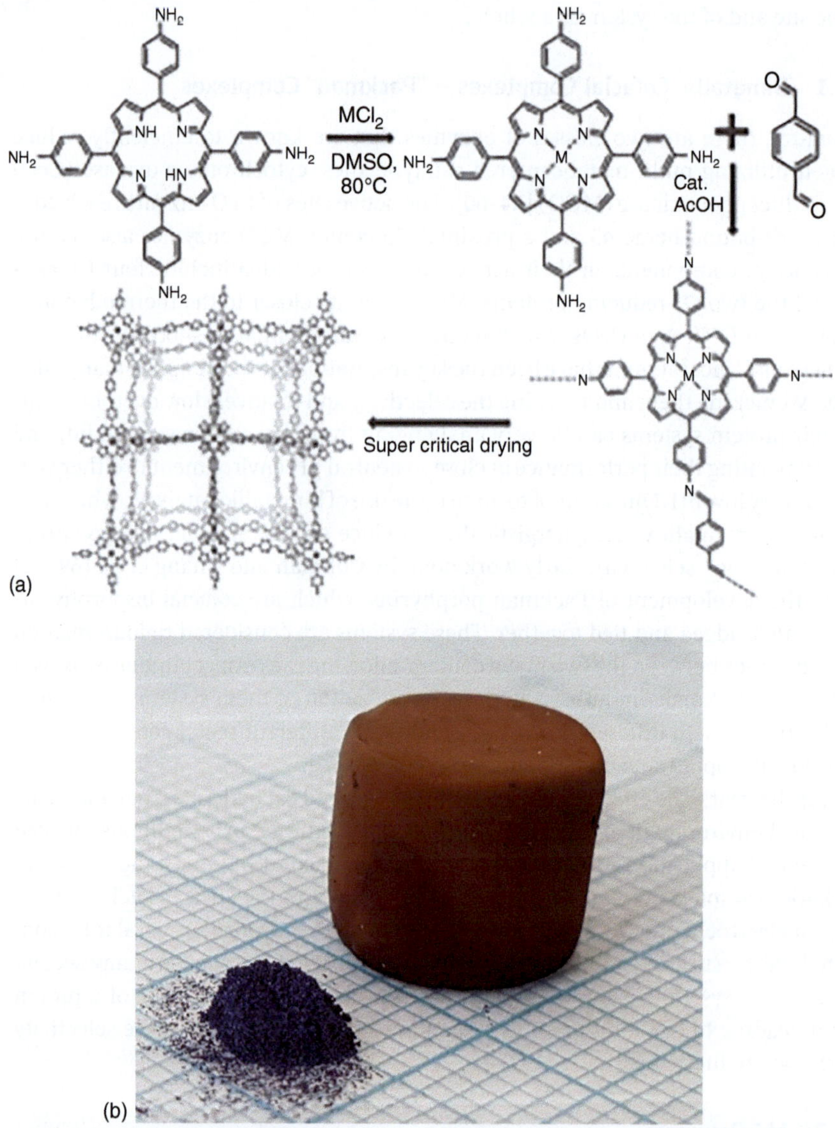

Figure 6.14 Description of the synthesis of porphyrin aerogels (a) and photograph of 30 mg of porphyrin powder and the porphyrin aerogel synthesized from the same amount of porphyrin (b). Source: Reproduced from Ref. [87] with permission of John Wiley & Sons.

One of the simplest ways known today to synthesize such materials is by a one-pot synthesis of MOFs, also known as coordination polymers [78–81]. In this synthesis, the metal centers are coordinated by organic or inorganic molecules, making them a stable framework with a high density of catalytic sites and usually with high surface area. Despite the ease of synthesis of these materials, MOFs are not considered as a good choice for ORR catalysis since these materials are usually not conductive enough by themselves. This can be changed if they are grown inside porous carbon supports [78, 80].

Polymerization and especially electropolymerization of aromatic complexes such as porphyrins and phthalocyanines have been used to answer both demands of high conductivity and high density of catalytic sites. Macor and Spiro [82] were the first to study electropolymerized metalloporphyrin films as catalysts for ORR. A more convenient method of polymerization was introduced by Bettelheim et al. [83, 84] using aminophenylporphyrins via a synthesis similar to the electropolymerization of aniline. The same method was also adopted by Friedman et al. [85, 86] to electropolymerize corroles. In all methods, a thin but dense polymeric film is deposited on a conductive support, enabling good electrical conductivity and good ORR activity with ultrahigh catalytic site density.

These polymers, although having answered the demand for high catalytic site density, have a great disadvantage in their accessibility to oxygen, and mass transfer in general, due to their ultradense structure. In order to make all the catalytic sites accessible to oxygen, the use of carbon aerogels became of great interest, and quickly followed by the preparation of polymeric aerogels from the catalysts themselves, giving both high density of catalytic sites and high accessibility to these sites (Figure 6.13 left) [13, 36, 87]. Figure 6.14 also shows pictures of porphyrin powder and the porphyrin aerogel synthesized from this powder.

6.4 The Physical and/or Chemical Interactions Between the Catalyst and Its Support Material

Choosing a catalyst for ORR and optimizing it to be as efficient as possible is unfortunately only the first step in preparation of cathodes for fuel cells. One other very important parameter influencing the catalysts' overall performance is their interaction/s with the support and the chemical and physical properties of the surface itself. There have been numerous papers on the preparation of ORR catalysts, where the most interesting part was the surface to which the catalyst is deposited on or attached to. Substrates, mostly high surface area, porous, carbonaceous supports, are used to incorporate MCs either via simple adsorption or covalently to the surface. Each of these interactions may have their sets of advantages and disadvantages.

CNTs have been very popular to covalently anchor MCs to their surface using functional groups such as imidazole and other similar molecules such as the axial ligands that were discussed earlier. Wei and Orellana [17, 47] showed how MCs, when attached to CNTs, can show much better ORR catalytic performance. Another surprising surface/catalyst assembly was seen in several cases where BP2000, a high

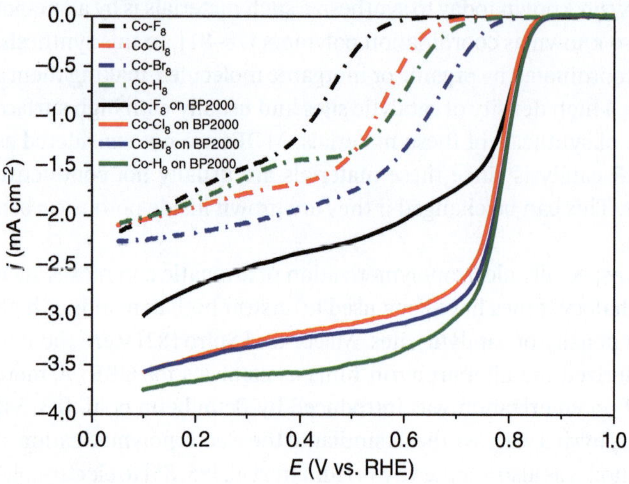

Figure 6.15 RDE measurements of β-halogenated Co-corroles deposited on a glassy carbon electrode (dash-dot line) and adsorbed on BP2000 (solid line). All measurements were conducted in a solution of 0.1 M KOH, 5 mV s^{-1} and at 900 rpm. Source: Reproduced from Ref. [24] with permission of The Royal Society of Chemistry.

surface area porous carbon, was used. In this case, a series of MCs deposited on a glassy carbon electrode and tested as ORR catalysts in both alkaline and acidic conditions showed distinct differences in their performance, but when the same MCs were adsorbed on BP2000, the performance was dramatically improved, and the differences between the MCs disappeared. An example for this phenomenon is given in Figure 6.15 [24]. This phenomena was thoroughly investigated, and it was concluded that the origin for the improved catalytic activity lies not in the carbon's chemical properties but rather in its unique porous structure. The pore size distribution of this material is such that it has a large quantity of certain size of pores (in this specific case, 20 nm in diameter). In these pores, the MCs are arranged in a specific distance from one another, enabling synergetic ORR catalysis, which in a way mimics cofacial MCs [88]. Hence, the use of BP2000 carbon has increased and became popular for the study of MCs today.

6.5 Effect of Pyrolysis

The ORR mechanism can result in the production of not only water molecules but also hydrogen peroxide in the case of acidic conditions or peroxide anion in the case of alkaline conditions, which tend to attack and degrade the ligand structure. In the early 1970s, Jahnke et al. [89] showed that heat treatment of phthalocyanines and other similar molecules can improve both the activity and durability of these catalysts. A decade later, after several pioneering studies were made on heat-treated MCs, Van Veen et al. [90] published a paper summarizing the ORR catalytic activity of a series of heat-treated porphyrins and phthalocyanines and compared them

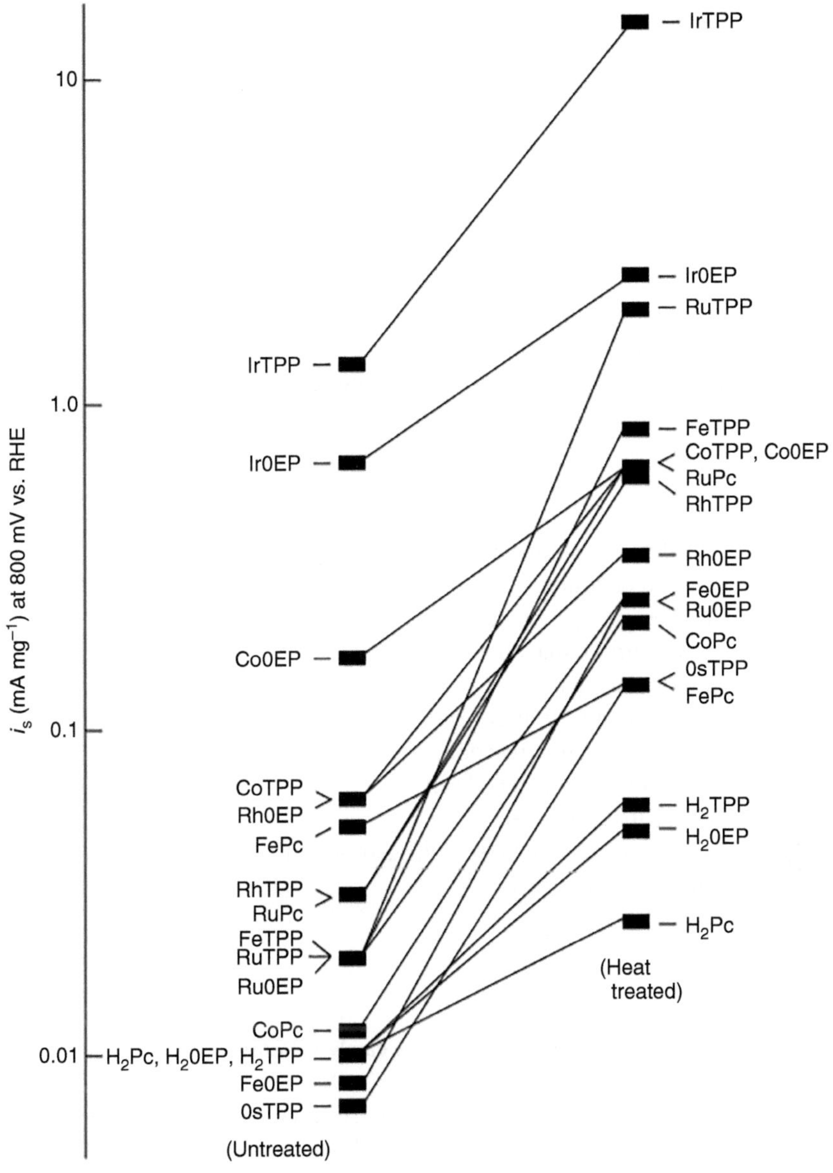

Figure 6.16 ORR activity of various porphyrins and phthalocyanines before and after heat treatment in an inert gas atmosphere. Source: Reproduced from Ref. [90] with permission of The Royal Society of Chemistry.

with their untreated analogues. The results are presented in Figure 6.16. It appears that in all cases, heat treatment improved the catalytic activity of the MCs. These findings opened a path for a whole new world of ORR catalytic materials that are all based on the presence of a catalytic center that is similar to that of MCs: a transition metal, centered within four nitrogen atoms, imbedded in a carbonaceous material. Some of the work done in this field is of heat treatment of MCs, but since these complexes are usually difficult to synthesize and hence quite expensive, much of it relates to the synthesis of what is called M–N–C (Metal–Nitrogen–Carbon) catalysts. These materials are synthesized from three main precursors – metal salt, nitrogen-containing precursor, and carbon precursor. One example for that is the work published by Zelenay et al. [91] who took cyanamide as the nitrogen precursor, $FeSO_4{\cdot}7H_2O$ as the metal precursor, and an activated carbon powder. These three were mixed together and heat treated at 1100 °C in nitrogen atmosphere. Although highly active, and although they show relatively (to MCs) good stability, heat treated PGM-free ORR catalysts are notoriously known for their structural uncertainty. The actual catalytic center and its exact surroundings are unknown in these catalytic systems. For this reason, in some cases, the use of MCs comes to simplify systems so that each parameter could be detected and thoroughly analyzed [92–95].

Acronyms

AEMFC	alkaline membrane electrolyte fuel cell
CNT	carbon nanotube
HOMO	highest occupied molecular orbital
LUMO	lowest unoccupied molecular orbital
MC	molecular catalyst
MOF	metal–organic frameworks
ORR	oxygen reduction reaction
PEFC	polymer electrolyte fuel cell
PGM	platinum group metal
SOHMO	singly occupied highest molecular orbital
TOF	turn-over frequencies

References

1 Jasinski, R. (1964). A new fuel cell cathode catalyst. *Nature* 201 (4925): 1212–1213.

2 Ramaswamy, N. and Mukerjee, S. (2012). Fundamental mechanistic understanding of electrocatalysis of oxygen reduction on Pt and non-Pt surfaces: acid versus alkaline media. *Advances in Physical Chemistry* 2012: 491604.

3 Venegas, R., Muñoz-Becerra, K., Candia-Onfray, C. et al. (2020). Experimental reactivity descriptors of M-N-C catalysts for the oxygen reduction reaction. *Electrochimica Acta* 332: 135340.

4 Tang, W., Qiu, Y., Li, X. et al. pH-dependent electrochemically catalyzed oxygen reduction behaviors of o-substituted Co(III) corroles. *Macroheterocycles* 13 (2): 156–162.

5 Chatterjee, S., Sengupta, K., Mondal, B. et al. (2017). Factors determining the rate and selectivity of $4e^-/4H^+$ electrocatalytic reduction of dioxygen by iron porphyrin complexes. *Accounts of Chemical Research* 50 (7): 1744–1753.

6 Dai, L., Xue, Y., Qu, L. et al. (2015). Metal-free catalysts for oxygen reduction reaction. *Chemical Reviews* 115 (11): 4823–4892.

7 Wang, L., Ambrosi, A., and Pumera, M. (2013). "Metal-free" catalytic oxygen reduction reaction on heteroatom-doped graphene is caused by trace metal impurities. *Angewandte Chemie* 125 (51): 14063–14066.

8 Masa, J., Zhao, A., Xia, W. et al. (2013). Trace metal residues promote the activity of supposedly metal-free nitrogen-modified carbon catalysts for the oxygen reduction reaction. *Electrochemistry Communications* 34: 113–116.

9 Wang, W., Jia, Q., Mukerjee, S., and Chen, S. (2019). Recent insights into the oxygen-reduction electrocatalysis of Fe/N/C materials. *ACS Catalysis* 9 (11): 10126–10141.

10 Blomberg, M.R. (2016). Mechanism of oxygen reduction in cytochrome c oxidase and the role of the active site tyrosine. *Biochemistry* 55 (3): 489–500.

11 Shpilman, J., Friedman, A., Zion, N. et al. (2019). A combined experimental and theoretical study of cobalt corroles as catalysts for oxygen reduction reaction. *The Journal of Physical Chemistry C* 123 (50): 30129–30136.

12 Chen, K., Liu, K., An, P. et al. (2020). Iron phthalocyanine with coordination induced electronic localization to boost oxygen reduction reaction. *Nature Communications* 11 (1): 4173.

13 Elbaz, L., Korin, E., Soifer, L., and Bettelheim, A. (2008). Electrocatalytic oxygen reduction by Co(III) porphyrins incorporated in aerogel carbon electrodes. *Journal of Electroanalytical Chemistry* 621 (1): 91–96.

14 Kosa, M., Levy, N., Elbaz, L., and Major, D.T. (2018). Theoretical study of the electrocatalytic reduction of oxygen by metallocorroles. *The Journal of Physical Chemistry C* 122 (31): 17686–17694.

15 Levy, N., Mahammed, A., Kosa, M. et al. (2015). Metallocorroles as nonprecious-metal catalysts for oxygen reduction. *Angewandte Chemie International Edition* 54 (47): 14080–14084.

16 Ramdhanie, B., Telser, J., Caneschi, A. et al. (2004). An example of O_2 binding in a cobalt(II) corrole system and high-valent cobalt–cyano and cobalt–alkynyl complexes. *Journal of the American Chemical Society* 126 (8): 2515–2525.

17 Orellana, W. (2011). Catalytic activity toward oxygen reduction of transition metal porphyrins covalently linked to single-walled carbon nanotubes: a density functional study. *Physical Review B* 84 (15): 155405.

18 Levy, N., Mahammed, A., Friedman, A. et al. (2016). Metallocorroles as non-precious metal electrocatalysts for highly efficient oxygen reduction in alkaline media. *ChemCatChem* 8 (17): 2832–2837.

19 Bazanov, M.I., Berezina, N.M., Karimov, D.R., and Berezin, D.B. (2012). Electrochemical and electrocatalytic properties of meso-triphenylcorrole and its complexes with Mn(III), Co(III), Cu(III), and Zn(II). *Russian Journal of Electrochemistry* 48 (9): 905–910.

20 Li, Y., Li, Q., Wang, H. et al. (2019). Recent progresses in oxygen reduction reaction electrocatalysts for electrochemical energy applications. *Electrochemical Energy Reviews* 2 (4): 518–538.

21 He, Q., Mugadza, T., Hwang, G., and Nyokong, T. (2012). Mechanisms of electrocatalysis of oxygen reduction by metal porphyrins in trifluoromethane sulfonic acid solution. *International Journal of Electrochemical Science* 7.

22 Pegis, M.L., Wise, C.F., Martin, D.J., and Mayer, J.M. (2018). Oxygen reduction by homogeneous molecular catalysts and electrocatalysts. *Chemical Reviews* 118 (5): 2340–2391.

23 Shi, Z. and Zhang, J. (2007). Density functional theory study of transitional metal macrocyclic complexes' dioxygen-binding abilities and their catalytic activities toward oxygen reduction reaction. *The Journal of Physical Chemistry C* 111 (19): 7084–7090.

24 Levy, N., Shpilman, J.S., Honig, H.C. et al. (2017). Surprising substituent effect in corroles on the electrochemical activation of oxygen reduction. *Chemical Communications* 53 (96): 12942–12945.

25 Nguyen, K.A., Day, P.N., Pachter, R. et al. (2002). Analysis of absorption spectra of zinc porphyrin, zinc meso-tetraphenylporphyrin, and halogenated derivatives. *The Journal of Physical Chemistry A* 106 (43): 10285–10293.

26 Mahammed, A., Mondal, B., Rana, A. et al. (2014). The cobalt corrole catalyzed hydrogen evolution reaction: surprising electronic effects and characterization of key reaction intermediates. *Chemical Communications* 50 (21): 2725–2727.

27 Abarca, G., Viera, M., Aliaga, C. et al. (2019). In search of the most active MN4 catalyst for the oxygen reduction reaction. The case of perfluorinated Fe phthalocyanine. *Journal of Materials Chemistry A* 7 (43): 24776–24783.

28 Govan, J., Abarca, G., Aliaga, C. et al. (2020). Influence of cyano substituents on the electron density and catalytic activity towards the oxygen reduction reaction for iron phthalocyanine. The case for Fe(II) 2,3,9,10,16,17,23,24-octa (cyano)phthalocyanine. *Electrochemistry Communications* 118: 106784.

29 Amanullah, S., Das, P.K., Samanta, S., and Dey, A. (2015). Tuning the thermodynamic onset potential of electrocatalytic O_2 reduction reaction by synthetic iron-porphyrin complexes. *Chemical Communications* 51 (49): 10010–10013.

30 Honig, H., Friedman, A., Zion, N., and Elbaz, L. (2020). Enhancement of the oxygen reduction reaction electrocatalytic activity of metallo-corroles using contracted cobalt(III) CF_3-corrole incorporated in high surface area carbon support. *Chemical Communications* 56 (61): 8627–8630.

31 Honig, H.C., Krishnamurthy, C.B., Borge-Durán, I. et al. (2019). Structural and physical parameters controlling the oxygen reduction reaction selectivity with carboxylic acid-substituted cobalt corroles incorporated in a porous carbon support. *The Journal of Physical Chemistry C* 123 (43): 26351–26357.

32 Dogutan, D.K., Stoian, S.A., McGuire, R. Jr., et al. (2011). Hangman corroles: efficient synthesis and oxygen reaction chemistry. *Journal of the American Chemical Society* 133 (1): 131–140.

33 Baran, J.D., Grönbeck, H., and Hellman, A. (2013). Analysis of porphyrines as catalysts for electrochemical reduction of O_2 and oxidation of H_2O. *Journal of the American Chemical Society* 136 (4): 1320–1326.

34 Tran, T.T.H., Chen, G.L., Hoang, T.K. et al. (2017). Effect of imidazole on the electrochemistry of zinc porphyrins: an electrochemical and computational study. *The Journal of Physical Chemistry A* 121 (37): 6925–6931.

35 Snitkoff, R.Z., Levy, N., Ozery, I. et al. (2019). Imidazole decorated reduced graphene oxide: a biomimetic ligand for selective oxygen reduction electrocatalysis with metalloporphyrins. *Carbon* 143: 223–229.

36 Elbaz, L., Korin, E., Soifer, L., and Bettelheim, A. (2010). Mediation at high potentials for the reduction of oxygen to water by cobalt porphyrin-quinone systems in porous aerogel carbon electrodes. *Journal of the Electrochemical Society* 157 (1): B27–B31.

37 Elbaz, L., Korin, E., Soifer, L., and Bettelheim, A. (2010). Evidence for the formation of cobalt porphyrin-quinone complexes stabilized at carbon-based surfaces toward the design of efficient non-noble-metal oxygen reduction catalysts. *Journal of Physical Chemistry Letters* 1 (1): 398–401.

38 Das, P.K. and Dey, A. (2014). Resonance Raman, electron paramagnetic resonance, and density functional theory calculations of a phenolate-bound iron porphyrin complex: electrostatic versus covalent contribution to bonding. *Inorganic Chemistry* 53 (14): 7361–7370.

39 Collman, J.P., Devaraj, N.K., Decréau, R.A. et al. (2007). A cytochrome c oxidase model catalyzes oxygen to water reduction under rate-limiting electron flux. *Science* 315 (5818): 1565.

40 Samanta, S., Das, P.K., Chatterjee, S. et al. (2013). O_2 reduction reaction by biologically relevant anionic ligand bound iron porphyrin complexes. *Inorganic Chemistry* 52 (22): 12963–12971.

41 Samanta, S., Das, P.K., Chatterjee, S., and Dey, A. (2015). Effect of axial ligands on electronic structure and O_2 reduction by iron porphyrin complexes: towards a quantitative understanding of the "push effect". *Journal of Porphyrins and Phthalocyanines* 19 (01n03): 92–108.

42 Pizarro, A., Abarca, G., Gutiérrez-Cerón, C. et al. (2018). Building pyridinium molecular wires as axial ligands for tuning the electrocatalytic activity of iron phthalocyanines for the oxygen reduction reaction. *ACS Catalysis* 8 (9): 8406–8419.

43 Ponce, I., Silva, J.F., Oñate, R. et al. (2012). Enhancement of the catalytic activity of Fe phthalocyanine for the reduction of O_2 anchored to Au(111) via conjugated self-assembled monolayers of aromatic thiols as compared to Cu phthalocyanine. *The Journal of Physical Chemistry C* 116 (29): 15329–15341.

44 Venegas, R., Recio, F.J., Riquelme, J. et al. (2017). Biomimetic reduction of O_2 in an acid medium on iron phthalocyanines axially coordinated to pyridine anchored on carbon nanotubes. *Journal of Materials Chemistry A* 5 (24): 12054–12059.

45 Cao, R., Thapa, R., Kim, H. et al. (2013). Promotion of oxygen reduction by a bio-inspired tethered iron phthalocyanine carbon nanotube-based catalyst. *Nature Communications* 4: 2076.

46 Kumar, S., Gonen, S., Friedman, A. et al. (2017). Doping and reduction of graphene oxide using chitosan-derived volatile N-heterocyclic compounds for metal-free oxygen reduction reaction. *Carbon* 120: 419–426.

47 Wei, P.-J., Yu, G.Q., Naruta, Y., and Liu, J.G. (2014). Covalent grafting of carbon nanotubes with a biomimetic heme model compound to enhance oxygen reduction reactions. *Angewandte Chemie International Edition* 53 (26): 6659–6663.

48 Tang, H., Yin, H., Wang, J. et al. (2013). Molecular architecture of cobalt porphyrin multilayers on reduced graphene oxide sheets for high-performance oxygen reduction reaction. *Angewandte Chemie International Edition* 52 (21): 5585–5589.

49 Wang, Z., Lei, H., Cao, R., and Zhang, M. (2015). Cobalt corrole on carbon nanotube as a synergistic catalyst for oxygen reduction reaction in acid media. *Electrochimica Acta* 171: 81–88.

50 Ni, C.L. and Anson, F.C. (1985). Relation between the potentials where adsorbed and unadsorbed cobalt(III) tetrakis (N-methylpyridinium-4-yl) porphyrin is reduced and those where it catalyzes the electroreduction of dioxygen. *Inorganic Chemistry* 24 (26): 4754–4756.

51 Elbaz, L., Korin, E., Soifer, L., and Bettelheim, A. (2009). Evidence for the formation of cobalt porphyrin–quinone complexes stabilized at carbon-based surfaces toward the design of efficient non-noble-metal oxygen reduction catalysts. *The Journal of Physical Chemistry Letters* 1 (1): 398–401.

52 Gasteiger, H.A. and Markovic, N.M. (2009). Just a dream-or future reality? *Science* 324 (5923): 48–49.

53 Wagner, F.T., Gasteiger, H., and Yan, S. (2003). What performance would non-Pt cathode catalysts need to achieve to be practical for transportation? *DOE Workshop on Non-Platinum Electrocatalysts,* New Orleans, US Department of Energy, 21 March 2003.

54 Boulatov, R., Collman, J.P., Shiryaeva, I.M., and Sunderland, C.J. (2002). Functional analogues of the dioxygen reduction site in cytochrome oxidase: mechanistic aspects and possible effects of CuB. *Journal of the American Chemical Society* 124 (40): 11923–11935.

55 Dey, S., Mondal, B., Chatterjee, S. et al. (2017). Molecular electrocatalysts for the oxygen reduction reaction. *Nature Reviews Chemistry* 1 (12): 0098.

56 Ferguson-Miller, S. and Babcock, G.T. (1996). Heme/Copper terminal oxidases. *Chemical Reviews* 96 (7): 2889–2908.

57 Kaila, V.R.I., Verkhovsky, M.I., and Wikström, M. (2010). Proton-coupled electron transfer in cytochrome oxidase. *Chemical Reviews* 110 (12): 7062–7081.

58 Wikström, M., Sharma, V., Kaila, V.R. et al. (2015). New perspectives on proton pumping in cellular respiration. *Chemical Reviews* 115 (5): 2196–2221.

59 Solomon, E.I., Heppner, D.E., Johnston, E.M. et al. (2014). Copper active sites in biology. *Chemical Reviews* 114 (7): 3659–3853.

60 Solomon, E.I., Sundaram, U.M., and Machonkin, T.E. (1996). Multicopper oxidases and oxygenases. *Chemical Reviews* 96 (7): 2563–2606.

61 Ramasamy, R.P., Luckarift, H.R., Ivnitski, D.M. et al. (2010). High electrocatalytic activity of tethered multicopper oxidase–carbon nanotube conjugates. *Chemical Communications* 46 (33): 6045–6047.

62 Cracknell, J.A. and Blanford, C.F. (2012). Developing the mechanism of dioxygen reduction catalyzed by multicopper oxidases using protein film electrochemistry. *Chemical Science* 3 (5): 1567–1581.

63 Dos Santos, L., Climent, V., Blanford, C.F., and Armstrong, F.A. (2010). Mechanistic studies of the 'blue' Cu enzyme, bilirubin oxidase, as a highly efficient electrocatalyst for the oxygen reduction reaction. *Physical Chemistry Chemical Physics* 12 (42): 13962–13974.

64 Blanford, C.F., Heath, R.S., and Armstrong, F.A. (2007). A stable electrode for high-potential, electrocatalytic O_2 reduction based on rational attachment of a blue copper oxidase to a graphite surface. *Chemical Communications* 17: 1710–1712.

65 Haas, A.S., Pilloud, D.L., Reddy, K.S. et al. (2001). Cytochrome c and cytochrome c oxidase: monolayer assemblies and catalysis. *The Journal of Physical Chemistry B* 105 (45): 11351–11362.

66 Burgess, J.D., Rhoten, M.C., and Hawkridge, F.M. (1998). Cytochrome c oxidase immobilized in stable supported lipid bilayer membranes. *Langmuir* 14 (9): 2467–2475.

67 Su, L., Kelly, J., and Hawkridge, F.M. (2007). Electroreduction of O_2 on cytochrome c oxidase modified electrode for biofuel cell. *ECS Transactions* 2 (11): 1.

68 Mukherjee, S., Mukherjee, A., Bhagi-Damodaran, A. et al. (2015). A biosynthetic model of cytochrome c oxidase as an electrocatalyst for oxygen reduction. *Nature Communications* 6 (1): 1–9.

69 Collman, J.P., Wagenknecht, P.S., and Hutchison, J.E. (1994). Molecular catalysts for multielectron redox reactions of small molecules: the "cofacial metallodiporphyrin" approach. *Angewandte Chemie International Edition in English* 33 (15–16): 1537–1554.

70 Chang, C.K., Liu, H.Y., and Abdalmuhdi, I. (1984). Electroreduction of oxygen by pillared cobalt(II) cofacial diporphyrin catalysts. *Journal of the American Chemical Society* 106 (9): 2725–2726.

71 Rosenthal, J. and Nocera, D.G. (2007). Role of proton-coupled electron transfer in O–O bond activation. *Accounts of Chemical Research* 40 (7): 543–553.

72 Loh, Z.-H., Miller, S.E., Chang, C.J. et al. (2002). Excited-state dynamics of cofacial pacman porphyrins. *The Journal of Physical Chemistry A* 106 (48): 11700–11708.

73 Le Mest, Y., Inisan, C., Laouénan, A. et al. (1997). Reactivity toward dioxygen of dicobalt face-to-face diporphyrins in aprotic media. Experimental and theoretical aspects. Possible mechanistic implication in the reduction of dioxygen. *Journal of the American Chemical Society* 119 (26): 6095–6106.

74 Liu, H.Y., Abdalmuhdi, I., Chang, C.K., and Anson, F.C. (1985). Catalysis of the electroreduction of dioxygen and hydrogen peroxide by an anthracene-linked dimeric cobalt porphyrin. *The Journal of Physical Chemistry* 89 (4): 665–670.

75 Peljo, P., Murtomäki, L., Kallio, T. et al. (2012). Biomimetic oxygen reduction by cofacial porphyrins at a liquid–liquid interface. *Journal of the American Chemical Society* 134 (13): 5974–5984.

76 Masa, J., Ozoemena, K.I., Schuhmann, W., and Zagal, J.H. (2013). Fundamental studies on the electrocatalytic properties of metal macrocyclics and other complexes for the electroreduction of O_2. In: *Electrocatalysis in Fuel Cells* (ed. M. Shao), 157–212. London: Springer.

77 Fukuzumi, S., Yamada, Y., and Karlin, K.D. (2012). Hydrogen peroxide as a sustainable energy carrier: electrocatalytic production of hydrogen peroxide and the fuel cell. *Electrochimica Acta* 82: 493–511.

78 Gonen, S. and Elbaz, L. (2018). Comparison of new metal organic framework-based catalysts for oxygen reduction reaction. *Data in Brief* 19: 281–287.

79 Gonen, S. and Elbaz, L. (2018). Metal organic frameworks as catalysts for oxygen reduction. *Current Opinion in Electrochemistry* 9: 179–188.

80 Gonen, S., Lori, O., Fleker, O., and Elbaz, L. (2019). Electrocatalytically active silver organic framework: Ag(I)-complex incorporated in activated carbon. *ChemCatChem* 11 (24): 6124–6130.

81 Zhao, C.-X., Li, B.Q., Liu, J.N. et al. (2019). Transition metal coordinated framework porphyrin for electrocatalytic oxygen reduction. *Chinese Chemical Letters* 30 (4): 911–914.

82 Macor, K.A. and Spiro, T.G. (1984). Oxidative electrochemistry of electropolymerized metalloprotoporphyrin films. *Journal of Electroanalytical Chemistry and Interfacial Electrochemistry* 163 (1): 223–236.

83 Bettelheim, A., White, B., and Murray, R.W. (1987). Electrocatalysis of dioxygen reduction in aqueous acid and base by multimolecular layer films of electropolymerized cobalt tetra(*o*-aminophenyl) porphyrin. *Journal of Electroanalytical Chemistry and Interfacial Electrochemistry* 217 (2): 271–286.

84 Yuan, S., Shui, J.L., Grabstanowicz, L. et al. (2013). A highly active and support-free oxygen reduction catalyst prepared from ultrahigh-surface-area porous polyporphyrin. *Angewandte Chemie* 125 (32): 8507–8511.

85 Friedman, A., Landau, L., Gonen, S. et al. (2018). Efficient bio-inspired oxygen reduction electrocatalysis with electropolymerized cobalt corroles. *ACS Catalysis* 8 (6): 5024–5031.

86 Friedman, A., Saltsman, I., Gross, Z., and Elbaz, L. (2019). Electropolymerization of PGM-free molecular catalyst for formation of 3D structures with high density of catalytic sites. *Electrochimica Acta* 310: 13–19.

87 Zion, N., Cullen, D.A., Zelenay, P., and Elbaz, L. (2020). Heat-treated aerogel as a catalyst for the oxygen reduction reaction. *Angewandte Chemie International Edition* 59 (6): 2483–2489.

88 Levy, N., Lori, O., Gonen, S. et al. (2020). The relationship of morphology and catalytic activity: a case study of iron corrole incorporated in high surface area carbon supports. *Carbon* 158: 238–243.

89 Jahnke, H., Schoenborn, M., and Zimmermann, G. (1974). Transition metal chelates as catalysts for fuel cell reactions. *Journal of Electrochemical Society* 121: 303–318.

90 Van Veen, J.R., van Baar, J.F., and Kroese, K.J. (1981). Effect of heat treatment on the performance of carbon-supported transition-metal chelates in the electrochemical reduction of oxygen. *Journal of the Chemical Society, Faraday Transactions 1: Physical Chemistry in Condensed Phases* 77 (11): 2827–2843.

91 Chung, H.T., Johnston, C.M., and Zelenay, P. (2009). Synthesis and evaluation of heat-treated, cyanamide-derived non-precious catalysts for oxygen reduction. *ECS Transactions* 25 (1): 485.

92 Chang, S.-T., Huang, H.C., Wang, H.C. et al. (2014). Effects of structures of pyrolyzed corrin, corrole and porphyrin on oxygen reduction reaction. *International Journal of Hydrogen Energy* 39 (2): 934–941.

93 Elbaz, L., Wu, G., and Zelenay, P. (2013). Heat-treated non-precious-metal-based catalysts for oxygen reduction. In: *Electrocatalysis in Fuel Cells* (ed. M. Shao), 213–246. London: Springer.

94 Widelöv, A. and Larsson, R. (1992). ESCA and electrochemical studies on pyrolysed iron and cobalt tetraphenylporphyrins. *Electrochimica Acta* 37 (2): 187–197.

95 Bezerra, C.W.B., Zhang, L., Lee, K. et al. (2008). A review of Fe–N/C and Co–N/C catalysts for the oxygen reduction reaction. *Electrochimica Acta* 53 (15): 4937–4951.

7

Recent Advances in Electrocatalysts for Hydrogen Oxidation Reaction in Alkaline Electrolytes

Indra N. Pulidindi[1] *and Meital Shviro*[2]

[1] *GSFC University, School of Science, Vigyan Bhavan, Fertilizernagar, Vadodara 391750, India*
[2] *National Renewable Energy Laboratory (NREL), Chemistry and Nanoscience Center, 15013 Denver West Parkway, Golden, CO 80401, USA*

7.1 Introduction

Anion-exchange membrane fuel cells (AEMFCs), as the next generation of cost-effective fuel cells, have attracted significant interest for more than a decade. The main advantage of the AEMFC technology is to overcome the cost of polymer electrolyte membrane fuel cells (PEMFCs) [1]. The alkaline environment allows the use of low-cost metal hardware and the design of less-expensive and non-platinum group metal (non-PGM)-based electrocatalysts. The main challenge in AEMFC is the slow kinetics of the hydrogen oxidation reaction (HOR). In alkaline media, the overall HOR process is represented as follows:

$$H_2 + 2OH^- \rightarrow 2H_2O + 2e^- \tag{7.1}$$

Three elementary steps are involved in the HOR mechanism:

1. Chemical dissociative adsorption of H_2 without electron transfer leads to two adsorbed H species (2H*). This reaction is called the Tafel reaction:

$$H_2 + 2^* \rightleftharpoons 2^*H_{ad} \tag{7.2}$$

2. Electrochemical dissociative adsorption of H_2 molecules resulting in the formation of an adsorbed hydrogen on the electrode surface is called the Heyrovsky step:

$$H_2 + {}^* + OH^- \rightleftharpoons {}^* - H_{ad} + e^- + H_2O \tag{7.3}$$

3. Adsorbed hydrogen atom combining with an OH^- leading to the formation of water and one electron is called the Volmer step:

$${}^* - H_{ad} + OH^- \rightleftharpoons {}^* + H_2O + e^- \tag{7.4}$$

Electrocatalysis for Membrane Fuel Cells: Methods, Modeling, and Applications, First Edition.
Edited by Nicolas Alonso-Vante and Vito Di Noto.

The symbol, "*" refers to the catalytic site needed for the binding of H_2 at the electrode surface.

Due to the involvement of OH^- in the Heyrovsky and Volmer steps, fundamental understanding of the HOR pathway in alkaline media is an issue of considerable debate. The opinions can be classified into two categories, HOR pathway with OH^- in the solution [1–6] and HOR pathway with OH_{ad} on the surface of the electrocatalysts (formed by OH^- and $OH_{ad} + e^-$) [1, 7–9].

In this chapter, we focus on the hydrogen oxidation reaction in alkaline conditions. One of the main efforts is to fulfill the potential of AEMFC by designing electrocatalysts with high performance, durability, and stability. Understanding the HOR mechanism, d-band character, size, morphology, alloying, and surface modification allows us to design the next-generation electrocatalysts for HOR [10].

7.2 Mechanism of the HOR in Alkaline Media

Strmcnik et al. [7] pioneered the fundamental HOR studies and proposed that the HOR kinetics, in an alkaline medium, is limited by the Volmer step $M-H_{ads} + OH^- \Rightarrow M + H_2O + e^-$. Reactive hydroxy species (OH_{ads}) is required to promote this step. A three-dimensional volcano plot was proposed wherein the HOR rate is codetermined by the binding of the hydrogen intermediate (H_{ads}) as well as of the OH_{ads} to the surface [11]. Zhao et al. [12] used first-principles calculations for understanding the mechanism of HOR on PtNi alloy catalysts. The enhanced HOR activity of platinum in alkaline by decorating it with nickel was studied to gain more insights into the HOR mechanism, $Pt_3Ni(111)$, $Pt_2Ni_2(111)$, and $PtNi_3(111)$, and its surface alloy $PtNi_{surf}(111)$ first-principles calculations were conducted. Free energy profiles suggested that the HOR on PtNi alloys proceeds through the **Tafel–Volmer mechanism** wherein gaseous H_2 is dissociatively adsorbed on the catalyst surface to yield two adsorbed H species (Tafel reaction), followed by its reaction with OH^- in the electrolyte (the rate-determining step) to form H_2O (Volmer reaction) [13]. The HOR activity of PtNi alloy catalyst is exclusively influenced by the adsorption of hydrogen rather than hydroxyl species, even though the oxophilic behavior of platinum is enhanced by alloying with nickel. The optimal catalytic active site for HOR should possess a moderate H adsorption free energy (ΔG_{H*}) of ~0.414 eV at a pH value of 13.

Alloying platinum with nickel elevated the d-band center (ε_d) from −2.61 to −2.08 eV, Figure 7.1, resulting in a ΔG_{H*} value close to 0.414 eV. As a result, the free energy barrier (E_a) for the rate-determining Volmer reaction (adsorbed H* reacting with OH^- of the electrolyte) lowered and leads to the highest HOR activity observed on the bulk alloy catalytic site of $PtNi_3(111)$. The corresponding values indicative of the HOR activity, namely HOR rate (k), current density (i_0), and transfer coefficient (α), are $9.42 \times 10^3\ s^{-1}$ per site, $1.42\ mA\ cm^{-2}$, and 0.07, respectively. These results provide theoretical guidance for designing electrocatalysts with improved activity for HOR in alkaline medium.

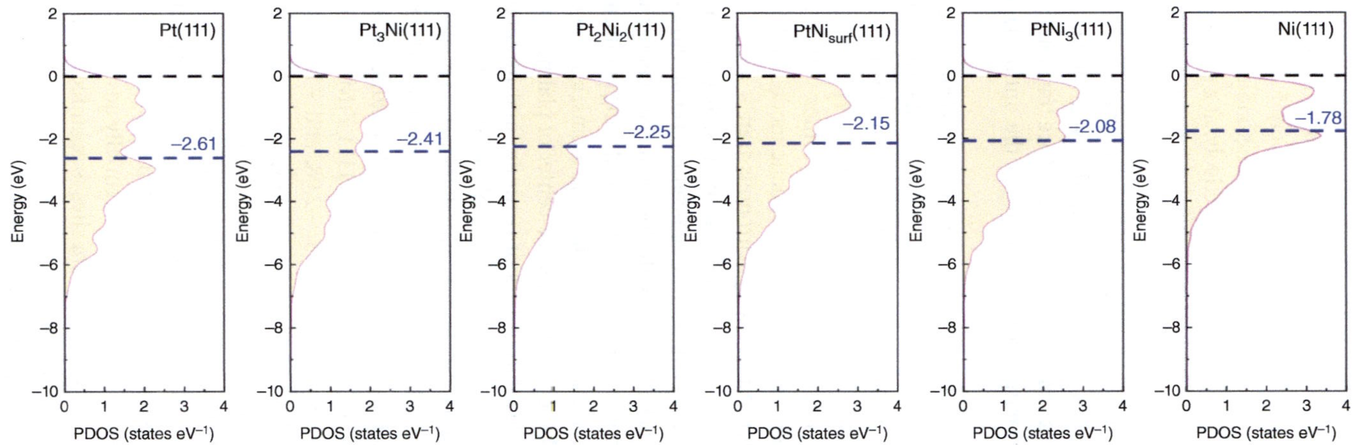

Figure 7.1 Projected density of states for the d-band of metal atoms around the FCC site for platinum, nickel, and PtNi alloys. The Fermi level marked with black dashed lines is set as the energy zero. Blue dashed lines show the d-band centers (in eV). The shadows indicate the integrating interval from the minimum energy to the Fermi level. Source: Reproduced from Ref. [12] with permission of the American Chemical Society.

In the alkaline environment, strongly oxophilic nickel will be oxidized and dissolve in the electrolyte medium as $Ni(OH)_2$. To understand the antioxidant capacity of PtNi alloys, the oxidation potential ($P_{oxidation}$) of nickel on the surface in pure Ni and PtNi alloys were calculated. The oxidation potential of nickel for Ni(111), $PtNi_3(111)$, $Pt_2Ni_2(111)$, $PtNi_{surf}(111)$, and $Pt_3Ni(111)$ were found to be 1.47, 1.05, 0.92, 0.81, and 0.74 V *vs.* SHE, respectively. The corresponding values relative to RHE were 0.70, 0.28, 0.16, 0.04, and −0.02 V, respectively, at a pH value of 13. With increasing platinum content, the oxidation potential decreases, and the corresponding antioxidant capacity increases for PtNi alloys. This is a result of the transfer of electrons from nickel to platinum in the alloys. Compared to pure nickel, the nickel atoms in the PtNi alloys have a reduced ability to provide electrons to oxygen-containing species, as some electrons on nickel are transferred to platinum. As a result, alloying nickel with platinum improves the antioxidant capacity of nickel.

Zhao et al. [12] proposed the HOR mechanism on the PtNi alloy catalyst, selected as a prototype of the highly active Pt-transition metal catalysts, for HOR in an alkaline medium at pH 13. pH 13 was chosen for modeling studies based on the experimental conditions (0.1 M KOH). The mechanism proposed is based on the first-principles DFT study yielding full free energy profiles with all the reaction barriers, hydrogen adsorption free energy, microkinetic model, and Tafel plot. The linear synchronous transition/quadratic synchronous transit method was used to locate the elementary reaction pathways and transition state (TS) and was further authenticated by the frequency calculations. Face-centered cubic close-packed (111) surface was modeled as this place possessed the lowest surface energy and is also the most abundant surface on most metal particles. Upon alloying platinum with nickel, some electrons transfer from nickel atoms to platinum atoms, leading to the positive charge center of nickel and the negative charge center of platinum atoms. This is due to the relative electronegativity values of platinum and nickel atoms, which are 2.28 and 1.91, respectively. Such formation of positive and negative charge centers is further revealed in the electron density-deformation maps of PtNi alloys. To understand the hydrogen adsorption activity of the metal surfaces, the d-band center (ε_d) of the surface metal atoms was calculated. The values of d-band centers (ε_d) of the metal surfaces Pt(111), $Pt_3Ni(111)$, $Pt_2Ni_2(111)$, $PtNi_{surf}(111)$, $PtNi_3(111)$, and Ni(111) were −2.61, −2.41, −2.25, −2.15, −2.08, and −1.78 eV, respectively. With an increase in the nickel content in the PtNi alloy, the d-band center shifted up toward the Fermi level that corresponds to the enhancement of adsorption activity of the metal surfaces with increasing content of nickel, Figure 7.1.

The H_2 molecule tends to horizontally adsorb on the top of the platinum in the PtNi alloy surfaces. The atop platinum on the PtNi alloy surface provides a stable adsorption site for H_2 while the H_2O species preferentially adsorb on the top of the nickel site. Adsorption of H_2 on atop platinum sites is regarded as physisorption as revealed from the weak adsorption energy (−0.19 to −0.37 eV) and by the elongation of bond in hydrogen in the adsorbed state (0.749–0.918 Å [H_2*] *vs.* 0.748 Å [free H_2]).

By increasing the platinum content, the interaction of H_2 with the alloy further strengthened. Adsorption of hydrogen on PtNi alloy surfaces is stronger than that on pure Pt(111) with an adsorption energy of −0.45 eV but weaker than that on Ni(111) wherein the adsorption energy is −0.72 eV. The hydrogen adsorption strength is dependent on the d-band center, ε_d, of the metal surfaces by a linear relationship, namely: $E_{ads} = -2.158 \times \varepsilon_d - 3.497$, $R^2 = 0.921\,53$. Ni(111) has a d-band center at relatively high energy ($\varepsilon_d = -1.78$ eV) compared to the d-band center of Pt(111) ($\varepsilon_d = -2.61$ eV) and, as a result, contains more empty antibonding states above the Fermi level resulting in a stronger bond with H, Figure 7.1. The d-band center (ε_d) of PtNi alloy is located between Pt(111) and Ni(111) shifted toward the Fermi level with increasing nickel content. As a result, the adsorption strength of H increases in the order: Pt(111) < Pt_3Ni(111) < Pt_2Ni_2(111) < $PtNi_{surf}$(111) < $PtNi_3$(111) < Ni(111). The most stable configuration of the first hydrogen adsorption is at the fcc site on the PtNi alloy surface. The most stable adsorption site for the second hydrogen is on a similar fcc site adjacent to the first adsorbed hydrogen. The second hydrogen is adsorbed at the fcc (NiNiNi) on the $PtNi_3$(111) surface. The adsorption energy of the second hydrogen is almost equal to the adsorption energy of the first hydrogen. The adsorption of the first hydrogen will have only a negligible effect on the adsorption of the second hydrogen atom. Even the defective surfaces, namely def-$PtNi_3$(111), prefer to adsorb H_2O and H_2 on the top nickel and platinum atoms, respectively.

At the anode in electrolyte membrane fuel cells, the activation of water leads to the formation of various surface intermediates, which are described as different surface phases. Different surface phases, namely hydrogen-covered surface, oxygen-covered surface, hydroxyl-covered surface, and water-covered surface, on different PtNi alloy surfaces were considered. The most stable surface phase in an aqueous environment can be described by the phase diagram depicting the change in free energy (ΔG^{water}) as a function of the electrode potential (E), as shown in Figure 7.2.

The free energy changes of the reactions that connect different surface phase Equations (7.5)–(7.7) were calculated using Eq. (7.8). The free energy change values (ΔG^{water}) were plotted as a function of the whole electrode potential (E) in the phase diagrams shown in Figure 7.2.

$$H_2O\,(\text{liquid}) + * \leftrightharpoons OH^* + e^- + H^+ \tag{7.5}$$

$$OH^* + * \rightleftharpoons O^* + e^- + H^+ \tag{7.6}$$

$$* + e^- + H^+ \rightleftharpoons H^* \tag{7.7}$$

$$\Delta G = \Delta E + \Delta ZPE - T\Delta S + \Delta G_E + \Delta G_{pH} \tag{7.8}$$

The change of free energy from the initial states (ISs) to final states (FSs) used for plotting the free-energy diagrams of the HOR is calculated from Eq. (7.8) wherein ΔE is the change in energy, and ΔZPE is the change in the zero-point energy. Zero-point energies and entropies were obtained from vibrational frequency computation at $T = 298$ K. ΔS is the change in entropy from the initial and final states. ΔG_E is the

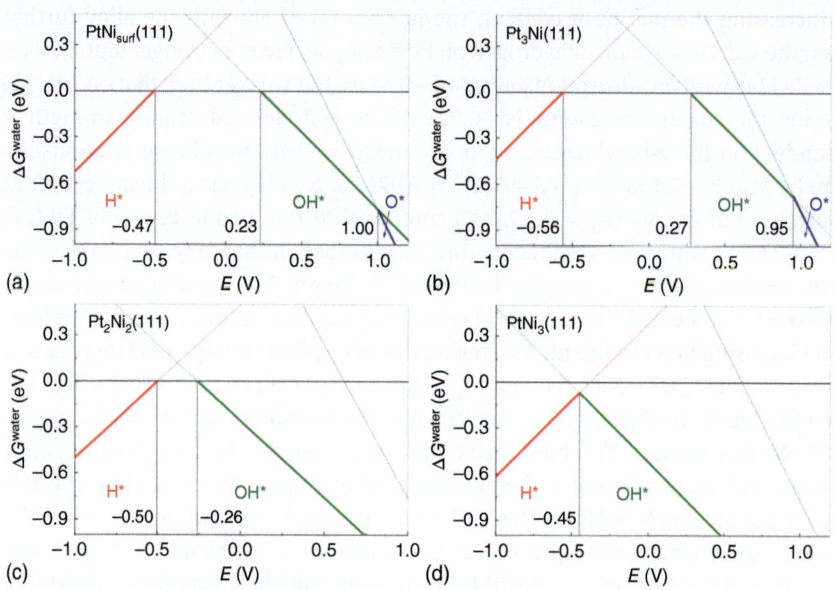

Figure 7.2 Phase diagram showing the free energy change for water in contact with different surfaces, (a) $PtNi_{surf}(111)$, (b) $Pt_3Ni(111)$, (c) $Pt_2Ni_2(111)$, and (d) $PtNi_3(111)$, as a function of whole electrode potential E. The free energy for liquid water at 298 K is defined as $\Delta G = 0$. $\Delta G_{H^*}^{Water}(E) = \Delta G_{H^*}^{Water}(0) + eE$; $\Delta G_{OH^*}^{Water}(E) = \Delta G_{OH^*}^{Water} - eE$; $\Delta G_{O^*}^{Water}(E) = \Delta G_{O^*}^{Water}(0) - 2eE$. Source: Reproduced from Ref. [12] with permission of the American Chemical Society.

bias effect on the free energy of the states involving an electron in the electrode $\Delta G_E = -neE$, where n is the number of the transferred electrons (e), and E is the whole electrode potential including the polarization effect (*vs.* SHE). U is set to be 0.059 V (*vs.* SHE), and ΔG_{pH} is the correction for the free energy of the H^+ ion at $pH \neq 0$. $\Delta G_{pH} = k_b T \cdot \ln 10 \cdot pH$, where k_b is the Boltzmann constant and pH = 13. In the phase diagram, Figure 7.2 shows the change in the free energy of water in contact with different surfaces as a function of whole electrode potential E, at 298 K and at a pH of 13. The lowest line corresponds to the thermodynamically most stable phase. The crossings of two bottom lines represent a phase change. At very negative potential, only hydrogen is stably adsorbed on all alloyed surfaces. Likewise, the surfaces of the alloy catalyst are covered with oxygen at high positive potentials. Upon alloying platinum with nickel, the potential of $H_2O^* \rightarrow OH^*$ phase transition is shifted downward (toward lower potential values), resulting in a reduction in the H_2O phase with an increase in the OH* phase as the nickel content in the alloy is increased. On $PtNi_3(111)$ surface, the clean surface disappears completely, and a transition from H* to OH* phase is observed at −0.45 and −0.46 V, respectively, and this is due to the increase in the oxophilic behavior of NiPt alloys with increasing nickel content since nickel is more oxophilic than platinum. This observation is further supported by the order of the OH adsorption energies of −0.10, −0.21, −0.25, −0.75, −1.01, and −1.14 eV on Pt(111), $Pt_3Ni(111)$, $PtNi_{surface}(111)$, $Pt_2Ni_2(111)$, $PtNi_3(111)$, and Ni(111) surfaces, respectively.

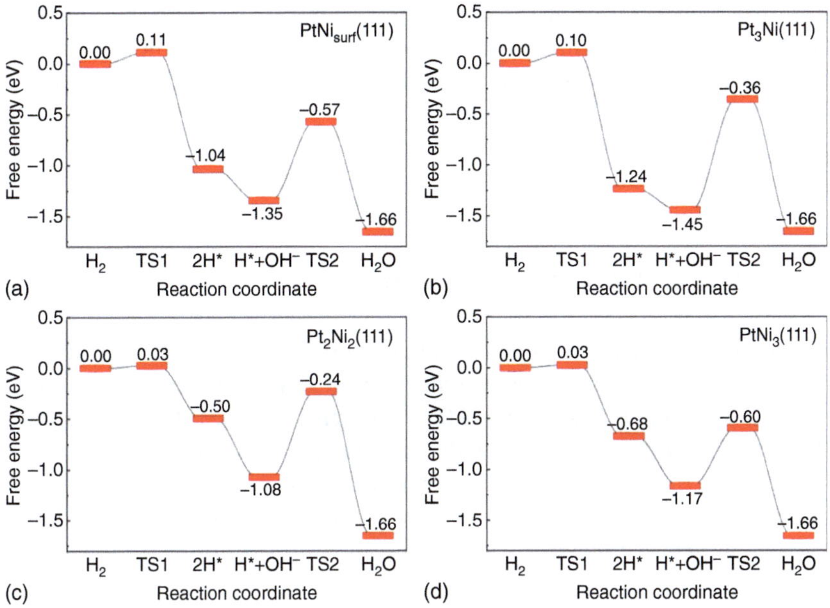

Figure 7.3 Free energy diagram of Tafel–Volmer mechanism involved in the HOR on (a) $PtNi_{surf}(111)$, (b) $Pt_3Ni(111)$, (c) $Pt_2Ni_2(111)$, and (d) $PtNi_3(111)$ at pH 13 and $T = 298$ K. Source: Reproduced from Ref. [12] with permission of the American Chemical Society.

For the alkaline HOR all possible elementary reactions were studied on $PtNi_{surf}(111)$, $Pt_3Ni(111)$, $Pt_2Ni_2(111)$, and $PtNi_3(111)$, and the calculated activation energy barriers of the reactions and the corresponding free energy changes as a function of reaction coordinate are depicted in Figure 7.3. Among the five possible elementary reactions of HOR, the Tafel reaction involving the dissociative adsorption of H_2 offers a low-energy pathway with an activation barrier in the range of 0.12–0.26 eV. On the contrary, for the Heyrovsky reactions, the H_2 reacts with either OH^- in solution or with adsorbed OH; the activation energy barrier is high in the range of 0.54–1.56 eV on all the PtNi alloy surfaces studied. The activation energy barrier for the subsequent Volmer reaction involving oxidative desorption of H* forming H_2O, Eq. (7.5), is 0.70–1.18 eV; the activation barrier is lower than that of the recombination of adsorbed OH and H, Eq. (7.6), 1.06–1.77 eV. However, the activation barrier for the Volmer reaction is still higher than that of Tafel reaction. Therefore, HOR on PtNi alloy surfaces prefers the Tafel–Volmer mechanism, with the Volmer reaction being the rate-determining step. The free energy diagrams, which represent the elementary processes of the most favorable Tafel–Volmer pathway on $PtNi_{surface}(111)$, $Pt_3Ni(111)$, $Pt_2Ni_2(111)$, and $PtNi_3(111)$, are shown in Figure 7.3. From the plots of free energy over the reaction coordinate, the activation energy barrier for the rate-determining reaction, i.e. oxidative desorption of H* to H_2O through the transitions state (TS2), is the lowest (−0.60 eV) on the surface of $PtNi_3(111)$ compared to any other PtNi alloy as the ideal catalyst with higher HOR activity. The optimal H adsorption free energy change for HOR (ΔG_{H*}) is

0.414 eV. The free-energy change, for the whole HOR (ΔG_{HOR}), was calculated to be −1.656 eV. $PtNi_3(111)$ exhibited the highest HOR activity with the lowest free energy barrier E_a, the smallest value of $|\Delta G_{H*} - 0.414|$, the highest rate constant, and the largest exchange current density.

To enhance the HOR activity of an electrocatalyst in an alkaline medium, it is essential to identify the alkaline HOR pathway and the rate-determining step.

7.3 Electrocatalysts for Alkaline HOR

In alkaline electrolytes, the HOR activity of the pure platinum-group metal (PGM) catalysts (Pt, Ir, and Pd) is significantly reduced by at least two orders of magnitude compared to the same PGMs under acidic conditions [2, 3, 14, 15]. In this regard, the development of electrocatalysts with comparable or much higher activity for alkaline HOR is the key to the overall performance enhancement of AEMFC. In this section, the recent development in PGM-based and non-PGM alkaline HOR electrocatalysts will be discussed.

7.3.1 Platinum Group Metal HOR Electrocatalysts

The activity of the catalysts for HOR depends on a combination of several parameters, such as the support, the morphology of electroactive catalyst species, composition, and size. The implementation of those parameters in the design of the electrocatalysts is crucial for achieving synergistic catalyst activity leading to enhanced overall fuel cell performance. Recently, the insertion of heteroatom into carbon has been used to improve the interaction between the nanoparticles and the support. For example, Kundu et al. [16] studied the effect of nitrogen-doped carbon as a support for Pt nanoparticles. In this study, they showed that the metallic platinum domains offered adsorption sites for the hydrogen atom; PtO_x promoted the adsorption of OH-species; and the N-doped carbon support acted as the highly conductive matrix favoring the charge transfer process. The presence of nitrogen, in the carbon matrix, leads to an improvement in the interaction between the hydrogen atoms and the catalyst surface. Bhowmik et al. [17] examined the interaction between palladium and carbon-nitride nanosheets, which led to improved activity, compared to commercial Pd/C, and attributed this phenomenon to the synergic effect between palladium nanoparticles and the support. Much research was also placed in using CeO_2 as support. The introduction of CeO_2 to palladium resulted in a fivefold increase in the performance and a 20-fold increase in the specific and mass activity compared to Pd/C [18]. The Pd—H bond interactions were weakened in the presence of CeO_2. The oxophilic nature of CeO_2 supplies reactive OH species that would remove the H_{ads} intermediate for the surface of palladium, resulting in improved HOR kinetics [18, 19]. The addition of 3d transition metal, such as nickel, leads to an enhancement of activity. For example, the Pd/Ni nanocomposite, which was synthesized by reducing the palladium complex on the nickel surface, in an aqueous solution, leads to the enhancement of HOR activity [8]. While evaluating the performance in a membrane electrode assembly for alkaline fuel cells a peak of 400 mW cm^{-2} was obtained, which is more than twice as much as that of a pure palladium catalyst. The high activity is attributed to the presence of OH_{ads} species on the surface of the oxophilic nickel that favor HOR kinetics. Since the distribution of palladium and nickel leads to the enhancement in HOR performance, three designed core–shell nanoparticulated catalysts of Pd@Ni, Ni@Pd, and NiPd alloys were tested as electrocatalysts for HOR [20]. The highest activity was measured on NiPd-alloyed nanoparticles and was attributed to the synergistic electrocatalytic effect between nickel and palladium surface atoms. Another example to enhance palladium activity is by decorating tungsten with palladium (Pd-d-W). The Pd-d-W electrocatalysts were designed by activating the surface of tungsten with palladium ions using a galvanic displacement reaction. The interaction between palladium and the oxophilic surface of tungsten oxide enhances the activity of palladium, leading to a reduction in the onset potential and an increase in specific current density. The effect on the enhanced HOR activity is attributed to the adsorption of the OH species near the surface of the palladium atoms [21]. Previous studies demonstrated that alloying is an effective way to enhance the HOR activity. For example, Pt–Ru alloy nanoparticles were found to improve drastically the HOR activity and the cell performance to about twice compared to platinum nanoparticle electrocatalysts [22]. Besides, Ru–Ir alloy nanoparticles increased the HOR activity to about four times compared to individual Ru, Ir, and Pt nanoparticle-based electrocatalysts [23]. Small metal nanoparticles of Ru–Ir (3.7 nm) supported on carbon were designed and tested to investigate the effect of alloying on the HOR activity. The Ru–Ir pair was found to be 19.5 and 3.5 times higher than the Ru–Ru and Ir–Ir pairs, respectively. The Tafel slope analysis suggested that the Ru–Ir pair promoted the Volmer step through the bifunctional mechanism. Ruthenium and iridium offer reactive OH^- and H species, resulting in an accelerated reaction between these species and thereby enhancing the HOR activity. Mao et al. [24] synthesized atomically dispersed nickel on ruthenium (Ni/Ru) which showed better HOR activity than bimetallic RuNi, pristine Ru, and commercial Pt/C catalysts. Isolation of nickel atoms on ruthenium nanocrystals optimized the hydrogen binding energy and decreased the free energy of the formation of water. The weakening of hydrogen binding energy, high electrochemically active surface area, and suitable oxophilic behavior are achieved by alloying platinum with ruthenium and supporting the alloy nanoparticles (3.0–3.8 nm) of $Pt_{0.25}Ru_{0.75}$ on nitrogen-doped carbon. The mass-specific activity of $Pt_{0.25}Ru_{0.75}$/N-doped carbon electrocatalyst was 1654 A g^{-1}(PtRu), and this value is 4.7 and 4 times higher than the activity of Pt/C and PtRu/C catalysts [25]. Mass activity superior to that of PtRu/C catalysts was achieved by confining the atomic clusters of ruthenium in the TiO_2 lattice. The electrocatalysts are CO tolerant and exhibited high HOR activity up to a potential of 0.9 V *vs.* RHE. Selective HOR is observed even with high CO content (10 vol%). Abundant Ru—Ti bond formations were observed along the TiO_2 lattice. Efficient electron transfer from electron-rich TiO_2 to Ru is attributed to the selective HOR and improved CO tolerance [26]. Alloying ruthenium with a 3d transition metal such as cobalt leads to a shift in the d-band and a weakening of H adsorption on ruthenium. Among RuCo, RuNi, and RuFe alloys, catalyst $Ru_{0.95}Co_{0.05}/C$ synthesized by impregnation method exhibited best HOR and ORR activities [27]. HOR activity of Pt/C modified with nickel was enhanced by the immersion of Pt/C in nickel chloride solution, leading to a weakening in the hydrogen adsorption. The specific mass current density was increased by 18% from 23.8 to 33.4 A g^{-1}. Modification of platinum with oxophilic metals was proposed as a way to improve the anodic electrocatalysts for HOR [28].

Table 7.1 summarizes the current state-of-the-art PGM-based HOR catalysts. As can be seen, different strategies were applied to enhance HOR activity and control the kinetics of the reaction. Such approaches are controlling the size, tuning the morphology, alloying with two or three metals, and improving the interaction between the metal and support. The main drawback is the use of precious metal. The idea of reducing them or not using them at all is needed and should further be explored by applying nonprecious metal electrocatalysts as a primary catalytic center.

7.3.2 Non-platinum Group Metal-based HOR Electrocatalysts

Compared to PEMFCs, the main advantage of AEMFCs is the fact that nonprecious metals offer good performance and could replace the use of platinum group metal as anode electrocatalysts [44–46]. H_2 is a chemically reactive gas, which adsorbs on most transition metal surfaces [47]. Among all those transition metals, nickel attracted much more attention but, unfortunately, has not been successful being used as a sole component for alkaline HOR electrocatalyst until now. Therefore, many efforts were implemented to enhance the alkaline HOR activity of nickel-based electrocatalysts to replace precious metals. Nickel nanoparticles supported on N-doped carbon nanotubes (Ni/N-CNT) have been synthesized by a wet chemical method [48]. When comparing the activity with pure nickel nanoparticles, Ni/N-CNT leads to a 21-fold increase in activity and is attributed to the synergetic effect between N-CNT and nickel. To further enhance the HOR activity, nickel-based catalysts supported on carbon matrices doped with diverse nonmetallic elements such as B, S, and N were investigated [49]. Those hybrid catalysts were designed using two steps: pyrolysis and reduction and compared to Ni/C. The activity order is as follows: Ni/SC > Ni/NC > Ni/BC ~ Ni/C, where SC, NC, and BC correspond to the S-, N-, and B-doped carbon support materials. Two main factors were proposed to improve activity: (i) the anchoring effect leading to smaller and more uniform particle sizes with large ECSA and (ii) the metal–support interaction that weakens the hydrogen bonding energy of nickel, leading to better specific activity [49]. To correlate the observed HOR activity of

Table 7.1 Summary of PGM-based electrocatalysts for alkaline HOR.

Electrocatalyst	Electrolyte concentration (M)/*T* (°C)	Mass activity @η = 50 mV (mA $mg^{-1}_{catalyst}$)	j_0 ECSA normalized (mA $cm^{-2}_{catalyst}$)	References
Pt/Au polycrystalline	0.1 M KOH/—	—	$1.580_{geometric}$	[29]
Pt/Au/C	0.1 M KOH/—	—	$2.750_{geometric}$	[29]
5% Pt/Vulcan XC72	0.1 M NaOH/40 °C	2052.00	1.710	[3]
10% Pd/Vulcan XC72	0.1 M NaOH/40 °C	115.50	0.110	[3]
10% Ir/Vulcan XC72	0.1 M NaOH/40 °C	448.40	0.760	[3]
Pt/C	1.0 M KOH/—	79.00	0.590	[30]
PdIr/C	1.0 M KOH/—	138.00	1.700	[30]
Pd/Au θ_{Pd}/Au = 0.39	0.1 M NaOH/20 °C	—	1.700	[31]
Pd/Au θ_{Pd}/Au = 0.23	0.1 M NaOH/20 °C	—	4.000	[31]
Pt/Cu nanowires	0.1 M KOH/—	330.00	1.010	[32]
20% Ir/C	0.1 M KOH/20 °C	292.30	0.470	[33]
20% Ir/C-300 °C	0.1 M KOH/20 °C	284.60	0.540	[33]
20% Ir/C-500 °C	0.1 M KOH/20 °C	260.40	0.620	[33]
20% Ir/C-600 °C	0.1 M KOH/20 °C	307.20	0.960	[33]
20% Ir/C-800 °C	0.1 M KOH/20 °C	187.70	1.360	[33]
1.5% Pd/Ni	0.1 M KOH	—	0.570	[34]
Ir_{NS}	0.1 M NaOH/25 °C	—	0.810	[35]
5% Pt/C	0.1 M KOH/20 °C	659.00	1.030	[4]
10% Pd/C	0.1 M KOH/20 °C	72.30	0.090	[4]
20% Ir/C	0.1 M KOH/20 °C	313.00	0.490	[4]
20% Rh/C	0.1 M KOH/20 °C	690.00	0.860	[4]
Pt nanowire	0.1 M KOH/—	99.00	$1.380_{geometric}$	[36]
PtRu nanowire	0.1 M KOH/—	157.00	$2.200_{geometric}$	[36]

(Continued)

Table 7.1 (Continued)

Electrocatalyst	Electrolyte concentration (M)/T (°C)	Mass activity @η = 50 mV (mA $mg^{-1}_{catalyst}$)	j_0 ECSA normalized (mA $cm^{-2}_{catalyst}$)	References
PtFe nanowire	0.1 M KOH/—	120.00	$1.680_{geometric}$	[36]
PtCo nanowire	0.1 M KOH/—	117.00	$1.640_{geometric}$	[36]
PtCu nanowire	0.1 M KOH/—	69.00	$0.970_{geometric}$	[36]
PtAu nanowire	0.1 M KOH/—	88.00	$1.23_{geometric}$	[36]
$PdNi_{NP}$	0.1 M KOH/—	3.90	$0.197_{geometric}$	[8]
Pd-CN_X	0.5 M KOH/—	—	$0.080_{geometric}$	[17]
20% Pd/C	0.1 M KOH/20 °C	91.90	0.120	[37]
20% Pd/C-300C	0.1 M KOH/20 °C	92.30	0.140	[37]
20% Pd/C-400C	0.1 M KOH/20 °C	76.20	0.230	[37]
20% Pd/C-500C	0.1 M KOH/20 °C	44.20	0.260	[37]
20% Pd/C-600C	0.1 M KOH/20 °C	17.10	0.240	[37]
Ir/CeO_2–C	0.1 M KOH/20 °C	73.5	$0.092_{geomtric}$	[38]
Pd-Ni/C	0.1 M KOH/—	104.00 (η = 100 mV)	1.380 (η = 100 mV)	[20]
Pt/C	0.1 M KOH/25 °C	403.00	0.226	[39]
IrNi/C	0.1 M KOH/25 °C	188.00	0.214	[39]
IrNi@PdIr/C	0.1 M KOH/25 °C	854.00	0.209	[39]
Ir_2Ru_1NWs/C	0.1 M KOH/—	994.00	0.099	[40]
Ir_1Ru_1NWs/C	0.1 M KOH/—	1416.00	0.126	[40]
Ir_1Ru_2NWs/C	0.1 M KOH/—	1813.00	0.0932	[40]
Ir_1Ru_3NWs/C	0.1 M KOH/—	3346.00	0.084	[40]
BCC PdCu/C-500 °C	0.1 M KOH/R.T.	522.00	$0.883_{geomtric}$	[41]
PtRu/C	0.1 M KOH/R.T.	1023.00	$1.463_{geomtric}$	[41]
IrNi@Ir	0.1 M KOH/R.T.	1.12	1.220 ± 0.100	[42]
$RuNi_1$	0.1 M KOH/R.T.	270.00	—	[24]
Pt/C	0.1 M KOH/R.T.	426.00	—	[24]
$IrMo_{0.59}$ nanoparticles	0.1 M KOH/30 °C	3850.00	1.150	[43]
Ir nanoparticles	0.1 M KOH/30 °C	690.00	0.320	[43]

ECSA, electrochemical surface area; j_0, exchange current density; θ_{Pd}, the palladium coverage; NW, nanowires.

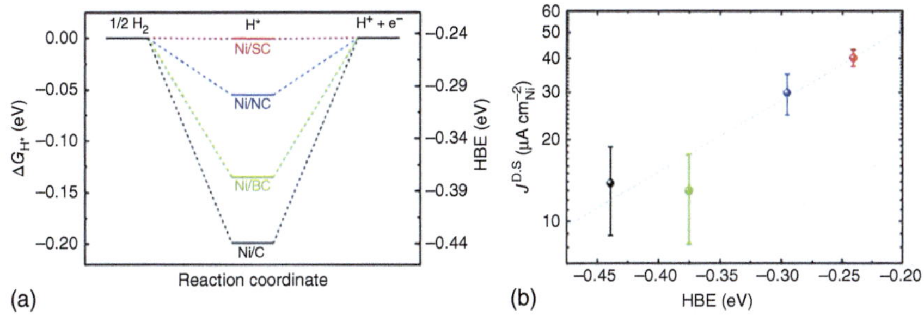

Figure 7.4 (a) Calculated adsorption free energy of H, ΔG_{H^*}, and HBE; (b) Dependence of exchange current density on the relative hydrogen binding energies (HBE). Source: Reproduced from Ref. [49] with permission of the Royal Society of Chemistry.

nickel supported on heteroatom-doped carbon supports to their Gibbs adsorption free energies of H (ΔG_{H^*}) or hydrogen binding energies (HBEs), density functional theory calculations were carried out. The values of Gibbs adsorption free energies of H, $|\Delta G_{H^*}|$, followed the order, Ni/SC < Ni/NC < Ni/BC < Ni/C, as shown in Figure 7.4a. The catalyst Ni/SC exhibited an optimal value of (~0 eV) Gibbs adsorption free energy of H (ΔG_{H^*}). As predicted theoretically from the DFT studies, Ni/SC exhibited 2.9 and 7.7 times higher specific activity and mass activity, respectively, than the dopant-free Ni/C catalyst. A linear relationship between the specific activity and the hydrogen binding energies of the four different catalysts studied, Ni/SC, Ni/NC, Ni/BC, and Ni/C, is observed, Figure 7.4b. The enhanced activity in the nickel-supported heteroatom-doped carbon supports, especially Ni/SC, is attributed to the strong metal–support interaction leading to enhance an HOR activity [49].

Another interesting catalyst is Ni_3N supported on carbon (Ni_3N/C) designed by reducing the nickel precursor under NH_3 atmosphere at elevated temperature [50]. Ni_3N/C presented the highest mass activity for HOR and breakdown potential for a PGM-free catalyst. The activity was attributed to the nickel d-band center downshifts and interfacial charge transfer from Ni_3N to the carbon support as factors leading to the weak binding of hydrogen and oxygen species in Ni_3N/C. The weak binding of hydrogen was suggested as the main reason for the high HOR activity. Bifunctional HOR mechanism, namely the dissociative adsorption of H and recombination of adsorbed intermediate species H* and OH* to form water, is not possible on pure nickel. This is due to the strong adsorption of hydrogen and a high activation barrier for the recombination reaction. The relatively strong adsorption of H (H_{ad}) is the main reason for the slow HOR kinetics on pure nickel. On the contrary, in the bimetallic NiCu catalyst with as low as 5% Cu, nearly 50% surface coverage of copper is observed, and therefore, weakening of (H_{ad}) hydrogen adsorption energy as well as decrease in the activation energy barrier for the recombination of H_{ad} and OH_{ad} was observed. Compared to nickel electrocatalyst, 10-fold enhancement in the Volmer step was achieved [51]. Wang et al. [52] prepared Ni–Cu binary alloy films by combinatorial magnetron co-sputtering method. A Volcano-type relationship was observed between HOR activity and copper content in the Ni–Cu bimetallic alloy film. With Ni–Cu binary alloy film containing 40 at.% Cu, the exchange current density was four times higher than that of pure nickel. Interfacial engineering resulted in a unique Ni_3N/Ni electrocatalyst which exhibits excellent HOR and HER activity [53]. The active sites are located at the interface of Ni_3N and nickel supported on Ni-foam. The electrocatalyst showed zero overpotential, long-term durability, 100% Faradaic efficiency, and excellent CO tolerance. The free-energy change for the adsorption of hydrogen at the interfacial sites was found to be almost zero. In the presence of CO, the maximum current density of Ni_3N/Ni/Nickel-foam decreased from 6.95 to 6.71 mA cm^{-2} (only 3.5% decrease). Another work that uses the strategy to construct the surface between metals and metal nitrides as an effective approach to design electrocatalysts for HOR is from Song et al. [54]. Low-cost Co_2N/Co electrocatalysts are designed through a facile electrodeposition and nitridation approach. The DFT calculation reveals the significant role of Co_2N/Co interfacial sites that possessed optimal hydrogen adsorption energy as well as facilitated water adsorption and dissociation on the catalyst surface. The electrocatalyst exhibited CO tolerance and long-term stability for HOR. Yang et al. [55] synthesized CeO_2-decorated nickel hybrid nanoparticles supported on carbon black (Vulcan XC-72) using solvothermal synthesis (CeO_2(r)-Ni/C-1). DFT calculations revealed that the improved HOR performance is associated with the promoted OH* adsorption energy and optimal ΔG_{H*} obtained from the synergistic effect between oxygen vacancy contained in CeO_2 and nickel species. The oxygen vacancies at the surface of the CeO_2 played an important role in accelerating the HOR kinetics in alkaline media.

The required demand for low-cost and earth-abundant electrocatalysts for the HOR led to the development of electrocatalysts using metal–organic frameworks (MOF). Yang et al. [56] designed Ni/NiO/C electrocatalyst protected by graphene using direct annealing of nickel-based MOF (Ni-BTC) and showed nearly zero onset potential, strong robustness, and great tolerance toward CO poisoning. Computational, electrochemical studies together with physical characterization show that the interfacing of Ni and NiO in Ni/NiO/C, with the presence of highly conductive graphene layers, leads to optimal binding energies of hydrogen and hydroxide species contributing to the overall HOR activity. Ni et al. [57] annealed $Ni_3(BTC)_2$ under nitrogen and low concentration of hydrogen (0, 1, 2, and 4 vol%), yielding carbon-supported nickel nanoparticles with the highest activity for HOR reported so far using non-PGM-based electrocatalysts.

Table 7.2 summarizes the non-PGM electrocatalysts designed in the past few years for HOR in alkaline conditions. As can be seen, the most common nickel-based catalysts take advantage of the oxophilic feature of Ni to tune their electronic properties. While adjusting the size, morphology, and composite with other transition metals and supports, it is possible to modulate the Ni activity and stability toward the best performing and low-cost electrocatalysts for HOR.

Table 7.2 Summary of non-PGM-based electrocatalysts for HOR.

Electrocatalyst	Electrolyte concentration (M)/*T* (°C)	Mass activity @η = 50 mV (mA $mg^{-1}_{catalyst}$)	j_0 ECSA normalized (mA $cm^{-2}_{catalyst}$)	References
CoNiMo	0.1 M KOH/20 °C	—	0.0150	[58]
$Ni_{0.95}Cu_{0.05}$/C	0.1 M NaOH/25 °C	1.5	0.0140	[29]
$Ni_{0.95}Cu_{0.05}$/Vulcan XC-72	0.1 M KOH/25 °C	2.5	0.0140	[59]
Ni/N-CNT	0.1 M KOH/R.T.	9.3	0.0280	[48]
Ni/CNT	0.1 M KOH/R.T.	1.9	0.0092	[48]
Ni	0.1 M KOH/R.T.	0.28	0.0013	[48]
Ni_3@h-$(BN)_1$/C-700NH_3	0.1 M KOH/—	—	0.0220	[60]
NiMo/KB	0.1 M NaOH/25 °C	4.5	0.0270	[61]
$Ni_{95}Cu_5$/KB	0.1 M KOH/—	—	0.0250	[62]
Ni/Ni_3N/NF	0.1 M KOH/25 °C	—	0.0030	[53]
Nonactivated Ni_{ED}/XC-72	0.1 M NaOH/25 °C	—	0.0062	[63]
Ch-activated NiED/XC72[a)]	0.1 M NaOH/25 °C	—	0.0560	[63]
EC-activated NiED/XC72[a)]	0.1 M NaOH/25 °C	—	0.0390	[63]
CeO_2(r)-Ni/C-1	0.1 M KOH/30 °C	12.28	0.0380	[64]
CeO_2-Ni/C-1	0.1 M KOH/30 °C	8.48	0.0260	[64]
Ni/C	0.1 M KOH/30 °C	5.89	0.0160	[64]
Ni/SC	0.1 M KOH/30 °C	11.0	0.0402	[49]
Ni/NC	0.1 M KOH/30 °C	—	0.0309	[49]
Ni/BC	0.1 M KOH/30 °C	—	0.0130	[49]
Ni/C	0.1 M KOH/30 °C	—	0.0148	[49]
Ni_3N/C	0.1 M KOH/—	24.38	0.0140	[49]
Ni_3N	0.1 M KOH/—	1.73	0.0170	[49]
Ni-Ref	0.1 M KOH/—	1.07	0.0020	[49]
Ni/NiO/C-700	0.1 M KOH/—	5.0	0.0260	[56]
Ni	0.1 M KOH/R.T.	—	0.0140	[65]
WNi_4	0.1 M KOH/R.T.	—	0.0680	[65]
$MoNi_4$	0.1 M KOH/R.T.	—	0.0650	[65]
Ni-H_2-2%	0.1 M KOH/—	50.4	0.028	[57]

j_0, exchange current density; CNT, carbon nanotube; h-BN, hexagonal boron nitride; KB, Ketjen black; ECSA, electrochemical surface area.

a) Nanostructured Ni nanoparticles supported on Vulcan XC-72 carbon (NiED/XC-72); The electrode subjected to oxidation either under contact with air (Ch activated) or by applying positive potential (EC activated); Reduction of Ni precursors via $NaBH_4$ in the presence of CeO_2 and XC-72 (CeO_2(r)-Ni/C-1); Heteroatoms-doped carbon supports (XC, X = S, N, or B).

7.4 Conclusions

This chapter provides fundamental information on the state-of-the-art PGM and non-PGM-based electrocatalysts for the HOR in alkaline conditions. One of the main challenges in the commercialization of AEMFCs is the slow kinetics of the HOR reaction, especially in the alkaline medium with PGM-based electrocatalysts compared to the acidic medium. The solution to overcome this problem depends on developing nickel-based systems with an emphasis on the downward shift of the d-band center of Ni(111) (−1.78 eV) away from the Fermi level. Strategies for shifting downward the d-band center (ε_d) of nickel were discussed (in the specific examples of $PtNi_3$(111) and Ni_3N systems). Such a strategy is crucial as the hydrogen adsorption strength varies linearly with the d-band center of the metal surfaces. Further, there exists a linear relationship between the hydrogen binding energies of the anode electrocatalyst and HOR specific activity. HOR reaction on Pt Ni alloys suggested proceeding through the Tafel–Volmer mechanism involving the dissociative adsorption of gaseous hydrogen on the catalyst surface yielding two adsorbed hydrogen species (Tafel reaction) and the subsequent reaction of adsorbed hydrogen with OH^- in the electrolyte to form H_2O (Volmer reaction, the rate-determining step). The optimal H adsorption free energy for HOR (ΔG_{H*}) and the free energy change for the whole HOR (ΔG_{HOR}) with the ideal HOR anode electrocatalyst $PtNi_3$ were found to be 0.414 and −1.656 eV. By a downward shift of the d-band center of nickel in the Ni_3N system, optimal hydrogen binding is achieved, resulting in a remarkable HOR activity (mass activity, 24.38 mA mg^{-1} catalyst). Another strategy for achieving the best HOR performance from nickel nanoparticles, via weakening of their hydrogen binding energy, is through a strong metal–support interaction. In a specific example, sulfur-doped carbon material support facilitated strong metal (Ni)–support interaction and has significantly weakened the hydrogen binding energy, leading to high HOR activity (mass activity, 11.0 mA mg^{-1} catalyst). Adopting a similar strategy of tuning the metal–support interaction to impose strain on the active catalytic sites (nickel nanoparticles in constrained geometry), Hu's group succeeded in designing the state-of-the-art anode electrocatalyst for HOR with outstanding activity (mass activity@η = 50 mV, 50.4 mA mg^{-1} catalyst) [57]. The nickel nanoparticle active sites in constrained geometries are achieved by choosing a peculiar Ni-metal–organic framework precursor and pyrolyzing the precursor in an N_2–H_2 (2 vol%) environment, leading to the formation of polycrystalline nickel nanoparticles (6.5 ± 2.9 nm). Charge transfer phenomenon from a nickel to carbon support was attributed to the downward shifting of the nickel d-band center away from the Fermi level, resulting in lower binding energies of H_{ads} and OH_{ads}. Overall, shifting the d-band center of nickel downward from the Fermi level leading to lowering of hydrogen binding energy appears to be the key to the success in finding the best nonnoble metal Ni-based anode electrocatalyst for HOR. This can be achieved by tuning the metal–support interaction and imposing strain on the geometry and configuration of nickel active sites. Such geometric and electronic effects between the active species (nickel nanoparticles) and extensively heteroatom-functionalized carbon supports (oxygen/sulfur/nitrogen) resulting from novel catalyst synthesis strategies are the path ahead to the groundbreaking invention in this field of research. The key parameters to be optimized are the location of the d-band center and the hydrogen binding energy of nickel nanoparticles in the Ni/C systems. This leaves much scope for the synthetic, electrochemical, and theoretical chemists to work on designing strategies for the nanoarchitecture of Ni/C systems for future anode electrocatalysts for HOR.

Acronyms

AEMFC	alkaline electrolyte membrane fuel cell
CNT	carbon nanotubes
DFT	density functional theory
ECSA	electrochemical surface area
FSs	final states
HBE	hydrogen energy binding
HOR	hydrogen oxidation reaction
ISs	initial states
MOF	metal-organic frameworks
N-CNT	N-doped carbon nanotube
PEMFC	polymer electrolyte membrane fuel cell
PGM	platinum group metal
RHE	reversible hydrogen electrode
SHE	standard hydrogen electrode
TS	transition state

References

1 Campos-Roldán, C.A. and Alonso-Vante, N. (2019). The hydrogen oxidation reaction in alkaline medium: an overview. *Electrochemical Energy Reviews* 2 (2): 312–331.

2 Sheng, W., Gasteiger, H.A., and Shao-Horn, Y. (2010). Hydrogen oxidation and evolution reaction kinetics on platinum: acid vs alkaline electrolytes. *Journal of the Electrochemical Society* 157 (11): B1529.

3 Durst, J., Siebel, A., Simon, C. et al. (2014). New insights into the electrochemical hydrogen oxidation and evolution reaction mechanism. *Energy & Environmental Science* 7 (7): 2255–2260.

4 Zheng, J., Sheng, W., Zhuang, Z. et al. (2016). Universal dependence of hydrogen oxidation and evolution reaction activity of platinum-group metals on pH and hydrogen binding energy. *Science Advances* 2 (3): 1–9.

5 Elbert, K., Hu, J., Ma, Z. et al. (2015). Elucidating hydrogen oxidation/evolution kinetics in base and acid by enhanced activities at the optimized Pt shell thickness on the Ru core. *ACS Catalysis* 5 (11): 6764–6772.

6 Rheinländer, P.J., Herranz, J., Durst, J., and Gasteiger, H.A. (2014). Kinetics of the hydrogen oxidation/evolution reaction on polycrystalline platinum in alkaline electrolyte reaction order with respect to hydrogen pressure. *Journal of the Electrochemical Society* 161 (14): F1448–F1457.

7 Strmcnik, D., Uchimura, M., Wang, C. et al. (2013). Improving the hydrogen oxidation reaction rate by promotion of hydroxyl adsorption. *Nature Chemistry* 5 (4): 300–306.

8 Alesker, M., Page, M., Shviro, M. et al. (2016). Palladium/nickel bifunctional electrocatalyst for hydrogen oxidation reaction in alkaline membrane fuel cell. *Journal of Power Sources* 304: 332–339.

9 Alia, S.M., Pivovar, B.S., and Yan, Y. (2013). Platinum-coated copper nanowires with high activity for hydrogen oxidation reaction in base. *Journal of the American Chemical Society* 135 (36): 13473–13478.

10 Ishikawa, K., Ohyama, J., Okubo, K. et al. (2020). Enhancement of alkaline hydrogen oxidation reaction of Ru-Ir alloy nanoparticles through bifunctional mechanism on Ru-Ir pair site. *ACS Applied Materials & Interfaces* 12 (20): 22771–22777.

11 Koper, M.T.M. (2013). Hydrogen electrocatalysis: a basic solution. *Nature Chemistry* 5 (4): 255–256.

12 Zhao, L., Liu, H., Liu, Y. et al. (2020). Mechanistic insights into the hydrogen oxidation reaction on PtNi alloys in alkaline media: a first-principles investigation. *ACS Applied Materials & Interfaces* 12 (36): 40248–40260.

13 Campos-Roldán, C.A., Calvillo, L., Granozzi, G., and Alonso-Vante, N. (2020). Alkaline hydrogen electrode and oxygen reduction reaction on Pt_xNi nanoalloys. *Journal of Electroanalytical Chemistry* 857: 113449.

14 Markovic, N.M., Gasteiger, H.A., and Ross, P.N. (1995). Oxygen reduction on platinum low-index single-crystal surfaces in sulfuric acid solution. Rotating ring – Pt(hkl) disk studies. *Journal of Physical Chemistry* 99 (11): 3411–3415.

15 Rheinländer, P., Henning, S., Herranz, J., and Gasteiger, H.A. (2013). Comparing hydrogen oxidation and evolution reaction kinetics on polycrystalline platinum in 0.1 M and 1 M KOH. *ECS Transactions* 50 (2): 2163–2174.

16 Kundu, M.K., Bhowmik, T., Mishra, R., and Barman, S. (2018). Platinum nanostructure/nitrogen-doped carbon hybrid: enhancing its base media HER/HOR activity through bi-functionality of the catalyst. *ChemSusChem* 11 (14): 2388–2401.

17 Bhowmik, T., Kundu, M.K., and Barman, S. (2016). Palladium nanoparticle-graphitic carbon nitride porous synergistic catalyst for hydrogen evolution/oxidation reactions over a broad range of pH and correlation of its catalytic activity with measured hydrogen binding energy. *ACS Catalysis* 6 (3): 1929–1941.

18 Miller, H.A., Lavacchi, A., Vizza, F. et al. (2016). A Pd/C-CeO_2 anode catalyst for high-performance platinum-free anion exchange membrane fuel cells. *Angewandte Chemie International Edition* 55 (20): 6004–6007.

19 Miller, H.A., Vizza, F., Marelli, M. et al. (2017). Highly active nanostructured palladium-ceria electrocatalysts for the hydrogen oxidation reaction in alkaline medium. *Nano Energy* 33: 293–305.

20 Shviro, M., Polani, S., Dunin-Borkowski, R.E., and Zitoun, D. (2018). Bifunctional electrocatalysis on Pd-Ni core–shell nanoparticles for hydrogen oxidation reaction in alkaline medium. *Advanced Materials Interfaces* 5 (9): 1–8.

21 Arulrajan, A.C., Subramanian, P., Singh, R.K., and Schechter, A. (2020). Pd-decorated tungsten as Pt-free bimetallic catalysts for hydrogen oxidation reaction in alkaline electrolyte. *Israel Journal of Chemistry* 60 (5–6): 563–569.

22 Wang, Y., Wang, G., Li, G. et al. (2015). Pt-Ru catalyzed hydrogen oxidation in alkaline media: oxophilic effect or electronic effect? *Energy & Environmental Science* 8 (1): 177–181.

23 Ohyama, J., Kumada, D., and Satsuma, A. (2016). Improved hydrogen oxidation reaction under alkaline conditions by ruthenium-iridium alloyed nanoparticles. *Journal of Materials Chemistry A* 4 (41): 15980–15985.

24 Mao, J., He, C.T., Pei, J. et al. (2020). Isolated Ni atoms dispersed on Ru nanosheets: high-performance electrocatalysts toward hydrogen oxidation reaction. *Nano Letters* 20 (5): 3442–3448.

25 Cong, Y., Chai, C., Zhao, X. et al. (2020). $Pt_{0.25}Ru_{0.75}$/N-C as highly active and durable electrocatalysts toward alkaline hydrogen oxidation reaction. *Advanced Materials Interfaces* 7 (11): 1–9.

26 Zhou, Y., Xie, Z., Jiang, J. et al. (2020). Lattice-confined Ru clusters with high CO tolerance and activity for the hydrogen oxidation reaction. *Nature Catalysis* 3 (5): 454–462.

27 Wang, H., Yang, Y., Disalvo, F.J., and Abrunã, H.D. (2020). Multifunctional electrocatalysts: Ru-M (M = Co, Ni, Fe) for alkaline fuel cells and electrolyzers. *ACS Catalysis* 10 (8): 4608–4616.

28 Huang, X., Long, C., Han, J. et al. (2020). Remarkably enhanced hydrogen oxidation reaction activity of carbon-supported Pt by facile nickel modification. *Chemical Research in Chinese Universities* 36 (1): 105–109.

29 Mahoney, E.G., Sheng, W., Yan, Y., and Chen, J.G. (2014). Platinum-modified gold electrocatalysts for the hydrogen oxidation reaction in alkaline electrolytes. *ChemElectroChem* 1 (12): 2058–2063.

30 Jervis, R., Mansor, N., Gibbs, C. et al. (2014). Hydrogen oxidation on PdIr/C catalysts in alkaline media. *Journal of the Electrochemical Society* 161 (4): F458–F463.

31 Henning, S., Herranz, J., and Gasteiger, H.A. (2015). Bulk-palladium and palladium-on-gold electrocatalysts for the oxidation of hydrogen in alkaline electrolyte. *Journal of the Electrochemical Society* 162 (1): F178–F189.

32 Alia, S.M. and Yan, Y. (2015). Palladium coated copper nanowires as a hydrogen oxidation electrocatalyst in base. *Journal of the Electrochemical Society* 162 (8): F849–F853.

33 Zheng, J., Zhuang, Z., Xu, B., and Yan, Y. (2015). Correlating hydrogen oxidation/evolution reaction activity with the minority weak hydrogen-binding sites on Ir/C catalysts. *ACS Catalysis* 5 (7): 4449–4455.

34 Bakos, I., Paszternák, A., and Zitoun, D. (2015). Pd/Ni synergestic activity for hydrogen oxidation reaction in alkaline conditions. *Electrochimica Acta* 176: 1074–1082.

35 Montero, M.A., De Chialvo, M.R.G., and Chialvo, A.C. (2016). Evaluation of the kinetic parameters of the hydrogen oxidation reaction on nanostructured iridium electrodes in alkaline solution. *Journal of Electroanalytical Chemistry* 767: 153–159.

36 Scofield, M.E., Zhou, Y., Yue, S. et al. (2016). Role of chemical composition in the enhanced catalytic activity of Pt-based alloyed ultrathin nanowires for the hydrogen oxidation reaction under alkaline conditions. *ACS Catalysis* 6 (6): 3895–3908.

37 Zheng, J., Zhou, S., Gu, S. et al. (2016). Size-dependent hydrogen oxidation and evolution activities on supported palladium nanoparticles in acid and base. *Journal of the Electrochemical Society* 163 (6): F499–F506.

38 Qin, B., Yu, H., Chi, J. et al. (2017). A novel Ir/CeO_2-C nanoparticle electrocatalyst for the hydrogen oxidation reaction of alkaline anion exchange membrane fuel cells. *RSC Advances* 7 (50): 31574–31581.

39 Qin, B., Yu, H., Jia, J. et al. (2018). A novel IrNi@PdIr/C core-shell electrocatalyst with enhanced activity and durability for the hydrogen oxidation reaction in alkaline anion exchange membrane fuel cells. *Nanoscale* 10 (10): 4872–4881.

40 Qin, B., Yu, H., Gao, X. et al. (2018). Ultrathin IrRu nanowire networks with high performance and durability for the hydrogen oxidation reaction in alkaline anion exchange membrane fuel cells. *Journal of Materials Chemistry A* 6 (41): 20374–20382.

41 Qiu, Y., Xin, L., Li, Y. et al. (2018). BCC-phased PdCu alloy as a highly active electrocatalyst for hydrogen oxidation in alkaline electrolytes. *Journal of the American Chemical Society* 140 (48): 16580–16588.

42 Liu, D., Lu, S., Xue, Y. et al. (2019). One-pot synthesis of IrNi@Ir core-shell nanoparticles as highly active hydrogen oxidation reaction electrocatalyst in alkaline electrolyte. *Nano Energy* 59 (January): 26–32.

43 Fu, L., Li, Y., Yao, N. et al. (2020). IrMo nanocatalysts for efficient alkaline hydrogen electrocatalysis. *ACS Catalysis* 10 (13): 7322–7327.

44 Cong, Y., Yi, B., and Song, Y. (2018). Hydrogen oxidation reaction in alkaline media: from mechanism to recent electrocatalysts. *Nano Energy* 44: 288–303.

45 Campos-Roldán, C.A., Calvillo, L., Boaro, M. et al. (2020). NiO-Ni/CNT as an efficient hydrogen electrode catalyst for a unitized regenerative alkaline microfluidic cell. *ACS Applied Energy Materials* 3 (5): 4746–4755.

46 Li, Y., Li, Z., Cong, H. et al. (2022). Boosting hydrogen oxidation performance of phase-engineered Ni electrocatalyst under alkaline media. *ACS Sustainable Chemistry & Engineering* 10 (11): 3682–3689.

47 Davydova, E.S., Mukerjee, S., Jaouen, F., and Dekel, D.R. (2018). Electrocatalysts for hydrogen oxidation reaction in alkaline electrolytes. *ACS Catalysis* 8 (7): 6665–6690.

48 Zhuang, Z., Giles, S.A., Zheng, J. et al. (2016). Nickel supported on nitrogen-doped carbon nanotubes as hydrogen oxidation reaction catalyst in alkaline electrolyte. *Nature Communications* 7: 1–8.

49 Yang, F., Bao, X., Zhao, Y. et al. (2019). Enhanced HOR catalytic activity of PGM-free catalysts in alkaline media: the electronic effect induced by different heteroatom doped carbon supports. *Journal of Materials Chemistry A* 7 (18): 10936–10941.

50 Ni, W., Krammer, A., Hsu, C.S. et al. (2019). Ni_3N as an active hydrogen oxidation reaction catalyst in alkaline medium. *Angewandte Chemie International Edition* 58 (22): 7445–7449.

51 Salmazo, D., Juarez, M.F., Oshchepkov, A.G. et al. (2019). On the feasibility of bifunctional hydrogen oxidation on Ni and NiCu surfaces. *Electrochimica Acta* 305: 452–458.

52 Wang, G., Li, W., Huang, B. et al. (2019). Exploring the composition-activity relation of Ni-Cu binary alloy electrocatalysts for hydrogen oxidation reaction in alkaline media. *ACS Applied Energy Materials* 2 (5): 3160–3165.

53 Song, F., Li, W., Yang, J. et al. (2018). Interfacing nickel nitride and nickel boosts both electrocatalytic hydrogen evolution and oxidation reactions. *Nature Communications* 9 (1): 4531.

54 Song, F., Li, W., Yang, J. et al. (2019). Interfacial sites between cobalt nitride and cobalt act as bifunctional catalysts for hydrogen electrochemistry. *ACS Energy Letters* 4 (7): 1594–1601.

55 Yang, F., Bao, X., Li, P. et al. (2019). Boosting hydrogen oxidation activity of Ni in alkaline media through oxygen-vacancy-rich CeO_2/Ni heterostructures. *Angewandte Chemie International Edition* 58 (40): 14179–14183.

56 Yang, Y., Sun, X., Han, G. et al. (2019). Enhanced electrocatalytic hydrogen oxidation on Ni/NiO/C derived from a nickel-based metal–organic framework. *Angewandte Chemie International Edition* 58 (31): 10644–10649.

57 Ni, W., Wang, T., Schouwink, P.A. et al. (2020). Efficient hydrogen oxidation catalyzed by strain-engineered nickel nanoparticles. *Angewandte Chemie* 132 (27): 10889–10893.

58 Sheng, W., Bivens, A.P., Myint, M. et al. (2014). Non-precious metal electrocatalysts with high activity for hydrogen oxidation reaction in alkaline electrolytes. *Energy & Environmental Science* 7 (5): 1719–1724.

59 Cherstiouk, O.V., Simonov, P.A., Oshchepkov, A.G. et al. (2016). Electrocatalysis of the hydrogen oxidation reaction on carbon-supported bimetallic NiCu particles prepared by an improved wet chemical synthesis. *Journal of Electroanalytical Chemistry* 783: 146–151.

60 Gao, L., Wang, Y., Li, H. et al. (2017). A nickel nanocatalyst within a h-BN shell for enhanced hydrogen oxidation reactions. *Chemical Science* 8 (8): 5728–5734.

61 Kabir, S., Lemire, K., Artyushkova, K. et al. (2017). Platinum group metal-free NiMo hydrogen oxidation catalysts: high performance and durability in alkaline exchange membrane fuel cells. *Journal of Materials Chemistry A* 5 (46): 24433–24443.

62 Roy, A., Talarposhti, M.R., Normile, S.J. et al. (2018). Nickel-copper supported on a carbon black hydrogen oxidation catalyst integrated into an anion-exchange membrane fuel cell. *Sustainable Energy & Fuels* 2 (10): 2268–2275.

63 Oshchepkov, A.G., Bonnefont, A., Pronkin, S.N. et al. (2018). Nanostructured nickel nanoparticles supported on vulcan carbon as a highly active catalyst for the hydrogen oxidation reaction in alkaline media. *Journal of Power Sources* 402 (September): 447–452.

64 Yang, F., Bao, X., Li, P. et al. (2019). Boosting hydrogen oxidation activity of Ni in alkaline media through oxygen-vacancy-rich CeO_2/Ni heterostructures. *Angewandte Chemie* 131 (40): 14317–14321.

65 Duan, Y., Yu, Z.Y., Yang, L. et al. (2020). Bimetallic nickel-molybdenum/tungsten nanoalloys for high-efficiency hydrogen oxidation catalysis in alkaline electrolytes. *Nature Communications* 11 (1): 1–10.

8

Membranes for Fuel Cells

Paolo Sgarbossa, Giovanni Crivellaro, Francesco Lanero, Gioele Pagot, Afaaf R. Alvi, Enrico Negro, Keti Vezzù, and Vito Di Noto

University of Padova, Section of Chemistry for Technologies (ChemTech), Department of Industrial Engineering, Via Marzolo 9, I-35131 Padova, Italy

8.1 Introduction

Nowadays, climate change is one of the biggest issues for humankind. The European Union (EU) is fighting this threat adopting a series of ambitious initiatives such as the European Green Deal, the European Climate Law, and the Fit for 55 package [1]. The objective of the EU is to reach a climate-neutral society by 2050. This transition involves all the aspects of daily life, from economy to politics, from the energy technologies to the social sciences. The climate-neutral target of the EU requires fundamental changes in the production and exploitation of energy, with a particular reference to the decarbonization of transport and the electricity conversion [2–4]. Similar policies are also under implementation elsewhere in the world (e.g. in the United States [5] and in Japan). To meet these ambitions requirements, a promising solution is to obtain energy from renewable sources (e.g. the sun, the wind, the tides, and geothermal sources) and to convert the excess of energy production into high-value fuels (e.g. hydrogen) by clean processes such as water electrolysis. The resulting "*green hydrogen*" can be stored for a long time and transported over long distances as the need arises, opening the door to the creation of a smart Energy Network within the EU. Finally, green hydrogen is used to obtain electricity with a high efficiency by means of suitable electrochemical energy conversion devices. Thus, the virtuous cycle of "*green*" electrical energy conversion is closed.

Fuel cells (FCs) are among the best systems to carry out the last step of this process. Indeed, FCs are electrochemical devices capable to convert the chemical energy of a fuel (typically hydrogen, but also hydrocarbons, alcohols, and hydrazine) directly into electrical energy with a very high efficiency [6]. FCs are generally classified depending on the type of fuel and electrolyte employed in the device. Among all FCs, polymer electrolyte fuel cells (PEFCs) are considered the best choice for electric vehicles and portable electronics applications [7] due to their moderate operation temperature (i.e. generally between 80 and 200 °C), high energy efficiency, and fast

Electrocatalysis for Membrane Fuel Cells: Methods, Modeling, and Applications, First Edition.
Edited by Nicolas Alonso-Vante and Vito Di Noto.

startup and shutdown times [8]. The heart of these devices comprises a polymer electrolyte (PE) formed into a membrane, which is one of the most important functional components of the FC [9]. The PE (i) acts as a separator for the feeds, thus avoiding short-circuits and (ii) allows for the selective transport of the ions required to maintain the charge neutrality in the cell during operation.

Excluding exotic applications [10, 11], PEFCs implement either cation-exchange membranes (CEMs) or anion-exchange membranes (AEMs), giving rise, respectively, to acidic or alkaline conditions [6, 12–14]. To become attracting for practical applications, the fundamental features that an ion-exchange membrane (IEM) must possess are (i) a high ionic conductivity; (ii) a low permeability toward undesired species; (iii) a high chemical and electrochemical stability; (iv) good mechanical properties; and (v) a low cost of production. The most commonly used CEMs in PEFCs are based on well-established perfluorinated proton-exchange ionomers such as Nafion® [15]. Perfluorinated CEMs generally exhibit (i) a high mechanical stability; (ii) a high chemical stability in a large variety of chemical environments (e.g. acidic, oxidizing, and reducing); (iii) a high ionic conductivity (is ca. $0.1\,S\,cm^{-1}$) in a considerably wide temperature range (from 0 to 80 °C); and (iv) quite a high production cost [16]. Aromatic- and polyimide-based CEMs such as sulfonated poly(ether ether)ketone (SPEEK) and polybenzimidazole (PBI) have recently been proposed as cost-effective and high-performing materials for CEMs [17, 18]. PEFCs working in alkaline conditions are of a great interest as they require significantly less platinum-group metals (PGMs) in the electrode configurations, curtailing costs and the risk to incur in supply bottlenecks. However, these FCs require AEMs which, up to now, do not match yet the performance and durability levels of CEMs [19]. Indeed, OH^- anions have a lower mobility in the bulk membrane with respect to H^+, and the activation energy for hydroxide transport is generally twice that of protons, resulting in a lower ionic conductivity [20, 21]. In addition, AEMs absorb easily the CO_2 found in the air, triggering a decrease in their performance. Thus, AEMs must be handled and stored in CO_2-free atmospheres [19]. Nevertheless, great leaps forward have been made in the past years for AEMs, progressively closing the performance and durability gaps with traditional CEMs.

This chapter is meant to provide the reader with a comprehensive overview of state-of-the-art and most recent breakthroughs in the PEs for PEFCs. The attention is focused on the main fundamental characteristics and features that PE must exhibit to be suitable for practical applications. A comprehensive overview of PEs is discussed, with an exhaustive description of their physicochemical properties and performance. The chapter ends with a description of the ion conduction mechanism and its correlation to the physicochemical features of the PEs, followed by a brief summary of the research perspectives.

8.2 Properties of the PE Separators

The separator plays a key role to set the overall performance and durability level of electrochemical energy conversion storage devices such as fuel cells (FCs), redox flow batteries (RFBs), and electrolyzers (ELs). Most separator membranes are made of PEs and are indicated hereinafter as "*ion-exchange membranes*" (IEMs). Thus, IEMs (that include both CEMs and AEMs) are the main focus of this chapter. There are some alternatives to IEMs (e.g. microporous membranes [22]), which will be referred explicitly when occurring in the following text. Several properties affect the conductivity of IEMs such as composition, type and degree of functionalization, and hydration level. Such properties affect the maximum power density of the device by reducing the ohmic drops [23] but also contribute to the durability imparting mechanical stability and impeding gas diffusion between the electrodes [24].

8.2.1 Benchmarking of IEMs

The United States Department of Energy (US DOE) has set highly authoritative technical targets for both IEMs and membrane-electrode assemblies (MEAs). The latter are the core component of PEFCs and are obtained by sandwiching the IEM between the electrode configurations. The US DOE targets are specifically intended for polymer electrolyte membrane fuel cell (PEMFC) power systems, stacks, and components for light-duty transportation applications (Table 8.1) [26]. They are the result of the inputs coming from the U.S. DRIVE Partnership, which includes automotive and energy companies, developers of stationary fuel cell power systems, and the Fuel Cell Technical Team of US DOE. Table 8.1 also includes technical targets suggested by other authoritative international institutions (i.e. the EU) [27].

Some of the technical targets for 2020 have already been achieved by state-of-the-art IEMs, while others have yet to be obtained. As an example, the Fumatech FS-715-RFS proton-exchange membrane exceeds the DOE targets for area resistance, which is lower than $0.02\,\Omega\,cm^{-2}$, and H_2 crossover, which is lower than $1.5\,mA\,cm^{-2}$. Among the most important parameters that are crucial for IEMs to reach the market are the cost, which is still an issue for state-of-the-art materials, and the chemical, electrochemical, and mechanical durability. For both new and developmental materials, the parameters of interest are usually benchmarked with a "*reference*" ion-exchange membrane of similar thickness [30]. In the case of proton-exchange membranes (PEMs) the benchmark is a membrane fabricated with a perfluorinated sulfonated ionomer such as Nafion [31], whereas in the case of AEMs, benchmarking is much more difficult, as there is no such well-assessed "*reference*" material (with the current possible exception of the Fumatech FAA-3 ionomer).

In the next few paragraphs, we present and discuss separately some specific parameters that are not reported in the US DOE targets list but are of paramount importance during the process of developing and optimizing new ionomers as they affect significantly the overall performance of the FC.

8.2.2 Ion-exchange Capacity (IEC)

The number of ion-exchange groups per membrane mass, or "*ion-exchange capacity*" (IEC), is directly correlated to the number of ions that are available for

Table 8.1 Targets for ion-exchange membranes and membrane electrode assemblies (MEAs).

	Key performance indicator [25]	Status 2015 (DOE) [26]	Target 2020 (DOE) [26]	Target 2020 (EU) [27]
Membrane	Area-specific resistance @ Max. operating T and 40–80 kPa water partial pressure/$\Omega\,cm^2$	0.072 (120 °C, 40 kPa)	0.02	n.a.
	Operating temperature/°C	<120	≤120	n.a.
	Conductivity ($S\,cm^{-1}$) at peak T, 20%RH	n.a.	n.a.	0.1
	Cost/$\$\cdot m^{-2}{}_{net}$	17	20	n.a.
	Key performance indicator	**Status 2015 (DOE) [26]**	**Target 2020 (DOE) [26]**	**Target 2024 (EU) [27]**
MEA with "*Low-PGM*" *ECs*	PGM loading/$mg\,cm^{-2}$	0.13 (cathode)	0.1 (cathode)	0.15 (total)
	Mass power/$g_{PGM}\,kW^{-1}$	0.16	0.125	0.08
	Durability/h	5000	5000	6000
	Conversion efficiency	>60%	>60%	>60%
	Key performance indicator	**Status 2021 [28]**	**Targets 2025 (DOE) [29]**	
MEA with "*PGM-free*" *ECs*	J @ 0.9 $V_{iR\text{-}free}$ (H_2–O_2, 1 atm)/$mA\,cm^{-2}$	44.2/30.2	>44	
	J @ 0.8 V (H_2-air, 1 atm)/$mA\,cm^{-2}$	151/85	>150	
	Power density @ 0.7 V (H_2-air, 1 atm)/$mW\,cm^{-2}$	600/535	>450	
	Loss in stability after 30k voltage cycling/%	NA/30	<40	

charge transport and greatly affects the overall conductivity of the IEM. Generally, the IEC is provided as the number of mmol or mEq of exchangeable ions per gram weight of dry membrane [32], thus indicating the density of ionic/highly polar moieties in the material. This figure will affect in turn the water uptake (WU, see also Section 8.2.3), the water content (i.e. the volume percentage of H_2O in the wet membrane), and the hydration number λ (i.e. the molar ratio between the molecules of H_2O in the IEM and the charged groups) [33, 34].

The IEC can be determined by various methods, including (i) simple and back-titration after ion exchange; (ii) measurement with ion-selective electrodes (e.g. glass pH probe) to quantify the concentration of H^+/OH^- ions in solution; and (iii) spectroscopic analysis of other ions of interest [35, 36]. As an example of application to a CEM, the membrane can be immersed in an aqueous 1.0 M H_2SO_4 solution to completely protonate the active sites, rinsed with doubly-distilled water to remove the excess acid, and then equilibrated in a 1.0 M NaCl water solution for 24 hours. The obtained acidic solution can then be titrated with a 5 mM NaOH solution with a suitable acid/base indicator (the titration must be carried out under an inert atmosphere to avoid carbonation of the titrant) [37, 38]. Finally, the IEC can be evaluated using Equation (8.1):

$$\mathrm{IEC} = \frac{n_{0,\mathrm{NaOH}} - n_{E,\mathrm{NaOH}}}{W} \tag{8.1}$$

where $n_{0,\mathrm{NaOH}}$ and $n_{E,\mathrm{NaOH}}$ are the moles of NaOH in the initial solution and at the equivalence point, respectively, and W is the mass of the dry membrane [39].

High IECs usually result in high ion conductivities, bestowing an improved current and power density to the final FC [40]. Nonetheless, the IEC and the number of polar/ionic moieties in the membrane are also often directly correlated to the latter's ability to absorb water (see also Section 8.2.3). Excessive hydration leads to structural modifications (e.g. swelling), which may result in a degradation of the mechanical properties and flooding of the electrocatalyst layers, hindering the transport of reactants, products, and active species [31, 41]. When considering AEMs, it is important to remember that the diffusion coefficient of the hydroxide ions is basically half of that of the protons in bulk water ($5.30 \cdot 10^{-5}$ cm^2 s^{-1} for OH^- and $9.31 \cdot 10^{-5}$ cm^2 s^{-1} for H^+ at 25 °C, respectively) [42]. On one side this makes it difficult to achieve AEMs with ion conductivities similar to those of state-of-the-art PEMs. On the other side, it requires a higher concentration of OH^- in AEMs to reach a comparable conductivity to PEMs. This results in materials for AEMs with a high IEC and, consequently, suffering a high risk of swelling and loss of mechanical stability [43]. We should point out that other ionomer features beyond IEC affect the conductivity (for more details, see also Section 8.4). Briefly, at a fixed IEC the ion conductivity is modulated by the microstructure (particularly the nanophase segregation) and the thermomechanical relaxations, which are heavily influenced by the chemical structure of the material (type of backbone, length of the macromolecular chains, cross-linking, etc.). These features, depending on the operational temperature, strongly contribute to the long-range ion charge transport mechanism [30, 41, 44].

8.2.3 Water Uptake (WU), Swelling Ratio (SR), and Water Transport

Water management is crucial in the operation of PEFCs [45–48]. In FCs the formation of water at the cathode (for PEMFCs) or at the anode (in AEMFCs) is fundamental to maintain the required hydration level of the membrane for optimal operation. An insufficient hydration will typically degrade the ion transport in the IEM, yielding a poor performance. On the other hand, an excess of water can hinder the transport of reactants and products in the electrode configurations, resulting in large mass transport overpotentials that degrade performance. Upon hydration, within the IEM the mobile ions are dissociated from the charged functional groups of the ionomer. Water also contributes to modulate the phase separation between

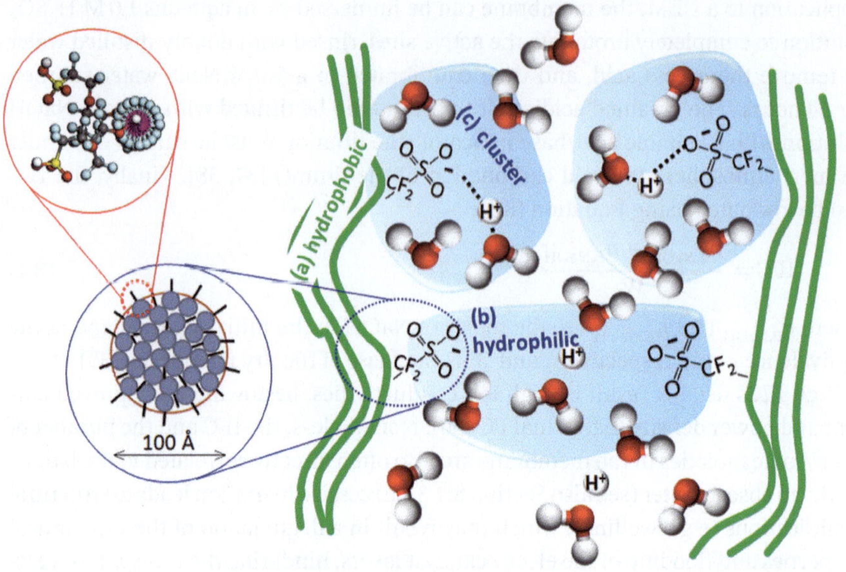

Figure 8.1 Structural regions in wet Nafion: (a) hydrophobic region (bundle of perfluorinated chains); (b) hydrophilic region; and (c) cluster (functional groups, water molecules, and cations).

hydrophilic and -phobic domains within the IEM, which plays a pivotal role to determine the conduction mechanism (see Figure 8.1 and Section 8.4) [44, 49]. Upon hydration, in most IEMs the ion conductivity (σ) raises significantly, by about two to three orders of magnitude [45, 46]. To quantify the amount of absorbed water per unit weight of dry membrane, water uptake (WU) is evaluated as in Eq. (8.2) [33, 34, 50]:

$$\mathrm{WU} = \frac{W_\mathrm{w} - W_\mathrm{d}}{W_\mathrm{d}} \times 100\% \tag{8.2}$$

where W_w and W_d are the "*wet*" and "*dry*" weights of the membrane after and before water absorption, respectively.

Upon hydration, the interconnection of hydrophilic domains within the hydrophobic polymer matrix is fundamental in IEMs to achieve a high conductivity [44, 51] (see also Section 8.4). In order to improve this interconnection, WU is often maximized by swelling the water domains (e.g. by immersing the IEM in hot water) [52, 53]. However, this approach may result in the disruption of the structure of the polymer matrix, producing mechanical instabilities and degrading ion selectivity [15]. Furthermore, a higher WU often results in the dilution of the mobile charged species, with a consistent reduction in ionic conductivity. As an example sulfonated polyimide membranes exhibit a WU of about 40% but present a much lower conductivity when compared with Nafion, which possesses a WU of around 22% [54].

In order to precisely quantify the amount of water present in the polymer matrix, the hydration number lambda (λ) is used. This number allows for a precise quantification of the relative number of water molecules per ion-exchange group (e.g. a sulfonic group in perfluorinated Nafion-like IEMs). The hydration number can be evaluated from the IEC (mmol g^{-1}) and WU (%) figures as in Eq. (8.3):

$$\lambda = \frac{n_{H_2O}}{n_{IEG}} = \frac{10 \cdot WU}{IEC \cdot M_{H_2O}} \tag{8.3}$$

As a consequence of the uptake of water, the structure at the mesoscale of IEMs undergoes significant changes; hydrophilic and hydrophobic regions are easily distinguished [45, 55]. The macroscopic structure of the IEM is affected as well [45, 51]. The swelling ratio (SR) is a way to assess these changes as it quantifies changes in its dimensions (i.e. length, width, and thickness) due to the presence of water [56]. SR is calculated as shown in Eq. (8.4):

$$\text{SR} = \frac{L_w - L_d}{L_d} \times 100\% \tag{8.4}$$

where L_w and L_d are the dimensions (typically, the length) of the wet and dry membrane, respectively. The mechanical stability and the ion selectivity of IEMs are strongly affected by the SR [31, 41]. These two features (WU and SR) correlate with the ability of IEMs to permeate water through them. The transport of water can be investigated with various methods such as effective drag coefficient [57], water balance experiments [58], and magnetic resonance imaging measurement [59].

8.2.4 Ionic Conductivity (σ)

In IEMs the ion conductivity is the capability to allow ions to migrate between the two sides of the membrane under the effect of a difference in electrical potential [60, 61]. Typically, the ionic conductivity is calculated by measuring the resistivity of the IEM against the flow of either alternating current (ac) or direct current (dc). Notwithstanding the apparent simplicity of the dc method chemical potential distribution, electrode reactions, and polarization have to be considered when measuring the resistivity with the current flow. Two- and four-terminal tests are the most common types of measurement methods. The former is mainly used to determine the resistivity of high-resistance materials (above $10^6\ \Omega$), while the latter is usually employed to measure the resistivity of ionic conductors with a low resistance due to a reduction in both the interface resistance and the polarization. Either method in conjunction with Eq. (8.5) will yield the ionic conductivity (σ):

$$\sigma = \frac{l}{R_S A} \tag{8.5}$$

where l is the distance between the probe electrodes, A is the cross-sectional area of the membrane sample, and R_s is the bulk resistance derived from an impedance analysis.

The ionic conductivity of low-conducting samples is typically measured by means of ac impedance or broadband electric spectroscopic techniques [62, 63].

Broadband electric spectroscopy (BES) is a powerful technique used to achieve an in-depth determination of the polarization phenomena and the dielectric relaxation events characterizing ion-conducting materials [62, 64]. In contrast to classical ac impedance measurements, BES experiments are usually carried out under a variety of different conditions in terms of temperature and/or degree of hydration [41, 62]; several cell geometries can be adopted (e.g. reflection and transmission) to better highlight different features of the electrical response of the material [62, 65, 66]. BES analyzes the electrical response of a sample exposed to a constant-amplitude ac potential with a frequency varying in a broad range [62, 67]. Thus, it is possible to probe the perturbations of relaxation and polarization phenomena in the material as alterations in the permittivity, conductivity, or other macroscopic properties of the sample. Two types of phenomena are the main promoters of the long-range charge migration in the materials. Polarization phenomena result from the accumulation of charge between domains with different values of permittivity [66, 68, 69], whereas dielectric relaxations are the result of the dynamics of the host matrix and are generally due to reorientations of the permanent or induced dipole moments characterizing regions of the molecular structure of the sample. Through the BES study of the electric response of a material, it is possible to elucidate in detail its long-range charge transport mechanism [67] (see also Section 8.4). The correlation of the physicochemical properties of an IEM with its long-range charge transport mechanism yields crucial information, which can be used to devise IEMs exhibiting improved features [62, 67]. A high ionic conductivity is crucial to decrease the resistance of the IEM, minimize ohmic losses, and obtain a high-power density.

In 1991, Springer, Zawodzinski, and Gottesfeld proposed a model correlating ion conductivity (σ) and λ in IEMs specifically based on Nafion 117 [70]. This model included transport through the IEM based on experimental transport parameters and ion conductivities measured as a function of the WU. The conductivity of the perfluorinated ionomer was measured for a range of water activities (relative humidity, RH) at 30 °C and for a fully hydrated membrane at 30 and 80 °C. The determined activation energy was assumed to apply at all values of λ. The fit of σ as a function of λ is reported in Eq. (8.6) for values of $\lambda > 1$, while for values below 1, the membrane ion conductivity was assumed constant:

$$\sigma_{30} = 0.005139\lambda - 0.00326$$
$$\sigma(T_{\text{cell}}) = \exp\left[1268\left(\frac{1}{303} - \frac{1}{273 + T_{\text{cell}}}\right)\right]\sigma_{30} \tag{8.6}$$

σ_{30} is the conductivity at 30 °C. This model has found wide application in most of the modeling works of IEMs until the present day [71–73].

8.2.5 Gas Permeability

The permselectivity of an IEM corresponds to the ability of the material to selectively allow the transport of specific charge carriers, while preventing the passage of other charged or neutral species [74, 75]. This feature is important in FCs especially concerning the permeation of H_2 and O_2. This phenomenon leads to losses in conversion efficiency and an excessive and localized increase in temperature due to a direct chemical reaction between H_2 and O_2, followed by possible burning of the cell components [76]. Generally speaking, gas diffusivity is dependent on the molecule size, the membrane porosity, the polymer structure, its polarity, gas concentration, and operating temperature. In the case of FCs, the gas permeability of IEMs is strictly dependent on the morphology of the hydrophilic domains and the percolation pathways. In general terms, the larger the WU is (and thus the size of the water domains), the higher the conductivity will be (with some exceptions, see also Section 8.2.3). However, at the same time it will be easier for neutral species (e.g. H_2 and O_2) to travel through the membrane and cause negative effects on the device performance. Another important aspect of IEMs that is strictly related to permeability is the degree of crystallinity. Whenever crystallinity is increased (due to aging, larger EW, annealing, etc.), gas permeability is reduced. This concept is straightforward if considered together with the tortuosity of percolation pathways: the increase in the distance between hydrophilic domains due to increased crystallinity leads to the reduction in gas diffusivity. Taken all together, gas permeability is strictly dependent on (i) the backbone length and conformation (affecting crystallinity); (ii) the chemistry of the side chains (influencing the amorphous regions and hydrophilic domains of the IEM); and (iii) the hydration level and the interconnection between domains as related to RH and temperature. Permselectivity/ionic selectivity can be evaluated by Equation (8.7) [30]:

$$S = \frac{\sigma}{P} \tag{8.7}$$

where σ is the ionic conductivity of the IEM, P corresponds to the total species permeability, and S represents the permselectivity/ionic selectivity. In FCs, gas permselectivity has been measured by a number of techniques such as gas chromatography, electrochemical methods, pressure-guided permeation, and time-lag studies [47].

8.2.6 Chemical Stability

In FCs, IEMs are required to withstand a series of highly aggressive conditions. The presence of H_2 and O_2 gases prompts the use of materials that must be very stable in both oxidizing and reducing conditions. Furthermore, the IEM materials must be stable in the temperature range between below freezing and 160 °C (depending on the application). The main factors affecting the chemical stability of an IEM, that is for the most part dependent on the PE used to fabricate it, are the type of mobile ions present in the PE formulation and the chemical species (ionic, radical, etc.) formed at the electrode configurations during FC operation. Usually, to obtain adequate conductivities and current densities in operating conditions, the IEM gives rise to either very acidic (in PEMFC) or very alkaline (in AEMFC) environments. This implies that very stable chemical structures are necessary to achieve durability and cyclability levels that comply with the requirements of FC for practical applications. The EU has set for the year 2024 a durability target for the entire FC system of light-duty vehicles at 6000 hours of operation with a loss of output power lower than

10%. To achieve this target, the contribution of the electrolyte membrane is crucial (see Table 8.1).

Usually, the chemical stability is determined experimentally by *ex situ* tests performed placing the membrane in a solution of either a strong acid ($[H_3O^+] = 0.1$ M from $HClO_4$, H_2SO_4, etc.) or a strong base ($[OH^-] = 0.1$ M from NaOH, KOH, etc.), measuring at definite times its conductivity. Other methods simulate more aggressive, strongly oxidizing conditions using the Fenton's reagent (4 ppm $FeSO_4$ in 3% H_2O_2) [77–79]. In fact, in operating conditions, radical species such as HOO· and HO· are formed at the electrodes [80, 81], the latter being the main source of oxidative stress and degradation in a PEMFC [82]. Although perfluorinated ionomers display an excellent chemical stability, maintaining a high ion conductivity also *in operando* and upon cycling in a corrosive environment, they are not immune to degradation [83, 84]. The hydroxyl radical can attack the weak points of the perfluorinated ionomer structure, leading to chain scission. In the case of hydrocarbon-based polymers the attack can trigger further oxidation processes, leading also to cross-linking and altered mechanical properties [85]. This irreversible aging can produce a deterioration of the integrity of the membrane and the failure of the entire FC [86].

The chemical stability of the membrane is more problematic in AEMFCs, where the presence of hydroxide anions, which are highly nucleophilic and interact strongly with the cationic groups on the polymer (ammonium, phosphonium, sulfonium, pyridinium, imidazolium, piperidinium, guanidinium, etc.), can trigger several possible degradation pathways involving both the ion-exchange moieties and the backbone. Some examples are represented in Figure 8.2 [87].

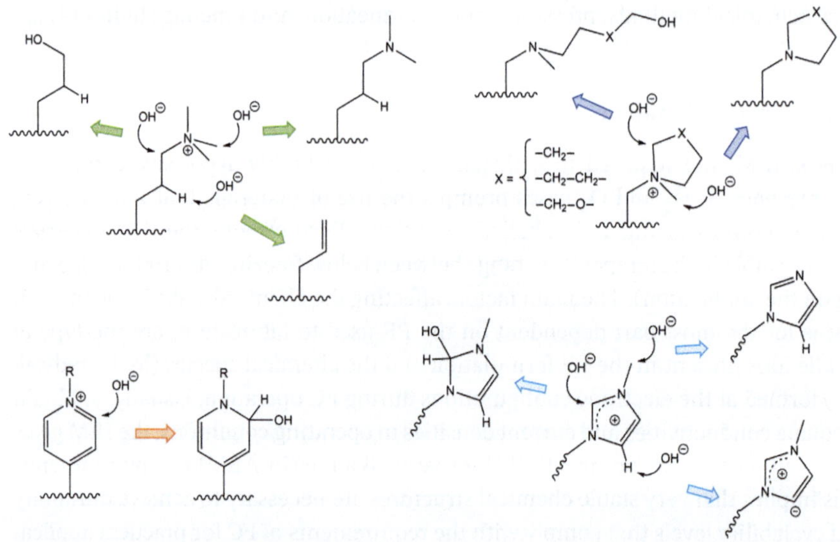

Figure 8.2 Examples of anion-exchange groups and their degradation products upon interaction with OH^- anions.

The same ionomer can undergo different degradation mechanisms depending on the pH and the temperature, making the comparison of *ex situ* stability tests difficult. As an example, tetraalkylammonium groups can yield either β-elimination at lower T and $[OH^-]$ or substitution at higher T and more basic conditions [88]. Up to now, in the literature there is still limited information on the chemical stability of AEMs. It is found that, with respect to phosphonium or sulfonium anion-exchange groups, nitrogen-based anion-exchange groups are more chemically stable. The best stability is displayed by heterocyclic anion-exchange groups (e.g. imidazolium or pyridinium) which are characterized by a conjugated structure and steric hindrance [89].

Research in both industrial and academic environments is constantly trying to enhance the chemical stability of IEMs mainly through the design of more stable backbones and/or ion-exchange functionalities. Among the possible stabilization approaches proposed in the literature, the most common are (i) cross-linking, to reduce the effects of chain scission; (ii) the addition of aromatic groups as side chains or repeating units in the backbone; (iii) the addition of bulky heteroatom-rich groups to increase the steric hindrance around the charged moiety; and (iv) the introduction of radical-scavenging additives.

8.2.7 Thermal and Mechanical Stability

Typically, chemically stable polymers also exhibit good thermal and mechanical properties (see also Section 8.2.6). The IEMs used in FCs require a wide temperature window of stability to operate. Sudden changes in the temperature or extreme conditions (e.g. start-up and shutdown, changes in power output, and local temperature increases) may cause the failure of the IEM [90]. Temperature has a strong influence over the most important physicochemical properties (e.g. ion conductivity, gas permeability, thermomechanical stability, and WU) as these latter are strongly influenced by the thermomechanical transitions taking place within the material [44]. Most importantly, thermomechanical properties are modulated by the hydrophobic/hydrophilic nanophase structure of the polymeric matrix, which is strongly affected by the hydration level. In turn, thermal and mechanical features are strictly related to each other since the changes produced by a temperature increase influence the mechanical behavior and vice versa. For example, PFSA ionomers face different thermal transitions as the temperature is changed. In detail, this family of ionomers usually presents four different thermomechanical relaxations referred to as γ, β, α, and α_{PC}. The γ relaxations are related to the local fluctuations of the polymer backbone; β relaxations are related to the reorientation of the perfluoroethereal side chains; α relaxations are associated with order–disorder conformational transitions occurring in hydrophobic PTFE-like domains, while the α_{PC} transition is associated with the long-range dynamics of both the backbone and the side chains due to the weakening of electrostatic interactions occurring within the hydrophilic domains. The "PC" subscript of α_{PC} indicates that this relaxation mode involves the dynamics of both the hydrophobic and -philic domains, unlike the α transition that is associated with the segmental motion of the perfluorinated backbone [91]. The temperatures of the β, α, and α_{PC} relaxations, which are correlated to their respective activation energies, change depending on the strength of these interactions. The specific features of these relaxations, which can also be revealed for non-PFSA materials, are strongly dependent on the chemistry of the ionomer. In some cases, most commonly in AEMs, the electrostatic interactions between the dipole moments located on the anion-exchange moieties trigger an additional disorder–order transition, referred to as δ. The studies on the mechanical properties highlight how these transitions influence the mechanical behavior of IEMs [67, 92]: γ and β tend to reduce the elastic modulus, resulting in less-stiff materials, while δ produces an increase in the elastic modulus (but will also decrease the conductivity, see also Section 8.4).

The presence of water also has a strong influence on the thermal and mechanical stability. In general terms, increasing the hydration level yields a decrease in the elastic modulus, the strength, and the break stress of the IEM due to a plasticizing effect triggered by a change in the interplay between hydrophobic and -philic domains. In particular, the presence of water allows for the reorganization of the polymer structure into a less-crystalline conformation, thus reducing the modulus. However, at extremely low values of hydration ($\lambda < 2$), the presence of water actually increases the elastic modulus due to the establishment of electrostatic interactions between charged ion-exchange groups [93].

Thermoanalytical techniques are extremely helpful to identify the structural changes of IEMs triggered by temperature, and they can provide valuable information also when studied as a function of other parameters (RH, WU, and others) [34, 37]. The most commonly used thermoanalytical techniques are thermogravimetry (TGA) and differential scanning calorimetry (DSC) [94]. The former consists of an analysis of the weight loss over temperature change, and it is able to identify the temperature and rate of degradation of materials in different conditions. The latter is used specifically to detect the temperature-driven changes inside the material, with a particular reference to the transitions that do not involve weight changes such as glass transitions or other kinds of reorganizational events. Another technique used to evaluate the thermal and mechanical stability of IEMs is dynamic mechanical analysis (DMA), which has been used to evaluate the viscoelastic properties of polymers and hybrid inorganic–organic materials as a function of the temperature [92]. DMA is based on the application of an oscillating force to the sample while the displacement is measured. This allows for the determination of the storage modulus E' (as related with the elastic properties) and the loss modulus E'' (as related with the dampening effects). At low temperatures, polymers tend to behave like glass with a high E' and a pronounced brittleness. At $T > T_g$ (T_g is the "*glass transition temperature*"), E' tends to change in accordance with the specific changes of polymer conformation [95, 96].

Several mechanical properties studies focus mainly on the study of the viscoelastic properties carried out by DMA. Though this technique is very suitable to analyze the viscoelastic response of a polymer as a function of temperature, it does not provide complete information about the material in its plastic deformation domain. For this reason, other techniques such as stress–strain behavior, time-dependent behavior, fatigue, and crack initiation have been considered. Stress–strain measurements allow to identify a series of important parameters such as the proportional limit stress, the break strain, and the maximum stress at the failure point. Stress–strain measurements are usually quantified depending on dimensional parameters (e.g. length and cross-sectional area); thus, it is important to define precisely the initial conditions of the material (hydration, temperature, and RH). The analysis of stress–strain behavior is generally parametrized as follows, Eq. (8.8):

$$\varepsilon = \frac{\Delta L}{L_{\text{ref}}} \tag{8.8}$$

where ε is the strain, and L_{ref} is the length of the sample. Most of the literature on mechanical analysis of IEMs refers to DMA and more in general to tensile deformation [47]. However, the effect of compression should receive the same attention, given the fact that in all practical FCs the MEA operates in compression. Compression studies attempt to quantify the swelling pressure under these conditions, showing that this parameter can reach up to 100 MPa depending on the compression ratio. However, it was found the dehydration effect of compression is negligible when membranes are compressed with a pressure below 10 MPa [47], which is significantly lower than that applied to a MEA as mounted in an FC.

8.2.8 Cost of the IEMs

The cost of the IEM is included in the figures of merit proposed by DOE (see Table 8.1). Indeed, the cost of 17 US$ m^{-2} already meets the required targets [97] when PFSA IEMs supported on electrospun ETFE (eETFE) are considered. However, this cost may vary in the future depending on (i) the availability of raw materials; and especially (ii) the process costs associated with the handling and recycling of the highly aggressive and toxic fluorinated compounds involved in the synthesis of PFSA. Thus, it is still relevant to look for viable IEMs that can meet the required targets for conductivity, durability, and stability as an alternative to PFSA-based systems.

Though the DOE target has already been met, the cost of the IEM is still a major contribution to the overall FC cost, at least at low or intermediate production volumes. If an increase in production is considered, the cost of the membrane would decrease more quickly than that of the PGM-based electrocatalysts. Considering an automotive PEMFC, going from an annual production volume of 1000–100,000 systems/year, the cost of the membrane (supported PFSA) would drop from US$ 1009 to 89 per system, while the PtCo-based electrocatalytic ink from US$ 1775 to 610 per system, respectively [98].

In conclusion, the cost of FCs can also be curtailed by reducing the cost of the IEM. To approach this issue, three viable substitutes to common PFSA membranes have been proposed, namely (i) composite/hybrid membranes with inorganic fillers or organic substrates [99]; (ii) functionalized polyaromatic IEMs (e.g. sulfonated polyether ether ketone, and polysulfone) [100, 101]; and (iii) modified perfluorinated IEMs [102]. More alternatives will be described in the following sections.

8.3 Classification of Ion-exchange Membranes

8.3.1 Cation-exchange Membranes (CEMs)

The majority of the studies in the development of IEMs has been focused on cation-exchange membranes (CEMs), usually proton-exchange membranes (PEMs), which act as the electrolyte in proton-exchange membrane fuel cells (PEMFCs). Several acidic functionalities such as sulfonic, phosphoric, or carboxylic groups are used in the preparation of PEMs, with those based on sulfonic groups being the most common, thanks to their thermal and chemical stability [89]. The concentration of sulfonic groups in CEMs affects remarkably the crucial properties of the membrane including ion-exchange capacity (IEC) and water uptake (WU). CEMs can be classified on the basis of the chemical structure of polymer backbone.

8.3.1.1 Perfluorinated Membranes

Perfluorinated IEMs are the most widely used electrolytes for application in FCs. Perfluorinated IEMs show (i) an exceptional chemical stability in a variety of chemical environments (e.g. acidic, oxidizing, and reducing); (ii) a high mechanical stability [114]; and (iii) a high ionic conductivity in a wide temperature range. For example, Nafion 117 shows proton conductivity of $0.015\,\text{S cm}^{-1}$ at 30 °C and $0.035\,\text{S cm}^{-1}$ at 60 °C [115] and Nafion 115 $0.0240\,\text{S cm}^{-1}$ at 20 °C and $0.0480\,\text{S cm}^{-1}$ at 60 °C [116]. Nafion (Table 8.2), which is produced by Dupont, was the first commercial perfluorinated membrane. Nowadays several companies offer similar ionomers; some examples include Solvay-Solexis (Aquivion and Hyflon-ion) and 3M. All the different commercially available perfluorinated ionomers are based on a poly(tetrafluoroethylene) (PTFE) backbone, which exhibits various types of perfluoroethereal side chains tipped with the ion-exchange sulfonic acid groups. The presence of both the fluorocarbon PTFE backbone and sulfonic acid groups in the ionomer triggers the formation of hydrophobic and -philic domains in the final IEM. Such domains play a crucial role to modulate the ion conduction mechanism.

The fluorocarbon PTFE backbone of Nafion has been shown to adopt mainly two different conformations, 15_7 and 10_3 [95, 117]. In general, the 15_7 and 10_3 conformations, Figure 8.3a,b, are associated with the crystalline and the amorphous hydrophobic PTFE domains of Nafion-based materials, respectively [37]. The relative abundance ratio of the 15_7 and 10_3 conformations can be modulated by the interactions between the Nafion macromolecules and "*guest*" species (e.g. different cations such as D^+ or Na^+, or inorganic nanofillers) [37, 50, 118]. In pristine dry Nafion ($\lambda \sim 0$) and "*Nafion* λ_2" (exhibiting $\lambda = 2$, a minimal hydration), the relative abundance of the 10_3 conformation is somewhat higher in comparison with Nafion Na and Nafion D_2O, both dry and λ_2 [45], and consequentially the dynamics of the side groups are strongly correlated to those of the fluorocarbon backbone chains, Figure 8.3c. Moreover, the degree of hydration and the nature of the cation associated with the sulfonic acid groups produce a significant effect on the water clusters in the hydrophilic domains, with the water present in the Nafion Na(λ_2)

Table 8.2 Selected examples of proton- and anion-exchange ionomers.

Proton-exchange membranes for PEMFCs			
Perfluorinated membranes	Nafion 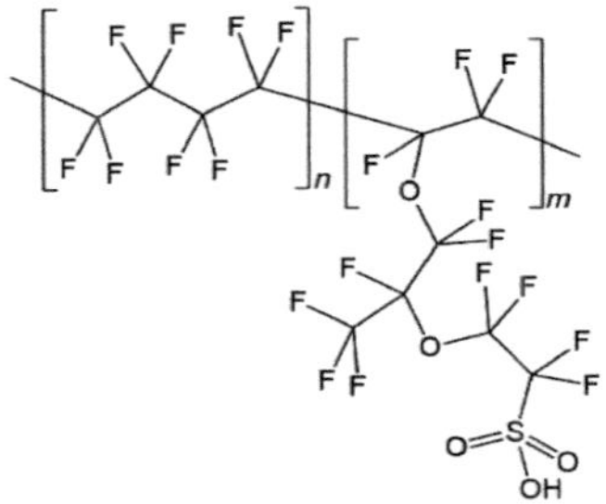	Other PFSA (perfluorosulfonic acid copolymers) 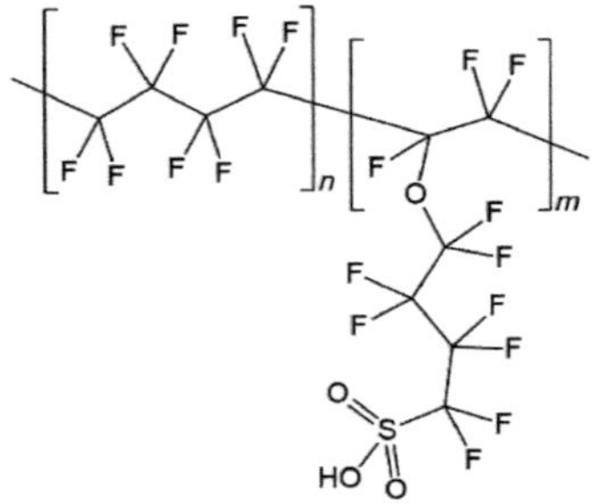	PFIA (perfluoro imide acid copolymer) [48] 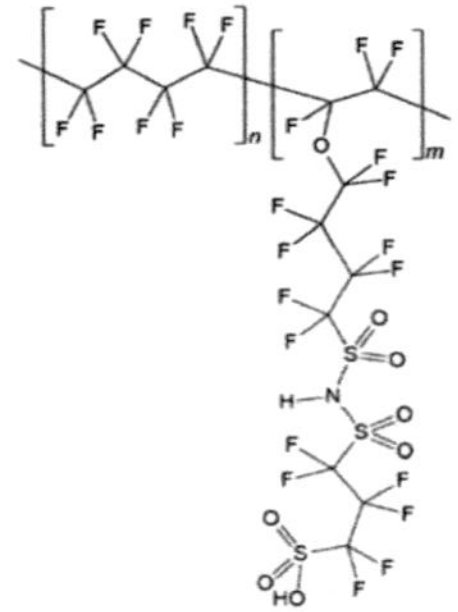
Aromatic membranes	SPEEK (sulfonated poly(ether ether ketone)) 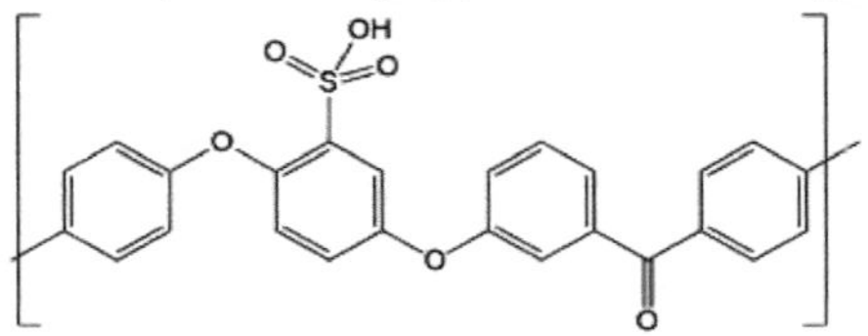	Sulfonated polysulfone [103, 104] 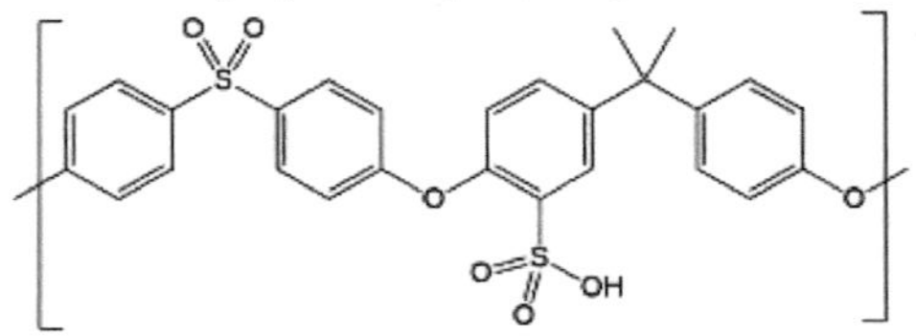	Sulfonated poly(phenylene) oxide [105, 106] 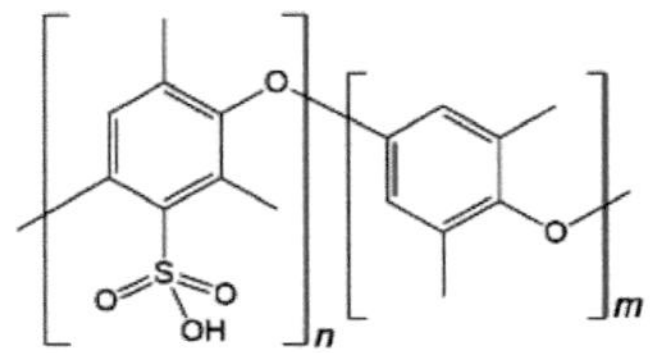

(Continued)

Table 8.2 (Continued)

Polyimidazoles	PBI4N (poly[2,2′-(m-phenylene)-5,5′-bibenzimidazole]) [96, 107–111] 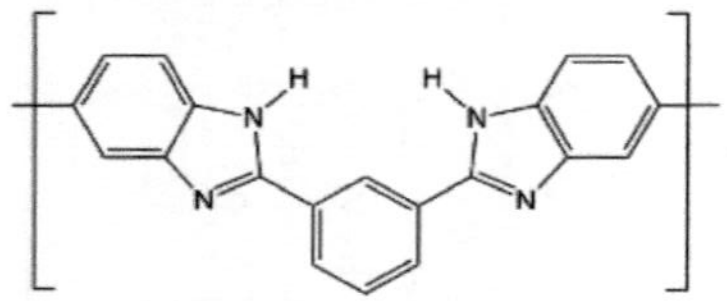	PBI5N (poly-2,2′(2,6-pyridine)-5,5′-bibenzimidazole) [110] 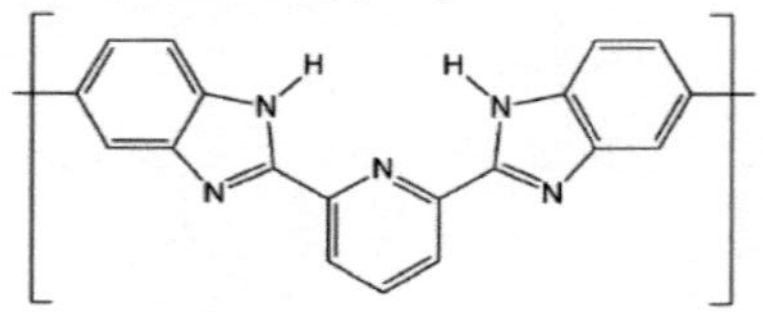	ABPBI (poly(2,5-benzimidazole)) 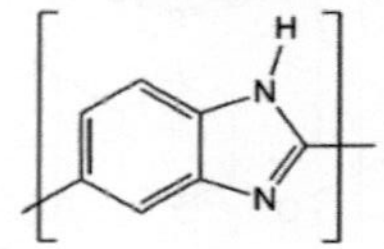
Anion-exchange membranes for AEMFCs			
Polyvinyl Benzyl Trimethyl Ammonium (PVBTMA)	EFTE-based radiation-grafted PVBTMA 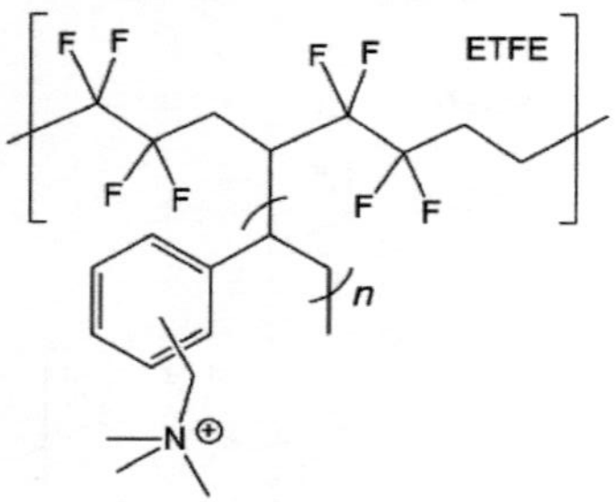	PVBTMA-co-PMB (Polyvinyl Benzyl Trimethyl Ammonium (PVBTMA) polymethyl butylene) 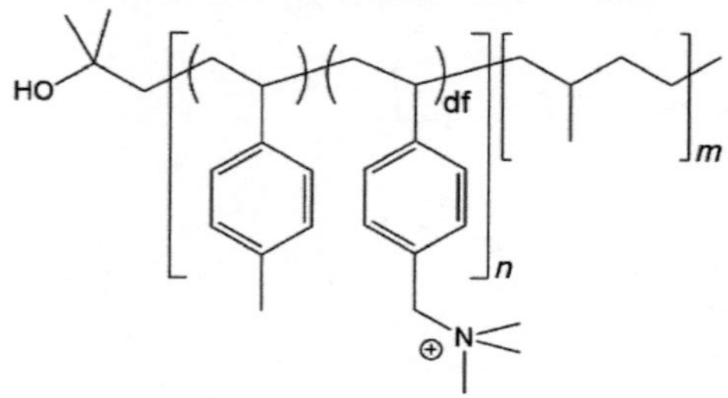	

Functionalized polymers	Functionalized poly(arylene ether ketone) 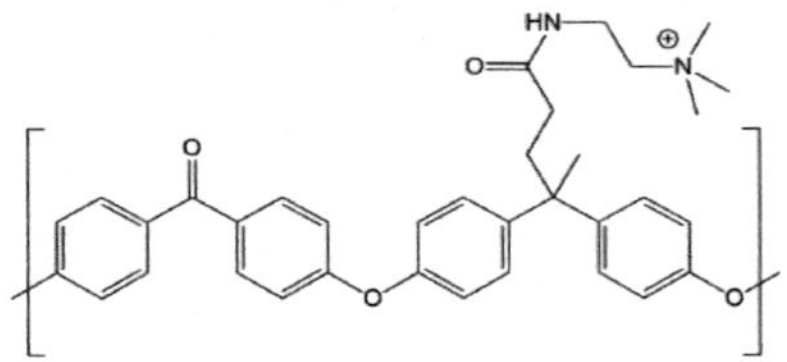	Functionalized poly(sulfone) 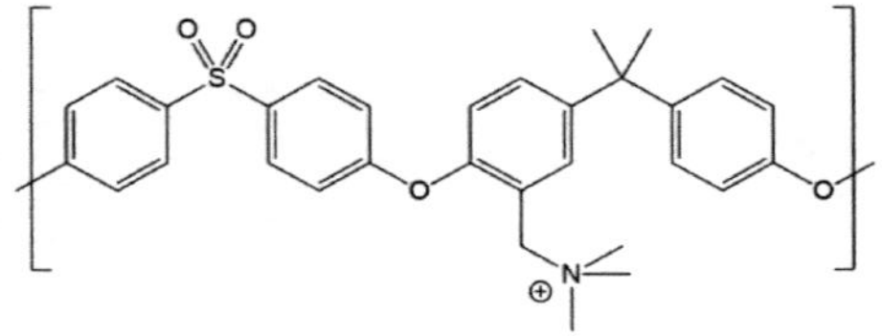	
	Functionalized poly(norbornene) [112] 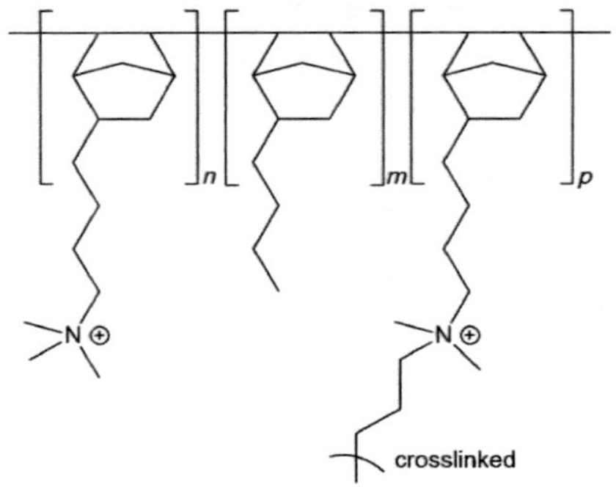	Functionalized poly(2,6-dimethyl-1,4-phenylene oxide) (PPO) 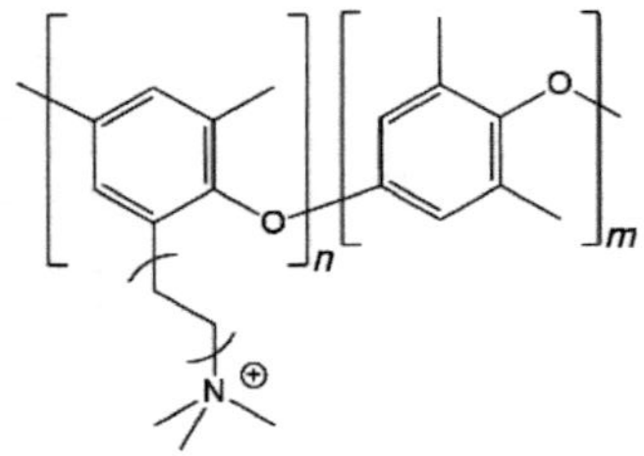	FPK (Functionalized Polyketone) [113] 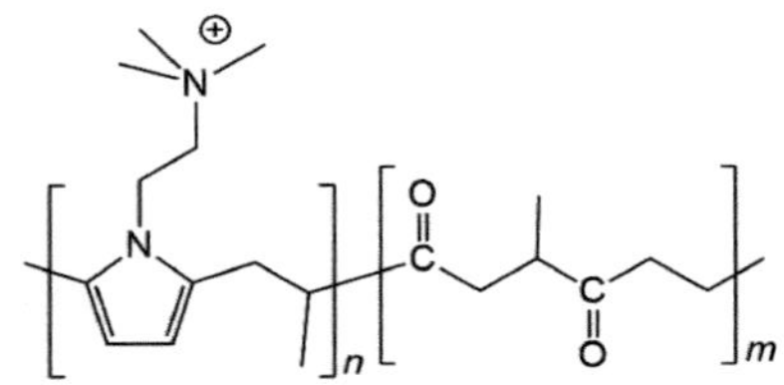

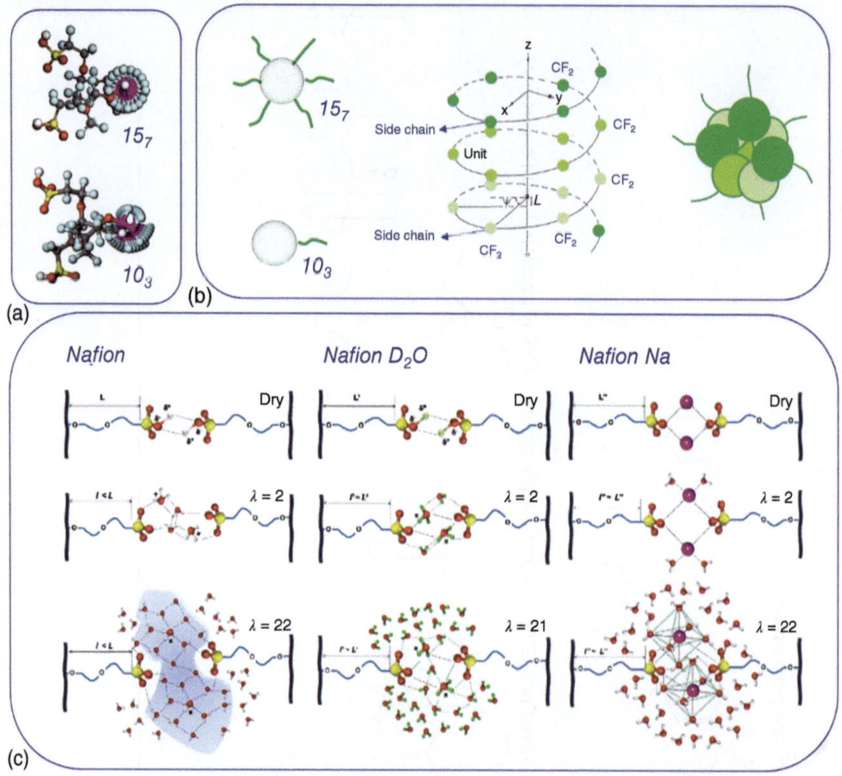

Figure 8.3 Perfluorinated membranes: (a) main conformational structures (15_7 and 10_3) present in perfluorinated materials; (b) 2D aspect for the structure of 15_7 and 10_3 chains; and (c) interaction among functional side groups, cations (H^+, D^+, and Na^+), and water molecules from dry to wet materials ($\lambda = 0$, 2, and 22). Source: Reproduced (adapted) from Refs. [45] and [48] with permissions of Elsevier and the American Chemical Society, respectively.

and Nafion Na(λ_{22}) ($\lambda = 22$ corresponding to a fully hydrated membrane) involved in the solvation of the Na^+ cations, yielding $[Na(H_2O)_n]^+$ complexes.

Among the perfluorinated ionomers is included the doubly functionalized perfluorosulfonic imide acid (PFIA) copolymer, which bears multiacid side chains. The PFIA has the same fluorocarbon backbone of conventional perfluorosulfonic acid (PFSA) ionomers, but the side chains present both a terminal sulfonic acid and a sulfonimide group (Table 8.2). By comparing PFIA with the simpler PFSA, it is interesting to focus on the modification of the polymer structure due to the different side chain: the fraction of crystalline domains, which is in this case associated with 10_3 and 2_1 conformations, increases from the PFSA to PFIA, while the abundance of amorphous domains in the 15_7 conformation decreases. In PFIA, the presence of two ion-exchange groups allows to increase dramatically the ion-exchange capacity. However, the ion conductivity of PFIA is lower than expected. If both protons were fully available for charge transport and the proton conduction mechanism remained the same as that of PFSA, at 100% RH PFIA should exhibit 1.5 times the conductivity of PFSA, while the experimental conductivities of PFIA and PFSA are very similar.

Higher conductivities are observed for PFIA at lower values of relative humidity (e.g. at 30°C and 40% RH, σ = 0.03 and 0.01 S cm^{-1} for PFIA and PFSA, respectively). This is ascribed to the nanophase segregation, as the transport enabled by the longer side chains in PFIA influences the overall proton conductivity more than the number of available protons [48].

8.3.1.2 Nonperfluorinated Membranes

The main drawback of perfluorinated IEMs is their high cost, pushing research toward the development of much cheaper nonperfluorinated polymer backbones for ionomers. Moreover, nonperfluorinated backbones exhibit potential advantages including easy functionalization, high oxidation resistance, high thermal and mechanical stability, and easy recycling.

Aromatic Backbone-based Membranes

Aromatic backbone ionomers are tested (i) to obtain IEMs with an improved chemical and thermal stability with respect to simple aliphatic polymers and (ii) for their ability to achieve the desired IEC by simple sulfonation. Several aromatic polymers are used to prepare CEMs: polysulfones [103, 104], poly(phenylene)oxides [105, 106], and polyether(ketones) [119, 120]. These materials show a good stability, thanks to the stronger C—H bond of aromatic carbons in comparison to the analogue aliphatic polymers [121]. Sulfonated poly(ether-ether)ketone (SPEEK) [30, 31, 41, 60, 122] is a highly promising aromatic ionomer to substitute PFSA ionomers such as Nafion. SPEEK has a good chemical and thermal stability; its precursors are easily available; and its processing is simple processing, curtailing the total cost of the final IEM compared to other ionomers [17]. The bulk conductivity of SPEEK increases with the degree of sulfonation from 9.5×10^{-4} to 3.9×10^{-3} S cm^{-1} at 25 °C for DS = 42% and 62%, respectively. This is not only due to the increase in the number of proton-exchange groups in the IEM but also to the increase in hydrophilicity. This raises the amount of absorbed water, facilitating proton transport [41].

Polyimide-based Membranes

Another family of aromatic polymer backbone ionomers used in CEMs are aromatic-based polyimides. This polymer shows a high mechanical and chemical stability and low crossover of active species. Thanks to its properties, CEMs based on aromatic-based polyimides display an excellent performance in electrochemical devices, comparable with that of Nafion, with a slightly lower thermal stability. A particular type of polymer based on this general structure that is very interesting for application in high-temperature proton-exchange membrane fuel cells (HT-PEMFCs) is polybenzimidazole (PBI). PBI-based CEMs exhibit a high chemical and thermal stability and are quite inexpensive. The most used polybenzimidazole for application in HT-PEMFCs is poly[2,2′-(m-phenylene)-5,5′-bibenzimidazole] (which is usually denoted as mPBI or PBI4N) [96, 107–111]. The very stable structure of mPBI makes it practically insoluble in classic solvents. Pristine mPBI includes no acidic functionalities (see Table 8.2). The high proton conductivity of PBI-based CEMs is obtained by swelling the PBI matrix with a suitable inorganic acid, typically H_3PO_4. The latter dissociates and releases the protons involved in the conduction mechanism [108]. For example, acid-swollen poly-2,2′-(2,6-pyridine)-5,5′-bibenzimidazole (PBI5N) membranes display conductivity values at $T = 135\,°C$ on the order of 10^{-2}–$10^{-3}\,S\,cm^{-1}$ [110].

8.3.2 Anion-exchange Membranes (AEMs)

Anion-exchange membranes (AEMs) are fundamental components for devices exploiting alkaline chemistries or requiring the exchange of anions to balance the charge. AEMs are the core of anion-exchange membrane fuel cells (AEMFCs). The charge carrier in AEMFCs is the OH^- ion rather than the H_3O^+ species exploited by PEMFCs. In AEMFCs, the electrooxidation of the fuel at the anode results in the generation of H^+ species, which react with the OH^- to produce water; the electrons are carried through an external load. The electrons then combine with oxygen and water at the cathode, generating OH^- [13].

In principle, AEMs would (i) transport selectively OH^- ions; (ii) exhibit no electronic conductivity; (iii) enable no fuel crossover; and (iv) display all the previous features also in the hydrated state. Failure to maintain stability will result in performance degradation. One of the main drawbacks of AEMFCs is the performance loss and the difficulty of maintaining the ionic conductivity of the AEM due to the presence of CO_2 contaminants in the feeds. In fact, when air is used instead of pure oxygen, a carbonation process occurs by reaction of OH^- with the CO_2, Equation (8.9). The main result of the carbonation process is a significant decrease in the anion conductivity and consequently a reduction in the AEMFCs power density and durability. Compared to the OH^- conductivity, the conductivity of CO_3^{2-} and HCO_3^- in AEMs is low due to the larger size of the carbonate and bicarbonate ions than the hydroxide ions, which slows down the transport mechanism [123]:

$$\begin{cases} O_2 + 2H_2O + 4e^- \rightarrow 4OH^- \\ OH^- + CO_2 \rightarrow HCO_3^- \\ HCO_3^- + OH^- \rightarrow CO_3^{2-} + H_2O \end{cases} \tag{8.9}$$

Another drawback of AEMFCs is the decrease in the ionic conductivity of the AEM as a consequence of the instability at high pH of both the anion-exchange groups and the polymer backbone due to the reactivity and high nucleophilicity of the OH^- anion (see also Section 8.2.6). Researchers are trying to improve the stability and conductivity of AEMs by increasing the IEC and water uptake through microstructure changes. This objective can be pursued by tuning (i) the chemistry of the backbone and (ii) the type and density of the functional groups [75]. These studies can also reduce the permeability to reactant gases [124, 125].

The main classes of AEMs currently under study bear positively charged anion-exchange side groups including quaternary ammonium, imidazolium, benzimidazolium, pyridinium, phosphonium, and sulfonium groups [87, 126]. Those groups can be linked to several polymeric backbones, including: poly(norbornene) [112], poly(arylene ether ketone) (PAEK) [127, 128], poly(arylene ether sulfone) [129, 130], poly(2,6-dimethyl-1,4-phenylene oxide) (PPO) [131, 132], and polyketone (PK) [113], after appropriate functionalization.

Among the most interesting AEMs are those based on functionalized poly (norbornene). Such ionomers are prepared as either homopolymers or block or random copolymers and show a high IEC (up to 4.55×10^{-3} Eq g^{-1}) [112]. The cross-linking with tetramethyl-1,6-hexanediamine allows to (i) avoid an excessive swelling due to a high WU and (ii) reach ionic conductivities around 200×10^{-3} S cm^{-1}, even at high temperature (80 °C) and without the separation of microphases in the polymer matrix. The alkaline stability tests proved a very low degradation (less than 1.35%) even after 1000 hours in 1 M KOH at 80 °C. Similarly, to proton-exchange PEs, the introduction of an aromatic moiety into the polymer backbone is thought to be one of the best ways to obtain high-performance AEMs. Poly(arylene ether ketone) (PAEK) backbone materials present good mechanical and thermal properties and show a high conductivity after functionalization with alkylammonium groups [127]. The effect of increasing the hydrophobicity of the ion-exchange side chains by increasing the length of the N-alkyl groups (from methyl to propyl) lowers the WU and the elongation but increases the tensile strength. In addition, the anion conductivity decreases from 21×10^{-3} S cm^{-1} to 1.4×10^{-3} S cm^{-1} at 80 °C.

Poly(arylene ether sulfone) IEMs functionalized with imidazolium groups [130] and stabilized by cross-linking show an IEC of up to 2.2×10^{-3} Eq g^{-1}, a high hydroxide conductivity (107×10^{-3} S cm^{-1}), good single cell performance (202 mA cm^2 at 0.6 V with 70% degree of functionalization) at 80 °C and 100% RH, and an improved durability and alkaline stability compared to the non-cross-linked versions [133]. Finally, IEMs based on functionalized poly(2,6-dimethyl-1,4-phenylene oxide) (PPO) display good thermal and mechanical properties, but their chemical stability and ion conductivity at high temperatures are still an open issue [131].

8.3.2.1 Functionalized Polyketones

Polyketones (PKs) are low-cost polymers obtained from the alternated copolymerization of carbon monoxide with ethylene or other olefins. PKs exhibit an outstanding mechanical and chemical stability, making them very promising as backbones for AEMs. The presence of the γ-diketonic repeating units allows the grafting of different side chains and functionalities through the Paal–Knorr cyclization reaction. By selecting a suitable primary amine, it is possible to introduce several different N-pendant groups, which can bestow different features to the polymer structure of the final IEM. The functionalized N-substituted polypyrrole–polyketone copolymers that can be prepared (that are more simply indicated as "*functionalized polyketones – FPKs*") demonstrate a range of excellent properties that are complying with the strict requirements of AEMs. The properties of FPKs are modulated by the amount and nature of the newly formed N-substituted pyrrole pendant side groups, which are introduced by selecting a suitable primary amine. Good thermal properties, easy processing, and high ion conductivity are demonstrated by functionalizing the PK backbone using 4-picolylamine, obtaining *N*-pyridylmethyl polypyrrole-polyketone [113].

At 80 °C, the quaternized pyridyl FPK yields ion conductivities of 8.6×10^{-3} and 10.5×10^{-3} S cm^{-1} in the iodide and hydroxide forms, respectively [113, 134]. Moreover, the hydroxide form of the AEM demonstrates a very good stability in alkaline conditions. No chemical decomposition is observed in the membrane after soaking it in KOH solution for 72 hours. Furthermore, the ion conductivity of the AEM is demonstrated to remain constant for at least 30 days at 80 °C [113]. Other FPK membranes were prepared by functionalizing the PK backbone with 1,2-diaminopropane. The resulting IEM shows a high ionic conductivity: 9×10^{-4} S cm^{-1} at 120 °C for the iodide form. In the hydroxide form, the IEM exhibits a conductivity of the same order of magnitude. Thermogravimetric analyses reveal that this FPK is thermally stable up to 200 °C [135].

8.3.2.2 Poly(Vinyl Benzyl Trimethyl Ammonium) (PVBTMA) Polymers

Several polymers based on the vinyl benzyl trimethyl ammonium (PVBTMA) structure have attracted interest. Radiation-grafted AEMs based on poly(ethylene-co-tetrafluoroethylene)[ETFE] films [136, 137] were the first membranes of this type demonstrating promising conductivities while maintaining good mechanical characteristics. A quaternary ammonium ETFE radiation-grafted (ETFE-g-PVBTMA) AEM in the hydroxide form shows the highest conductivity in comparison with the same AEM exchanged with Cl^-, CO_3^{2-}, and HCO_3^-. The conductivity of the (ETFE-g-PVBTMA) AEM in the hydroxide form at 60 and 80 °C is equal to $112\pm20\times10^{-3}$ and $132\pm20\times10^{-3}$ S cm^{-1}, respectively [138]. The IEC of the same (ETFE-g-PVBTMA) AEM is 1.78 ± 0.02 mEq g^{-1}. This AEM exhibits a performance level similar to that of PFSA CEMs such as Nafion.

The poly(vinyl benzyl trimethyl ammonium bromide)-block-poly(methylbutylene) ([PVBTMA][Br]-*b*-PMB) AEM consists of a block copolymer including both hydrophilic and -phobic moieties. The hydrophilic “A” blocks comprise a statistical distribution of repeat units based on both vinylbenzyltrimethylammonium bromide ([VBTMA][Br]) and 4-methylstyrene (4MS). The hydrophobic “B” blocks comprise repeat units of methylbutylene. The hydrophobic B blocks are larger than the hydrophilic A blocks, allowing for the modulation of the structural features and for the chemical, mechanical, and thermal stabilization of the ionomer. The [PVBTMA][Br]-*b*-PMB AEM exhibits a high WU (ca. 125%) and an IEC of 2.2 mmol g^{-1}. BES studies report the analysis of the electric response of the membrane and show that its ionic conductivity is strongly dependent on temperature. Indeed, in the dry state, the conductivity of the [PVBTMA][Br]-*b*-PMB AEM increases from 2.5×10^{-17} S cm^{-1} at 5 °C to 2.0×10^{-11} S cm^{-1} at 115 °C [139].

The AEM fabricated with the [poly(2,6-dimethyl-1,4-phenylene oxide)-*b*-poly (vinyl benzyl trimethyl ammonium)] copolymer shows (i) an OH^- conductivity of 138×10^{-3} S cm^{-1} and 106×10^{-3} S cm^{-1} at 60 °C [140]; (ii) a low water uptake; and (iii) a reduced swelling; and (iv) no chemical degradation over two weeks. These features suggest that the AEM is highly promising for electrochemical applications.

8.3.2.3 Poly(sulfones) (PS)

The introduction of an aromatic moiety into the polymer backbone is one of the best ways to obtain AEMs exhibiting a high performance and a long durability. Polysulfones (PS) include such aromatic moieties and present a high chemical inertness. Thanks to this property, PS is suitable for application in FC membranes [133]. In addition, PS presents a very high tolerance to high temperatures, and it is very easy to fabricate into a membrane with reproducible properties. AEMs based on PS backbones were developed, demonstrating a high alkaline stability when soaked in 40 wt% aqueous solutions of NaOH at 70–80 °C for over 300 hours. Quaternary ammonium groups were introduced on PS backbones through chloromethylation and subsequent quaternization. The resulting PS-based membrane exhibited an IEC between 0.5 and 1.8×10^{-3} Eq g^{-1}, an electrical resistance between 0.5 and 5 Ω cm^{-2} [141], and conductivities in the range between 10 and 20×10^{-3} S cm^{-1} at 30 °C [142].

8.3.3 Hybrid Ion-exchange Membranes

The utilization of hybrid ion-exchange membranes in energy conversion and storage devices has gained much attention over the past few decades due to the growing interest of the research community in innovative power generation technologies. There are still some technical drawbacks in reference PFSA ionomers (e.g. Nafion) toward their implementation in FCs. In an FC operating at high temperatures (e.g. above the boiling point of water, 100 °C), the IEM undergoes dehydration, and its conductivity drops by orders of magnitude. As an example, for conventional Nafion membranes, the conductivity value reaches 0.1 S cm^{-1} at RH > 90% at $T \approx 80$ °C. However, there is a significant drop in conductivity value to <0.01 S cm^{-1} at 40% relative humidity. Additionally, dehydration (i) causes shrinkage of the IEM, resulting in a poor contact with the electrode configurations and (ii) raises gas crossover due to an increased accessibility of gas flow pathways [143]. New strategies are constantly needed to devise IEMs able to overcome these limitations [144].

Four classes of CEMs have been proposed to address the above drawbacks: (i) partially fluorinated membranes; (ii) composite membranes; (iii) aromatic polymer membranes, and (iv) hybrid inorganic–organic membranes. In this regard, the incorporation of nanoparticles (e.g. SiO_2, TiO_2, ZrO_2, zeolites, and zirconium phosphates) in IEMs, yielding hybrid inorganic–organic composite membranes, has proven to be a promising strategy yielding membranes with improved properties (e.g. reduced fuel crossover and high conductivity at temperatures above 100 °C under low RH conditions). There is also a high fundamental scientific interest to develop hybrid IEMs to study how the interactions between the various components affect the overall features of the final system [92].

Hybrid inorganic–organic PEs are classified into two groups, Figure 8.4. Class A consists of multiphase PEs, while class B comprises monophase polymer electrolytes. Class A is further divided into two groups: Class A1 includes composite polymer electrolytes in which oxide nanopowders are incorporated within the PEs to enhance the chemical, mechanical, and electrochemical properties.

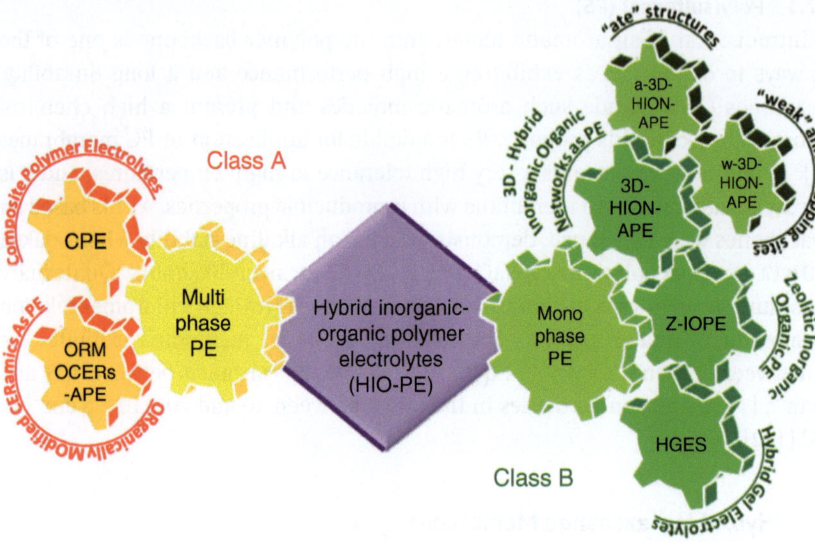

Figure 8.4 Classification of hybrid inorganic–organic polymer electrolytes (HIO-PE) [145]. Source: Reproduced (adapted) from Ref. [9] with permission of Elsevier.

Class A2 involves organically modified ceramics. Organic and inorganic nanodomains are formed through a sol–gel reaction using organometallic silicon precursors and basic components of the PEs. Class B is divided into three subclasses. B1 includes three-dimensional hybrid inorganic–organic network PEs (3D-HION-APE), wherein metal or nonmetal atoms act as bridges between polymer chains. B1 comprises two types of hybrid polymer electrolytes as follows. B1a ("ate" 3D-HION-APE) includes the hybrid PE formed via bridging anionic metal or nonmetal species fixed in the polymer backbone; the mobile cations are the main participant in the resultant ionic conductivity. B1b (w-3D-HION-APE) is the PE with weak anion-trapping sites. This class includes the incorporation of metal or nonmetal atoms with weak Lewis acid characteristics into the skeleton of 3D-HION-APE. Class B2 includes zeolitic-inorganic organic PEs (z-IOPE), wherein inorganic clusters (obtained by the reaction of two or more inorganic coordination complexes) are linked through organic macromolecules. Class B3 makes up "*hybrid gel electrolytes (HGEs)*" conventionally synthesized by the sol–gel process using inorganic or organometallic precursors carrying ion-conducting cation and low-molecular or macromolecular liquids [145].

8.3.3.1 Hybrid Membranes with Single Ceramic Oxoclusters [$P/(M_xO_y)_n$]

The incorporation of inorganic nanoparticles into the polymer matrix is a promising strategy to achieve IEMs exhibiting (i) an enhanced chemical and mechanical stability; (ii) an improved ionic conductivity at high temperatures; and (iii) a more facile water management. Furthermore, the inclusion of inorganic nanofillers leads to a significant increase in the interfacial area between the inorganic nanoparticles and the polymer matrix, which creates a considerable volume fraction with unique properties. Various types of metal oxide nanofillers can be incorporated in the polymer matrix including SiO_2 [92], HfO_2 [117], ZrO_2, TiO_2, Ta_2O_5, and WO_3 [37, 49]. It has been demonstrated that the addition of inorganic additives in a perfluorosulfonic acid matrix prompts the formation of "*dynamic cross-links*" R-SO_3H···[M_xO_y] ··HSO_3-R in the system. The properties of the resulting composite IEMs are modulated by these dynamics, which are typically established within the polar domains. In particular, there is a considerable improvement in the mechanical properties of composite IEMs based on a PFSA matrix. Additionally, the acidity of the M_xO_y filler affects the concentration and strength of the dynamic cross-links in the hydrophilic domains and plays a key role to regulate the hydration, conductivity, and thermal stability of the IEM.

The formation of dynamic cross-links between acid side groups and inorganic nanofillers modifies the properties of the nanocomposite IEM by affecting the macromolecular structure and dynamics of the host. In the case of Nafion (N)-based IEMs it has been demonstrated that out of four possible helical conformations of the PTFE backbone (i.e. 15_7 ($D14\pi/15$), 2_1 ($D2h$), 10_3 ($D6\pi/10$), and 4_1 ($D4$)), only two (i.e. 15_7 and 10_3) are detected for nanocomposite [N/(M_xO_y)] IEMs. The 15_7 conformation is predominant and gives rise to highly ordered polymeric bundles, while the 10_3 conformation is mostly found as a disordered structure. The type of metal oxide used in the preparation of nanocomposite [N/(M_xO_y)] IEMs determines the ratio between these two conformational components. Indeed, the secondary structure of host ionomer in hydrophobic PTFE domains of Nafion plays a crucial role in modulating the amount and type of distributed water domains in nanocomposite [N/(M_xO_y)] IEMs [37, 92]. For instance, the WU increases along the IVB group in the order $Ti < Zr < Hf$ and decreases along the 6° period in the order $Hf > Ta > W$ of the periodic table (see inset of Figure 8.5a) [37]. Four different species of water have been detected in composite membranes: (i) water dimers $(H_2O)_2$ located in the bulk environment of the membrane where dielectric constant is low, i.e. near the hydrophobic fluorocarbon domains; (ii) acid water $[H_3O^+\cdots(H_2O)_n]$; (iii) water molecules solvating oxonium ion clusters directly interacting with the sulfonic anion groups, $[H_3O^+\cdots SO_3^-](H_2O)_n$; and (iv) water molecules interacting with metal oxoclusters and forming their primary hydration shell $[M_xO_y\cdots(H_2O)_n]$. The concentration of each type of water domain embedded in the bulk of [N/(M_xO_y)] IEMs depends on the acidity of M_xO_y oxoclusters, Figure 8.5a. Nanofillers with acidic coordination sites give rise to localized and dipolar interactions; the density of cross-links is high, but the strength of each cross-link is low. In this case, the conductivity mainly occurs in the bulk water nanodomains embedded in the [N/(M_xO_y)] IEMs. In general, nanofillers with acidic coordination sites result in IEMs capable of holding more water, which leads to an increased WU and a more facile proton transport. However, an excessive concentration of nanofiller may trigger the formation of nanofiller aggregates; in this case the density of charge carriers is decreased, and the overall conductivity of the IEM is degraded. The incorporation of inorganic nanofillers into an ionomeric host may also improve the conductivity of the resulting hybrid IEM as compared to its nanofiller-free counterpart. For example, the conductivities of [N/(M_xO_y)] IEMs

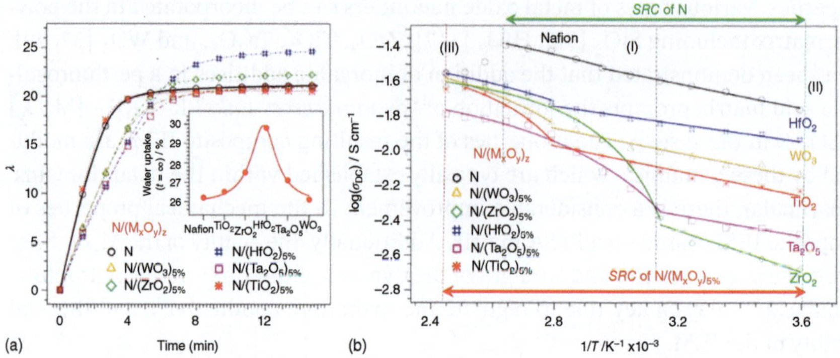

Figure 8.5 (a) Water uptake of $N/(M_xO_y)_z$ membranes, with $z = 0$ and 5 wt% and M = W, Zr, Hf, Ta, and Ti, expressed as λ vs. time. The inset reports the maximum WU for $t = \infty$ as a function of M. (b) Dependence of $\log(\sigma_{DC})$ vs. $1/T$ for $N/(M_xO_y)_z$ membranes, with $z = 0$ and 5 wt% and M = W, Zr, Hf, Ta, and Ti. I, II, and III indicate the three conductivity regions, and SRC the Stability Range of Conductivity [37, 49, 92, 117]. Source: Reproduced (adapted) from Refs. [37] and [49] with permission of Elsevier.

doped with SiO_2, HfO_2, and WO_3 are ca. $60 \times 10^{-3}\,S\,cm^{-1}$ at 105 °C, $28 \times 10^{-3}\,S\,cm^{-1}$ at 135 °C, and $25 \times 10^{-3}\,S\,cm^{-1}$ at 135 °C, respectively, Figure 8.5b. The Stability Range of Conductivity (SRC) is defined as the temperature range where Eq. (8.10) holds true [49, 92]:

$$\frac{\partial(\ln \sigma_{DC})}{\partial\left(\frac{1}{T}\right)} < 0 \tag{8.10}$$

The SRC is a measure of the capability of the IEM to operate at high temperatures. The upper limit of the SRC is directly correlated to the maximum temperature at which the IEM can operate. Nanocomposite Nafion (N)-based IEMs, $N/(SiO_2)_z$, $N/(HfO_2)_z$, and $N/(WO_3)_z$ are characterized by a wider Stability Range of Conductivity (SRC) as compared to pristine Nafion. The SRC of pristine Nafion, $N/(SiO_2)_{5\%}$, $N/(HfO_2)_{5\%}$, and $N/(WO_3)_{5\%}$ is equal to $5\,°C < T < 105\,°C$, $5\,°C < T < 115\,°C$, $5\,°C < T < 135\,°C$, and $5\,°C < T < 135\,°C$, respectively. Thus, $N/(HfO_2)_{5\%}$ and $N/(WO_3)_{5\%}$ are promising for application in PEMFCs operating at temperatures higher than that of reference Nafion.

Another strategy to devise acidic FCs able to operate at high T is to design IEMs wherein proton transport is not mediated by water. Acid–base polymer membranes have been investigated as promising alternative to Nafion that can maintain a high conductivity (on the order of $0.1\,S\,cm^{-1}$) at high temperatures ($T > 120\,°C$) and under low RH humidity conditions. The addition of acidic moieties into a basic polymer matrix promotes proton conduction via acid–base interactions (ionic cross-linking and hydrogen bonding bridges) [146]. One of the most successful high-temperature acid–base PE developed so far is phosphoric acid-doped poly[2,2′-(*m*-phenylene)-5,5′-bibenzimidazole] (PBI4N) [147]. PBI4N is an amorphous heterocyclic basic polymer and is characterized by excellent thermal stability (T_g = 425–436 °C), chemical resistance, and high mechanical strength [148]. The potential of acid-doped PBI4N to be utilized in HT-PEMFCs at low RH was first demonstrated by Wainright et al. in 1995 [147]. Among various inorganic acids (sulfuric acid, nitric acid, phosphoric acid [PA], perchloric acid, hydrochloric acid, etc.), doping of PBI4N with phosphoric acid is widely studied because of the promising properties of H_3PO_4 such as a high thermal stability, a high boiling point, and a high proton conductivity in anhydrous conditions [149]. The conductivity of PBI4N/PA membranes increases as the acid doping level, temperature, and humidification are raised. The acid doping level is usually given as the number of moles of acid per repeat unit of polymer and plays a key role in the determination of the conductivity and the mechanical properties of the IEMs. Various methods have been explored to increase the acid uptake for improved electrochemical properties. Among them, the incorporation of inorganic nanofillers into the polymer matrix is a promising strategy [96, 107].

The inclusion of inorganic nanofillers into phosphoric acid-doped PBI4N composite membranes $[PBI4N(M_xO_y)_x](H_3PO_4)_y$, where M = Zr, Hf, demonstrated that the presence of nanofillers tends to disrupt the hydrogen bonding interactions PBI4N··· PBI4N between PBI4N chains. Hence, the nanofiller acts as a plasticizer. It has been shown that the increase in inorganic filler concentration reduces the biphenyl torsional angle (φ) of bibenzimidazole groups, which improves the planarity of the repeat units (i.e. the aromatic rings in the repeat unit become more coplanar). As a result, the ring stacking becomes less staggered, and interplanar distances along the neighboring planes are reduced. Furthermore, the level of acid increases with the increase in nanofiller concentration, which is responsible for the higher conductivity value of the composite membranes as compared to PBI4N. At a high doping level there is a significant amount of acid within the membrane, which is dispersed homogeneously (i) along the interfaces between the inorganic nanoparticles and polymer domains and (ii) along the polymer chains ($PBI4N^{n+}/H_2PO_4^-/H_3PO_4$, $PBI4N^{n+}/M_xO_y/H_2PO_4^-/H_3PO_4$, PBI4N/PBI4N, $PBI4N^{n+}/PBI4N^{n+}$, $PBI4N/M_xO_y$, and $PBI4N^{n+}/M_xO_y$ interfaces). Therefore, the increase in concentration of nanofiller notably increases the density of inorganic-rich domains, leading to a dramatic increase in acid uptake. At first, acid molecules are aligned along the surface of the nanofillers. However, once all the basic sites on the polymer host become occupied, the excess acid groups form clusters like those that are postulated to be evenly distributed along the polymer chains, Figure 8.6 [107].

Moreover, polarization and dielectric phenomena are facilitated by the new chemistry formed at the interfaces between the nanofiller and the PBI host (as described above) as a result of acid doping. The latter leads to a sharp increase in conductivity values in comparison with undoped analogs. For example, the $[PBI4N(HfO_2)_{0.04}](H_3PO_4)_{12}$ sample at 215 °C presents a conductivity value which exceeds $90 \times 10^{-3}\ S\ cm^{-1}$. This value is higher than that observed for pristine PBI4N under the same conditions ($48 \times 10^{-3}\ S\ cm^{-1}$) [111].

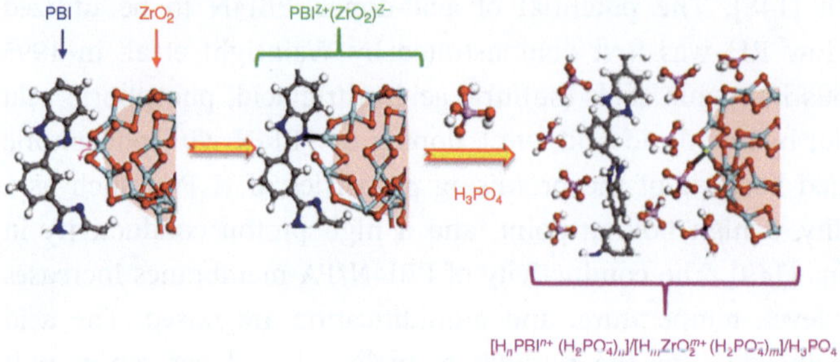

Figure 8.6 Interactions among ZrO_2 nanoparticles, PBI chains, and H_3PO_4 molecules in H_3PO_4-doped nanocomposite [PBI4N$(ZrO_2)_x$] membranes, which lead to the formation of protonated PBI4NH$^+$ and zirconium dihydrogen phosphate $(H_mZrO_2^{m+}(H_2PO_4^-)_m)$. Source: Reproduced (adapted) from Ref. [96] with permission of John Wiley & Sons.

8.3.3.2 Hybrid Membranes Comprising Surface-functionalized Nanofillers

One additional avenue to devise high-performing hybrid IEMs is to modulate the surface chemistry of the nanofiller by suitable functionalization. This approach tailors the interactions between the nanofiller and the macromolecular host, affecting crucial physicochemical features of the resulting hybrid IEM such as the nanophase segregation and the establishment of dynamic cross-links, with the ultimate purpose to improve properties such as ion conductivity, chemical stability, and mechanical stability. In this respect, sulfonation of metal oxide nanoparticles yields acidic moieties on their surface, which modulates the density of mobile protons and influences the percolation pathways in the bulk membrane. In addition to functionalizing nanoparticles with proton-exchange groups, metal oxide nanofillers can also be functionalized with hydrophobic moieties. Highly hydrophobic nanofillers suppress the water uptake of hybrid IEMs, at the same time maintaining a high conductivity. For instance, the fabrication of a hybrid [N/Si_{80}F] IEM comprising silica nanoparticles functionalized with fluoroalkyloxirane is reported [118]. The bulky, nonpolar perfluorinated chains on the surface of the nanoparticles hinder the agglomeration of the filler and act as a plasticizer for the hydrophobic domains of the Nafion host. Hydrophobic nanofillers preferentially interact with the fluorocarbon backbone chains of the host material and trigger the conformational transition $15_7 \leftrightarrow 10_3$ in the hydrophobic PTFE-like domains of Nafion host. Furthermore, cross-linking interactions between the sulfonic acid group and silanol group on the surface of functionalized Si_{80}F nanoparticles led to the mechanical stabilization of the IEMs. The composite membrane [N/$(Si_{80}F)_{0.7}$] showed a maximum conductivity of $83 \times 10^{-3}\,S\,cm^{-1}$ at 135 °C at a low water uptake [118].

8.3.3.3 Hybrid Membranes Doped with Hierarchical *"Core–Shell"* Nanofillers

Hybrid nanocomposite IEMs doped with "*core-shell*" nanofillers of formula $[(M1_mO_n)(M2_xO_y)_z]$, where a "*core*" of an oxide with a high Mohs hardness $(M1_mO_n)$ (M1 = Zr, Ti, Hf) is covered by a thin layer of an oxide characterized by a lower Mohs hardness $(M2_xO_y)$ (M2 = Si, Hf, Ta, W), demonstrated improved

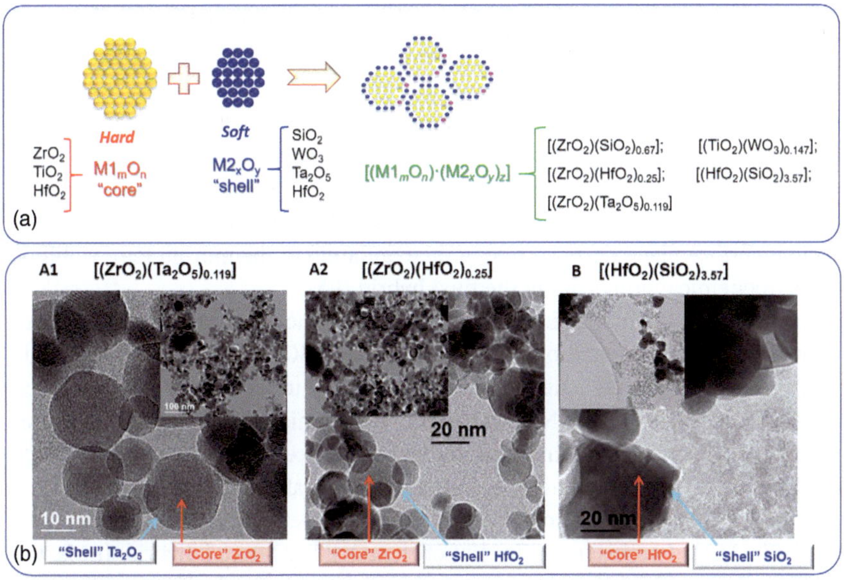

Figure 8.7 "*Core-shell*" nanoparticles: (a) a few examples of $[(M1_mO_n)(M2_xO_y)_z]$ systems; (b) HR-TEM image of $[(ZrO_2)(Ta_2O_5)_{0.119}]$ (class A1), $[(ZrO_2)(HfO_2)_{0.25}]$ (class A2), and $[(HfO_2)(SiO_2)_{3.57}]$ (class B) obtained by ball milling [34, 150]. Source: Reproduced from Ref. [34] with permission of Elsevier.

mechanical properties and proton conductivity as compared to pristine Nafion and $[Nafion(M_xO_y)_n]$ membranes, Figure 8.7 [34, 95, 151].

"*Core-shell*" nanofillers are classified into two categories as follows:

(i) In **Type A** "*core-shell*" systems chemical bonding interactions are established between the "*core*" and the "*shell*" materials. Such materials are characterized by a compatible crystal structure and chemical behavior. Type A is further divided into A1 and A2. Type A1, for example $[(ZrO_2)(Ta_2O_5)_{0.119}]$, are "*core-shell*" nanofillers characterized by strong chemical interactions between two components originated by the formation of metal ion complexes partially coordinated by oxygen atoms on the surface of ZrO_2 "*core.*" The increased surface energy of hard ZrO_2 nanoparticles facilitates the coordination between the surface Zr ion complexes and the surface terminal oxygen atoms of soft Ta_2O_5. This leads to the formation of strong chemical bonds between monoclinic "*core*" ZrO_2 nanoparticles and monoclinic Ta_2O_5 "*shell*" through oxygen bridges, resulting in an isomorphous structure [150, 152]. Type A2 consists of nanoparticles based on materials with a different crystal structure chemical bonded to each other, such as in $[(ZrO_2)(HfO_2)_{0.25}]$, wherein a triclinic HfO_2 "*shell*" is chemically bonded to the surface of a monoclinic ZrO_2 "*core*" [153].

(ii) **Type B** "*core-shell*" systems are obtained by a physical adhesion of the "*soft*" phase characterized by nanoparticles with a low Mohs hardness onto the surface of nanoparticles with a high Mohs hardness. In this case, the "*shell*" is physically interacting with the "*core*." These physical "*core–shell*" nanoparticles are formed when the two starting materials exhibit very different and incompatible crystal structures [34].

Dynamic cross-links are established between the "*core–shell*" nanofiller and the ionomer host (e.g. Nafion). The properties of the resulting nanocomposite IEMs are largely modulated by the interactions developed between the nanofillers and perfluoroethereal side chains of the Nafion host. Thus, the acid–base character of nanofiller surface plays a crucial role in the regulation of the morphology, stability, and electrochemical properties of the final hybrid IEM.

It has been demonstrated that nanofillers with basic coordination sites trigger nondirectional ionic interactions. In this case, the $-SO_3H$ groups of Nafion side chains protonate the basic sites distributed on the surface of nanofillers, which increases the strength of the Nafion-nanofiller dynamic cross-links. Upon hydration, protons bonded by coordination sites on the surface of the nanofillers undergo delocalization. As a result, Nafion behaves as a neutralized polymer matrix, thus improving the thermal, mechanical, and chemical properties of the resulting hybrid IEMs [150, 153]. For instance, in Nafion/$[(ZrO_2)(Ta_2O_5)_{0.119}]$ hybrid IEMs the cross-linking interactions taking place between the "*core–shell*" nanofiller and polymer matrix $RSO_3H\cdots[(ZrO_2)(Ta_2O_5)_{0.119}]\cdots SO_3HR$ improve the mechanical modulus of hybrid IEM in two ways: (i) The presence of $RSO_3H\cdots$(filler) dynamic cross-links promotes the $15_7 \rightarrow 10_3$ conformational transition of PTFE chains of Nafion, which disrupts the crystalline order along the polymer chains and (ii) the electrostatic interactions present in the polar ionic aggregate domains are weakened, thus facilitating the mechanical mobility of PTFE chains of Nafion [150]. Furthermore, an increase in the number density and strength of nanofiller-Nafion dynamic cross-links decreases the swelling ability by reducing the free volume within the hybrid IEMs. The latter indicates that a lower amount of water is required for the solvation of the hydrophilic groups of Nafion involved in dynamic cross-links, thus maintaining the mechanical integrity of the membrane, Figure 8.8 [153]. The hybrid membranes doped with $[(ZrO_2)(Ta_2O_5)_{0.119}]$ "*core–shell*" nanofillers demonstrated elastic modulus of 1 MPa up to 200 °C, while pristine Nafion undergoes an irreversible elongation at 150 °C, Figure 8.8 [154].

The distribution of water in the hybrid IEM comprising the "*core–shell*" nanofillers is strongly influenced by the cross-linking interactions inside its polar cages, similarly to what was revealed in hybrid IEMs comprising conventional oxoclusters (see also Section 8.3.3.1). For instance, six different water domains have been investigated for hybrid IEMs doped with $(M1_mO_n)(M2_xO_y)_z$ "*core–shell*" nanofillers, as follows: (i) H_2O domains solvating the cross-links $[(M1_mO_n)(M2_xO_y)_z \cdots H_3O^+\cdots SO_3^-\text{-}](H_2O)_n$; (ii) H_2O solvating oxonium ions directly interacting with $-SO_3^-$ anion groups: $[H_3O^+\cdots SO_3^-\text{-}](H_2O)_n$; (iii) H_2O domains solvating cation–anion interacting species $[(M1_mO_n)(M2_xO_y)_zH^+\cdots SO_3^-\text{-}]\cdot(H_2O)_n$; (iv) water domains hydrating oxonium ions $[H_3O^+\cdots(H_2O)_n]$, acidic water; (v) bulk water not associated with oxonium ions $(H_2O)_n$; and (vi) water forming the primary hydration shell of $[(M1_mO_n)(M2_xO_y)_z]$ oxoclusters [151]. However, the water is mainly distributed as bulk water $(H_2O)_n$ and water solvating the cross-links

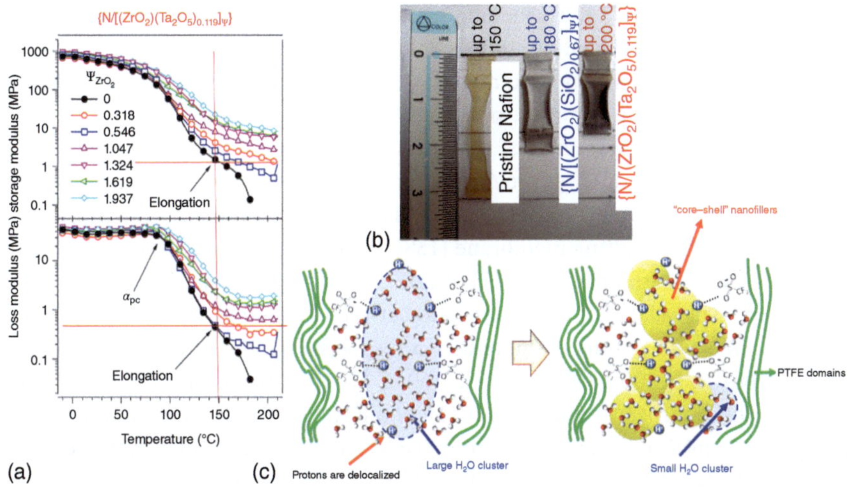

Figure 8.8 N/[(M1$_m$O$_n$)(M2$_x$O$_y$)$_z$] inorganic–organic hybrid membranes: (a) temperature spectra of storage modulus and loss modulus of {Nafion/[(ZrO$_2$)·(Ta$_2$O$_5$)$_{0.119}$]$_\psi$} nanocomposite membranes with $0 \leq \psi \leq 1.937$; (b) photo of membranes after DMA measurements; and (c) scheme of interactions in hydrophobic and -philic domain of Nafion before and after the incorporation of nanofillers.

([(M1$_m$O$_n$)(M2$_x$O$_y$)$_z$ ⋯ H_3O^+ ⋯ SO_3^-)(H_2O)$_n$, which plays a crucial role in the modulation of the conductivity of the nanocomposite membranes.

Thus, the incorporation of nanofillers into the polymer matrix better channelizes the bulk water, thus facilitating the charge migration process owing to an efficient delocalization of protons in the membrane at low hydration and high temperatures.

8.3.4 Porous Membranes

Porous membranes are physical separators characterized by the presence of micro/nanopores in their structure. The main advantages of porous membranes with respect to common IEMs are (i) lower cost and ease of production [75] and (ii) higher chemical and mechanical stability [155]. Several types of porous membranes can be devised, as they can be fabricated from inorganic, organic, hybrid, or natural compounds. In general, the pores of these membranes have a diameter of ca. 50–100 Å [156]. Conventionally, these materials are employed to take advantage of their ability to select the species to be transferred on the basis of their size. In principle, porous membranes are not the best choice for application in FCs due to the high permeability of reactants, which are often gaseous species. Nevertheless, porous membranes are used as (i) host for phosphoric acid (PA) or KOH doping, offering an improved uptake; (ii) support layer, in order to ameliorate the mechanical properties of the final membrane; and (iii) unconventional separators.

8.3.4.1 Porous Membranes as Host Material

The presence of pores in the membrane structure enhances the uptake of PA or other conducting species (e.g. KOH), thus improving the ion conductivity of the separator

[157]. Indeed, the pores in these membranes act as acid/basic reserve voids, offering more room for functionalization. In this field, the polybenzimidazole (PBI) porous membrane is the most widely used material. It is mentioned that these membranes are not the same as the compact, pore-free PBI membranes described above (see also "Polyimide-based Membranes" section). The porous PBI must be doped with a suitable component in order to provide either proton or OH^- conduction (e.g. with PA or KOH, respectively). Typically, the PA doping level is proportional to the ion conductivity of the resulting membrane [158]; thus, a high doping level is required to enhance the performance of the HT-PEMFC during operation. For example, the acid doping level (ADL, i.e. the molar ratio between PA and PBI) can be improved by a factor of 3 in a porous membrane. Correspondingly, the proton conductivity is raised from 70 to $190 \times 10^{-3}\,S\,cm^{-1}$ at 160 °C [159]. In terms of HT-PEMFC performance, the improved proton conductivity allows to obtain a maximum power density of $595\,mW\,cm^{-2}$, compared to $480\,mW\,cm^{-2}$ demonstrated by the device including the dense PBI membrane [159]. Pores in PBI can be obtained by adding a low-molecular-weight compound (porogen) during the preparation of the membrane [160], which is then easily removed by washing with appropriate solvents [161]. Examples of porogen compounds are (i) glucose and saccharose [160, 162]; (ii) ionic liquids [163]; and (iii) dibutyl phthalate [161, 164]. A porous structure can also be formed by vapor-induced phase separation (i.e. the solvent casting step to obtain the membrane is carried out in a constant climate chamber kept at 50 °C and 90% RH) [165] or by means of a non-solvent-induced phase separation (i.e. the PBI solution is added into a PBI-insoluble solvent) [159, 166].

This approach is also used to obtain porous membranes for application in AEMFCs [167]. Indeed, PBI can be doped with KOH, thus obtaining an OH^- ion conductor. In these membranes, the porosity results to be proportional to the amount of porogen (dibutyl phthalate) added during the synthesis [167]. In parallel, the KOH uptake raises from 24% to 58% for dense PBI and porous PBI, respectively [167]. As a result, the ion conductivity of dense PBI is ca. $19 \times 10^{-3}\,S\,cm^{-1}$ at 65 °C, while that of the porous membrane is $78 \times 10^{-3}\,S\,cm^{-1}$ at the same temperature [167]. When implemented in an AEMFC, the maximum power density obtained for porous PBI is $72\,mW\,cm^{-2}$, which is 1.8 times higher than that determined using a dense PBI AEM (i.e. $41\,mW\,cm^{-2}$) [167].

8.3.4.2 Porous Membranes as Support Layer

Despite the high conductivity and chemical stability of reference PFSA IEMs, these systems suffer from a low mechanical strength, especially at high temperatures [168]. This drawback limits the long-term stability of an FC and can be addressed by the use of porous membranes endowed with strong mechanical properties. Indeed, pores allow to easily exchange active ions (e.g. H^+ and OH^-) even though the material making up the membrane is dielectric. At the same time, the porous membrane can act as a strong skeleton for a second ion-conducting component. To achieve this objective, polymers such as polytetrafluoroethylene (PTFE) [169–171], polyethylene (PE) [168], polyether ether ketone (PEEK) [172], and polysulfone (PS) [173] are used. When a PTFE porous membrane is used as support for sulfonated-PEEK (SPEEK), it is observed that the PEMFC mounting the resulting separator displays a higher resistance with respect to that of a cell assembled using a pure SPEEK membrane [171]. Nevertheless, these disadvantages are compensated by the possibility to use a thinner membrane, thanks to its higher mechanical strength and lower gas permeability with respect to a pure SPEEK separator [171]. In particular, porous separators with a thickness of 45 μm can be prepared with the composite material and are stable over cycling, compared to the 75 μm of the pure SPEEK IEM [171]. At 0.6 V, the PEMFC employing the PTFE-SPEEK composite membrane loses only 70 mA cm^{-2} with respect to that using pure SPEEK, reaching a value of 730 mA cm^{-2} [171].

8.3.4.3 Porous Membranes as Unconventional Separators

Pichonat et al. demonstrated the possibility to use a porous silicon-based separator in miniaturized FC applications [174, 175]. The silicon support consists of either a phosphorus-doped 0.016–0.024 Ω cm n^+-type (1 0 0)-oriented silicon wafer [174] or a boron-doped 0.005 Ω cm n^+-type (1 0 0)-oriented silicon wafer [175]. In both cases, the porosity is obtained by means of an electrochemical anodization process, which gives rise to a porosity of 50% and pores with a diameter of 6–30 nm [174, 175]. The porous silicon separator is impregnated with Nafion in order to provide a proton conductivity of ca. 40×10^{-3} S cm^{-1} [174]. A maximum power density of 20 mW cm^{-2} is demonstrated, which is still far from practical uses, but it is promising for this unconventional application [174]. The porous silicon wafer can also be converted into proton conductors by chemically grafting silane functionalities bearing acid groups to mimic the structure of an ionomer (e.g. Nafion) [175]. An ionic conductivity of 1.1×10^{-3} S cm^{-1} is demonstrated, which allows to assemble a miniaturized FC yielding a maximum power density of 17 mW cm^{-2} [175]. This cell shows a slightly higher permeability toward gases with respect to that implementing the Nafion-impregnated separator, since the pores of the former system are not filled and compact like those of Nafion [175].

8.4 Mechanism of Ion Conduction

The study of the ion conduction mechanism plays a pivotal part to (i) optimize the application of known IEMs and (ii) design and develop new ionomers. The mechanism of ion conduction is associated with the long-range migration of charges within the IEM at the nanoscale. We saw in the previous sections how hydration is a crucial parameter to modulate the conductivity mechanism in several families of IEMs. It is possible to determine such mechanism from the macroscopic features of the electric response of a material coupled with its structural and compositional characterization.

In this respect, broadband electrical spectroscopy (BES) is a technique that can deliver a wealth of information on the electrical behavior of a material and its relaxation phenomena [62], complementing and supporting the observations coming from spectroscopic, thermomechanical, and structural techniques such as FT-IR, Raman, DSC, DMA, and XRD. BES is used to define the phenomena contributing to the overall conductivity of ionomers, but it also allows for the extraction of fundamental thermodynamic information including activation energies and average distance of charge migration. In perfluorinated ionomers, but more generally in PEs where domains with different polarity/hydrophilicity are present, either pristine or mixed with nanofillers in hybrid IEMs, BES has been profitably used to elucidate the transport dynamics and conduction mechanism [91, 92, 176].

BES measurements, carried out over a wide range of frequencies (10^{-2} to 10^7 Hz) and temperatures (123–423 K) on Nafion membranes at different levels of hydration (Figure 8.9 shows the relative permittivity, ε_r, of Nafion in both dry, $\lambda = 0$, and wet, $\lambda = 22$, conditions), highlight the different relaxation modes that contribute to charge transport.

In the dry Nafion, which behaves as a dielectric material ($\varepsilon_r \sim 2.3$), up to three dielectric relaxation modes are revealed, Figure 8.10. They can be classified based on the involved moieties at molecular level as α-, β-, and γ-mode, in order of increasing frequency. The α-mode corresponds to the diffusion of conformational states of 15_7 and 10_3 PTFE backbone chains (segmental motions). The β-modes, which correspond to the motion of the polyether side chains, can be further associated with the dipole moment fluctuation of (i) the ether side chains directly bound to the PTFE backbone and (ii) the ether side chains bearing the sulfonic acid groups. Finally, the γ-mode corresponds to the short-range motion of CF_2 units of PTFE backbone chains. The α-transition, which is difficult to reveal in wet ion-conducting membranes, can be easily detected in perfectly dry Nafion membranes.

There are two observed polarization modes, respectively, named σ_{EP}, or electrode polarization, and σ_{IP}, or interdomain polarization. The former polarization mode corresponds to the accumulation of charge at the membrane/electrode interface, while the latter mode derives from the accumulation of charge at the interface

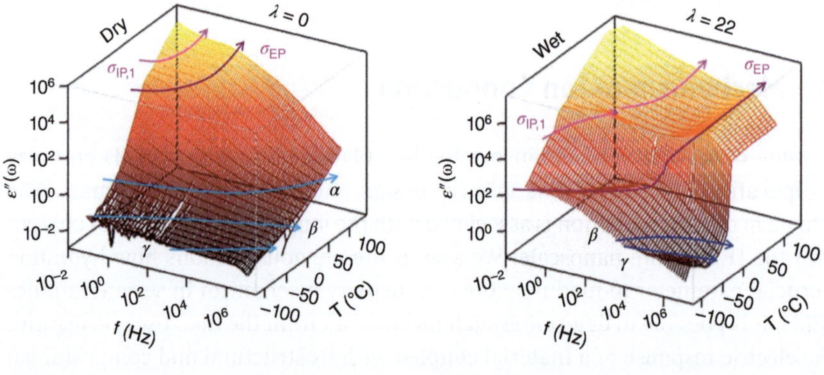

Figure 8.9 3D spectra of the imaginary permittivity (ε'') of dry Nafion ($\lambda = 0$, on the left) and wet Nafion ($\lambda = 22$, on the right) as function of temperature and frequency. The spectra reveal the presence of one electrode polarization (σ_{EP}), up to two interdomain polarizations ($\sigma_{IP,i}$), and three different dielectric relaxations: α (segmental motion of the fluorocarbon chains), β (fluctuation of the dipole moment of polyether side chains), and γ modes (short-range motion of CF_2 units of PTFE backbone chains).

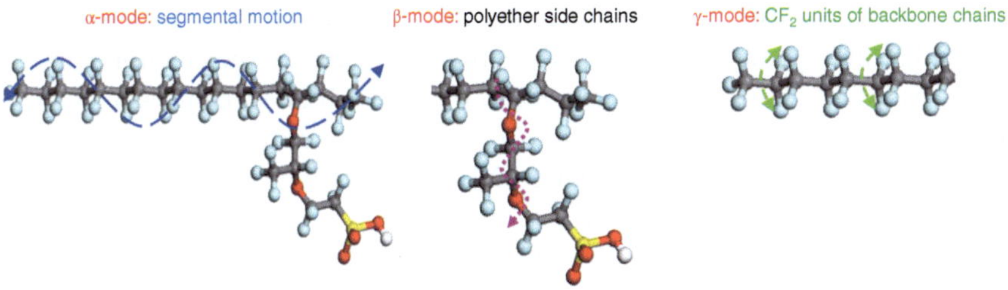

Figure 8.10 Dielectric relaxations detected in perfluorinated IEMs: α-mode is attributed to the segmental motion of the PTFE backbone chains, β-mode to the fluctuation of the dipole moment of polyether side chains, and γ mode to the short-range motion of CF_2 units of PTFE backbone chains. Source: Reproduced (adapted) from Ref. [177] with permission of the American Chemical Society.

between domains with different permittivity where the protons are easily delocalized. The presence of a single interdomain polarization event suggests that there are only two types of domains, corresponding to (i) the hydrophilic sulfonated perfluoroether domains and (ii) the hydrophobic PTFE backbone.

Bearing in mind the microscopic structure of the perfluorinated ionomers, in their dry form (WU = 0) the motion of the proton (H^+) is possible only by hopping from one ion-exchange site to the next, thus forming percolation pathways (σ_{IP}) at the interface between the hydrophilic and -phobic domains, Figure 8.11. Being limited to this mechanism, which is much slower if compared to the fast exchange of H^+ between water molecules, the ion conductivity is extremely low, below 10^{-7} S cm^{-1} [67].

In wet conditions PFSA shows a splitting in the β relaxations since two distinct phenomena are segregated at low T, ascribed to the 10_3 and 15_7 conformations of the side chains. This transposes to four different components of σ_{IP} due to the different nature of the hydrophilic and -phobic domains occurring at different helicoidal

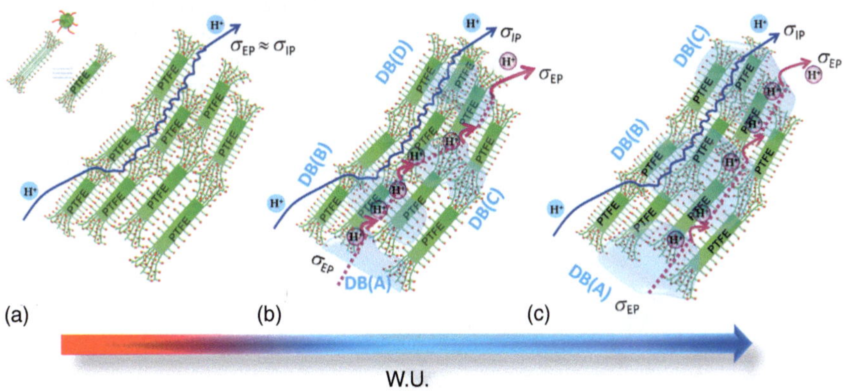

Figure 8.11 Conductivity mechanisms in perfluorinated membranes as function of WU values: (a) dry; (b) partially hydrated ($2 \leq \lambda \leq 6$); and (c) fully hydrated (wet, i.e. $\lambda = 22$ for Nafion) membranes. PTFE: bundle of perfluorinated chains, σ_{EP} = electrode polarization conductivity, and σ_{IP} = interdomain conductivity.

conformations (i.e. the two conformations 15_7 and 10_3 and their corresponding hydrophilic and hydrophobic domains). As the WU increases, the amount of water molecules close to the ion-exchange sites in the hydrophilic domains is increased. In this condition, long-range conduction (σ_{IP}) occurs primarily via proton exchange between different delocalization bodies (DB(A) → DB(B)), which is facilitated by water (or H_3PO_4) and the segmental motion of the hydrophobic domains (peristaltic mechanism). A "*delocalization body – DB*" is defined as the wet volume of the IEM, where the H^+ is delocalized with respect to the time scale of the overall conduction mechanism, Figure 8.12. The long-range migration of protons occurs through the contact of contiguous DBs favored by the dynamics of the host polymer at a progressively higher rate, becoming the main contribution to the ion conduction. Increasing the equivalent weight (EW) of the ionomer corresponds to raise the length of the PTFE sections between ion-exchange groups, reducing the charge density and IEC, and also increasing the crystalline fraction of the polymer matrix. Thus, the larger crystalline domains lead to longer percolation pathways for proton migration between DBs, reducing both the probability of charge transfer and the overall conductivity.

Similar mechanisms can be invoked, with due adaptation, for the materials presenting a similar permittivity domain segregation in their structure, such as PFIA and PFSA membranes. In terms of proton conductivity, PFIA has a similar behavior to PFSA at extreme water contents (0% or 100%) but shows higher conductivity values at intermediate hydration levels (see also Section 8.3.1.1). Thanks to the smaller water clusters in hydrophilic domains (as proved by the smaller conductivity gap at 0 °C), protons in PFIA are more efficiently exchanged between DBs. Again, the interdomain polarization is diagnostic of the number of domains with different permittivity; from the four σ_{IP} detected in the dry material (two crystalline conformations both with hydrophilic and -phobic domains), we move to six components in the

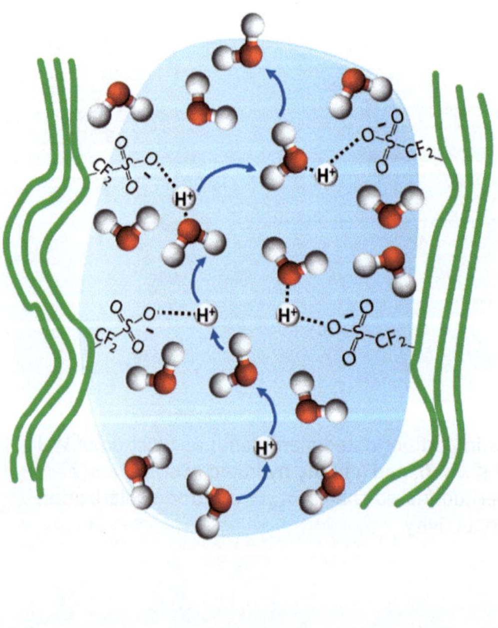

Figure 8.12 Schematic depiction of a delocalization body, i.e. a wet polar domain of a perfluorinated membrane, showing the possible fast proton-exchange processes responsible for the proton delocalization within this volume.

wet form (three crystalline conformations with two domains, each with a different permittivity). If σ_{IP} dominates the overall conductivity below the freezing point of water (0 °C), then charges migrate along interdomain migration pathways. At high T, σ_{EP} contributes progressively more and more, becoming dominant in the overall conductivity above 50 °C. In these conditions charge migration through DBs is responsible for the overall conductivity.

In SPEEK both σ_{EP} and σ_{IP} rise with the degree of sulfonation (DS) (see also "Aromatic Backbone-based Membranes" section), which corresponds to a higher density of $-SO_3H$ groups. A higher DS will raise the concentration of mobile protons, but also increase the hydrophilicity of the IEM, modulating the WU and the proton mobility. This can be easily visualized by determining the average proton migration distance, $<r_{ED}>$, between hydrophilic domains in the bulk membrane, Figure 8.13, that can be determined from the conductivity data by applying the Nernst–Einstein equation. In fact, by increasing the DS and, correspondingly, the density of ion-exchange groups, we allow for the formation of a higher number of hydrophilic domains close to one another, reducing the distance between DBs and streamlining the conduction pathways through the IEM.

The most common anion-exchange ionomers are also structurally characterized by (i) regions with a higher hydrophilicity, where the anion-exchange groups are present as side chains and (ii) other more hydrophobic regions where the nonfunctionalized polymer units (hydrocarbon-, polyether-, or polyketone-based) bestow chemical stability to the IEM in the alkaline environment. In AEM we can expect a similar conduction behavior as that described for PEMs, but we must remember that (i) the hydroxide anion (OH^-) is far less mobile than the proton, leading to a reduced

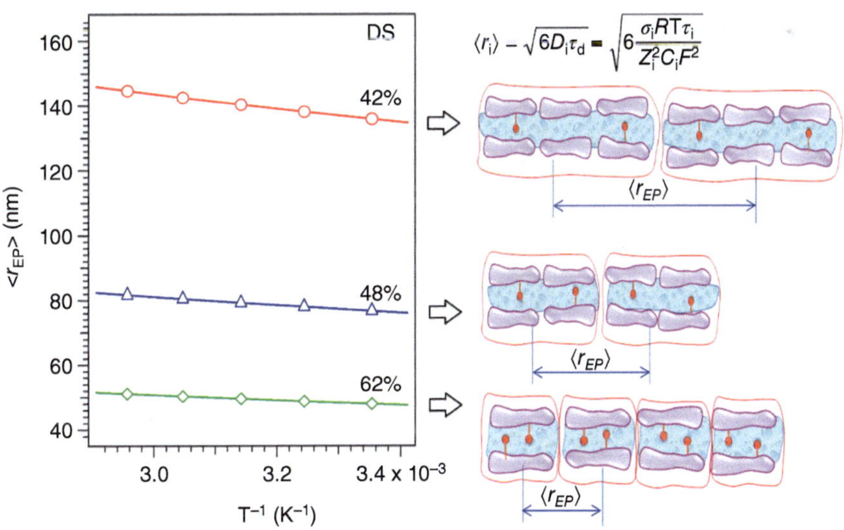

Figure 8.13 SPEEK-based membranes: average migration distance ($<r_{EP}>$) attributed to σ_{EP} as a function of Sulfonation Degree (SD) and temperature (on the left) and schematic representation of the size of the Delocalization Bodies (DBs) as a function of DS in the wet IEMs. Source: Reproduced (adapted) from Ref. [41] with permission of Elsevier.

conductivity at similar IEC values and (ii) OH^- (and other anions, in general) tends to form stronger and closer ion pairs with the countercation in the ionomer structure. This gives rise to dipole interactions that can hinder both the relative chains motion of the host matrix and the anion transport through the membrane.

In the block copolymer [PVBTMA][X]-*b*-PMB (poly(vinylbenzyltrimethylammonium)-*b*-poly(methylbutylene) bromide, with X = Br, or hydroxide, with X = OH) the long-range anion migration process occurs when X^- is exchanged between different delocalization bodies (DBs).

This process is modulated by the local dipole fluctuations of the benzyl trimethyl ammonium side groups (β_2 relaxation mode, Figure 8.14) supported by the H_2O solvation processes and the relaxations of the host matrix. The effect of the relative motion of the polymer host is evident at a higher temperature (80 °C in the OH^- form with 41.7% degree of functionalization, DF%) with a sudden drop in ion conductivity, Figure 8.15. At this temperature, indicated with T_δ, a disorder–order transition in the hydrophilic domains is observed (see also Section 8.2.7). At T_δ the randomly oriented local dipole moments of the R-TMA $^+X^-$ ion-pair side groups of different backbone chains can couple in an ordered arrangement due to a strong dipolar interaction that limits the mobility of the charge carriers and lowers the overall conductivity.

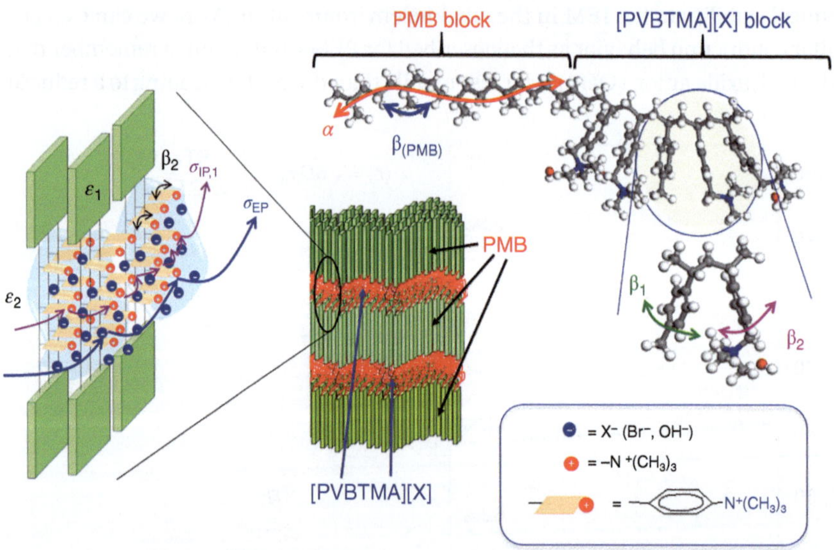

Figure 8.14 Conductivity mechanism and dielectric relaxations for [PVBTMA][X]-*b*-PMB diblock copolymer membrane with X = Br or OH. σ_{EP} and $\sigma_{IP,1}$ are the conductivity pathways as described in the text. ε_1 and ε_2 are the permittivity of PMB and [PVBTMA][X] blocks, respectively. Dielectric relaxations: α-mode (segmental motion of the PMB chains); $\beta_{(PMB)}$-mode (fluctuation of the local dipole moment of MB repeat units); β_1-mode (fluctuation of the dipole moment of the tolyl side groups); and β_2-mode (fluctuation of the dipole moment of the benzyl trimethyl ammonium side groups) [139, 178]. Source: Reproduced (adapted) from Ref. [139] with permission of the Royal Society of Chemistry.

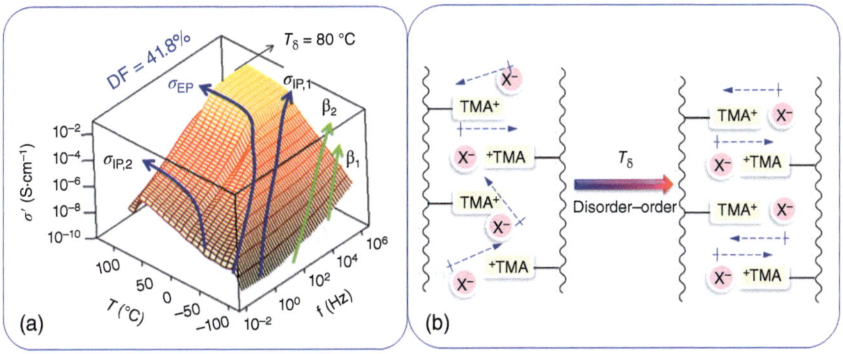

Figure 8.15 (a) 3D plot of real conductivity for [PVBTMA][OH]-*b*-PMB diblock copolymer membrane ($DF\% = \frac{n_{[BTMA][X]}}{n_{[BTMA][X]}+n_{4MS}} \cdot 100 = 41.8\%$, where $n_{[BTMA][X]}$ and n_{4MS} are the average moles of vinyl benzyl trimethyl ammonium and tolyl groups, respectively) as function of frequency (0.01 ÷ 10^7 Hz) and temperature (−105 ÷ 150 °C) [139, 178]; (b) Disorder–order transition, T_δ, in hydrophilic domains. At T_δ the randomly oriented ion-pair local dipole moments (ip-μ) of R-TMA^+X^- side groups are ordered together. Source: Reproduced (adapted) from Ref. [178] with permission of the American Chemical Society.

In AEMs consisting of poly[*N*-(4-methyl-methylpyridinium)ethylenepyrrole]- or poly[*N*-(2-trimethylammonium-ethyl)ethylenepyrrole]-polyketone block copolymer, three conductivity regions are observed. The latter are delimited by the main thermal transitions of the IEM (i.e. T_g, water melting and T_I). The electric response is dominated by the two polarization events σ_{IP} and σ_{EP}. The four dielectric relaxations are associated either to the segmental motions of the poly(ethylenepyrrole) (α_2) and polyketone chains (α_1) or to the local fluctuations of the dipole moments of the carbonyl groups along the backbone (β_1) or those of the N-functionalized pyrrole units (β_2). The α_1-mode is dominant when the hydrophobic domains prevail as in the dry membrane. When the membrane is hydrated, the relaxation of the hydrophilic polyethylene pyrrole (PEP) backbone chains (α_2) prevails, and the segmental motion of the polymer matrix modulates the long-range anion transfer. The overall conductivity is again the superposition of (i) the exchange of X^- between DBs and (ii) the migration of the anion along the interdomain percolation pathways (IPPs) at the interfaces between PK domains with low ε and PEP domains with higher permittivity.

Hybrid membranes can be classified by the type of nanofiller used (see also Section 8.3.3). The nature of the filler interaction with the ionomer host has an important effect on the structure of the final IEM at the nanoscale and thus on the phenomena involved in the conduction mechanism. For instance, when considering the nanocomposite [Nafion/$(SiO_2)_x$] membranes, they show two molecular relaxation phenomena [92]. The former is associated with the α-relaxation mode of PTFE fluorocarbon domains, and the latter to the β-relaxation mode of the sulfonic acid side groups of Nafion. The two relaxation modes are strongly coupled and indicate the interactions between the polar groups of the host polymer and the surface atoms of the $(SiO_2)_x$ nanofillers, Figure 8.16. The study suggests that charge migration occurs along interconnecting channels and polar hydrophilic

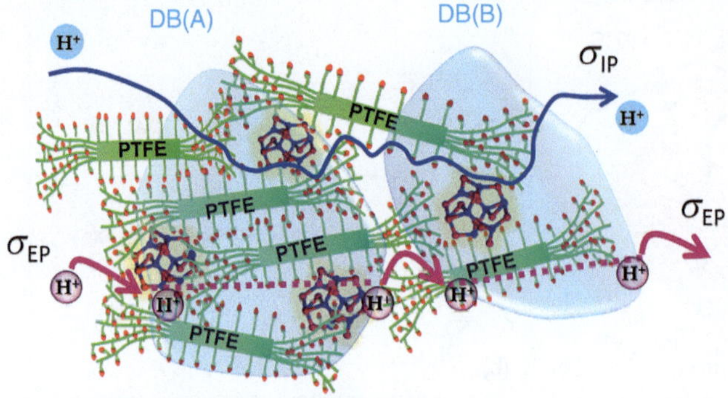

Figure 8.16 Conductivity mechanisms in hybrid membranes ($N/(M_xO_y)_n$ and $N/[(M1_mO_n)(M2_xO_y)_z]$): PTFE: bundle of perfluorinated chains, σ_{EP} = electrode polarization conductivity, and σ_{IP} = interdomain conductivity.

clusters, thanks to proton-exchange events by hopping between fluctuating water species in different hydrophilic domains. The process is then modulated by the different types of interstitial water domains, their amount in the IEM, the density of $SiO_2\cdots(HSO_3)$-cross-links, and the segmental motions of the PTFE fluorocarbon backbone.

The extended study on the electrical behavior of a series of metal oxide nanohybrids [49] [Nafion/$(M_xO_y)_n$] (with M = Ti, Zr, Hf, Ta, W; $n = 5$ wt%) confirmed that conduction is bottlenecked by proton hopping between the polar hydrated clusters. This process is mediated by the several factors observed before for [Nafion/$(SiO_2)_x$] materials. In brief, the conductivity in these membranes takes place through a *peristaltic*-like mechanism, where the long-range charge migration takes place when different water domains (DBs) intersect with one another allowing the hopping of the protons.

There are examples where the filler itself can contribute to the proton conductivity as in the case of the hybrid of Nafion with sulfated-ZrO_2 as the filler. In this system, the acidic functionalities on the surface of the filler allow the deprotonation of the acid moieties of the polymer host even in dry conditions, resulting in a conductivity of 3×10^{-3} S cm^{-1} at 120 °C [179]. At temperatures below the melting point of water, the ionic conductivity of the wet membrane is like that of the dry membrane, with the long-range charge transport occurring via a hopping mechanism. At higher temperatures, proton conduction is facilitated by the motion of the host matrix in the dry system, while it is largely driven by the water clusters in the wet membrane, with a smaller contribution from the motion of the host matrix. When the temperature approaches 0 °C in wet conditions, the proton-rich interfaces provide a percolation pathway in the polar domains that are sterically hindered by the presence of solid water.

It is interesting to note that the use of nanofillers functionalized with hydrophobic pendant groups can also enhance the ion conductivity of Nafion. In fact, the introduction of silica-based nanoparticles functionalized with heptadecafluorononyl surface groups [118] (see also Section 8.3.3.2) is responsible for (i) a plasticizing effect on the fluorocarbon domains of perfluorinated ionomer and (ii) the formation of cross-links between the residual silanol groups on the surface of the nanofiller's silica "*core*" and the sulfonic acid groups of the host polymer. These produce a reduced WU of the hybrid IEM but also a wider stability range of conductivity ($5 \leq T \leq 135\,°C$), with σ comparable to that of the pristine recast Nafion (up to $0.083\,S\,cm^{-1}$ at 135 °C), see also Section 8.3.3.1.

The strength of the interactions between nanofiller particles and Nafion host are directly correlated with the amount of 10_3 helical conformation in the host ionomer, which allows for more efficient cross-linking. Nevertheless, the strength of these interactions mainly depends on the surface properties and basicity of the nanofiller as it can be seen comparing the "*core–shell*" hybrid IEMs [Nafion/$(SiO_2)(HfO_2)_x$] and [Nafion/$(ZrO_2)(HfO_2)_x$]. In the former IEM the H^+ of the sulfonic groups of Nafion are localized in direct dipolar interactions with the nanofiller particles. The opposite phenomenon occurs in the latter IEM, which comprises a more basic filler, where the ion conductivity is enhanced by the efficient delocalization of protons [50, 153] despite the increase in WU and diffusion coefficients. These phenomena depend mainly on the general increase in polymer chains disorder due to the presence of the nanofiller particles.

Nafion membranes incorporating "*core–shell*" $(ZrO_2)(Ta_2O_5)_{0.119}$ nanofillers (see also Section 8.3.3.3) show an improved conductivity and mechanical stability with respect to pristine Nafion [152]. The BES analysis of the hybrid IEMs indicates that the long-range ion transport is closely related to the segmental motions of the host matrix, with a stabilizing effect of the $RSO_3^-\cdots[H\text{-filler-}H]^{2+}\cdots SO_3R$ dynamic cross-links on the stability of both hydrophilic and -phobic domains. The reduced WU of the hybrid IEM, when compared to pristine Nafion in the same conditions, suggests that the proton conduction occurs at the interfaces between the hydrophilic domains of the polymer and the surface of the filler's nanoparticles. The presence of the nanofiller (i) contributes to reduce the size of the water clusters in the wet IEM (DBs) but (ii) also enhances the ion conduction (see also Section 8.3.3.2 – Figure 8.7c).

The plasticizing effect of nanofiller is also observed in PBI-based hybrid IEMs. The nanocomposite polybenzimidazole membranes doped with zirconia [PBI4N $(ZrO_2)_x$]$(H_3PO_4)_y$ proved that the nanofiller can affect the interchain dipole interactions of the ionomer, modulating the overall electrical response of the IEM [180]. Acid doping yields an increase in ion conductivity by modifying the chemistry at the interfaces between the nanofiller and the PBI host (see also Section 8.3.3.1 – Figure 8.8), facilitating the dipolar fluctuations and the segmental motion of the polymer chains. Ion conductivity can proceed by proton hopping between the binding sites of the polymer host and hopping between $H_nPBI4N^{n+}/H_nPBI4N^{n+}$ and $H_nPBI4N^{n+}/H_mZrO_2^{m+}$ sites at the interfaces with the nanofiller. The amount of nanofiller has an important effect on the mechanical and ion conduction properties of the IEM. At a content of ZrO_2 up to 8 wt% the thermal and mechanical properties of the hybrid IEM increase as the acid uptake is raised [96]. In this case the protonation of PBI4N chains forms $H_2PO_4^-$ species, which coordinate the surface of the nanoparticles of the filler increasing both (i) the number of polymer–nanofiller dynamic cross-links and (ii) the delocalization of the mobile protons responsible for the long-range charge migration. A further increase in the amount of nanofiller beyond 8 wt% yields a decrease in conductivity triggered by (i) the segregation of the nanofiller within the host polymer matrix and (ii) the disruption of the polymer nano- and microstructure.

The effect of nanofiller content is also evident in $[PBI4N(HfO_2)_x](H_3PO_4)_y$ hybrid IEMs (see also Section 8.3.3.1) [107], where the disruption of the interchain hydrogen bonding allows the aromatic units to adopt a more coplanar structure. This facilitates a less-staggered structure of the aromatic planes, but only up to a nanofiller fraction of $x = 0.08$. Indeed, when the content of HfO_2 exceeds this value, a phase separation occurs forming different domains: (i) PBI4N nanofiller domains and (ii) pure nanofiller domains. Coherently, ion conductivity increases up to $x = 0.1$ (with a maximum at $x = 0.04$) owing to the introduction of defects into the morphology of the IEM. This facilitates charge transfer and the relaxation modes of the host PBI but then slightly decreases as the amount of HfO_2 nanofiller is raised. Taken all together, the conduction mechanism in hybrid PBI-based IEMs can be associated with the formation of conductivity pathways that involve either interdomain proton migration through hopping events or proton exchange between different delocalization bodies [111].

A comparison of the ion conductivities displayed by common ionomers and some of the hybrid IEMs obtained from them are reported in Table 8.3.

8.5 Summary and Perspectives

The rational development of IEMs that are suitable for application in low-temperature FCs is fraught with difficulties, especially due to the highly demanding and often conflicting requirements that must be satisfied. A high ionic conductivity (typically more than $0.1\,S\,cm^{-1}$) is desirable to maximize FC power density but must also correspond to (i) a very high selectivity in the transport of the species of interest and (ii) a minimized permeability to reactant species (e.g. H_2 and O_2). The IEM must be highly durable, to comply with the requirements of applications that mandate lifetimes of several thousands of hours at least. This is particularly hard to achieve by an IEM forced to operate inside an FC in very harsh conditions in terms of extreme pH values and broadly different chemical potentials. The IEM must also be inexpensive, free of strategic elements, and easy to dispose/recycle, to facilitate large-scale deployment. This conflicts directly with the complex processes that are often necessary to prepare and upscale the IEM.

The research has devised several possible solutions to address the complex issues outlined above. To obtain viable IEMs it is necessary to finely tune the chemical composition, morphology, and structure of very complex materials, often consisting of a polymer matrix wherein the backbone is functionalized with suitable ion-exchange moieties. In several instances further components must be added, with a particular reference to (i) species promoting the selective transport of ions (e.g. water and phosphoric acid); (ii) additives meant to raise conductivity especially at low hydration

Table 8.3 Comparative conductivity of selected IEMs.

Class	Electrolyte group		Membrane	Conductivity[a)] (S cm^{-1})	T_{max} of SRC[b)] (°C)	References
PEMs	Perfluorinated	Nafion	Nafion	$3.0 \cdot 10^{-2}$	105	[150]
			Nafion/ APNB-10%	$8.2 \cdot 10^{-2}$ (20 °C)	/	[11]
			PVDF/Nafion (PVDF/$N_{0.9}$)	$4.2 \cdot 10^{-2}$	80	[177]
			{Nafion/[(ZrO_2) (Ta_2O_5)$_{0.119}$]$_{13}$}	$7.1 \cdot 10^{-2}$	160	[150]
		PFSA	PFSA 1100 EW	$8.3 \cdot 10^{-2}$	120	[91]
		PFIA	PFIA	$6.5 \cdot 10^{-2}$	120	[48]
		Gore	Gore	$9.6 \cdot 10^{-2}$		[13]
	Aromatic	SPEEK	SPEEK, DS = 62%	$3.5 \cdot 10^{-3}$	110	[41]
			Me-SPEEK	$1.6 \cdot 10^{-1}$ (100 °C)	/	[120]
			Ph-SPEEK	$1.8 \cdot 10^{-1}$ (100 °C)	/	[120]
			[SPEEK/ (WO_3)$_{0.09}$]	$4.9 \cdot 10^{-3}$ (20 °C)	110	[30]
		PS	Polysulfone (triazole-grafted polysulfone)	$1.2 \cdot 10^{-1}$	160	[157]
		PPO	PPO-TBD	$1.3 \cdot 10^{-3}$ (80 °C)	190	[106]
			PPO-TDB-Me	$1.1 \cdot 10^{-2}$ (80 °C)	190	[106]
	Polyimide	PBI	PBI/H_3PO_4	$1.0 \cdot 10^{-2}$	160	[18]
			PBI4N/(H_3PO_4)$_{11}$	$5.4 \cdot 10^{-2}$	225	[180]
			PBI5N/H_3PO_4	$1.0 \cdot 10^{-2}$	220	[110]
			[PBI4N/ (HfO_2)$_{0.13}$]/ (H_3PO_4)$_{13}$	$9.0 \cdot 10^{-2}$	225	[96, 107]
			[PBI4N/ (ZrO_2)$_{0.21}$]/ (H_3PO_4)$_{13}$	$1.1 \cdot 10^{-1}$	225	[180]
			PBI-RGO/ PPBI-80/ PBI-RGO	$1.1 \cdot 10^{-1}$	170	[162]
		Porous PBI	Asym PBI67	$6.0 \cdot 10^{-2}$	160	[163]
			m-PBI350-2	$1.8 \cdot 10^{-1}$	160	[159]
		ABPBI	ABPBI-grafted montmorillonite (15PBMPP)	$1.57 \cdot 10^{-1}$ (140 °C)	/	[181]

(Continued)

Table 8.3 (Continued)

Class	Electrolyte group		Membrane	Conductivity[a] (S cm^{-1})	T_{max} of SRC[b] (°C)	References
AEMs	PVBTMA	EFTE based	ETFE-g-PVBTMA	$1.32 \cdot 10^{-1}$ (80 °C)	/	[138]
		PMB based	[PVBTMA][Br]-*b*-PMB	$6.7 \cdot 10^{-5}$	120	[139]
			[PVBTMA][OH]-*b*-PMB; DF = 80.7%	$4.4 \cdot 10^{-3}$	120	[178]
		Poly (norbornene)	Random copolymer (GTR75-5)	$1.94 \cdot 10^{-3}$ (80 °C)	/	[112]
			Block copolymer (GT75-5)	$2.01 \cdot 10^{-3}$ (80 °C)	/	[112]
		PS	PSf135-ImOH	$4 \cdot 10^{-2}$ (20 °C)	150	[133]
		FPK	FPKm(I) (4-Pic)	$3.3 \cdot 10^{-2}$	120	[113, 134]
			FPKm(OH) (4-Pic)	$2.1 \cdot 10^{-2}$	120	
			FPKm/$(Ta_2O_5)_{0.0066}$ (I) (*t*MAE)	$1.96 \cdot 10^{-2}$	80	[135]
	Porous	qPVBz/OHL	PTFE/qPVBz/OHL (20 μm)	$2.2 \cdot 10^{-2}$ (60 °C)	200	[169]

a) The conductivity of all the membrane is measured in wet conditions, with the exception of PBI-based materials that are imbibed with H_3PO_4.
b) SRC – stability range of conductivity.

(e.g. inorganic nanofillers) by modulating the interactions and the nanophase segregation within the IEM; and (iii) means to improve the mechanical and dimensional stability (e.g. porous polymer layers).

A detailed investigation of the physicochemical properties and the electric response of such complex systems provides critical tools to guide the preparation of IEMs exhibiting improved properties. In this regard, elucidation of the long-range ion conduction mechanism by advanced techniques, such as DSC, DMA, and BES, has played a key role in ensuring that research in this field is not carried out by a slow, tedious, and cumbersome "*trial-and-error*" approach.

In the case of CEMs for application in PEMFCs and HT-PEMFCs, the most promising systems still remain PFSA-based IEMs and phosphoric-acid-doped PBI-based IEMs, respectively. After decades of research such IEMs are mostly able to achieve the technical targets for large-scale applications. However, there is still room to explore different systems (e.g. IEMs with aromatic hydrocarbon backbones) especially if we consider the high cost and environmental impact of PFSA membranes due to the difficulty to synthesize and dispose complex perfluorinated chemicals and materials.

In comparison with CEMs, the development status of AEMs is not as advanced. It proved particularly difficult to develop ionomers exhibiting simultaneously (i) a very sturdy backbone able to bestow a sufficient level of mechanical and thermal stability (e.g. polyethylene and polyketones) and (ii) a high density of stable anion-exchange moieties (e.g. aromatic and highly encumbered N-based groups) able to provide a high conductivity and tolerate the degradation triggered by species such as OH^- anions and radicals such as ·OH and ·OOH formed upon AEMFC operation. Despite these drawbacks, great progresses have been achieved with functionalized polyketone AEMs and radiation-grafted ETFE- and PE-based AEMs. The latter are quickly closing the performance and durability gaps with state-of-the-art perfluorinated CEMs.

Taken all together, the development of membranes for low-temperature fuel cells such as PEMFCs, AEMFCs, and HT-PEMFCs has reached a sufficient level to warrant performance and durability levels complying with the requirements for large-scale application. This will allow the FCs to fulfill their promise as one of the cornerstones of the "*hydrogen economy*," for a not-too-distant future when energy technologies will be fully compatible with the environment.

Acronyms

ADL	acid-doping level
AEMs	anion-exchange membranes
BES	broadband electric spectroscopy
CEMs	cation-exchange membranes
DB	delocalization body
DF	functional degree
Els	electrolyzes
FCs	fuel cells
IEC	ion-exchange capacity
IEM	ion-exchange membrane
PA	phosphoric acid
PBI	polybenzimidazole
PE	polymer electrolyte
PE	polyethylene
PEFC	polymer electrolyte fuel cells
PEMs	proton-exchange membranes
PFSA	perfluorosulfonic acid
PGMs	platinum-group metals
PSU	polysulfone
[PVBTMA][X]-*b*-PMB	poly(vinylbenzyltrimethylammonium)-*b*-poly (methylbutylene)
RFBs	redox flow batteries

SPEEK	sulfonated poly(ether ether)ketone
SR	swelling ratio
WU	water uptake

Symbols

A	surface area
L_w	"*wet*" length of the membrane
L_d	"*dry*" length of the membrane
n_{H_2O}	mole of water
n_{IEC}	mole of IEC
$n_{0,NaOH}$	moles of NaOH in the initial solution
$n_{E,NaOH}$	the moles of NaOH at the equivalent point
R_s	bulk resistance
W	weight
W_d	"*dry*" weight of the membrane
W_w	"*wet*" weight of the membrane
λ	hydration number

References

1 Bäckstrand, K. (2022). Towards a climate-neutral union by 2050? The European green deal, climate law, and green recovery. In: *Routes to a Resilient European Union: Interdisciplinary European Studies* (ed. A.B. Engelbrekt et al.), 39–61. Cham: Springer International Publishing.

2 Madurai Elavarasan, R., Pugazhendhi, R., Irfan, M. et al. (2022). State-of-the-art sustainable approaches for deeper decarbonization in Europe – an endowment to climate neutral vision. *Renewable and Sustainable Energy Reviews* 159: 112204.

3 Liu, W., Song, M.S., Kong, B. et al. (2017). Flexible and stretchable energy storage: recent advances and future perspectives. *Advanced Materials* 29 (1): 1603436.

4 Patterson, M., Macia, N.F., and Kannan, A.M. (2015). Hybrid microgrid model based on solar photovoltaic battery fuel cell system for intermittent load applications. *IEEE Transactions on Energy Conversion* 30 (1): 359–366.

5 Danilovic, N.Z. and Iryna, Z. (2021). Hydrogen's big shot. *The Electrochemical Society Interface* 30 (4): The Electrochemical Society.

6 Mauritz, K.A., Nalawade, A., and Hassan, M.K. (2012). Proton exchange membranes for H_2 fuel cell applications, in sol-gel processing for conventional and alternative energy. In: *Editors. Sol-Gel Processing for Conventional and Alternative Energy* (ed. M. Aparicio, A. Jitianu, and L.C. Klein), 73–98. Boston, MA: Springer US.

7 Cullen, D.A., Neyerlin, K.C., Ahluwalia, R.K. et al. (2021). New roads and challenges for fuel cells in heavy-duty transportation. *Nature Energy* 6 (5): 462–474.

8 Itoh, T. (2010). Chapters 13 – Hyperbranched polymer electrolytes for high temperature fuel cells. In: *Polymer Electrolytes* (ed. C. Sequeira and D. Santos), 524–549. Woodhead Publishing.

9 Di Noto, V., Lavina, S., Giffin, G.A. et al. (2011). Polymer electrolytes: present, past and future. *Electrochimica Acta* 57: 4–13.

10 Zuo, Z., Zhao, X., and Manthiram, A. (2013). High-performance blend membranes composed of an amphoteric copolymer containing supramolecular nanosieves for direct methanol fuel cells. *RSC Advances* 3 (19): 6759–6762.

11 Liu, L., Wang, C., He, Z. et al. (2021). Revisiting Nafion membranes by introducing ammoniated polymer with norbornene to improve fuel cell performance. *Journal of Power Sources* 506: 230164.

12 Dekel, D.R. (2018). Review of cell performance in anion exchange membrane fuel cells. *Journal of Power Sources* 375: 158–169.

13 Das, G., Choi, J.H., Nguyen, P.K.T. et al. (2022). Anion exchange membranes for fuel cell application: a review. *Polymers* 14 (6): 1197.

14 Pan, M., Pan, C., Li, C. et al. (2021). A review of membranes in proton exchange membrane fuel cells: transport phenomena, performance and durability. *Renewable and Sustainable Energy Reviews* 141: 110771.

15 Jiang, B., Yu, L., Wu, L. et al. (2016). Insights into the impact of the Nafion membrane pretreatment process on vanadium flow battery performance. *ACS Applied Materials & Interfaces* 8 (19): 12228–12238.

16 Pineri, M., Gebel, G., Davies, R.J. et al. (2007). Water sorption–desorption in Nafion® membranes at low temperature, probed by micro X-ray diffraction. *Journal of Power Sources* 172 (2): 587–596.

17 Mu, D., Yu, L., Liu, L. et al. (2017). Rice paper reinforced sulfonated poly(ether ether ketone) as low-cost membrane for vanadium flow batteries. *ACS Sustainable Chemistry & Engineering* 5 (3): 2437–2444.

18 Aili, D., Yang, J., Jankova, K. et al. (2020). From polybenzimidazoles to polybenzimidazoliums and polybenzimidazolides. *Journal of Materials Chemistry A* 8 (26): 12854–12886.

19 Gottesfeld, S., Dekel, D.R., Page, M. et al. (2018). Anion exchange membrane fuel cells: current status and remaining challenges. *Journal of Power Sources* 375: 170–184.

20 Slade, R.C.T. and Varcoe, J.R. (2005). Investigations of conductivity in FEP-based radiation-grafted alkaline anion-exchange membranes. *Solid State Ionics* 176 (5-6): 585–597.

21 Yan, J. and Hickner, M.A. (2010). Anion exchange membranes by bromination of benzylmethyl-containing poly(sulfone)s. *Macromolecules* 43 (5): 2349–2356.

22 Lu, W., Yuan, Z., Zhao, Y. et al. (2016). High-performance porous uncharged membranes for vanadium flow battery applications created by tuning cohesive and swelling forces. *Energy & Environmental Science* 9 (7): 2319–2325.

23 Shi, M., Dai, Q., Li, F. et al. (2020). Membranes with well-defined selective layer regulated by controlled solvent diffusion for high power density flow battery. *Advanced Energy Materials* 10 (34): 2001382.

24 Shin, H.Y., Cha, M.S., Hong, S.H. et al. (2017). Poly(p-phenylene)-based membrane materials with excellent cell efficiencies and durability for use in vanadium redox flow batteries. *Journal of Materials Chemistry A* 5 (24): 12285–12296.

25 Hydrogen and Fuel Cell Technologies Office, doe multi-year research, development, and demonstration plan, 3.4 fuel cells, 2016. https://www.energy.gov/sites/prod/files/2017/05/f34/fcto_myrdd_fuel_cells.pdf. 2017; https://www.energy.gov/sites/prod/files/2017/05/f34/fcto_myrdd_fuel_cells.pdf (accessed 03 June 2023).

26 Office, U.H.a.F.C.T. DOE technical targets for polymer electrolyte membrane fuel cell components. https://www.energy.gov/eere/fuelcells/doe-technical-targets-polymer-electrolyte-membrane-fuel-cell-components (accessed 03 June 2023).

27 EU fuel cells and hydrogen 2 joint undertaking (FCH 2 JU), addendum to the multi – annual work plan 2014 – 2020. https://www.fch.europa.eu/sites/default/files/MAWP%20final%20version_endorsed%20GB%2015062018%20%28ID%203712421%29.pdf. 2015; https://www.fch.europa.eu/sites/default/files/MAWP%20final%20version_endorsed%20GB%2015062018%20%28ID%203712421%29.pdf (accessed 03 June 2023).

28 Lister S., Advanced PGM-free cathode engineering for high power density and durability. https://www.hydrogen.energy.gov/pdfs/review21/fc171_litster_2021_o.pdf (accessed 03 June 2023).

29 He, Y. and Wu, G. (2022). PGM-free oxygen-reduction catalyst development for proton-exchange membrane fuel cells: challenges, solutions, and promises. *Accounts of Materials Research* 3 (2): 224–236.

30 Sun, C., Negro, E., Vezzù, K. et al. (2019). Hybrid inorganic-organic proton-conducting membranes based on SPEEK doped with WO_3 nanoparticles for application in vanadium redox flow batteries. *Electrochimica Acta* 309: 311–325.

31 Xi, J., Li, Z., Yu, L. et al. (2015). Effect of degree of sulfonation and casting solvent on sulfonated poly(ether ether ketone) membrane for vanadium redox flow battery. *Journal of Power Sources* 285: 195–204.

32 Drioli, E. and Giorno, L. (2018). *Encyclopedia of Membranes.* Springer.

33 Nawn, G., Vezzù, K., Negro, E. et al. (2019). Structural analyses of blended Nafion/PVDF electrospun nanofibers. *Physical Chemistry Chemical Physics* 21 (20): 10357–10369.

34 Di Noto, V., Boaretto, N., Negro, E. et al. (2012). Inorganic–organic membranes based on Nafion, $[(ZrO_2)\cdot(HfO_2)_{0.25}]$ and $[(SiO_2)\cdot(HfO_2)_{0.28}]$. Part I: Synthesis, thermal stability and performance in a single PEMFC. *International Journal of Hydrogen Energy* 37 (7): 6199–6214.

35 Karas, F., Hnát, J., Paidar, M. et al. (2014). Determination of the ion-exchange capacity of anion-selective membranes. *International Journal of Hydrogen Energy* 39 (10): 5054–5062.

36 Moukheiber, E., De Moor, G., Flandin, L. et al. (2012). Investigation of ionomer structure through its dependence on ion exchange capacity (IEC). *Journal of Membrane Science* 389: 294–304.

37 Di Noto, V., Gliubizzi, R., Negro, E. et al. (2007). Hybrid inorganic–organic proton conducting membranes based on Nafion and 5wt.% of M_xO_y (M=Ti, Zr, Hf, Ta and W): Part I. Synthesis, properties and vibrational studies. *Electrochimica Acta* 53 (4): 1618–1627.

38 Sun, C., Negro, E., Nale, A. et al. (2021). An efficient barrier toward vanadium crossover in redox flow batteries: the bilayer [Nafion/$(WO_3)_x$] hybrid inorganic-organic membrane. *Electrochimica Acta* 378: 138133.

39 Hwang, G.-J., Ohya, H., and Nagai, T. (1999). Ion exchange membrane based on block copolymers. Part III: preparation of cation exchange membrane. *Journal of Membrane Science* 156 (1): 61–65.

40 Luo, X., Holdcroft, S., Mani, A. et al. (2011). Water, proton, and oxygen transport in high IEC, short side chain PFSA ionomer membranes: consequences of a frustrated network. *Physical Chemistry Chemical Physics* 13 (40): 18055–18062.

41 Di Noto, V., Piga, M., Giffin, G.A. et al. (2012). Broadband electric spectroscopy of proton conducting SPEEK membranes. *Journal of Membrane Science* 390-391: 58–67.

42 Lee, S.H. and Rasaiah, J.C. (2011). Proton transfer and the mobilities of the H^+ and OH^- ions from studies of a dissociating model for water. *The Journal of Chemical Physics* 135 (12): 124505.

43 Agel, E., Bouet, J., and Fauvarque, J.F. (2001). Characterization and use of anionic membranes for alkaline fuel cells. *Journal of Power Sources* 101 (2): 267–274.

44 Sun, C., Zlotorowicz, A., Nawn, G. et al. (2018). [Nafion/$(WO_3)_x$] hybrid membranes for vanadium redox flow batteries. *Solid State Ionics* 319: 110–116.

45 Negro, E., Kuleshova, L., D'Aguanno, B. et al. (2013). The influence of the cationic form and degree of hydration on the structure of Nafion™. *Solid State Ionics* 252: 84–92.

46 Hofmann, D.W.M., Kuleshova, L., D'Aguanno, B. et al. (2009). Investigation of water structure in Nafion membranes by infrared spectroscopy and molecular dynamics simulation. *The Journal of Physical Chemistry B* 113 (3): 632–639.

47 Kusoglu, A. and Weber, A.Z. (2017). New insights into perfluorinated sulfonic-acid ionomers. *Chemical Reviews* 117 (3): 987–1104.

48 Kusoglu, A., Vezzù, K., Hegde, G.A. et al. (2020). Transport and morphology of a proton exchange membrane based on a doubly functionalized perfluorosulfonic imide side chain perflourinated polymer. *Chemistry of Materials* 32 (1): 38–59.

49 Di Noto, V., Lavina, S., Negro, E. et al. (2009). Hybrid inorganic–organic proton conducting membranes based on Nafion and 5wt% of M_xO_y (M=Ti, Zr, Hf, Ta and W). Part II: Relaxation phenomena and conductivity mechanism. *Journal of Power Sources* 187 (1): 57–66.

50 Giffin, G.A., Lavina, S., Pace, G. et al. (2012). Interplay between the structure and relaxations in selemion AMV hydroxide conducting membranes for AEMFC applications. *Journal of Physical Chemistry C* 116 (45): 23965–23973.

51 Lin, J., Wu, P.H., Wycisk, R. et al. (2008). Properties of water in prestretched recast Nafion. *Macromolecules* 41 (12): 4284–4289.

52 Hinatsu, J.T., Mizuhata, M., and Takenaka, H. (1994). Water uptake of perfluorosulfonic acid membranes from liquid water and water vapor. *Journal of the Electrochemical Society* 141 (6): 1493–1498.

53 Zawodzinski, T.A., Springer, T.E., Davey, J. et al. (1993). A comparative study of water uptake by and transport through ionomeric fuel cell membranes. *Journal of the Electrochemical Society* 140 (7): 1981–1985.

54 Yang, P., Long, J., Xuan, S. et al. (2019). Branched sulfonated polyimide membrane with ionic cross-linking for vanadium redox flow battery application. *Journal of Power Sources* 438: 226993.

55 Zawodzinski, T., Derouin, C., Radzinski, S. et al. (1993). Water uptake by and transport through Nafion® 117 membranes. *Journal of the Electrochemical Society* 140: 1041.

56 Li, Z., Dai, W., Yu, L. et al. (2014). Properties investigation of sulfonated poly (ether ether ketone)/polyacrylonitrile acid-base blend membrane for vanadium redox flow battery application. *ACS Applied Materials & Interfaces* 6 (21): 18885–18893.

57 Janssen, G.J.M. and Overvelde, M.L.J. (2001). Water transport in the proton-exchange-membrane fuel cell: measurements of the effective drag coefficient. *Journal of Power Sources* 101 (1): 117–125.

58 Yan, Q., Toghiani, H., and Wu, J. (2006). Investigation of water transport through membrane in a PEM fuel cell by water balance experiments. *Journal of Power Sources* 158 (1): 316–325.

59 Teranishi, K., Tsushima, S., and Hirai, S. (2006). Analysis of water transport in PEFCs by magnetic resonance imaging measurement. *Journal of the Electrochemical Society* 153 (4): A664.

60 Yin, B., Yu, L., Jiang, B. et al. (2016). Nano oxides incorporated sulfonated poly(ether ether ketone) membranes with improved selectivity and stability for vanadium redox flow battery. *Journal of Solid State Electrochemistry* 20 (5): 1271–1283.

61 Mai, Z., Zhang, H., Li, X. et al. (2011). Nafion/polyvinylidene fluoride blend membranes with improved ion selectivity for vanadium redox flow battery application. *Journal of Power Sources* 196 (13): 5737–5741.

62 Di Noto, V., Giffin, G.A., Vezzù, K. et al. (2012). *Broadband Dielectric Spectroscopy: A Powerful Tool for the Determination of Charge Transfer Mechanisms in Ion Conductors, in Solid State Proton Conductors: Properties and Applications in Fuel Cells*, 109–183. Chichester: Wiley.

63 Sone, Y., Ekdunge, P., and Simonsson, D. (1996). Proton conductivity of Nafion 117 as measured by a four-electrode AC impedance method. *Journal of the Electrochemical Society* 143 (4): 1254–1259.

64 Di Noto, V. (2002). Electrical spectroscopy studies of lithium and magnesium polymer electrolytes based on PEG400. *The Journal of Physical Chemistry B* 106 (43): 11139–11154.

65 Bertasi, F., Giffin, G.A., Vezzù, K. et al. (2017). A lipophilic ionic liquid based on formamidinium cations and TFSI: the electric response and the effect of

CO_2 on the conductivity mechanism. *Physical Chemistry Chemical Physics* 19 (38): 26230–26239.

66 Piccolo, M., Giffin, G.A., Vezzù, K. et al. (2013). Molecular relaxations in magnesium polymer electrolytes via GHz broadband electrical spectroscopy. *ChemSusChem* 6 (11): 2157–2160.

67 Di Noto, V., Piga, M., Pace, G. et al. (2008). Dielectric relaxations and conductivity mechanism of Nafion: studies based on broadband dielectric spectroscopy. *ECS Transactions* 16: 1183.

68 Di Noto, V., Longo, D., and Münchow, V. (1999). Ion-oligomer interactions in poly(ethylene glycol)400/$(LiCl)_x$ electrolyte complexes. *The Journal of Physical Chemistry B* 103 (14): 2636–2646.

69 Vezzù, K., Zago, V., Vittadello, M. et al. (2006). Effect of subcritical CO_2 on ionic conductivity of $\{Al[O(CH_2CH_2O)_{8.7}]_\rho/(LiClO_4)_z\}_n$ hybrid inorganic-organic networks. *Electrochimica Acta* 51 (8): 1592–1601.

70 Springer, T., Zawodzinski, T.A., and Gottesfeld, S. (1991). Polymer electrolyte fuel cell model. *Journal of the Electrochemical Society* 138: 2334.

71 Shah, A.A., Yu, F., Xing, W.W. et al. (2022). Machine learning for predicting fuel cell and battery polarisation and charge–discharge curves. *Energy Reports* 8: 4811–4821.

72 Kravos, A., Kregar, A., Penga, Ž. et al. (2022). Real-time capable transient model of liquid water dynamics in proton exchange membrane fuel cells. *Journal of Power Sources* 541: 231598.

73 Corda, G., Fontanesi, S., and d'Adamo, A. (2022). Methodology for PEMFC CFD simulation including the effect of porous parts compression. *International Journal of Hydrogen Energy* 47 (32): 14658–14673.

74 Gubler, L. (2019). Membranes and separators for redox flow batteries. *Current Opinion in Electrochemistry* 18: 31–36.

75 Shi, Y., Eze, C., Xiong, B. et al. (2019). Recent development of membrane for vanadium redox flow battery applications: a review. *Applied Energy* 238: 202–224.

76 Ye, D., Tu, Z., Yu, Y. et al. (2014). Hydrogen permeation across super-thin membrane and the burning limitation in low-temperature proton exchange membrane fuel cell. *International Journal of Energy Research* 38 (9): 1181–1191.

77 Thayumanasundaram, S., Piga, M., Lavina, S. et al. (2010). Hybrid inorganic-organic proton conducting membranes based on Nafion, SiO_2 and triethylammonium trifluoromethanesulfonate ionic liquid. *Electrochimica Acta* 55 (4): 1355–1365.

78 Hao, J., Gao, X., Jiang, Y. et al. (2018). Crosslinked high-performance anion exchange membranes based on poly (styrene-b-(ethylene-co-butylene)-b-styrene). *Journal of Membrane Science* 551: 66–75.

79 Zeng, Q.H., Liu, Q.L., Broadwell, I. et al. (2010). Anion exchange membranes based on quaternized polystyrene-block-poly (ethylene-ran-butylene)-block-polystyrene for direct methanol alkaline fuel cells. *Journal of Membrane Science* 349 (1-2): 237–243.

80 Danilczuk, M., Coms, F.D., and Schlick, S. (2009). Visualizing chemical reactions and crossover processes in a fuel cell inserted in the ESR resonator: detection by spin trapping of oxygen radicals, Nafion-derived fragments, and hydrogen and deuterium atoms. *The Journal of Physical Chemistry B* 113 (23): 8031–8042.

81 Ohguri, N., Nosaka, A.Y., and Nosaka, Y. (2010). Detection of OH radicals as the effect of Pt particles in the membrane of polymer electrolyte fuel cells. *Journal of Power Sources* 195 (15): 4647–4652.

82 Coms, F.D. (2008). The chemistry of fuel cell membrane chemical degradation. *ECS Transactions* 16 (2): 235.

83 De Bruijn, F., Dam, V., and Janssen, G. (2008). Durability and degradation issues of PEM fuel cell components. *Fuel Cells* 8 (1): 3–22.

84 Yu, T.H., Sha, Y., Liu, W.G. et al. (2011). Mechanism for degradation of Nafion in PEM fuel cells from quantum mechanics calculations. *Journal of the American Chemical Society* 133 (49): 19857–19863.

85 Schlick, S. (2018). *The Chemistry of Membranes Used in Fuel Cells: Degradation and Stabilization*. Wiley.

86 Gubler, L., Nauser, T., Coms, F.D. et al. (2018). Perspective—prospects for durable hydrocarbon-based fuel cell membranes. *Journal of the Electrochemical Society* 165 (6): F3100.

87 Varcoe, J.R., Atanassov, P., Dekel, D.R. et al. (2014). Anion-exchange membranes in electrochemical energy systems. *Energy and Environmental Science* 7 (10): 3135–3191.

88 Park, E.J., Maurya, S., Hibbs, M.R. et al. (2019). Alkaline stability of quaternized Diels–Alder polyphenylenes. *Macromolecules* 52 (14): 5419–5428.

89 Ran, J., Wu, L., He, Y. et al. (2017). Ion exchange membranes: new developments and applications. *Journal of Membrane Science* 522: 267–291.

90 Xi, J., Jiang, B., Yu, L. et al. (2017). Membrane evaluation for vanadium flow batteries in a temperature range of −20–50°C. *Journal of Membrane Science* 522: 45–55.

91 Giffin, G.A., Haugen, G.M., Hamrock, S.J. et al. (2013). Interplay between structure and relaxations in perfluorosulfonic acid proton conducting membranes. *Journal of the American Chemical Society* 135 (2): 822–834.

92 Di Noto, V., Gliubizzi, R., Negro, E. et al. (2006). Effect of SiO_2 on relaxation phenomena and mechanism of ion conductivity of [Nafion/$(SiO_2)_x$] composite membranes. *The Journal of Physical Chemistry B* 110 (49): 24972–24986.

93 Bauer, F., Denneler, S., and Willert-Porada, M. (2005). Influence of temperature and humidity on the mechanical properties of Nafion® 117 polymer electrolyte membrane. *Journal of Polymer Science Part B: Polymer Physics* 43 (7): 786–795.

94 Hussein, I.A. and Zaidi, S.M.J. (2009). Thermal and mechanical properties of fuel cell polymeric membranes: structure-property relationships. In: *Polymer Membranes for Fuel Cells* (ed. S.M.J. Zaidi and T. Matsuura), 235–252. Boston, MA: Springer US.

95 Di Noto, V., Piga, M., Lavina, S. et al. (2010). Structure, properties and proton conductivity of Nafion/[$(TiO_2)\cdot(WO_3)_{0.148}]_{\psi}TiO_2$ nanocomposite membranes. *Electrochimica Acta* 55 (4): 1431–1444.

96 Nawn, G., Pace, G., Lavina, S. et al. (2015). Nanocomposite membranes based on polybenzimidazole and ZrO_2 for high-temperature proton exchange membrane fuel cells. *ChemSusChem* 8 (8): 1381–1393.

97 James, B. (2018). Cost projections of PEM fuel cell systems for automobiles and medium-duty vehicles Office of Energy Efficiency & Renewable Energy: Washington, DC, USA 44.

98 James, B.D., Huya-Kouadio, J.M., and Houchins, C. et al. (2018). Mass production cost estimation of direct H_2 PEM fuel cell systems for transportation applications: 2018 update. Strategic Analysis Inc. p. 355.

99 Wannek, C., Glüsen, A., and Stolten, D. (2010). Materials, manufacturing technology and costs of fuel cell membranes. *Desalination* 250 (3): 1038–1041.

100 Miyake, J. and Miyatake, K. (2017). Fluorine-free sulfonated aromatic polymers as proton exchange membranes. *Polymer Journal* 49 (6): 487–495.

101 Kerres, J., Tang, C.M., and Graf, C. (2004). Improvement of properties of poly(ether ketone) ionomer membranes by blending and cross-linking. *Industrial and Engineering Chemistry Research* 43 (16): 4571–4579.

102 Yee, R.S.L., Rozendal, R.A., Zhang, K. et al. (2012). Cost effective cation exchange membranes: a review. *Chemical Engineering Research and Design* 90 (7): 950–959.

103 Gindt, B.P., Abebe, D.G., Tang, Z.J. et al. (2016). Nanoporous polysulfone membranes via a degradable block copolymer precursor for redox flow batteries. *Journal of Materials Chemistry A* 4 (11): 4288–4295.

104 Ren, J., Dong, Y., Dai, J. et al. (2017). A novel chloromethylated/quaternized poly(sulfone)/poly(vinylidene fluoride) anion exchange membrane with ultra-low vanadium permeability for all vanadium redox flow battery. *Journal of Membrane Science* 544: 186–194.

105 Di Vona, M.L. and Knauth, P. (2020). Electrochemical synthesis of ion exchange polymers: comparison between hydroxide and proton conductors. *Solid State Ionics* 352.

106 Narducci, R., Ercolani, G., Becerra-Arciniegas, R.A. et al. (2019). "Intrinsic" anion exchange polymers through the dissociation of strong basic groups: PPO with grafted bicyclic guanidines. *Membranes* 9 (5).

107 Nawn, G., Pace, G., Lavina, S. et al. (2015). Interplay between composition, structure, and properties of new H_3PO_4-doped $PBI_4N\text{–}HfO_2$ nanocomposite membranes for high-temperature proton exchange membrane fuel cells. *Macromolecules* 48 (1): 15–27.

108 Conti, F., Majerus, A., Di Noto, V. et al. (2012). Raman study of the polybenzimidazole–phosphoric acid interactions in membranes for fuel cells. *Physical Chemistry Chemical Physics* 14 (28): 10022–10026.

109 Giffin, G.A., Conti, F., Lavina, S. et al. (2014). A vibrational spectroscopic and modeling study of poly(2,5-benzimidazole) (ABPBI) – Phosphoric acid interactions in high temperature PEFC membranes. *International Journal of Hydrogen Energy* 39 (6): 2776–2784.

110 Di Noto, V., Piga, M., Giffin, G.A. et al. (2011). Structure–property interplay of proton conducting membranes based on PBI_5N, SiO_2–Im and H_3PO_4 for high temperature fuel cells. *Physical Chemistry Chemical Physics* 13 (26): 12146–12154.

111 Nawn, G., Vezzù, K., Bertasi, F. et al. (2017). Electric response and conductivity mechanism in H_3PO_4-doped polybenzimidazole-4N– HfO_2 nanocomposite membranes for high temperature fuel cells. *Electrochimica Acta* 228: 562–574.

112 Mandal, M., Huang, G., Hassan, N.U. et al. (2020). Poly(norbornene) anion conductive membranes: homopolymer, block copolymer and random copolymer properties and performance. *Journal of Materials Chemistry A* 8 (34): 17568–17578.

113 Nawn, G., Vezzù, K., Cavinato, G. et al. (2018). Opening doors to future electrochemical energy devices: the anion-conducting polyketone polyelectrolytes. *Advanced Functional Materials* 28 (29): 1706522.

114 Mauritz, K.A. and Moore, R.B. (2004). State of understanding of Nafion. *Chemical Reviews* 104 (10): 4535–4586.

115 Sigwadi, R., Dhlamini, M.S., Mokrani, T. et al. (2019). The proton conductivity and mechanical properties of Nafion(R)/ZrP nanocomposite membrane. *Heliyon* 5 (8): e02240.

116 Lu, J., Tang, H., Xu, C. et al. (2012). Nafion membranes with ordered mesoporous structure and high water retention properties for fuel cell applications. *Journal of Materials Chemistry* 22 (12): 5810.

117 Vittadello, M., Negro, E., Lavina, S. et al. (2008). Vibrational studies and properties of hybrid inorganic–organic proton conducting membranes based on Nafion and hafnium oxide nanoparticles. *The Journal of Physical Chemistry B* 112 (51): 16590–16600.

118 Di Noto, V., Boaretto, N., Negro, E. et al. (2010). New inorganic–organic proton conducting membranes based on Nafion and hydrophobic fluoroalkylated silica nanoparticles. *Journal of Power Sources* 195 (23): 7734–7742.

119 Zhang, B., Zhang, S., Weng, Z. et al. (2016). Quaternized adamantane-containing poly(aryl ether ketone) anion exchange membranes for vanadium redox flow battery applications. *Journal of Power Sources* 325: 801–807.

120 Liu, B., Robertson, G.P., Kim, D.S. et al. (2007). Aromatic poly(ether ketone)s with pendant sulfonic acid phenyl groups prepared by a mild sulfonation method for proton exchange membranes. *Macromolecules* 40 (6): 1934–1944.

121 Laberty-Robert, C., Valle, K., Pereira, F. et al. (2011). Design and properties of functional hybrid organic-inorganic membranes for fuel cells. *Chemical Society Reviews* 40 (2): 961–1005.

122 Liu, S., Wang, L., Li, D. et al. (2015). Novel amphoteric ion exchange membranes by blending sulfonated poly(ether ether ketone)/quaternized poly(ether imide) for vanadium redox flow battery applications. *Journal of Materials Chemistry A* 3 (34): 17590–17597.

123 Grew, K.N., Ren, X., and Chu, D. (2011). Effects of temperature and carbon dioxide on anion exchange membrane conductivity. *Electrochemical and Solid-State Letters* 14 (12): B127–B131.

124 Mallinson, S.L., Varcoe, J.R., and Slade, R.C.T. (2014). Examination of amine-functionalised anion-exchange membranes for possible use in the all-vanadium redox flow battery. *Electrochimica Acta* 140: 145–151.

125 Zeng, L., Zhao, T.S., Wei, L. et al. (2019). Anion exchange membranes for aqueous acid-based redox flow batteries: current status and challenges. *Applied Energy* 233-234: 622–643.

126 Merle, G., Wessling, M., and Nijmeijer, K. (2011). Anion exchange membranes for alkaline fuel cells: a review. *Journal of Membrane Science* 377 (1-2): 1–35.

127 Tuan, C.M. and Kim, D. (2016). Anion-exchange membranes based on poly(arylene ether ketone) with pendant quaternary ammonium groups for alkaline fuel cell application. *Journal of Membrane Science* 511: 143–150.

128 Liu, D., Lin, L., Xie, Y. et al. (2021). Anion exchange membrane based on poly (arylene ether ketone) containing long alkyl densely quaternized carbazole derivative pendant. *Journal of Membrane Science* 623: 119079.

129 Kim, J.H., Vinothkannan, M., Kim, A.R. et al. (2020). Anion exchange membranes obtained from poly (arylene ether sulfone) block copolymers comprising hydrophilic and hydrophobic segments. *Polymers* 12 (2): 325.

130 Lee, K.H., Cho, D.H., Kim, Y.M. et al. (2017). Highly conductive and durable poly (arylene ether sulfone) anion exchange membrane with end-group cross-linking. *Energy & Environmental Science* 10 (1): 275–285.

131 Liu, L., Chu, X., Liao, J. et al. (2018). Tuning the properties of poly (2, 6-dimethyl-1, 4-phenylene oxide) anion exchange membranes and their performance in H_2/O_2 fuel cells. *Energy & Environmental Science* 11 (2): 435–446.

132 Ran, J., Wu, L., Ru, Y. et al. (2015). Anion exchange membranes (AEMs) based on poly (2, 6-dimethyl-1, 4-phenylene oxide)(PPO) and its derivatives. *Polymer Chemistry* 6 (32): 5809–5826.

133 Zhang, F.X., Zhang, H.M., and Qu, C. (2011). Imidazolium functionalized polysulfone anion exchange membrane for fuel cell application. *Journal of Materials Chemistry* 21 (34): 12744–12752.

134 Vezzù, K., Nawn, G., Pagot, G. et al. (2019). Relaxation phenomena and conductivity mechanisms in anion-exchange membranes derived from polyketone. *Electrochimica Acta* 319: 253–263.

135 Alvi, A.R., Vezzù, K., Pagot, G. et al. (2022). Inorganic-organic hybrid anion conducting membranes based on ammonium-functionalized polyethylene pyrrole-polyethylene ketone copolymer. *Macromolecular Chemistry and Physics* 223 (8): 2100409.

136 Herman, H., Slade, R.C.T., and Varcoe, J.R. (2003). The radiation-grafting of vinylbenzyl chloride onto poly(hexafluoropropylene-co-tetrafluoroethylene) films with subsequent conversion to alkaline anion-exchange membranes:

optimisation of the experimental conditions and characterisation. *Journal of Membrane Science* 218 (1): 147–163.

137 Varcoe, J.R., Slade, R.C.T., Yee, E.L.H. et al. (2007). Poly(ethylene-co-tetrafluoroethylene)-derived radiation-grafted anion-exchange membrane with properties specifically tailored for application in metal-cation-free alkaline polymer electrolyte fuel cells. *Chemistry of Materials* 19 (10): 2686–2693.

138 Pandey, T.P., Maes, A.M., Sarode, H.N. et al. (2015). Interplay between water uptake, ion interactions, and conductivity in an e-beam grafted poly(ethylene-co-tetrafluoroethylene) anion exchange membrane. *Physical Chemistry Chemical Physics* 17 (6): 4367–4378.

139 Di Noto, V., Giffin, G.A., Vezzù, K. et al. (2015). Interplay between solid state transitions, conductivity mechanisms, and electrical relaxations in a [PVBTMA][Br]-b-PMB diblock copolymer membrane for electrochemical applications. *Physical Chemistry Chemical Physics* 17 (46): 31125–31139.

140 Pandey, T.P., Sarode, H.N., Yang, Y. et al. (2016). A highly hydroxide conductive, chemically stable anion exchange membrane, poly(2,6 dimethyl 1,4 phenylene oxide)-b-Poly(vinyl benzyl trimethyl ammonium), for electrochemical applications. *Journal of the Electrochemical Society* 163 (7): H513–H520.

141 Zschocke, P. and Quellmalz, D. (1985). Novel ion exchange membranes based on an aromatic polyethersulfone. *Journal of Membrane Science* 22 (2): 325–332.

142 Lu, S., Pan, J., Huang, A. et al. (2008). Alkaline polymer electrolyte fuel cells completely free from noble metal catalysts. *Proceedings of the National Academy of Sciences of the United States of America* 105 (52): 20611–20614.

143 Vielstich, W., Lamm, A., and Gasteiger, H. (New York, 2003). *Handbook of Fuel Cells. Fundamentals, Technology, Applications*. John Wiley & Sons.

144 Pabby, A.K., Rizvi, S.S., and Requena, A.M.S. (2008). *Handbook of Membrane Separations: Chemical, Pharmaceutical, Food, and Biotechnological Applications*. CRC Press.

145 Di Noto, V., Lavina, S., Giffin, G.A. et al. (2011). *Polymer Electrolytes: Present, Past and Future*, 4–13. Elsevier.

146 Savadogo, O. (2004). Emerging membranes for electrochemical systems: Part II. High temperature composite membranes for polymer electrolyte fuel cell (PEFC) applications. *Journal of Power Sources* 127 (1-2): 135–161.

147 Wainright, J., Wang, J.T., Weng, D. et al. (1995). Acid-doped polybenzimidazoles: a new polymer electrolyte. *Journal of the Electrochemical Society* 142 (7): L121.

148 Samms, S., Wasmus, S., and Savinell, R. (1996). Thermal stability of proton conducting acid doped polybenzimidazole in simulated fuel cell environments. *Journal of the Electrochemical Society* 143 (4): 1225.

149 Ma, Y.-L., Wainright, J.S., Litt, M.H. et al. (2003). Conductivity of PBI membranes for high-temperature polymer electrolyte fuel cells. *Journal of the Electrochemical Society* 151 (1): A8.

150 Di Noto, V., Piga, M., Giffin, G.A. et al. (2012). Interplay between mechanical, electrical, and thermal relaxations in nanocomposite proton conducting membranes based on Nafion and a [(ZrO_2)·(Ta_2O_5) 0.119] core–shell nanofiller. *Journal of the American Chemical Society* 134 (46): 19099–19107.

151 Di Noto, V., Boaretto, N., Negro, E. et al. (2008). New inorganic–organic proton conducting membranes based on Nafion® and [(ZrO_2)·(SiO_2) 0.67] nanoparticles: synthesis vibrational studies and conductivity. *Journal of Power Sources* 178 (2): 561–574.

152 Di Noto, V., Piga, M., Negro, E. et al. (2013). New nanocomposite proton conducting membranes based on a core–shell nanofiller for low relative humidity fuel cells. *RSC Advances* 3 (41): 18960–18969.

153 Noto, V., Boaretto, N., Negro, E. et al. (2012). Inorganic-organic membranes based on Nafion,[(ZrO_2)–(HfO_2) 0.25] and [(SiO_2)–(HfO_2) 0.28] nanoparticles. Part II: relaxations and conductivity mechanism. *International Journal of Hydrogen Energy* 37: 6215–6227.

154 Di Noto, V., Piga, M., Negro, E. et al. (2012). New hybrid inorganic-organic proton conducting membranes based on Nafion and a [(ZrO_2)·(Ta_2O_5) 0.119] oxide core-shell nanofiller. *MRS Online Proceedings Library (OPL)* 1384.

155 Qiao, L., Zhang, H., Lu, W. et al. (2018). Advanced porous membranes with slit-like selective layer for flow battery. *Nano Energy* 54: 73–81.

156 Arora, P. and Zhang, Z. (2004). Battery separators. *Chemical Reviews* 104 (10): 4419–4462.

157 Dong, C., Xu, X., Zhang, J. et al. (2022). Proton transport of porous triazole-grafted polysulfone membranes for high temperature polymer electrolyte membrane fuel cell. *International Journal of Hydrogen Energy* 47 (13): 8492–8501.

158 Krishnan, N.N., Joseph, D., Duong, N.M.H. et al. (2017). Phosphoric acid doped crosslinked polybenzimidazole (PBI-OO) blend membranes for high temperature polymer electrolyte fuel cells. *Journal of Membrane Science* 544: 416–424.

159 Lee, S., Nam, K.H., Seo, K. et al. (2020). Phase inversion-induced porous polybenzimidazole fuel cell membranes: an efficient architecture for high-temperature water-free proton transport. *Polymers* 12 (7): 1604.

160 Li, J., Li, X., Yu, S. et al. (2014). Porous polybenzimidazole membranes doped with phosphoric acid: preparation and application in high-temperature proton-exchange-membrane fuel cells. *Energy Conversion and Management* 85: 323–327.

161 Wang, S., Zhao, C., Ma, W. et al. (2012). Preparation and properties of epoxy-cross-linked porous polybenzimidazole for high temperature proton exchange membrane fuel cells. *Journal of Membrane Science* 411-412: 54–63.

162 Cai, Y., Yue, Z., Teng, X. et al. (2018). Radiation grafting graphene oxide reinforced polybenzimidazole membrane with a sandwich structure for high temperature proton exchange membrane fuel cells in anhydrous atmosphere. *European Polymer Journal* 103: 207–213.

163 Jheng, L.-C., Hsu, S.L.C., Tsai, T.Y. et al. (2014). A novel asymmetric polybenzimidazole membrane for high temperature proton exchange membrane fuel cells. *Journal of Materials Chemistry A* 2 (12): 4225–4233.

164 Shen, C.-H., Jheng, L., Hsu, S.L. et al. (2011). Phosphoric acid-doped cross-linked porous polybenzimidazole membranes for proton exchange membrane fuel cells. *Journal of Materials Chemistry* 21 (39): 15660–15665.

165 Geng, K., Tang, H., Ju, Q. et al. (2021). Symmetric sponge-like porous polybenzimidazole membrane for high temperature proton exchange membrane fuel cells. *Journal of Membrane Science* 620: 118981.

166 Wang, P., Peng, J., Yin, B. et al. (2021). Phosphoric acid-doped polybenzimidazole with a leaf-like three-layer porous structure as a high temperature proton exchange membrane for fuel cells. *Journal of Materials Chemistry A* 9 (46): 26345–26353.

167 Zarrin, H., Jiang, G., Lam, G.Y.Y. et al. (2014). High performance porous polybenzimidazole membrane for alkaline fuel cells. *International Journal of Hydrogen Energy* 39 (32): 18405–18415.

168 Saleem, J., Gao, P., Barford, J. et al. (2013). Development and characterization of novel composite membranes for fuel cell applications. *Journal of Materials Chemistry A* 1 (45): 14335–14343.

169 Cao, Y.-C., Scott, K., and Wang, X. (2012). Preparation of polytetrafluoroethylene porous membrane based composite alkaline exchange membrane with improved tensile strength and its fuel cell test. *International Journal of Hydrogen Energy* 37 (17): 12688–12693.

170 Cao, Y.-C., Xu, C., Zou, L. et al. (2015). A polytetrafluoroethylene porous membrane and dimethylhexadecylamine quaternized poly (vinyl benzyl chloride) composite membrane for intermediate temperature fuel cells. *Journal of Power Sources* 294: 691–695.

171 Xing, D.M. et al. (2005). Characterization of sulfonated poly(ether ether ketone)/polytetrafluoroethylene composite membranes for fuel cell applications. *Fuel Cells* 5 (3): 406–411.

172 Zhang, C., Yue, X., Yang, Y. et al. (2020). Thin and methanol-resistant reinforced composite membrane based on semi-crystalline poly (ether ether ketone) for fuel cell applications. *Journal of Power Sources* 450: 227664.

173 Zhang, X., Glüsen, A., and Garcia-Valls, R. (2006). Porous lignosulfonate membranes for direct methanol fuel cells. *Journal of Membrane Science* 276 (1): 301–307.

174 Pichonat, T., Gauthier-Manuel, B., and Hauden, D. (2004). A new proton-conducting porous silicon membrane for small fuel cells. *Chemical Engineering Journal* 101 (1): 107–111.

175 Pichonat, T. and Gauthier-Manuel, B. (2005). Development of porous silicon-based miniature fuel cells. *Journal of Micromechanics and Microengineering* 15 (9): S179–S184.

176 Di Noto, V., Negro, E., Sanchez, J.Y. et al. (2010). Structure-relaxation interplay of a new nanostructured membrane based on tetraethylammonium trifluoromethanesulfonate ionic liquid and neutralized Nafion 117 for high-temperature fuel cells. *Journal of the American Chemical Society* 132 (7): 2183–2195.

177 Vezzù, K., Nawn, G., Negro, E. et al. (2020). Electric response and conductivity mechanism of blended polyvinylidene fluoride/Nafion electrospun nanofibers. *Journal of the American Chemical Society* 142 (2): 801–814.

178 Vezzù, K., Maes, A.M., Bertasi, F. et al. (2018). Interplay between hydroxyl density and relaxations in poly(vinylbenzyltrimethylammonium)-b-poly (methylbutylene) membranes for electrochemical applications. *Journal of the American Chemical Society* 140 (4): 1372–1384.

179 Giffin, G.A., Piga, M., Lavina, S. et al. (2012). Characterization of sulfated-zirconia/Nafion® composite membranes for proton exchange membrane fuel cells. *Journal of Power Sources* 198: 66–75.

180 Nawn, G., Vezzù, K., Bertasi, F. et al. (2018). Electric response and conductivity mechanism reciprocity in H_3PO_4-doped Polybenzimidazole-4N-ZrO_2 nanocomposite membranes. *Solid State Ionics* 320: 172–176.

181 Altaf, F. et al. (2021). Synthesis and electrochemical investigations of ABPBI grafted montmorillonite based polymer electrolyte membranes for PEMFC applications. *Renewable Energy* 164: 709–728.

9

Supports for Oxygen Reduction Catalysts: Understanding and Improving Structure, Stability, and Activity

Iwona A. Rutkowska, Sylwia Zoladek, and Pawel J. Kulesza

University of Warsaw, Faculty of Chemistry, Pasteura 1, PL-02-093, Warsaw, Poland

9.1 Introduction

There has been continuous progress in the development of electrocatalytic systems for oxygen reduction reaction (ORR), particularly with respect to potential applications in low-temperature fuel cells [1–11]. On practical grounds, the most important issues concern the improvement of stability and activity, as well as diminishing the formation of undesirable hydrogen peroxide intermediate, H_2O_2 and HO_2^- in acid and alkaline media, respectively. Despite intensive research to replace highly expensive platinum-rich materials with alternative electrocatalysts, such as the noble-metal-free, e.g. with heteroatom-doped, iron- or cobalt-modified, totally metal-free-functionalized, or derivatized carbon-based (including graphene-type), nanomaterials [11–18], the practical ORR catalysts are still based on platinum. Thus, there is a need of better utilization of catalytic sites and significant lowering of the noble metal loadings. Among numerous fundamental problems to be resolved, a problem of the choice of a suitable support or carrier for highly dispersed Pt-metal-based nanoparticles still exists.

While exhibiting long-term stability and having large electrochemically active surface area, a useful support should prevent agglomeration of catalytic centers, exhibit activating interactions with them, facilitate oxygen mass transfer and water removal, and assure good electrical conductivity to active sites at the electrocatalytic interface. In this respect, nanostructured active carbon supports [19, 20], which are characterized by large surface areas and good conductivity, have been largely considered as supports for Pt electrocatalysts, and their application significantly contributed to the progress in fuel cell technology.

For immobilization of catalytic nanoparticles, e.g. Pt, activated carbon, carbon black Vulcan XC-72R, acetylene black, black pearls, graphite, and graphitized materials are good candidates due to their high conductivities, appropriate pore structures, and high surface areas. Graphitization of carbons leads typically to their better corrosion resistance [21]. In particular, the Vulcan XC-72R carbon has attracted much attention due to its good compromise among adequate

Electrocatalysis for Membrane Fuel Cells: Methods, Modeling, and Applications, First Edition.
Edited by Nicolas Alonso-Vante and Vito Di Noto.

Brunauer–Emmett–Teller isotherm (BET) surface area (~250 $m^2 g^{-1}$), mesopore and macropore percentage equal to 54%, and high electric conductivity (~2.77 $S cm^{-1}$) [19, 20]. Among other carbon materials, Black Pearls 2000, Acetylene Black, Ketjen Black, and Macsorb have also been considered, but their performance as supports for Pt during ORR has been modest relative to that of Vulcan XC-72R [20].

Despite efforts to improve stability, the interfacial corrosion of carbon carriers, particularly, when combined with their weak interaction with Pt nanoparticles, decreases the practical durability of carbon-supported Pt electrocatalysts [22]. The carbon corrosion through interfacial oxidation is the most pronounced during start-up of the fuel cell operation [23]. Consequently, the Pt particles tend to agglomerate into larger particles or collapse into caves following detachment from the carbon surfaces [24]. To minimize the problems mentioned above, approaches such as surface functionalization, introduction of additives and graphitization of carbon blacks, and application of carbon nanotubes (CNTs) or graphene have been considered. Among other important issues is to maximize the electrochemically active surface area and to utilize carbons of sufficiently high porosity to immobilize platinum typically deposited in the form of the highly dispersed nanoparticles of diameters from 2 to 5 nm [25–27]. The so-called particle proximity effect often influences the observed weakening of the OH adsorption strength [28].

9.2 Carbon Black Supports

Carbon blacks are produced by the incomplete combustion of heavy petroleum products and constitute paracrystalline carbons that are characterized by high surface-area-to-volume ratios and negligible contents of aromatic hydrocarbons. While being color pigments of high tinting strength and stability, carbon blacks are employed as inks and paints. Because they are also good conductors of electricity, carbon blacks have been widely used as support materials or carriers for fuel cell catalysts [29]. Under oxidative conditions, they suffer from some corrosion, thus leading to performance degradation and durability problems, particularly at high potentials [30, 31]. Care must be exercised to assure good mass (reactant) transport capabilities [32]. Carbon blacks consist primarily of pores not exceeding 8 nm [33] that are enough to accommodate catalytic nanoparticles of comparable (or lower) sizes, Figure 9.1. On the other hand, the limited interior surface areas and pore volumes in carbon black particles tend to hinder the mass transport during electrode reactions [34]. It has been postulated that admixing or combining the electrocatalytic systems with mesoporous carbons, such as carbon nanofibers (CNFs) or CNTs, could be useful in this respect.

Figure 9.1 Cartoon illustrating immobilization of platinum nanoparticles in pores existing on surfaces of carbon supports.

The active functional groups (e.g. –CO, –COOH, and –CN) existing on surfaces of carbon nanostructures can interact with metal precursors during the catalyst synthesis, thus affecting the size and degree of dispersion of obtained nanoparticles [28, 29]. It has been demonstrated that weak interaction does not support formation and immobilization of small (below 1 nm) metal nanoparticles. The nitrogen-doped carbons not only provide anchoring sites during preparation (deposition) of metal nanoparticles, but they are also capable of promoting catalytic reactions. Indeed, the 1-nm-thick layer of pyridinic-N-doped carbons tends to improve distribution and performance of the supported Pt, both in terms of the ORR activity and the catalyst's durability [35].

9.3 Decoration and Modification with Metal Oxide Nanostructures

Recent studies [36, 37] demonstrate the feasibility of deposition of certain metal oxides, e.g. CeO_x [38], on carbon supports, to provide nucleation sites for Pt nanoparticles and to produce electrocatalytic systems characterized by high electrochemically active surface areas (ECSAs) and homogeneous distribution of small-size Pt nanoparticles, Figure 9.2a. Such systems seem to be capable of tuning the activity and durability of Pt sites and to combine catalytic proficiencies of both components. In this respect, the nanostructured metal oxide additive can be considered as cocatalyst and the support component. The improved activity and durability during ORR have also been attributed to strong interactions of metal oxides with catalytic (Pt) centers. In addition to possible changes of the electron

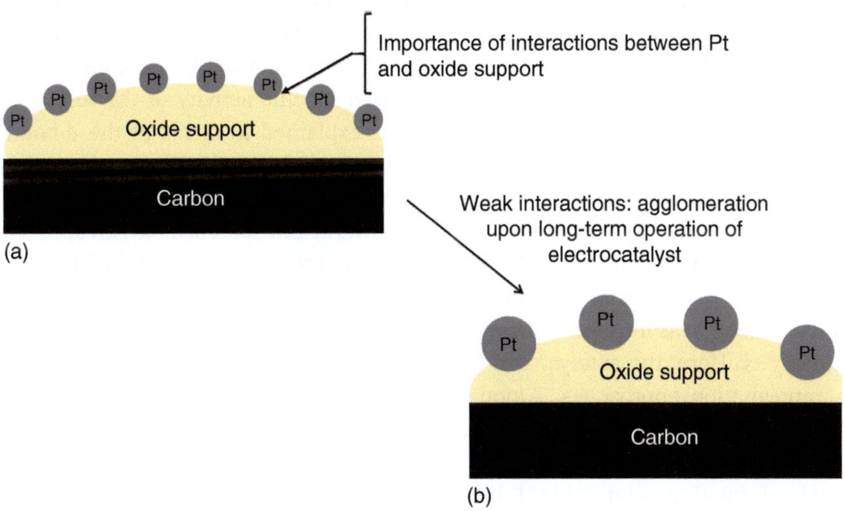

Figure 9.2 Schematic diagram illustrating (a) importance of the strength of interactions between the catalytic Pt nanoparticles, and (b) the metal oxide supports existing on carbon carriers.

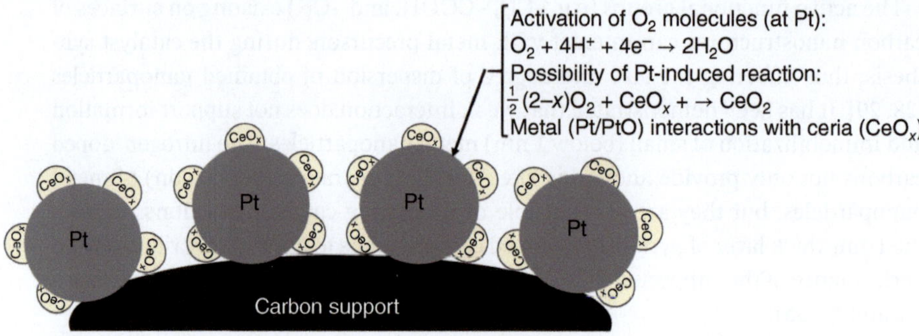

Figure 9.3 Hybrid (functionalized) electrocatalytic system composed of ceria (CeO_x)-decorated Pt nanoparticles deposited onto carbon support (carriers).

density on the surface of Pt, such interactions are likely to diminish agglomeration of Pt-nanostructures upon long-term operation (ORR) of electrocatalysts, Figure 9.2. On the contrary, weak metal support interactions would not preclude undesirable sintering of Pt sites, Figure 9.2b. Historically, the activating interactions of the metal oxide clusters with noble metal nanoparticles were postulated and demonstrated during electrooxidation of small organic molecules [39, 40]. The fact that certain metal oxides exhibit higher resistance to corrosion, relative to common carbon supports, could also explain improvement in the long-term operation of catalysts [41, 42]. Modification of the electronic structure of Pt surfaces may promote adsorption of O_2, activation of the molecule, and cleavage of O=O bond, Figure 9.3 [43]. A great deal of research has been devoted to searching for suitable metal oxide materials for polymer electrolyte membrane fuel cell (PEMFC) applications.

In the search for suitable metal oxide nanomaterials for decoration of carbon supports, tantalum oxide (Ta_2O_5) was considered: the catalyst was synthesized by deposition of the oxide on carbon black (Vulcan), followed by immobilization of Pt from platinum colloids [36]. The increased specific activity of the resulting hybrid system toward the ORR (at 0.9 V) was explained in terms of the d-band approach explored together with density functional theory (DFT). It was rationalized that tantalum should bring down the d-band center of platinum, thus resulting in weakening of bonding between platinum and adsorbed oxygen species. Furthermore, the preferential adsorption of –OH groups on the oxide, relative to platinum (Pt^0), surfaces was also postulated [36]. In another example, ceria, or more precisely CeO_x, was introduced to improve performance of carbon-black-supported Pt catalysts. Here, the catalyst was prepared by combining the precipitation and co-impregnation approaches. Among the crucial issues was the suppression of the Pt-oxide formation in the presence of nonstoichiometric $Ce^{IV,III}O_x$, Figure 9.4; it was postulated that Ce^{III} sites, rather than Pt, were oxidized to Ce^{IV} at the Pt-oxide-formation potential [37]. Indeed, the inhibition of the Pt-oxide formation seems to be of primary importance in elucidating the enhancement effect, since the ORR activity at Pt-oxide is much lower than on the bare Pt surface.

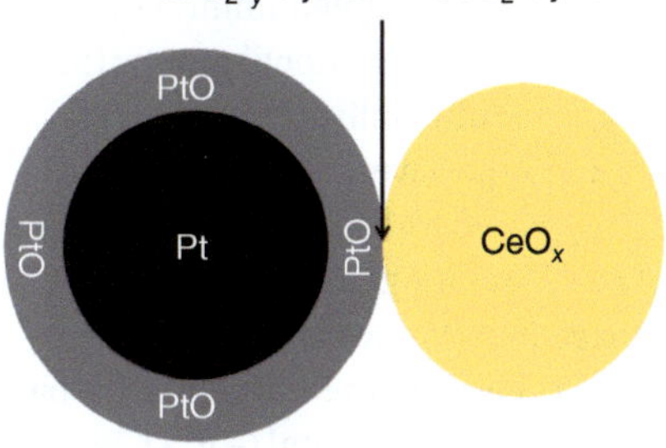

Figure 9.4 Diminishing of the formation of platinum oxo-species on catalytic Pt through mutual platinum–ceria interactions.

9.4 Carbon Nanotube as Carriers

CNTs are allotropes made of cylindrical carbon nanostructures with diameters typically at the nanometer level [44, 45]. They are typically categorized into single-walled carbon nanotubes (SWCNTs) and multiwalled carbon nanotubes (MWCNTs). The latter nanostructures consist of multiple rolled layers (concentric tubes) of graphene. CNTs can exhibit remarkable electrical conductivity, and, due to existence of strong bonds between carbon atoms, they show exceptional mechanical tensile strength and thermal conductivity. In addition, they can be chemically modified or functionalized to attain desired properties that can be used in a wide variety of applications, including catalysis. In comparison to carbon blacks, CNTs have fewer impurities, e.g. sulfur that can poison platinum electrocatalysts. Because immobilization and anchoring of Pt nanoparticles on smooth and chemically inert surfaces is often difficult, CNTs are commonly subjected to harsh treatments involving interfacial acid oxidation and functionalization to provide oxygen-binding groups for the metal catalytic sites [46]. Both SWCNT and MWCNT have been effectively used as support materials for dispersed metal catalysts of importance to the ORR electrocatalysis [46, 47].

An important advantage coming from use of CNTs in electrocatalysis is their high resistance toward corrosion, when compared to carbon-black-based systems. Consequently, the CNT-supported systems exhibit good electrochemical stability [30, 48]. Representative examples include application of SWCNTs or MWCNTs onto which Pt nanoparticles [49, 50] have been immobilized or MWCNT-supported flowerlike Pt-nanostructures [48]. All these systems are characterized by high Pt-mass activities during ORR in acid media. But in comparison to pristine CNTs, the exceptionally high activity of Pt (efficient ORR according to the four-electron mechanism both during diagnostic and fuel cell measurements) was found when it was deposited and dispersed over defect-free highly purified CNTs. Here, Pt centers were generated by reducing the hexachloroplatinate complex in nitrogen atmosphere. Furthermore, the stability of the support (achieved by heat treatment) and the stability of the deposited Pt sites (resistance to agglomeration) were noted. It was proposed that interactions between the crystalline Pt nanoparticles and

CNTs, which involved redistribution of charge between 2p orbitals of carbon and 5d orbitals of Pt, facilitated immobilization of platinum on carbon surface without involvement of any functional groups [51].

Finally, the bimetallic Pt-based catalysts, e.g. MWCNT-supported PtNi [52] and PtRu [53], have been considered for the ORR and compared to other carbon-supported Pt-based electrocatalytic systems. It should be remembered, however, that CNTs are generally quite expensive, and their rigidity and "sharpness" can damage membranes in PEMFCs.

As in the case of the carbon-black-based systems [36], application of CNTs decorated with certain metal (e.g. Ti, W, V, and Nb) oxide nanostructures as support for the immobilized Pt nanoparticles may promote their electrocatalytic activity due to their better dispersion and the possibility of metal–support interactions leading to changes in the electronic structure of Pt, in addition to the preferential adsorption of the –OH groups ($-OH_{ads}$) on the oxide rather than Pt surfaces. For example, it has been demonstrated that formation of the triple-junction interface between Pt and CNTs decorated with CeO_2 (in reality substoichiometric CeO_x) has a positive effect on the electrocatalytic ORR at platinum [54]. Here, good dispersion of platinum has been correlated with the feasibility of keeping Pt catalytic sites in contact with both the carbon and the metal oxide nanostructures. Furthermore, as in the case of ORR at Pt supported onto CeO_x-decorated carbon black [37], it is reasonable to expect that due to the existence of oxygen vacancies in substoichiometric CeO_x the formation of inhibiting Pt-oxides at neighboring Pt sites is largely suppressed, Figure 9.3, and consequently, the ORR currents are increased, and the onset potential of the ORR is shifted toward more positive values. Strong interfacial interactions between Pt and ceria, as well as correlation between changes in the electronic structure of platinum and the degree of substoichiometry of CeO_x, namely the content of Ce^{III} (relative to Ce^{IV}) ionic sites and the related population of oxygen vacancies, have been confirmed by XPS analysis. Moreover, the application of atomic layer deposition method has facilitated formation of highly and uniformly dispersed Pt nanoparticles, thus leading to high ECSA values. On the whole, the ORR performance of Pt supported onto CNTs decorated with ceria has been superior in comparison to the behavior of commercial carbon-black (Vulcan XC-72)-supported platinum at the same loading [54–56]. In addition, combination of Pt electrocatalyst with CNTs and cerium oxide nanostructures results in the enhanced durability due to better resistance of Pt coexisting and interacting with substoichiometric CeO_x (existing on robust CNTs) to the oxidative corrosion.

Among other examples of the systems utilizing CNTs decorated with transition metal oxides, the system based on Pt nanocrystals (*ca.* 2 nm size), which were well dispersed over Mn_3O_4-coated CNTs conductive network [57], should be mentioned here. The advantages of the reported system concerned simplicity of the synthetic approach, which was performed under mild conditions with no need of any capping agent for platinum or additional reducing agent because the prereduced manganese oxide induced generation of platinum sites. Furthermore, a synergistic cocatalytic effect, originating from the existence of heterojunction interfaces formed between Pt and Mn_3O_4, was postulated, leading to the enhanced electrocatalytic activity of ORR.

9.5 Doping, Modification, and Other Carbon Supports

It has been established that doping of the originally inert carbon nanostructures, including graphene, may lead to formation of catalytically active sites by replacement of the lattice carbons with heteroatoms [58]. In general, fairly high level of activity toward ORR has been demonstrated for the N-doped carbons in alkaline media, contrary to the relatively poorer performance and stability in acidic environments.

Strong Pt-support interaction should lead to improvement in the electrocatalytic ORR performance and, thus, would permit designing low Pt-loading catalysts. It has been reported that the phosphorus-doped carbon nanotubes (P-CNTs) were prepared and used as the support of low Pt-loading catalyst to design low-cost ORR electrocatalyst in acidic medium. Based on rotating disk voltammetric electrochemical diagnostic experiments as well as X-ray photoelectron spectroscopic (XPS) analysis and DFT calculations, it was postulated that P-CNT-supported platinum (0.85 wt% Pt) showed not only the exceptionally high electrocatalytic activity but also the long-term stability during the ORR in acid medium.

While CNTs are composed of graphene layers wrapped into perfect cylinders, CNFs are cylindrical nanostructures with graphene layers arranged as stacked cones, cups, or plates. Thus, the main difference between CNTs and CNFs concerns the exposure of active edge planes. The unique surface morphology, thermal properties, and crystalline features of the porous CNFs should be correlated with their large surface areas, existence of small pores, and disordered graphene sheets [59]. Consequently, CNFs have been considered for electrochemical energy conversion and storage and as carrier materials for various catalysts. CNFs provide an active and well-defined high surface area material for immobilization of catalytic nanocenters including Pt nanoparticles. In this respect, CNF-supported Pt have been reported to show an improved electrocatalytic activity toward the ORR, relative to the carbon black-based systems [30].

9.6 Graphene as Catalytic Component

Graphene is another allotrope of carbon consisting of a single layer of atoms chemically bonded in the hexagonal pattern that is characteristic of graphite, i.e. arranged in a two-dimensional honeycomb lattice. While each carbon atom is connected there to three nearest neighbors by a σ-bond, it contributes one electron to a conduction band extended over the whole sheet what makes graphene semimetallic with remarkable electron mobility at room temperature. The corresponding resistivity of graphene sheets is on the level of $10^{-6}\ \Omega\,cm$ [60], which is lower than that characteristic of silver at room temperature. Obviously, the charge propagation dynamics would be affected by adsorption of "contaminants" (e.g. traces of metals or dopants) and molecules such as water and oxygen. On the whole, graphene has the ability to meet expectations for good supports of platinum nanoparticles. In other words, the graphene-based systems can meet criteria such as high surface area, electronic conductivity, electrochemical stability, and the ability to distribute uniformly Pt catalytic sites [11, 17, 61]. Because graphene constitutes two-dimensional planar sheets with an open structure, both sides of graphene sheets can be utilized for supporting catalytic centers [62].

Graphene-type supports are certainly capable of anchoring noble metal nanoparticles [11, 12, 14, 16] including platinum electrocatalysts [11, 17, 46, 61]. It should be remembered that graphene can exist in different forms. A single sheet of sp^2-hybridized carbon atoms can be derived from bulk graphite material by mechanical exfoliation or via chemical procedures. The chemical approach may lead to additional doping of the carbon lattice or introduction of functional groups. The chemical oxidation to graphene oxide results in the increased hydrophilicity due to formation of oxygen-containing groups, typically carboxyl or epoxide ones. Having in mind potential application as support for catalytic nanoparticles, high porosity of the oxidized form (graphene oxide) seems to be advantageous. But the oxidized system's overall stability (resistance to oxidative corrosion) and conductivity are decreased. A reduction step, that is typically performed chemically and is irreversible, produces so-called reduced graphene oxide. In comparison to the oxidized form, the latter system is more hydrophobic, but it still contains hydrophilic domains [63]. Among important diagnostic experiments, spectroscopic monitoring of surface oxygen-containing functional groups, such as –C–OH, –COOH, –C=O, and –C–O–C–, has been achieved using Fourier Transformed Infrared Spectroscopy (FTIR), and relative intensities of G and D bands at about 1600 and 1350 cm^{-1} (using Raman) should be mentioned. Typically, reduced graphene oxide (RGO) surfaces are characterized by the low degree of organization of graphitic structure and by the presence of defects. When these interfacial features are combined with some hydrophilicity, the increased adsorptive capabilities and enhanced cocatalytic activity of RGO can be envisioned. Furthermore, RGO can be viewed as robust and conductive catalytic material or carrier [63, 64]. Thus, RGO seems to be the optimum form of graphene for electrocatalytic applications, including oxygen reduction.

A classical procedure for deposition of Pt nanoparticles onto graphene sheets involved reduction of H_2PtCl_6, using $NaBH_4$, in the suspension of graphene oxide (GO) [64]. The resulting partially reduced GO-Pt catalyst was subjected to further reduction with hydrazine at elevated temperature (300 °C). The observed enhancement of the electrochemically active surface area was correlated with a spatial morphology of the catalytic film in which Pt nanoparticles (anchored to single graphene sheets) induced separation between graphene flakes. Under such conditions, transport of reactants (oxygen and water) was feasible.

Graphene support could facilitate electron transport through the system's basal planes to edges. Faster electron transfers at edges resulted in lower contact resistances when particles are deposited therein. Indeed, the ball-milled graphene supports exhibited higher activity than those based on graphene flakes during ORR [65]. Obviously, the ball-milling procedure tended to produce the defect-rich oxidized surface functionalities. A synthetic Chemical Vapor Deposition (CVD) methodology was explored for fabrication of electrodes of controlled morphology [66]. Such systems permitted controlled arrangement of deposited metal nanoparticles.

Among practical limitations appearing during preparation of the graphene-based catalytic inks are the poor dispersion of graphene in solvents and its strong tendency to agglomerate in aqueous solutions. But following deposition of catalytic films, care must be exercised to diminish the aggregation of Pt nanoparticles on graphene surfaces and to avoid the restacking of graphene layers caused by their attraction due to van der Waals forces.

Some improvement can be achieved through intentional modification of graphene surfaces [67], for example, by doping with a single element (N, S, P, and B) or bielement doping (N and S or N and B) or by chemical functionalization, e.g. to obtain derivatized systems with metal oxides, polymers, metal phthalocyanines, or sulfonates.

In the case of graphene supports doped with heteroatom species, the foreign heteroatoms are integrated directly into the graphitic planes [11, 17, 68]. Based on DFT studies, strong hybridization between platinum orbitals and sp^2 dangling bonds at the defective N-doped graphene sites (thus facilitating attachment of platinum) can be postulated [69]. Obviously, the carbon network of graphene is somewhat damaged by doping. On the other hand, chemical functionalization involving either covalent or noncovalent approaches could not have any destructive effect on the carbon network [67].

Both basal and edge planes of graphene could interact with the metal electrocatalytic sites [30], whereas only basal planes could be exposed in the case of CNTs. An interface based on basal panes only is certainly more homogeneous in comparison to that built up of both basal and edge planes. The latter heterogeneous surfaces are believed to facilitate stronger dispersion and stabilization of metal catalytic sites.

The interfacial functional groups on the surfaces of graphene oxide species facilitate their dispersion and "dissolution" in different solvents [70]. Obviously, population of the functional groups can be varied on surfaces by changing the extent of oxidation. As already mentioned, the functional groups can serve as the anchors or nucleating centers for the controlled growth of catalytic nanoparticles [62]. There are also specific strategies to functionalize graphene surfaces, e.g. by covalent grafting with *p*-phenyl SO_3H or *p*-phenyl NH_2 groups [70]. The functional groups were found to not only facilitate the homogeneous distribution of Pt nanoparticles on the surfaces of graphene supports but diminishing the Pt average particle size but also to strengthen interaction between Pt atoms and functional groups of the support, thus minimizing the migration and coalescence of Pt nanoparticles during the ORR experiments or in the course of accelerated durability tests.

Nitrogen is known to efficiently create defects on surfaces of carbon materials, thus increasing the edge plane exposure and improve the ORR activity [71]. It is also noteworthy that utilization of graphene facilitates the fast electron transfer rate during ORR, particularly under conditions of the fuel cell operation [30].

Incorporation of sulfur to graphene surfaces would lead to poisoning or contamination of the Pt catalytic surfaces. On the other hand, immobilization of sulfur within the graphitic framework of a support material should result in strong catalyst–support interactions [72]. To fabricate the sulfur-doped graphene, typically phenyl disulfide and graphene oxide are mixed together and then heated in a furnace at 1000 °C for 30 minutes under the atmosphere of high-purity argon. While the heterogeneous dopant atoms seem to serve as nucleation sites permitting dense surface coverage, the uniform nucleation of Pt on the sulfur-doped graphene can be attributed to strong Pt–S interactions [68, 72]. High system's catalytic activity toward ORR, relative to the performance of the commercial Pt/C catalyst, has been postulated [72].

Other types of nanostructured carbons, such as mesoporous carbons, carbon nanocoils, or carbon aerogels, have also been considered as support materials for the ORR catalysts [31].

9.7 Metal Oxide-containing ORR Catalysts

Important alternative approaches include development of metal oxide composite supports for catalytic metal nanoparticles. Metal oxides, such as CeO_2, ZrO_2, WO_3, TiO_2, and SnO_2, have also been studied as components of the ORR catalysts, Figure 9.5, both in basic and acidic media [30, 37, 73–75]. Nanostructured metal oxides can act as supports for catalytic nanoparticles, such as Pt, Figure 9.5a, or they can be introduced onto particles to modify their catalytic properties, Figure 9.5b. In principle, the metal oxide supports could also be combined or decorated with traces of another "activating" metal oxide, Figure 9.5c [76, 77]. Having in mind possible enhancements effects, metal oxide supports usually suffer from the limited electron conductivity, but they often offer very good corrosion resistance and improve overall stability [30, 75]. In practice, metal oxides are used in combination with carbon supports to obtain desirable properties, such as high surface area and high conductivity. The resulting metal-oxide-containing systems tend to improve the stability and the activity of the noble-metal-based electrocatalytic materials. Routinely, the surfaces of carbon carriers are functionalized with metal oxides before they are used as supports for catalysts, e.g. Pt nanoparticles. Depending on the nature of metal oxide, such functionalization approaches could delay corrosion of carbon carrier; but, in some cases, they may accelerate the degradation process of the carbon material, e.g. oxidation.

Representative examples include TiO_2-supported Pt and Pt dispersed over TiO_2 existing on the CNT carriers [78]. The latter system was electrocatalytically more

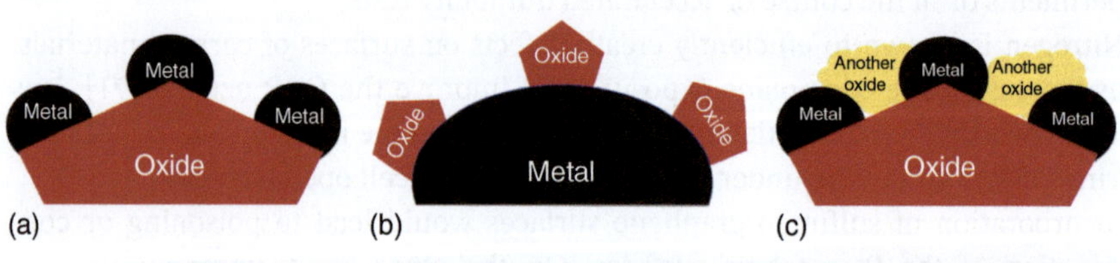

Figure 9.5 Nanostructured metal-oxide-containing electrocatalytic systems: (a) metal oxide acts as support for catalytic metal (e.g. Pt) sites; (b) catalytic metal (e.g. Pt) is decorated with metal-oxide nanostructures; and (c) metal-oxide-supported catalytic (e.g. Pt) sites are further decorated with additional (distinct) metal oxide nanostructures.

active, in comparison to the simple CNT-free one. While incorporation of highly conductive CNTs improved charge distribution at the electrocatalytic interface, introduction of robust TiO_2 nanostructures boosted the catalytic system's stability. Also, the analogous system utilizing carbon black (instead of CNTs) showed improvement in activity for ORR and stability upon introduction of TiO_2 [79]. The influence of two distinct oxides, titania and molybdenum oxide, with and without various carbon carriers, on the activity and stability of Pt and bimetallic-PtRu-based catalytic systems was also addressed [53]. The durability tests confirmed the stabilizing role of TiO_2.

Strong interactions between WO_3 and platinum were realized during both electrooxidations, e.g. of methanol, and electroreductions, e.g. of oxygen, in acid media [80, 81]. First, the Pt-induced formation of nonstoichiometric hydrogen tungsten oxide bronzes (H_xWO_3, where $0 < x < 1$) with strong reduction properties was found to induce electroreduction of H_2O_2 (to H_2O), the undesirable O_2-reduction intermediate. Furthermore, WO_3 was demonstrated to act as robust matrix-stabilizing Pt particles and preventing their agglomeration. A unique method was proposed to generate and immobilize Pt catalytic sites within WO_3 at low loadings such as $5\,\mu g\,cm^{-2}$ through anodic dissolution of the Pt counter electrode [82]. The bifunctional activity of the resulting system in terms of promoting both reduction of oxygen (at traces of Pt) and reductive decomposition of the H_2O_2 intermediate (at mixed-valent W(VI,V)) was postulated. The concept was later extended to the formation of Pt-free WO_3-modified carbon-supported $RuSe_x$ nanoparticles exhibiting potent electrocatalytic properties toward oxygen reduction [83] and tolerance to contamination by methanol crossing-over through the membrane in fuel cell. Although the overall efficiency of the proposed system was lower compared to the Pt-based systems, modification of $RuSe_x$ with ultrathin layers of WO_3 led to the enhancement effect (increased electrocatalytic current, lower production of hydrogen peroxide, and higher heterogeneous electron transfer rate constant) as well as resulted in a shift of the oxygen reduction potential toward more positive values. It was apparent from the XPS data that WO_3 was chemisorbed and interacted with $RuSe_x$ and tended to increase the metallic behavior of Se (and indirectly of Ru) [12]. Such features were likely to increase activity and better selectivity of the catalyst. In general, films of partially reduced WO_3 [80, 84, 85], in which hydrogen tungsten(VI,V) oxide bronzes, H_xWO_3, coexist with substoichiometric oxygen-deficient tungsten(VI,IV) oxides, WO_{3-y} ($0 < y < 1$), or forms H_xWO_{3-y}, were characterized by fast electron transfers, good proton mobility, and high porosity as well as by high reactivity toward reductions of such inert reactants as bromate or hydrogen peroxide [82, 85, 86].

Representative examples of the activation of carbon-supported Pt nanoparticles involved their combination or codeposition with nonstoichiometric transition metal oxides, such as MoO_x (with Mo-oxidation state below VI) and TiO_x (with Ti-oxidation state below IV). Despite the partial blocking of the electrochemically active Pt-surface area with the neighboring metal oxide agglomerates, the composite catalysts of carbon-supported Pt admixed with MoO_x and TiO_x tended to exhibit higher catalytic activity toward ORR, in comparison to the simple carbon-supported Pt catalyst [87]. This phenomenon was explained in terms of the increased activity of catalytic sites for the ORR due to the synergetic effect originating from possible activating interactions between platinum and the oxide materials. Under such conditions, the possibility, of using smaller amounts of a noble metal catalyst (Pt), was postulated [88]. To assure the intimate contact between Pt and TiO_2 (anatase), the metallic Pt nanosites were formed during reduction of platinum cationic species on the oxide surface by photogenerated (UV-irradiation) electrons. The observed lattice contraction of TiO_2, as demonstrated using X-ray diffraction (XRD), was assigned to the interfacial charge transfer between the oxide and metal atoms.

Application of TiO_2 at the electrocatalytic interface facilitates the uptake of water, thus permitting the membrane electrode assemblies to be operated in fuel cells with dry gases without any external humidification [89]. The ability of the oxide layer to anchor Pt nanoparticles and to prevent them from agglomeration or coalescence during the thermal treatment should be emphasized. When it comes to the potential application in polymer electrolyte fuel cells, it is also worth mentioning the increased durability and the ability to inhibit the formation of the undesirable hydrogen peroxide intermediate (responsible for the degradation of the polymer electrolyte membrane, PEM). It has been postulated that the photochemically deposited platinum on titania exhibits superior stability relative to platinum supported onto pristine carbon supports [90].

Strong interactions between Pt and nonstoichiometric cerium(IV,III) oxide, CeO_x, and enhanced activity of the nanocomposite system of CeO_x-supported Pt and CNTs during the ORR were further explored [56, 91]. Here, the cerium-containing metal organic framework (MOF) served as a sacrificial precursor during the synthesis of the composite catalyst. It was apparent from the powder XRD and transmission electron microscopic (TEM) data that the utilization of MOF facilitated formation of CeO_x-nanostructures capable of being in the close contact with surfaces of Pt nanoparticles, thus preventing or diminishing their agglomeration. The XPS data were consistent with the view that the interactions between CeO_x and Pt tended to stabilize Pt^0 and Ce^{3+} oxidation states in the nanocomposite system. Additional experiments involving the CO-stripping (oxidation) analysis and Raman spectroscopy were consistent with the existence of strong interaction between Pt and CeO_x, Figures 9.3 and 9.4. The enhanced activity of the composite system was also demonstrated when it was utilized as cathode catalyst in the hydrogen–oxygen microlaminar flow fuel cell.

Among other oxide-based composite systems, platinized carbon-supported SnO_2 and ZnO were also considered [79]. In all cases, the enhancement of electrocatalytic activity toward the ORR in acid medium was observed. The interfacial electronic changes (e.g. shifting of valence d states of platinum) due to strong interaction between Pt centers and metal oxides were postulated. Such phenomena were further confirmed by the CO-stripping voltammetry (oxidation) as well as by *in situ* infrared reflection absorption spectroscopy. The fact that the oxidative desorption (stripping) of CO (to CO_2) was shifted to lower (less-positive) potentials in the presence of metal oxide species could be of importance not only to the oxidation of simple organic fuels (e.g. methanol) but also when it comes to the removal of CO poisoning species which may appear during oxidative corrosion of carbon supports [92–95].

Strong metal (Pt)–support (WO_3 or TiO_2) interactions were described in terms of changes in electronic properties of platinum, namely in the binding energy of Pt 4f and increase of the Pt 4f peak asymmetry [96] leading to the increased electron density on platinum. Consequently, upon deposition of platinum on the oxide support, the Pt surface became more metallic and apparently contained less surface oxides (Pt oxo species). Having in mind that metallic platinum (Pt^0) acts as heterogeneous electrocatalyst for ORR, the enhancement effects can be envisioned under such conditions. Changes in the metal oxide parameters, including lattice parameters due to the possibility of alloy formation, were also postulated.

A nanocomposite system of Pt centers dispersed over silica was proposed using chemical reduction method [97]. At low Pt populations (with Pt-to-SiO_2 ratios equal to 1 : 1 or 2 : 1), the dispersions were quite uniform and formed the interconnected particle network. Features such as high electrochemically active surface area and improved charged transfer kinetics were mentioned with respect to fuel cell applications.

Deposition of very small amounts of platinum on carbon-supported niobium oxide nanoparticles (NbO_2 or Nb_2O_5) was explored toward fabrication of highly active catalysts toward ORR [43]. While Nb_2O_5 nanoparticles were prepared by heating a peroxo niobic acid sol, NbO_2 nanoparticles were obtained by annealing the Nb_2O_5 nanoparticles in the hydrogen atmosphere. Rotating disk electrode tests demonstrated that the proposed system exhibited the approximately three times higher Pt-mass activity for the ORR in comparison to the commercial Pt/C catalyst. The observed high activity of Pt deposit was interpreted in terms of the reduced OH adsorption caused by the lateral repulsions between PtOH and oxide surface species. The utility of niobium oxide (NbO_2) nanoparticles as supports for Pt was postulated, and the possibility of avoiding problems related to the substrate oxidation and corrosion was emphasized.

9.8 Photodeposition of Pt on Various Oxide–Carbon Composites

Pt deposited on the corrosion-resistant metal oxide-based supports has shown better ORR activity and durability due to the strong metal–support interactions between the Pt and the metal oxide nanostructures [96, 98, 99]. However, selective deposition of Pt nanoparticles on the metal oxide component, rather than the carbon carrier of the composite support, is a difficult task to achieve. In this respect, photoelectrochemical approach can be employed as a useful tool for the single-step selective deposition of Pt nanoparticles on the semiconductor metal oxides (with larger band gaps), incorporated in carbon-based support materials [79, 88, 100–103]. Superior electrocatalytic ORR activity and durability of photodeposited Pt nanoparticles on various oxide–carbon composites has been demonstrated [79, 101, 103]. For example, it has been apparent from XRD analysis that photodeposition of platinum on composite supports made of carbon carriers and TiO_2 or WO_3 leads to the formation of Pt nanoalloys at the interface [101]. Furthermore, the support-induced changes in the electronic structure of the photodeposited Pt have been correlated with the observed improvement of the ORR electrocatalytic activity. Based on XPS analysis [96], the binding energy (BE) values corresponding to Pt $4f_{7/2}$ for $Pt/WO_3/C$ and $Pt/TiO_2/C$ catalysts were 70.94 and 70.90 eV, respectively, whereas the BE value characteristic of the simple carbon supported Pt is equal to 71.19 eV. Similar downshift in the BE of Pt photodeposited on the composite of SnO_2 with carbon carriers has also been observed [100]. This downshift in the BE is consistent with the oxide-support-induced increase in the interfacial electronic density of the photodeposited Pt due to support effect. Such electronic modifications can reflect charge transfers from the metal oxide to Pt atoms or changes in the lattice parameters due to possibility of the formation of alloys at the interface. The results of XRD analyses of Pt photodeposited platinum on the composites of carbon carriers with TiO_2, SnO_2, and ZnO_2 imply crystalline structure of Pt in such catalysts [79]. While the chemical fabrication of platinum nanoparticles on the oxide–carbon composites typically results in the uneven distribution and in agglomeration of the metal centers, the photodeposition leads to the formation of dendritic Pt nanoclusters of various sizes and shapes because the photogenerated electron–hole pair in the metal oxide (unevenly dispersed) acts as a nucleation site for Pt deposition.

While considering supports made of metal oxides deposited onto carbon carriers, the distribution of the metal oxide is also of importance, and it can be improved by pretreatment of the carbon surfaces with acids that tends to increase the interfacial density of active sites available for nucleation of metal oxide nanostructures [104]. In this respect, the system utilizing TiO_2 and carbon carriers has acted as promising support permitting uniform dispersion of Pt nanoparticles (size, 3.2 ± 0.5 nm). The fact that the half-wave potential for the electroreduction of oxygen has been observed at 0.88 V, i.e. at *ca.* 50 mV more positive potential, in comparison to that determined for ORR at conventional carbon-supported Pt in 0.1 mol dm^{-3} $HClO_4$ (0.83 V vs. Reversible Hydrogen Electrode, RHE) [104], is consistent with the existence of specific metal (Pt)–support (TiO_2) interactions affecting the electronic nature of Pt surfaces and partially suppressing, or shifting, the formation of the inhibiting Pt-oxides. The TiO_2-containing catalytic systems have also exhibited very good durability described in terms of the decrease of only 1 mV in the half-wave potential value after 10 000 potential cycles between 0.6 and 1.0 V vs. RHE. CNTs can also be successfully utilized to prepare TiO_2-containg supports for Pt highly active during ORR [105].

Among different oxides, SnO_2 can be used to decorate carbon carriers as well as to serve as template for photodeposition of platinum [100]. The systems containing 40 wt% SnO_2 have been characterized by fairly large electrochemically active surface area (69 ± 1 m^2 g^{-1} Pt) and exhibited promising ORR activity (half-wave potential close to 0.9 V). Although the long-term stability has been postulated, its durability, in particular, the resistance of SnO_2 to oxidative corrosion in acid (e.g. 0.1 mol dm^{-3} $HClO_4$ electrolyte), requires further systematic investigation. Once more, the improved performance has been attributed to strong and specific metal–support interactions inducing electronic modification of the catalytic centers as evidenced. This view has been supported with the results of diagnostic experiments, such as voltammetric stripping of CO adsorbates and X-ray studies using XPS and XRD analyses [100]. It is noteworthy that the CO stripping (voltammetric oxidation) pattern obtained at the photodeposited Pt dispersed onto the hybrid support of SnO_2-decorated carbon carriers resembles the characteristic voltammogram for CO oxidation at PtSn alloy [106]. Similar CO voltammetric stripping results have also been observed at the electrocatalytic system utilizing TiO_2 [107]. The sensitivity of the Pt-photodeposition approach to the physicochemical (heterogeneous) identity (presence of distinct nanodomains) of surfaces existing on carbon-containing metal oxide supports [100, 108] should also be taken into consideration during fabrication of platinum nanoclusters. Further research is expected and needed along this line.

9.9 Other Supports

Although carbon supports have been reported to improve performance of catalytic sites during ORR, parasite site effects such as the oxidation (corrosion) of carbon in the presence of O_2 at high electrode potential have been identified as one of the major degradation problems when it comes to their application in fuel cells [75]. The representative examples of alternative noncarbon supports include conducting polymers (e.g. polyaniline and polypyrrole), hybrid organic–inorganic systems [109], and mesoporous silica. However, issues such as limited conductivity or stability may limit true applications of silica or conducting polymers, respectively. No major breakthroughs have yet been achieved with these carbon-free supports [30].

During long-term operation of fuel cells, the agglomeration of metal nanoparticles and the electrochemical corrosion of supports lead to the degradation of catalysts. The agglomeration of metal nanoparticles can be reduced by choosing a support exhibiting strong interactions with the metal catalytic sites. Ideally, it should also be much less corrosive than common carbon supports. An interesting alternative originates from the possibility of utilizing gold core, instead of carbon, covered by platinum shell [110]. An ultimate goal of this research has been to prepare systems characterized by reasonably high catalytic activity but with small amount of expensive platinum [111]. Ideally, the Pt shell should be as thin as possible is thin enough, whereas the ORR electrocatalytic activity can be optimized by choosing the appropriate core material, possibly, composed of less inexpensive components. In this respect, the electronic nature and morphology of the Pt surface, as well as mutual interactions between core and shell, are of primary importance. Obviously, the position of the d-band center of platinum within the thin shell, which depends on the strain (geometric effects), is crucial. Furthermore, the possible enhancement effects may also reflect the electronic interaction between the top Pt-layer and the core material that have effect on reactants' adsorption energies. Special attention has been paid to utilization of the gold core [110]. In addition to its high electronic conductivity, the Au core is characterized by the larger lattice parameter, when compared to platinum, thus introducing tensile stresses within the Pt monolayer. Under such conditions, the ORR enhancement effect has been correlated with the decrease in the formation of inhibiting PtOH [110]. Finally, it has been demonstrated that platinum electrocatalytic nanoparticles can be stabilized against dissolution (corrosion) by their modification with gold clusters [112]. The superior stability of Au-modified Pt, relative to pristine Pt, has been particularly apparent from the electrochemical diagnostic experiments involving 30 000 voltammetric potential cycles between 0.6 and 1.1 V under oxidizing conditions (in the presence of O_2). Evidence is also provided that the carbon-supported bimetallic Au-Pt/C (20%, 20%) catalysts tend to drive the ORR according to the four-electron reduction mechanism transfer pathway to H_2O and to shift the half-wave potential for ORR approximately 100 mV more positive when compared to the performance of conventional carbon-supported Pt [113]. In addition to the abovementioned interfacial (structural and electronic) changes of the Pt catalyst, the bifunctional mechanism involving Au-site promotion and Pt-site activation by –OH adsorption (formation of Au–OH) has been postulated [113]. The importance of this bifunctional mechanism has been emphasized when it comes to the ORR in alkaline media where platinum is responsible for the dissociative chemisorption of O_2 via formation of HO_2^- which is subject to further decomposition via two-electron reduction on Au.

9.10 Alkaline Medium

The catalytic systems for ORR, under alkaline (rather than acidic) conditions, are typically more stable and kinetically more favorable. The research along this line is of importance to the development of anion-exchange membrane fuel cells, Zn-air batteries, and alkaline-based Li-air batteries. Here, significant progress in the design and synthesis of efficient ORR catalysts from earth-abundant elements, instead of precious metals, for operation in alkaline media should be noted. Special attention has been paid to carbon composite materials because of their reasonable catalytic activity, durability, and cost. As mentioned, by intentional doping with heteroatom (e.g. N, S, B, or P), the electronic and geometric properties of carbon can be adjusted, thus providing more catalytically active sites and enhancing the mutual interaction between carbon structure ("support") and active sites [11, 114]. To achieve high catalytic activity and improved durability, transition metals are considered to catalyze carbonization of nitrogen/carbon precursors and to form highly graphitized carbon nanostructures with favorable nitrogen doping. The idea of intentional functionalization toward the enhancement of activity of the carbons' catalytic sites during ORR has been pursued through deliberate decoration metal oxides or sulfides. Rational designs of advanced alkaline ORR catalysts, which are characterized by good stability and promising performance, refer to the concepts of carbon composite systems.

For instance, to prepare uniformly distributed Pt_2Pd-alloyed nanoparticles immobilized within nitrogen-rich nanopores of graphene supports, platinum phthalocyanine (PtPc) and palladium phthalocyanine (PdPc) were used (following calcination at high temperatures) as precursors of both Pt and Pd metals and the source of nitrogen dopant forming the nitrogen sites within the graphene skeleton [115]. Strong interactions between Pt_2Pd-alloy nanoparticles and N-derivatized graphene units (in nanopores) were postulated to explain the observed ORR enhancement effects. The appearance of the characteristic peak at 399.4 eV in the system's N1s XPS spectrum implied possibility of strong interactions of the type N–Pt or N–Pd, namely between Pt_2Pd nanoparticles and N-doped site within nanopores. These features were correlated with the increased electrocatalytic activity (ORR in 0.1 mol dm^{-3} KOH) and the system's good stability preventing agglomeration of catalytic Pt_2Pd nanocenters. Indeed, the pyridinic-N and pyrrolic-N could work as metal coordination sites because of the capability of donating lone-pair electrons. The feasibility of stronger adherence of Pt_2Pd nanoparticles to N-derivatized, rather than undoped, graphene sites was supported with DFT calculations.

In another approach, the ultrathin adsorbate-type layers of polyoxometalates (namely Keggin-type phospho-dodeca-molybdates) were demonstrated not only to adsorb strongly on both metal (Au or Ag) and carbon surfaces [12, 116] but also to act as capping ligands, thus stabilizing metal nanoparticles, linking them to the reduced graphene oxide supports and facilitating their dispersion. During experiments in alkaline medium, polyoxometalate adsorbates were sacrificed and removed to obtain bare, highly active Au nanostructures immobilized on carbon supports. Obviously, the presence of pores and structural defects, e.g. on poorly organized graphitic structures or reduced graphene oxide [12, 116], supports facilitated dispersion of metal nanoparticles. For the same mass loadings (e.g. 30 μg cm^{-2}) and comparable morphologies of catalytic metal nanoparticles, the system utilizing the reduced-graphene-oxide support produced, relative to the unsupported Au and the Vulcan-carbon-black-supported Au-catalysts, lower amounts of the undesirable HO_2^- intermediate [12]. Furthermore, application of RGO as support for Au evidently affected the onset potential for ORR and shifted it toward more positive values (0.9 V). The observed synergism could be interpreted in terms of activating interactions between the catalytic Au nanoparticles and the graphene support [117–119], leading to the decrease in the dissociation energy of the O_2 molecule adsorbed on the Au-metal in the presence of the carbon support or to the reduction in the stability of HO_2^--intermediate species. The ability of the reduced-graphene-oxide-supported Au to induce decomposition of hydrogen peroxide intermediate should also be mentioned here.

High activity of the reduced-graphene-oxide-supported Au alloyed with Pd was also demonstrated for ORR in alkaline medium [120]. It has been apparent from the C 1s XPS data that, following solvothermal formation and deposition of bimetallic AuPd nanoparticles, the C—C bonds, rather than the C—O bonds (surface oxygenous groups), were becoming predominant on the graphene surfaces. Furthermore, the Au 4f XPS signals of the RGO-supported Au—Pd nanoparticles were consistent with the existence of metallic Au^0 as the main species (known to be active toward O_2-electroreduction in alkaline media). The system's electrocatalytic activity was also promoted by the presence of Pd^0 and Pd^{2+} sites. Under such conditions the electron transfer numbers could reach 3.95 during ORR; i.e. they were very close to theoretical four-electron pathway consistent with the direct reduction to H_2O.

9.11 Toward More Complex Hybrid Systems

Surprisingly high catalytic activity toward ORR (in alkaline medium) is exhibited by the hybrid system formed by Co_3O_4 nanocrystals grown and deposited on carbon supports such as the mildly reduced graphene oxide [121]. Separately, Co_3O_4 and graphene species exhibit low ORR catalytic activity when considered as single components under analogous experimental conditions. The catalyst's performance can further be improved by N-doping of the carbon support. The voltammetric data of the rotating disk are consistent with the behavior comparable to that of the commercial Pt/C catalyst, i.e. the involvement of almost four electrons during the reduction process in the potential range from 0.75 to 0.6 V (vs. RHE). The half-wave potential has been determined to be equal to 0.83 V at 1600 rpm (0.86 V has been found for Vulcan-supported Pt). The rotating ring-disk electrode (RRDE) measurements have been employed to monitor the formation of peroxide-type species (HO_2^-) during the ORR process and have yielded values not exceeding 6% over the potential range of 0.8–0.45 V with the effective electron transfer number of 3.9. On the contrary, the bare (Co_3O_4-free) N-doped reduced graphene oxide support has exhibited low ORR activity involving predominantly only two electrons.

The experimental results and theoretical considerations clearly indicate that N-doped sites in carbon supports form stable centers for attachment of noble and nonprecious metal nanoparticles. In the case of graphene, both the active sites at edges and in the basal plane are of importance [122]. But simple N-doping seems to be not enough for effective breaking of the O—O bond in O_2 or HOO*. The situation changes following Co-doping, particularly when the Co–N_4 complex is formed (e.g. within the basal plane of graphene): it activates the O_2 dissociation and facilitates the H_2O desorption. Thus, the ORR can proceed on the Co–N_4-functionalized surfaces with lower overpotentials. Special attention needs to be paid to uniform deposition and optimization of the coupling between transition-metal (e.g. Co) and nonmetallic dopants [123]. Upon subjecting to pyrolysis at 600 °C, the catalyst reaches the performance comparable to that known for the Vulcan-supported platinum in alkaline electrolyte. Further hybridization of catalytic systems composed of N-doped carbons (especially graphene) and Co nanoparticles by introducing N-doped graphitic shells could improve their performance and lead to powerful ORR catalysts [124]. Here, the four-electron ORR pathway has been dominant in alkaline media.

Further functionalization of hybrid systems involving immobilization of not only cobalt but also zinc as cometal on N_x/carbon nanosheets should also be mentioned [11]. The resulting catalytic system is characterized by highly positive onset potential for oxygen reduction (0.96 V) competitive to that characteristic of the commercial Vulcan-supported platinum studied under analogous conditions (0.94 V). It is noteworthy that the latter system exhibits reasonable ORR electrocatalytic activity in acidic medium (with the onset potential of 0.85 V; nevertheless, somewhat lower than that for commercial Vulcan-supported Pt, 0.93 V). Features, such as stability and good methanol tolerance, are of importance here. With respect to ORR, the cobalt-based system seems to exhibit the relatively highest electrocatalytic activity [125].

Some attention should be paid to carbon, including graphene, supports onto which iron nanostructures can be immobilized for ORR in alkaline medium. A representative example utilizes iron nanoparticles (5.0 wt% Fe) generated within nitrogen moieties following thermal decomposition (N content from 4.0% to 12.0%, depending on temperature) of graphitic carbon nitride immobilized on graphene sheets containing iron salts, e.g. $FeCl_3$ [126]. In addition to the performance of this nitride–graphene composite resembling electrocatalytic behavior of the commercially available Pt-based electrodes (30 wt% Pt on Vulcan XC-72), the system's long-term stability and tolerance to methanol crossover effect should be mentioned. In another example of the efficient catalyst for oxygen electroreduction, both Fe and Co have been coordinated to N-doped active edge sites created by controlled oxidative etching of porous graphene. Here, the coexistence of N-coordinated bimetallic (Co/Fe) centers with nitrogen sites locked at the pore openings tends to reduce substantially the overpotential for ORR in alkaline medium [126, 127]. Due to its good stability, the catalytic system has been considered as cathode material in the anion-exchange membrane fuel cell: here, a power density of *ca.* 35 $mW\,cm^{-2}$ has been achieved in relation to the 60 $mW\,cm^{-2}$ shown by the Pt-based system. Finally, the concept of the carbon-dual-doping, namely, the three-dimensional (3-D) graphene with both nitrogen and sulfur to produce a support for catalytic $CoFe_2O_4$ nanoparticles, should be mentioned here [128]. An important feature of the graphene with 3-D hierarchical porous structure is that it provides not only a good conducting network for electrons but also promotes the efficient mass transport of the reactant (O_2 and H_2O) molecules and electrolyte species during the redox processes. Under such conditions, more catalytic sites seem to be utilized because stacking of graphene layers seems to be suppressed.

Hybrid systems composed of the reduced graphene oxide-supported platinum and multiwalled CNT-supported iridium (both noble metals utilized at low loadings on the level of 15 and $\leq 2\,\mu g\,cm^{-2}$, respectively) have been proposed as catalytic materials for the reduction of oxygen in 0.5 $mol\,dm^{-3}$ H_2SO_4 [129]. The role of CNTs is to improve charge distribution at the electrocatalytic interface and facilitate the transport of oxygen and electrolyte in the catalytic systems by lowering the extent of reduced graphene oxide restacking during solvent evaporation. The enhancement effect originating from the addition of traces of iridium (supported onto CNTs) to platinum, utilized at low loading, has been attributed to the high ability of iridium to induce decomposition of the undesirable hydrogen peroxide intermediate.

Preparation of hierarchical, particularly noble-metal free, catalytic systems is of particular importance for the ORR process. Core–shell-type Pt-free highly active electrocatalysts have been proposed [130]. They utilize metal ($FeSn_{0.5}$)-based nanoparticles embedded in a carbon nitride shell templated on the graphene nanoplatelet core support. Here, instead of conventional metal, metal oxide, or alloyed metal nanoparticles, subnanometric metal (Fe and Sn) clusters, which are bound within coordination nests formed by carbon and nitrogen ligands existing on surfaces of the carbon nitride shells, have been demonstrated to act as the catalytically active sites during ORR in alkaline medium. Diagnostic experiments with use of RRDE voltammetry and gas-diffusion electrode measurements point on ORR overpotential of only *ca.* 70 mV higher than that observed at the XC-72R carbon black-supported platinum. Among important issues are high stability, selectivity toward the four-electron reduction of oxygen to water, high tolerance to methanol poisoning, and possibility of application in anion-exchange membrane fuel cells.

9.12 Stabilization Approaches

Practical ORR electrocatalysts are often exposed to potentials higher than 0.9 V vs. RHE. During long-term operation under such conditions, oxidative corrosion of virtually all carbons, including graphene-type materials, takes place. Degradation of carbon supports results typically in loss of catalytic centers or in their poisoning (e.g. in a case of Pt with –CO, –CHO, or –COH adsorbates). A reasonable approach is to introduce traces of nanostructured metal oxides in the vicinity of Pt catalytic sites: the hydroxyl groups existing on the oxide surfaces could induce desorption of passivating CO-type species coming from the carbon support corrosion, Figure 9.6. An interesting example of the highly durable ORR catalyst includes a hybrid composite material which was prepared by carefully mixing carbon black with reduced-graphene-oxide-supported Pt [131]. Here, by inserting carbon black nanostructures between the platinized RGO, its durability was significantly improved. It was postulated that the two-dimensional reduced graphene oxide functioned as a "mesh" preventing leaching of dissolved Pt species into the electrolyte, whereas carbon black nanostructures could serve as active sites for recapture or renucleation of small Pt clusters. Similar stabilization effects were postulated for the platinized sandwich structure in which Vulcan XC-72 is placed in between graphene nanosheets [132].

Stabilities of different supports for Pt nanoparticles, such as carbon black, reduced graphene oxide, and multiwalled CNTs, have been compared in 0.1 mol dm^{-3} $HClO_4$ [133]. It has been found that, while the latter substrate was stable, the first one tended to degrade more readily. For comparison, Vulcan XC-72 can be considered as the moderately stable substrate. In practice, stability reflects many experimental parameters including the material's porosity and the degree of formation of the hydrogen peroxide intermediate.

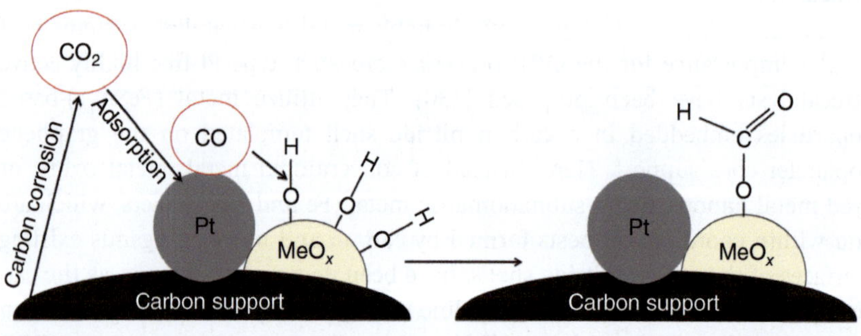

Figure 9.6 Cartoon illustrating interactions of hydroxyl groups on ceria with CO adsorbates existing on platinum (as a result of the oxidative corrosion of carbon support) leading to their removal.

9.13 Conclusions and Perspectives

Although the application of well-conducting high-surface-area nanostructured carbons as supports for Pt catalytic centers has significantly contributed to the progress in fuel cell technology, interfacial corrosion of carbon carriers and degradation of active Pt nanoparticles through their agglomeration and detachment, which decreases practical durability and long-term activity of the ORR catalysts, still requires attention and some improvement when it comes to practical applications. Surface functionalization (e.g. with certain nanostructured metal oxides), introduction of additives and doping with elements, and graphitization of carbon blacks have been considered to minimize problems mentioned above. Some substoichiometric metal oxides, such as CeO_x, TiO_{2-x}, H_xWO_{3-y}, or NbO_x, or hydroxyl-group-rich metal oxides (ZrO_2 or SiO_2) are capable of tuning the activity and durability of not only Pt catalytic sites but of carbon supports as well. Among important issues is promotion of adsorption and activation of oxygen molecule (weakening and splitting of O=O double bond) through modification of the interfacial electronic structure of catalytic platinum as a result of strong interactions of metal oxides with catalytic Pt. It is reasonable to expect that certain functionalized, particularly N-, P-, or B-doped, carbon catalysts, in which the lattice carbons are replaced with heteroatoms (also available for further coordination with metals or cocatalysts), would be of particular interest to the future fuel cell research and ORR studies not only in alkaline media but in acid media as well. A concept of preparation of hierarchical core–shell-type systems, in which nonnoble metal nanoparticles are embedded in carbon nitride shells covering graphene-nanoplatelet core supports, can be viewed as a promising approach to stabilization, activation, and nanostructuring of carbon-supported ORR catalysts. However, it should be remembered that, while developing novel hybrid systems, care should be exercised to assure good mass transport of reactants (oxygen and water). More attention has to be paid to requirements coming from the ORR pathways, i.e. to mechanistic and kinetic details. Formation of hydrogen peroxide-type intermediates would have to be further decreased in practical systems. In this respect, additives of certain metal (e.g. tungsten, cerium, or traces of iridium) oxides could promote decomposition of H_2O_2, both through reduction (to H_2O) or disproportionation (to O_2 and H_2O). Consequently, new advanced ORR catalysts are likely to be multicomponent and multifunctional with distinct catalytic sites capable of promoting four-electron reduction of oxygen and inducing decomposition of any hydrogen peroxide. In other words, certain combinations of metal oxides or mixed-metal oxides, which exhibit desired stability activity or porosity, are likely to be considered as novel support materials or for admixing or decoration of the electrocatalytic ORR systems. Further progress in the area of intentional doping with single elements (N, S, P, or B), bielemental, or multielemental doping, as well as in chemical functionalization toward improved derivatized systems, can be expected. Nevertheless, all specific strategies will have to take into account the requirements of high conductivity in addition to the long-term physicochemical stability and catalytic durability.

Acknowledgment

This work has been supported by the National Science Center (NCN, Poland) under Opus Project 2018/29/B/ST5/02627. We also acknowledge support from the European Union EIT Raw Materials ALPE 19247 Project (Specific Grant Agreement No. EIT/RAW MATERIALS/SGA2020/1).

Acronyms

BE	binding energy
BET	Brunauer–Emmett–Teller isotherm
CNF	carbon nanofiber
CNT	carbon nanotube
CVD	chemical vapor deposition
DFT	density functional theory
FTIR	Fourier-transformed infrared spectroscopy
GO	graphene oxide
ORR	oxygen reduction reaction
MWCNT	multiwall carbon nanotube
PEMFC	polymer electrolyte membrane fuel cell
RGO	reduced graphene oxide
RHE	reversible hydrogen electrode
RRDE	rotating ring disk electrode
SWCNT	single-wall carbon nanotube
TEM	transmission electron microscopy
XPS	X-ray photoelectron spectroscopy

References

1 Shao, M., Chang, Q., Dodelet, J.-P., and Chenitz, R. (2016). Recent advances in electrocatalysts for oxygen reduction reaction. *Chemical Reviews* 116: 3594–3657.

2 Wang, Y.-J., Zhao, N., Fang, B. et al. (2015). Carbon-supported Pt-based alloy electrocatalysts for the oxygen reduction reaction in polymer electrolyte membrane fuel cells: particle size, shape, and composition manipulation and their impact to activity. *Chemical Reviews* 115: 3433–3467.

3 Long, N.V., Yang, Y., Thi, C.M. et al. (2013). The development of mixture, alloy, and core-shell nanocatalysts with nanomaterial supports for energy conversion in low-temperature fuel cells. *Nano Energy* 2: 636–676.

4 Lewera, A., Miecznikowski, K., Hunger, R. et al. (2010). Electronic-level interactions of tungsten oxide with unsupported Se/Ru electrocatalytic nanoparticles. *Electrochimica Acta* 55: 7603–7609.

5 Katsounaros, I., Cherevko, S., Zeradjanin, A.R., and Mayrhofer, K.J.J. (2014). Oxygen electrochemistry as a cornerstone for sustainable energy conversion. *Angewandte Chemie International Edition* 53: 102–121.

6 Bashyam, R. and Zelenay, P. (2006). A class of non-precious metal composite catalysts for fuel cells. *Nature* 443: 63–66.

7 Masa, J., Zhao, A., Xia, W. et al. (2014). Metal-free catalysts for oxygen reduction in alkaline electrolytes: influence of the presence of Co, Fe, Mn and Ni inclusions. *Electrochimica Acta* 128: 271–278.

8 Hickner, M.A., Herring, A.M., and Coughlin, E.B. (2013). Anion exchange membranes: current status and moving forward. *Journal of Polymer Science Part B: Polymer Physics* 51: 1727–1735.

9 Ramaswamy, N. and Mukerjee, S. (2011). Influence of inner- and outer-sphere electron transfer mechanisms during electrocatalysis of oxygen reduction in alkaline media. *Journal of Physical Chemistry C* 115: 18015–18026.

10 Qu, L., Liu, Y., Baek, J.B., and Dai, L. (2010). Nitrogen-doped graphene as efficient metal-free electrocatalyst for oxygen reduction in fuel cells. *ACS Nano* 4: 1321–1326.

11 Kulesza, P.J., Zak, J.K., Rutkowska, I.A. et al. (2018). Elucidation of role of graphene in catalytic designs for electroreduction of oxygen. *Current Opinion in Electrochemistry* 9: 257–264.

12 Zoladek, S., Rutkowska, I.A., Blicharska, M. et al. (2017). Evaluation of reduced-graphene-oxide-supported gold nanoparticles as catalytic system for electroreduction of oxygen in alkaline electrolyte. *Electrochimica Acta* 233: 113–122.

13 Li, J.-C., Hou, P.-X., and Liu, C. (2017). Heteroatom-doped carbon nanotube and graphene-based electrocatalysts for oxygen reduction reaction. *Small* 13 (45): 1702002.

14 Perivoliotis, D.K. and Tagmatarchis, N. (2017). Recent advancements in metal-based hybrid electrocatalysts supported on graphene and related 2D materials for the oxygen reduction reaction. *Carbon* 118: 493–510.

15 Fazio, G., Ferrighi, L., Perilli, D., and Di Valentin, C. (2016). Computational electrochemistry of doped graphene as electrocatalytic material in fuel cells. *International Journal of Quantum Chemistry* 116: 1623–1640.

16 Raj, C.R., Samanta, A., Noh, S.H. et al. (2016). Emerging new generation electrocatalysts for the oxygen reduction reaction. *Journal of Materials Chemistry A* 4: 11156–11178.

17 Higgins, D., Zamani, P., Yu, A., and Chen, Z. (2016). The application of graphene and its composites in oxygen reduction electrocatalysis: a perspective and review of recent progress. *Energy & Environmental Science* 9: 357–390.

18 Huang, X., Qi, X., Boey, F., and Zhang, H. (2012). Graphene-based composites. *Chemical Society Reviews* 41: 666–686.

19 Bezerra, C.W.B., Zhang, L., Liu, H. et al. (2007). A review of heat-treatment effects on activity and stability of PEM fuel cell catalysts for oxygen reduction reaction. *Journal of Power Sources* 173: 891–908.

20 Tang, S., Sun, G., Qi, J. et al. (2010). Review of new carbon materials as catalyst supports in direct alcohol fuel cells. *Chinese Journal of Catalysis* 31: 12–17.

21 Shim, J., Lee, C.R., Lee, H.K. et al. (2001). Electrochemical characteristics of Pt–WO_3/C and Pt–TiO_2/C electrocatalysts in a polymer electrolyte fuel cell. *Journal of Power Sources* 102: 172–177.

22 Sui, S., Wang, X., Zhou, X. et al. (2017). A comprehensive review of Pt electrocatalysts for the oxygen reduction reaction: nanostructure, activity, mechanism and carbon support in PEM fuel cells. *Journal of Materials Chemistry A* 5: 1808–1825.

23 Huang, S.Y., Ganesan, P., Park, S., and Popov, B.N. (2009). Development of a titanium dioxide-supported platinum catalyst with ultrahigh stability for polymer electrolyte membrane fuel cell applications. *Journal of the American Chemical Society* 131: 13898–13899.

24 Sasaki, K., Marinkovic, N., Isaacs, H.S., and Adzic, R.R. (2016). Synchrotron-based in situ characterization of carbon-supported platinum and platinum monolayer electrocatalysts. *ACS Catalysis* 6: 69–76.

25 Nørskov, J.K., Rossmeisl, J., Logadottir, A. et al. (2004). Origin of the overpotential for oxygen reduction at a fuel-cell cathode. *The Journal of Physical Chemistry. B* 108: 17886–17892.

26 Roy, S.C., Harding, A.W., Russell, A.E., and Thomas, K.M. (1997). Spectroelectrochemical study of the role played by carbon functionality in fuel cell electrodes. *Journal of the Electrochemical Society* 144: 2323–2328.

27 Auer, E., Freund, A., Pietsch, J., and Tacke, T. (1998). Carbons as supports for industrial precious metal catalysts. *Applied Catalysis, A* 173: 259–271.

28 Speder, J., Spanos, I., Zana, A. et al. (2015). From single crystal model catalysts to systematic studies of supported nanoparticles. *Surface Science* 631: 278–284.

29 Jung, N., Chung, D.Y., Ryua, J. et al. (2014). Pt-based nanoarchitecture and catalyst design for fuel cell applications. *Nano Today* 9: 433–456.

30 Shurma, S. and Pollet, B.G. (2012). Support materials for PEMFC and DMFC electrocatalysts—a review. *Journal of Power Sources* 208: 96–119.

31 Yaldagard, M., Jahanshahi, M., and Seghatoleslami, N. (2013). Carbonaceous nanostructured support materials for low temperature fuel cell electrocatalysts—a review. *World Journal of Nano Science and Engineering* 3: 121–153.

32 Li, D., Lv, H., Kang, Y. et al. (2016). Progress in the development of oxygen reduction reaction catalysts for low-temperature fuel cells. *Annual Review of Chemical and Biomolecular Engineering* 7: 509–532.

33 Uchida, M., Fukuoka, Y., Sugawara, Y. et al. (1996). Effects of microstructure of carbon support in the catalyst layer on the performance of polymer-electrolyte fuel cells. *Journal of the Electrochemical Society* 143: 2245–2252.

34 Park, Y.-C., Tokiw, H., Kakinuma, K. et al. (2016). Effects of carbon supports on Pt distribution, ionomer coverage and cathode performance for polymer electrolyte fuel cells. *Journal of Power Sources* 315: 179–191.

35 Kim, H.S., Lee, Y., Lee, J.G. et al. (2016). Platinum catalysts protected by N-doped carbon for highly efficient and durable polymer-electrolyte membrane fuel cells. *Electrochimica Acta* 193: 191–198.

36 Baturina, O.A., Garsany, Y., Zega, T.J. et al. (2008). Oxygen reduction reaction on platinum/tantalum oxide electrocatalysts for PEM fuel cells. *Journal of the Electrochemical Society* 155: B1314–B1321.

37 Masuda, T., Fukumitsu, H., Fugane, K. et al. (2012). Role of cerium oxide in the enhancement of activity for the oxygen reduction reaction at Pt–CeO_x nanocomposite electrocatalyst – an in situ electrochemical X-ray absorption fine structure study. *Journal of Physical Chemistry C* 116: 10098–10102.

38 Xu, F., Wang, D., Sa, B. et al. (2017). One-pot synthesis of Pt/CeO_2/C catalyst for improving the ORR activity and durability of PEMFC. *International Journal of Hydrogen Energy* 42: 13011–13019.

39 Ueda, A., Yamada, Y., Ioroi, T. et al. (2003). Electrochemical oxidation of CO in sulfuric acid solution over Pt and PtRu catalysts modified with TaO_x and NbO_x. *Catalysis Today* 84: 223–229.

40 Kulesza, P.J., Pieta, I.S., Rutkowska, I.A. et al. (2013). Electrocatalytic oxidation of small organic molecules in acid medium: enhancement of activity of noble metal nanoparticles and their alloys by supporting or modifying them with metal oxides. *Electrochimica Acta* 110: 474–483.

41 Chalgin, A., Song, C., Tao, P. et al. (2020). Effect of supporting materials on the electrocatalytic activity, stability and selectivity of noble metal-based catalysts for oxygen reduction and hydrogen evolution reactions. *Progress in Natural Science: Materials International* 30: 289–297.

42 Debe, M.K. (2012). Electrocatalyst approaches and challenges for automotive fuel cells. *Nature* 486: 43–51.

43 Sasaki, K., Zhang, L., and Adzic, R.R. (2008). Niobium oxide-supported platinum ultra-low amount electrocatalysts for oxygen reduction. *Physical Chemistry Chemical Physics* 10: 159–167.

44 Tasis, D., Tagmatarchis, N., Bianco, A., and Prato, M. (2006). Chemistry of carbon nanotubes. *Chemical Reviews* 106: 1105–1136.

45 Rathinavel, S., Priyadharshini, K., and Panda, D. (2021). A review on carbon nanotube: an overview of synthesis, properties, functionalization, characterization, and the application. *Materials Science and Engineering B* 268: 115095.

46 Sevjidsuren, G., Zils, S., Kaserer, S. et al. (2010). Effect of different support morphologies and Pt particle sizes in electrocatalysts for fuel cell applications. *Journal of Nanomaterials* 2010: 852786.

47 Yuan, X.Z., Song, C., Wang, H., and Zhang, J. (2010). EIS diagnosis for PEM fuel cell performance. In: *Electrochemical Impedance Spectroscopy in PEM Fuel Cells: Fundamentals and Applications*, 193–262. London: Springer.

48 Ghosh, S. and Raj, C.R. (2010). Facile in situ synthesis of multiwall carbon nanotube supported flowerlike Pt nanostructures: an efficient electrocatalyst for fuel cell application. *Journal of Physical Chemistry C* 114: 10843–10849.

49 Kongkanand, A., Kuwabata, S., Girishkumar, G., and Kamat, P. (2006). Single-wall carbon nanotubes supported platinum nanoparticles with improved electrocatalytic activity for oxygen reduction reaction. *Langmuir* 22: 2392–2398.

50 Wang, S., Jiang, S.P., Whit, T.J. et al. (2009). Electrocatalytic activity and interconnectivity of Pt nanoparticles on multiwalled carbon nanotubes for fuel cells. *Journal of Physical Chemistry C* 113: 18935–18945.

51 Zhou, J., Zhou, X., Sun, X. et al. (2007). Interaction between Pt nanoparticles and carbon nanotubes – an X-ray absorption near edge structures (XANES) study. *Chemical Physics Letters* 437: 229–232.

52 Du, S., Lu, Y., Malladi, S.K. et al. (2014). A simple approach for PtNi–MWCNT hybrid nanostructures as high performance electrocatalysts for the oxygen reduction reaction. *Journal of Materials Chemistry A* 2: 692–698.

53 Khotseng, L., Bangisa, B., Modibedi, M., and Linkov, V. (2016). Electrochemical evaluation of Pt-based binary catalysts on various supports for the direct methanol fuel cell. *Electrocatalysis* 7: 1–12.

54 Chen, J., Li, Z., Chen, Y. et al. (2020). An enhanced activity of Pt/CeO_2/CNT triple junction interface catalyst prepared by atomic layer deposition for oxygen reduction reaction. *Chemical Physics Letters* 755: 137793.

55 Li, Y., Zhang, X., Wang, S., and Sun, G. (2018). Durable platinum-based electrocatalyst supported by multiwall carbon nanotubes modified with CeO_2. *ChemElectroChem* 5: 2442–2448.

56 Luo, Y., Calvillo, L., Daiguebonne, C. et al. (2016). A highly efficient and stable oxygen reduction reaction on Pt/CeO_x/C electrocatalyst obtained via a sacrificial precursor based on a metal-organic framework. *Applied Catalysis B: Environmental* 189: 39–50.

57 Xiao, Y.P., Jiang, W.J., Wan, S. et al. (2013). Self-deposition of Pt nanocrystals on Mn_3O_4 coated carbon nanotubes for enhanced oxygen reduction electrocatalysis. *Journal of Materials Chemistry A* 1: 7463–7468.

58 Zhu, C. and Dong, S. (2013). Recent progress in graphene-based nanomaterials as advanced electrocatalysts towards oxygen reduction reaction. *Nanoscale* 5: 1753–1767.

59 Ji, L. and Zhang, X. (2009). Fabrication of porous carbon nanofibers and their application as anode materials for rechargeable lithium-ion batteries. *Nanotechnology* 20: 155705.

60 Sagade, A.A., Neumaier, D., Schall, D. et al. (2015). Highly air stable passivation of graphene based field effect devices. *Nanoscale* 7: 3558–3564.

61 Zaman, S., Wang, M., Liu, H. et al. (2022). Carbon-based catalyst supports for oxygen reduction in proton-exchange membrane fuel cells. *Trends in Chemistry* 4: 886–906.

62 Vinayan, B.P., Nagar, R., and Ramaprabhu, S. (2012). Synthesis and investigation of mechanism of platinum–graphene electrocatalysts by novel co-reduction techniques for proton exchange membrane fuel cell applications. *Journal of Materials Chemistry* 22: 25325–25334.

63 Stobinski, L., Lesiak, B., Malolepszy, A. et al. (2014). Graphene oxide and reduced graphene oxide studied by the XRD, TEM and electron spectroscopy methods. *Journal of Electron Spectroscopy and Related Phenomena* 195: 145–154.

64 Seger, B. and Kamat, P.V. (2009). Electrocatalytically active graphene-platinum nanocomposites. Role of 2-D carbon support in PEM fuel cells. *Journal of Physical Chemistry C* 113: 7990–7995.

65 Jeon, I.Y., Choi, H.J., Jung, S.M. et al. (2013). Large-scale production of edge-selectively functionalized graphene nanoplatelets via ball milling and their use as metal-free electrocatalysts for oxygen reduction reaction. *Journal of the American Chemical Society* 135: 1386–1393.

66 Stamatin, S.N., Hussainova, I., Ivanov, R., and Colavita, P.E. (2016). Quantifying graphitic edge exposure in graphene-based materials and its role in oxygen reduction reactions. *ACS Catalysis* 6: 5215–5221.

67 Goth, K., Kawabata, K., Fujii, E. et al. (2009). The use of graphite oxide to produce mesoporous carbon supporting Pt, Ru, or Pd nanoparticles. *Carbon* 47: 2120–2124.

68 Jiao, Y., Zheng, Y., Jaroniec, M., and Qiao, S.Z. (2014). Origin of the electrocatalytic oxygen reduction activity of graphene-based catalysts: a roadmap to achieve the best performance. *Journal of the American Chemical Society* 136: 4394–4403.

69 Tian, Y., Liu, Y.-J., Zhao, J.-X., and Ding, Y.-H. (2015). High stability and superior catalytic reactivity of nitrogen-doped graphene supporting Pt nanoparticles as a catalyst for the oxygen reduction reaction: a density functional theory study. *RSC Advances* 5: 34070–34077.

70 Paredes, J.I., Villar-Rodil, S., Martınez-Alonso, A., and Tascon, J.M.D. (2008). Graphene oxide dispersions in organic solvents. *Langmuir* 24: 10560–10564.

71 Li, W., Xin, Q., and Yan, Y. (2010). Nanostructured Pt–Fe/C cathode catalysts for direct methanol fuel cell: the effect of catalyst composition. *International Journal of Hydrogen Energy* 35: 25–30.

72 Wang, R., Higgins, D.C., Hoque, M.A. et al. (2013). Controlled growth of platinum nanowire arrays on sulfur doped graphene as high performance electrocatalyst. *Scientific Reports* 3: 2431.

73 Khotseng, L. (2018). Chapter 2 – Electrocatalysts for fuel cells and hydrogen evolution – theory to design. In: *Oxygen Reduction Reaction* (ed. A. Ray, I. Mukhopadhyay, and R.K. Pati), 25–50. IntechOpen.

74 Rutkowska, I.A. and Kulesza, P.J. (2018). Metal oxide cluster and polyoxometallate supports for noble metal nanoparticles in efficient electrocatalysis. In: *Encyclopedia of Interfacial Chemistry: Surface Science and Electrochemistry*, vol. 5 (ed. K. Wandelt), 207–216. Elsevier.

75 Wang, Y.J., Wilkinson, D.P., and Zhang, J. (2011). Noncarbon support materials for polymer electrolyte membrane fuel cell electrocatalysts. *Chemical Reviews* 111: 7625–7651.

76 Hornberger, E., Bergmann, A., Schmies, H. et al. (2018). In situ stability studies of platinum nanoparticles supported on ruthenium–titanium mixed oxide (RTO) for fuel cell cathodes. *ACS Catalysis* 8: 9675–9683.

77 Baker, A.M., Stewart, S.M., Ramaiyan, K.P. et al. (2021). Doped ceria nanoparticles with reduced solubility and improved peroxide decomposition activity for PEM fuel cells. *Journal of the Electrochemical Society* 168: 024507.

78 Montero-Ocampo, C., Garcia, J.R.V., and Estrada, E.A. (2013). Comparison of TiO_2 and TiO_2-CNT as cathode catalyst supports for ORR. *International Journal of Electrochemical Science* 8: 12780–11280.

79 Ruiz Camacho, B., Morais, C., Valenzuela, M.A., and Alonso-Vante, N. (2013). Enhancing oxygen reduction reaction activity and stability of platinum via oxide-carbon composites. *Catalysis Today* 202: 36–43.

80 Kulesza, P.J. and Faulkner, L.R. (1989). Electrodeposition and characterization of three-dimensional tungsten(VI,V)-oxide films containing spherical Pt microparticles. *Journal of the Electrochemical Society* 136: 707–713.

81 Rutkowska, I.A., Wadas, A., Zoladek, S. et al. (2018). Activation of reduced-graphene-oxide supported Pt nanoparticles by aligning with WO_3-nanowires toward oxygen reduction in acid medium: diagnosis with rotating-ring-disk voltammetry and double-potential-step chronocoulometry. *Journal of the Electrochemical Society* 165: J3384–J3391.

82 Kulesza, P.J., Grzybowska, B., Malik, M.A., and Galkowski, M. (1997). Tungsten oxides as active supports for highly dispersed platinum microcenters: electrocatalytic reactivity toward reduction of hydrogen peroxide and oxygen. *Journal of the Electrochemical Society* 110: 1911–1917.

83 Kulesza, P.J., Miecznikowski, K., Baranowska, B. et al. (2006). Tungsten oxide as active matrix for dispersed carbon-supported $RuSe_x$ nanoparticles: enhancement of the electrocatalytic oxygen reduction. *Electrochemistry Communications* 8: 904–908.

84 Kulesza, P.J. and Faulkner, L.R. (1988). Electrochemical preparation of electrodes modified with non-stoichiometric mixed-valent tungsten(VI,V) oxides. *Journal of Electroanalytical Chemistry* 248: 305–320.

85 Kulesza, P.J. and Faulkner, L.R. (1988). Electrocatalysis at a novel electrode coating of nonstoichiometric tungsten(VI,V) oxide aggregates. *Journal of the American Chemical Society* 110: 4905–4913.

86 Roland, J.F. and Anson, F.C. (1992). Incorporation of redox-active cations into tungsten oxide coatings on electrodes: enhancement of coating stability and electrocatalytic activity. *Journal of Electroanalytical Chemistry* 336: 245–261.

87 Elezovic, N.R., Babic, B.M., Radmilovic, V.R. et al. (2009). Synthesis and characterization of MoO_x-Pt/C and TiO_x-Pt/C nano-catalysts for oxygen reduction. *Electrochimica Acta* 54: 2404–2409.

88 Timperman, L., Feng, Y.J., Vogel, W., and Alonso-Vante, N. (2010). Substrate effect on oxygen reduction electrocatalysis. *Electrochimica Acta* 55: 7558–7563.

89 De Tacconi, N.R., Chenthamarakshan, C.R., Rajeshwar, K. et al. (2008). Photocatalytically generated Pt/C–TiO_2 electrocatalysts with enhanced catalyst dispersion for improved membrane durability in polymer electrolyte fuel cells. *Journal of the Electrochemical Society* 155: B1102–B1109.

90 Vogel, W., Timperman, L., and Alonso-Vante, N. (2010). Probing metal substrate interaction of Pt nanoparticles: structural XRD analysis and oxygen reduction reaction. *Applied Catalysis A: General* 377: 167–173.

91 Kostuch, A., Rutkowska, I.A., Dembinska, B. et al. (2021). Enhancement of activity and development of low Pt content electrocatalysts for oxygen reduction reaction in acid media. *Molecules* 26: 5147.

92 Roen, L.M., Paik, C.H., and Jarvi, T.D. (2004). Electrocatalytic corrosion of carbon support in PEMFC cathodes. *Electrochemical and Solid-State Letters* 7: A19–A22.

93 Jang, S.-E. and Kim, H. (2010). Effect of water electrolysis catalysts on carbon corrosion in polymer electrolyte membrane fuel cells. *Journal of the American Chemical Society* 132: 14700–14701.

94 Oh, J.-G., Lee, W.H., and Kim, H. (2012). The inhibition of electrochemical carbon corrosion in polymer electrolyte membrane fuel cells using iridium nanodendrites. *International Journal of Hydrogen Energy* 37: 2455–2461.

95 Lee, W.H. and Kim, H. (2014). The electrochemical society, find out more optimization of electrode structure to suppress electrochemical carbon corrosion of gas diffusion layer for unitized regenerative fuel cell. *Journal of the Electrochemical Society* 161: F729–F733.

96 Lewera, A., Timperman, L., Roguska, A., and Alonso-Vante, N. (2011). Metal–support interactions between nanosized Pt and metal oxides (WO_3 and TiO_2) studied using X-ray photoelectron spectroscopy. *Journal of Physical Chemistry C* 115: 20153–20159.

97 Seger, B., Kongkanand, A., Vinodgopal, K., and Kamat, P.V. (2008). Platinum dispersed on silica nanoparticle as electrocatalyst for PEM fuel cell. *Journal of Electroanalytical Chemistry* 621: 198–204.

98 Estudillo-Wong, L.A., Ramos-Sanchez, G., Calvillo, L. et al. (2017). Support interaction effect of platinum nanoparticles on non-, Y-, Ce-doped anatase and its implication on the ORR in acid and alkaline media. *ChemElectroChem* 4: 3264–3275.

99 Hussain, S., Erikson, H., Kongi, N. et al. (2020). Oxygen reduction reaction on nanostructured Pt-based electrocatalysts: a review. *International Journal of Hydrogen Energy* 45: 31775–31797.

100 Hussain, S., Kongi, N., Erikson, H. et al. (2019). Platinum nanoparticles photo-deposited on SnO_2-C composites: an active and durable electrocatalyst for the oxygen reduction reaction. *Electrochimica Acta* 316: 162–172.

101 Timperman, L., Lewera, A., Vogel, W., and Alonso-Vante, N. (2010). Nanostructured platinum becomes alloyed at oxide-composite substrate. *Electrochemistry Communications* 12 (12): 1772–1775.

102 Ma, J.W., Habrioux, A., and Alonso-Vante, N. (2013). Enhanced HER and ORR behavior on photodeposited Pt nanoparticles onto oxide–carbon composite. *Journal of Solid State Electrochemistry* 17: 1913–1921.

103 Estudillo-Wong, L.A., Luo, Y., Diaz-Real, J.A., and Alonso-Vante, N. (2016). Enhanced oxygen reduction reaction stability on platinum nanoparticles photo-deposited onto oxide-carbon composites. *Applied Catalysis B: Environmental* 187: 291–300.

104 Wang, J., Xu, M., Zhao, J. et al. (2018). Anchoring ultrafine Pt electrocatalysts on TiO_2-C via photochemical strategy to enhance the stability and efficiency for oxygen reduction reaction. *Applied Catalysis B: Environmental* 237: 228–236.

105 Kong, J., Qin, Y.-H., Wang, T.-L., and Wang, C.-W. (2020). Photodeposition of Pt nanoparticles onto TiO_2@CNT as high-performance electrocatalyst for oxygen reduction reaction. *International Journal of Hydrogen Energy* 45 (3): 1991–1997.

106 Borbath, I., Guban, D., Paszti, Z. et al. (2013). Controlled synthesis of Pt_3Sn/C electrocatalysts with exclusive Sn–Pt interaction designed for use in direct methanol fuel cells. *Topics in Catalysis* 56 (11): 1033–1046.

107 Timperman, L., Luo, Y., and Alonso-Vante, N. (2016). On the availability of active sites for the hydrogen peroxide and oxygen reduction reactions on highly dispersed platinum nanoparticles. *ChemElectroChem* 3 (10): 1705–1712.

108 Campos-Roldán, C.A., Ramos-Sánchez, G., Gonzalez-Huerta, R.G. et al. (2016). Influence of sp^3–sp^2 carbon nanodomains on metal/support interaction, catalyst durability, and catalytic activity for the oxygen reduction reaction. *ACS Applied Materials & Interfaces* 8: 23260–23269.

109 Kulesza, P.J., Chojak, M., Karnicka, K. et al. (2004). Network films composed of conducting polymer-linked and polyoxometalate-stabilized platinum nanoparticles. *Chemistry of Materials* 16: 4128–4134.

110 Sarkar, A., Kerr, J.B., and Cairns, E.J. (2013). Electrochemical oxygen reduction behavior of selectively deposited platinum atoms on gold nanoparticles. *ChemPhysChem* 14: 2132–2142.

111 Gong, K., Su, D., and Adzic, R.R. (2010). Platinum-monolayer shell on $AuNi_{0.5}Fe$ nanoparticle core electrocatalyst with high activity and stability for the oxygen reduction reaction. *Journal of the American Chemical Society* 132: 14364–14366.

112 Zhang, J., Sasaki, K., Sutter, E., and Adzic, R.R. (2007). Stabilization of platinum oxygen-reduction electrocatalysts using gold clusters. *Science* 315: 220–222.

113 Kumar, S.S. and Phani, K.L.N. (2009). Exploration of unalloyed bimetallic Au–Pt/C nanoparticles for oxygen reduction reaction. *Journal of Power Sources* 187: 19–24.

114 Zak, J.K., Negro, E., Rutkowska, I.A. et al. (2018). Graphene-based nanostructures in electrocatalytic oxygen reduction. In: *Encyclopedia of Interfacial Chemistry: Surface Science and Electrochemistry*, vol. 5 (ed. K. Wandelt), 651–659. Elsevier.

115 Zhong, X., Qin, Y., Chen, X. et al. (2017). PtPd alloy embedded in nitrogen-rich graphene nanopores: high-performance bifunctional electrocatalysts for hydrogen evolution and oxygen reduction. *Carbon* 114: 740–748.

116 Zoladek, S., Blicharska-Sobolewska, M., Krata, A.A. et al. (2020). Heteropolytungstate-assisted fabrication and deposition of catalytic silver nanoparticles on different reduced graphene oxide supports: electroreduction of oxygen in alkaline electrolyte. *Journal of Electroanalytical Chemistry* 875: 114694.

117 Yin, H., Tang, H., Wang, D. et al. (2012). Facile synthesis of surfactant-free Au cluster/graphene hybrids for high-performance oxygen reduction reaction. *ACS Nano* 6: 8288–8297.

118 Lim, D.H. and Wilcox, J. (2011). DFT-based study on oxygen adsorption on defective graphene-supported Pt nanoparticles. *Journal of Physical Chemistry C* 115: 22742–22747.

119 Slanac, D.A., Hardin, W.G., Johnston, K.P., and Stevenson, K.J. (2012). Atomic ensemble and electronic effects in Ag-rich AgPd nanoalloy catalysts for oxygen reduction in alkaline media. *Journal of the American Chemical Society* 134: 9812–9819.

120 Lv, J.J., Li, S.S., Wang, A.J. et al. (2014). Monodisperse Au-Pd bimetallic alloyed nanoparticles supported on reduced graphene oxide with enhanced electrocatalytic activity towards oxygen reduction reaction. *Electrochimica Acta* 136: 521–528.

121 Liang, Y., Li, Y., Wang, H. et al. (2011). Co_3O_4 nanocrystals on graphene as a synergistic catalyst for oxygen reduction reaction. *Nature Materials* 10: 780–786.

122 Li, F., Shu, H., Hu, C. et al. (2015). Atomic mechanism of electrocatalytically active Co–N complexes in graphene basal plane for oxygen reduction reaction. *ACS Applied Materials & Interfaces* 7: 27405–27413.

123 Liu, R., Jin, Y., Xu, P. et al. (2016). A hybrid-assembly approach towards nitrogen-doped graphene aerogel supported cobalt nanoparticles as high performance oxygen reduction electrocatalysts. *Journal of Colloid and Interface Science* 464: 83–88.

124 Zhang, G., Lu, W., Cao, F. et al. (2016). N-doped graphene coupled with Co nanoparticles as an efficient electrocatalyst for oxygen reduction in alkaline media. *Journal of Power Sources* 302: 114–125.

125 Liu, X., Liu, W., Ko, M. et al. (2015). Metal (Ni, Co)-metal oxides/graphene nanocomposites as multifunctional electrocatalysts. *Advanced Functional Materials* 25: 5799–5808.

126 Parvez, K., Yang, S., Hernandez, Y. et al. (2012). Nitrogen-doped graphene and its iron-based composite as efficient electrocatalysts for oxygen reduction reaction. *ACS Nano* 6: 9541–9550.

127 Palaniselvam, T., Kashyap, V., Bhange, S.N. et al. (2016). Nanoporous graphene enriched with Fe/Co-N active sites as a promising oxygen reduction electrocatalyst for anion exchange membrane fuel cells. *Advanced Functional Materials* 26: 2150–2162.

128 Yan, W., Cao, X., Tian, J. et al. (2016). Nitrogen/sulfur dual-doped 3D reduced graphene oxide networks-supported $CoFe_2O_4$ with enhanced electrocatalytic activities for oxygen reduction and evolution reactions. *Carbon* 99: 195–202.

129 Dembinska, B., Zlotorowicz, A., Modzelewska, M. et al. (2020). Low-noble-metal-loading hybrid catalytic system for oxygen reduction utilizing reduced-graphene-oxide-supported platinum aligned with carbon-nanotube-supported iridium. *Catalysts* 10: 689.

130 Negro, E., Delpeuch, A.B., Vezzù, K. et al. (2018). Toward Pt-free anion-exchange membrane fuel cells: Fe–Sn carbon nitride–graphene core–shell electrocatalysts for the oxygen reduction reaction. *Chemistry of Materials* 30: 2651–2659.

131 Li, Y., Li, Y., Zhu, E. et al. (2012). Stabilization of high-performance oxygen reduction reaction Pt electrocatalyst supported on reduced graphene oxide/carbon black composite. *Journal of the American Chemical Society* 134: 12326–12329.

132 He, D., Cheng, K., Peng, T. et al. (2013). Graphene/carbon nanospheres sandwich supported PEM fuel cell metal nanocatalysts with remarkably high activity and stability. *Journal of Materials Chemistry A* 1: 2126–2132.

133 Schonvogel, D., Hulstede, J., Wagner, P. et al. (2017). Stability of Pt nanoparticles on alternative carbon supports for oxygen reduction reaction. *Journal of the Electrochemical Society* 164: F995–F1004.

Part IV

Physical–Chemical Characterization

10

Understanding the Electrocatalytic Reaction in the Fuel Cell by Tracking the Dynamics of the Catalyst by X-ray Absorption Spectroscopy

Ditty Dixon[1], Aiswarya Bhaskar[2], and Aswathi Thottungal[2]

[1] *Mahatma Gandhi University, School of Chemical Sciences, PD Hills, Kottayam, Kerala 686560, India*
[2] *Central Electrochemical Research Institute (CSIR-CECRI), Electrochemical Power Source Division, Karaikudi, Tamil Nadu 630003, India*

10.1 Introduction

Several X-ray as well as electron-based characterization techniques, such as X-ray diffraction (XRD) [1], X-ray photoelectron spectroscopy (XPS) [2], and transmission electron microscopy (TEM) [3], are routinely used in the laboratory to obtain structural as well as electronic information about the material under investigation. X-ray diffraction can provide information regarding, e.g., the bulk crystal structure, lattice strain, phase fraction, etc. of the material under *ex situ* as well as *in operando* conditions. One of the prerequisite conditions for XRD is that the material under investigation should have long-range order. Hence, nanomaterials such as electrocatalysts are not the ideal system that can be studied by XRD. On the other hand, XPS can probe the electronic state (valence as well as spin state) of an element under investigation. Although in XPS an X-ray source (Al K-alpha) is used to get the electronic information, it is necessary to probe the binding energy (BE) of the electrons that are emitted from the sample due to the interaction with X-rays. In general, to carry out any technique where electrons are used as a source such as TEM or scanning electron microscopy (SEM) or as a probe (XPS, TEM, and SEM), it is necessary to have an ultrahigh vacuum (UHV) environment. This makes any *operando* measurements using these techniques extremely difficult or nearly impossible. Therefore, to investigate any electrocatalyst and its electrocatalytic activity or mechanism under fuel cell operating conditions, it is necessary to have unique techniques that can probe nanoparticles (1–12 nm) with short-range order exposed or surrounded by gaseous as well as the aqueous environment. Since all the characterization techniques use either a specific wavelength (XPS, XRD, and Raman spectroscopy) [4] or a narrow window of wavelengths (infrared spectroscopy) from the electromagnetic spectrum, the choice of the characterization technique is dictated by the extent of attenuation of the photons by the electrocatalyst and the fuel cell environment. In this context, conventional infrared spectroscopy (IR) cannot be used under *operando* conditions to characterize fuel cell catalyst since

Electrocatalysis for Membrane Fuel Cells: Methods, Modeling, and Applications, First Edition.
Edited by Nicolas Alonso-Vante and Vito Di Noto.

the carbon/water present in the electrode/interface will completely attenuate the IR radiation. In principle, the light must penetrate several layers, such as the bipolar plate, liquid methanol/water, and gas diffusion layer (GDL), to reach the electrode surface. Depending on the geometry of the measurement, that is transmission mode or reflectance/emission, the light has to travel through either the whole membrane electrode assembly (MEA) or half of the MEA. Thus, the light has to penetrate a few millimeters to get the information about the electrocatalyst. In the electromagnetic spectrum, only X-rays with energy above 5000 eV (hard X-ray) can penetrate through the bulk of MEA [5]. Since fuel cell hardware is a nonpermeable and leak-proof system, incorporating a suitable X-ray window into the hardware of the fuel cell enables *operando* measurements. Figure 10.1 shows the attenuation of X-rays of different energy by the different components of the fuel cell.

From the various aspects discussed above, hard X-ray absorption spectroscopy (XAS) and X-ray emission spectroscopy (XES) can overcome many of the limitations faced by other characterization techniques such as IR and Raman spectroscopy [6]. Moreover, UHV conditions are not necessary for XAS or XES measurements as the absorption edges of many noble metal catalysts used in fuel cells fall in a hard X-ray regime. The absorption edge of Pt, the most commonly used catalyst in the fuel cell, is at 11.6 keV (L_3-edge). In any case, to carry out *operando* measurements at the corresponding catalyst absorption edges (Pt L_3, 11.6 keV; Pd K, 24.3 keV; or Ru K, 22.1 keV), it is necessary to modify the fuel cell in such a way that X-rays can penetrate through the catalyst layer. This is usually achieved by drilling the

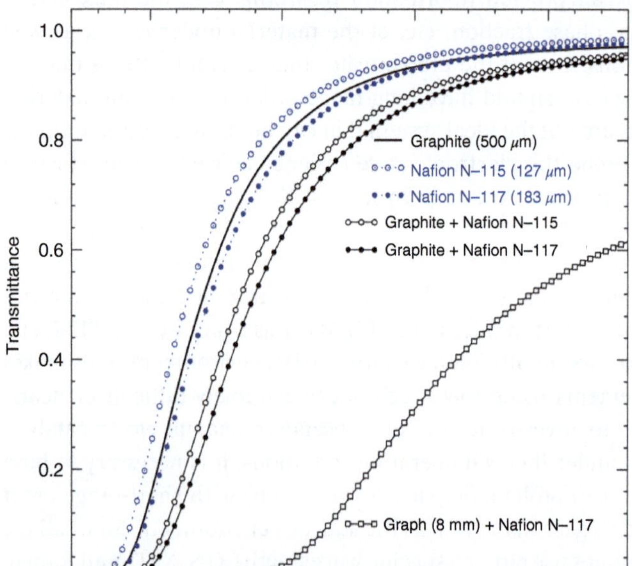

Figure 10.1 Extent of transmittance or attenuation of X-ray photons by different components of a fuel cell. Source: Reproduced from Ref. [5] with the permission of International Union of Crystallography.

metallic end plate with a geometry that allows the collection of fluorescence signals from the electrode (large solid angle). Often, the position of this window drilled in the endplate will be aligned with the channel region of the graphite flow field. This geometry enables the acquisition of data with relatively less background noise since the thickness of the graphite will be higher on land compared to the channel. Moreover, the catalyst in contact with the channel region is always exposed to a higher concentration of reacting species such as H_2, O_2, or methanol compared to the land. Further reduction in the background noise could be obtained by thinning down the graphite bipolar plate or drilling a part of the graphite flow field and sealing it with a Kapton window. More or less, these strategies are used in different *operando* fuel cell designs reported in the literature [5, 7–11].

10.2 A Short Introduction to XAS

X-ray absorption spectroscopy has been widely used by both catalytic and the electrocatalytic community to study the activity of various materials such as metallic nanoparticles, metal oxide nanoparticles, and zeolites. To carry out XAS measurements on these materials, absorption edges corresponding to the atom of interest are chosen, i.e. K-, L_1-, L_2-, L_3-edges. Mainly in catalysis or electrocatalysis, feasibility to carry out *operando* or *in situ* measurements determines the selection of absorption edge. For instance, the Ru K-edge which is at 22.1 keV is preferred over Ru L-edges, as all the L-edges of Ru fall below 4000 keV where the *operando* measurements are difficult to be performed. Again, the L_3-edge of Pt (11.6 keV) is preferred over its K-edge which is at 78.3 keV. Such high energy can be detrimental to the samples. Moreover, not every synchrotron facility can provide sufficient photon flux at such high energy. Once the absorption edge is selected, the sample or the *operando* cell is scanned with an X-ray beam with energy lower (−150 eV) and greater (1000 eV) than that of the absorption edge. Then the linear X-ray absorption coefficient (μE) of the sample is obtained from the Beer–Lambert's law using Eq. (10.1):

$$\ln (I_0/I_t) = \mu x \tag{10.1}$$

where I_0 is the initial intensity of the photon, I_t is the intensity of the photon after absorption by the sample, μ is the linear absorption coefficient, and x is the distance traveled by the photon (usually the sample thickness).

When the linear X-ray absorption coefficient is plotted against the X-ray energy, the X-ray absorption spectrum is obtained. X-ray absorption spectroscopy spectrum (Mn K-edge) obtained for MnO_2 is shown in Figure 10.2.

X-ray absorption coefficient (μ) should decrease with an increase in X-ray energy; however, when the energy of the photon ($h\nu$) is equivalent to the BE or the energy required to remove an electron from a particular shell (K, L, M, etc.) to the continuum, the μ increases dramatically. This translates into the absorption edge as shown in Figure 10.2. Thus, the position of the absorption edge is characteristic of an element as well as its oxidation state. Any transition of the electron within the absorbing atom (below the absorbing edge) or its interference with surrounding

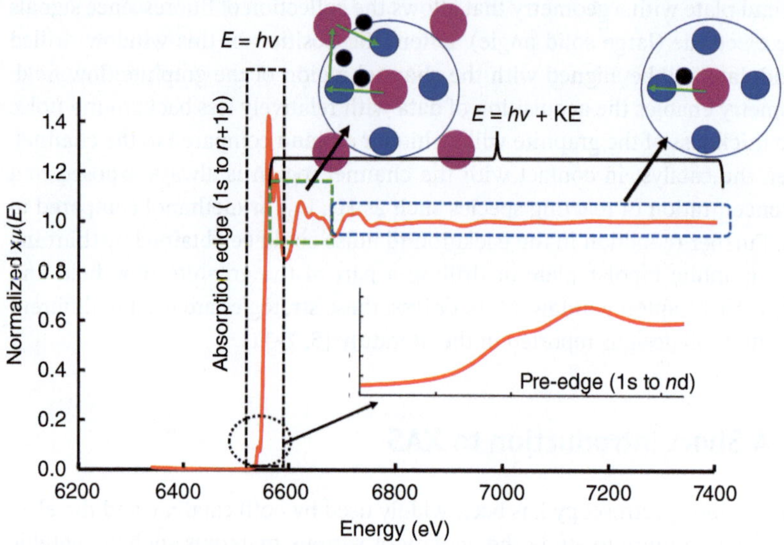

Figure 10.2 Scheme depicting various X-ray absorption events, occurring before and after Mn K-edge in the XAS spectra obtained for the MnO_2 pellet sample.

elements (above the absorbing edge) will be reflected in the absorption coefficient (or XAS spectra) and appears as peaks or oscillations. The pre-edge peaks observed below the absorption edge in the Mn K-edge, Figure 10.2, are due to the transition of an electron from 1s to 3d orbital. Although 1s to 3d transition is electric dipole forbidden in a centrosymmetric molecule, 4p–3d orbital mixing due to distortion can give electric dipole character and thus can result in the appearance of pre-edge peaks. Hence, shape and intensity of pre-edge peaks can be used to get the electronic structure as well as the local symmetry of the transition metal. Once the X-ray energy surpasses the edge energy, the electron ejected from the shell will have some kinetic energy (KE), $h\nu = \text{BE} + \text{KE}$, and it starts interacting with the surrounding atom. This reflects in the oscillation beyond the absorption edge and is due to the constructive or destructive interference of the ejected photoelectron wave with the electron wave of the surrounding atom. Further, the nature of the interference (constructive or destructive) depends on the bond distance between the absorbing atom and the scattering atom, and the amplitude of the corresponding oscillation depends on the atomic weight as well as the number of the scattering atom. The high-energy region beyond the absorption edge can be resolved into two distinct regions, namely (i) X-ray absorption near-edge structure (XANES) and (ii) extended X-ray absorption fine structure (EXAFS). This distinction is mainly based on the difference in photoelectron scattering length due to the difference in the KE acquired by the photoelectron. When the KE of the photoelectron is lower (0–150 eV above the absorption edge), the inelastic mean free path of the photoelectron will be higher. That is, the electron can travel longer, and multiple scattering events can occur. This region is called XANES, where information on average long-range order can be obtained. When the KE is larger (150 eV above the absorption edge), the inelastic mean free path will be smaller and single scattering events will be dominated. This region is called EXAFS, and it gives information about the local coordination of the absorbing atom [12]. Detailed interpretation of these regions is carried out *by ab initio* multiple scattering codes such as FEFF [13–16]. Nevertheless, features of XANES, considered as fingerprints, are often compared with standard materials to get information about the oxidation state as well as the coordination environment.

10.3 Application of XAS in Electrocatalysis

In the field of electrocatalysis, XAS has been used either to understand the structure of the catalyst and its influence on the electrocatalytic activity or to elucidate the electrocatalytic mechanism by monitoring the adsorbate species on the catalyst surface [17–23]. To study the structure of the catalyst, experiments are carried out under different conditions such as *ex situ*, *in situ,* and *operando*. To track the adsorbate, XAS measurements need to be carried out under *in situ* or *operando* conditions, and thus the overall electrocatalytic mechanism can be elucidated. The main difference between *in situ* and *operando* measurements is that in the latter one, the catalyst is under real fuel cell operating conditions, i.e. with mass (flow of gas or methanol etc.) as well as current flow. On the other hand, XAS measurement on catalyst powder under flow of H_2, CO, or O_2 gas can be called as an "*in situ* measurement" because the nature of the catalyst is completely different from the catalyst in a fuel cell. Moreover, electrochemical potential or current is not generated in such a system [24]. Similarly, XAS measurements in an electrochemical cell may also be categorized as *in situ* measurements as the system is in an environment containing a large volume of acidic/alkaline electrolyte at the electrode interface with limited mass as well as current flow, a different situation from that found in thc fuel cell.

10.3.1 *Ex Situ* Characterization of Electrocatalyst

From the *ex situ, in situ*, and *operando* XAS studies, information such as (i) the amount of amorphous metal oxide phase relative to the metallic phase [25–27], (ii) the extent of alloying in bimetallic nanoparticle systems [28], and (iii) relative particle size [29, 30] can be obtained. Dedicated *ex situ* analysis is sometimes preferred over *in situ* or *operando* analysis as the quality of spectra obtained from the *ex situ* measurements is by far superior to those obtained from any *operando* measurements. Usually, background scattering from the X-ray window, GDL, membrane liquid, and gaseous phase, present during the *operando* measurement, increases the noise of the XAS spectra. Most often, the noise is manifested mainly in the high-energy region of XAS spectra (EXAFS region). This often leads to shorter k range (Å^{-1}) values since higher k range features cannot be resolved from noise. A shorter k range of EXAFS leads to less information content which further leads to larger uncertainties in the EXAFS analysis [31]. Therefore, to get precise information such as extent of alloying and particle size, high-quality *ex situ* data are preferred.

Qualitative and quantitative determination of the amorphous metal oxide phase is important since the metal oxide phase can be the active center for electrocatalysis. For example, the ruthenium oxide in Pt/Ru can facilitate the bifunctional mechanism by providing the –OH groups to the Pt that is to the –CO group during the methanol oxidation process or during CO-poisoning [32–34]. However, the oxide phase if present is also susceptible to leaching and can initiate particle growth, leading to an overall decrease in the electrochemically active surface area [35–37]. Usually, the metal oxide phase is formed during the synthesis procedure or due to atmospheric oxidation. Determining the oxidation state or the nature of oxide phase from XAS K-edge spectra is very straightforward if the oxide is more than 5% (the limit for XAS quantification) relative to the metallic phase. When the metallic catalyst system has a high amount of oxide phase, the corresponding XAS K-edge spectra will be shifted to higher energy compared to its metal foil spectra. For example, when comparing Ru K-edge spectrum obtained from a pristine electrode fabricated with a commercial Pt/Ru (HiSPEC 10000, Johnson Mathey [JM])catalyst with the XAS spectrum of an Ru metal foil, it is observed that the edge position of the spectrum, corresponding to the commercial catalyst, was shifted to a higher energy compared to that of the metal spectrum, Figure 10.3.

This indicates that the Ru fraction in the commercial catalyst is oxidized. When the linear combination fit (LCF) was carried out on the XAS spectrum of the commercial sample using the XAS spectra of Ru metal and RuO_2 as standards, it was found that 50% of the Ru, in the commercial sample, was either oxidized or contained ruthenium oxide [25]. By LCF, it was also found that only 20% of the oxide fraction remained, when this catalyst was subjected to direct methanol fuel cell (DMFC) cycling. As mentioned before, the oxide fraction is prone to leaching under DMFC operating conditions and can migrate to the cathode. This was characterized using *ex situ* XAS by analyzing the cathodes of aged DMFC stack, where the obtained Ru K-edge XAS spectra had similar features to that of RuO_2 [38]. Shyam et al. [27] conducted a detailed EXAFS investigation of different commercial

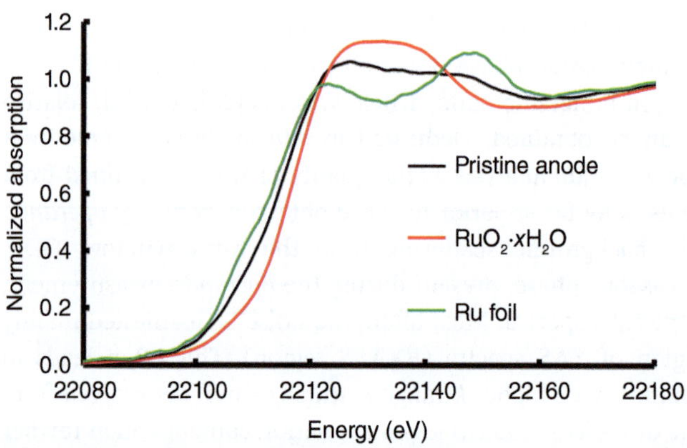

Figure 10.3 Ru K-edge spectra obtained from pristine Pt/Ru anode catalyst, ruthenium oxide, and ruthenium metal foil. Source: Reproduced from Ref. [25] with the permission of Elsevier.

catalysts and concluded that the activity, as well as the stability of the $Pt/Ru/RuO_x$ catalyst system, is attributed to the size of the Ru-oxide islands. They found that the commercial Tanaka catalyst (TEC90110) with a larger Ru oxide island was more active and stable for CO oxidation compared to the JM catalyst (HiSPEC 9000) with smaller Ru oxide islands. Smaller Ru oxide islands in JM catalyst are more prone to dissolution, and therefore, the catalyst showed less electrochemical activity for CO oxidation. They used the coordination number (CN) ratio Ru–Ru/Ru–Pt obtained from EXAFS fitting to determine the size of the Ru oxide islands [27].

Determination of oxide phase or oxidation state from L-edges (L_3-edge for Pt) is slightly different from K-edge as L_3-edge arises due to the transition of $2p_{3/2}$ electron to 5d orbital. The absorption edge corresponding to the L_3-edge spectra of Pt is called a white-line, and the intensity of the white-line gives the occupancy of the 5d orbital. Therefore, the white-line intensity of the L_3-edge XAS spectra corresponding to metallic Pt should be lower than its oxidized form. Figure 10.4 shows the Pt L_3-edge spectra obtained from commercial JM catalyst Pt/C (HiSPEC 9000), the catalyst scratched from MEA made with this catalyst, the catalyst scratched from aged MEA, and the Pt foil.

The commercial catalyst powder has a higher white-line intensity compared to other catalyst samples, indicating a higher amount of oxidized Pt. During the MEA fabrication process, which involves dispersing the catalyst powder in alcohol media, a fraction of this oxide phase may get reduced to its metallic state, as reflected in the lowering of the white-line intensity in comparison with the fresh catalyst. The aged cathode catalyst (after 5000 hours of DMFC operation) had the lowest white-line intensity and was overlapping with that of the Pt foil. Hence, it can be concluded that with aging, the oxidized phase present in the commercial or fresh MEA gets reduced and forms metallic Pt. When the Pt-oxide is reduced, it facilitates the growth of Pt nanoparticles (increase in particle size) which can be elucidated from the EXAFS region of the XAS spectra of the above samples. Fourier transform (FT)

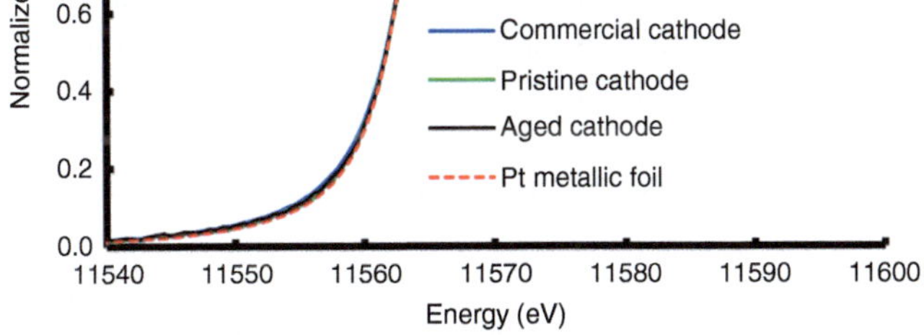

Figure 10.4 Pt L_3-edge spectra obtained for commercial Pt/C, pristine and aged catalyst prepared out of commercial catalyst, and Pt metallic foil.

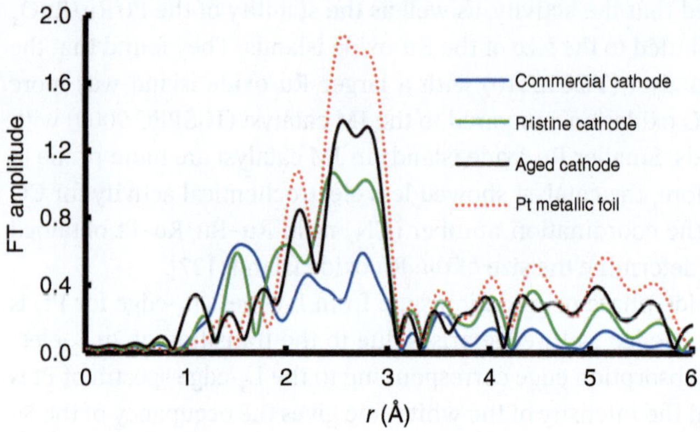

Figure 10.5 EXAFS FT obtained from Pt L_3-edge spectra for commercial Pt/C, pristine and aged catalyst prepared out of commercial catalyst, and Pt metallic foil.

amplitude obtained from the EXAFS region of the above XAS spectra is shown in Figure 10.5.

The peak at 2.7 Å corresponds to the scattering of Pt–Pt, and the amplitude of this peak directly correlates with the particle size or CN. It can be seen from the figure that the Pt metal foil (bulk Pt) has the highest amplitude, and when the EXAFS fitting is done on this FT, it yields a maximum first shell CN of 12. Hence, any nanoparticle system of Pt will have a lower first shell CN or lower EXAFS FT amplitude. So now it can be concluded that the commercial catalyst with lower EXAFS FT amplitude contains smaller Pt particles. As explained before from the XANES analysis, the commercial catalysts are relatively more oxidized. This is again reflected in the EXAFS FT as the FT that corresponds to commercial catalyst has an additional peak at *ca.* 1.5 Å, corresponding to the Pt–O coordination characteristic of the platinum oxide. As mentioned before, when this commercial catalyst is subjected to the MEA fabrication process, the oxide fraction gets reduced to metallic state, which is reflected by a decrease in the amplitude of the peak corresponding to the Pt–O signal in the EXAFS FT plot. Concomitantly, there is an increase in the amplitude of the Pt–Pt shell, indicating an increase in particle size of Pt. With aging, the Pt–O shell almost disappears, and an increase in the amplitude of the Pt–Pt shell is observed, indicating a further increase in particle size. The EXAFS results were supported by both TEM and powder XRD studies where particle size and crystallite size were found to increase with aging [38]. Extended X-ray absorption fine structure fitting results for the abovementioned samples and the crystallite size calculated from XRD are summarized in Table 10.1. Note that XRD gives ideally the average crystallite size and not particle size, but in selected cases like for single crystals, the crystallite size can be the same as particle size.

A more precise way to calculate the nanoparticle size from CN was developed by Frenkel et al. [28] by fitting the EXAFS up to the fourth coordination shell, and this fitting process is more rigorous and involves several multiple scattering paths. A graph showing the relationship between the CN as well as the nanoparticle

Table 10.1 Crystallite size calculated from XRD and EXAFS fitting results obtained for the for-cathode catalysts.

	Crystallite size (XRD) (nm)	Coordination number (CN) Pt–Pt	Distance *r*Pt–Pt (Å)
Commercial cathode	2.0	3.1	2.74
Pristine cathode	4.0	7.2	2.75
Aged cathode	5.0	8.2	2.76
Pt metallic foil	—	12	2.77

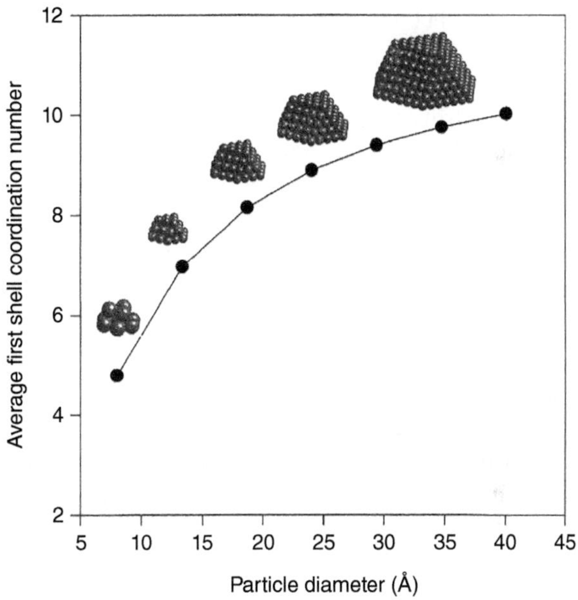

Figure 10.6 Correlation between particle size and coordination number obtained from EXAFS fitting. Source: Reproduced from Ref. [28] with the permission of the American Chemical Society.

size (particle diameter) is shown in Figure 10.6. Therefore, the CN obtained from EXAFS fitting can be used as a complementary technique to XRD/TEM for obtaining crystallite/particle size.

Since EXAFS is highly sensitive to local coordination, it can be a powerful tool to characterize bimetallic nanoparticle systems, especially to distinguish if the two metals present in the system exist as an alloy or segregated. The Pt L_3-edge EXAFS FT obtained for Pt/C as well as Pt/Ru is shown in Figure 10.7. As explained before, a shell corresponding to Pt/Pt is visible for metallic Pt at 2.7 Å. However, for Pt/Ru, although it should contain a coordination shell corresponding to Pt/Pt and Pt/Ru (if alloyed), a splitting was observed in the FT corresponding to metallic coordination shell [21, 30]. This happens due to the alloying of Pt and Ru, i.e. when the Pt is replaced by Ru from its face-centered cubic (fcc) lattice, or when the Ru is replaced by Pt from its hexagonally closed packed (hcp) lattice (or alloying) the bond

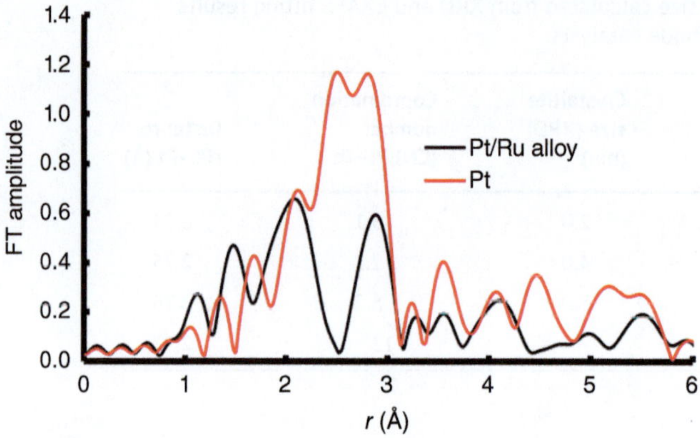

Figure 10.7 Pt L_3-edge EXAFS FT obtained for pure Pt nanoparticles as well as Pt/Ru alloy nanoparticles.

distance intrinsic to pure metal in either of the abovementioned close packing will be distorted. Hence, the new alloyed fcc or hcp lattice will have Pt–Pt and Pt–Ru, coordination, with two different bond lengths. Due to the relatively smaller size of the Ru atom, the bond length corresponding to Pt–Ru will be slightly smaller compared to that of Pt–Pt. This slight difference in the bond distance leads to dampening of oscillation for alloy-based systems and will be reflected in the much larger splitting of the metallic shell in the EXAFS FT.

10.3.2 *Operando* XAS Studies

One of the most important advantages of XAS is that it can provide information about material of any state (solid, liquid, or gas) under every condition. As an extreme case, to understand the coordination or the structural chemistry at the earth's core, XAS measurements are carried out on a sample contained in a diamond anvil cell (DAC) in which earth core conditions (very high temperature and pressure) are generated [39]. Hence, it is relatively straightforward to carry out *operando* measurement of fuel cell once the fuel cell hardware is modified as described in the former section. The *operando* fuel cell setup to record XAS as well as XRD data is shown in Figure 10.8 [40].

Operando XAS measurements are carried out on almost every fuel cell system operated with various fuels such as H_2 gas, reformate gas (both low and high temperature), liquid methanol, gaseous methanol, and liquid ethanol. Mainly the change in XANES features, as well as the EXAFS fitting results, correlates to elucidate the electrocatalytic mechanism. Primarily, the changes in XANES features of the catalytic centers during the *operando* measurement can be due to the nature of the adsorbate, geometry of its coordination with the metal surface, and its coverage. On the other hand, changes in the EXAFS fitting parameters, such as distance between the nearest neighbor, CN, and EXAFS Debye–Waller factor (DW), could be linked to the structural changes associated with catalyst during the electrocatalytic process. In general,

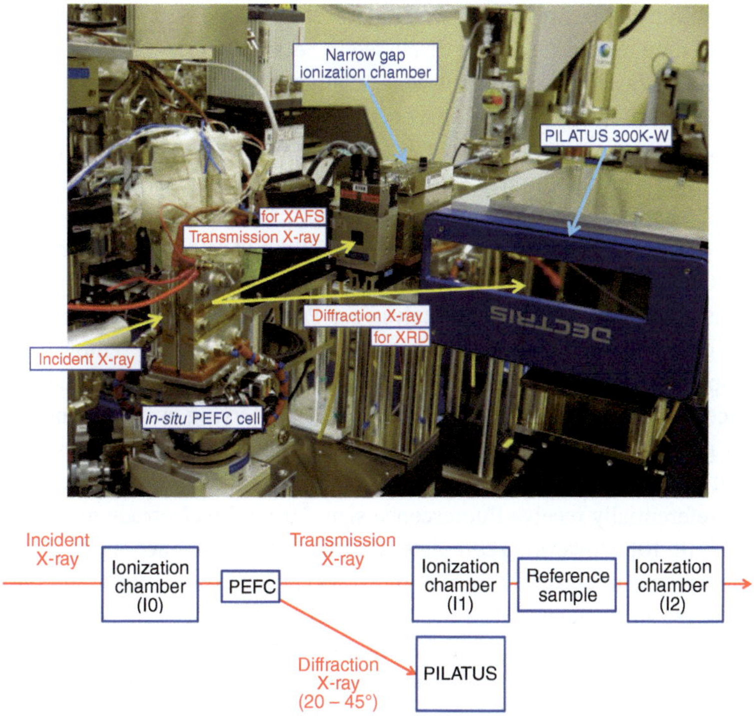

Figure 10.8 Photograph of fuel cell *operando* setup at Spring 8. The scheme of X-ray optics in the experimental hutch is shown in the bottom. Source: Reproduced from Ref. [40] with the permission of the American Chemical Society.

dedicated experiments are not carried out to get exclusive information about XANES or EXAFS, but rather spectrum is recorded over the entire range, −150 eV below and 1000 eV above the edge. However, when time resolution is critical or information is required in a very short time (seconds or milliseconds) the energy range might be limited within the XANES, i.e. to 150 eV above the edge. Nevertheless, with the advent of fourth-generation synchrotron light sources, with high photon flux coupled with either fast-moving monochromator or bend crystals (dispersive XAS) the acquisition of the whole XAS data (both XANES and EXAFS) is possible within millisecond time resolution [40–42].

Mainly *operando* measurements involve a combination of both time- and potential/current-dependent modes of operation of the fuel cell. Usually, XAS spectra are recorded at a specific voltage (potentiostatic mode) or current (galvanostatic mode) during fuel cell operation. The fuel cell is maintained at this particular potential or current till the complete or desired number of spectra are recorded. Then the subsequent potential step or current step is applied, and the XAS spectra are recorded at the new potential or current. The key factor that determines the time interval between each potential step or current step is the scan rate or the time required to record a spectrum. Scan rate can be in the range of hours for conventional XAS scan (high resolution) to a few minutes for quick-EXAFS scan or even millisecond for dispersive XAS scan. Time-resolved measurements recorded with millisecond time resolution will be discussed in Section 10.4.

Many *operando* studies in fuel cells analyze the changes in the white-line intensity (Pt L_3-edge), changes in XANES line shape, or shift in edge position to correlate the activity of the catalyst for various electrocatalytic reactions such as oxygen reduction reaction (ORR), hydrogen oxidation reaction (HOR), or methanol oxidation reaction (MOR). Since Pt-based catalysts are often used on both anode and cathode of the fuel cell, to get precise information about the activity of the Pt catalyst for a particular reaction (HOR, MOR, or ORR), often the counter electrode (electrode not under investigation) is substituted by a non-Pt catalyst, e.g. Pd/C [43]. Sometimes, a small portion of the counter electrode (Pt based) in the X-ray beam path is removed to get information exclusive from the working electrode [8]. However, both techniques could influence the overall performance of the fuel cell. Carrying out experiments in fluorescent geometry could reduce the signal from the counter electrode since the fluorescent detector is placed almost 45° to the X-ray window, and hence, the detector will preferentially receive fluorescence signal from the electrode which is directly facing it [8, 10]. However, a complete screening of fluorescence signal from the counter electrode is not possible since hard X-rays can penetrate deeper, and the situation gets worse when the working electrode is very thin. Using a thicker working electrode often leads to an X-ray self-absorption effect (reduction in absorption amplitude) which can be corrected to some extent by proper data reduction process. Dedicated *operando* fuel cell hardware is also developed to screen the entire fluorescent signal from the counter electrode, and in this type of design, a second X-ray window is drilled to collect the fluorescent signal coming from the working electrode [44].

In almost all the *operando* investigations irrespective of the element (Pt, Ru, Pd, Co, etc.) used as catalyst, a reduction in oxidation state or oxide phase corresponding to the element was observed as soon as the electrode is brought under fuel cell operating conditions [45]. This is mostly reflected as the decrease in the white-line intensity (Pt L_3-edge) or decrease in the edge position/change in XANES shape (Co and Ru K-edges). During the *operando* studies with Pt/Co (1 : 1) as cathode catalyst (Pd/C as the anode), the white-line intensity of both Co K-edge and Pt L_3-edge was found to decrease [46]. Further, with the reduction in the cell voltage, subtle decrease in white-line intensity was observed for both Co K- and Pt L_3-edge. As far as the EXAFS analysis of the above system was concerned, in agreement with XANES observation, the Co–O as well as Pt–O scattering, i.e. intensity of the peak around 1.5 Å, decreased when the catalyst was exposed to fuel cell operating conditions. This indicates that in the as-prepared catalyst, some portion of both Pt and Co was oxidized. *Operando* XAS was also used to study the CO oxidation reaction at Pt as well as at Ru-modified Pt surfaces. The EXAFS analysis gave direct insight into the coordination of CO with both Pt and Ru [18, 47]. For the pure Pt system, the *r*Pt–CO scattering distance was found to be 1.85 Å, which indicates a linearly bonded CO. The Ru–CO scattering path obtained from Ru K-edge EXAFS analysis of the Ru-modified Pt catalyst was found to be 2.04 Å, indicating bridge-type coordination of CO with Ru. It is worth mentioning that in this study, the CN corresponding to Pt–Pt, Pt–Ru or Ru–Ru remains unchanged in the presence of CO and after stripping of CO. This also indicates that CO is adsorbed onto the surface of Pt or Ru. More information about the coordination of Ru on the surface of the Pt was obtained from *in situ* studies (carried out in electrochemical cells). It was found that the Ru formed on the surface of the Pt mainly exists as oxy/hydroxide phase. During the fuel cell cycling (0.7–0.05 V) the oxy/hydroxide species was reduced and alloyed with Pt. When the cell voltage was brought back to 0.7 V, it was found that the reduced Ru was reoxidized, and an oxide phase consisting of Ru^{3+} and Ru^{4+} is formed. Additional information regarding the dissolution of the Ru from the anode was obtained by quantitatively tracking the RuO_2 phase found in the Pt/Ru catalyst spatially (methanol inlet, middle, and outlet region) during the DMFC operation [25]. When the data from spatially resolved *operando* Ru K-edge XAS measurements were analyzed, it was found that, throughout the DMFC operation voltage (0–0.7 V) the methanol inlet region had a higher amount of Ru-oxide phase compared to the methanol outlet region (Figure 10.9).

This was mainly attributed to the higher concentration of methanol in the inlet region, resulting in higher anodic potential. This inhomogeneous distribution of RuO_2 arising from the difference in concentration of methanol leads to the inhomogeneous distribution of current in DMFC. In the event of CO_2 bubble formation, the uniform methanol flow can be hampered and fuel starvation happens, leading to cell voltage reversal (especially in the stack). Cell voltage reversal is highly detrimental as it increases the anode potential to almost 1.2 V vs. standard hydrogen electrode (SHE), and such high voltage can induce carbon corrosion leading to irreversible damage of MEA. The effect of fuel starvation on Pt/Ru catalyst was studied by carrying out *operando* XAS measurements on a DMFC, which was forced to operate in fuel starvation mode by stopping the methanol flow. The Ru K-edge XANES obtained for the corresponding anode catalyst (Pt/Ru, JM HiSPEC 10000) during the fuel starvation process is shown in Figure 10.10a.

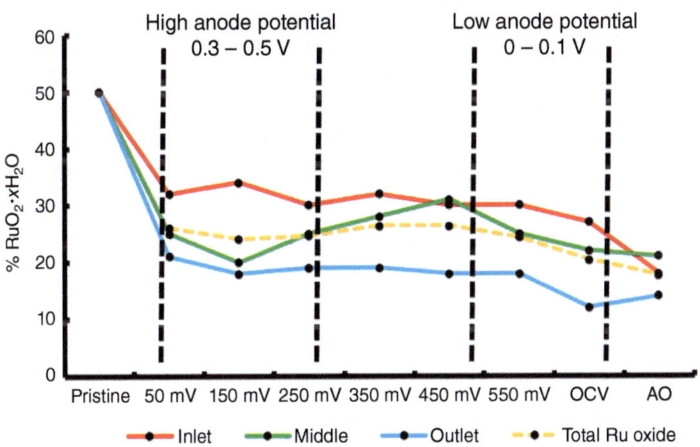

Figure 10.9 Amount of $RuO_2 \cdot xH_2O$ formed/remained at different regions of DMFC in accordance with the cell voltage. The amount of Ru-oxide was quantified by LCF fitting of Ru K-edge XANES.

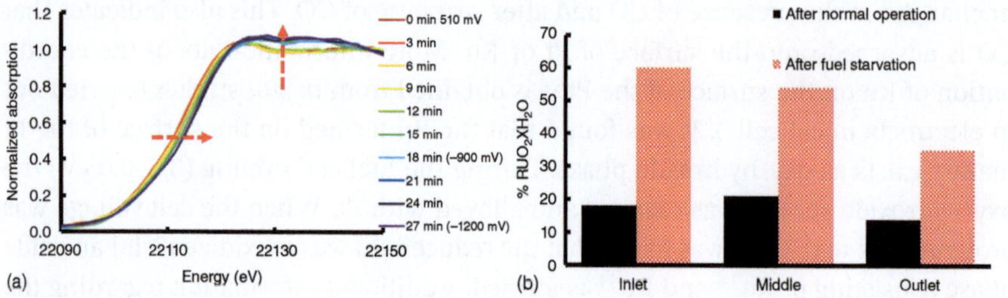

Figure 10.10 (a) XANES Ru K-edge obtained during fuel starvation operation; (b) amount of RuO_2 after normal and fuel starvation operations. Source: Reproduced from Ref. [25] with the permission of Elsevier.

After the fuel starvation process, the amount of RuO_2 phase in the anode increased, and it had a higher phase fraction of oxide compared to the pristine MEA. This indicates that some of the initially alloyed Ru is also oxidized during the fuel starvation process. Therefore, the fuel starvation process can degrade the catalyst as the excess ruthenium oxide, RuO_2, formed during the starvation process can increase both Pt and Ru dissolution. Moreover, among the different regions of DMFC, the methanol inlet showed a much higher RuO_2 phase fraction compared to the methanol outlet region after the fuel starvation process. The amount of RuO_2 phase remained after the normal operation and fuel starvation operation is shown in Figure 10.10b, which indicates that catalyst degradation was much more severe in the methanol inlet region. This is because once the methanol flow has stopped, the methanol concentration decreases slowly. Since the flow of methanol has completely stopped, with time the residual water starts accumulating more at the methanol inlet region (bottom of the cell). However, since the cell is forced to operate galvanostatically, the anode potential increases, and the current is compensated by the water oxidation which will be more prevalent in the methanol inlet region. This leads to an excess oxidation of Ru at the methanol inlet region. On the other hand, water does not accumulate at the methanol outlet region, located at the top of the cell hardware, and since the fuel cell is heated externally, the outlet region dries up quickly compared to other regions. Thus, the methanol outlet region becomes electrochemically inactive, and hence, the catalyst will no longer get oxidized. Therefore, in the event of fuel starvation, the catalyst at the electrochemically active regions will be degraded faster compared to inactive regions, and so fuel starvation should be minimized as possible.

10.4 $\Delta\mu$ XANES Analysis to Track Adsorbate

During the very early *operando* XAS measurements on fuel cells, the full potential of XAS to track the adsorbate has not been explored. In this context, the very first *operando* XAS measurement, carried out by Viswanathan et al. [9], concluded that any improvement observed in the fuel cell performance, after the conditioning period, was due to the improvement in the membrane hydration as well as electrode wetting. The authors reached this conclusion since they did not observe any significant changes in the XANES. However, the adsorption of any molecule (H_2, CO, and O_2) on the surface of any catalyst can induce subtle changes in the XANES features [48]. Thus, a fingerprint related to the adsorbed species is embedded in the XAS data, and with careful background subtraction and other data reduction process (normalization, alignments, etc.) it is possible to extract, from the XAS data, the information about adsorbates. Pioneering work related to extracting the information about adsorbates from the XANES, during *in situ* as well *operando* conditions, was developed by Ramaker et al. [49, 50]. The so-called $\Delta\mu$ technique is simply a process in which a background-corrected, normalized XAS spectrum is subtracted from a reference spectrum. The reference spectrum used for $\Delta\mu$ analysis is measured during the *in situ/operando* measurement at a specific potential/condition at which the catalyst surface has no adsorbate. For instance, in the double layer region, the catalyst surface does not have any chemisorbed adsorbate, and therefore, XAS spectra measured at a potential that falls in the double layer region can be called a reference spectrum. Any XAS spectra measured outside the double layer region will have a contribution from adsorbates. Therefore, if a XANES of the catalyst obtained from the double layer is subtracted from a XANES obtained from any other potential outside the double layer, the difference spectrum obtained corresponds to the adsorbate signature. To correlate this experimental adsorbate signature, theoretical adsorbate signatures are generated from a small platinum cluster (Pt_6-Janin cluster) using FEFF8. The unique property of this small cluster is that it can coordinate with adsorbate species in all possible sites, i.e. atop, bridged, fcc, and hcp [51]. Janin clusters with adsorbate coordinated at different sites are shown in Figure 10.11.

Similar to the experimentally generated adsorbate signature, theoretical signatures can be generated by subtracting the XANES obtained for the adsorbate-free

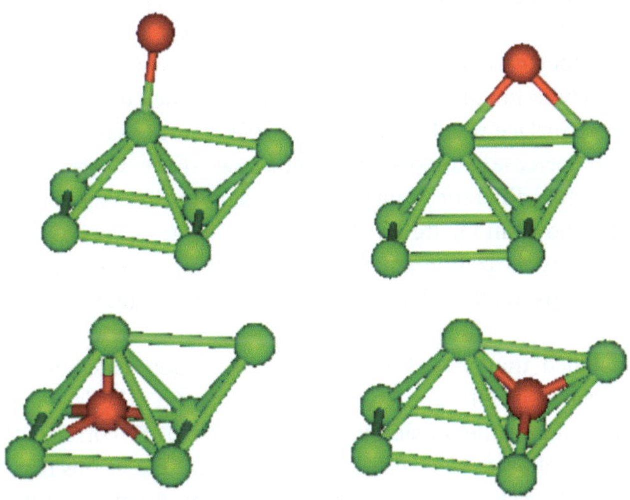

Figure 10.11 Janin clusters showing different coordination sites, top left: atop, top right: bridged, bottom left: fcc, and bottom right: hcp.

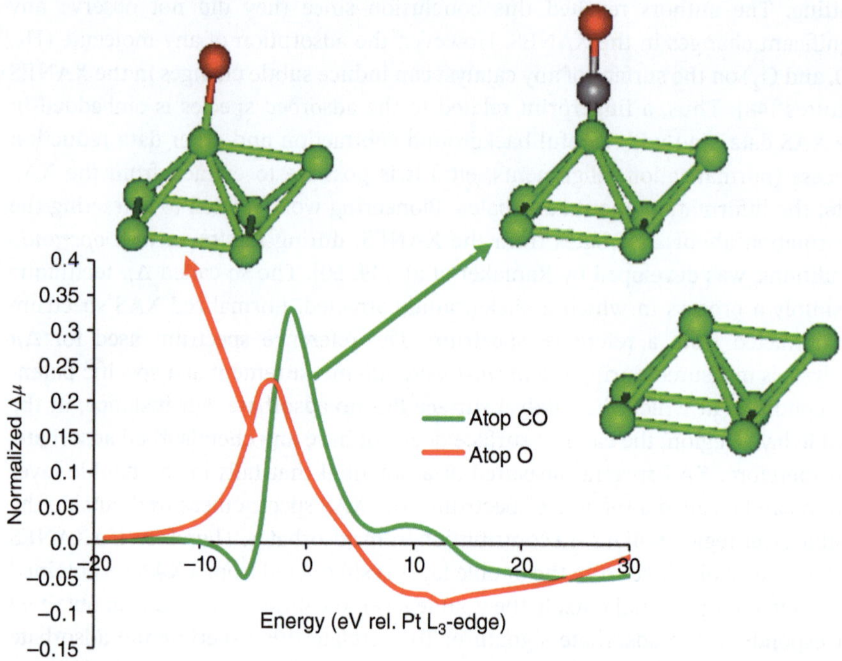

Figure 10.12 Theoretical $\Delta\mu$ signature obtained for atop CO and atop O from FEFF8 calculation.

Janin cluster and one with the adsorbate. The theoretical $\Delta\mu$ signature obtained for the oxygen and carbon monoxide in atop coordination is shown in Figure 10.12.

One of the earliest works employing the $\Delta\mu$ technique under fuel cell *operando* conditions was reported by Roth et al. [52]. Using the $\Delta\mu$ technique, these authors tracked both OH and CO coverages on Pt/Ru catalyst at the anode of a DMFC (MOR). From the $\Delta\mu$ signature, a strong correlation between OH and CO coverages on the Pt surface was observed. It was found that with an increase in the current density, OH coverage on the Pt surface was found to increase, and at the same time, CO coverage was found to decrease. Moreover, they found two distinct threshold regions for OH coverage at two different estimated anode polarization potentials. The detailed $\Delta\mu$ analysis is shown in Figure 10.13. The OH coverage or the $\Delta\mu$ signature observed at the anode polarization potential below 400 mV or preignition region was correlated to OH adsorbate on Pt/Ru which are near to hydrated ruthenium oxide ($RuO(H)_n$) island.

Therefore, it was concluded that at this potential, MOR is dominated by a bifunctional mechanism where the OH group is provided by the $RuO(H)_n$ islands. The OH signature observed above the anode bias potential of 450 mV correlated with the OH adsorbed on Ru-alloyed Pt. Hence, it was concluded that at this bias potential, methanol oxidation occurs via the ligand mechanism. Similar results were also obtained for *operando* measurements using reformate gas (H_2 gas with 150 ppm CO) [53]. In this study, a composition of physically mixed Pt/C and Ru/C

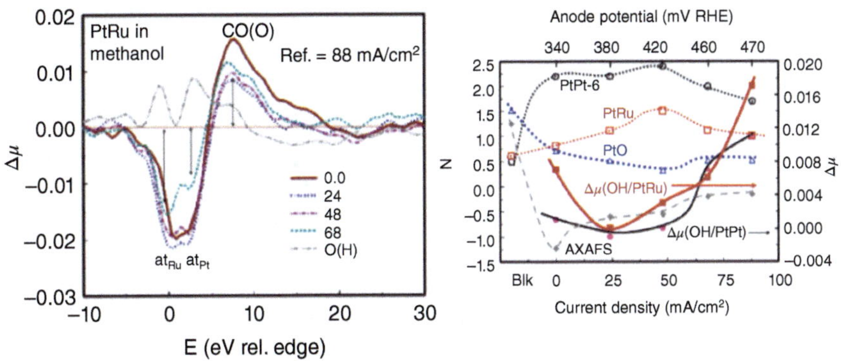

Figure 10.13 Left: $\Delta\mu$ signature obtained during methanol operation; the OH adsorbates at Ru (at_{Ru}) and Pt (at_{Pt} near to Ru islands) are marked. Right: $\Delta\mu$ magnitude obtained plotted against current density and estimated anode potential values. Source: Reproduced from Ref. [52] with the permission of the American Chemical Society.

catalysts showed superior CO tolerance compared to Pt/Ru alloy catalyst. The small Ru (metallic) islands generated on the Pt surface in the physically mixed Pt/Ru catalyst can facilitate CO oxidation through an efficient bifunctional mechanism. The authors attributed the CO oxidation observed for the Pt/Ru alloy catalyst at low current predominantly to the ligand mechanism, as the bifunctional mechanism may be inefficient due to the large RuO_x island present in the Pt/Ru alloy catalyst. With additional work carried out in the electrochemical cell, it was finally concluded that the CO removal by the OH groups, generated on either Ru or Pt, greatly depends on the cluster size and morphology of the Pt/Ru catalyst. Especially, it was found that the relative size of the Ru island dictates the relative magnitude of the bifunctional or ligand mechanism. Similar findings were also observed for ethanol oxidation on Pt/Ru and Pt/Sn catalysts [54]. Again, it was observed that OH adsorption increases with increasing current, and with increasing OH coverage simultaneously decreases the C species adsorbed on the Pt surface. Although the $\Delta\mu$ could not differentiate the different intermediate C species formed during ethanol oxidation (CO, $COCH_x$, and CH_x) adsorbed on Pt accurately, the correlation among the relative coverage of C (overall), OH, and O with the electrochemical potential changes is still observed.

CO oxidation can also take place on the Pt surface by a direct mechanism at a temperature of 170 °C, which is the basis of the high-temperature polymer electrolyte membrane fuel cell (HT-PEMFC). For HT-PEM fuel cells, polybenzimidazole (PBI) membranes are used instead of Nafion. To enhance the proton conductivity of this membrane, phosphoric acid is also used as an electrolyte. This makes the HT-PEM fuel cell susceptible to phosphoric acid poisoning. When the $\Delta\mu$ signature of both CO and PO_4 was monitored in an operating HT-PEMFC, the synergistic interplay between CO and PO_4 adsorption was found [55]. That is, when the concentration of CO in the gas feed was increased (with a maximum of 3% CO), a corresponding increase in PO_4 adsorption was also observed. It was proposed that when the CO concentration increases, the PO_4 adsorption gets more localized above the monolayer of threefold hydrogen (hydrogen underpotential deposition, H_{upd}). Nevertheless, when the CO concentration was increased above 2%, the CO poisoning increased dramatically and decreased the HOR considerably. When the $\Delta\mu$ signature of PO_4 adsorbate at the cathode was followed at different operating temperatures (50, 110, and 170 °C), it was found that the amount of PO_4 adsorbed strongly decreased with increasing the cell temperature [56]. Moreover, when the PO_4 adsorbate was monitored with respect to the cell voltage (at 170 °C), it was found that at the intermediate cell voltage a maximum PO_4 coverage happened, and at the same time H as well as OH adsorption dominated at low and high cell voltage, respectively.

The $\Delta\mu$ analysis was exclusively used for *operando* space-resolved studies by Dixon et al. [57] where spectra were recorded at a different area of DMFC. $\Delta\mu$ signatures of both the anode and the cathode were analyzed in the regions where concentration of reactants varied considerably. In this work, adsorbate coverages at methanol inlet (high methanol concentration) and middle and outlet (low methanol concentration) regions as well as oxygen inlet (high oxygen concentration) and middle and outlet regions (high water content) were monitored by $\Delta\mu$ analysis. The space-resolved performed studies, at both high and low oxygen flow rates, revealed a "cross-talk" between the anode and the cathode. It was found that when the cell was operated at high oxygen flow, the amount of OH coverage on the cathode (say oxygen outlet) is dictated by the amount of H^+ crossing the membrane from the anode, which is geometrically opposite to the methanol inlet region. On the other hand, when the cell was operated at a low oxygen flow rate, methanol/CO oxidation on the anode, especially at the methanol inlet region, was dictated by the oxygen outlet where now methanol is oxidized and H^+ is crossing from the cathode to anode. This is because, at a very low oxygen flow rate, the DMFC operates in bifunctional mode, especially at high cell voltage (close to open circuit voltage [OCV]). During bifunctional operation, the oxygen inlet which is geometrically opposite to the methanol outlet region acts like a galvanic cell, and the oxygen outlet region which is geometrically opposite to the methanol inlet region acts like an electrolytic cell. In the galvanic region, normal DMFC operation persists due to a relatively high concentration of oxygen. However, at the oxygen outlet region (bottom of the cell), due to relatively low oxygen concentration compared to the oxygen inlet, a high amount of methanol crossover happens since it is geometrically opposite to the methanol inlet region. This results in methanol oxidation at the cathode, and the region behaves like an electrolytic cell, where the current is consumed. $\Delta\mu$ signature while operating DMFC with a low oxygen flow rate revealed the presence of CO at the cathode, supporting the bifunctional operation.

10.5 Time-resolved *Operando* XAS Measurements in Fuel Cells

To study the dynamics events such as particle size changes and intermediate adsorbate coverage and finally to elucidate the kinetics of the fuel cell catalytic process, time-resolved *operando* XAS measurements are extremely important. A dedicated

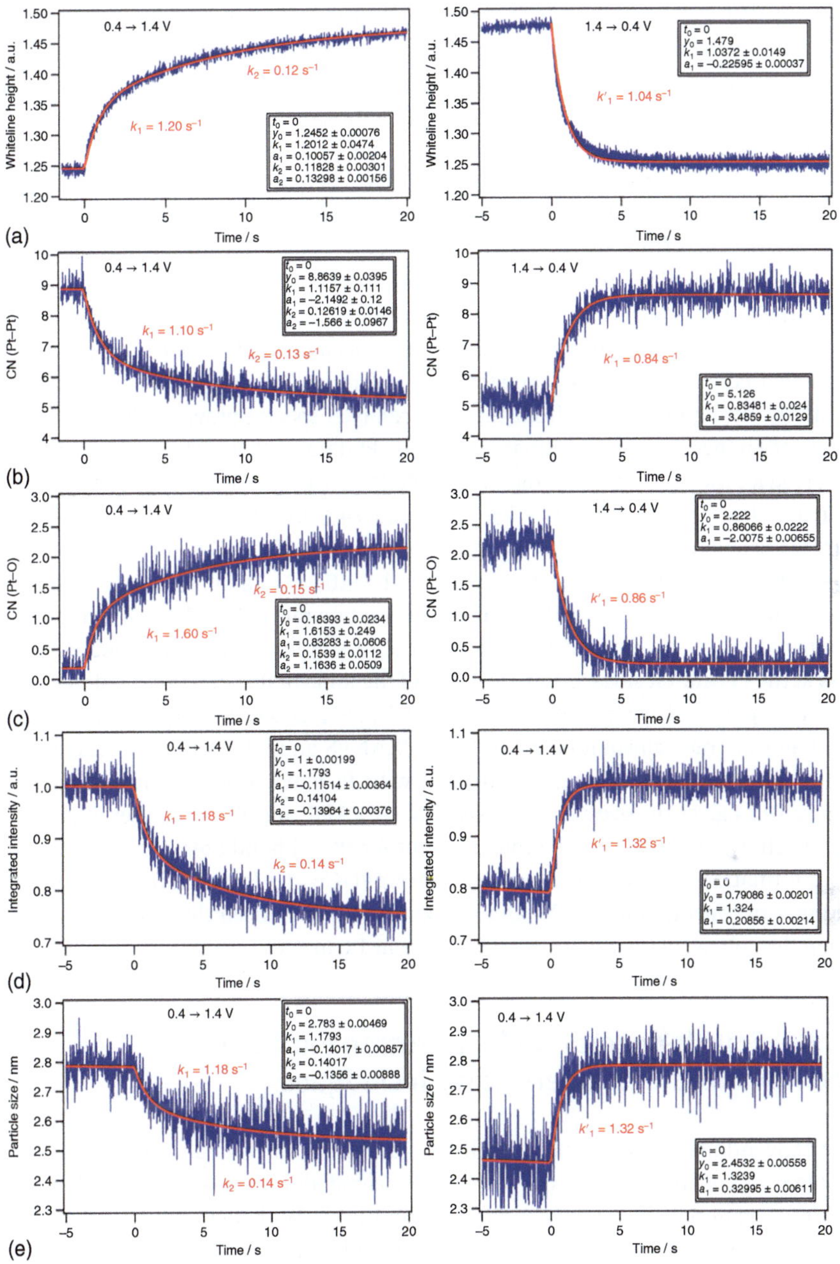

Figure 10.14 XAS analysis results from *operando* time-resolved measurements when cell voltage was switched from 0.4 to 1.4 V (left) and from 1.4 to 0.4 V (right). (a) White-line intensity; (b) Coordination number (CN) Pt–Pt; (c) CN Pt–O; (d) Integrated intensity corresponding to (220) reflection; (e) Crystallite size changes from XRD. Source: Reproduced from Ref. [40] with the permission of the American Chemical Society.

beamline (BL36XU) exclusively for carrying out time-resolved XAS measurements in fuel cells was commissioned in Spring-8, Japan. The details of the beamline can be found elsewhere [58]. With the QXAFS monochromator installed at this beamline, XAS spectra could be recorded with a time resolution of 20 milliseconds [59]. In one of the measurements carried out at this beamline, Sekizawa et al. [40] studied the structural kinetics of the Pt the Pt nanoparticles when the electrochemical voltage of the fuel cell was switched back and forth from 0.4 to 1.0 V. Within the experiment time of 40 seconds (time to reach 0.4 or 1.0 V), XANES spectra with 100 milliseconds time resolution and a complete EXAFS region with 500 milliseconds were recorded. From the curve-fitting analysis of the time-resolved XANES spectra, the rate constant for Pt oxidation, that is when cell voltage was increased from 0.4 to 1.0 V, was estimated as $k = 0.20 \pm 0.02\ s^{-1}$, and for the Pt reduction (1.0–0.4 V) $k' = 0.8 \pm 0.1\ s^{-1}$. Similarly, fitting the EXAFS from the time-resolved spectra provides the rate constant for changes in the coordination of both Pt–Pt and Pt–O. The rate constants observed for the Pt—O bond formation and Pt–Pt breaking when the cell voltage was increased from 0.4 to 1.0 V were $1.3 \pm 0.6\ s^{-1}$ and $0.4 \pm 0.1\ s^{-1}$, respectively. Likewise, during the potential decrement cycle, the rate constants of Pt—O bond breaking and Pt—Pt bond formation were estimated to be $1.5 \pm 0.9\ s^{-1}$ and $0.5 \pm 0.2\ s^{-1}$, respectively. Finally, the authors attributed the slow reaction rate of Pt—Pt bond formation compared to the Pt—O bond formation as the reason for the reversible redox process at the Pt catalyst surface and the cause of the unfavorable dissolution of Pt with repetitive potential cycling. In a similar work by Sekizawa et al. [60], by combining XRD and the rate constant obtained from QXANES and QEXAFS measurements, the relative size/phase changes associated with Pt nanoparticles were estimated. Figure 10.14 gives the overview of QXANES, QEXAFS, and XRD analyses obtained for the study. Estimation of the rate constant for Pt—O bond breaking and Pt—Pt bond formation on bimetallic Pt alloy catalyst (Pt–Co and Pt–Ni) revealed that the addition of transition metal (Ni and Co) kinetically enhanced the reduction of Pt when the voltage of the cell was decreased from 1.0 to 0.4 V.

10.6 Fourth-generation Synchrotron Facilities and Advanced Characterization Techniques

The conventional XAS technique itself has addressed many fundamental research problems involving the operating as well as degradation mechanisms of different electrode materials in various fuel cell chemistries. With the advent of fourth-generation synchrotron facilities with brighter and more coherent beams, the electrocatalytic mechanism could be studied with much less time resolution (less than milliseconds). Most importantly, complementary techniques such as Resonance X-ray Emission (RXES) as well as diffraction techniques, e.g. pair distribution function, could be coupled along with XAS measurements. Since the future will witness nonnoble metal-based catalyst involving first-row transition metal, RXES measurements of the first-row transition metal at K-edge are expected to give complementary information (valence state) as that of corresponding to soft XAS at

L-edges. Nevertheless, the information from *operando* RXES could be considered much more accurate since soft XAS measurements are carried out under ultrahigh vacuum conditions, and often sample preparation involves the dismantling of the electrochemical cells.

10.6.1 Total-reflection Fluorescence X-ray Absorption Spectroscopy

As explained before, most of the catalyst systems (Pt/C or Pt/Ru/C, etc.) also contain the corresponding metal oxide. Since the surface of the nanoparticle is more vulnerable to the atmospheric or chemical oxidation process, the concentration of oxide will be higher on the surface. Using depth profiling techniques one can differentiate the electrocatalytic processes taking place at the surface oxide layer as well as on metallic nanoparticles. Moreover, the information also gives insight into the degradation of catalysts, such as metal dissolution and particle growth. Total reflection fluorescence XAS using hard X-ray can be a powerful tool to study the catalyst surface. Moreover, by varying the incident angle of the beam, depth profiling concerning thickness of the electrode can be achieved. Since the sample volume is relatively low (a few nanometers) for this kind of measurement, a high flux focused beam, such as the fourth-generation synchrotron facilities, will be of great advantage.

10.6.2 Resonant X-ray Emission Spectroscopy (RXES)

To understand the electrocatalytic mechanism of the electrode, it is important to get information about the unoccupied "d states" of the transition metals involved. The information about the unoccupied d states of the transition metals is usually obtained from their corresponding L-edges [61]. As far as Pt (the most commonly used catalyst) is concerned, the L-edges fall in the hard X-ray region, and it is easy to carry out *operando* measurements. However, once catalysts based on first-row (Co, Ni, etc.) or second-row (Pd, Ru, etc.) transition metals are used, the *operando* investigations at L-edges become highly challenging since they fall in the soft X-ray regime, and all the measurements need to be carried out under high vacuum conditions. Transition metal 1s2p RXES involves both photon "in" (X-ray absorption) and photon "out" (X-ray emission) processes in a hard X-ray regime. However, the experiment often requires relatively high photon flux (compared to conventional XAS) as well as a wavelength-dispersive spectrometer. Nevertheless, electrochemical cycling associated with the fuel cells needs to be optimized as RXES data acquisition time is in the order of several minutes (>10 minutes). However, with high photon flux, the acquisition time can be reduced. In the same detector setup, one can also carry out high-energy resolution fluorescence detection (HERFD) by tuning the wavelength-dispersive spectrometer to the maximum of the Kα emission and the incident energy through the absorption edge. Both RXES and HERFD are quite useful, especially when local symmetry around the transition metal is broken by the ligand-to-metal charge transfer. In such a situation, estimation of the oxidation state from the conventional XAS spectra will become more complex, and RXES and HERFD could give a straightforward solution.

10.6.3 Combined XRD and XAS

Several novel nanomaterials are proposed as catalysts mainly as an alternative to Pt-based catalysts for the fuel cell, for instance transition metal–organic frameworks and transition metal oxide. Since these nanomaterials lack long-range order, XAS is the ideal tool to investigate them. Due to the complexity of EXAFS analysis (fitting beyond first shell; approx. 3.5 Å) as well as due to the high background noise involved during an *operando* measurement (especially with fluorescence measurements), a complete picture of local coordination might not be available. Therefore, total scattering experiments are proposed in conjunction with XAS measurements [62, 63]. Although it is not an element-specific technique, the pair distribution function (PDF) analysis associated with total scattering experiments can provide information about coordination beyond the first shell or long-range orders. Fast automated switching between XAFS and PDF measurements is essential for such combined measurements. With the high flux and coherent beam available at the fourth-generation synchrotron facilities, *operando* measurements involving fuel cell systems should see a dramatic change as the users might utilize advanced techniques such as RXES and X-ray Raman Scattering (XRS) as well as the combination of different techniques such as XAS and XRD. Moreover, grazing incident measurements that give depth profiles of the electrode might witness a routine usage as explained before. However, the highly focused beam can be detrimental to the sample as it can cause beam damage of the sample. The beam damage not only induces structural changes in the electrocatalyst, but it also can be hazardous as it could compromise the structural integrity of the membrane (especially the region in the beam path) and could cause a short circuit (mixing of H_2 and O_2). The highly focused beam also demands a highly homogeneous sample, since the small sample volume through which the beam passes should represent the whole sample. On the other hand, with the development of modern laboratory XAS machines, the *operando* XAS as well as XES measurements will be accessible to a much larger electrocatalysis devoted community. Although laboratory XAS/XES require a longer acquisition time, the data quality is sufficient to get information such as oxidation state and adsorbate. Hence, modern laboratory XAS/XES machines can help in the development of next-generation fuel cells by identifying novel, advanced, and low-cost catalyst systems.

10.7 Conclusions

From the above discussion, it can be concluded that XAS is a unique technique by which the catalyst in a fuel cell under operating conditions can be investigated. By interpreting the XAS data, information regarding the electronic state of the metal nanoparticle, its coordination geometry, and the nature of adsorbed molecule or atom on its surface can be obtained. The XANES region of the operando XAS spectra gives insight into the adsorbed species on the surface of the catalyst. Depending on the type of fuel cell, the corresponding adsorbate or intermediate species bound to the catalyst surface can be tracked and correlated to the voltage. Thus, one can elucidate the electrocatalytic mechanism involved between the catalyst and the respective oxidant/reductant molecule. This gives the possibility to scrutinize or develop novel catalysts with superior electrocatalytic activity. XAS can also qualitatively and quantitatively estimate the amorphous or short-range ordered phase in the catalyst system. For instance, the size of the amorphous RuO_2 determines the CO oxidation mechanism, and the amount of amorphous RuO_2 phase susceptibility of Ru leaching and crossover can be predicted. The EXAFS interpretation can give detailed insight into the local coordination changes or particle growth. This information is highly relevant for understanding the catalyst degradation process. Dedicated time-resolved measurements can provide a rate constant related to bond formation between the metal catalyst and adsorbing species, and hence kinetics of the electrocatalytic process can be estimated. Finally, with the advent of the fourth-generation synchrotron light sources, advanced measurements such as RXES and XRS which give the precise electronic states for first- and second-row transition metals will be explored.

Acronyms

CN	coordination number
DAC	diamond anvil cell
DMFC	direct methanol fuel cell
DW	Debye–Waller factor
EXAFS	extended X-ray absorption fine structure
FT	Fourier transform
GDL	gas diffusion layer
HERFD	high-energy resolution fluorescence detection
HOR	hydrogen oxidation reaction
HT-PEM	high-temperature polymer electrolyte membrane
HT-PEMFC	high-temperature polymer electrolyte membrane fuel cell
IR	infrared spectroscopy
KE	kinetic energy
LCF	linear combination fit
MEA	membrane electrode assembly
MOR	methanol oxidation reaction
OCV	open circuit voltage
ORR	oxygen reduction reaction
PBI	polybenzimidazole
PDF	pair distribution function
RXES	resonance X-ray Emission
SEM	scanning electron microscopy
TEM	transmission electron microscopy
UHV	ultrahigh vacuum

XANES	X-ray absorption near-edge structure
XAS	X-ray absorption spectroscopy
XES	X-ray emission spectroscopy
XPS	X-ray photoelectron spectroscopy
XRD	X-ray diffraction
XRS	X-ray Raman scattering

References

1 Warren, B.E. (1990). *X-Ray Diffraction*, 402. Courier Corporation.

2 van der Heide, P. (2011). *X-Ray Photoelectron Spectroscopy: An Introduction to Principles and Practices*, 220. Wiley.

3 Williams, D.B. and Carter, C.B. (2009). *Transmission Electron Microscopy: A Textbook for Materials Science*, 804. Springer Science & Business Media.

4 Larkin, P. (2017). *Infrared and Raman Spectroscopy: Principles and Spectral Interpretation*, 288. Elsevier.

5 Principi, E., Di Cicco, A., Witkowska, A., and Marassi, R. (2007). Performance of a fuel cell optimized for in situ X-ray absorption experiments. *Journal of Synchrotron Radiation* 14: 276–281.

6 van Bokhoven, J.A. and Lamberti, C. (2016). *X-Ray Absorption and X-Ray Emission Spectroscopy: Theory and Applications*, 940. Wiley.

7 Wiltshire, R.J.K., King, C.R., Rose, A. et al. (2005). A PEM fuel cell for in situ XAS studies. *Electrochimica Acta* 50 (25–26): 5208–5217.

8 Roth, C., Benker, N., Mazurek, M. et al. (2005). Development of an in-situ cell for X-ray absorption measurements during fuel cell operation. *Advanced Engineering Materials* 7 (10): 952–956.

9 Viswanathan, R., Hou, G., Liu, R. et al. (2002). *In-situ* XANES of carbon-supported Pt–Ru anode electrocatalyst for reformate-air polymer electrolyte fuel cells. *The Journal of Physical Chemistry B* 106 (13): 3458–3465.

10 Dixon, D., Schröder, A., Schökel, A. et al. (2010). Design of in-situ experimentation for the study of fuel cells with X-rays and neutrons. *Materials Testing* 52 (10): 725–735.

11 Russell, A.E. and Rose, A. (2004). X-ray absorption spectroscopy of low temperature fuel cell catalysts. *Chemical Reviews* 104 (10): 4613–4636.

12 Bunker, G. (2010). *Introduction to XAFS: A Practical Guide to X-Ray Absorption Fine Structure Spectroscopy*. Cambridge University Press.

13 Rehr, J.J., Kas, J.J., Prange, M.P. et al. (2009). Ab initio theory and calculations of X-ray spectra. *Comptes Rendus Physique* 10 (6): 548–559.

14 Rehr, J.J., Kas, J.J., Vila, F.D. et al. (2010). Parameter-free calculations of X-ray spectra with FEFF9. *Physical Chemistry Chemical Physics* 12 (21): 5503–5513.

15 Rehr, J.J. and Albers, R.C. (2000). Theoretical approaches to x-ray absorption fine structure. *Reviews of Modern Physics* 72 (3): 621–654.

16 Newville, M. (2001). IFEFFIT: interactive XAFS analysis and FEFF fitting. *Journal of Synchrotron Radiation* 8: 322–324.

17 Principi, E., Witkowska, A., Dsoke, S. et al. (2009). An XAS experimental approach to study low Pt content electrocatalysts operating in PEM fuel cells. *Physical Chemistry Chemical Physics* 11: 9987.

18 Maniguet, S., Mathew, R.J., and Russell, A.E. (2000). EXAFS of carbon monoxide oxidation on supported Pt fuel cell electrocatalysts. *The Journal of Physical Chemistry B* 104 (9): 1998–2004.

19 Stoupin, S., Rivera, H., Li, Z. et al. (2008). Structural analysis of sonochemically prepared PtRu versus Johnson Matthey PtRu in operating direct methanol fuel cells. *Physical Chemistry Chemical Physics* 10 (42): 6430.

20 Via, G.H., Drake, K.F., Meitzner, G. et al. (1990). Analysis of EXAFS data on bimetallic clusters. *Catalysis Letters* 5 (1): 25–33.

21 McBreen, J. and Mukerjee, S. (1995). In situ X-ray absorption studies of a Pt-Ru electrocatalyst. *Journal of the Electrochemical Society* 142 (10): 3399–3404.

22 Mathew, R.J. and Russell, A.E. (2000). XAS of carbon supported platinum fuel cell electrocatalysts: advances towards real time investigations. *Topics in Catalysis* 10 (3/4): 231–239.

23 Hwang, B.J., Chen, C.H., Sarma, L.S. et al. (2006). Probing the formation mechanism and chemical states of carbon-supported Pt–Ru nanoparticles by in situ X-ray absorption spectroscopy. *The Journal of Physical Chemistry B* 110 (13): 6475–6482.

24 Roth, C., Benker, N., Mazurek, M. et al. (2007). Pt–Ru fuel cell catalysts subjected to H_2, CO, N_2 and air atmosphere: an X-ray absorption study. *Applied Catalysis A: General* 319: 81–90.

25 Dixon, D., Schoekel, A., and Roth, C. (2019). Spatially resolved quantification of ruthenium oxide phase in a direct methanol fuel cell operated under normal and fuel starved conditions. *Electrochimica Acta* 298: 52–58.

26 Liu, D.G., Lee, J.F., and Tang, M.T. (2005). Characterization of Pt Ru/C catalysts by X-ray absorption spectroscopy and temperature-programmed surface reaction. *Journal of Molecular Catalysis A: Chemical* 240 (1–2): 197–206.

27 Shyam, B., Arruda, T.M., Mukerjee, S., and Ramaker, D.E. (2009). Effect of RuO_xH_y island size on PtRu particle aging in methanol. *The Journal of Physical Chemistry C* 113 (45): 19713–19721.

28 Frenkel, A.I., Hills, C.W., and Nuzzo, R.G. (2001). A view from the inside: complexity in the atomic scale ordering of supported metal nanoparticles. *The Journal of Physical Chemistry B* 105 (51): 12689–12703.

29 Yevick, A. and Frenkel, A.I. (2010). Effects of surface disorder on EXAFS modeling of metallic clusters. *Physical Review B* 81 (11): 115451.

30 Alayoglu, S., Zavalij, P., Eichhorn, B. et al. (2009). Structural and architectural evaluation of bimetallic nanoparticles: a case study of Pt–Ru core–shell and alloy nanoparticles. *ACS Nano* 3 (10): 3127–3137.

31 Vlaic, G., Andreatta, D., and Colavita, P.E. (1998). Characterisation of heterogeneous catalysts by EXAFS. *Catalysis Today* 41 (1–3): 261–275.

32 Saida, T., Sugimoto, W., and Takasu, Y. (2010). Enhanced activity and stability of Pt/C fuel cell anodes by the modification with ruthenium-oxide nanosheets. *Electrochimica Acta* 55 (3): 857–864.

33 Rolison, D.R., Hagans, P.L., Swider, K.E., and Long, J.W. (1999). Role of hydrous ruthenium oxide in Pt–Ru direct methanol fuel cell anode electrocatalysts: the importance of mixed electron/proton conductivity. *Langmuir* 15 (3): 774–779.

34 Lasch, K., Jorissen, L., Friedrich, K.A., and Garche, J. (2003). The function of ruthenium oxides in Pt–Ru catalysts for methanol electro-oxidation at low temperatures. *Journal of Solid State Electrochemistry* 7 (9): 619–625.

35 Piela, P., Eickes, C., Brosha, E. et al. (2004). Ruthenium crossover in direct methanol fuel cell with Pt–Ru black anode. *Journal of the Electrochemical Society* 151 (12): A2053.

36 Schoekel, A., Melke, J., Bruns, M. et al. (2016). Quantitative study of ruthenium cross-over in direct methanol fuel cells during early operation hours. *Journal of Power Sources* 301: 210–218.

37 Gancs, L., Hult, B.N., Hakim, N., and Mukerjee, S. (2007). The impact of Ru contamination of a Pt/C electrocatalyst on its oxygen-reducing activity. *Electrochemical and Solid-State Letters* 10 (9): B150–B154.

38 Dixon, D., Wippermann, K., Mergel, J. et al. (2011). Degradation effects at the methanol inlet, outlet and center region of a stack MEA operated in DMFC. *Journal of Power Sources* 196 (13): 5538–5545.

39 Kuramochi, K., Ishimatsu, N., Sakai, T. et al. (2020). An application of NPD to double-stage diamond anvil cells: XAS spectra of rhenium metal under high pressures above 300 GPa. *High Pressure Research* 40 (1): 119–129.

40 Sekizawa, O., Uruga, T., Higashi, K. et al. (2017). Simultaneous operando time-resolved XAFS–XRD measurements of a Pt/C cathode catalyst in polymer electrolyte fuel cell under transient potential operations. *ACS Sustainable Chemistry & Engineering* 5 (5): 3631–3636.

41 Yoshida, H., Kinumoto, T., Iriyama, Y. et al. (2007). XAS study for degradation mechanism of Pt/C catalyst during potential cycling test. *ECS Transactions* 11 (1): 1321.

42 Leach, A.S., Hack, J., Amboage, M. et al. (2021 Jun). A novel fuel cell design for *operando* energy-dispersive x-ray absorption measurements. *Journal of Physics: Condensed Matter* 33 (31): 314002.

43 Russell, A.E., Maniguet, S., Mathew, R.J. et al. (2001). In situ X-ray absorption spectroscopy and X-ray diffraction of fuel cell electrocatalysts. *Journal of Power Sources* 96 (1): 226–232.

44 Petrova, O., Kulp, C., van den Berg, M.W.E. et al. (2011). A spectroscopic proton-exchange membrane fuel cell test setup allowing fluorescence x-ray absorption spectroscopy measurements during state-of-the-art cell tests. *Review of Scientific Instruments* 82 (4): 044101.

45 Stoupin, S., Chung, E.H., Chattopadhyay, S. et al. (2006). Pt and Ru X-ray absorption spectroscopy of PtRu anode catalysts in operating direct methanol fuel cells. *The Journal of Physical Chemistry B* 110 (20): 9932–9938.

46 Witkowska, A., Dsoke, S., Principi, E. et al. (2008). Pt–Co cathode electrocatalyst behaviour viewed by in situ XAFS fuel cell measurements. *Journal of Power Sources* 178 (2): 603–609.

47 Wiltshire, R.J.K., King, C.R., Rose, A. et al. (2009). Effects of composition on structure and activity of PtRu/C catalysts. *Physical Chemistry Chemical Physics* 11 (13): 2305.

48 Bus, E. and van Bokhoven, J.A. (2007). Hydrogen chemisorption on supported platinum, gold, and platinum-gold-alloy catalysts. *Physical Chemistry Chemical Physics* 9 (22): 2894.

49 Teliska, M., O'Grady, W.E., and Ramaker, D.E. (2005). Determination of O and OH adsorption sites and coverage in situ on Pt electrodes from Pt L23 X-ray absorption spectroscopy. *The Journal of Physical Chemistry B* 109 (16): 8076–8084.

50 Ramaker, D.E. and Koningsberger, D.C. (2010). The atomic AXAFS and $\Delta\mu$ XANES techniques as applied to heterogeneous catalysis and electrocatalysis. *Physical Chemistry Chemical Physics* 12 (21): 5514.

51 Janin, E., von Schenck, H., Göthelid, M. et al. (2000). Bridge-bonded atomic oxygen on Pt(110). *Physical Review B* 61 (19): 13144.

52 Roth, C., Benker, N., Buhrmester, T. et al. (2005). Determination of O[H] and CO coverage and adsorption sites on PtRu electrodes in an operating PEM fuel cell. *Journal of the American Chemical Society* 127 (42): 14607–14615.

53 Scott, F.J., Roth, C., and Ramaker, D.E. (2007). Kinetics of CO poisoning in simulated reformate and effect of Ru island morphology on PtRu fuel cell catalysts as determined by operando X-ray absorption near edge spectroscopy. *The Journal of Physical Chemistry C* 111 (30): 11403–11413.

54 Melke, J., Schoekel, A., Dixon, D. et al. (2010). Ethanol oxidation on carbon-supported Pt, PtRu, and PtSn catalysts studied by operando X-ray absorption spectroscopy. *The Journal of Physical Chemistry C* 114 (13): 5914–5925.

55 Caldwell, K.M., Kaserer, S., Roth, C., and Ramaker, D.E. (2015). Following adsorbate coverage on anodes of high-temperature polymer electrolyte membrane fuel cells in the presence of CO and H_2O by using in operando X-ray absorption spectroscopy. *ChemElectroChem* 2 (10): 1502–1509.

56 Kaserer, S., Caldwell, K.M., Ramaker, D.E., and Roth, C. (2013). Analyzing the influence of H_3PO_4 as catalyst poison in high temperature PEM fuel cells using in-operando X-ray absorption spectroscopy. *The Journal of Physical Chemistry C* 117 (12): 6210–6217.

57 Dixon, D., Habereder, A., Farmand, M. et al. (2012). Space resolved, in operando X-ray absorption spectroscopy: investigations on both the anode and cathode in a direct methanol fuel cell. *The Journal of Physical Chemistry C* 116 (13): 7587–7595.

58 Sekizawa, O., Uruga, T., Tada, M. et al. (2013). New XAFS beamline for structural and electronic dynamics of nanoparticle catalysts in fuel cells under operating conditions. *Journal of Physics: Conference Series* 430: 012020.

59 Tada, M., Murata, S., Asakoka, T. et al. (2007). In situ time-resolved dynamic surface events on the Pt/C cathode in a fuel cell under operando conditions. *Angewandte Chemie International Edition* 46 (23): 4310–4315.

60 Ishiguro, N. and Tada, M. (2018). Structural kinetics of cathode events on polymer electrolyte fuel cell catalysts studied by operando time-resolved XAFS. *Catalysis Letters* 148 (6): 1597–1609.

61 Rovezzi, M. and Glatzel, P. (2014). Hard x-ray emission spectroscopy: a powerful tool for the characterization of magnetic semiconductors. *Semiconductor Science and Technology* 29 (2): 023002.

62 Lin, F., Liu, Y., Yu, X. et al. (2017). Synchrotron X-ray analytical techniques for studying materials electrochemistry in rechargeable batteries. *Chemical Reviews* 117 (21): 13123–13186. https://pubs.acs.org/doi/10.1021/acs.chemrev.7b00007.

63 Bak, S.M., Shadike, Z., Lin, R. et al. (2018). In situ/operando synchrotron-based X-ray techniques for lithium-ion battery research. *NPG Asia Materials* 10: 563–580. https://www.nature.com/articles/s41427-018-0056-z.

Part V

Modeling

11

Unraveling Local Electrocatalytic Conditions with Theory and Computation

Jun Huang[1], *Mohammad J. Eslamibidgoli*[1], *and Michael H. Eikerling*[1,2,3]

[1] *Institute of Energy and Climate Research, Theory and Computation of Energy Materials (IEK-13), Forschungszentrum Jülich GmbH, Wilhelm-Johnen-Straße, 52425 Jülich, Germany*
[2] *Jülich Aachen Research Alliance, JARA Energy, Wilhelm-Johnen-Straße, 52425 Jülich, Germany*
[3] *RWTH Aachen University, Theory and Computation of Energy Materials, Faculty of Georesources and Materials Engineering, Intzestraße, 52062 Aachen, Germany*

11.1 Local Reaction Conditions: Why Bother?

This chapter centers around challenges in theoretical and computational electrocatalysis. The overarching theme, as the title suggests, is concerned with reaction conditions – not the global, externally provided parameters under which an electrochemical cell is operated (cell potential, temperature, reactant partial pressures, and stoichiometries) but those that prevail locally in the nanoscale interfacial region close to the surface of the electrocatalyst material, i.e. the electric potential, pH, ion concentrations, and solvent configuration at the reaction plane. These are the conditions that matter for mechanistic studies of reaction mechanisms and pathways and for the comparative evaluation of materials in terms of their electrocatalytic activity, selectivity, and stability. Local reaction conditions are a result of a complex interplay of electronic structure effects, surface processes, and electrolyte phenomena at the microscale. The reward for solving this puzzle is huge: precise knowledge of the electronic, electrostatic, and ionic conditions at the interface is the key to the development of electrocatalytic materials that optimize the performance of electrochemical energy devices.

Local or microscopic reaction conditions determine the electrocatalytic activity of the catalyst for the target reaction as the key comparative metric of catalyst development. Another important selection criterion of electrocatalyst materials is their selectivity to form desired chemical products while suppressing the formation of others that are unwanted. Moreover, catalyst stability to degradation under harsh operating conditions, as well as economic viability (i.e. cost and materials abundance) and compatibility with other components in the electrochemical cell are further considerations to make.

Reaction conditions are essential to solving the activity–stability conundrum in polymer electrolyte fuel cells and electrolysis cells. As is well known [1–3], both the activity and stability of Pt group metal catalysts, e.g., Pt for the oxygen reduction

Electrocatalysis for Membrane Fuel Cells: Methods, Modeling, and Applications, First Edition.
Edited by Nicolas Alonso-Vante and Vito Di Noto.

reaction (ORR) in fuel cells and Ir based catalysts for the oxygen evolution reaction (OER) in electrolysis cells, are sensitive to the pH at the surface of the catalyst, which must be finely tuned to optimize cell performance and lifetime. The activity and selectivity of CO_2 reduction cells are highly sensitive to the pH and the buffer equilibrium in the near-surface region [4, 5]. Surface deprotonation and cation adsorption are known to significantly modify the surface condition in Ni oxide based catalysts for the ORR in alkaline membrane fuel cells [6, 7].

Phenomena in the near-surface region of the catalyst, i.e. within a few nanometers from the surface, are extremely challenging to grasp and difficult to calculate with theoretical and computational approaches. The fundamental challenges have their origin in the complexity of the interface and the enormous potential variations in the near-surface region that correspond to huge electric fields, caused by a complex interplay of surface charging phenomena and polarization effects. These effects are highly sensitive to changes in surface termination, type and abundance (stability) of chemisorbed species, and structure and composition of the solvent. A recent computational study revealed the high sensitivity of the interfacial electric potential distribution at an electrochemical interface to the distance between the metal region and the electrolyte region used in the computational model [8].

Theory and computation have become quintessential tools to navigate the parameter space that controls the performance and longevity of electrocatalytic materials and electrochemical devices. Viable methods process as input details on materials structure, composition, and reaction conditions. Their output includes mechanistic insights into reaction mechanisms and pathways as well as tools or metrics to evaluate materials in terms of stability, selectivity, and activity and, based thereon, select, modify, or design materials. The core challenge, connecting input to output, is a self-consistency problem that couples the electrode potential to variables for the electronic structure of the solid electrode, solvent properties, and ion distributions in the electrolyte as well as specific properties of the boundary region in between. In this chapter, we discuss theoretical and computational approaches that strive to accomplish this feat.

The chapter is organized as follows. Section 11.2 introduces the basic concepts of electrochemical cells and interfaces, followed by an illustration of the key characteristics of electrocatalytic interfaces in Section 11.3. Section 11.4 summarizes three types of surface charge effects in shaping the local reaction condition. Section 11.5 discusses challenges in first-principles simulations of electrochemical interfaces. Section 11.6 introduces a concerted theoretical–computational approach to model local reaction conditions in electrocatalysis, which is exemplified for a case study in Section 11.7. We give a brief outlook on challenges and perspectives of the field in Section 11.8.

11.2 From Electrochemical Cells to Interfaces: Basic Concepts

An electrochemical cell consists of two electrodes and an electrolyte solution in between, as schematically shown in Figure 11.1. The electric potential difference,

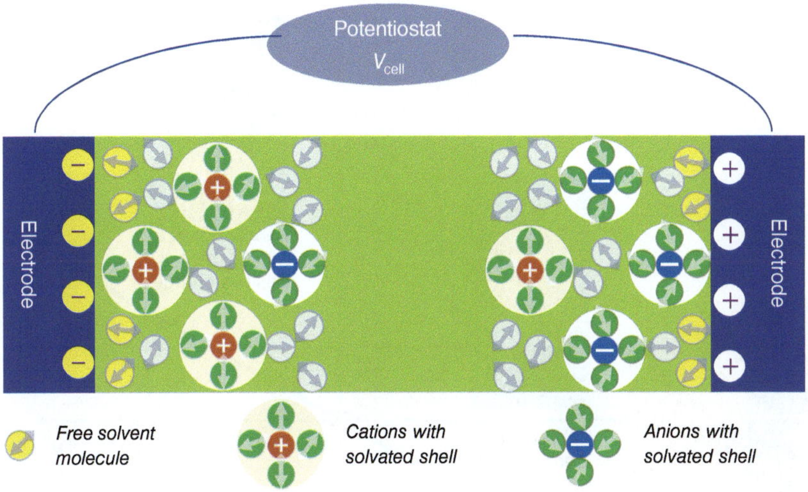

Figure 11.1 An electrochemical cell with two charged electrode/electrolyte interfaces.

V_{cell}, between the two electrodes can be modulated with a potentiostat. Variation in V_{cell} controls the difference in the electrochemical potential of electrons between the two electrodes. Electrons flow from the electrode with the higher electrochemical potential to the other side. The resultant electron flow may or may not cross the electrode–electrolyte interfaces, depending on the reactivity of the electrode. For the simplest case of ideally polarizable electrodes, the electron flow is terminated at the electrode–electrolyte interface regardless of V_{cell}. Excess/deficient electrons, corresponding to a net negative/positive surface charge, are distributed over a skin layer of several angstrom thickness on the electrode surface [9]. Counterions are attracted to, and coions are repelled from the electrode surface via Coulomb forces, as illustrated in Figure 11.1. In addition, the orientation of solvent molecules responds to the strong electric field that is generated by surface charges. These coupled phenomena lead to the formation of a nonelectroneutral region of a few nanometer thickness in the vicinity of the electrified electrode, which is termed as electric diffuse layer (see Chapter 3).

Ultimately, the distributions of the electric potential, ion concentration, and solvent orientation in the diffuse layer are dictated by the excess surface charge density, denoted σ_M, which is a function of the electrode potential, ϕ_M. The relation between σ_M and ϕ_M is called the surface charging relation. Determining the $\sigma_M - \phi_M$ relation is an essential prerequisite for understanding local reaction conditions. Figure 11.2 schematically illustrates the surface charging relation of ideally polarizable interfaces. The electric potential in the solution bulk, ϕ_S, is taken as the reference. The electric potential in the electrode bulk, ϕ_M, is tunable. A potential drop, χ, that is caused by electron spillover from the metal phase into the electrolyte phase and extends over several angstroms, exists on the metal surface [9]. χ is related to the total electron density at the electrode surface [10]. Variations in the excess electron density modulated by ϕ_M will only negligibly change χ [9]. It is thus customary and a useful first approximation to assume that χ is constant.

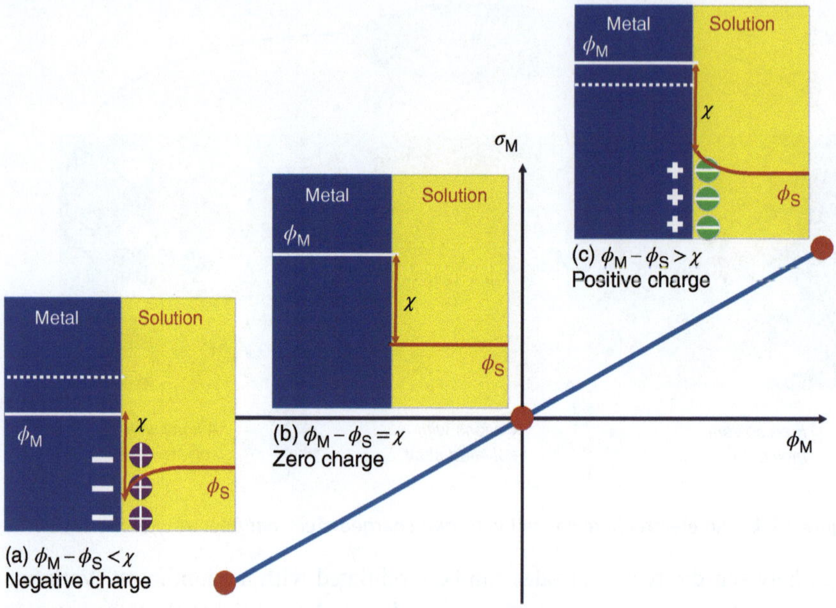

Figure 11.2 Surface charging relation of an ideally polarizable interface.

For the case of $\phi_M - \phi_S < \chi$, Figure 11.2a, the electric potential in the diffuse layer is negative relative to ϕ_S. Therefore, cations are accumulated and anions depleted, resulting in a positive net charge in the diffuse layer. Correspondingly, there will be excess electrons on the metal surface and $\sigma_M < 0$. If ϕ_M is increased, the electric potential in the diffuse layer becomes more positive while χ remains constant, meaning that less cations are accumulated and less anions repelled. Therefore, σ_M becomes less negative and shifts toward zero. For the case of $\phi_M - \phi_S = \chi$, as illustrated in Figure 11.2b, the distribution of electric potential is uniform in the electrolyte phase, and there is no excess charge in the diffuse layer or on the electrode surface, i.e. $\sigma_M = 0$. This particular value of ϕ_M is the potential of zero charge (pzc) [11]. For the case of $\phi_M - \phi_S > \chi$, shown in Figure 11.2c, anions are accumulated and cations depleted in the diffuse layer, i.e. $\sigma_M > 0$.

A quantitative determination of the $\sigma_M = f(\phi_M)$ relation requires the relation between ion concentrations and electric potential in the electrolyte phase. The rudimentary starting point is the Boltzmann relation, viz. $c_i = c_i^0 \exp(-z_i f(\phi - \phi_S))$, which is valid in the dilute electrolyte limit, where c_i^0 is the bulk concentration, ϕ the local electric potential, z_i the charge number, and $f = F/RT$. Substituting the Boltzmann relation into the Poisson equation results in the Poisson–Boltzmann equation, which, closed with boundary conditions, allows calculating the distributions of electric potential and ion concentrations in the diffuse layer. Thereafter, the net charge in the diffuse layer and σ_M can be obtained as functions of ϕ_M [12]. The Poisson–Boltzmann relation can be modified to account for more complex situations, including higher surface charge densities or concentrated electrolytes, that cause ion–ion interaction and correlation effects, such as steric repulsion and ion saturation, or polarization effects due to solvent ordering [13, 14]. These complications will modify the slope of the $\sigma_M = f(\phi_M)$ relation, but they will preserve the monotonicity of the $\sigma_M = f(\phi_M)$ relation [13].

11.3 Characteristics of Electrocatalytic Interfaces

The crucial difference of electrocatalytic interfaces compared to ideally polarizable interfaces is that the electron flow, driven by V_{cell}, can cross the electrode–electrolyte boundary. These electrons are transferred between the metal electrode and redox species in the electrolyte phase. Let us first consider a prototypical electrocatalytic process, namely chemisorption [15, 16]. One of the most studied example of chemisorption is adsorption of halogen ions onto metal surfaces [17–19]. During this process, the electron that previously resided on the electronic orbital of the halogen ion in the electrolyte phase is gradually transferred to the electronic orbital of the electrode as the halogen ion approaches the electrode surface [19]. When the halogen ion arrives at its energetically most favorable position, the electron is shared by both the electrode and the halogen ion due to strong electronic interactions between them. In other words, the electron is only partially transferred from the halogen ion to the electrode, resulting in a partially charged halogen ion instead of a halogen atom [16, 19]. The charged adsorbates, together with the compensating charge localized on metal surface atoms, create a surface dipole moment, which is termed chemisorption-induced surface dipole moment, μ_{chem}, as shown in Figure 11.3a [12, 19].

We assume that the chemisorbates are lined up rigidly in the inner Helmholtz plane (IHP) at a distance δ_{IHP} away from the electrode. These chemisorbates possess

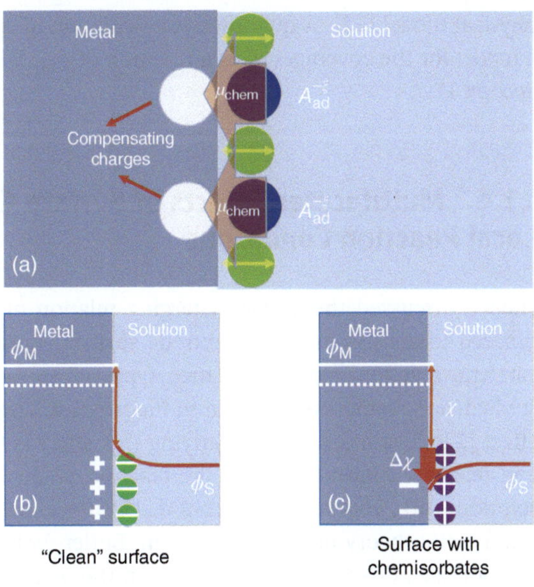

Figure 11.3 How chemisorption modifies the electrode–electrolyte interface.

a net charge of ξ_i and a coverage of θ_i, with the subscript i indexing the type of chemisorbates. Each charged chemisorbate and the compensating charge that is located at metal surface atoms contribute a dipole moment of $\mu_i = 2\delta_{\mathrm{IHP}} e_0 \xi_i$ with e_0 the elementary charge. Here, μ_i is positive when the dipole moment is parallel to the outward-pointing normal vector of the electrode surface. Therefore, the total chemisorption-induced surface dipole moment can be written as

$$\mu_{\mathrm{chem}} = \sum_i N\theta_i\mu_i = \sum_i N\theta_i\delta_{\mathrm{IHP}} e_0 \xi_i \tag{11.1}$$

where N is the total number density of adsorption sites.

The chemisorption-induced surface dipole moment introduces an additional contribution to the potential drop on the electrode surface, $\Delta\chi$, Figure 11.3c. Having approximated the adlayer as a parallel-plate capacitor, we can readily calculate $\Delta\chi$ as

$$\Delta\chi = \frac{\mu_{\mathrm{chem}}}{\varepsilon_{\mathrm{IHP}}} = \sum_i \frac{Ne_0\delta_{\mathrm{IHP}}}{\varepsilon_{\mathrm{IHP}}}\theta_i\xi_i \tag{11.2}$$

Equation (11.2) indicates that $\Delta\chi$ is negative for negatively charged chemisorbates, e.g., chemisorbed halogen atoms, and positive for positively charged chemisorbates, e.g. chemisorbed hydrogen atoms. In Figure 11.3, we compare the surface charging relation for the ideally polarized surface, Figure 11.3b, and the electrocatalytic surface with chemisorbates, Figure 11.3c. At a high potential $\phi_M - \phi_S > \chi$, the clean surface is positively charged, but the surface with chemisorbates could be negatively charged due to the additional negative potential drop $\Delta\chi$. Consequently, the surface charging relation could be nonmonotonic with a second pzc in the high potential region for electrocatalytic interfaces [12, 13]. A second pzc was measured by Frumkin and Petrii in the 1970s [20]. Only until very recently has this nonmonotonic surface charging relation been rationalized by a physical model [12]. A quantitative calculation of $\Delta\chi$ as a function of ϕ_M requires a model for the coverages θ_i as a function of ϕ_M. Such an example will be given in Section 11.7.

11.4 Multifaceted Effects of Surface Charging on the Local Reaction Conditions

Having discussed the surface charging relation of electrocatalytic interfaces, we proceed to discuss how the electrode surface charge influences the local electrocatalytic reaction condition. Three types of surface charging effects have been studied, as schematically shown in Figure 11.4. The first type is the diffuse charge effect [22]. The basic idea underlying this effect is that the local electric potential and reactant concentration at the reaction plane deviate from the values in the solution bulk, Figure 11.4a [22–24]. Diffuse charge effects can be incorporated into kinetic theory using the Frumkin–Butler–Volmer theory. For a general redox reaction, i.e., $O + ne \rightleftharpoons R$, in which both the oxidant and reductant are solvable in

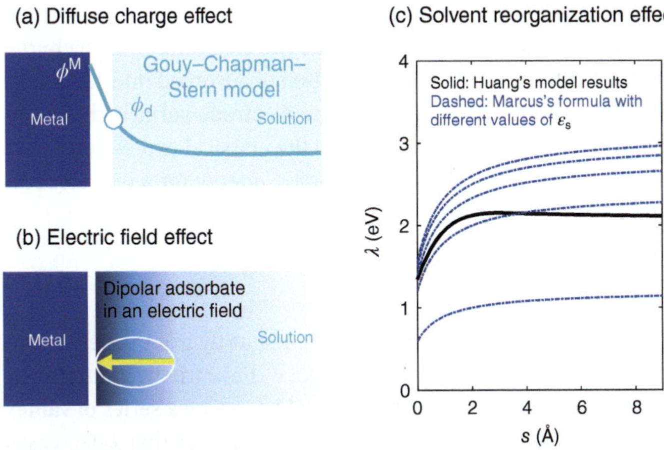

Figure 11.4 Multifaceted double-layer effects on the local reaction condition: (a) diffuse charge effect, (b) electric field effect, and (c) solvent reorganization effect. Source: Reproduced from Ref. [21] with permission of AIP Publishing. The dashed blue lines correspond to different values of ε_s used in Eq. (11.6).

the electrolyte phase, the Frumkin–Butler–Volmer theory expresses the current density, positive-defined for the oxidation direction, as [25, 26].

$$j = Fk_O c_R^0 \exp\left(\left(1-\alpha_R^*\right) nf(\phi_M - \phi_S)\right) - Fk_R c_O^0 \exp\left(-\alpha_R^* nf(\phi_M - \phi_S)\right) \quad (11.3)$$

where k_R and k_O are the rate constants of the reduction and oxidation, respectively, and we have $k_R/k_O = \exp(f(E^0 - E_{pzc}))$, with $f = F/RT$, F Faraday constant, R gas constant, T the absolute temperature, E^0 the standard equilibrium potential, and E_{pzc} the potential of zero charge. c_R^0 and c_O^0 denote the bulk solution concentrations of the reductant and oxidant, respectively, n the number of transferred electrons, and ϕ_S the electric potential in the bulk solution. The apparent transfer coefficient with correction for the diffuse charge effect is given by [26],

$$\alpha_R^* = \alpha_R \frac{C_{DL} + \frac{z_O}{n\alpha_R} C_{Stern}}{C_{DL} + C_{Stern}} \quad (11.4)$$

with $\alpha_R = 1/2$ being the original transfer coefficient, z_O the valence of the oxidant, C_{Stern} the capacitance of the Stern layer, and C_{DL} the capacitance of the diffuse layer. C_{DL} is dependent on the surface charge density, and so is α_R^*. Zhang and Huang [25] analyzed the condition under which α_R^* becomes negative.

The second type of effect consists in the impact of the interfacial electric field on the adsorption energy of reaction intermediates bearing internal dipole moments or forming surface dipoles with their image charges, Figure 11.4b [27]. The electrostatic Gibbs energy contribution due to interaction of adsorbed intermediates that possess a dipole moment of μ_i with the local electric field E is written as [28],

$$\Delta\Delta G_i = \mu_i E \quad (11.5)$$

where we have neglected the second and higher order terms related to electronic polarization. The local electric field E is modulated by the electrode surface charge and can contribute to the Gibbs free energy of adsorption of adsorbed intermediates.

The third type of effect concerns the impact of the electrode surface charge on the local solvent reorganization energy. According to the classical expression of the standard Marcus theory, the local solvent reorganization energy for a one-electron transfer reaction at the electrode surface is calculated as [21],

$$\lambda = \frac{e_0^2}{8\pi}\left(\frac{1}{a} - \frac{1}{2d}\right)\left(\frac{1}{\varepsilon_\infty} - \frac{1}{\varepsilon_s}\right) \tag{11.6}$$

with a being the radius of the ion, d the distance of the ion away from the electrode surface, ε_∞ the optical permittivity, and ε_s the local total dielectric permittivity. In Figure 11.4c, the dashed lines are calculated from Eq. (11.6) using a series of values for ε_s ranging from $3\varepsilon_0$ to $15\varepsilon_0$ (every $3\varepsilon_0$). Marcus theory indicates that λ decreases as the ion approaches closer to the interface and as ε_s decreases. The local total dielectric permittivity ε_s exhibits a more pronounced decrease as the surface charge density increases, because less free solvent molecules are left in the diffuse layer due to the accumulation of more counterions, and simultaneously because the orientational polarization of these interfacial solvent molecules approach a more saturated and thus more rigid configuration [29]. Consequently, λ will decrease as the surface charge density becomes larger [29].

11.5 The Challenges in Modeling Electrified Interfaces using First-principles Methods

First-principles computational methods in electrochemistry employ density functional theory (DFT) at their core to provide atomic-scale insights into the structure, energetics, and reaction kinetics of electrochemical processes [30]. The overarching objectives are to understand how the composition and the structure of electrode material and electrolyte impact the energetics and dynamics of adsorption and charge transfer processes; how surface impurities and chemisorbed species, including reaction intermediates, affect pathways of multistep reactions; and how solvent species and ions modulate interfacial properties and local reaction conditions.

The principal conceptual challenge of theoretical and computational electrocatalysis is to account in a self-consistent manner for the effects of electrode potential, pH, and ion concentrations in the electrolyte on the structure and charging properties of the electrified interface [31]. The ideal approach to accomplish this is to design and implement a fully *ab initio* computational algorithm to explicitly control the value of the metal phase (or electrode) potential; additionally, in this approach the simulation should be performed in the relevant length and time scales to properly sample the representative statistical configurations in the electrolyte region [32, 33]. This ideal approach would allow for a fully self-consistent simulation – i.e. without empirical input needed – of the potential distribution between the metal bulk and the electrolyte bulk at a given electrode potential.

Although it is difficult to realize this approach in practice for the near future, assuming the electrode potential will be successfully addressed with the grand-canonical DFT methods [34], *ab initio* molecular dynamics (AIMD) may be accelerated with domain-specific hardware designs [35] (e.g. multi-GPU processing) for large-scale and longer simulations. Moreover, employing machine-learned interatomic potentials may open new avenues to bridge the length/time scales [36–38].

However, considering the current computational limitations to self-consistently account for the complexities of the electrode/electrolyte system, DFT calculations are mainly employed in a hierarchy of methods, e.g. in combination with thermodynamics and microkinetic models, or kinetic Monte Carlo simulations, to gain insights into the interface properties [30, 39–41]. For example, one approach to simulate the local reaction condition in DFT calculations is to try to pin the potential to a control variable that uniquely defines the state or surface configuration of the electrode. A suitable control variable for this is, for example, the oxide coverage of a Pt electrode in the potential range for the ORR; the relation of this variable to the electrode potential is well studied in experiment [6]. This control variable might be more difficult to identify for other materials of interest, and reliable experimental data might not always be available.

Various first-principles schemes were developed to properly simulate the local reaction condition at electrified interfaces, each with their own capabilities, achievements, and limitations. These include (i) step-by-step calculations of Gibbs free energies of adsorption of reaction intermediates, with steps linked together with the help of the computational hydrogen electrode (CHE) [42, 43], (ii) extrapolation of the unit cell in periodic slab-type calculations to infinite size to eliminate finite-size effects on activation and reaction energies of charge transfer reactions [44, 45], (iii) compensating charge and explicit consideration of a reference electrode to simulate the applied potential [46], (iv) explicit treatment of electrified interface with *ab initio* molecular dynamics simulations [47], (v) effective screening medium combining electronic DFT with mean field theories and continuum solvation for the electrolyte region [48–50], and (vi) grand-canonical density functional theory (GC-DFT) that combines electronic and classical DFT for different regions [51, 52]. In the following sections, we provide a brief account of these schemes.

11.5.1 Computational Hydrogen Electrode

The CHE, as the most widely used approach in computational electrochemistry, is a combination of periodic DFT calculations for the metal electrode region and the grand-canonical approach to formulate the Gibbs free energy of adsorption of reaction intermediates formed in each step of the overall electrochemical process [42, 43, 53, 54]. The electrode and electrolyte phases are treated as infinite reservoirs for electrons and protons, respectively. The setup assumes a reference state in which molecular hydrogen in the gas phase is in equilibrium with the solvated proton and the electron $\left(\frac{1}{2}H_2(g) \rightleftarrows H^+(aq) + e^-\right)$ under standard condition ($T = 298$ K,

$P = 1$ atm, and pH 0). Referencing the potential to that of the standard hydrogen electrode (SHE), E_{SHE}, the imprecise calculation of the proton solvation energy in water is avoided, and instead the DFT-calculated energy of $\frac{1}{2}H_2(g)$ is used to calculate the Gibbs free energy of adsorption, ΔG_{CHE},

$$\Delta G_{CHE}(n, E_{SHE}, \text{pH}) = \Delta E_{DFT} - T\Delta S + \Delta \text{ZPE} - n\left(\frac{1}{2}\mu_{H_2} - eE_{SHE}\right) - K_B T \ln(10)\text{pH} \tag{11.7}$$

Here, n is the number of protons (and electrons) transferred, ΔE_{DFT} is the binding energy of the intermediate calculated using DFT, and $T\Delta S$ and ΔZPE are corrections for entropy and zero-point energy contributions for adsorbates and solvent molecules. In the CHE scheme the terms for potential and pH are added explicitly in the equation. The approach has been applied to find stable surface structures and generate Pourbaix diagrams of various systems, including Pt(111), Ni(111), and Ag(111) under ORR conditions [55]; study the adsorption of halides on Pt and Cu(111) [56]; and explore the coadsorption of sulfate, bisulfate, and hydrogen on Pt and Au(111) [57]. It has also been employed to study reaction pathways and predict activity trends and perform comparative materials analyses [58–60]. The CHE concept, however, makes significant approximations in the treatment of the complex local reaction conditions and interfacial structure. It assumes zero net charge at all potentials, and the impact of the solvent reorganization on interface properties and kinetic processes is completely neglected. This leads to shortcomings in the description of pH effects, which is currently the subject of active research [45, 61, 62].

11.5.2 Unit-cell Extrapolation, Explicit Solvated Protons, and Excess Electrons

In the atomistic approach originally proposed by Rossmeisl, Nørskov, and coworkers [44, 63], the potential is tuned by adding excess hydrogen atoms to the explicit water layer at the electrode surface. They spontaneously form solvated protons and excess electrons at the electrode surface. The resulting shift in the work function is then related to the potential scale of the normal hydrogen electrode [44, 64], where the experimental reference of the absolute standard hydrogen electrode potential (ASHEP) was pinpointed at 4.42 V by Tripkovic et al. [65]. This value is in close agreement with the ASHEP value of 4.44 V recommended by Trasatti [66]. With this explicit treatment of solvated protons and excess electrons at the metal surface, the elementary processes involved in the hydrogen evolution reaction were evaluated [63]. By calculating the activation energy as a function of electrode potential, a Brønsted–Evans–Polanyi-type relationship between activation energy and reaction energy was obtained [67].

In this scheme, cell extrapolation to infinite size is proposed to address the finite-size effects in periodic slab-type calculations on the activation and reaction energies of charge transfer reactions [45, 68]. Methods based on linear extrapolation make two assumptions: (i) the chemical and electrostatic contributions to reaction energetics are separable; (ii) the interface behaves as a simple capacitor with constant capacitance (or with a linear charging relation). These methods do not account for polarization effects induced by chemisorbed partially charged adsorbates and charge delocalization. Moreover, away from the potential of zero charge, the electrostatic charging and polarization properties of the boundary region respond nonlinearly to changes in electrode potential, invalidating linear potential extrapolation. These nonlinear charging effects modify surface electronic states, short-range electronic interactions with near-surface species, adsorption strength, or orientational ordering of polar solvent molecules [69].

11.5.3 Counter Charge and Reference Electrode

To simulate an applied potential by adding (or subtracting) electrons, Lozovoi et al. proposed a grand-canonical scheme for electrons [46]. To this end, they introduced a virtual "reference electrode," where to maintain electroneutrality of the periodically repeated structure, compensating charges are added to the unit cell. The supercell calculations are performed at constant electron chemical potential, rather than at a constant number of electrons. The countercharge is added in different ways; Lozovoi and Alavi employed a Gaussian charge sheet, i.e. a localized charged plane defined as reference electrode [70]. Kastlunger et al. presented an approach based on changing the total number of electrons in the self-consistent cycle, while enforcing charge neutrality through the introduction of a jellium countercharge dispersed in an implicit solvent region above the slab [71]. Taylor et al. proposed a double reference electrode approach [72]. It uses a compensating background charge that is distributed homogeneously over the unit cell to maintain overall charge neutrality, whereas explicit water molecules are introduced into the vacuum region between the metal slabs to model the solution/metal interface [73]. The approach was exemplified for the structure of water molecules as a function of potential at Cu(111) and to study the reaction energetics for the dehydrogenation of methanol at Pt(111), indicating the critical role of interfacial charge polarization in rationalizing the structure and reactivity of the interface as a function of potential. In the original paper [72], however, it is indicated that ensemble averaging, and the choice of the water model, is critical in this approach, and the authors proposed *ab initio* molecular dynamics simulations of the electrode/electrolyte system as a required step for advancing this general methodology [47, 74].

11.5.4 Effective Screening Medium and mPB Theory

In the effective screening medium (ESM) approach proposed by Otani and Sugino [48], Green's function technique was employed to modify the Poisson solver part of the standard pseudopotential plane-wave scheme. In this way, the periodic boundary condition requirement is removed along the surface normal direction resolving the corresponding restrictions in modeling potential distribution for the charged surfaces/interfaces. To model the electrolyte ions, the modified Poisson–Boltzmann (mPB) theory is considered. It was discussed that the computational time required to obtain the electronic structures and atomic force is almost equal to that of the standard DFT scheme, making the approach appropriate for performing AIMD simulations [49]. The approach was used to apply a bias potential to water/Pt(111) interface to predict the configuration of water as a function of applied potential [49].

Later, this approach was refined by Jinnouchi and Anderson [50] as well as Dabo et al. [75] combining DFT and mPB theory through a smooth and flexible dielectric model function to more accurately describe the regions handled by DFT and mPB and was employed to obtain the potential of zero charge (PZC) of Pt(111) [50]. This periodic continuum solvation method based on mPB has later been applied to calculate the potential of zero charge and differential capacitance of several metal and CO-covered Pt surfaces [76], as well as the electrochemical reduction of CO_2 on Cu(100) [77], leading to agreement with the experimental studies of the PZC.

11.5.5 Grand-canonical DFT

This approach introduces a grand-canonical ensemble of electrons at fixed potential directly implemented in the self-consistent field procedure for Kohn–Sham DFT calculations. It uses joint density functional theory [51, 52, 78], which combines electronic DFT for the metal region and classical DFT for the electrolyte region. To solve the Kohn–Sham equation at fixed potential, Sundararaman et al. [34] implemented two algorithms based on grand-canonical self-consistent field (GC-SCF) and grand-canonical auxiliary Hamiltonian (GC-AuxH) for free energy minimization. This approach directly mimics the experimental condition in a way that the number of electrons varies to preserve the system at constant electron chemical potential, while ionic response in a continuum solvation model of electrolyte keeps the system neutral. As a case study, grand-canonical DFT was applied to analyze the underpotential deposition of copper on platinum in an electrolyte containing chloride ions, explaining the origin of the second underpotential voltammetric peak for this system [34].

11.6 A Concerted Theoretical–Computational Framework

In the past few years, we have developed a concerted theoretical–computational framework for modeling interface properties and electrocatalytic reactions, as schematically shown in Figure 11.5 [25, 79–81]. The framework combines DFT-based first-principles calculations for parameterization of specific interface properties, a mean-field model of the double layer, and a microkinetic model for multistep kinetics. DFT calculations are used to determine the reaction pathways and the chemisorption-induced surface dipole moments. The mean-field model of the double layer considers dipolar effects due to chemisorption of oxygen species, solvent orientational polarization, and ionic effects in the electrolyte. From the mean-field double-layer model, we can calculate the local reaction conditions

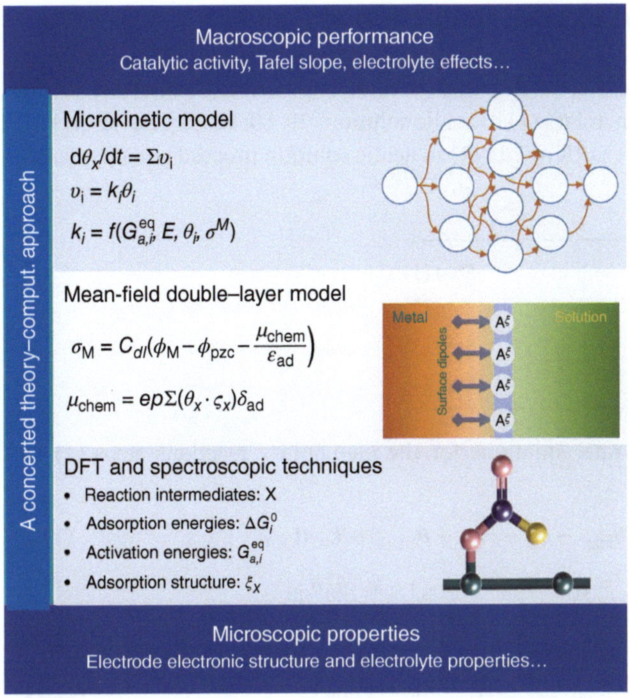

Figure 11.5 A concerted theory–computation framework for modeling electrocatalytic reactions.

required for the microkinetic model. In the microkinetic model, the reaction rate of each elementary reaction step is formulated using the Frumkin–Butler–Volmer theory, where the reactant concentration, electrolyte-phase potential, and solvent reorganization energy are employed at the reaction plane. The reaction free energy of each elementary reaction step is obtained from DFT calculations, with proper modifications such as considering lateral interactions between reaction intermediates.

There are several important couplings among the microkinetic model, the double-layer model, and the elementary electron transfer model, which ought to be solved in a self-consistent manner. Firstly, the coverages of the reaction intermediates are solved from the microkinetic model under the steady-state condition, from which we calculate the chemisorption-induced surface dipole moment that is then implanted into the double-layer model as a boundary condition at the inner Helmholtz plane. Secondly, the local reaction conditions, obtained as the output of the double-layer model, are the input of the microkinetic model, which in turn defines the boundary condition for the double-layer model. Thirdly, the surface charge density calculated from the double-layer model impacts the binding energies of reaction intermediates. Solving the whole model at a series of electrode potentials, we can build up the map between the microscopic parameter space of catalyst composition and interfacial properties and the macroscopic parameter space of electrode performance.

11.7 Case Study: Oxygen Reduction at Pt(111)

We demonstrate the above concerted theoretical–computational framework for the case of oxygen reduction at Pt(111) in acidic solution [79, 80, 82, 83]. Based on DFT calculations [42, 84], the ORR at Pt(111) in acidic solution proceeds mainly via the following pathway:

$O_2 + H^+ + e \rightleftharpoons OOH_{ad}$	Step (1)
$OOH_{ad} \rightleftharpoons O_{ad} + OH_{ad}$	Step (2)
$O_{ad} + H^+ + e \rightleftharpoons OH_{ad}$	Step (3)
$OH_{ad} + H^+ + e \rightleftharpoons H_2O$	Step (4)

The complete set of rate equations for the elementary reactions steps (1)–(4) above is

$$v_1 = k_1[O_2][H^+](\theta_{max} - \theta_O - \theta_{OH} - \theta_{OOH}) - k_{-1}\theta_{OOH} \quad (11.8)$$

$$v_2 = k_2\theta_{OOH}(\theta_{max} - \theta_O - \theta_{OH} - \theta_{OOH}) - k_{-2}\theta_O\theta_{OH} \quad (11.9)$$

$$v_3 = k_3[H^+]\theta_O - k_{-3}\theta_{OH} \quad (11.10)$$

$$v_4 = k_4[H^+]\theta_{OH} - k_{-4}(\theta_{max} - \theta_O - \theta_{OH} - \theta_{OOH}) \quad (11.11)$$

Here, θ_i is the coverage of intermediate i, and θ_{max} is the maximum coverage which is usually taken to be unity. Note that $[H^+]$ is employed at the reaction plane, which is taken to be the inner boundary layer of the double-layer model. Below, we adopt the assumption of $\theta_{OOH} = 0$ because OOH is highly unstable at Pt(111) [85].

$k_{\pm i}$ are rate constants

$$k_{\pm i} = k_i^0 \exp\left(-\frac{\Delta G_{a,\pm i}}{RT}\right) \quad (11.12)$$

where $k_i^0 = k_B T/h$ is a prefactor assuming a transmission coefficient of unity, with k_B being Boltzmann constant and h Planck constant. $\Delta G_{a,\pm i}$ is the activation barrier in either forward (+) or reverse (−) direction. According to the Brønsted–Evans–Polanyi (BEP) relation, $\Delta G_{a,\pm i}$ is linearly related to the corresponding reaction Gibbs energy of the elementary step, ΔG_i, with the proportionality factor being the symmetry factor β_i for the forward reaction and $(1 - \beta_i)$ for the reverse reaction.

ΔG_i ($i = 1, 3, 4$), the reaction Gibbs energies, are given by

$$\Delta G_1 = \Delta G^{\ominus}_{OOH} - \Delta G^{\ominus}_{O_2} + F(\phi_M - \phi_{RP}) \quad (11.13)$$

$$\Delta G_3 = \Delta G^{\ominus}_{OH} - \Delta G^{\ominus}_{O} + F(\phi_M - \phi_{RP}) + \delta G_3^{OX} \quad (11.14)$$

$$\Delta G_4 = -\Delta G^{\ominus}_{OH} + F(\phi_M - \phi_{RP}) + \delta G_4^{OX} \quad (11.15)$$

where $\Delta G^{\ominus}_{OX}$ refers to the adsorption free energy under the standard state (0 V, pH 0, and $\theta_{OX} = 0.25$ ML [monolayer]), ϕ_{RP} is the local electric potential at the reaction plane, and δG_i^{OX} describes the variation in the reaction free energy caused by lateral interactions of the intermediates. Under the Frumkin adsorption condition, we have

$$\delta G_3^{\mathrm{OX}} = \xi_{\mathrm{OH}}\left(\theta_{\mathrm{OH}} - \theta_{\mathrm{OH}}^{\ominus}\right) - \xi_{\mathrm{O}}\left(\theta_{\mathrm{O}} - \theta_{\mathrm{O}}^{\ominus}\right) \tag{11.16}$$

$$\delta G_4^{\mathrm{OX}} = -\xi_{\mathrm{OH}}\left(\theta_{\mathrm{OH}} - \theta_{\mathrm{OH}}^{\ominus}\right) \tag{11.17}$$

where ξ_{OX} is the lateral interaction coefficient for oxygen-containing species OX, and we use $\theta_{\mathrm{OX}}^{\ominus} = 0.25$ according to the definition of standard state. Field-dependent effects expressed in Eq. (11.5) can be incorporated into the δG_i^{OX} terms.

The local reaction condition is determined from the mean-field double-layer model which can be developed on varying levels. For the simplest version, we have [80, 81]

$$\sigma_{\mathrm{M}} = C_{\mathrm{dl}}(\phi_{\mathrm{M}} - \phi_{\mathrm{S,bulk}} - \chi - \Delta\chi) \tag{11.18}$$

where C_{dl} is the double-layer capacitance, χ is the potential drop due to electron spillover, and $\Delta\chi$ is the additional potential drop created by the chemisorption-induced surface dipole moment, given in Eq. (11.2). If we use the Poisson–Boltzmann equation for the diffuse layer, σ_{M} is solved from [12]

$$\frac{2RT}{F}\operatorname{arcsinh}\left(\frac{\lambda_{\mathrm{D}}F\sigma_{\mathrm{M}}}{2\varepsilon_{\mathrm{s}}RT}\right) + \sigma_{\mathrm{M}}\frac{\delta_{\mathrm{IHP}}}{\varepsilon_{\mathrm{IHP}}} = (\phi_{\mathrm{M}} - \phi_{\mathrm{S,bulk}}) - \chi - \Delta\chi \tag{11.19}$$

where ε_{s} is the dielectric permittivity of the solution phase. The double-layer model can be further complicated by considering the ion size effect [13] and the polarization of solvent molecules [29].

Given σ_{M}, ϕ_{RP} and $[\mathrm{H^+}]_{\mathrm{RP}}$ can be calculated (using the solution of the Poisson–Boltzmann equation),

$$\phi_{\mathrm{RP}} = \frac{2RT}{F}\operatorname{arcsinh}\left(\frac{\lambda_{\mathrm{D}}F\sigma_{\mathrm{M}}}{2\varepsilon_{\mathrm{s}}RT}\right)$$

$$[\mathrm{H^+}]_{\mathrm{RP}} = c_{\mathrm{H^+,b}}\exp\left(-\frac{F\phi_{\mathrm{RP}}}{RT}\right)$$

where $c_{\mathrm{H^+,b}}$ is the proton concentration in the bulk solution.

Under steady-state conditions, the reaction rates of elementary steps are interrelated. As long as the reaction pathway in steps (1)–(4) is considered, the reaction rates must fulfill the following balance: $v_{\mathrm{ORR}} = v_1 = v_2 = v_3 = v_4/2$ [19]. Using this relation, solutions for θ_i and v_{ORR} can be obtained [79]:

$$\frac{\theta_{\mathrm{OH}}}{\theta_{\max}} = \frac{\left(K_4^*k_1^*\right)^{-1} + \left(k_4^*/2\right)^{-1}}{\left(k_1^*\right)^{-1} + \left(k_3^*\right)^{-1} + \left(k_4^*/2\right)^{-1} + \left(K_4^*k_1^*\right)^{-1} + \left(K_3^*k_4^*/2\right)^{-1} + \left(K_3^*K_4^*k_1^*\right)^{-1}} \tag{11.20}$$

$$\frac{\theta_{\mathrm{O}}}{\theta_{\max}} = \frac{\left(k_3^*\right)^{-1} + \left(K_3^*k_4^*/2\right)^{-1} + \left(K_3^*K_4^*k_1^*\right)^{-1}}{\left(k_1^*\right)^{-1} + \left(k_3^*\right)^{-1} + \left(k_4^*/2\right)^{-1} + \left(K_4^*k_1^*\right)^{-1} + \left(K_3^*k_4^*/2\right)^{-1} + \left(K_3^*K_4^*k_1^*\right)^{-1}} \tag{11.21}$$

$$\frac{\theta_{\max}}{v_{\mathrm{ORR}}} = (k_1^*)^{-1} + (k_3^*)^{-1} + (k_4^*/2)^{-1} + (K_4^* k_1^*)^{-1} + (K_3^* k_4^*/2)^{-1} + (K_3^* K_4^* k_1^*)^{-1} \tag{11.22}$$

where $k_1^* = k_1[O_2][H^+], k_3^* = k_3[H^+]$, and $k_4^* = k_4[H^+]$ are the rate constants that are dependent on the reactant concentrations, and $K_3^* = \frac{k_3^*}{k_{-3}}$, and $K_4^* = k_4^*/k_{-4}$ are corresponding equilibrium constants. The adsorbate coverages θ_{OH} and θ_{O} are substituted into Eq. (11.1) to close the equation set of the model, which is summarized in Figure 11.5.

The potential-dependent θ_{OH} and θ_{O} were validated against results of X-ray photoelectron spectroscopy (XPS) measurements on a Pt(111) electrode in 0.1 M N_2-saturated HF solution, Figure 11.6a [86, 87]. Nonmonotonic surface charging behavior is seen in Figure 11.6b, with σ_{M} inverting toward negative values above 0.78 V due to chemisorption-induced surface dipoles [79]. The nonmonotonic surface charging relation on Pt(111) was corroborated by Martínez-Hincapié et al. [88]. Accordingly, interfacial reaction conditions ϕ_{RP} and $[H^+]_{\mathrm{RP}}$ are strongly potential dependent, Figure 11.6b. The ORR rate expression was parameterized according to the polarization curves at three pHs measured using the rotating disk electrode (RDE) technique (2500 rpm) on a Pt(111) electrode in NaF/HClO$_4$

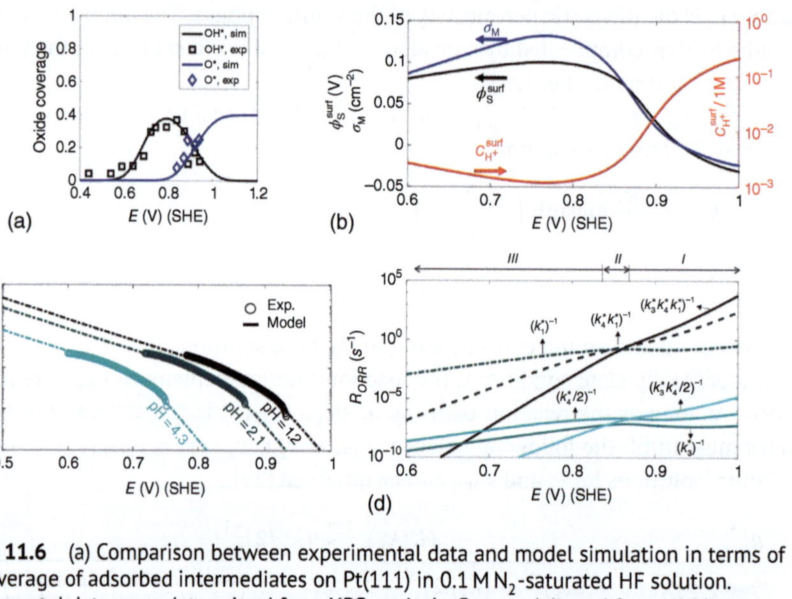

Figure 11.6 (a) Comparison between experimental data and model simulation in terms of the coverage of adsorbed intermediates on Pt(111) in 0.1 M N_2-saturated HF solution. Experimental data were determined from XPS analysis. Source: Adapted from Wakisaka et al. [86, 87]. (b) Surface charging relation and local reaction conditions as functions of ϕ_{M} calculated by the model at pH 1.2. (c) Model validation according to the polarization curves at three pHs measured using the RDE technique (2500 rpm) on a Pt(111) electrode in NaF/HClO$_4$ electrolyte saturated with O_2 of 1 bar at room temperature [1]. The ORR current density is normalized to the value at 0.95 V (RHE, reversible hydrogen electrode).
(d) Rate-determining resistance term in the ORR rate expression. Source: Reproduced from Ref. [79] with permission of Royal Society of Chemistry.

electrolyte saturated with O_2 of 1 bar at room temperature, as shown in Figure 11.6c [1]. The separate resistive terms of the ORR rate expression in Eq. (11.22) are shown in Figure 11.6d. The rate-determining resistance term varies among $R_1 = (K_3^* K_4^* k_1^*)^{-1}$ for $E > 0.86\ V$, $R_2 = (K_4^* k_1^*)^{-1}$ for $0.84\ \text{V} < E < 0.86\ V$, and $R_3 = (k_1^*)^{-1}$ for $E < 0.84\ V$. These consecutive transitions provide conclusive explanation of the potential-dependent Tafel slope. As the rate-determining resistance terms share the form, $(K_3^{*\nu} K_4^{*\mu} k_1^*)^{-1}$, with the exponents ν, $\mu = 0$ or 1, the intrinsic differential Tafel slope reads

$$b = \frac{2.3RT}{F} \frac{1}{\nu + \mu + 0.5} \tag{11.23}$$

It is readily seen that (i) $\nu = \mu = 1$, $b \cong 24$ mV dec^{-1} when $E > 0.86$ V; (ii) $\nu = 0$, $\mu = 1$, $b \cong 40$ mV dec^{-1} when $0.84\ \text{V} < E < 0.86\ V$; and (iii) $\nu = \mu = 0$, $b \cong 120$ mV dec^{-1} when $E < 0.84$ V.

11.8 Outlook

Precise knowledge of the local reaction conditions is a crucial prerequisite for the assessment of the activity, selectivity, and durability of electrocatalysts. The core task in this realm is to develop a model for the reactive electrochemical interfaces. Given the fact that the thickness of the electric double layer (EDL) is on the orders of tens of nanometers, it is prohibitively expensive to adopt the fully first-principles computation route. A viable approach is to combine first-principles computation and theoretical models into a multiscale frame. Specifically, first-principles computation is employed to obtain a set of fundamental parameters characterizing the microscopic structure of the EDL and the short-range interactions between the electrocatalyst materials and the solution species. These parameters are then passed to a mean-field theoretical model for the EDL encompassing both the electrocatalyst materials and the electrolyte solution. A particularity of the electrocatalytic EDL is that the EDL structure is transformed by adsorbed intermediates, whose coverages are dependent on the local reaction conditions and which in turn shape the local reaction conditions via regulating the surface electrostatics. The development of such a concerted theoretical–computational approach is still at an early stage.

Acronyms

AIMD	*ab initio* molecular dynamics
ASHEP	absolute standard hydrogen electrode potential
BEP	Brønsted–Evans–Polanyi
CHE	computational hydrogen electrode
DFT	density functional theory
EDL	electric double layer
GC-AuxH	grand-canonical auxiliary Hamiltonian

GC-DFT	grand-canonical density functional theory
GC-SCF	grand-canonical self-consistent field
IHP	inner Helmholtz plane
mPB	modified Poisson–Boltzmann
ORR	oxygen reduction reaction
PZC	potential of zero charge
RDE	rotating disk electrode
RHE	reversible hydrogen electrode
XPS	X-ray photoelectron spectroscopy

Symbols

α	transfer coefficient
δ_{IHP}	distance for IHP
$\varepsilon_{\mathrm{IHP}}$	dielectric constant of the space between the metal surface and the IHP
θ_i	coverage
μ_{chem}	chemisorption-induced surface dipole moment
μ_i	dipole moment
ξ_i	net charge
σ_{M}	excess surface charge density
ϕ_{M}	electrode potential
ϕ_{S}	electric potential in the solution bulk
$\Delta\chi$	potential drop due to μ_{chem}
χ	potential drop due to electron spillover
b	Tafel slope
c_i^0	bulk concentration
F	Faraday constant
k_{O}	rate constants of the oxidation reaction
k_{R}	rate constants of the reduction reaction
R	gas constant
T	absolute temperature
V_{cell}	electric potential difference
z_i	charge number

References

1 Briega-Martos, V., Herrero, E., and Feliu, J.M. (2017). Effect of pH and water structure on the oxygen reduction reaction on platinum electrodes. *Electrochimica Acta* 241: 497–509.

2 Nishimoto, T., Shinagawa, T., Naito, T., and Takanabe, K. (2020). Microkinetic assessment of electrocatalytic oxygen evolution reaction over iridium oxide in unbuffered conditions. *Journal of Catalysis* 391: 435–445.

3 Zhang, M.-K., Chen, W., Xu, M.-L. et al. (2021). How buffers resist electrochemical reaction-induced pH shift under a rotating disk electrode configuration. *Analytical Chemistry* 93: 1976–1983.

4 Bohra, D., Chaudhry, J.H., Burdyny, T. et al. (2019). Modeling the electrical double layer to understand the reaction environment in a CO_2 electrocatalytic system. *Energy & Environmental Science* 12: 3380–3389.

5 Ringe, S., Morales-Guio, C.G., Chen, L.D. et al. (2020). Double layer charging driven carbon dioxide adsorption limits the rate of electrochemical carbon dioxide reduction on gold. *Nature Communications* 11: 33.

6 Diaz-Morales, O., Ferrus-Suspedra, D., and Koper, M.T.M. (2016). The importance of nickel oxyhydroxide deprotonation on its activity towards electrochemical water oxidation. *Chemical Science* 7: 2639–2645.

7 Garcia, A.C., Touzalin, T., Nieuwland, C. et al. (2019). Enhancement of oxygen evolution activity of nickel oxyhydroxide by electrolyte alkali cations. *Angewandte Chemie International Edition* 58: 12999–13003.

8 Fernandez-Alvarez, V.M. and Eikerling, M.H. (2019). Interface properties of the partially oxidized Pt(111) surface using hybrid DFT–solvation models. *ACS Applied Materials & Interfaces* 11: 43774–43780.

9 Huang, J., Li, P., and Chen, S. (2020). Potential of zero charge and surface charging relation of metal-solution interphases from a constant-potential jellium-Poisson-Boltzmann model. *Physical Review B* 101: 125422.

10 Lang, N.D. and Kohn, W. (1970). Theory of metal surfaces: charge density and surface energy. *Physical Review B* 1: 4555–4568.

11 Trasatti, S. and Lust, E. (1999). The potential of zero charge. In: *Modern Aspects of Electrochemistry* (ed. R.E. White, J.O.M. Bockris, and B.E. Conway), 1–215. Boston, MA: Springer US.

12 Huang, J., Malek, A., Zhang, J., and Eikerling, M.H. (2016). Non-monotonic surface charging behavior of platinum: a paradigm change. *Journal of Physical Chemistry C* 120: 13587–13595.

13 Huang, J., Zhou, T., Zhang, J., and Eikerling, M. (2018). Double layer of platinum electrodes: non-monotonic surface charging phenomena and negative double layer capacitance. *The Journal of Chemical Physics* 148: 044704.

14 Kornyshev, A.A. (2007). Double-layer in ionic liquids: paradigm change? *The Journal of Physical Chemistry. B* 111: 5545–5557.

15 Muscat, J.P. and Newns, D.M. (1978). Chemisorption on metals. *Progress in Surface Science* 9: 1–43.

16 Schmickler, W. and Guidelli, R. (2014). The partial charge transfer. *Electrochimica Acta* 127: 489–505.

17 Roman, T. and Groß, A. (2013). Periodic density-functional calculations on work-function change induced by adsorption of halogens on Cu(111). *Physical Review Letters* 110: 156804.

18 Gossenberger, F., Roman, T., and Groß, A. (2016). Hydrogen and halide co-adsorption on Pt(111) in an electrochemical environment: a computational perspective. *Electrochimica Acta* 216: 152–159.

19 Ávila, M., Juárez, M.F., and Santos, E. (2020). Role of the partial charge transfer on the chloride adlayers on Au(100). *ChemElectroChem* 7: 4269–4282.

20 Frumkin, A.N. and Petrii, O.A. (1975). Potentials of zero total and zero free charge of platinum group metals. *Electrochimica Acta* 20: 347–359.

21 Marcus, R.A. (1965). On the theory of electron-transfer reactions. VI. Unified treatment for homogeneous and electrode reactions. *The Journal of Chemical Physics* 43: 679–701.

22 Frumkin, A. (1933). Wasserstoffüberspannung und Struktur der Doppelschicht. *Zeitschrift für Physikalische Chemie* 164A: 121–133.

23 Delahay, P. (1965). *Double Layer and Electrode Kinetics*. Wiley.

24 Parsons, R. (1969). The effect of specific adsorption on the rate of an electrode process. *Journal of Electroanalytical Chemistry and Interfacial Electrochemistry* 21: 35–43.

25 Zhang, L. and Huang, J. (2020). Understanding surface charge effects in electrocatalysis. Part I: peroxodisulfate reduction at Pt(111). *The Journal of Physical Chemistry C* 124: 16951–16960.

26 van Soestbergen, M. (2012). Frumkin-Butler-Volmer theory and mass transfer in electrochemical cells. *Russian Journal of Electrochemistry* 48: 570–579.

27 Nørskov, J.K., Studt, F., Abild-Pedersen, F., and Bligaard, T. (2014). Poisoning and promotion of catalysts. In: *Fundamental Concepts in Heterogeneous Catalysis*, 150–154. Wiley.

28 Hammer, B. and Nørskov, J.K. (1995). Electronic factors determining the reactivity of metal surfaces. *Surface Science* 343: 211–220.

29 Huang, J. (2020). Mixed quantum-classical treatment of electron transfer at electrocatalytic interfaces: theoretical framework and conceptual analysis. *The Journal of Chemical Physics* 153: 164707.

30 Calle-Vallejo, F. and Koper, M.T.M. (2012). First-principles computational electrochemistry: achievements and challenges. *Electrochimica Acta* 84: 3–11.

31 Eslamibidgoli, M.J. and Eikerling, M.H. (2018). Approaching the self-consistency challenge of electrocatalysis with theory and computation. *Current Opinion in Electrochemistry* 9: 189–197.

32 Calegari Andrade, M.F., Ko, H.-Y., Zhang, L. et al. (2020). Free energy of proton transfer at the water–TiO_2 interface from ab initio deep potential molecular dynamics. *Chemical Science* 11: 2335–2341.

33 Groß, A., Gossenberger, F., Lin, X. et al. (2014). Water structures at metal electrodes studied by ab initio molecular dynamics simulations. *Journal of the Electrochemical Society* 161: E3015–E3020.

34 Sundararaman, R., Goddard, W.A., and Arias, T.A. (2017). Grand canonical electronic density-functional theory: algorithms and applications to electrochemistry. *The Journal of Chemical Physics* 146: 114104.

35 Dally, W.J., Turakhia, Y., and Han, S. (2020). Domain-specific hardware accelerators. *Communications of the ACM* 63: 48–57.

36 Bartók, A.P., Payne, M.C., Kondor, R., and Csányi, G. (2010). Gaussian approximation potentials: the accuracy of quantum mechanics, without the electrons. *Physical Review Letters* 104: 136403.

37 Behler, J. and Parrinello, M. (2007). Generalized neural-network representation of high-dimensional potential-energy surfaces. *Physical Review Letters* 98: 146401.

38 Wang, H., Zhang, L., Han, J., and Weinan, E. (2018). DeePMD-kit: a deep learning package for many-body potential energy representation and molecular dynamics. *Computer Physics Communications* 228: 178–184.

39 Reuter, K. (2016). Ab initio thermodynamics and first-principles microkinetics for surface catalysis. *Catalysis Letters* 146: 541–563.

40 Stamatakis, M. and Vlachos, D.G. (2012). Unraveling the complexity of catalytic reactions via kinetic Monte Carlo simulation: current status and Frontiers. *ACS Catalysis* 2: 2648–2663.

41 Andersen, M., Plaisance, C.P., and Reuter, K. (2017). Assessment of mean-field microkinetic models for CO methanation on stepped metal surfaces using accelerated kinetic Monte Carlo. *The Journal of Chemical Physics* 147: 152705.

42 Nørskov, J.K., Rossmeisl, J., Logadottir, A. et al. (2004). Origin of the overpotential for oxygen reduction at a fuel-cell cathode. *The Journal of Physical Chemistry B* 108: 17886–17892.

43 Nørskov, J.K., Bligaard, T., Logadottir, A. et al. (2005). Trends in the exchange current for hydrogen evolution. *Journal of the Electrochemical Society* 152: J23.

44 Rossmeisl, J., Skúlason, E., Björketun, M.E. et al. (2008). Modeling the electrified solid–liquid interface. *Chemical Physics Letters* 466: 68–71.

45 Chan, K. and Nørskov, J.K. (2015). Electrochemical Barriers Made Simple. *The Journal of Physical Chemistry Letters* 6: 2663–2668.

46 Lozovoi, A.Y., Alavi, A., Kohanoff, J., and Lynden-Bell, R.M. (2001). Ab initio simulation of charged slabs at constant chemical potential. *The Journal of Chemical Physics* 115: 1661–1669.

47 Sakong, S., Forster-Tonigold, K., and Groß, A. (2016). The structure of water at a Pt(111) electrode and the potential of zero charge studied from first principles. *The Journal of Chemical Physics* 144: 194701.

48 Otani, M. and Sugino, O. (2006). First-principles calculations of charged surfaces and interfaces: a plane-wave nonrepeated slab approach. *Physical Review B* 73: 115407.

49 Otani, M., Hamada, I., Sugino, O. et al. (2008). Structure of the water/platinum interface––a first principles simulation under bias potential. *Physical Chemistry Chemical Physics* 10: 3609–3612.

50 Jinnouchi, R. and Anderson, A.B. (2008). Electronic structure calculations of liquid-solid interfaces: combination of density functional theory and modified Poisson-Boltzmann theory. *Physical Review B* 77: 245417.

51 Sundararaman, R., Letchworth-Weaver, K., Schwarz, K.A. et al. (2017). JDFTx: software for joint density-functional theory. *SoftwareX* 6: 278–284.

52 Letchworth-Weaver, K. and Arias, T.A. (2012). Joint density functional theory of the electrode-electrolyte interface: application to fixed electrode potentials, interfacial capacitances, and potentials of zero charge. *Physical Review B* 86: 075140.

53 Oberhofer, H. (2018). Electrocatalysis beyond the computational hydrogen electrode. In: *Handbook of Materials Modeling: Applications: Current and Emerging*

Materials (ed. W. Andreoni and S. Yip), 1–33. Cham: Springer International Publishing.

54 Groß, A. (2021). Grand-canonical approaches to understand structures and processes at electrochemical interfaces from an atomistic perspective. *Current Opinion in Electrochemistry* 27: 100684.

55 Hansen, H.A., Rossmeisl, J., and Nørskov, J.K. (2008). Surface Pourbaix diagrams and oxygen reduction activity of Pt, Ag and Ni(111) surfaces studied by DFT. *Physical Chemistry Chemical Physics* 10: 3722–3730.

56 Gossenberger, F., Roman, T., and Groß, A. (2015). Equilibrium coverage of halides on metal electrodes. *Surface Science* 631: 17–22.

57 Gossenberger, F., Juarez, F., and Groß, A. (2020). Sulfate, bisulfate, and hydrogen co-adsorption on Pt(111) and Au(111) in an electrochemical environment. *Frontiers in Chemistry* 8: 634–634.

58 Karlberg, G.S., Jaramillo, T.F., Skúlason, E. et al. (2007). Cyclic voltammograms for H on Pt(111) and Pt(100) from first principles. *Physical Review Letters* 99: 126101.

59 Eslamibidgoli, M.J. and Eikerling, M.H. (2015). Electrochemical formation of reactive oxygen species at Pt (111)—a density functional theory study. *ACS Catalysis* 5: 6090–6098.

60 Rossmeisl, J., Nørskov, J.K., Taylor, C.D. et al. (2006). Calculated phase diagrams for the electrochemical oxidation and reduction of water over Pt(111). *The Journal of Physical Chemistry B* 110: 21833–21839.

61 Andreussi, O. and Marzari, N. (2014). Electrostatics of solvated systems in periodic boundary conditions. *Physical Review B* 90: 245101.

62 Quaranta, V., Hellström, M., and Behler, J. (2017). Proton-transfer mechanisms at the water–ZnO interface: the role of presolvation. *The Journal of Physical Chemistry Letters* 8: 1476–1483.

63 Skúlason, E., Karlberg, G.S., Rossmeisl, J. et al. (2007). Density functional theory calculations for the hydrogen evolution reaction in an electrochemical double layer on the Pt(111) electrode. *Physical Chemistry Chemical Physics* 9: 3241–3250.

64 Fawcett, W.R. (2008). The ionic work function and its role in estimating absolute electrode potentials. *Langmuir* 24: 9868–9875.

65 Tripkovic, V., Björketun, M.E., Skúlason, E., and Rossmeisl, J. (2011). Standard hydrogen electrode and potential of zero charge in density functional calculations. *Physical Review B* 84: 115452.

66 Trasatti, S. (1986). The absolute electrode potential: an explanatory note (recommendations 1986). *Pure and Applied Chemistry* 58: 955–966.

67 Skúlason, E., Tripkovic, V., Björketun, M.E. et al. (2010). Modeling the electrochemical hydrogen oxidation and evolution reactions on the basis of density functional theory calculations. *The Journal of Physical Chemistry C* 114: 18182–18197.

68 Chan, K. and Nørskov, J.K. (2016). Potential dependence of electrochemical barriers from ab initio calculations. *The Journal of Physical Chemistry Letters* 7: 1686–1690.

69 Schnur, S. and Groß, A. (2009). Properties of metal–water interfaces studied from first principles. *New Journal of Physics* 11: 125003.

70 Lozovoi, A.Y. and Alavi, A. (2003). Reconstruction of charged surfaces: general trends and a case study of Pt(110) and Au(110). *Physical Review B* 68: 245416.

71 Kastlunger, G., Lindgren, P., and Peterson, A.A. (2018). Controlled-potential simulation of elementary electrochemical reactions: proton discharge on metal surfaces. *The Journal of Physical Chemistry C* 122: 12771–12781.

72 Taylor, C.D., Wasileski, S.A., Filhol, J.-S., and Neurock, M. (2006). First principles reaction modeling of the electrochemical interface: consideration and calculation of a tunable surface potential from atomic and electronic structure. *Physical Review B* 73: 165402.

73 Filhol, J.-S. and Neurock, M. (2006). Elucidation of the electrochemical activation of water over Pd by first principles. *Angewandte Chemie International Edition* 45: 402–406.

74 Eslamibidgoli, M.J., Groß, A., and Eikerling, M. (2017). Surface configuration and wettability of nickel(oxy)hydroxides: a first-principles investigation. *Physical Chemistry Chemical Physics* 19: 22659–22669.

75 Dabo, I., Li, Y., Bonnet, N., and Marzari, N. (2010). Ab initio electrochemical properties of electrode surfaces. In: *Fuel Cell Science* (ed. A. Wieckowski and J. Norskov), 415–431. Wiley.

76 Fang, Y.-H., Wei, G.-F., and Liu, Z.-P. (2013). Theoretical modeling of electrode/electrolyte interface from first-principles periodic continuum solvation method. *Catalysis Today* 202: 98–104.

77 Goodpaster, J.D., Bell, A.T., and Head-Gordon, M. (2016). Identification of possible pathways for C–C bond formation during electrochemical reduction of CO_2: new theoretical insights from an improved electrochemical model. *The Journal of Physical Chemistry Letters* 7: 1471–1477.

78 Petrosyan, S.A., Briere, J.-F., Roundy, D., and Arias, T.A. (2007). Joint density-functional theory for electronic structure of solvated systems. *Physical Review B* 75: 205105.

79 Huang, J., Zhang, J., and Eikerling, M. (2018). Unifying theoretical framework for deciphering the oxygen reduction reaction on platinum. *Physical Chemistry Chemical Physics* 20: 11776–11786.

80 Zhou, D., Wei, J., He, Z.-D. et al. (2020). Combining single crystal experiments and microkinetic modeling in disentangling thermodynamic, kinetic, and double-layer factors influencing oxygen reduction. *The Journal of Physical Chemistry C* 124: 13672–13678.

81 Zhu, X. and Huang, J. (2019). Modeling electrocatalytic oxidation of formic acid at platinum. *Journal of the Electrochemical Society* 167: 013515.

82 Zhang, Y., Zhang, J., and Huang, J. (2019). Potential-dependent volcano plot for oxygen reduction: mathematical origin and implications for catalyst design. *The Journal of Physical Chemistry Letters* 10: 7037–7043.

83 Huang, J. and Eikerling, M. (2019). Modeling the oxygen reduction reaction at platinum-based catalysts: a brief review of recent developments. *Current Opinion in Electrochemistry* 13: 157–165.

84 Tripković, V., Skúlason, E., Siahrostami, S. et al. (2010). The oxygen reduction reaction mechanism on Pt(111) from density functional theory calculations. *Electrochimica Acta* 55: 7975–7981.

85 Rossmeisl, J., Karlberg, G.S., Jaramillo, T., and Nørskov, J.K. (2009). Steady state oxygen reduction and cyclic voltammetry. *Faraday Discussions* 140: 337–346.

86 Wakisaka, M., Suzuki, H., Mitsui, S. et al. (2009). Identification and quantification of oxygen species adsorbed on Pt(111) single-crystal and polycrystalline Pt electrodes by photoelectron spectroscopy. *Langmuir* 25: 1897–1900.

87 Wakisaka, M., Udagawa, Y., Suzuki, H. et al. (2011). Structural effects on the surface oxidation processes at Pt single-crystal electrodes studied by X-ray photoelectron spectroscopy. *Energy & Environmental Science* 4: 1662–1666.

88 Martínez-Hincapié, R., Climent, V., and Feliu, J.M. (2018). Peroxodisulfate reduction as a probe to interfacial charge. *Electrochemistry Communications* 88: 43–46.

Part VI

Protocols

12

Quantifying the Activity of Electrocatalysts

Karla Vega-Granados and Nicolas Alonso-Vante

University of Poitiers, IC2MP-UMR-CNRS 7285, 4 rue Michel Brunet, F-86073 Poitiers Cedex 9, France

12.1 Introduction: Toward a Systematic Protocol for Activity Measurements

Electrochemical generators, ranging from metal–air batteries to fuel cells, require electrode materials with good electrocatalytic activity and low cost, which are durable, essentially for the O_2 reduction reaction (ORR), which is a multielectron charge transfer reaction that limits the efficiency of fuel cells [1]. Therefore, such ORR electrocatalysts are essential for proton-exchange membrane fuel cells (PEMFCs). PEMFCs efficiently produce electricity by electrochemical oxidation of H_2 fuel at their anodes and reduction of O_2 from air at their cathodes, using Pt nanoparticles [2] or Pt alloys [3, 4] as catalysts, usually supported on high surface area carbons, such as Vulcan carbon or Ketjen black. Therefore, ORR, in acidic or basic electrolyte, is a slow four-electron transfer reaction resulting in high overpotentials at the cathodes of PEMFCs, requiring high Pt catalyst loadings, which increases costs. These parameters must be optimized in the design of advanced materials.

The findings reported in the literature for the study of electrocatalytic reactions show a wide range of experimental conditions as well as data processing. Reported variations include parameters such as the catalyst mass loading on the electrode, the components of the electrocatalytic inks, the choice of the cell material, and the purity of the electrolytes. All these variations in parameters lead to difficulties and inefficiencies when it comes to making comparisons between materials [5]. This often results in retesting the same materials under different conditions to ensure the reported results. To overcome this, several benchmarking studies have been done by different research groups, varying the mentioned conditions aiming at standardizing the testing of half-cells [6–9]. Once methods and configurations for benchmark catalysts have been established, further adaptations to different systems can be made following the same line.

The compiled results presented herein demonstrate the difference in the data. This chapter devoted to protocol should enable more researchers to determine critical electrochemical evaluation parameters: (i) electrochemical surface area,

Electrocatalysis for Membrane Fuel Cells: Methods, Modeling, and Applications, First Edition.
Edited by Nicolas Alonso-Vante and Vito Di Noto.

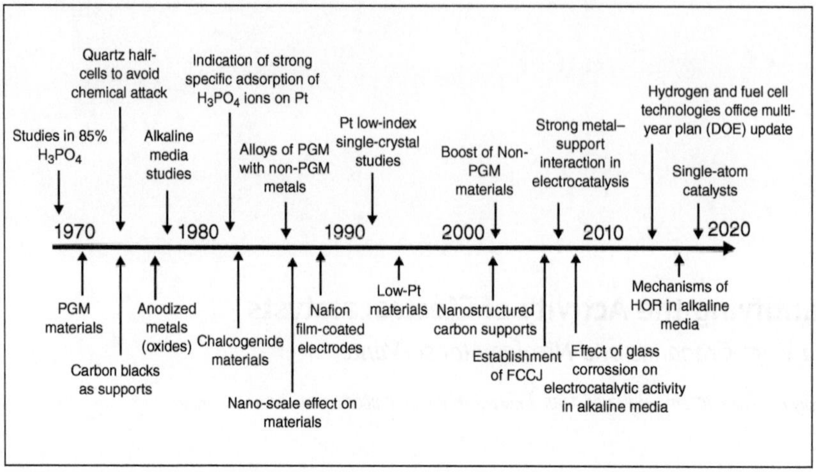

Figure 12.1 Milestones that have influenced the protocols of H_2-O_2 fuel cell reactions in half-cell for over the past five decades. Source: Adapted from Refs. [10–21].

(ii) Pt mass specific activity, j_m, (iii) Pt area specific activity, j_s, which are indispensable parameters of true ORR catalyst activity, further, (iv) difference in the activity of the electrolytic medium, and (v) the consequences of electrode potential determination referenced to the same scale, whatever the pH to make comparisons with data reported in different laboratories around the world.

Half-cell protocols have evolved as new information is acquired and new devices are developed. Figure 12.1 summarizes some key points that have been made and impacted on fuel cell reactions, in the half-cell configuration, over the past five decades [10–21]. The objective of this chapter is to unify the protocols reported in the half-cell tests to properly evaluate the different types of materials toward electrocatalytic reactions involved in hydrogen fuel cells. Different approaches will be addressed ranging from the selection of materials, important aspects of the experimental setup, to the importance of stability and durability tests in half-cells.

12.2 Materials Consideration

12.2.1 PGM Group

A wide variety of materials have been studied for the electrocatalytic reactions occurring in polymer electrolyte membrane (PEM) fuel cells: the hydrogen oxidation reaction (HOR) and the oxygen reduction reaction (ORR). The most studied and efficient materials for many electrocatalytic reactions, over several decades, are the so-called platinum group metals, or PGM, e.g. Ru, Rh, Pd, Os, Ir, and Pt [22]. This group of metals is known for its high chemical stability in various hostile environments, and also for its high costs and low natural availability. Among the elements in this group, Pt has been categorized as the benchmark electrocatalyst for low-temperature fuel cell reactions, achieving fast electrode kinetics on the oxidation of hydrogen with

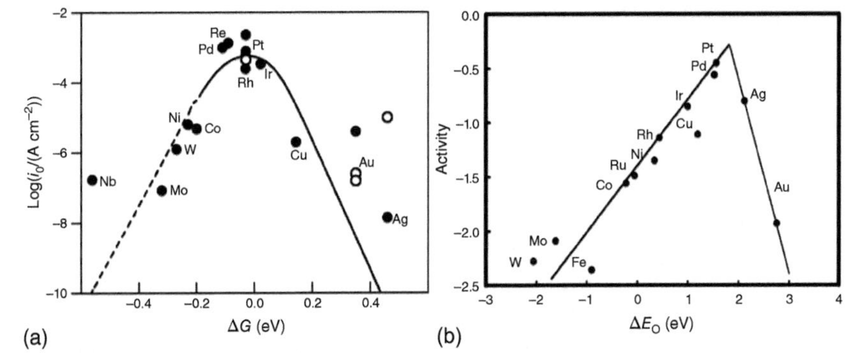

Figure 12.2 (a) Volcano plot of the hydrogen-exchange current density as a function of the hydrogen adsorption free energy obtained by DFT calculations. Source: Reproduced from Ref. [25] with permission of American Chemical Society. (b) Oxygen reduction activity plotted as a function of the oxygen binding energy obtained by DFT calculations. Source: Reproduced from Ref. [27] with permission of American Chemical Society.

a low Pt content, say 0.05 mg_{Pt} cm^{-2} [23]. On the other hand, in the ORR, Pt still presents high overpotentials to achieve good activity which translates to voltage losses in the fuel cell device. This is one of the main drawbacks for the ORR activity of PEM fuel cells. The origin of the activity of electrocatalytic materials has been discussed by several authors, and commonly, the energies of adsorption of hydrogen [24, 25] and oxygen [26, 27] with metallic surfaces have been used as a descriptor, recognizing Sabatier's principle [28]. Figure 12.2 shows volcano-shaped plots relating the activity of different metals for the hydrogen-electrode and oxygen reduction reaction to the corresponding element adsorption energy.

From Figure 12.2, it can be seen that the PGM remains at the top of the volcano curve; however, its electrocatalytic activity still leaves room for improvement, which is why new materials are being developed.

12.2.2 Low PGM and PGM-free Approaches

It was established by the USA Department of Energy (DOE) that over 40% of the cost of PMFCs is attributed to the need of Pt materials to obtain good electrocatalytic activity, specifically at the cathode for the ORR [29], due to the reaction's sluggish kinetics. Following the DOE's guidelines for 2020 targets, in fuel cells for transportation applications, the PGM mass loading of catalysts should be lower than 0.125 mg cm^{-2} with a mass activity of 0.44 A mg_{PGM}^{-1} @ 0.9 V [30]. To achieve this, in the past decades the use of so-called low-PGM and PGM-Free materials has been studied by several groups [31–35], for the H_2–O_2 fuel-cell reactions.

A common strategy that achieves the low-PGM approach is to reduce the size of Pt particles and make them disperse on conductive supports to create a larger electrocatalytic surface. The influence of particle size has been studied by different research groups over the past decades [2, 36–38]. In this context, it was assessed that the particle size of Pt catalysts influences the ORR activity in acid medium [38]. Another strategy to reduce the mass loading of Pt is to prepare it with other component to yield nanoalloy particles, where primarily the first row of the transition metals has been thoroughly studied [39]. This approach can cause modifications to the surface of active sites, as well as generate improvements in the strength of the adsorption energy of reactive species and intermediates species by means of geometric factors (changing Pt–Pt interatomic distances), and electronic effects (d-band modifications) [40, 41].

For example, Toda et al. [42] reported an important ORR activity enhancement in 0.1 M $HClO_4$ using nonsupported Pt–Ni alloys with 30 at.% of Ni content, and observed that after electrochemical experiments, a Pt skin was formed on the surface of the catalyst. Similar effects were later reported with Pt–Fe (50 at.% Fe), and Pt–Co (40 at.% Co) alloys by the same group [43], and the results were explained by the increase in the surface 5d vacancies, since Ni, Co, and Fe have more 5d vacancies than Pt, thus leading to an increase in 2π electron donation from O_2 to the surface atoms of Pt, enhancing the O_2 adsorption and weakening the O—O bond, Figure 12.3. A similar improvement in activity was reported for HOR, as well as improvements in CO tolerance with Pt—Fe, Pt—Co, and Pt—Ni alloys with Pt skins.

Regarding the PGM-free catalysts for H_2–O_2 fuel cell reactions, a group of materials, known as M–N–C catalysts, consisting of nitrogen and transition metals, M = Fe, Co, Mn, codoped carbons that form a carbon-embedded MN_4 moiety, have shown interesting results as PGM-free materials in the ORR in both acid and alkaline media [45]. The synthesis of these materials is based on the high-temperature pyrolysis of a metal precursor, nitrogen precursor, and carbon support or precursors to avoid agglomeration. Metal organic frameworks (MOFs) have commonly been used as precursors for the synthesis of these materials [44, 46]. Figure 12.3b shows the ORR curves in 0.1 M $HClO_4$ and 0.1 M KOH of an Fe–N–C catalyst obtained using a Zn-based MOF (ZIF-8) as precursor (FePhen@MOF-ArNH_3), showing comparable performance with respect to Pt/C.

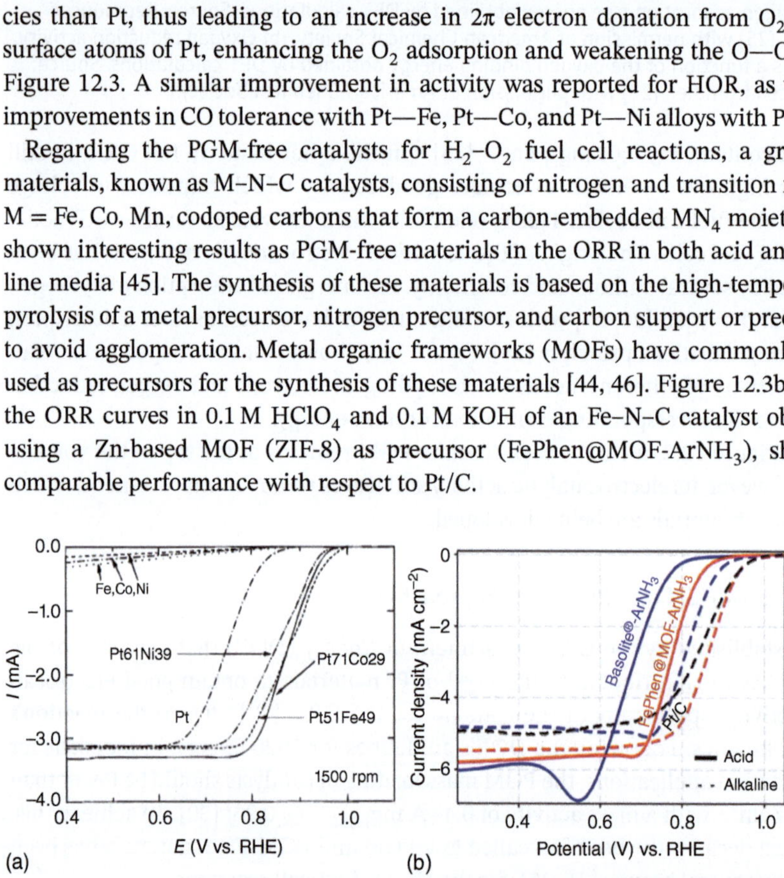

Figure 12.3 (a) ORR RDE curves for Pt alloys at 1500 rpm obtained in 0.1 M $HClO_4$. Source: Reproduced from Ref. [43] with permission of The Electrochemical Society. The Electrochemical Society. (b) ORR curves in 0.1 M $HClO_4$ and 0.1 M KOH of an Fe–N–C catalyst (FePhen@MOF-ArNH_3) compared to commercial Pt/C from Tanaka Kikinzoku International KK (Japan) and commercial Basolite (ZIF-8). Curves obtained at 20 mV s^{-1} and 1600 rpm. Source: Reproduced from Ref. [44] with permission of Springer Nature/CC BY 4.0.

Another alternative for the implementation of PGM-free electrocatalysts is the study of alkaline fuel cells, or anion-exchange membrane fuel cells (AEMFCs), as it has been reported for the activity of nonprecious metals, for both H_2–O_2 fuel cell reactions and metal-free materials, can be as good as Pt or even higher for ORR [47].

12.2.3 Impact of Support Effects on Catalytic Sites

The support of electrocatalysts plays a significant role in many aspects, as it contributes to the proper distribution of nanoparticles of the active phase along high surface areas. It can also modulate the activity toward the reaction through the metal–support interactions, as well as the stability and durability of the catalyst in different environments. Commonly, carbon black supports are chosen as support due to their high specific surface area and conductivity. Some examples are Vulcan XC-72 Carbon with a surface area of $250\,m^2\,g^{-1}$, Ketjen Black with a surface area of up to $1270\,m^2\,g^{-1}$, and Black Pearls 2000 with surface area of up to $1500\,m^2\,g^{-1}$ [48].

Recently, nanostructured carbon materials, such as carbon nanotubes (CNTs), graphene (as reduced graphene oxide, rGO), carbon nanofibers (CNFs), and carbon nanohorns (CNHs), have been used as electrocatalysts supports, being a substitute of the typical carbon blacks, as they can provide an enhanced electrical conductivity and surface area, and even be more tolerant to corrosion [49]; however, carbon corrosion cannot be avoided completely when using carbon supports. In this matter, hybrid carbon supports have been studied to provide better stability, which can be achieved by doping or functionalizing nanostructured carbon supports or combining with other materials, such as metal oxides to form composites, or even using noncarbon materials as supports.

An important support effect is the so-called strong metal support interaction (SMSI), which was first reported in 1978 [50, 51] to describe an effect in the chemisorption properties of noble metals supported in a number of metal oxides. This phenomenon has also been observed electrochemically for pure carbon supports, induced by the photodeposition method onto preferred graphitic carbon domains [52, 53]. In the past decades it has been taken advantage of this effect in electrocatalysis, using oxide-carbon composites as supports to provide sites for the selective deposition of metals (via, e.g. photodeposition) onto the oxide sites while having a conductive support at the same time [54, 55]. In this matter, there is evidence that deposited metals become alloyed with the metal of the oxide of the support, enhancing electrocatalytic activity [4].

This interaction can be probed by the CO stripping process, observing the negative potential shift of the CO oxidation peaks, see Figure 12.4. The induced electronic effect has been studied using X-ray photoelectron spectroscopy (XPS). In the case of Pt on oxide-carbon composites, and on highly graphitized carbon, a downshift in the binding energy (BE) of Pt 4f electrons is shown, owed to a local increase in electron density on Pt [56, 57].

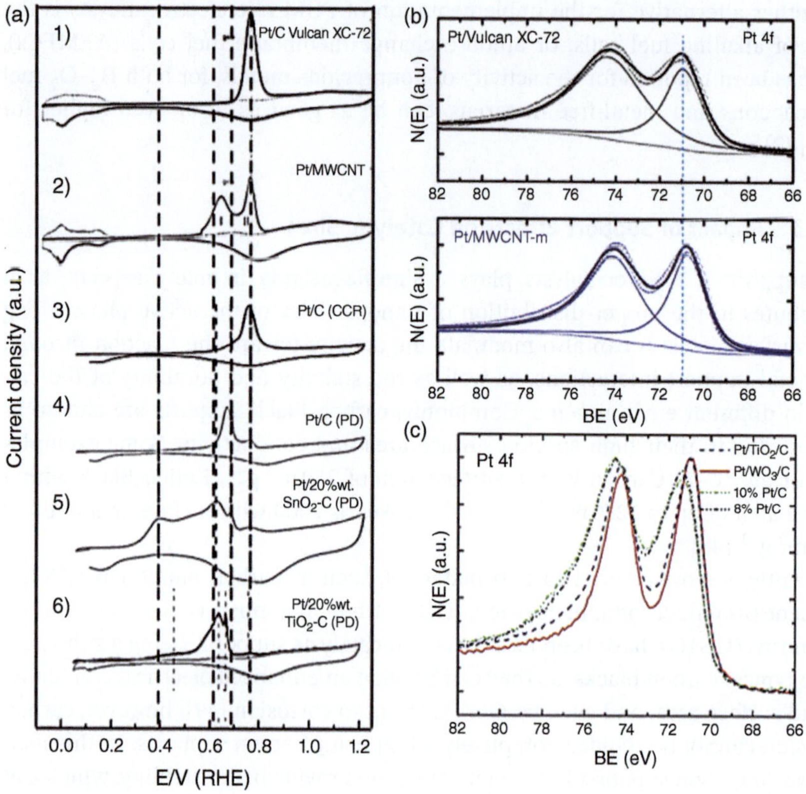

Figure 12.4 (a) Electrochemical probe of the SMSI effect using the CO stripping technique on different supports with the same electrocatalytic center. (b) XPS Pt 4f region comparison between Pt/Vulcan C and Pt/MWCNT. (c) XPS Pt 4f region comparison between commercial 10 wt% Pt/C, chemically prepared by Pt/C (carbonyl chemical route, [CCR]), and Pt photodeposited (PD) onto oxide-carbon composites. Source: From Ref. [56] with permission of American Chemical Society. Figures a1, a2 and b: From Ref. [57] with permission of American Chemical Society. Figures a3, a4 and a5: From Ref. [58] with permission of Elsevier. Figure a6: From Ref. [54] with permission of Elsevier.

12.3 Electrochemical Cell Considerations

12.3.1 Cell Configuration and Material

Fuel cells are two-electrode devices, where one electrochemical reaction takes place on each electrode (the fuel oxidation at the anode and the reduction of oxygen at the cathode) in a simultaneous matter. The response of the fuel cell parameters reflects the sum of the contributions from each component of the device (the anode, the cathode, the electrolyte, the cell configuration, etc.). However, before testing electrode materials in the fuel cell configuration, which is not a simple task, it is important to evaluate their performance in a half-cell configuration, where only one of the reactions is monitored and the response is directly attributable to the electrode material if the measurements are made carefully. For these preliminary studies, a three-electrode setup is commonly used, each one being the working electrode (WE), containing the catalytic material being tested for the specific electrochemical reaction; the counter electrode (CE), which closes a circuit with the working electrode allowing the flow of electrons in the system; and the third electrode is the reference electrode, which is used to measure the potential of the WE.

The appropriate material of the electrochemical cell depends on the electrolyte of choice. While a glass cell is appropriate for measurements in acid and neutral electrolytes, in the case of experiments being carried out in alkaline medium, the effect of glass corrosion could be significant if a glass cell is being used or if there are glass components as a part of the electrochemical setup. As it is well known and studied since the past century, glass corrodes in the presence of alkaline solutions [59], resulting in soluble ions that interfere with surface reactions in electrocatalysis. To avoid the presence of considerable amounts of contaminants such as traces of silicon, boron, aluminum, and lead, the electrochemical cell must be constructed of a chemically inert material for use in alkaline environments, such as polymers like tetrafluoroethylene (Teflon). The effect of glass corrosion is still underestimated in some studies, even with the knowledge that most catalytic surfaces are overly sensitive to the presence of spectator ions coming from the electrolyte. Studies by Mayrhofer et al. [60, 61] and Campos-Roldan et al. [62] proved that there is a negative influence in the evaluation of electrocatalytic reactions on polycrystalline Pt due to the presence of glass corrosion products, such as silicates and trace amounts of boron, aluminum, and lead, confirmed by induced coupled plasma (ICP) when glass is present in controlled amounts, even when measuring in electrochemical cells made of polymeric materials. These effects are shown in Figure 12.5.

Besides the corrosion of glass, there are other sources of contaminants that could disturb the electrochemical experiments in half-cell testing, such as low purity of the chemical regents used to prepare the supporting electrolyte, an inappropriate choice of reference electrode, and an incomplete cleaning of the electrochemical cell.

A good starting point, after choosing the cell material, is its correct cleaning, to decrease the sources of possible impurities that interfere with the experiments. Different procedures have been proposed for this task, with the aim of oxidizing any trace of metallic ions or organic residues from the walls of the cell and its components. The reported methods include filling glass cells with concentrated H_2SO_4 and leaving it to stand for several hours and then rinsing with Milli-Q water (18 MΩ cm); the use of some glassware surfactants and then soaking in concentrated H_2SO_4; or a mixture of H_2SO_4 with HNO_3 or H_2SO_4 with Nochromix® to be rinsed afterward with boiling Milli-Q water [63].

Other methods imply the use of stronger oxidants, such as a acidified potassium permanganate solution (obtained by mixing potassium permanganate with dilute sulfuric acid) by submerging the material for several hours, or the use of piranha solution (3 : 1 v/v of concentrated H_2SO_4 with 30% H_2O_2), either by soaking the glass or polymer cell and its components and then rinsing with water [8] or using it as a second step after using an acidified permanganate solution procedure to properly remove the remains of manganese oxide or permanganate anions.

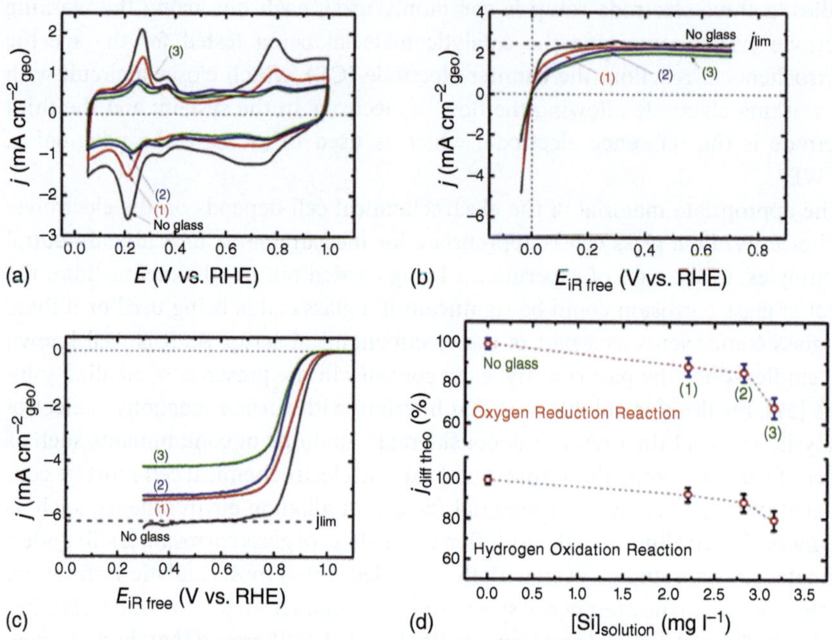

Figure 12.5 (a) Cyclic voltammograms of 20 wt% Pt/C (JM) performed on a Teflon cell in N_2-sat. 0.1 M KOH, with increasing quantities of glass (from curves 1–3) and with no glass in the system. (b) HOR curves of the same material and glass quantities as on figure (a) performed in H_2-sat. (c) 0.1 M KOH ORR of the same material and glass quantities as on figure curves performed in O_2-sat. KOH. Polarization curves were performed at 1600 rpm and 5 mV s^{-1}. (d) Limiting diffusional current density at $E = 0.6$ V vs. RHE for ORR (top) and at 0.3 V vs. RHE for HOR (bottom). Source: Reproduced from Ref. [62] with permission of IOP Publishing/CC BY 4.0.

Table 12.1 describes reports of electrochemical cell cleaning procedures prior to experiments.

The choice of RE must be done carefully, since their equilibrium must remain stable on common electrolyte conditions, which are harsh environments (e.g. 0.1 M $HClO_4$, $[H^+] = 0.1$ M, pH 1 and 0.1 M KOH $[OH^-] = 0.1$ M, pOH $= -\log [OH^-] = 1$, pH 14–1 = 13).

Some RE can be susceptible to the precipitation of their ionic salts in certain electrolyte environments, or their ions can leach to the electrolyte medium, like the Ag/AgCl electrode, where the high chloride concentrations can leach into the solution and interfere with the electrokinetic reactions taking place in the WE [65]. Also, the exposure of the Ag/AgCl electrode to strong alkaline solutions can shift its potential due to the formation of AgO_x species, making its use less reliable in this environment [62].

To avoid possible sources of contamination of the system or the shift of the equilibrium potential, the recommended RE for developing studies on acid electrolytes is the reversible hydrogen electrode (RHE), while its use is not recommended directly in alkaline electrolytes, since the OH^- species disturb the Pt/H_2 equilibrium [62]; instead, the Hg/HgO electrode is highly recommended for studies on alkaline media [66]. More details and conventions regarding RE are discussed in Section 12.3.3.

Table 12.1 Chemical treatments reported for the cleaning of electrochemical cells to avoid organic residues and metallic contaminations.

Chemicals	Procedure	References
Conc. H_2SO_4 Milli-Q water	1. Fill the cell with H_2SO_4 coating the whole interior. 2. Soak cell overnight. 3. Rinse with Mill-Q water 8×.	[63]
Sparkleen™ glassware surfactant Mill-Q water	1. Wash with surfactant. 2. Rinse with Milli-Q water. 3. Follow either one of the acid treatments: (a) Soak in conc. H_2SO_4 for two-four hours. (b) 1 : 1 mixture of $H_2SO_4 + HNO_3$ for four hours. (c) 1 : 1 mixture of H_2SO_4 + Nochromix for four hours. 4. Rinse with Milli-Q water and boil twice.	[63]
Conc. NaOH Piranha solution (98% H_2SO_4, 30% H_2O_2 3 : 1 v/v) Milli-Q water	1. Soak in concentrated NaOH. 2. Soak in piranha solution for 24 hours. 3. Soak in Milli-Q water and boil, changing water five times each 30 min.	[8]
$KMnO_4$ Conc. H_2SO_4 Piranha solution (98% H_2SO_4, 30% H_2O_2 3 : 1 v/v) Milli-Q water	1. Soak in a mixture of $KMnO_4$ with H_2O-H_2SO_4 1 : 1 (v/v) overnight. 2. Rinse with piranha solution (bubbles must be observed). 3. Rinse with Milli-Q water. 4. Boil in Milli-Q water.	[64]
Conc. HNO_3 Conc. HCl Milli-Q water	1. Immerse in a mixture of conc. HNO_3 and conc. HCl (1 : 2 v/v) for 30 min. 2. Rinse with Milli-Q water.	[64]

12.3.2 Electrolyte

12.3.2.1 Purity

The sources of electrolyte contamination are often ignored. As it was discussed, the electrochemical cell and the reference electrode could be a source of such contaminations. In addition, the electrolyte itself can alter the expected activity of a catalyst. The use of high-purity chemicals is always recommended, including the water for electrolyte preparation. In the early 1980s, McNicol showed the effect of using different grades of water purification processes in measuring the electrochemical surface activity (ECSA) of Pt, finding that if the electrodes are immersed and polarized in

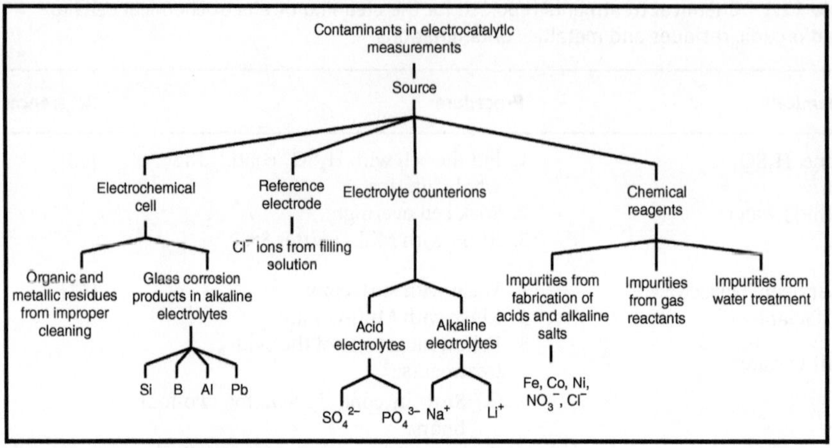

Figure 12.6 Possible contaminants in electrochemical measurements and their common sources.

the double-layer region for long periods, contaminants can block the surface sites of Pt [67]. Nowadays, Milli-Q water is the choice for the electrolyte preparation as well as ultrahigh purity (UHP) gasses and chemicals. Effects of chemical contamination coming from the chemical reagents used to prepare electrolytes have been also reported [68, 69]. Figure 12.6 schematizes the source of different contaminants that interfere with electrochemical studies.

12.3.2.2 Protons vs. Hydroxide Ions

The reactions involved in PEM fuel cells can be read in the Pourbaix diagram of water, Figure 12.7, noticing that the potential at which they can be carried out thermodynamically is a function of pH, according to the standard hydrogen electrode (SHE) scale. For example, the ORR goes from 1.23 V/SHE at pH 0 to 0.4 V/SHE at pH 14. With the pH variation, the free energy of the surface increases, and, during this process, the adsorption free energy of oxygen and its intermediates decrease. This phenomenon is commonly used as a descriptor for ORR activity. In this matter, a greater number of surfaces have a sufficiently high O_2 adsorption energy to carry out the ORR in alkaline rather than in acid media [71].

However, other factors must be taken into account, such as the nature of the ORR itself, which is an inner-sphere electron transfer reaction, where O_2 is expected to be chemisorbed directly at the active sites of the electrode. An outer-sphere mechanism has been proposed for alkaline conditions, Figure 12.8b, since specifically adsorbed hydroxyl ions, water dipoles from the solvent (acting as source of protons), and chemisorbed O_2 coexist in the inner Helmholtz plane (IHP) of the double layer, whereas solvated hydroxyl ions are located in the outer Helmholtz plane (OHP) along with solvated O_2 molecules. In this matter, solvated O_2 molecules interact with the hydroxyl species adsorbed on the surface of the electrode via a hydrogen bond, having a lower bond energy than the one expected for direct O_2 chemisorption ($<35\ kJ\ mol^{-1}$ vs. $>300\ kJ\ mol^{-1}$). Therefore, solvated oxygen molecules, stabilized

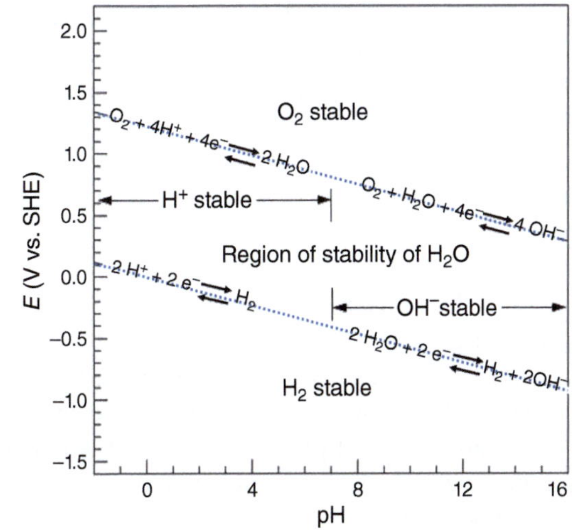

Figure 12.7 Pourbaix diagram of water. Source: From Ref. [70] with permission of Springer Nature.

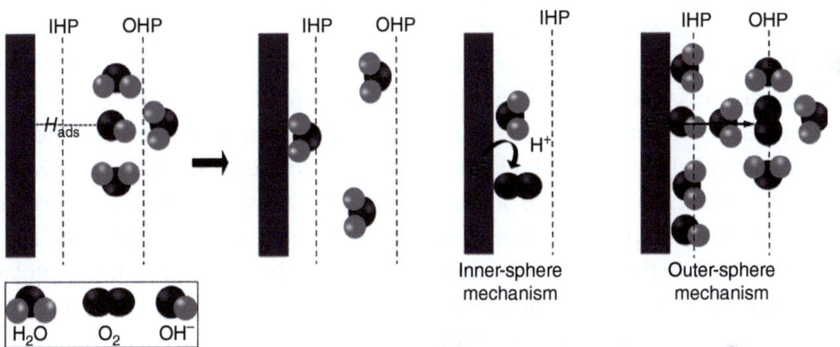

(a) (b)

Figure 12.8 (a) Double-layer structure of the HOR process at Pt catalyst in alkaline electrolyte, (b) Double-layer structure of the ORR proceeding via inner- and outer-sphere mechanisms, suggested for alkaline electrolytes, based on the mechanisms reported in Refs. [72] and [73].

in the OHP, provide an outer sphere electron transfer. This mechanism has been proposed as an alternative concept to understand why several materials, including PGM, non-PGM, and metal oxides, can serve as good fuel cell cathodes [72].

A different scenario occurs with the HOR. A correct assessment of this reaction is complicated in acid medium since its kinetics are very fast with Pt electrodes. Voltage losses in the cells below 5 mV have been reported when using electrode with a Pt mass loadings of $0.05\,mg_{Pt}\,cm^{-2}$. In this respect, the high HOR/HER exchange current density makes its precise quantification complicated when using the conventional RDE setup due to the mass transport limitations [74].

On the other hand, it is known that HOR exchange current density, in alkaline medium on Pt, proceeds with two orders of magnitude lower than on acid medium. Several hypotheses have been formulated to explain this phenomenon. Analogous to the double-layer modification for the ORR, a similar mechanism in alkaline media has been proposed [73], Figure 12.8a, in which water acts as the H_{ads} precursor via $Pt + H_2O + e^- \rightarrow Pt - H_{ads} + OH^-$. OH^- does not dissociate completely from the water molecule, but instead it remains stabilized in the OHP via a hydrogen bond with the H_{ads} species. When the potential is negative enough, the OH^- species are further cleaved completely from the water molecule and diffuse to the bulk of the solution. Other authors suggest that a change in the hydrogen binding energy (HBE) is responsible for the HOR activity in alkaline media, as HBE increases along with pH [75].

12.3.2.3 Influence of Counterions

Supporting electrolytes are key components of the half-cell electrochemical studies, since their main roles include reducing the resistance of the solution for electrochemical reactions to take place and to eliminate the contribution of migration of the mass transport [76]. According to the IUPAC, a supporting electrolyte is a solution with a high concentration of ionic species that are nonelectroactive in the studied potential range [77]. However, several reports have shown that some ionic species, in commonly used electrolytes, are not only spectators to electrochemical reactions. For example, the effect of sulfate ions on Pt surfaces has been studied by several groups, showing that SO_4^{2-} and HSO_4^- species strongly adsorb [78, 79] on well-defined surfaces, Figure 12.9a, blocking the sites for the O_2 adsorption [20], thus reducing the ORR activity of the electrocatalyst. This negative effect is enhanced in the presence of phosphate [84], Figure 12.9b, and chloride ions [85], Figure 12.9c,d. In this way, several studies have proven that the ORR activity decreases in the order $HClO_4 > H_2SO_4 > H_3PO_4 > HCl$ on Pt surfaces due to surface blocking via the anion adsorption [86]. In the case of some non-PGM metals, the anion adsorption is not an important issue, for example on Fe-based electrocatalysts [86, 87]. However, these materials are still not as active as their PGM counterparts.

The influence of electrolytes also exists when alkaline electrolytes are used. In this case this is due to noncovalent interactions, rather than specific adsorption of ions. To explain why Pt electrocatalytic activity is affected, this mechanism was suggested by Strmcnik et al. [83], Figure 12.9e. In this matter, it has been discussed that the activity trend for ORR and HOR over Pt follows $Cs^+ > K^+ > Na^+ > Li^+$, as shown in Figure 12.9g. The electrocatalytic activity is inversely proportional to the charge density of the alkali metal ion and directly proportional to hydration energy of the cation. This phenomenon occurs via the hydrated-cations interaction with adsorbed OH species over the Pt surface, stabilizing the OH adsorption, making electrocatalytic processes more endothermic. The effect has been supported by different authors [62, 88, 89].

12.3.3 Electrode Potential Measurements

The primary standard in electrochemical measurements, having a potential of zero by definition, is the standard hydrogen electrode (SHE), which consists of a platinized platinum surface (wire or a sheet) in contact with a hydronium-containing solution while hydrogen gas is supplied continuously [90]. The activities of the components must remain in standard conditions, which makes the use of this

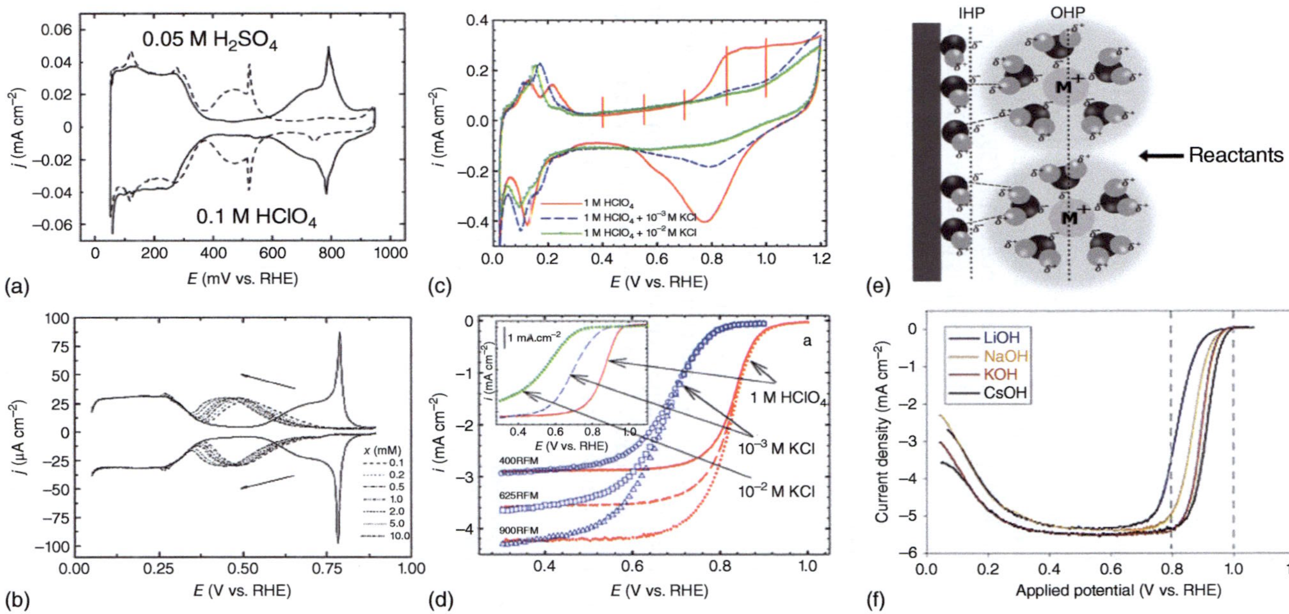

Figure 12.9 (a) Voltammograms of a Pt(111) electrode in N_2 sat. acid electrolytes. Source: From Ref. [80] with permission of American Chemical Society. (b) Comparison between cyclic voltammograms of a Pt(111) electrode in 0.1 M $HClO_4$ and increasing concentrations of NaH_2PO_4. Source: From Ref. [81] with permission of Elsevier. (c) Cyclic voltammograms of a commercial Pt/C electrode (E-TEK) in Ar sat. 1 M $HClO_4$ with increasing KCl concentrations. (d) ORR polarization curves of the Pt/C electrode performed with the same conditions as (c). Source: Reproduced from Ref. [82] with permission of American Chemical Society. (e) Mechanism of metal cation stabilization influencing surface reactivity for electrocatalysis. (f) Effect of alkali metal cations on the ORR of a Pt(111) electrode. Source: Reproduced from Ref. [83] with permission of Springer Nature.

electrode not so practical. Thus, if any of the activities of the SHE differs from the unity, it becomes the reversible hydrogen electrode (RHE), and the potential may shift according to the Nernst equation:

$$H^+(aq) + e^- \leftrightarrows \frac{1}{2}H_2(g) \tag{12.1}$$

$$E = E^\circ - \frac{RT}{F}\ln\left(\frac{\sqrt{f(H_2)/p^\circ}}{a(H^+)}\right) \tag{12.2}$$

where E is the electrode potential, E° is the standard potential of the $H^+(aq)/H_2(g)$ couple, $f(H_2)$ is the fugacity of H_2, p° is the standard pressure, and $a(H^+)$ is the activity of H^+ (aq) [91]. By fixing the fugacity of the gas to the unitary value, the activity of H^+(aq) can be varied, and remembering the definition of pH (pH = −log $[a_{H^+}]$) one can understand the RHE potential dependence on pH (59.2 mV decrease per unit of pH increase at 298.15 K). This relationship is illustrated in Figure 12.10. The variation of the potential by fixing the activity of H^+(aq) can also be represented; however, in laboratory conditions, the difference from the unit value is considered to be negligible, so it is not usually taken into account.

There are certain cases in which the use of the RHE is not practical due to incompatibilities with the nature of the electrolyte, as it has been mentioned in Section 12.2.1, or its availability in some laboratories. For this reason, there are different kinds of available RE that can be used depending on the needs of the study. It is known that metal/metal oxide reference electrodes such as Hg/HgO and Ag/AgCl electrode are compatible in alkaline medium if care is taken on not using glass-body or Vycor®-containing electrodes and if leaching of Cl^- does not occur in the case of the latter [90]. An important practice for the use of these electrodes

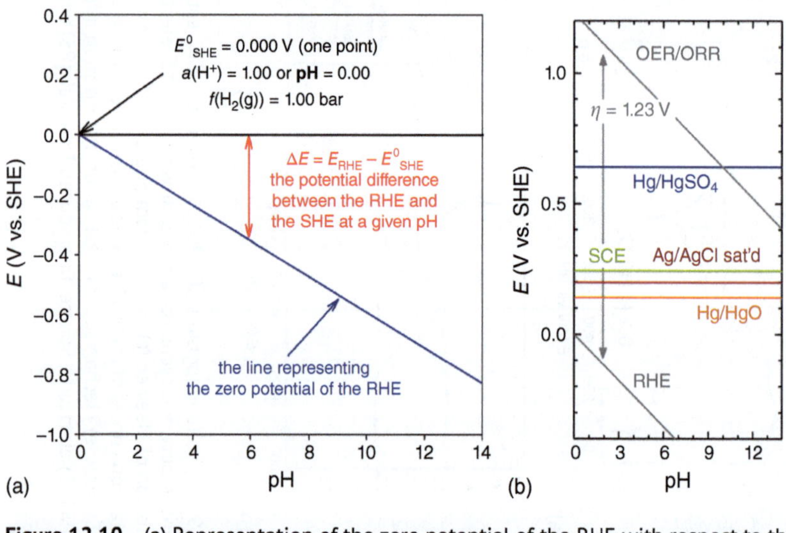

Figure 12.10 (a) Representation of the zero potential of the RHE with respect to the SHE. Source: Reproduced from Ref. [91] with permission of American Chemical Society. (b) Identification of commonly used RE, in the SHE, scale. Source: Reproduced from Ref. [8] with permission of John Wiley & Sons.

Table 12.2 Potentials of commonly employed reference electrodes at 25 °C [76, 90].

Electrode	Equilibrium	Filling Solution and concentration	Potential vs. SHE @ 25 °C
RHE[a)]	$2\,H^+ + 2\,e^- \leftrightarrows H_2$	0.1 M $HClO_4$	−0.059 V
		0.5 M H_2SO_4	−0.018 V
		0.1 M H_3PO_4	−0.059 V
SCE	$Hg_2Cl_2 + 2\,e^- + \leftrightarrows 2Hg^\circ + 2Cl^-$	0.1 M KCl	0.337 V
		1.0 M KCl	0.280 V
		Sat. KCl	0.244 V
Ag/AgCl	$AgCl + e^- \leftrightarrows Ag^\circ + Cl^-$	1 M KCl	0.236 V
		3 M KCl	0.207 V
		Sat. KCl	0.197 V
Hg/HgO[b)]	$HgO + H_2O + 2e^- \leftrightarrows Hg^\circ + 2OH^-$	1 M $[OH^-]$	0.098 V
		0.1 M $[OH^-]$	0.165 V
$Hg/HgSO_4$	$Hg_2SO_4 + 2\,e^- \leftrightarrows 2Hg^\circ + SO_4^{2-}$	1 M H_2SO_4	0.615 V
		Sat. K_2SO_4	0.651 V

a) Recommended for acid electrolytes.
b) Recommended for alkaline electrolytes.

is the calibration prior to making any electrochemical studies. This can be done by performing a voltammetry scan on the HER/HOR potential region using a Pt electrode in $H_{2(g)}$ saturated electrolyte, where the intercept on the potential axis at zero current corresponds to an offset value that can be converted to the RHE scale using:

$$E_{RHE} = E_{measured} - E_{offset} \tag{12.3}$$

where $E_{measured}$ corresponds to the measured potential in the used RE scale [8].

In an equal matter, measured potentials can be converted to the RHE scale, which is a common convention for reporting potential since the SHE depends on the pH. This is done by employing

$$E_{RHE} = E_{measured} + 0.059\ \mathrm{pH} + E^\circ{}_{RE} \tag{12.4}$$

where $E^\circ{}_{RE}$ is the potential, in standard condition of the used RE.

Table 12.2 presents some of the more commonly used reference electrodes and their potential with respect to the SHE at 25 °C.

12.3.4 Preparation of Electrodes

A common practice is to disperse electrocatalytic materials in powder form in an appropriate solvent medium (commonly water–alcohol mixtures at different proportions, using isopropyl alcohol (IPA) or ethyl alcohol) to obtain suspensions of the electrocatalytic materials. These suspensions are known as electrocatalytic inks, from which an aliquot is further deposited on a suitable disk electrode made of, e.g. a conductive substrate such as glassy carbon (GC). As for the electrocatalytic ink preparation, several different recipes can be found. Typically, a catalyst ink contains water, an organic solvent such as IPA, and a small quantity of an ionomer which, in this case, acts like a binder to favor the adhesion of the catalyst to the substrate and maintain its mechanical stability during the analysis.

The pretreatment of GC electrodes usually consists in mechanically polishing the surface with the aid of silicon carbide paper, diamond pastes, or alumina slurries [92], in the latter case, using different particle sizes going from coarser to smallest (e.g. 1, 0.3, and 0.05 μm) [93] on a polishing cloth, ending with an ultrasonic bath treatment in water and isopropanol for a few minutes to remove impurities from the surface [94] or even with electrochemical treatments [95], depending on the desired finish according to the study to be performed. However, aiming at standardizing protocols, alumina polishing followed by ultrasonication is the generally accepted method [8, 63].

The most used ionomer for electrochemical studies in acid media is Nafion®, which was developed by DuPont company and first reported in 1966. It consists of a copolymer of tetrafluoroethylene (TFE) and the perfluoro(vinyl ether) monomer with a sulfonyl fluoride ($-SO_2F$) terminal group which is hydrolyzed and acidified to make an acid-containing side-chain polymer [96]. Its use in fuel cells relies on the difference in polarity between the fluorocarbon backbone and the terminal sulfonate groups, causing hydrophobic and hydrophilic regions.

The second ones are highly hygroscopic, allowing the hydration of the sulfonate groups, creating nanoscopic "channels," which promote the diffusion of protons from the anode to the cathode [99]. An important effect, which is commonly neglected, is the influence of the use of Nafion in alkaline electrochemical studies. As it has been stated, Nafion is permeable to protons in solution, meaning a low affinity for the diffusion of hydroxide species of the alkaline media. In this regard, Jervis et al. proved the importance of using an alkaline ionomer for electrocatalytic studies in alkaline media by comparing Pt/C electrode performance for the HOR using inks prepared with acid (Nafion) and alkaline (AS-4, Tokuyama corporation) ionomers. A blocking effect of the catalyst surface by the acidic ionomer was shown, reflected in the ECSA measurements and HOR activity as shown in Figure 12.11f [98].

Different studies have been made to investigate the role of the components of the catalyst inks. For example, Shinozaki et al. [100] prepared Pt/C catalyst inks without IPA and Nafion, finding out that the obtained results for the ORR in Pt materials were not reproducible. In other study, Takahashi and Kocha [6] found the importance of adding an optimal quantity of IPA for Pt/C catalysts, contributing to a good dispersion of the inks since most of the catalysts are highly hydrophobic. In this matter, reproducibility was promoted for Pt/C catalysts at 24 wt% of isopropanol content. It is important to note that the best ink formulation for different catalysts could be different, depending on the properties of the solid, so the proportions of each component should be optimized. To obtain reproducible and valid results, the inks

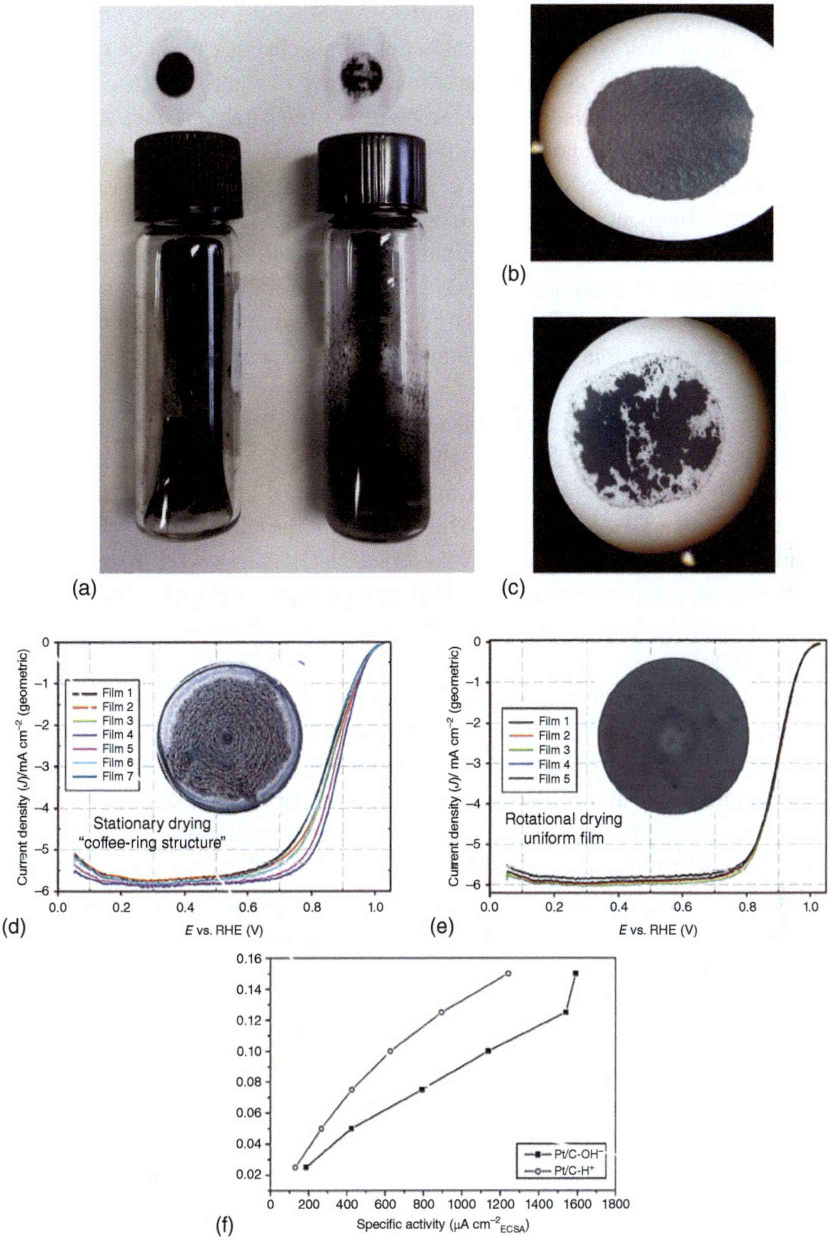

Figure 12.11 (a) Comparison between a well-dispersed ink vs. a badly dispersed one. (b) Drop of a well-dispersed ink over paper observed with a 2× magnification with an optical microscope. (c) Drop of a badly dispersed ink in paper observed with a 2× magnification with an optical microscope. (d) ORR polarization curves of a commercial 40 wt% Pt/C (JM) using an RDE and stationary drying. (e) ORR polarization curves of a commercial 40 wt% Pt/C (JM) using an RDE and rotational drying. Source: (d, e) Reproduced from Ref. [97] with permission of IOP Publishing. (f) Comparison of the specific activity of a Pt/C electrode in 0.1 M KOH using a proton-conducting ionomer (Nafion) and a hydroxide-conductive ionomer (AS-4, Tokuyama). Source: Reproduced from Ref. [98] with permission of The Electrochemical Society.

must be dispersed until a homogeneous mixture is observed and no sedimentation of the particles occurs. Any surface that is not homogeneous cannot be quantified for activity. The effect of a badly dispersed ink is shown in Figure 12.11a–c. The time of dispersion may vary depending on the nature of the catalyst. Pollet et al. [101] reported that for commercial Pt/C (E-Tek) electrodes, ECSA measurements were reproducible after at least 30 minutes of ultrasonication, and by probing different ultrasonication frequencies (20 kHz, 3.3–12.23 W and 40 kHz, 1.82 W) they found that ECSA values were consistently higher at higher ultrasonication frequencies, but not using higher power, and an ECSA decrease after 30 minutes of sonication due to possible detachment of the Pt particles from the support, or Pt sonochemical dissolution, as well as fomenting agglomeration by weakening the metal–support interaction.

Nowadays, plenty of electrocatalyst benchmarking studies for ink preparation are available, mostly for PGM materials, where different ink compositions have been reported with associated MA and SA, mainly for the ORR [102]. There are scarce studies regarding non-PGM materials. One example of the latter is the work reported by Osmieri et al. [103] where different inks for Fe–N–C catalysts were manufactured for a membrane–electrode assembly (MEA) configuration, noting that the presence of water (82 wt%) in combination with IPA is key for a good dispersion and interaction with the ionomer.

After depositing an aliquot of the ink on the polished electrode, the drying process is the next step. Different methods have been reported, such as leaving the electrode on a stationary position to air dry, the use of an oven to accelerate the process, using a flow of an inert gas, and rotational drying [97]. The first mentioned methods are known to produce films of variable qualities, of which several cases result on a non-homogeneous distribution of the catalysts, since the flux of evaporation is higher at the edges of the droplet rather than in the center. This generates a liquid depletion zone at the edges that is followed by a liquid flow carrying particles from the center in an outward direction to compensate this loss, creating particle aggregations on the edges, a phenomenon resulting in the "coffee ring structure" [104]. For this reason, the rotational drying method was proposed by Garsany et al. [104], based on the spin-coating technology. The method consists of mounting the electrode on an inverted RDE rotator and rotating at 700 rpm, in a matter that the air flows from the center of the film toward the outside, helping to create a homogeneous surface with the purpose of avoiding any mass transport limitations due to nonuniform catalyst layers. The benefits of the rotational drying procedures are shown in Figure 12.11d,e by contrasting with stationary drying.

Once the electrode has dried, an activation of the electrochemical surface is of importance to compare among different experiments performed on similar surfaces to assure that the measured surface is the same. This protocol usually involves the application of cycling potentials between the oxygen and hydrogen evolution reactions. For Pt, for example, the positive potential scanning aids to oxidize impurities on the surface so they can be desorbed and activate the surface for hydrogen adsorption [105].

12.3.5 Well-defined and Nanoparticulated Objects

The study of well-defined surfaces is a useful approach at understanding the link among surface structure, electrocatalytic activity, and reactivity, as it has been the base for more complex systems [106].

The achievement of well-defined surfaces is not an easy task, since first, a single crystal must be prepared, then the identification of the direction of the crystallographic plane must be done, and finally, the crystal must be cut and cleaned to expose a desired specific macroscopic surface. The latter procedure is commonly done by hydrogen flame annealing at nearly 1110 °C, as reported by Clavilier et al. [78] for the preparation of Pt electrodes.

In regard to the fuel cell reactions, the ORR has been proved to be dependent on the crystallographic structure of the electrode surface, as shown in Figure 12.12c for the ORR on different well-oriented surfaces of Pt. For this reason, well-defined surfaces are fundamental for the understanding of real-world materials.

Besides single-crystal materials, shape and surface-controlled nanoparticles have been explored, since it has been proved that it is possible to modulate the reactivity and selectivity of metal nanoparticles, e.g. Pt, by controlling its morphology, exposing different surface sites according to their shape [109]. Different synthesis methods have been studied with the aim of shaping and controlling the surface of nanoparticles for electrocatalytic reactions; e.g. Au nanoparticles have been

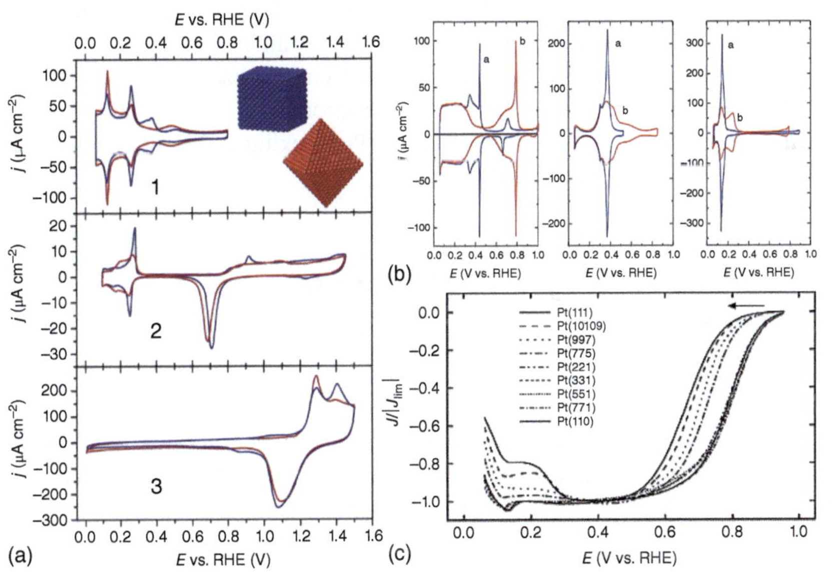

Figure 12.12 (a) Representative cyclic voltammetry curves of clean cubic and octahedral Pt(1), Pd(2), and Au(3) in 0.5 M H_2SO_4 for 1 and 2 and 0.1 M NaOH for 3. Source: From Ref. [107] with permission of Elsevier. (b) Cyclic voltammograms of the three Pt basal planes in the order (111), (100), and (110) in 0.5 M H_2SO_4 (blue lines) and 0.1 M $HClO_4$ (red lines). Source: Reproduced from Ref. [106] with permission of John Wiley & Sons. (c) ORR curves of well-oriented surfaces of Pt in 0.1 M $HClO_4$. Source: Reproduced from Ref. [108] with permission of Elsevier.

obtained in a microemulsion setup, exposing different crystal facets depending on added adsorbates, showing a higher ratio of (100) and (111) planes when iodine is present as a part of the synthesis method, and by contrary they are almost absent when sulfide is used as an adsorbate [110]. The effects on surface-sensitive reactions correlated with single-crystal studies have shown that the Au(100) electrode is the most active for the ORR in alkaline electrolytes among other single-crystalline Au surfaces, and the results are in good agreement with the ones obtained by nanoparticles when iodine is present due to the exposure of this preferred facet. The approach of exposing highly active preferred facets of different metals has been reported for electrocatalytic reactions, e.g. gold nanocubes exposing (100) facets for the ORR [111] and tetrahedral and octahedral Pt exposing (100) and (111) facets for the oxidation of formic acid and methanol [112]. It is not a simple task to electrochemically characterize well-defined nanoparticles and obtain correlations with their exposed facets, since their highly sensitive surfaces need to be as clean as possible. For this purpose, residues of the synthesis method must be removed without changing their initial surface structure. There is no universal surface-cleaning method for all nanoparticles, but some of them have been listed in recent studies [107]. Figure 12.12a shows electrochemical surface differences between clean cubic and octahedral Pt, Pd, and Au nanoparticles obtained in 0.5 M H_2SO_4 for Pt and Pd and in 0.1 M NaOH for Au.

Another surface-sensitive process is the adsorption of the so-called spectator species from the supporting electrolyte. This effect has been studied for different surfaces, Pt being the most studied one, showing the differences between anion adsorption from two acids (H_2SO_4 and $HClO_4$) on the basal planes of Pt in Figure 12.12b. It has been discussed that the adsorption of the SO_4^{-2} and HSO_4^{-1} anions has a greater effect on the surfaces of Pt blocking active sites, which is prejudicial for electrocatalytic activity [106].

12.4 Parameters Diagnostic of Electrochemical Performance

12.4.1 Surface Area

The determination of the surface area is of great importance, because it is related to the number of surface-active sites participating in the electrocatalytic reactions, useful to quantify electrocatalytic activity. It should be considered that electrode surfaces are not always smooth; they often present microirregularities that make the real surface area of the material several orders of magnitude larger than the considered macroscopic geometric area [113]. This effect is enhanced when nanostructured materials are studied as well with the high surface area supports. The ratio between the real surface area (RSA) and the geometric surface area is called roughness factor (R_F) [114]. These effects are schematized in Figure 12.13.

The real surface area of electrodes can be determined by numerous methods, including methods for quantifying the area of liquid electrodes such as Hg, Ga,

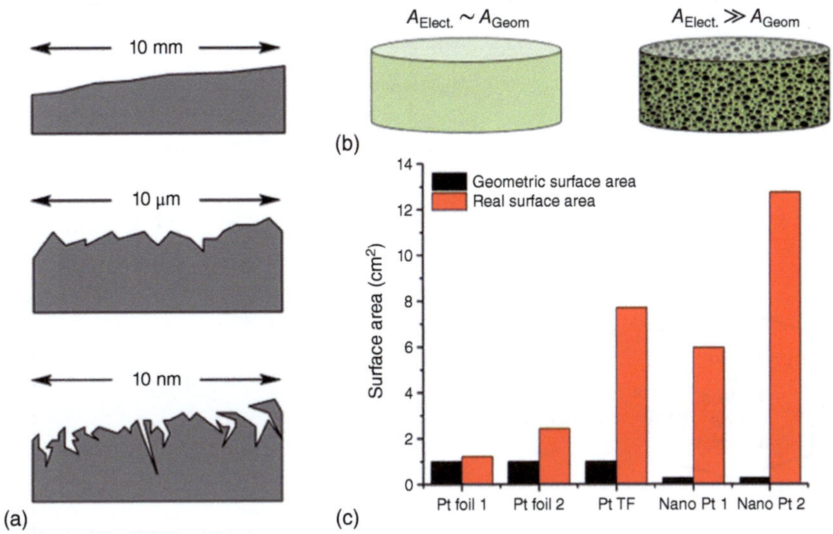

Figure 12.13 (a) Differences of the electrode surface at different scales, revealing surface topographies that influence the real surface area. Source: Reprinted from Ref. [115] with permission of American Chemical Society. (b) Representation of the surface of a conventional flat electrode vs. a nanostructured electrode. Source: Reproduced from Ref. [116] with permission of American Chemical Society. (c) Comparison between geometric surface area of different Pt electrodes: foil 1: Butane-annealed Pt foil, foil 2: Aqua regia-etched Pt foil, TF: thin film, nano 1 : 20 wt% Pt/C E-Tek at 30 μg cm^{-2}, nano 1 : 20 wt% Pt/C E-Tek at 60 μg cm^{-2}. Source: Adapted from Ref. [117].

and some liquid alloys and different methods for solid electrodes, as it has been proposed by Trasatti and Petri in 1991 [113] for the Physical Chemistry Division Commission on Electrochemistry of the International Union of Pure and Applied Chemistry (IUPAC). Also, the RSA can be expressed in terms of the Electrochemically Active Surface Area (ECSA), a parameter that is useful to compare among different electrodes. Figure 12.14 shows a guide to choose an appropriate method for RSA determination from the ones described herein.

12.4.2 Hydrogen Underpotential Deposition Integration

This method is valid for transition metals that present reversible hydrogen adsorption at more positive potentials than the hydrogen evolution potential on the surface, process known as the underpotential deposition of hydrogen, H_{UPD}. The metals that present this phenomenon are Pt, Rh, Ir, and Pd. In the case of the latter, the process becomes complicated, as the simultaneous absorption of H in the bulk of the metal occurs, making it difficult to separate both phenomena, which is why different methods have been proposed for the RSA determination of Pd electrodes [118].

The principle of this method assumes that the charge under the H_{UPD} region corresponds to the adsorption of one hydrogen atom per each metal atom of the surface, and then the charge that corresponds to a one-to-one hydrogen-metal complex

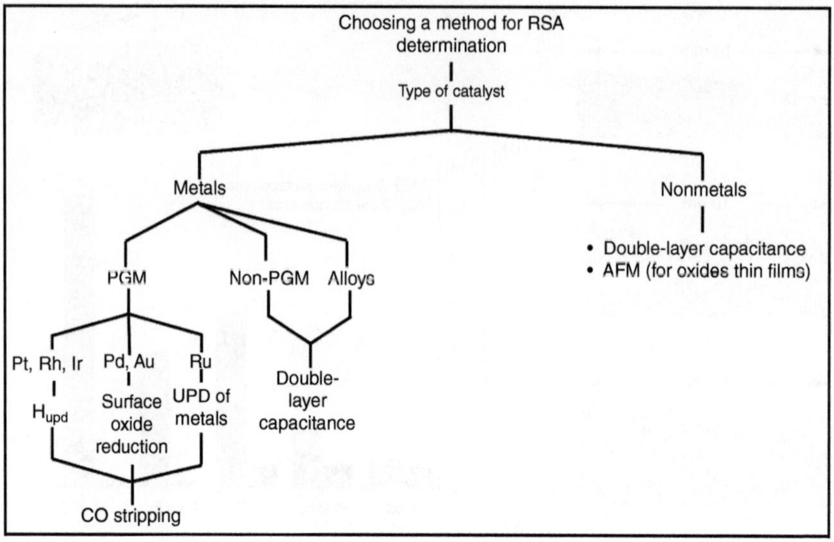

Figure 12.14 Guide for choosing a method for determining RSA of electrocatalysts according to the studied material.

per unit surface area can be calculated based on the distribution of surface metal atoms [113].

For the applicable abovementioned metals, the integration of both hydrogen adsorption and desorption regions, previously corrected by extracting the double-layer capacitance charge, corresponds to charge density of the adsorption/desorption of one hydrogen atom per each metal surface atom q_H. It is more common to use the charge associated with the hydrogen desorption region, instead of the adsorption charge, Figure 12.15. Both processes should have the same charge, but some differences have been observed, since the cathodic process could be influenced by the onset of the hydrogen evolution reaction. An average of the charge of both regions can also be reported. The RSA can be determined by Eq. (12.5):

$$\text{RSA} = \frac{q_H}{q_H^*} \tag{12.5}$$

where q_H is the measured charge from the hydrogen adsorption/desorption region, and q_H^* is the charge that corresponds to the oxidation of a monolayer of adsorbed hydrogen. q_H^* is obtained by the distribution of the atom density on the surface of the metal. This value is well defined for single-crystal structures, and for polycrystalline surface it is an average of the three lower index planes (100), (111), and (110). The accepted averaged values for polycrystalline Pt, Rh, and Ir are 210, 221, and 218 $\mu C\ cm^{-2}$, respectively [114], based on the assumption that the density of atoms on the surface is 1.31×10^{15} atoms cm^{-2} [113].

12.4.2.1 Surface Oxide Reduction

This method is used for metals that present defined oxygen adsorption regions at potentials less negative than the onset potential of the oxygen evolution and their

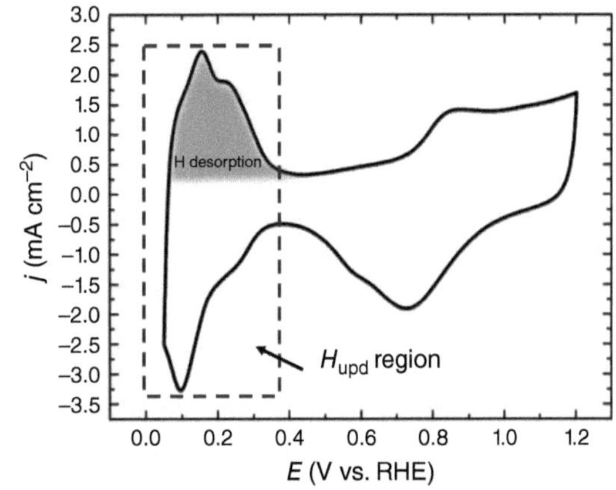

Figure 12.15 Cyclic voltammogram of a 20 wt% Pt/C (JM) catalyst in 0.1 M $HClO_4$ showing hydrogen desorption region (shaded) used for RSA calculation.

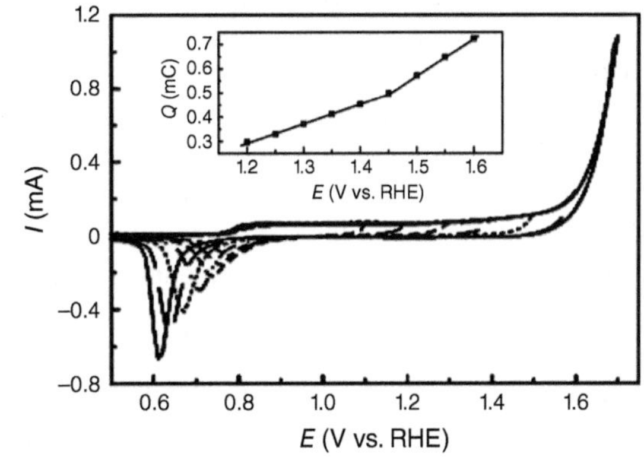

Figure 12.16 Curves obtained from a Pd sheet in 0.5 M H_2SO_4 with different upper potential limits for the oxide reduction method for RSA measurement. The insert shows the integration of the oxide reduction peak after double-layer correction, where the change of slope represents the formation of a complete PdO monolayer. Source: Adapted from Ref. [118] with permission of Chinese Physical Society.

corresponding reductions. The upper potential limit at which a monolayer of oxygen is chemisorbed with a one-to-one correspondence to the surface of the metal before oxygen evolution is taken into account [118]. This is done by varying the upper potential limit of the voltammogram and plotting the integrated cathodic charge, after the double-layer correction, against it, as shown in Figure 12.16. A straight line is obtained, and the coordinates at which it changes its slope correspond to the complete formation of a metal-O monolayer. The concentration of oxygen at the surface is considered in terms of density of atoms, assuming that every atom adsorbs a maximum of two O atoms, as it has been described for metals such as Pt, Au, and Pd [119].

It is preferable to take into account the cathodic charge rather than the charge of the anodic region because more than one process can be involved with the anodic branch, such as the electrochemical dissolution of the electrode material and oxygen evolution giving place to an overestimation of the RSA [114]. This method is commonly used for Pd- and Au-based electrodes, as the H_{UPD} method is not accurate or absent for these metals. The concentration of oxygen at the surface is considered in terms of density of atoms. The q^* values are 434 μC cm^{-2} and 390 ± 10 μC cm^{-2} for Pd and Au, respectively, with the potentials to achieve a monolayer of Pd—O and Au—O being 1.45 and 1.70 V vs. RHE, respectively [114].

12.4.2.2 CO Monolayer Oxidation (CO Stripping)

Carbon monoxide is a molecule with high affinity to some metals, including the PGM. On one hand, it is considered as a catalyst poison since it adsorbs to the surface of the metal and blocks the active sites for electrochemical reactions. On the other hand, this property can be useful for determining the RSA, since the adsorbed CO can be completely oxidized to CO_2 and be desorbed from the surface at an applied potential (CO stripping). Normally, the electrode in study is polarized at a specific potential (where no oxidation occurs) while CO is bubbled in the electrochemical cell to obtain an adsorbed monolayer of CO on the surface of the electrode; then, while the electrode is still polarized, CO bubbling is switched by N_2 to eliminate the excess CO from the solution for at least 20 minutes. The adsorbed CO monolayer is further oxidized during a cyclic voltammetry scan, and a second cyclic scan is applied to obtain a baseline for the oxidation peak to be integrated and obtain the CO oxidation charge, as shown in Figure 12.17. Then, the surface area can be obtained by normalizing the charge by the specific charge of an ideal two-electron transfer ($420\,\mu C\,cm^{-2}$) [120].

The reactions involved in this process proceed via a Langmuir–Hinshelwood mechanism, where CO and oxygen-containing molecules are the reactants, the latter are generally ascribed as OH species. Thus, the reaction rate is dependent on the surface coverage of the species and its mobility [121]. On Pt, the mechanism starts with the adsorption of the oxygenated species on the surface [122, 123]:

$$H_2O + Pt \rightleftarrows Pt - OH_{ads} + H^+ + e^- \quad (12.6)$$

And in order for the reaction to be fulfilled, CO must be adsorbed at sites adjacent to the oxygenated species. Under such conditions, a transition state involving an adsorbed carboxylic group has been reported [121] to be further oxidized to CO_2:

$$Pt - CO_{ads} + Pt - OH_{ads} \rightarrow Pt - COOH_{ads} \quad (12.7)$$

$$Pt - COOH_{ads} \rightarrow CO_2 + H^+ + e^- + 2\,Pt \quad (12.8)$$

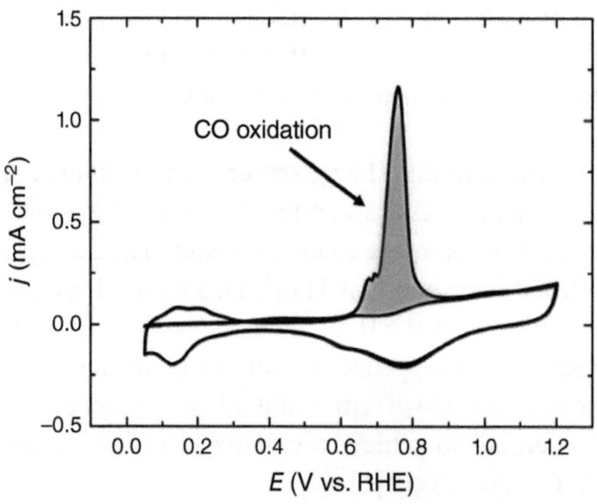

Figure 12.17 CO stripping curves of a 20 wt% Pt/C (JM) catalyst in 0.1 M $HClO_4$ showing the CO oxidation region (shaded) used for the RSA calculation.

12.4.2.3 Underpotential Deposition of Metals

This technique is applicable when the surface of the metals does not present well-defined hydrogen and oxygen desorption/adsorption regions. The underpotential deposition (UPD) of metals occurs at more positive potentials than the potential of the M^+/M redox couple, avoiding a bulk deposition of the metal phase. It has been proposed that this phenomenon occurs when there is a higher affinity of the adsorbing metal to the substrate [114].

This method was originally proposed by Franklin et al. [124] using the UPD of Pb on silver to quantify the surface area of an electrode, with the premise that the observed UPD peaks of the metals that present this phenomenon correspond to a monolayer of metal in the surface. The RSA can be obtained by Eq. (12.9):

$$\mathrm{RSA} = \frac{q_M}{q_M^*} \tag{12.9}$$

where q_{M} is the measured charge of the monolayer of deposited metal, and q_{M}^* is the charge associated with a monolayer of deposited metal. The most commonly used metals are Cu, Ag, and Pb, with the corresponding values of q_{M}^*being 420, 210, and 300 $\mu C\, cm^{-2}$, respectively [114]. An example of the RSA determination of Pt using Cu is displayed in Figure 12.18.

For this method, it is important to identify the potential where a monolayer of the metal is completed, even though in some cases it is not possible to complete a monolayer, having a multilayer deposition before the full monolayer is completed. It is reported that for Pt, the arrangement of Pt and ad-atoms can be 1 : 1 when using Cu and Ag, but for other metals such as Cd, Sn, and Pb, this ratio can vary [114].

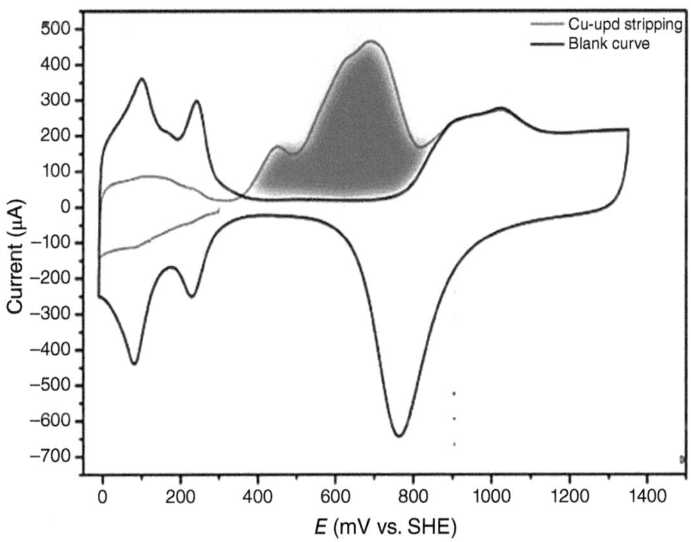

Figure 12.18 Cyclic voltammetry of polycrystalline Pt performed in 0.5 M H_2SO_4 solution and Cu stripping curve obtained in 0.4 M H_2SO_4 0.5 M containing 4.5 mM of $CuSO_4$. Source: Adapted from Ref. [114].

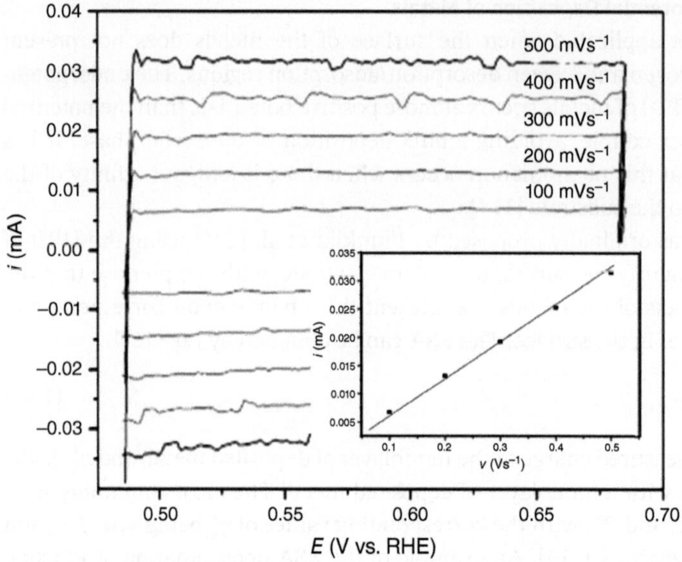

Figure 12.19 Curves obtained for the double-layer capacitance method for determining RSA of a polycrystalline Pt electrode in 0.5 M H_2SO_4. The insert shows the double-layer charging current at the potential of 0.6 V vs. RHE with respect to the scan rate. Source: Adapted from Ref. [114].

12.4.2.4 Double-layer Capacitance

Herein, the RSA can be obtained by performing cyclic voltammetry curves on the double layer capacitance region of electrodes at various scan rates, and then plotting the capacitive current with respect to the scan rate. If the only process occurring in the selected potential range for cycling is the double layer charging, a straight line is obtained from the plot with a slope that corresponds to the double layer differential capacitance, shown on Figure 12.19. Then, the RSA can be calculated with Eq. (12.10).

$$\mathrm{RSA} = \frac{C}{C_{\mathrm{ref}}} \tag{12.10}$$

where, C_{ref}, is a reference capacitance per unit area. This parameter is commonly unknown for several catalytic materials, as it depends on the electrode potential, the structure of the surface, and the electrolyte. For this, only the obtained capacitance is generally reported by several groups instead of an RSA value.

This method presents some advantages when compared to the others, even though it is less precise. There is no need for polarization at specific potentials that could change the studied surface; it is not destructive; and it gives a value that involves all of the components of the surface, not only the metal contributions.

12.4.3 Electrocatalysts Site Density

This term is self-explanatory, as it refers to the amount of active electrocatalytic sites per surface area of a catalyst; however, putting values to it is not a trivial task. It is

Table 12.3 Activity descriptors for electrocatalytic activity.

Activity descriptor	Definition	Units
Geometric activity	Current at a given potential normalized by the geometric area.	$mA\ cm_{geo}^{-2}$
Mass activity (MA)	Current per unit loading mass of electrocatalyst	$mA\ g_{electrocatalyst}^{-1}$
Electrochemically active surface area (ECSA)	RSA per gram of catalyst	$m^2\ g_{cat}^{-1}$
Turnover frequency (TOF)	Number of electrons produced or consumed per active site per second	$e\text{-}site^{-1}\ s^{-1}$
Specific activity (SA)	Current at a unit potential normalized by the real surface area of the electrocatalyst	$\mu A\ cm_{cat}^{-2}$

common to refer to the term turn-over frequency (TOF), which is a concept that has been adapted from enzymatic catalysis to heterogeneous catalysis, representing the molecules reacting per active site per unit time [125]. This, in turn, has been translated into electrocatalysis as the number of electrons produced or consumed per active site per second [126].

Thus, activity can be reported in a number of different forms, taking into account specific parameters, such as the geometric area, the mass content of the catalyst, or the overpotential to reach a certain current magnitude. Table 12.3 summarizes different activity parameters commonly used in electrocatalysis where some of the activity descriptors are related as described in Eq. (12.11) [127]:

$$MA = SA \times ECSA \tag{12.11}$$

The mentioned activity parameters present limitations; for example, geometric activity can be useful for describing the performance of a device, but it does not reflect the intrinsic activity of a catalyst. It is generally reported as the overpotential to reach a given current density (normalized by the geometric surface area of the electrode) or as the half-wave potential (common for ORR) [126].

In the case of MA, this parameter is dependent on the size of the catalyst particles, since these are related to the number of active sites. For surfaces processes, MA is not representative of the intrinsic activity of the catalyst, since it assumes that all the atoms on the electrode are active sites. However, this parameter can be useful to reflect the cost efficiency of the material, which is an important parameter at a greater scale.

Although important efforts have been made to describe the intrinsic activity of electrocatalysts, it is still a challenging task. To report TOF, one must know the obtained current density as well as the number of active sites. The first task can be done with a potentiostat and then convert the current to number of electrons; in the case of the active site determinations, it is not that simple. The method to obtain the TOF is converting the surface area to active sites using a coefficient that defines the number of surface atoms per unit surface area. This coefficient is commonly obtained by literature data, which can make it inaccurate, since the site density can change depending on the specific material and the orientation of their surfaces. TOF can be expressed as it follows [127]:

$$\mathrm{TOF} = \frac{\mathrm{SA}}{q} \times \frac{\mathrm{ECSA}\, M_{\mathrm{cat}}}{N_A} = \frac{\mathrm{MA}M_{\mathrm{cat}}}{qN_A} \tag{12.12}$$

where q is the elemental charge, M_{cat} refers to the molar mass of the catalyst, and N_A is the Avogadro number.

Other methods are reported to measure the number of active sites for specific materials; for example, in non-PGM materials, techniques such as nitrite electrochemical stripping and CO cryoadsorption have been used to estimate surface density (SD) of specific N-sites [128]. In this context, SA is a useful representation of the catalyst's intrinsic activity. This parameter is obtained by the normalization of the current density by the real surface area of the electrode, obtained by electrochemical techniques.

12.4.4 Data Evaluation (Half-Cell Reactions)

Since both the HOR and the ORR are mass transport limited, the activity of catalysts toward these reactions is commonly studied using the rotating disk electrode method (RDE), which is a technique that helps to control diffusion limiting currents [129]. It is required to extract the kinetic data of the experimental results obtained by the RDE technique before quantifying the electrocatalytic activities of the employed catalysts. This is done by applying the Koutecký–Levich equation (12.13), where it is shown that the inverse of the measured current density, j_{meas}, is composed of the sum of a kinetic contribution, j_{k}, and a mass transport diffusional contribution, j_{d}, which is defined by the Levich equation (12.14), where n represents the number of transferred electrons in the reaction, D is the diffusion coefficient of the electroactive species in the specific electrolyte (H_2 or O_2), ω is the rotation rate of the RDE in rad/s, v is the kinematic viscosity of the electrolyte, and C^* is the concentration of the electroactive species in the bulk of the solution. By substituting Eq. (12.14) into (12.13), Eq. (12.15) is obtained [76].

$$\frac{1}{j_{\mathrm{meas}}} = \frac{1}{j_k} + \frac{1}{j_d} \tag{12.13}$$

$$j_{\mathrm{d}} = 0.62nFAD^{2/3}\omega^{1/2}v^{-1/6}C^* \tag{12.14}$$

$$\frac{1}{j_{\mathrm{meas}}} = \frac{1}{j_k} + \frac{1}{0.62nFAD^{2/3}\omega^{1/2}v^{-1/6}C^*} \tag{12.15}$$

Since diffusional limiting current is dependent on the rotation rate of the electrode, a series of linear voltammetry scans are applied at different rotation speeds. The theoretical value of the limiting current can be obtained using Eq. (12.14). The experimental value should be more than 90% of this current density to assure that there are no implications of contaminants in the system as well as to validate the

Table 12.4 Koutecký–Levich oxygen parameters in common electrolyte media.

Electrolyte	D_{O_2}/ 10^5 cm^2 s^{-1}	$C_{O_2}^*$/ 10^6 mol cm^{-3}	References
0.5 M H_2SO_4	1.4	1.1	[131]
0.1 M $HClO_4$	1.93	1.26	[132]
0.1 M KOH	1.9	1.2	[133]

experimental operation [8]. Then, the Koutecký–Levich plot is constructed, with a slope of $0.62nFAD^{2/3}\nu^{-1/6}C^*$ which is further used to extract the kinetic current density with a modification of Eq. (12.13):

$$\frac{1}{j_{\mathrm{k}}} = \frac{1}{j_{\mathrm{meas}}} - \frac{1}{j_{\mathrm{d}}} = \frac{j_{\mathrm{d}} - j_{\mathrm{meas}}}{j_{\mathrm{meas}} \times j_{\mathrm{d}}} \tag{12.16}$$

For both reactions it is recommended for the mass transport correction to be restricted to the region below 80% of the diffusional limiting current to avoid errors in the analysis [74, 130].

Once the pertinent corrections have been performed, one can calculate the specific activity, SA, by dividing the kinetic current at a specific potential by the calculated ECSA, and the mass activity, MA, by dividing the kinetic current over the amount of active phase mass loading of the catalyst, as described by Eq. (12.11), which are common parameters for the ORR activity [97].

Different values have been reported for D and C^* terms of Eq. (12.14), which are dependent on the nature of the electrolyte, its concentration, pressure, and temperature of the system. Table 12.4 provides generally accepted values at 25 °C used in the literature.

12.4.5 The $E_{1/2}$ and E (j_{Pt} (5%)) Parameters

The ORR is the most hindered kinetic electrochemical process taking place in the low-temperature fuel cells, resulting in large overpotentials and a significant energy efficiency loss of ECs. Therefore, research effort is focused on the development of new electrocatalysts with low Pt-loading or non-PGM-based metals which exhibit a performance similar to that of the state-of-art Pt electrocatalysts. A comparison of the performance of electrocatalysts with that of the Pt-based reference electrocatalysts is not a simple job, especially for non-PGM electrocatalysts. In this scenario, the development of fast, easy, and accurate methods to test the performance of these materials is mandatory. As discussed above, cyclic voltammetry with thin-film rotating ring disk electrode (CV-TF-RRDE) method is one of the most powerful and effective techniques for the determination of electrochemical parameters diagnostic of the elementary events responsible for the overall ORR processes [134]. In the literature, two methods are proposed to determine the activation potential of an electrocatalyst in the ORR process by means of the CV-TF-RRDE technique. The activation potential is the overpotential at which an electrocatalyst starts performing

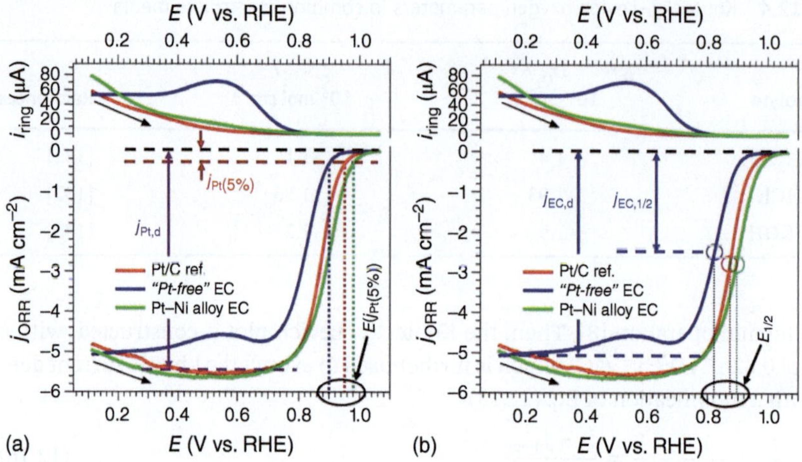

Figure 12.20 (a) Determination of $j_{Pt}(5\%)$ ($j(5\%) = \left(\frac{5}{100}\right) \cdot j$) and E ($j_{Pt}(5\%)$) values for the reference and the selected ORR electrocatalysts. (b) Measurement of the $E_{1/2}$ values for the reference and selected ORR electrocatalysts. j_{Pt_d} is the limiting current density of a reference electrocatalyst. Source: Reproduced from Ref. [137] with permission of Elsevier.

the ORR process. In principle, it is a pure kinetic parameter which is correlated with the activation energy barrier of the electrocatalyst. The first method consists in accepting that the half-wave potential ($E_{1/2}$) [135, 136] determined in the ORR profile of CV-TF-RRDE measurements is a sufficiently accurate parameter to represent the activation potential of an electrocatalyst. Unfortunately, as elsewhere demonstrated [137], this parameter appears to be correlated with the superposition of the kinetic, diffusion, and transport phenomena [138] of the electrocatalyst.

Recently, a new simple, general, and accurate method for the determination of the ORR activation potential of the electrocatalysts has been proposed [137]. This second method, which is based on a solid electrochemical analysis, is rigorous and provides an activation potential which is directly correlated with the activation energy barrier of the ORR kinetic process. Experimentally, the method consists of the following steps: (a) acquisition of accurate CV-TF-RRDE profiles of a reference Pt electrocatalyst and the different electrocatalysts under analysis both in acid and alkaline conditions (Figure 12.20a); (b) determination of the limiting current density (j_{Pt_d}) of a reference electrocatalyst in the CV-TF-RRDE profile at 0.35 V vs. RHE (Figure 12.20a); (c) evaluation of $j_{Pt}(5\%)$, i.e. the current density corresponding to 5% of j_{Pt_d} value of the reference electrocatalyst ($j_{Pt}(5\%) = \left(\frac{5}{100}\right) \cdot j_{Pt\_d}$); (d) measurement of E ($j(5\%)$), which is the activation potential of the electrocatalyst under analysis in correspondence to the $j_{Pt}(5\%)$ value of the reference electrocatalyst (Figure 12.20b).

E ($j_{Pt}(5\%)$) is a descriptor which:

1. allows to determine accurately the performance in the kinetic regime of any electrocatalyst by fixing in the ORR a suitable yield in current for a specific amount of consumed O_2. In this way, an accurate comparison of the ability of each type

Table 12.5 Comparison between $E\ (j_{Pt}(5\%))$ and $E_{1/2}$ for the "model" ORR electrocatalysts shown in Figure 12.20.

Electrocatalyst	$E\ (j_{Pt}(5\%))$/ mV vs. RHE	$E_{1/2}$/ mV vs. RHE
Pt/C ref.	955	873
"Pt-free" EC	903	822
Pt-Ni alloy EC	983	895

EC, electrocatalyst; ORR, oxygen reduction reaction.
Voltammograms measured at 20 mV s^{-1}; temperature: 298 K; support electrolyte: 0.1 M KOH; rotating speed: 1600 rpm; E_{ring}: 1.2 V vs. RHE.
Source: This result has been taken from ref. [137] with the permission from Elsevier.

of electrocatalyst (PGM or non-PGM) to convert chemical into electrochemical energy at the same conditions can be obtained. The most suitable method, to determine in the kinetic regime of this condition, is to fix a yield in current for a specific amount of consumed O_2. As demonstrated in Ref. [137], this is possible by evaluating for a reference EC the current density at 5% of its limiting value (j_{Pt_d}) measured at 0.35 V vs. RHE;
2. is directly correlated with the activation energy barrier and the symmetry factor of the electrocatalytic kinetic process under investigation; permits to monitor the status of the kinetic activation electrochemical energy barrier of each electrocatalyst as it is converting a fixed chemical into electrochemical energy, at the same conditions of a reference electrocatalyst.

Results shown in Table 12.5 demonstrate that with respect to the most commonly adopted $E_{1/2}$ parameter, $E\ (j_{Pt}(5\%))$ is a reliable quantitative and accurate descriptor to investigate and compare the performance of any type of electrocatalyst in the ORR kinetic regime [137].

12.5 Stability Tests

A common problem in the study and development of PEMFC is the degradation of the electrodes and the membrane. The origin of this situation can be related to the intrinsic physicochemical properties of the catalyst materials, the design and assembly of the device itself, or the generation of chemical radicals in the catalytic process interfering with the reactions.

An important tool that helps to evaluate the stability in the early stages of development of devices is the use of accelerated degradation tests (ADT), which are experiments that simulate realistic operation conditions of fuel cells (MEAs and fuel cell stacks), with the aim of accelerating and understanding the degradation mechanisms involved in their functions. The results of these tests could aid to make a prediction on the lifetime of the specific device assembly on its application. A variant of these experiments is their performance in *ex situ* conditions, meaning, the testing of half-cells, which is a faster and cheaper way to study the performance of the catalyst materials and their aging [139]. Currently, there are many reports that propose different settings for the ASTs in half-cells, such as variable temperatures, use of rotation, number of cycles, and different gaseous environments; all of these variants make it difficult to have a fair comparison between the results of each. For this reason, in 2001, two organizations, the United States Department of Energy (DOE) and the Fuel Cell Commercialization Conference of Japan (FCCCJ), proposed similar standard procedures for the ADT involving several aspects of importance in the use and development of fuel cells, including half-cell testing protocols that mimic the main degradation mechanisms occurring in the devices [140].

As a part of the US Department of Energy (DOE), the Hydrogen and Fuel Cell Technologies Office has a Durability Working Group that meets twice a year to develop tools for the study of the degradation mechanisms of the PEMFCs in realistic conditions, such as in the presence of impurities in the fuel and air, freezing temperatures, and with start–stop cycles, conditions known for causing stresses in the chemical and mechanical performance of devices.

Two setups of experiments are commonly proposed to resemble the operating conditions of fuel cell vehicles. These are start/stop cycles, where carbon corrosion in the catalytic support can be the main problem in the cathode due to the exposure to high potentials [141] and the load cycles, where Pt dissolution can be a problem in benchmark Pt catalysts. Both degradation mechanisms can be monitored via the loss of ECSA.

In the case of the FCCCJ, a triangular wave potential cycle from 1.0 to 1.5 V vs. RHE is proposed to test start/stop cycles with a 30-seconds initial potential hold at 500 mV s^{-1} in an N_2 saturated 0.1 M $HClO_4$ solution making measurements of the electrochemical surface area via cyclic voltammetry and to test ORR activity every 10–1000 cycles. In the case where the ECSA decreases below half of its initial value, the test should be stopped; otherwise, it should continue until 60,000 cycles. On the other hand, to test load cycles, a rectangular wave potential cycle is proposed, from 0.6 to 1.0 V vs. RHE in the same experimental conditions as the latter, making each cycle last six seconds and testing the same parameters as in the start/stop cycles but continuing up to 400,000 cycles [142]. The DOE proposed the same procedure to simulate start/stop cycles with the change in the scan rate, making it 100 mV s^{-1} and decreasing the number of cycles to 30,000. The proposed test, at lower potentials, is also very similar with the change in the cycle time to 16 seconds per cycle [143]. Some authors prefer to adhere to the tests proposed by the FCCJ to decrease experiment times as it is difficult to maintain optimal electrolyte conditions for prolonged time without changing it [139]. It is important to consider the design of the ADT or accelerated stability tests (AST) with congruence to the information that it is desired to be extracted from the results.

As it is mentioned, a part of the ADT is the evaluation of the ECSA and ORR activity in between potential cycling. This is done in N_2- and O_2- saturated electrolytes, respectively, which can make the procedure even longer, or involve the change in electrolyte setup, which can also affect catalyst performance. An optimization of this procedure has been proposed by Nagai et al. [144], making it easier to obtain intermediate cycle evaluations for catalysts that are less stable than Pt, like some Pt alloys. This is made by measuring the ECSA via H_{UPD} in a O_2-saturated electrolyte using the RDE at low rotation speeds as 50 rpm to avoid any implications in the hydrogen region.

The catalyst candidates that fulfill the requirements of the screening tests for activity and durability in half-cell mode can be further upscaled to produce larger quantities and be tested in more realistic conditions, with the final test being a full-size MEA in PEMFC short stacks having an assembly of 15–30 individual cells [6].

12.6 Data Evaluation (Auxiliary Techniques)

Nowadays, it is common for electrochemical studies to complete the results with other available techniques that can be coupled with the electrochemical cells. Herein, some of these techniques will be mentioned.

The differential electrochemical mass spectrometry (DEMS) measures the rate of formation of gaseous species at the electrochemical interface by detecting the mass/charge produced as a function of the applied electrode potential, as it was described in 1984 by Wolter and Heitbaum [145]. This technique was based on the electrochemical mass spectrometry technique, developed by Bruckenstein and Gadde in the 1970s, which consisted in coupling the electrochemical system with a mass spectrometry device. It was then possible to measure the quantity of gaseous product at time increments of about 20 seconds [146]. The DEMS technique consists of an electrochemical cell, a mass spectrometer, and a membrane inlet system, the latter being a hydrophobic porous membrane onto which the electrode was deposited, and directly attached to the ionization chamber of the mass spectrometer, so that every volatile product is directly ionized and detected after its formation. The volume of the ionization chamber is pumped continuously, producing a mass signal that is proportional to the rate of the incoming flow of the products that are formed at the electrode [145]. Several electrochemical studies have been reported using this technique, generally for the detection and quantification of reaction intermediates for the oxidation of small organics such as alcohols [147] and formic acid [148]. Interestingly, results on the corrosion of carbon supports by detecting CO_2 in half-cell [149] and MEA configurations were obtained [150].

Another interesting technique that helps to complete complex electrochemical studies is the in-situ Fourier Transform Infrared Spectroscopy (FT-IR). It is important to obtain molecular information on species adsorbed at the electrode surfaces and on the species in solution that participate in the electrochemical reactions [151]. The set-up for these kinds of experiments is designed to overcome different obstacles that could perturb the signal-to-noise-ratio, such as the adsorption of the IR-beam from the electrolyte, partial energy loss due to reflection at the surface of the electrode, and weak absorption from the adsorbates on the surface of the electrode [151]. This technique is also commonly used for identifying product intermediates from alcohol oxidation reactions and in CO_2 reduction, although interesting results have been obtained to highlight the adsorption of anions on Pt electrode surfaces in different acid electrolytes and their correlation with the reduction activity of molecular oxygen [152]. The effect of CO adsorption and tolerance for the HOR has also been studied with this technique [153].

The development of scanning electrochemical microscopy (SECM) in 1989 by Bard et al. [154] allowed the acquirement of topographic information of surfaces immersed in an electrolyte. It uses an ultramicroelectrode with a tip diameter of 10 μm or less that can be moved in the *x*, *y*, and *z* directions with reference to a substrate. The current that flows through the tip is measured to obtain the surface information. This method has been compared to scanning tunneling microscopy (STM); however, in SECM the distance from the tip to the studied surface is larger (1 nm to 10 μm) than on STM (in the order of 1 nm). Also, in STM, the tunneling current is a non-Faradaic process, while on SECM the current is the product of a redox process on the tip and the substrate controlled by electron transfer kinetics and mass transfer processes in the solution. SECM can operate in different modes, the most common being the imaging mode, in which chemical reactivity maps can be obtained as they are operating. Two types of imaging techniques can be distinguished, being the topographic (where a redox mediator is added to be oxidized or reduced at the SECM tip, to measure the response of the opposite reaction on the substrate) and the reactivity imaging mode (the generated products of the substrate reach the electrode tip to be oxidized or reduced) [155]. Other modes that are non–imaging can be used to perform electrochemical experiments in the "nanogap" range, determined by the placement of the tip relative to the substrate, avoiding mass transfer limitations, for example, to measure electron transfer rate kinetics [155].

Besides *in situ* and operando techniques, *ex situ* methods can also provide important information about the properties of the electrodes. For example, Arenz and Zana [139] have reported the use of transmission electron microscopy to evaluate electrode stability by depositing electrocatalysts on transmission electron microscopy (TEM) grids to be used as working electrodes as well as the use of identical location TEM (IL-TEM) where the same area of the electrode can be traced before and after the AST. The latter technique was also used by Arán-Ais et al. [156], to put in evidence the effect of the surface activation methodology by cyclic voltammetry on shape-defined Pt electrodes, observing drastic change in the NPs by particle migration and changes in the shape and morphology.

12.6.1 Surface Atoms vs. Bulk

Electrocatalysts are single center element or composed from different elements. As it is known, electrocatalysis is a surface-sensitive process, and a number of different strategies to generate active surfaces have been described such as the use of alloys, which is a strategy to modify electrocatalytic activity toward a specific reaction, and reduce the amount of the precious component, e.g. in Pt-M alloys. In these cases, segregation to the surface of one of the metals can occur, generating a specific element-enriched surface. This effect can be characterized by the use of surface-sensitive techniques such as low energy ion scattering spectroscopy (LEIS) and Auger electron spectroscopy (AES), as reported by Stamenkovic et al. [157, 158] in the study of Pt alloys with Co and Ni for the ORR. They found that a Pt "skin" was generated on the prepared alloys, where the outermost surface layer consisted of only Pt atoms, improving the ORR activity by the electronic modification of a Co-enriched subsurface layer.

An important effect that has been studied is the particle size, in relation with electrocatalytic activity, where it has been shown that mass and specific activities for the ORR are dependent on the particle size of Pt, as it has been mentioned previously [36]. In this regard, the SA increases by fourfold going from 1.3 to 2.2 nm, and it starts to slowly decrease as the particle size increases from there [2]. In the nanoscale, as it has been mentioned, two factors play an important role that reflect on electrochemical activity: electronic and geometric factors [37]. It is of importance to remark that, as the particle size decreases, the amount of exposed surface atoms increases, which in turn modify the atomic and electronic structure of the surface, along with the surface defects [159].

The interesting effects that have been found for electrocatalysis in the nanoscale have led to another approach at enhancing electrochemical activity in regard to the shapes and surface characteristics of the catalysts. This emerging field is the single-atom electrocatalysis, where the single-atom electrocatalysts (SAC) are defined as isolated metal atoms dispersed on supports with no interaction between them [159]. Their study is focused on how their specific characteristics such as no metal bonds present on them and unique electronic structures can enhance interactions between their supports and reactive molecules.

12.7 Conclusions

This chapter described the key aspects on the evaluation of electrocatalytic materials in half-cell configuration, taking into account the importance of unifying protocols in order to replicate the measurements and acquire valid results for comparisons, e.g. on platinum and platinum-based materials.

Some of the mentioned aspects apply equally on all systems, such as the choice for electrocatalytic cell material depending on its stability on the pH of the electrolyte, the measurement of electrode potentials, sources of contaminants in the cell that interfere with the results, and the descriptors for electrocatalytic activity. However, it is important to remark that there are no universal solutions for working in electrochemical setups, since several aspects depend on the type of material in terms of electronic properties, shapes, interaction with the support, and electrolytic medium. However, procedures on benchmark materials act as a systematic guide for further studies, e.g. the measurement of electrochemically active surface area and stability tests.

It is still a challenge to standardize protocols and avoid retesting materials on specific conditions to properly compare them; however, as time has passed and new knowledge has been gathered, experiments in electrocatalysis are pointing toward that direction.

Acknowledgments

K. Vega-Granados thanks Conacyt-Mexico for financial support through scholarship no. 744006 and from the MyDCI program of the Autonomous University of Baja California. The authors acknowledge financial support from the European Union (ERDF) "Région Nouvelle Aquitaine."

Acronyms

ADT	accelerated degradation test
AEMFC	anion-exchange membrane fuel cell
AES	Auger electron spectroscopy
AST	accelerated stability test
BE	binding energy
CCR	carbonyl chemical route
CE	counter electrode
CNF	carbon nanofibers
CNH	carbon nanohorns
CNTs	carbon nanotubes
CV-TF-RRDE	cyclic voltammetry with thin-film rotating ring disk electrode
DEMS	differential electrochemical mass spectrometry
DOE	Department of Energy of the United States
ECSA	electrochemically active surface area
FCCJ	fuel cell commercialization conference of Japan
FT-IR	Fourier transform infrared spectroscopy
GC	glassy carbon
HBE	hydrogen-binding energy
HER	hydrogen evolution
HOR	hydrogen oxidation reaction
ICP	inductively coupled plasma
IHP	inner Helmholtz plane
IL-TEM	identical location transmission electron microscopy
IPA	isopropyl alcohol
IUPAC	International Union of Pure and Applied Chemistry
JM	Johnson Matthey
LEIS	low-energy ion-scattering spectroscopy
MA	mass activity
MEA	membrane–electrode assembly
MOF	metal–organic frameworks
MWCNT	multiwalled carbon nanotube
NPs	nanoparticles
OHP	outer Helmholtz plane
ORR	oxygen reduction reaction
PD	photodeposited

PEM	polymer electrolyte membrane
PEMFC	polymer electrolyte membrane fuel cell
PGM	platinum group metals
RDE	rotating disk electrode
RE	reference electrode
R_F	roughness factor
RHE	reversible hydrogen electrode
rpm	revolutions per minute
RSA	real surface area
SA	specific activity
SAC	single-atom catalysts
SD	surface density
SECM	scanning electrochemical microscopy
SHE	standard hydrogen electrode
SMSI	strong metal–support interaction
STM	scanning tunneling microscopy
TEM	transmission electron microscopy
TFE	tetrafluoroethylene
TOF	turnover frequency
UHP	ultrahigh purity
WE	working electrode
XPS	X-ray photoelectron spectroscopy
ZIF	zeolitic imidazole framework

Symbols

$a(H^+)$	activity of H^+ (aq)
C	capacitance
C^*	concentration of the electroactive species in the bulk of the solution
C_{ref}	reference capacitance per unit area
D	diffusion coefficient of the electroactive species
$E^\circ{}_{RE}$	potential, in standard condition of the used RE
E_{RHE}	potential of RHE
E	electrode potential
E°	standard potential of the H^+(aq)/ H_2(g) couple
$E_{1/2}$	half-wave potential
$E\ (j_{Pt}(\mathbf{5}\ \%))$	potential measured at $\boldsymbol{j_{Pt}}(5\%)$
$E_{measured}$	measured potential in the used RE scale
E_{ring}	potential of the ring
$f(H_2)$	fugacity of H_2
H_{UPD}	underpotential deposition of hydrogen
j_{meas}	measured current density
j_k	kinetic current density
j_d	mass transport diffusional contribution

$\boldsymbol{j}_{\mathbf{Pt}}(5\%)$	current density corresponding to 5% of $j_{\mathrm{Pt\_d}}$ value
$j_{\mathrm{Pt\_d}}$	limiting current density of the reference electrocatalyst
M_{cat}	molar mass of the catalyst
n	number of transferred electrons
N_{A}	Avogadro number
$p°$	standard pressure
Q	elemental charge
q_{H}	to charge density of the adsorption/desorption of one hydrogen atom per each metal surface atom
q_{H}^{*}	obtained by the distribution of the atom density on the surface of the metal
q_{M}	measured charge of the monolayer of deposited metal
q_{M}^{*}	charge associated with a monolayer of deposited metal
ν	kinematic viscosity of the electrolyte
ω	rotation rate of the RDE

References

1 Alonso-Vante, N. (2010). Platinum and non-platinum nanomaterials for the molecular oxygen reduction reaction. *ChemPhysChem* 11 (13): 2732–2744. https://doi.org/10.1002/cphc.200900817.

2 Shao, M., Peles, A., and Shoemaker, K. (2011). Electrocatalysis on platinum nanoparticles: particle size effect on oxygen reduction reaction activity. *Nano Letters* 11 (9): 3714–3719. https://doi.org/10.1021/nl2017459.

3 Yang, H., Vogel, W., Lamy, C., and Alonso-Vante, N. (2004). Structure and electrocatalytic activity of carbon-supported Pt–Ni alloy nanoparticles toward the oxygen reduction reaction. *The Journal of Physical Chemistry B* 108 (30): 11024–11034. https://doi.org/10.1021/jp049034+.

4 Timperman, L., Lewera, A., Vogel, W., and Alonso-Vante, N. (2010). Nanostructured platinum becomes alloyed at oxide-composite substrate. *Electrochemistry Communications* 12 (12): 1772–1775. https://doi.org/10.1016/j.elecom.2010.10.020.

5 Masa, J., Andronescu, C., and Schuhmann, W. (2020). Electrocatalysis as the nexus for sustainable renewable energy: the gordian knot of activity, stability, and selectivity. *Angewandte Chemie International Edition* 59 (36): 15298–15312. https://doi.org/10.1002/anie.202007672.

6 Takahashi, I. and Kocha, S.S. (2010). Examination of the activity and durability of PEMFC catalysts in liquid electrolytes. *Journal of Power Sources* 195 (19): 6312–6322. https://doi.org/10.1016/j.jpowsour.2010.04.052.

7 Čolić, V., Tymoczko, J., Maljusch, A. et al. (2015). Experimental aspects in benchmarking of the electrocatalytic activity. *ChemElectroChem* 2 (1): 143–149. https://doi.org/10.1002/celc.201402295.

8 Wei, C., Rao, R.R., Peng, J. et al. (2019). Recommended practices and benchmark activity for hydrogen and oxygen electrocatalysis in water splitting and

fuel cells. *Advanced Materials* 31 (31): 1806296. https://doi.org/10.1002/adma.201806296.

9 Kocha, S.S., Shinozaki, K., Zack, J.W. et al. (2017). Best practices and testing protocols for benchmarking ORR activities of fuel cell electrocatalysts using rotating disk electrode. *Electrocatalysis* 8 (4): 366–374. https://doi.org/10.1007/s12678-017-0378-6.

10 Kunz, H.R. and Gruver, G.A. (1975). The catalytic activity of platinum supported on carbon for electrochemical oxygen reduction in phosphoric acid. *Journal of the Electrochemical Society* 122 (10): 1279–1287. https://doi.org/10.1149/1.2134000.

11 Appleby, A.J. (1970). Oxygen reduction on active platinum in 85% orthophosphoric acid. *Journal of the Electrochemical Society* 117 (5): 641. https://doi.org/10.1149/1.2407593.

12 Appleby, A.J. (1970). Evolution and reduction of oxygen on oxidized platinum in 85% orthophosphoric acid. *Journal of Electroanalytical Chemistry and Interfacial Electrochemistry* 24 (1): 97–117. https://doi.org/10.1016/S0022-0728(70)80011-0.

13 Bagotzky, V.S., Shumilova, N.A., and Khrushcheva, E.I. (1976). Electrochemical oxygen reduction on oxide catalysts. *Electrochimica Acta* 21 (11): 919–924. https://doi.org/10.1016/0013-4686(76)85066-9.

14 Hsueh, K.L., Gonzalez, E.R., Srinivasan, S., and Chin, D.T. (1984). Effects of phosphoric acid concentration on oxygen reduction kinetics at platinum. *Journal of the Electrochemical Society* 131 (4): 823–828. https://doi.org/10.1149/1.2115707.

15 Paffett, M.T., Beery, J.G., and Gottesfeld, S. (1988). Oxygen reduction at $Pt_{0.65}Cr_{0.35}$, $Pt_{0.2}Cr_{0.8}$ and roughened platinum. *Journal of the Electrochemical Society* 135 (6): 1431–1436. https://doi.org/10.1149/1.2096016.

16 Lawson, D.R., Whiteley, L.D., Martin, C.R. et al. (1988). Oxygen reduction at nafion film-coated platinum electrodes: transport and kinetics. *Journal of the Electrochemical Society* 135 (9): 2247–2253. https://doi.org/10.1149/1.2096247.

17 Couturier, G., Kirk, D.W., Hyde, P.J., and Srinivasan, S. (1987). Electrocatalysis of the hydrogen oxidation and of the oxygen reduction reactions of Pt and some alloys in alkaline medium. *Electrochimica Acta* 32 (7): 995–1005. https://doi.org/10.1016/0013-4686(87)90024-7.

18 Peuckert, M., Yoneda, T., Betta, R.A.D., and Boudart, M. (1986). Oxygen reduction on small supported platinum particles. *Journal of the Electrochemical Society* 133 (5): 944–947. https://doi.org/10.1149/1.2108769.

19 Alonso-Vante, N., Jaegermann, W., Tributsch, H. et al. (1987). Electrocatalysis of oxygen reduction by chalcogenides containing mixed transition metal clusters. *Journal of the American Chemical Society* 109 (11): 3251–3257. https://doi.org/10.1021/ja00245a013.

20 Markovic, N.M., Gasteiger, H.A., and Ross, P.N. (1995). Oxygen reduction on platinum low-index single-crystal surfaces in sulfuric acid solution: rotating ring-Pt(hkl) disk studies. *The Journal of Physical Chemistry* 99 (11): 3411–3415. https://doi.org/10.1021/j100011a001.

21 Nicole, J., Tsiplakides, D., Pliangos, C. et al. (2001). Electrochemical promotion and metal–support interactions. *Journal of Catalysis* 204 (1): 23–34. https://doi.org/10.1006/jcat.2001.3360.

22 ElGuindy, M. (2001). Platinum group metals: alloying, properties, and applications. In: *Encyclopedia of Materials: Science and Technology* (ed. K.H.J. Buschow, R.W. Cahn, M.C. Flemings, et al.), 7117–7121. Oxford: Elsevier https://doi.org/10.1016/B0-08-043152-6/01260-2.

23 Gasteiger, H.A., Panels, J.E., and Yan, S.G. (2004). Dependence of PEM fuel cell performance on catalyst loading. *Journal of Power Sources* 127 (1–2): 162–171.

24 Nørskov, J.K., Bligaard, T., Logadottir, A. et al. (2005). Trends in the exchange current for hydrogen evolution. *Journal of the Electrochemical Society* 152 (3): J23–J26.

25 Skúlason, E., Tripkovic, V., Björketun, M.E. et al. (2010). Modeling the electrochemical hydrogen oxidation and evolution reactions on the basis of density functional theory calculations. *The Journal of Physical Chemistry C* 114 (42): 18182–18197. https://doi.org/10.1021/jp1048887.

26 Rossmeisl, J., Karlberg, G.S., Jaramillo, T., and Nørskov, J.K. (2009). Steady state oxygen reduction and cyclic voltammetry. *Faraday Discussions* 140: 337–346.

27 Nørskov, J.K., Rossmeisl, J., Logadottir, A. et al. (2004). Origin of the overpotential for oxygen reduction at a fuel-cell cathode. *Journal of Physical Chemistry B* 108 (46): 17886–17892.

28 Sabatier, P. (1911). Hydrogénations et déshydrogénations par catalyse. *Berichte der Deutschen Chemischen Gesellschaft* 44 (3): 1984–2001. https://doi.org/10.1002/cber.19110440303.

29 Wang, X.X., Swihart, M.T., and Wu, G. (2019). Achievements, challenges and perspectives on cathode catalysts in proton exchange membrane fuel cells for transportation. *Nature Catalysis* 2 (7): 578–589. https://doi.org/10.1038/s41929-019-0304-9.

30 United States Department of Energy (2017). Multi-Year Research, Development, and Demonstration Plan Planned program activities for 2011-2020. https://www.energy.gov/sites/prod/files/2017/05/f34/fcto_myrdd_fuel_cells.pdf (accessed 5 June 2023).

31 Du, L., Prabhakaran, V., Xie, X. et al. (2020). Low-PGM and PGM-free catalysts for proton exchange membrane fuel cells: stability challenges and material solutions. *Advanced Materials* 33 (6): e1908232. https://doi.org/10.1002/adma.201908232.

32 Zhong, H., Estudillo-Wong, L.A., Gao, Y. et al. (2020). Cobalt-based multi-component oxygen reduction reaction electrocatalysts generated by melamine thermal pyrolysis with high performance in an alkaline hydrogen/oxygen microfuel cell. *ACS Applied Materials & Interfaces* 12 (19): 21605–21615. https://doi.org/10.1021/acsami.0c02884.

33 Feng, Y. and Alonso-Vante, N. (2008). Nonprecious metal catalysts for the molecular oxygen-reduction reaction. *Physica Status Solidi B* 245 (9): 1792–1806. https://doi.org/10.1002/pssb.200879537.

34 Li, J.-C., Cheng, M., Li, T. et al. (2019). Carbon nanotube-linked hollow carbon nanospheres doped with iron and nitrogen as single-atom catalysts for the oxygen reduction reaction in acidic solutions. *Journal of Materials Chemistry A* 7 (24): 14478–14482. https://doi.org/10.1039/C9TA00508K.

35 Zhuang, Z., Giles, S.A., Zheng, J. et al. (2016). Nickel supported on nitrogen-doped carbon nanotubes as hydrogen oxidation reaction catalyst in alkaline electrolyte. *Nature Communications* 7: 10141. https://doi.org/10.1038/ncomms10141.

36 Mukerjee, S. and McBreen, J. (1998). Effect of particle size on the electrocatalysis by carbon-supported Pt electrocatalysts: an in situ XAS investigation. *Journal of Electroanalytical Chemistry* 448 (2): 163–171. https://doi.org/10.1016/S0022-0728(97)00018-1.

37 Mayrhofer, K.J.J., Blizanac, B.B., Arenz, M. et al. (2005). The impact of geometric and surface electronic properties of Pt-catalysts on the particle size effect in electrocatalysis. *The Journal of Physical Chemistry. B* 109 (30): 14433–14440.

38 Kinoshita, K. (1990). Particle size effects for oxygen reduction on highly dispersed platinum in acid electrolytes. *Journal of the Electrochemical Society* 137 (3): 845–848. https://doi.org/10.1149/1.2086566.

39 Stamenkovic, V.R., Mun, B.S., Arenz, M. et al. (2007). Trends in electrocatalysis on extended and nanoscale Pt-bimetallic alloy surfaces. *Nature Materials* 6 (3): 241–247. https://doi.org/10.1038/nmat1840.

40 Wang, Y.-J., Zhao, N., Fang, B. et al. (2015). Carbon-supported Pt-based alloy electrocatalysts for the oxygen reduction reaction in polymer electrolyte membrane fuel cells: particle size, shape, and composition manipulation and their impact to activity. *Chemical Reviews* 115 (9): 3433–3467. https://doi.org/10.1021/cr500519c.

41 Mukerjee, S., Srinivasan, S., Soriaga, M.P., and McBreen, J. (1995). Role of structural and electronic properties of Pt and Pt alloys on electrocatalysis of oxygen reduction: an in situ XANES and EXAFS investigation. *Journal of the Electrochemical Society* 142 (5): 1409–1422. https://doi.org/10.1149/1.2048590.

42 Toda, T., Igarashi, H., and Watanabe, M. (1998). Role of electronic property of Pt and Pt alloys on electrocatalytic reduction of oxygen. *Journal of the Electrochemical Society* 145 (12): 4185–4188. https://doi.org/10.1149/1.1838934.

43 Toda, T., Igarashi, H., Uchida, H., and Watanabe, M. (1999). Enhancement of the electroreduction of oxygen on Pt alloys with Fe, Ni, and Co. *Journal of the Electrochemical Society* 146 (10): 3750–3756. https://doi.org/10.1149/1.1392544.

44 Strickland, K., Miner, E., Jia, Q. et al. (2015). Highly active oxygen reduction non-platinum group metal electrocatalyst without direct metal–nitrogen coordination. *Nature Communications* 6 (1): 7343. https://doi.org/10.1038/ncomms8343.

45 Li, J., Ghoshal, S., Liang, W. et al. (2016). Structural and mechanistic basis for the high activity of Fe–N–C catalysts toward oxygen reduction. *Energy & Environmental Science* 9 (7): 2418–2432. https://doi.org/10.1039/C6EE01160H.

46 He, Y., Tan, Q., Lu, L. et al. (2019). Metal-nitrogen-carbon catalysts for oxygen reduction in PEM fuel cells: self-template synthesis approach to enhancing catalytic activity and stability. *Electrochemical Energy Reviews* 2 (2): 231–251. https://doi.org/10.1007/s41918-019-00031-9.

47 Campos-Roldán, C.A. and Alonso-Vante, N. (2019). The hydrogen oxidation reaction in alkaline medium : an overview. *Electrochemical Energy Reviews* 2 (2): 312–331. https://doi.org/10.1007/s41918-019-00034-6.

48 Xing, W. (2008). Catalyst layer composition optimization. In: *PEM Fuel Cell Electrocatalysts and Catalysts Layers* (ed. J. Zhang), 1003–1040. Springer.

49 Anwar, M.T., Yan, X., Asghar, M.R. et al. (2019). Recent advances in hybrid support material for Pt-based electrocatalysts of proton exchange membrane fuel cells. *International Journal of Energy Research* 43 (7): 2694–2721. https://doi.org/10.1002/er.4322.

50 Tauster, S.J., Fung, S.C., and Garten, R.L. (1978). Strong metal-support interactions. Group 8 noble metals supported on titanium dioxide. *Journal of the American Chemical Society* 100 (1): 170–175. https://doi.org/10.1021/ja00469a029.

51 Tauster, S.J. and Fung, S.C. (1978). Strong metal-support interactions: occurrence among the binary oxides of groups IIA–VB. *Journal of Catalysis* 55 (1): 29–35. https://doi.org/10.1016/0021-9517(78)90182-3.

52 Campos-Roldán, C.A., González-Huerta, R.G., and Alonso-Vante, N. (2018). The oxophilic and electronic effects on anchored platinum nanoparticles on sp^2 carbon sites: the hydrogen evolution and oxidation reactions in alkaline medium. *Electrochimica Acta* 283: 1829–1834. https://doi.org/10.1016/j.electacta.2018.07.104.

53 Campos-Roldan, C.A., Ramos-Sanchez, G., Gonzalez-Huerta, R.G. et al. (2016). Influence of sp^3-sp^2 carbon nanodomains on metal/support interaction, catalyst durability, and catalytic activity for the oxygen reduction reaction. *ACS Applied Materials & Interfaces* 8 (35): 23260–23269. https://doi.org/10.1021/acsami.6b06886.

54 Estudillo-Wong, L.A., Luo, Y., Díaz-Real, J.A., and Alonso-Vante, N. (2016). Enhanced oxygen reduction reaction stability on platinum nanoparticles photo-deposited onto oxide-carbon composites. *Applied Catalysis B: Environmental* 187: 291–300. https://doi.org/10.1016/j.apcatb.2016.01.030.

55 Timperman, L. and Alonso-Vante, N. (2011). Oxide substrate effect toward electrocatalytic enhancement of platinum and ruthenium–selenium catalysts. *Electrocatalysis* 2 (3): 181–191. https://doi.org/10.1007/s12678-011-0052-3.

56 Lewera, A., Timperman, L., Roguska, A., and Alonso-Vante, N. (2011). Metal–support interactions between nanosized Pt and metal oxides (WO_3 and TiO_2) studied using X-ray photoelectron spectroscopy. *The Journal of Physical Chemistry C* 115 (41): 20153–20159. https://doi.org/10.1021/jp2068446.

57 Ma, J., Habrioux, A., Morais, C. et al. (2013). Spectroelectrochemical probing of the strong interaction between platinum nanoparticles and graphitic domains of carbon. *ACS Catalysis* 3 (9): 1940–1950. https://doi.org/10.1021/cs4003222.

58 Hussain, S., Kongi, N., Erikson, H. et al. (2019). Platinum nanoparticles photo-deposited on SnO_2-C composites: an active and durable electrocatalyst for the oxygen reduction reaction. *Electrochimica Acta* 316: 162–172. https://doi.org/10.1016/j.electacta.2019.05.104.

59 Molchanov, V.S. and Prikhidko, N.E. (1957). Corrosion of silicate glasses by alkaline solutions. *Bulletin of the Academy of Sciences of the USSR, Division of chemical science* 6 (10): 1179–1184. https://doi.org/10.1007/BF01167384.

60 Mayrhofer, K.J.J., Wiberg, G.K.H., and Arenz, M. (2008). Impact of glass corrosion on the electrocatalysis on Pt electrodes in alkaline electrolyte. *Journal of the Electrochemical Society* 155 (1): P1–P5. https://doi.org/10.1149/1.2800752.

61 Analysis of the impact of individual glass constituents on electrocatalysis on Pt electrodes in alkaline solutions (2008). *Journal of the Electrochemical Society* (155). https://doi.org/10.1149/1.2904882

62 Campos-Roldán, C.A., González-Huerta, R.G., and Alonso-Vante, N. (2018). Experimental protocol for HOR and ORR in alkaline electrochemical measurements. *Journal of the Electrochemical Society* 165 (15): J3001–J3007. https://doi.org/10.1149/2.0011815jes].

63 Garsany, Y., Baturina, O.A., Swider-lyons, K.E., and Kocha, S.S. (2010). Experimental methods for quantifying the activity of platinum electrocatalysts for the oxygen. *Analytical Chemistry* 82 (15): 6321–6328.

64 Napporn, T.W., Holade, Y., Kokoh, B. et al. (2018). Chapter 9 – electrochemical measurement methods and characterization on the cell level. In: *Fuel Cells and Hydrogen* (ed. V. Hacker and S. Mitsushima), 175–214. Elsevier https://doi.org/10.1016/B978-0-12-811459-9.00009-8.

65 Mayrhofer, K., Ashton, S., Kreuzer, J., and Arenz, M. (2009). An electrochemical cell configuration incorporating an ion conducting membrane separator between reference and working electrode. *International Journal of Electrochemical Science* 4: 1–9.

66 Nickell, R.A., Zhu, W.H., Payne, R.U. et al. (2006). Hg/HgO electrode and hydrogen evolution potentials in aqueous sodium hydroxide. *Journal of Power Sources* 161 (2): 1217–1224. https://doi.org/10.1016/j.jpowsour.2006.05.028.

67 McNicol, B.D., Miles, R., and Short, R.T. (1983). The influence of electrolyte purity on electrochemical measurements using platinum electrodes. *Electrochimica Acta* 28 (9): 1285–1290. https://doi.org/10.1016/0013-4686(83)85018-X.

68 Subbaraman, R., Danilovic, N., Lopes, P.P. et al. (2012). Origin of anomalous activities for electrocatalysts in alkaline electrolytes. *The Journal of Physical Chemistry C* 116 (42): 22231–22237. https://doi.org/10.1021/jp3075783.

69 Strmcnik, D., Li, D., Lopes, P.P. et al. (2015). When small is big: the role of impurities in electrocatalysis. *Topics in Catalysis* 58 (18): 1174–1180. https://doi.org/10.1007/s11244-015-0492-8.

70 McCafferty, E. (2010). Thermodyamics of corrosion: Pourbaix diagrams. In: *Introduction to Corrosion Science* (ed. E. McCafferty), 95–117. New York, NY: Springer.

71 Firouzjaie, H.A. and Mustain, W.E. (2020). Catalytic advantages, challenges, and priorities in alkaline membrane fuel cells. *ACS Catalysis* 10 (1): 225–234. https://doi.org/10.1021/acscatal.9b03892.

72 Ramaswamy, N. and Mukerjee, S. (2012). Fundamental mechanistic understanding of electrocatalysis of oxygen reduction on Pt and non-Pt surfaces: acid versus alkaline media. *Advances in Physical Chemistry* 2012: 491604. https://doi.org/10.1155/2012/491604.

73 Ramaswamy, N., Ghoshal, S., Bates, M.K. et al. (2017). Hydrogen oxidation reaction in alkaline media: relationship between electrocatalysis and electrochemical double-layer structure. *Nano Energy* 41: 765–771. https://doi.org/10.1016/j.nanoen.2017.07.053.

74 Sheng, W., Gasteiger, H.A., and Shao-Horn, Y. (2010). Hydrogen oxidation and evolution reaction kinetics on platinum: acid vs alkaline electrolytes. *Journal of the Electrochemical Society* 157 (11): B1529. https://doi.org/10.1149/1.3483106.

75 Sheng, W., Zhuang, Z., Gao, M. et al. (2015). Correlating hydrogen oxidation and evolution activity on platinum at different pH with measured hydrogen binding energy. *Nature Communications* 6 (1): 5848. https://doi.org/10.1038/ncomms6848.

76 Bard, A.J. and Faulkner, L.R. (2001). *Electrochemical Methods*. Wiley.

77 Meites, L., Zuman, P., and Nurnberg, H.W. (1985). Recommended terms, symbols, and definitions for electroanalytical chemistry (recommendations 1985). *Pure and Applied Chemistry* 57 (10): 1491–1505. https://doi.org/10.1351/pac198557101491.

78 Clavilier, J., Faure, R., Guinet, G., and Durand, R. (1980). Preparation of monocrystalline Pt microelectrodes and electrochemical study of the plane surfaces cut in the direction of the {111} and {110} planes. *Journal of Electroanalytical Chemistry and Interfacial Electrochemistry* 107 (1): 205–209. https://doi.org/10.1016/S0022-0728(79)80022-4.

79 Gamboa-Aldeco, M.E., Herrero, E., Zelenay, P.S., and Wieckowski, A. (1993). Adsorption of bisulfate anion on a Pt(100) electrode: a comparison with Pt(111) and Pt(poly). *Journal of Electroanalytical Chemistry* 348 (1): 451–457. https://doi.org/10.1016/0022-0728(93)80151-7.

80 Wang, J.X., Markovic, N.M., and Adzic, R.R. (2004). Kinetic analysis of oxygen reduction on Pt(111) in acid solutions: intrinsic kinetic parameters and anion adsorption effects. *The Journal of Physical Chemistry B* 108 (13): 4127–4133. https://doi.org/10.1021/jp037593v.

81 Mostany, J., Martínez, P., Climent, V. et al. (2009). Thermodynamic studies of phosphate adsorption on Pt(111) electrode surfaces in perchloric acid solutions. *Electrochimica Acta* 54 (24): 5836–5843. https://doi.org/10.1016/j.electacta.2009.05.040.

82 Arruda, T.M., Shyam, B., Ziegelbauer, J.M. et al. (2008). Investigation into the competitive and site-specific nature of anion adsorption on Pt using in situ X-ray absorption spectroscopy. *The Journal of Physical Chemistry C* 112 (46): 18087–18097. https://doi.org/10.1021/jp8067359.

83 Strmcnik, D., Kodama, K., van der Vliet, D. et al. (2009). The role of non-covalent interactions in electrocatalytic fuel-cell reactions on platinum. *Nature Chemistry* 1 (6): 466–472. https://doi.org/10.1038/nchem.330.

84 He, Q., Yang, X., Chen, W. et al. (2010). Influence of phosphate anion adsorption on the kinetics of oxygen electroreduction on low index Pt(hkl) single crystals. *Physical Chemistry Chemical Physics* 12 (39): 12544–12555. https://doi.org/10.1039/C0CP00433B.

85 Stamenkovic, V.M., Markovic, N., and Ross, P.N. (2001). Structure-relationships in electrocatalysis: oxygen reduction and hydrogen oxidation reactions on Pt(111) and Pt(100) in solutions containing chloride ions. *Journal of Electroanalytical Chemistry* 500 (1): 44–51. https://doi.org/10.1016/S0022-0728(00)00352-1.

86 Holst-Olesen, K., Reda, M., Hansen, H.A. et al. (2018). Enhanced oxygen reduction activity by selective anion adsorption on non-precious-metal catalysts. *ACS Catalysis* 8 (8): 7104–7112. https://doi.org/10.1021/acscatal.8b01584.

87 Strickland, K., Pavlicek, R., Miner, E. et al. (2018). Anion resistant oxygen reduction electrocatalyst in phosphoric acid fuel cell. *ACS Catalysis* 8 (5): 3833–3843. https://doi.org/10.1021/acscatal.8b00390.

88 Danilovic, N., Subbaraman, R., Strmcnik, D. et al. (2012). The effect of noncovalent interactions on the HOR, ORR, and HER on Ru, Ir, and $Ru_{0.50}Ir_{0.50}$ metal surfaces in alkaline environments. *Electrocatalysis* 3 (3): 221–229. https://doi.org/10.1007/s12678-012-0100-7.

89 Angelucci, C.A., Varela, H., Tremiliosi-Filho, G., and Gomes, J.F. (2013). The significance of non-covalent interactions on the electro-oxidation of alcohols on Pt and au in alkaline media. *Electrochemistry Communications* 33: 10–13. https://doi.org/10.1016/j.elecom.2013.03.039.

90 Kahlert, H. (2010). Reference electrodes. In: *Electroanalytical Methods: Guide to Experiments and Applications* (ed. F. Scholz, A.M. Bond, R.G. Compton, et al.), 291–308. Berlin, Heidelberg: Springer Berlin Heidelberg https://doi.org/10.1007/978-3-642-02915-8_15.

91 Jerkiewicz, G. (2020). Standard and reversible hydrogen electrodes: theory, design, operation, and applications. *ACS Catalysis* 10 (15): 8409–8417. https://doi.org/10.1021/acscatal.0c02046.

92 Kamau, G.N. (1988). Surface preparation of glassy carbon electrodes. *Analytica Chimica Acta* 207: 1–16. https://doi.org/10.1016/S0003-2670(00)80777-1.

93 Weisshaar, D.E. and Kuwana, T. (1985). Considerations for polishing glassy carbon to a scratch-free finish. *Analytical Chemistry* 57 (1): 378–379. https://doi.org/10.1021/ac00279a090.

94 Hu, I.-F., Karweik, D.H., and Kuwana, T. (1985). Activation and deactivation of glassy carbon electrodes. *Journal of Electroanalytical Chemistry and Interfacial Electrochemistry* 188 (1): 59–72. https://doi.org/10.1016/S0022-0728(85)80050-4.

95 Kiema, G.K., Aktay, M., and McDermott, M.T. (2003). Preparation of reproducible glassy carbon electrodes by removal of polishing impurities. *Journal of Electroanalytical Chemistry* 540: 7–15. https://doi.org/10.1016/S0022-0728(02)01264-0.

96 Yandrasits, M. and Hamrock, S. (2012). 10.33 – poly(Perfluorosulfonic acid) membranes. In: *Polymer Science: A Comprehensive Reference* (ed.

K. Matyjaszewski and M. Möller), 10: 601–619. Amsterdam: Elsevier https://doi.org/10.1016/B978-0-444-53349-4.00283-1.

97 Garsany, Y., Ge, J., St-Pierre, J. et al. (2014). Analytical procedure for accurate comparison of rotating disk electrode results for the oxygen reduction activity of Pt/C. *Journal of the Electrochemical Society* 161 (5): F628–F640. https://doi.org/10.1149/2.036405jes.

98 Jervis, R., Mansor, N., Sobrido, A.J. et al. (2017). The importance of using alkaline ionomer binders for screening electrocatalysts in alkaline electrolyte. *Journal of the Electrochemical Society* 164 (14): F1551–F1555. https://doi.org/10.1149/2.0441714jes.

99 Moilanen, D.E., Spry, D.B., and Fayer, M.D. (2008). Water dynamics and proton transfer in nafion fuel cell membranes. *Langmuir* 24 (8): 3690–3698. https://doi.org/10.1021/la703358a.

100 Shinozaki, K., Zack, J.W. et al. (2015). Oxygen reduction reaction measurements on platinum electrocatalysts utilizing rotating disk electrode technique: II. Influence of ink formulation, catalyst layer uniformity and thickness oxygen reduction reaction measurements on platinum II influence. *Journal of the Electrochemical Society* 162 (12): F1384–F1396. https://doi.org/10.1149/2.0551512jes.

101 Pollet, B.G. and Goh, J.T.E. (2014). The importance of ultrasonic parameters in the preparation of fuel cell catalyst inks. *Electrochimica Acta* 128: 292–303. https://doi.org/10.1016/j.electacta.2013.09.160.

102 Kocha, S.S., Zack, J.W., Alia, S.M. et al. (2013). Influence of ink composition on the electrochemical properties of Pt/C electrocatalysts. *ECS Transactions* 50 (2): 1475–1485. https://doi.org/10.1149/05002.1475ecst.

103 Osmieri, L., Wang, G., Cetinbas, F.C. et al. (2020). Utilizing ink composition to tune bulk-electrode gas transport, performance, and operational robustness for a Fe–N–C catalyst in polymer electrolyte fuel cell. *Nano Energy* 75: 104943. https://doi.org/10.1016/j.nanoen.2020.104943.

104 Garsany, Y., Singer, I.L., and Swider-Lyons, K.E. (2011). Impact of film drying procedures on RDE characterization of Pt/VC electrocatalysts. *Journal of Electroanalytical Chemistry* 662 (2): 396–406. https://doi.org/10.1016/j.jelechem.2011.09.016.

105 Chen, Q.-S., Solla-Gullón, J., Sun, S.-G., and Feliu, J.M. (2010). The potential of zero total charge of Pt nanoparticles and polycrystalline electrodes with different surface structure: the role of anion adsorption in fundamental electrocatalysis. *Electrochimica Acta* 55 (27): 7982–7994. https://doi.org/10.1016/j.electacta.2010.03.050.

106 Climent, V. and Feliu, J.M. (2017). Surface electrochemistry with Pt single-crystal electrodes. *Advances in Electrochemical Science and Engineering* 1–57. https://doi.org/10.1002/9783527340934.ch1.

107 Montiel, M.A., Vidal-Iglesias, F.J., Montiel, V., and Solla-Gullón, J. (2017). Electrocatalysis on shape-controlled metal nanoparticles: progress in surface cleaning methodologies. *Current Opinion in Electrochemistry* 1 (1): 34–39. https://doi.org/10.1016/j.coelec.2016.12.007.

108 Kuzume, A., Herrero, E., and Feliu, J.M. (2007). Oxygen reduction on stepped platinum surfaces in acidic media. *Journal of Electroanalytical Chemistry* 599 (2): 333–343. https://doi.org/10.1016/j.jelechem.2006.05.006.

109 Vidal-Iglesias, F.J., Arán-Ais, R.M., Solla-Gullón, J. et al. (2012). Electrochemical characterization of shape-controlled Pt nanoparticles in different supporting electrolytes. *ACS Catalysis* 2 (5): 901–910. https://doi.org/10.1021/cs200681x.

110 Hernández, J., Solla-Gullón, J., and Herrero, E. (2004). Gold nanoparticles synthesized in a water-in-oil microemulsion: electrochemical characterization and effect of the surface structure on the oxygen reduction reaction. *Journal of Electroanalytical Chemistry* 574 (1): 185–196. https://doi.org/10.1016/j.jelechem.2003.10.039.

111 Hernández, J., Solla-Gullón, J., Herrero, E. et al. (2007). Electrochemistry of shape-controlled catalysts: oxygen reduction reaction on cubic gold nanoparticles. *The Journal of Physical Chemistry C* 111 (38): 14078–14083. https://doi.org/10.1021/jp0749726.

112 Solla-Gullón, J., Vidal-Iglesias, F.J., López-Cudero, A. et al. (2008). Shape-dependent electrocatalysis: methanol and formic acid electrooxidation on preferentially oriented Pt nanoparticles. *Physical Chemistry Chemical Physics* 10 (25): 3689–3698. https://doi.org/10.1039/B802703J.

113 Trasatti, S. and Petri, O.A. (1991). Real surface area measurements in electrochemistry. *Pure and Applied Chemistry* 63 (5): 711–734.

114 Łukaszewski, M., Soszko, M., and Czerwiński, A. (2016). Electrochemical methods of real surface area determination of noble metal electrodes – an overview. *International Journal of Electrochemical Science* 11: 4442–4469. https://doi.org/10.20964/2016.06.71.

115 Doña Rodríguez, J.M., Herrera Melián, J.A., and Pérez Peña, J. (2000). Determination of the real surface area of Pt electrodes by hydrogen adsorption using cyclic voltammetry. *Journal of Chemical Education* 77 (9): 1195. https://doi.org/10.1021/ed077p1195.

116 Voiry, D., Chhowalla, M., Gogotsi, Y. et al. (2018). Best practices for reporting electrocatalytic performance of nanomaterials. *ACS Nano* 12 (10): 9635–9638. https://doi.org/10.1021/acsnano.8b07700.

117 Chen, D., Tao, Q., Liao, L.W. et al. (2011). Determining the active surface area for various platinum electrodes. *Electrocatalysis* 2 (3): 207–219. https://doi.org/10.1007/s12678-011-0054-1.

118 Fang, L.-l., Tao, Q., Li, M.-f. et al. (2010). Determination of the real surface area of palladium electrode. *Chinese Journal of Chemical Physics* 23 (5): 543–548. https://doi.org/10.1088/1674-0068/23/05/543-548.

119 Rand, D.A.J. and Woods, R. (1971). The nature of adsorbed oxygen on rhodium, palladium and gold electrodes. *Journal of Electroanalytical Chemistry and Interfacial Electrochemistry* 31 (1): 29–38. https://doi.org/10.1016/S0022-0728(71)80039-6.

120 Rudi, S., Cui, C., Gan, L., and Strasser, P. (2014). Comparative study of the electrocatalytically active surface areas (ECSAs) of Pt alloy nanoparticles evaluated by H_{upd} and CO-stripping voltammetry. *Electrocatalysis* 5 (4): 408–418. https://doi.org/10.1007/s12678-014-0205-2.

121 Iwasita, T. and Ciapina, E.G. (2010). Mechanistic aspects of carbon monoxide oxidation. *Handbook of Fuel Cells* https://doi.org/10.1002/9780470974001.f500014.

122 Maillard, F., Eikerling, M., Cherstiouk, O.V. et al. (2004). Size effects on reactivity of Pt nanoparticles in CO monolayer oxidation: the role of surface mobility. *Faraday Discussions* 125: 357–377. https://doi.org/10.1039/B303911K.

123 Ciapina, E.G., Santos, S.F., and Gonzalez, E.R. (2018). Electrochemical CO stripping on nanosized Pt surfaces in acid media: a review on the issue of peak multiplicity. *Journal of Electroanalytical Chemistry* 815: 47–60. https://doi.org/10.1016/j.jelechem.2018.02.047.

124 Franklin, T.C. and Franklin, N.F. (1976). The use of underpotential deposition to measure the surface area of metals. *Surface Technology* 4: 431–440.

125 Burwell, R.L. (1976). Manual of symbols and terminology for physicochemical quantities and units-appendix II. *Pure and Applied Chemistry* 46 (1): 71–90. https://doi.org/10.1351/pac197646010071.

126 Wei, C., Sun, S., Mandler, D. et al. (2019). Approaches for measuring the surface areas of metal oxide electrocatalysts for determining their intrinsic electrocatalytic activity. *Chemical Society Reviews* 48 (9): 2518–2534. https://doi.org/10.1039/C8CS00848E.

127 Alonso-Vante, N. (2018). *Chalcogenide Materials for Energy Conversion: Pathways to Oxygen and Hydrogen Reactions*. Switzerland: Nanostructure Science and Technology. Springer International Publishing. https://doi.org/10.1007/978-3-319-89612-0_2.

128 Luo, F., Choi, C.H., Primbs, M.J.M. et al. (2019). Accurate evaluation of active-site density (SD) and turnover frequency (TOF) of PGM-free metal–nitrogen-doped carbon (MNC) electrocatalysts using CO cryo adsorption. *ACS Catalysis* 9 (6): 4841–4852. https://doi.org/10.1021/acscatal.9b00588.

129 Shinozaki, K., Zack, J.W., Richards, R.M. et al. (2015). Oxygen reduction reaction measurements on platinum electrocatalysts utilizing rotating disk electrode technique. I impact of impurities, measurement protocols and applied corrections. *Journal of the Electrochemical Society* 162 (10): F1144–F1158. https://doi.org/10.1149/2.1071509jes.

130 Mayrhofer, K.J.J., Strmcnik, D., Blizanac, B.B. et al. (2008). Measurement of oxygen reduction activities via the rotating disc electrode method: from Pt model surfaces to carbon-supported high surface area catalysts. *Electrochimica Acta* 53 (7): 3181–3188. https://doi.org/10.1016/j.electacta.2007.11.057.

131 Xing, W., Yin, M., Lv, Q. et al. (2014). Chapter 1 – oxygen solubility, diffusion coefficient, and solution viscosity. In: *Rotating Electrode Methods and Oxygen Reduction Electrocatalysts* (ed. W. Xing, G. Yin, and J. Zhang), 1–31. Amsterdam: Elsevier. https://doi.org/10.1016/B978-0-444-63278-4.00001-X.

132 Zhang, L., Wei, M., Wang, S. et al. (2015). Highly stable PtP alloy nanotube arrays as a catalyst for the oxygen reduction reaction in acidic medium. *Chemical Science* 6 (5): 3211–3216. https://doi.org/10.1039/C5SC00124B.

133 Davis, R.E., Horvath, G.L., and Tobias, C.W. (1967). The solubility and diffusion coefficient of oxygen in potassium hydroxide solutions. *Electrochimica Acta* 12 (3): 287–297. https://doi.org/10.1016/0013-4686(67)80007-0.

134 Di Noto, V., Negro, E., Nale, A. et al. (2021). Hidden in plain sight: unlocking the full potential of cyclic voltammetry with the thin-film rotating (ring) disk electrode studies for the investigation of oxygen reduction reaction electrocatalysts. *Current Opinion in Electrochemistry* 25. https://doi.org/10.1016/j.coelec.2020.08.008.

135 Sarapuu, A., Hussain, S., Kasikov, A. et al. (2019). Electroreduction of oxygen on Nafion®-coated thin platinum films in acid media. *Journal of Electroanalytical Chemistry* 848: 113292. https://doi.org/10.1016/j.jelechem.2019.113292.

136 Curnick, O.J., Pollet, B.G., and Mendes, P.M. (2012). Nafion®-stabilised Pt/C electrocatalysts with efficient catalyst layer ionomer distribution for proton exchange membrane fuel cells. *RSC Advances* 2 (22): 8368–8374. https://doi.org/10.1039/C2RA21071A.

137 Di Noto, V., Pagot, G., Negro, E. et al. (2022). A formalism to compare electrocatalysts for the oxygen reduction reaction by cyclic voltammetry with the thin-film rotating ring-disk electrode measurements. *Current Opinion in Electrochemistry* 31: 100839. https://doi.org/10.1016/j.coelec.2021.100839.

138 Wang, J., Zhao, C.-X., Liu, J.-N. et al. (2021). Quantitative kinetic analysis on oxygen reduction reaction: a perspective. *Nano Materials Science* 3 (3): 313–318. https://doi.org/10.1016/j.nanoms.2021.03.006.

139 Arenz, M. and Zana, A. (2016). Fuel cell catalyst degradation: identical location electron microscopy and related methods. *Nano Energy* 29: 299–313. https://doi.org/10.1016/j.nanoen.2016.04.027.

140 Rodgers, M.P., Brooker, R.P., Mohajeri, N. et al. (2012). Comparison of proton exchange membranes degradation rates between accelerated and performance tests service comparison of proton exchange membranes degradation rates between accelerated and performance tests. *Journal of the Electrochemical Society* 159 (7): F338. https://doi.org/10.1149/2.040207jes.

141 Tang, H., Qi, Z., Ramani, M., and Elter, J.F. (2006). PEM fuel cell cathode carbon corrosion due to the formation of air/fuel boundary at the anode. *Journal of Power Sources* 158 (2): 1306–1312. https://doi.org/10.1016/j.jpowsour.2005.10.059.

142 Ohma, A., Shinohara, K., Iiyama, A. et al. (2019). Membrane and catalyst performance targets for automotive fuel cells by FCCJ membrane, catalyst, MEA WG. *ECS Transactions* 41 (1): 775–784. https://doi.org/10.1149/1.3635611.

143 DOE Durability Working Group (2011). Rotating Disk-Electrode Aqueous Electrolyte Accelerated Stress Tests for PGM Eletrocatalyst/Support Durability Evaluation. https://www.energy.gov/sites/prod/files/2015/08/f25/fcto_dwg_pgm_electrocatalyst_support_aqueous_ast.pdf (accessed 5 June 2023).

144 Nagai, T., Jahn, C., and Jia, H. (2019). Improved accelerated stress tests for ORR catalysts using a rotating disk electrode. *Journal of the Electrochemical Society* 166 (7): F3111–F3115. https://doi.org/10.1149/2.0161907jes.

145 Wolter, O. and Heitbaum, J. (1984). Differential electrochemical mass spectroscopy (DEMS) — a new method for the study of electrode processes. *Berichte der Bunsengesellschaft für Physikalische Chemie* 88 (1): 2–6. https://doi.org/10.1002/bbpc.19840880103.

146 Bruckenstein, S. and Gadde, R.R. (1971). Use of a porous electrode for in situ mass spectrometric determination of volatile electrode reaction products. *Journal of the American Chemical Society* 93 (3): 793–794. https://doi.org/10.1021/ja00732a049.

147 González Pereira, M., Dávila Jiménez, M., Elizalde, M.P. et al. (2004). Study of the electrooxidation of ethanol on hydrophobic electrodes by DEMS and HPLC. *Electrochimica Acta* 49 (22): 3917–3925. https://doi.org/10.1016/j.electacta.2004.01.080.

148 Wolter, O., Willsau, J., and Heitbaum, J. (1985). Reaction pathways of the anodic oxidation of formic acid on Pt evidenced by 18O labeling—a DEMS study. *Journal of the Electrochemical Society* 132 (7): 1635–1638. https://doi.org/10.1149/1.2114179.

149 Ashton, S.J. and Arenz, M. (2012). Comparative DEMS study on the electrochemical oxidation of carbon blacks. *Journal of Power Sources* 217: 392–399. https://doi.org/10.1016/j.jpowsour.2012.06.015.

150 Li, W. and Lane, A.M. (2009). Investigation of Pt catalytic effects on carbon support corrosion of the cathode catalyst in PEM fuel cells using DEMS spectra. *Electrochemistry Communications* 11 (6): 1187–1190. https://doi.org/10.1016/j.elecom.2009.04.001.

151 Ye, J.-Y., Jiang, Y.-X., Sheng, T., and Sun, S.-G. (2016). In-situ FTIR spectroscopic studies of electrocatalytic reactions and processes. *Nano Energy* 29: 414–427. https://doi.org/10.1016/j.nanoen.2016.06.023.

152 Nesselberger, M. and Arenz, M. (2016). In situ FTIR spectroscopy: probing the electrochemical Interface during the oxygen reduction reaction on a commercial platinum high-surface-area catalyst. *ChemCatChem* 8 (6): 1125–1131. https://doi.org/10.1002/cctc.201501193.

153 Ogihara, Y., Yano, H., Matsumoto, T. et al. (2016). In situ FTIR analysis of CO-tolerance of a Pt-Fe alloy with stabilized Pt skin layers as a hydrogen anode catalyst for polymer electrolyte fuel cells. *Catalysts* 7 (12): 8–8. https://doi.org/10.3390/catal7010008.

154 Bard, A.J., Fan, F.R.F., Kwak, J., and Lev, O. (1989). Scanning electrochemical microscopy. Introduction and principles. *Analytical Chemistry* 61 (2): 132–138. https://doi.org/10.1021/ac00177a011.

155 Jantz, D.T. and Leonard, K.C. (2018). Characterizing electrocatalysts with scanning electrochemical microscopy. *Industrial & Engineering Chemistry Research* 57 (22): 7431–7440. https://doi.org/10.1021/acs.iecr.8b00922.

156 Aran-Ais, R.M., Vidal-Iglesias, F.J., Farias, M.J.S. et al. (2017). Understanding CO oxidation reaction on platinum nanoparticles. *Journal of Electroanalytical Chemistry* 793: 126–136. https://doi.org/10.1016/j.jelechem.2016.09.031.

157 Stamenković, V., Schmidt, T.J., Ross, P.N., and Marković, N.M. (2002). Surface composition effects in electrocatalysis: kinetics of oxygen reduction on well-defined Pt_3Ni and Pt_3Co alloy surfaces. *The Journal of Physical Chemistry B* 106 (46): 11970–11979. https://doi.org/10.1021/jp021182h.

158 Stamenkovic, V.R., Mun, B.S., Mayrhofer, K.J.J. et al. (2006). Effect of surface composition on electronic structure, stability, and electrocatalytic properties of Pt-transition metal alloys: Pt-skin versus Pt-skeleton surfaces. *Journal of the American Chemical Society* 128 (27): 8813–8819. https://doi.org/10.1021/ja0600476.

159 Zhu, C., Fu, S., Shi, Q. et al. (2017). Single-atom electrocatalysts. *Angewandte Chemie International Edition* 56 (45): 13944–13960. https://doi.org/10.1002/anie.201703864.

13

Durability of Fuel Cell Electrocatalysts and Methods for Performance Assessment

Bianca M. Ceballos and Piotr Zelenay

Los Alamos National Laboratory, Materials Physics and Applications Division, Bikini Atoll Road, Los Alamos, NM 87545, USA

13.1 Introduction

Advancements in alternative energy are necessary to overcome the rising worldwide energy demand and the negative environmental impacts of fossil fuel consumption. Hydrogen fuel is an attractive energy source that can be consumed in a fuel cell to generate electricity with water and heat as innocuous byproducts [1, 2]. Hydrogen fuel cells are a promising alternative technology compared to current internal combustion engines due to their high efficiency, the renewable feedstock available to generate hydrogen fuel, and the elimination of harmful exhaust byproducts [3]. For these reasons, fuel cells, together with battery-powered vehicles, are likely to replace existing technologies in automotive transportation, portable power, and other applications.

Previous investigations of polymer electrolyte fuel cells (PEFCs) and platinum group metal-free (PGM-free) oxygen reduction reaction (ORR) catalysts have been successful in advancing toward practical application in vehicles [4–8]. In 2014, the Toyota Mirai became the first commercial fuel cell vehicle on the market with a price of c. US$ 57,000. To date, the price of the Mirai has not changed significantly due to the high production costs of fuel cell stacks which rely on high loadings of platinum. Improvements in cost reduction and durability are necessary to support the growth and feasibility of fuel cell vehicle technologies, especially for heavy-duty transportation applications.

PEFCs rely on the production and transfer of protons and electrons via electrochemical reactions. At the anode, a catalyst facilitates hydrogen oxidation reaction (HOR), described by Eq. (13.1),

$$H_2 \rightarrow 2H^+ + 2e^- \tag{13.1}$$

to generate protons and electrons that are transferred through the proton exchange membrane (red arrows in Figure 13.1) to the cathode. The electrons and protons

Electrocatalysis for Membrane Fuel Cells: Methods, Modeling, and Applications, First Edition.
Edited by Nicolas Alonso-Vante and Vito Di Noto.

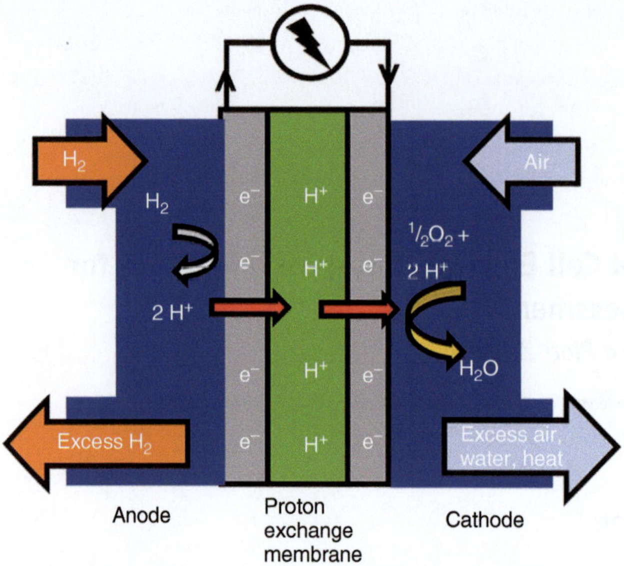

Figure 13.1 Schematic of a polymer electrolyte fuel cell (PEFC) where H_2 oxidation occurs at the anode to generate H^+. The proton then permeates through the proton exchange membrane and participate in oxygen reduction reaction (ORR) at the cathode.

supplied at the cathode participate in ORR, described by Eq. (13.2),

$$\tfrac{1}{2}O_2 + 2H^+ + 2e^- \rightarrow H_2O \tag{13.2}$$

to produce water by using a catalyst, Figure 13.1 (yellow arrow).

Current state-of-the-art PEFCs rely on highly dispersed Pt-based nanoparticle catalysts on high surface-area carbon supports to reach high catalytic rates of both anodic and cathodic reactions. Desirable fuel cell performance is dictated by catalyst loading and reaction kinetics of HOR and ORR [9]. At Pt-based catalysts, HOR is fast, requiring low loadings of Pt at the anode [10]. However, the ORR kinetics are sluggish and require higher Pt catalyst loadings [10, 11]. The high Pt loadings, totaling as much as 0.3 $mg_{Pt}\ cm^{-2}$ of the fuel cell membrane electrode assembly (MEA), account for the large cost of fuel cell stacks. Breakthroughs in cost reduction of the PEFC require lower loadings of Pt-based catalysts or replacement of PGM catalysts with catalysts based on earth-abundant materials.

Advancements made utilizing PGM-free catalysts have focused on Fe–N–C electrocatalysts, benefiting from the low cost of iron, nitrogen, and carbon precursor materials [6, 7, 12]. In order for a PGM-free catalyst to replace Pt in a fuel cell the catalyst must (1) achieve high ORR activity approaching necessary kinetic benchmarks, (2) demonstrate suitable stability over thousands of hours, and (3) offer significant cost advantage in fuel cell stack production [13, 14]. Present-day Fe–N–C catalysts can reach high ORR activities but suffer from suboptimal durability compared to their PGM counterparts [5, 15–17].

Current advancements in PGM-free electrocatalysis focus on improving durability via identifying active site type, quantifying active site density, and studying degradation pathways. The active sites in PGM-free catalysts are different from those in Pt-based catalysts because they are embedded in the carbon matrix, whereas Pt-based catalysts rely on Pt nanoparticles that are distinct from the carbon support. In PGM-free catalysts, active sites are formed simultaneously with the carbon matrix material at high temperatures. Gaining a better understanding of the active site structure of PGM-free catalysts could help direct catalyst development to increase active site densities and/or turnover frequency (TOF) to improve overall performance in PEFCs. Current understanding of degradation pathways for PGM-free catalysts includes: (1) micropore flooding, (2) anion adsorption and protonation of active site, (3) site demetallation, and (4) carbon oxidation (corrosion) [16–18]. Degradation of PGM-free catalysts likely includes combinations of these pathways that lead to catalyst deactivation.

Overall, comparison of catalyst durability in the literature is difficult due to inconsistent use of electrochemical techniques under various experimental conditions. Methods used to evaluate catalyst durability range from chronoamperometric holds at various potentials and different durations to voltammetric cycling under inert or O_2 atmosphere. Standardizing methods used to assess catalyst durability would positively impact the field by allowing for easier comparison and identification of catalysts suitable for PEFC applications [16]. Herein, we aim to address these challenges by describing a set of experimentally validated protocols to be used as a common testing platform for PGM-free and PGM ORR electrocatalysts to assess durability and performance for low-temperature PEFCs for automotive applications [19]. Current understanding of PGM-free degradation pathways and an overview of approaches to mitigate durability issues resulting from carbon corrosion will also be discussed.

13.2 Fuel Cell PGM-free Electrocatalysts for Low-temperature Applications

Recent advancements in PGM-free electrocatalyst and electrode development have focused on improving ORR catalyst performance and durability in MEAs for PEFC applications. PGM-free catalyst synthetic approaches have progressed from hard-template metal-nitrogen-carbon (M–N–C) precursors (i.e. metal porphyrins) [8, 20–22] to metal-nitrogen-carbon rich metal organic framework (MOF) materials [16, 23, 24].

In fact, MOF-derived electrocatalysts have shown superior ORR performance, where the high activity exhibited by these catalysts is thought to be related to the high content of MN_x sites originating from the MOF precursor [16]. Therefore, achieving high M–N–C content by increasing metal-loadings is desirable for generating active electrocatalysts. However, rising metal-loadings frequently results in metal atom aggregation that hinders MN_x site formation. Utilizing synthesis strategies that rely on precursors which include coordination frameworks such as MOFs or porous organic polymers (POPs) is an effective method to circumvent this challenge and achieve atomically dispersed metal species, increase metal-loadings, and prevent metal aggregation. In addition, pyrolysis or high-temperature treatment of M–N–C catalysts leads to important changes of material characteristics that are linked to better ORR performance. For example, an increase in both pyridinic and graphitic nature of Fe–N–C sites after pyrolysis has correlated to a significant increase in ORR activity [7, 25–27]. Since bulk iron species or iron nanoparticles can form at high temperatures, post treatment of pyrolyzed materials via acid leaching or subsequent heat treatment are important to remove these non-reactive metal species.

Despite the high initial ORR performance observed with many *M*–N–C catalysts, long-term catalyst stability remains a crucial challenge. To date, very few reported PGM-free electrocatalysts have shown stability for over 200 hours of operation in a PEFC [7, 28]. An apparent correlation between initial high activity and long-term instability of *M*–N–C ORR catalysts is not well understood. Differences in catalyst testing conditions, electrochemical methods, and evaluation of performance make understanding these differences even more difficult to recognize.

Herein, we summarize the current understanding of PGM-free catalyst degradation pathways, as well as report a set of protocols, performance benchmarks, and test conditions for PGM-free catalysts in MEAs. These protocols have been developed and validated by four individual national laboratories participating in the DOE Electrocatalysis Consortium (ElectroCat), involving Los Alamos National Laboratory (LANL), Argonne National Laboratory (ANL), National Renewable Energy Laboratory (NREL), and Oak Ridge National Laboratory (ORNL) [19].

13.3 PGM-free Electrocatalyst Degradation Pathways

One of the largest barriers to widespread use of PGM-free catalysts is their rapid performance loss under operating conditions [17]. Initial catalyst performance loss in the MEA is often fast, occurring within the first few hours of testing, while a slower decline in activity is observed over longer test times. Decreases in activity are thought to be due to a combination of mechanisms including (i) Fe loss or demetallation and (ii) carbon oxidation via hydrogen peroxide attack, whereas contributions from mechanisms like (iii) competing protonation events at the active site or neighboring N-atoms or (iv) micropore flooding are heavily debated in the literature [15, 17, 18, 29–32].

While significant improvements in ORR activity have allowed PGM-free catalysts to narrow the performance gap to Pt-based catalysts, their durability and stability remain insufficient for their application in automotive transportation fuel cells. Thus, the importance of investigating the causes of catalyst deactivation and instability is key to generating catalysts with extended utility lifetimes for fuel cell applications.

13.3.1 Demetallation

The nature of ORR active sites in PGM-free catalysts is the focus of ongoing research [6, 23, 33–37]. Some studies support the hypothesis that metal-containing active sites are responsible for ORR catalysis, while others focus on metal-free active sites centered on the extended C–N structure as sites for ORR catalysis [25, 34, 38–43].

Surface-exposed free metal atoms, metal clusters, or nanoparticles, are not stable in a PEFC acidic environment [44, 45]. In Fe–N–C catalysts, iron loss is believed to proceed via loss of FeN_x species, nanoparticles, iron carbide, or iron sulfide species through direct Fe leaching [46] or loss of FeN_x moieties due to carbon corrosion [47]. Although dissolved bulk-iron species are unlikely to participate in ORR catalysis, they have been found to have detrimental effects on the overall PEFC performance by participating in other degenerative pathways, which are discussed in the carbon oxidation section [48].

A recent density functional theory (DFT) study by Holby et al. [45] utilized DFT descriptors to model metal dissolution (or demetallation) of FeN_4 in a graphene monolayer and bilayer as a function of pH and potential. The proposed stability diagram predicts Fe^{2+} dissolution from unligated FeN_4 sites at pH values <2 (Figure 13.2). This predicted instability of bare FeN_4 sites has significant implications for practical applications in PEFC operation and testing via rotating ring disk electrode (RRDE) in acidic environments. When considering O_2 vs N_2 or Ar purged environments, the authors find that ligation of the FeN_4 site leads to better site stability. Where under O_2 free environments the FeN_4 site remains ligated and is stable from dissolution. In contrast, in the presence of O_2, FeN_4 site degradation is accelerated due to several possible factors including the generation of bare FeN_4 sites during ORR catalyst turnover and possible H_2O_2 production. However, the authors also find that binding of O_2 to FeN_4 sites generates a proposed hydroxyl terminal surface species (HO-FeN_4) that is predicted to be stable to dissolution at low pH conditions (Figure 13.2). This suggests that ligation to FeN_4 sites may play an important role for increased stability toward Fe dissolution.

Bulk Fe dissolution to Fe^{2+} is favored in acids at potentials above −0.66 V vs. standard hydrogen electrode (SHE) [49]. Under similar acidic conditions, Holby et al. [45] calculate an equilibrium potential for dissolution from a carbon-hosted FeN_4 site to be +0.65 V vs. computational hydrogen electrode (CHE). The large difference

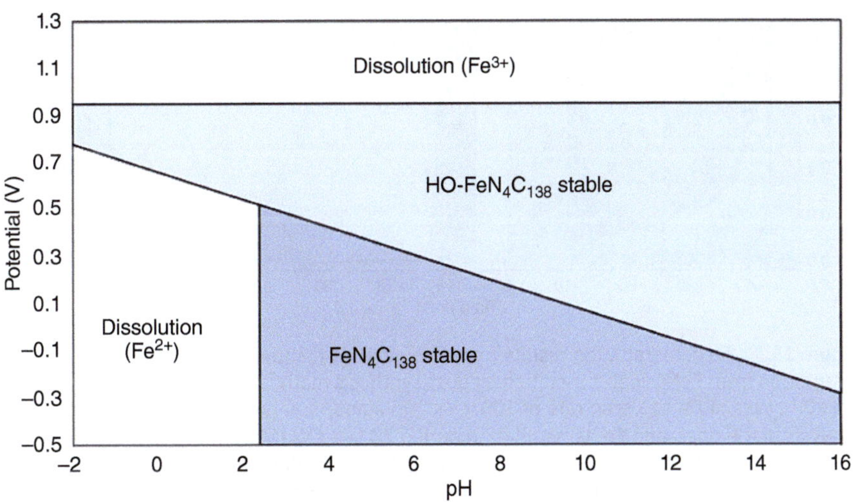

Figure 13.2 Stability diagram describing proposed Fe dissolution as a function of pH and potential for FeN_4C_{138}, HO-FeN_4C_{138}, $Fe_{aq}{}^{2+}$ (1 M), and $Fe_{aq}{}^{3+}$ (1 M) phases at 298 K. Source: Reproduced from Ref. [45] with the permission of the American Chemical Society.

in potential for iron dissolution to Fe^{2+} is thought to be due to the stability of the surrounding carbon or graphene-N_4 sites that result from iron loss. In the case of iron from graphene-N_4 structures, residual N sites can be stabilized via protonation.

Choi et al. [44, 47] conducted cycling studies with inductively coupled plasma mass spectrometry (ICP-MS) and differential mass spectroscopy (DEMS) in a scanning flow cell (SFC) to investigate the effects of iron demetallation on catalyst performance under RRDE and fuel cell conditions. The authors found that acid washing does not fully remove inactive Fe particles due to the high open circuit potential (*ca.* 0.9 V vs. RHE) and the stable phases of iron that form at low pH values. Both a solvated Fe^{2+} species and insoluble $Fe^{III}(OH)_3$ or $Fe^{III}O(OH)$ species were hypothesized, with the insoluble Fe species remaining intact after acid washing. However, continuous iron leaching occurred when cycling at low potentials (<0.7 V vs. RHE), consistent with the expected redox potential for Fe^{II}/Fe^{III} of 0.77 V vs. SHE. At the same time, the observed loss of iron from the catalysts did not significantly influence catalyst fuel cell durability over the first 50 hours when comparing FeNC materials with various post-treatments (acid washing, CV cycling, and chemical reducing agent) intended to remove inactive iron. The Fe–N–C catalyst sample without post-treatment, which also yielded the highest dissolved iron content detected in solution, showed similar durability to post-treated catalysts when comparing polarization curves, Figure 13.3. This suggests that early iron demetallation is not a major contributor to catalyst durability. Instead, the authors found the measured performance loss for the Fe–N–C materials correlated well with production of H_2O_2 and might include more complex degradation pathways such as Fe-centered reactions with reactive oxygen species (ROS).

Yang et al. [32] recently reported a DFT study on degradation mechanisms for Fe-N-C materials. Therein, they show that binding of two hydroxyl groups to form

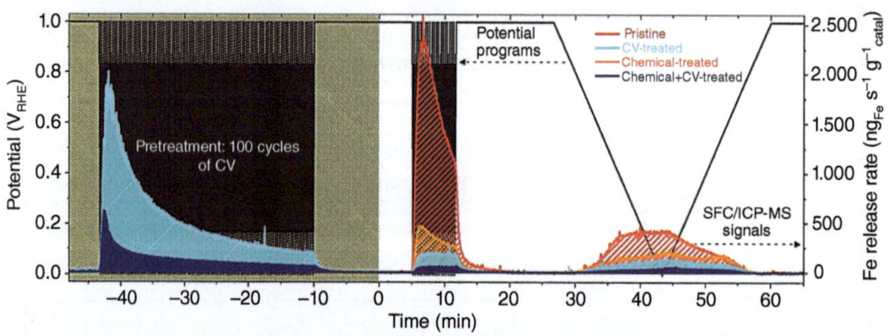

Figure 13.3 Fe demetallation results from Choi et al. [44] showing operando Fe dissolution from FeNC-wet-1 modified by post-synthesis methods. FeNC-wet-1 was treated via 100 cycles of CV at a scan rate of 100 mV s^{-1} in a potential range of 1.0–0.0 V_{RHE} before analysis of the operando Fe demetallation during 20 fast CVs (100 mV s^{-1}) and 1 slow CV (1 mV s^{-1}) in a potential range of 1–0 V_{RHE} (cyan). Fe dissolution during this CV pretreatment is indicated (the section below the time scale of zero minutes). Chemical modification of FeNC-wet-1 progressed in 10 mM $SnCl_2$ + 0.1 M HCl solution for two hours under Ar bubbling, and then Fe demetallation was investigated (orange). Combined post-synthesis methods, i.e. chemical treatment followed by CV treatment, were performed on FeNC-wet-1, and its Fe demetallation was also examined (navy). Source: Reproduced from Ref. [44] with the permission of the American Chemical Society.

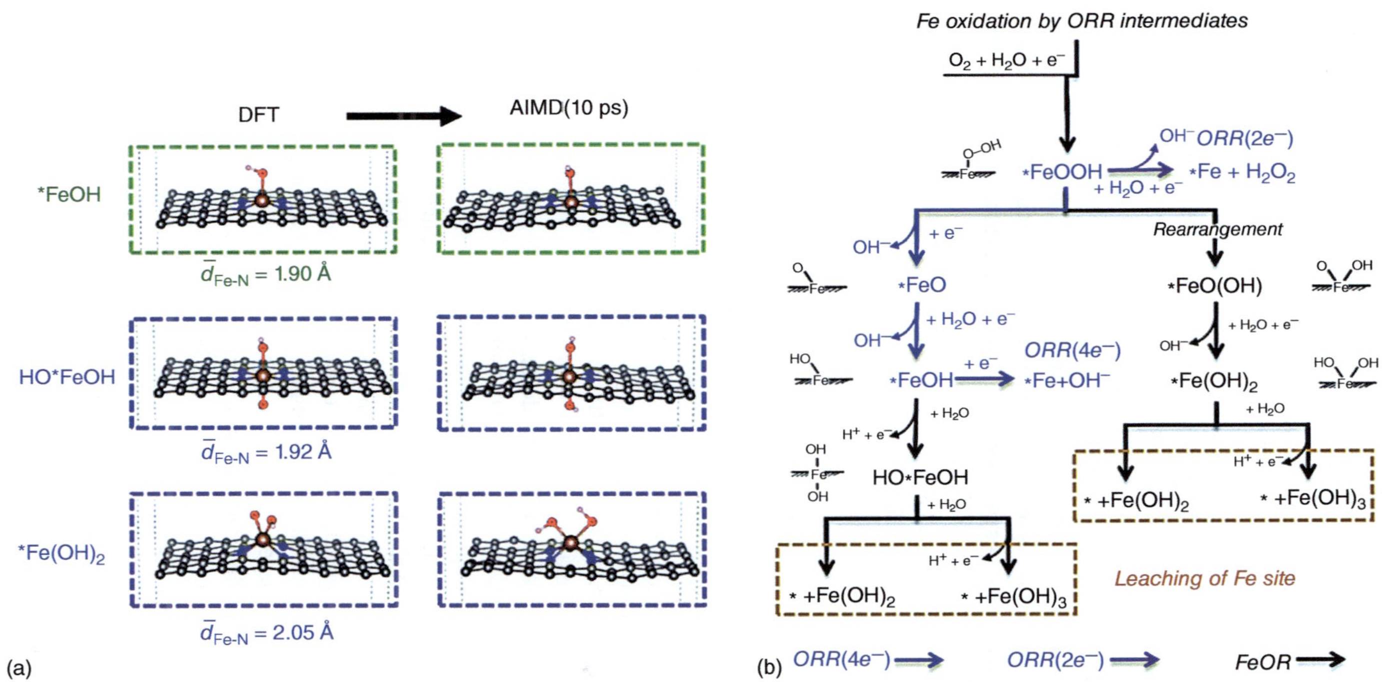

Figure 13.4 DFT study by Yang et al. [32] depicting the (a) density functional theory (DFT)- and *ab initio* molecular dynamics (AIMD)-optimized configurations of *FeOH, HO*FeOH, and *$Fe(OH)_2$ intermediates in an Fe–N–C catalyst, and (b) a schematic detailing the proposed Fe site oxidation mechanisms involving ORR intermediates at an Fe–N–C catalyst. Source: Reproduced from Ref. [32] with the permission of the Royal Society of Chemistry.

an $Fe(OH)_2$ moiety, where both OH^- groups are bound on the same side of the Fe atom, can occur via ORR intermediates (Figure 13.4). Binding of these two hydroxyl groups causes a geometric distortion around the Fe and allows for the protonation of neighboring N-sites. Subsequent N-protonation events trigger leaching of Fe. The authors also found a competitive relationship between –OH desorption at the Fe site, necessary for catalyst turnover for ORR, and the binding of a second –OH group at the Fe in a *trans* configuration, which is predicted to block the activity at the Fe site. At potentials above 0.6 V vs. RHE, the predicted ΔG of OH desorption is expected to be larger than that of further OH adsorption. Thus, blockage of the Fe site is more likely to occur at potentials above 0.6 V vs. RHE than active-site turnover, resulting in activity losses of the Fe–N–C catalyst.

Further studies on demetallation through the loss of FeN_x moieties have focused on the effects of acidic H_2O_2 washes to mimic electrochemical Fenton reaction conditions. Two recent studies by Choi et al. [15] and Xia et al. [50] have described conflicting results. Choi et al. [15] found that acidic H_2O_2 treatments of an Fe–N–C single-atom catalyst synthesized at 1050 °C resulted in no significant Fe demetallation, but did observe the formation of oxygen functional groups at the carbon. The resulting oxygen functional groups led to loss of activity at the retained Fe-centered sites. However, these oxygen functional groups could be electrochemically removed, resulting in partial catalyst activity recovery. In contrast, Xia et al. [50] observed iron demetallation of Fe–N–C catalysts treated at lower temperatures (600 or 900 °C) and corroborated the loss in performance, as well as the formation of oxygen functional groups on the carbon observed by Choi et al. The difference in stability of proposed FeN_x moieties by Xia et al. may stem from intrinsic differences in the type of sites formed at different temperatures, where more stable sites are formed at high temperatures (>1000 °C). In the study by Xia et al. [50], the authors synthesized an Fe–N–C material at 1100 °C and tested the effect of acidic H_2O_2 treatment. The authors reported a slight decrease in TOF value of the FeN_x sites and a large decrease in active site density of more than 60% (via *in situ* molecular probe stripping) suggesting activity losses are possibly from Fe demetallation. Consequently, it is likely that the presence of oxygen functional groups at the carbon is not solely responsible for the loss of iron observed in these materials since researchers did not universally observe iron loss.

The loss of iron through either direct Fe leaching or the loss of FeN_x sites has important implications on catalyst durability, which makes the understanding of active site structure necessary to advance the field. Activity losses from iron demetallation are well documented, but evidence suggests that iron leaching does not account for early performance losses [44, 46, 47, 50]. Instead, some iron demetallation can be attributed to inactive Fe species that do not participate in ORR but are most likely involved in downstream reactions resulting in catalyst degradation via carbon oxidation.

13.3.2 Carbon Oxidation

Under acidic conditions, ORR can proceed through two pathways: direct $4e^-$ and $4H^+$ transfer to generate water (reaction 13.3) or a $2e^-$ and $2H^+$ transfer that results in hydrogen peroxide (reaction 13.4). Hydrogen peroxide can be further electrochemically reduced or disproportionate to form water (reactions 13.5 and 13.6, respectively).

$$O_2 + 4H^+ + 4e^- \rightarrow 2H_2O \qquad E^\circ = 1.23\,V \tag{13.3}$$

$$O_2 + 2H^+ + 2e^- \rightarrow H_2O_2 \qquad E^\circ = 0.67\,V \tag{13.4}$$

$$H_2O_2 + 2H^+ + 2e^- \rightarrow 2H_2O \qquad E^\circ = 1.77\,V \tag{13.5}$$

$$H_2O_2 \rightarrow 2H_2O + O_2 \tag{13.6}$$

Together with the hydrogen peroxide formation, the alternative peroxide-driven degenerative pathways become likely. As a result, several investigations into the effects of peroxide on catalyst performance loss have been reported [15, 20, 48, 51, 52]. As described in the previous section, oxidation of carbon or nitrogen sites is thought to lead to loss of activity due to the loss of FeN_x sites.

Iron cations are known to catalyze the decomposition of peroxide to reactive oxygen species (ROS), such as hydroxyl ($^{\bullet}OH$) and hydroperoxyl ($^{\bullet}OOH$) radicals in acidic medium, via Fenton reactions [51]. The reactions proceed via the oxidation of Fe^{2+} ions by hydrogen peroxide to catalyze the disproportionation of H_2O_2 to generate ROS in solution. The ROS generated in this process, like the hydroxyl radical, are strong oxidants that can participate in non-selective oxidation of organic compounds. Thus, it is hypothesized that the indiscriminate oxidation by ROS generated in solution can initiate carbon oxidation via secondary reactions at the electrode surface.

Goellner et al. [48] found that degradation of M–N–C type catalysts (M = Fe, Co, or Cr) can be induced by adding various amounts of H_2O_2. Increased amounts of H_2O_2 led to decreased ORR activity of the studied catalysts, with Cr-N-C showing the largest degradation effect followed by Fe–N–C and Co–N–C (Figure 13.5). Changes in structure were monitored via X-ray absorption spectroscopy (XAS) and Mössbauer spectroscopy before and after degradation with H_2O_2. The authors found minor changes in both XAS and Mössbauer spectra, indicating a retention of most Fe–N–C and Co–N–C moieties in the pristine and degraded catalyst. However, the measured loss of Fe and Co from the materials was 17% and 27%, respectively. The treatment of these catalyst with H_2O_2 resulted in a loss of metal content, as well as decreased activity without large changes in the observed material structure monitored by XAS and Mössbauer spectroscopy.

Lefevre et al. [51] investigated the production of peroxide at Fe–N–C catalysts and the influence of peroxide on catalyst activity. The authors treated a series of Fe–N–C catalysts with 5–30 vol% peroxide in H_2SO_4 at pH 1. From these experiments, they found that the catalyst activity loss correlates with the Fe content. The catalyst that showed the greatest performance loss also contained the least amount of iron. However, Fe content alone was not the only contributing factor determining stability. The authors also demonstrated that catalysts treated at a high temperature of 900 °C were more stable than those prepared under lower pyrolysis temperatures, despite having the same relative abundance of FeN_2/C to FeN_4/C sites.

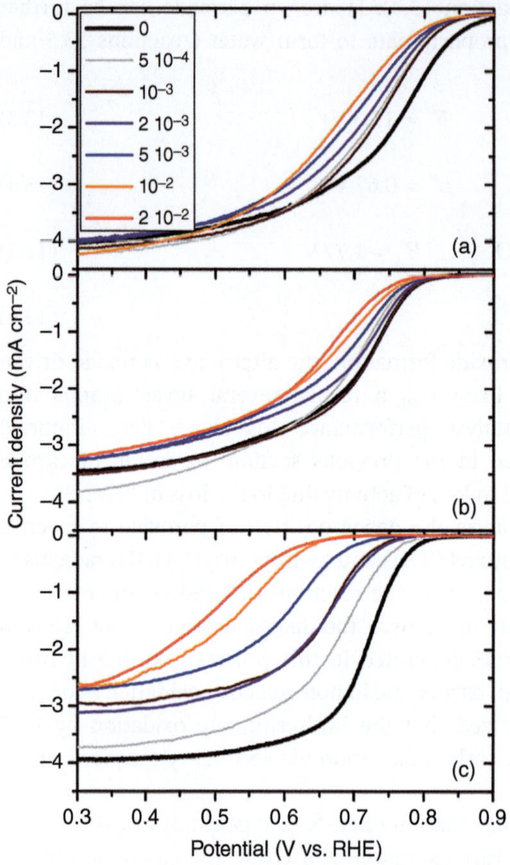

Figure 13.5 Study of the effect of different H_2O_2 concentrations (inset) on (a) Fe–N–C; (b) Co–N–C; (c) Cr–N–C catalysts ORR performance. Source: Reproduced from Ref. [48] with the permission of IOP Publishing.

Recently, Xia et al. [50] observed a similar dependence of Fe-N-C catalyst stability to peroxide conditions and catalyst pyrolysis temperature (900 or 1100 °C), with the catalyst pyrolyzed at the higher temperature losing less Fe after exposure to peroxide. The authors exposed the catalysts to different accelerated stress tests (ASTs) to further investigate the influence of surface oxidation on catalyst durability. AST-1 included 10k square cycles between 0.6 and 1.0 V vs. RHE (holding time of three seconds) under O_2, while AST-2 involved 1k or 2k cycles between 1.0 and 1.5 V vs. RHE at a scan rate of 500 mV s^{-1} under N_2. For both catalysts, AST-2 resulted in considerably higher carbon oxidation than AST-1. The level of surface carbon oxidation in catalysts exposed to AST-2 was similar despite large differences in catalyst stability to Fe loss. This suggests that carbon oxidation is not a dominant factor influencing catalyst stability.

Choi et al. [15] studied peroxide-treated catalysts using extended X-ray absorption fine structure (EXAFS) and Fe Mössbauer spectroscopy. They showed that after peroxide exposure Fe–N–C sites remained in similar electronic and coordination structure, but the catalyst yielded lower TOF values. They also found that peroxide treatment resulted in an increase at oxygen content at the carbon in the material when analyzed by XPS and proposed that adventitious ROS are participating in carbon oxidation via Fenton reactions, which do not result in active site decomposition [15]. Instead, they proposed that carbon oxidation results in weakened O_2 binding at Fe-centered sites, leading to lower TOF values. In addition, the authors found that electrochemical reduction of the oxidized catalyst allowed for the recovery of catalytic performance.

The effects of peroxide and resulting ROS on catalyst stability have been shown to result in an increase of surface carbon oxidation with varying influence on Fe loss [44, 47, 50]. Overall, the literature suggests that carbon oxidation is a relatively minor catalyst degradation pathway. The effects of ROS on other factors such as ionomer could play a significant role on fuel cell performance, but are beyond the scope of this text [52].

13.3.3 Micropore Flooding

Dodelet and coworkers [53] invoked micropore flooding to explain the catalyst performance decay in the first 15 hours of PEFC testing of Fe-porphyrin- and ZIF-8-derived catalysts. In the proposed degradation path, micropores are assumed to have hydrophobic properties that protect active sites hosted within the microporous channels. The performance loss occurs when pore hydrophobicity is gradually lowered via oxidation by peroxide generated during fuel cell operation, causing pore flooding, limiting oxygen transport, and decreasing overall performance.

This hypothesis was disputed by Banham and coworkers [54], who found no difference in the double-layer capacitance before and after fuel cell testing, which suggested that micropores were partially or fully wet at the beginning of life (BOL) for these catalysts. Polarization curves showed that catalysts suffered mostly kinetic losses rather than mass transport losses that would support the micropore flooding hypothesis. Kinetic losses were attributed to the loss and/or deactivation of active sites. Lastly, attempts to recover catalyst activity by heating at 80 °C for five hours under N_2 did not help to recover catalyst performance. Thus, the micropore flooding is probably not a large contributing factor to the initial catalyst performance losses compared to demetallation.

13.3.4 Nitrogen Protonation and Anionic Adsorption

Herranz et al. [55] proposed that N protonation could facilitate subsequent adsorption of sulfate-type anions that lead to lower TOF values for catalysts. However, recent studies have shown that Fe–N–C catalyst deactivation does not depend on anion adsorption. Later, Dodelet and coworkers [53] speculated that fluorination of carbon and iron sites could lead to active site stability. However, they found that anion adsorption has no effect on initial ORR performance loss and suggested that the theory of anion neutralization should be abandoned as a contributing factor to the early decay of ORR catalyst performance.

In a recent report [32], Yang et al. investigated the susceptibility of N protonation in Fe–N–C catalysts using DFT. The authors found that protonation is most likely to occur at the Fe site rather than at the neighboring N sites in a pathway that ultimately produces molecular hydrogen. It was also found that protonation at the Fe sites does not result in a significant change in structure of the FeN_x site.

The current understanding of the field suggests that neither the protonation of N sites nor anionic adsorption is significant enough to explain initial catalyst performance losses observed during testing of Fe–N–C catalysts.

13.4 PGM-free Electrocatalyst Durability and Metrics

A generally accepted standard testing procedure for PGM-free electrocatalyst durability and performance assessment would facilitate the development of durable ORR catalysts. Current testing often relies on chronoamperometry and cyclic voltammetry tests developed for PGM catalysts rather than considering the important differences required to develop PGM-free specific testing protocols. In particular, AST protocols for PGM catalysts use a square-wave from 0.6 to 0.95 V under an inert-gas atmosphere to speed up metal dissolution rates and losses in the electrochemically active surface areas (ECSA). These test conditions cause very little performance losses in PGM-free catalysts and often misrepresent the actual durability of PGM-free catalysts.

Instead, test conditions that more closely simulate conditions of PEFC cathode operation, including cycling under oxygen or air, are needed for the assessment of the true PGM-free catalyst durability [56, 57]. The dissimilarities in catalyst degradation conditions for PGM and PGM-free catalysts point toward inherent differences in catalyst active site structure and ORR mechanism. As such, active sites in PGM-free catalysts are more susceptible to degradation pathways induced under an O_2 atmosphere, thus highlighting the importance and need for independent testing protocols for PGM-free class of materials.

Herein, we discuss durability test protocols and catalyst performance assessment for PGM-free ORR electrocatalysts for the PEFC cathode reported by Zhang et al. [19]. The protocol validation was performed in a cross-laboratory study with two independently synthesized ORR electrocatalysts at different research centers. The results of the protocol validation suggest that the developed protocol is effective and reliable for comparing durability and activity of various materials, with concurring outcomes.

13.4.1 Performance and Durability Evaluation in Air-supplied Fuel Cell Cathode

Assessment of a PGM-free electrocatalyst durability in an MEA involves performance determination using multiple electrochemical techniques, including (i) square-wave cycling, (ii) polarization curves, and (iii) cyclic voltammetry under standardized conditions. Variations in experimental conditions such as the length of voltage step, hold time, and gas flow rate have made comparison of catalyst performance in MEAs challenging. Table 13.1 summarizes the proposed testing conditions, metrics used to assess performance, frequency of measurements, and target activity.

Table 13.1 PGM-free MEA durability testing in air protocol.

Test conditions	Square-wave cycle in air: steps between 0.60 V (3 s) and the OCV (but no higher than 0.925 V to minimize carbon corrosion occurring between 0.925 V and 0.95 V) (3 s); a rise time of 0.5 s or less; a single differential cell ≥5 cm^2; maintaining a minimum stoichiometry of ~10; a non-stabilized membrane; report fuel cell polarization curve and cathode cyclic voltammetry in N_2 (or another inert gas) at time intervals and conditions specified below.
Total number of cycles	Until the OCV drops below 0.80 V (up to 30,000)
Cycle time	$6\,s < t \leq 7\,s$
Temperature	80 °C
Anode/cathode RH	100%/100%
Anode gas/cathode gas	H_2/air
Pressure	100 kPa total partial pressure of $N_2 + O_2$

Metric	Frequency	Target
Catalytic activity[a)]	At beginning and end of test minimum	≤40% loss of initial catalytic activity
H_2-air performance at 0.80 V[b),c)]	After 0, 100, 1,000, 5,000, 10,000, and 30,000 cycles	0.300 A cm^{-2} at beginning of test
H_2-air performance at high current density[c)–e)]	After 0, 100, 1,000, 5,000, 10,000, and 30,000 cycles	≤30 mV loss at 0.80 A cm^{-2}

a) Current density (A cm^{-2}) and mass activity (A $mg_{catalyst}^{-1}$) at 0.90 V (*iR*-corrected) in a H_2-O_2 fuel cell, 100 kPa partial pressure of O_2, 100% relative humidity (anode and cathode), 80 °C. Average and individual values are reported for three consecutive curves (the value for each curve is an average of negative- and positive-trending plots). H_2–O_2 polarization curves were acquired from OCV to 0.60 V in 25 mV steps (which were smaller if needed), with a hold time of 45 seconds per point. Constant anode/cathode gas flows of 0.7/0.35 SLPM were used for a 5 cm^2 differential cell, corresponding to ~10/10 stoichiometry at 2 A cm^{-2} for H_2/O_2 (proportionately larger flow rates were used for larger cells). The average current during the last 15 seconds was reported. Multiple points were required at high voltage. An *iR*-free voltage of 0.90 V should be determined on the basis of these points.

b) Current density (A cm^{-2}) at 0.80 V in H_2-air fuel cell, 100 kPa total pressure of $N_2 + O_2$, 100% relative humidity, 80 °C. The average values, total hold time, and average current are reported as in footnote a). H_2-air polarization curves were acquired from 0.20 to 0.70 V in 50 mV steps and from 0.70 V to OCV in 25 mV steps (which were smaller if needed). Constant anode/cathode gas flows of 0.7/1.7 SLPM were used for a 5 cm^2 differential cell, corresponding to ~10/10 stoichiometry at 2 A cm^{-2} for H_2/air (proportionately larger flow rates were used for larger cells).

c) The MEA of the same type, as used in activity testing (membrane, electrode, catalyst loading, and so on).

d) H_2-air fuel cell, 100 kPa total partial pressure of $N_2 + O_2$, 100% RH, 80 °C. The average values, total hold time, and average current are reported as in footnote a).

e) Before durability testing and after 30,000 cycles: polarization curve at 250 kPa, 75% RH, 95 °C (these conditions are subject to updates on the basis of the initial testing results). The average values, total hold time, and average current are reported as in footnote a).

Source: Reproduced from Ref. [19] with the permission of Springer Nature.

Zhang et al. [19] recommend utilizing a differential cell $\geq 5\,cm^2$ for testing in order to minimize effects associated with non-uniform O_2 delivery across the cathode, enhance mass transport, and maintain constant pressure at high flow rates. They also suggest using a non-stabilized membrane that is free of additives, such as cerium, to clearly assess actual catalyst degradation.

Square-wave voltage cycling should be performed in air, with steps between 0.60 V and open circuit voltage (OCV) at −0.1 V and rise times of 0.5 seconds or less. Experimenters should beware of sampling at high potentials, between 0.925 V and 0.95 V, to prevent carbon corrosion. Cycling should be carried out until the measured OCV drops below 0.80 V or up to 30k cycles as a benchmark for induced changes in the catalyst material. For low-temperature applications, these cycling experiments should be performed at 80 °C at anode/cathode relative humidity (RH) of 100%/100% and a 100 kPa total partial pressure of N_2 and O_2.

Polarization curves should be recorded under both (1) H_2/air with constant anode/cathode gas flows of 0.7/1.7 standard liters per minute (SLPM) for a $5\,cm^2$ cell and (2) H_2/O_2 conditions with constant anode/cathode gas flows of 0.7/0.35 SLPM. Polarization curves should be acquired from 0.20 to 0.70 V in 50 mV steps from 0.70 V to OCV in 25 mV steps (or using smaller steps if needed). These flow rates correspond to *ca.* 10/10 stoichiometric ratio at $2\,A\,cm^{-2}$ for H_2/air, which can be adjusted proportionately for larger cells.

Catalyst activity can be assessed via polarization curve at the beginning and end of test, at a minimum by considering current density ($A\,cm^{-2}$) and mass activity ($A\,mg_{catalyst}^{-1}$) at 0.90 V vs. RHE (*iR*-corrected) in H_2/O_2 fuel cell. Experiments should be conducted at 80 °C with 100 kPa partial pressure of O_2 at 100% RH for anode and cathode. Reported values should include the average and individual measurements corresponding to three consecutive curves, with the reported curve being an average of negative- and positive-going polarization curves. Hold times at each voltage should be 45 seconds per point, where the reported current is the average current measured during the last 15 seconds of the hold. Successful catalyst durability would show an activity loss that is ≤40% of initial activity.

Performance of the MEA electrocatalysts should be assessed at both 0.8 V and high current densities. The tests should be acquired after 0, 100, 1k, 5k, 10k, and 30k cycles to monitor catalyst activity changes over time. Current density ($A\,cm^{-2}$) measurements at 0.80 V should be collected using the same MEA type as used in activity testing in a H_2/air fuel cell at 80 °C, 100 kPa total partial pressure of N_2 and O_2, and 100% RH. Target H_2/air performance at 0.80 V is $0.300\,A\,cm^{-2}$ at the beginning of testing (BOT) [13, 14]. For H_2/air performance at high current densities, polarization curves should be collected before durability testing and after 30k cycles. The polarization curve should be acquired at 250 kPa, 75% RH at 95 °C. Target durability performance for H_2/air would show ≤30 mV loss at $0.80\,A\,cm^{-2}$ after testing at 0, 100, 1k, 5k, 10k, and 30k cycles [13, 14].

Lastly, it is recommended to record cyclic voltammetry (CV) under N_2 after 0, 100, 1k, 5k, 10k, and 30k cycles. The proposed protocol suggests collection of CV experiments in two potential windows (i) from 0.20 to 0.70 V vs. RHE and (ii) from 0.70 V to OCV vs. RHE with a scan rate of $20\,mV\,s^{-1}$ at 80 °C with 100% RH. These CV experiments are important diagnostic tools to monitor variations in total double-layer capacitance, catalyst surface functionalization, or other electrochemical behavior that could be insightful for determining causes for performance loss during cycling experiments. Unlike PGM catalysts, the estimated surface-area changes cannot be used to predict ORR activity. For PGM-free catalysts, it is common to observe an increase in double-layer capacitance over time. This increase is often associated with roughening of the surface, for example, due to carbon corrosion, which results in a loss active sites and catalytic activity.

Another recommendation includes electrochemical impedance spectroscopy (EIS) for monitoring changes in proton transport resistance across the membrane. This study is important to distinguish performance losses of the catalyst from losses caused by the ionomer degradation.

13.4.2 Assessment of Carbon Corrosion in Nitrogen-purged Cathode

Carbon corrosion testing is valuable for understanding performance losses due to changes in the carbon support. The testing includes: (i) triangle sweep cycling to simulate and accelerate carbon corrosion during start-up/shut-down operation, (ii) polarization curves to evaluate the effect of induced carbon corrosion on catalyst performance, and (iii) CV as a diagnostic tool for discriminating between corrosion and other possible phenomena influencing catalyst performance. Details for catalyst testing conditions, performance metrics, and maximum allowable carbon corrosion losses are summarized in Table 13.2 [19].

Triangle sweep cycling in N_2 (or in another inert gas) between 1.0 and 1.5 V at 500 mV s^{-1} has been suggested to accelerate carbon corrosion phenomena [19]. Cycling should continue until OCV drops below 0.80 V or up to 5k cycles. Polarization curves should be acquired within two different potential windows: (i) from 0.20 to 0.70 V in 50 mV steps and (ii) from 0.70 V to OCV in 25 mV steps or smaller, if necessary, with constant anode/cathode gas flows of 0.7/1.7 SLPM for both H_2/air and H_2/O_2 fuel cells.

Loss in H_2/air performance should be assessed at 1.5 A cm^{-2} after 0, 100, 1k, 2k, and 5k cycles at 250 kPa total pressure (sum of the partial pressures) at both the cathode and anode, 75% RH and at 80 °C. Individual and average values for three consecutive curves should be reported for a total hold time of 45 seconds per point and an average value for the measured current after 15 seconds. The durability target is ≤30 mV of voltage loss after cycling.

Cyclic voltammetry can be used to assess changes in electrochemical surface area under N_2 and conducted after 0, 100, 1k, 2k, and 5k cycles. Cyclic voltammetry scans should be acquired between 0.20 and 0.70 V vs. RHE with a scan rate of 20 mV s^{-1} at 80 °C with 100% RH. Target losses due to carbon corrosion should be no more than 40% of double-layer charging current value.

13.4.3 Determination of Performance Loss upon Cycling Cathode Catalyst in Nitrogen

Assessment of a PGM-free electrocatalyst durability in MEAs includes interrogating performance under N_2 using multiple electrochemical experiments: (i) square-wave cycling, (ii) polarization curves, and (iii) cyclic voltammetry utilizing the same

Table 13.2 PGM-free MEA carbon corrosion in N_2 protocol.

Test conditions	Triangle sweep cycle in N_2 (or another inert gas): 500 mV s^{-1} between 1.0 and 1.5 V; single differential cell ≥5 cm^2; non-stabilized membrane; report fuel cell polarization curve and cathode cyclic voltammetry in N_2 (or another inert gas) at time intervals and conditions specified below.
Total number of cycles	Until OCV drops below 0.80 V (up to 5k)
Cycles time	2 s
Temperature	80 °C
Anode/cathode RH	100%/100%
Anode gas/cathode gas	H_2/N_2 (or another inert gas)
Pressure	Atmospheric

Metric	**Frequency**	**Target**
Loss in H_2-air performance at 1.5 A cm^{-2} [a)]	After 0, 100, 1k, 2k, and 5k cycles	≤30 mV

a) Voltage (V) at 1.5 A cm^{-2} in H_2-air fuel cell, 250 kPa total pressure, 75% RH, 95 °C (these conditions are subject to updating based on the results of initial testing); acquire polarization curve from 0.20 to 0.70 V in 50 mV steps and from 0.70 V to OCV in 25 mV (smaller if needed) steps. Record H_2-air polarization curves at constant anode/cathode gas flows of 0.7/1.7 SLPM for a 5 cm^2 cell corresponding to *ca.* 10/10 stoic at 2 A cm^{-2} for H_2/air (use proportionately larger flow rates for larger cells). Report average and individual values for three consecutive curves (value for each curve should be an average of negative- and positive-going plots); total hold time of 45 seconds per point, report average value during the last 15 seconds.

Source: Reproduced from Ref. [19] with the permission of Springer Nature.

potential range, hold times, and scan rates for testing as described in Section 13.4.1 under air. Following the same standard protocol to assess catalyst durability and performance after cycling is important to understand potential catalyst deactivation pathways under specific conditions. Suggested test conditions, metrics used to assess performance, frequency of measurements, and target activity are detailed in Table 13.3.

Electrochemical cycling tests are recommended for durability testing to mimic the actual application conditions. In this case, testing under N_2 does not effectively influence catalyst durability; instead, it is useful for establishing baseline electrochemical activity in the absence of an O_2 substrate. Tracking the changes of the catalyst before and after exposure to the durability protocol under air can be used to better understand how changes in catalyst performance over time influence different features of the cyclic voltammetry scan.

Zhang et al. [19] suggest using square-wave cycling of MEAs under N_2 (or another inert gas), with steps between 0.60 V vs. RHE and OCV with rise times of 0.5 seconds or less to simulate drive cycle operation of PEFCs. Experimenters should beware of sampling at high voltages between 0.925 and 0.95 V to prevent carbon corrosion. Cycling experiments ought to conclude when OCV drops below 0.80 V vs. RHE or

Table 13.3 PGM-free MEA durability testing in N_2 protocol.

Test conditions	Square-wave cycle in N_2: steps between 0.60 V (3 s) and OCV (but no higher than 0.925 V to minimize carbon corrosion occurring between 0.925 and 0.95 V) (3 s); rise time of 0.5 s or less; single differential cell $\geq 5\ cm^2$; non-stabilized membrane; report fuel cell polarization curve and cathode cyclic voltammetry in N_2 (or another inert gas) at time intervals and conditions specified below.
Total number of cycles	Until OCV drops below 0.80 V (up to 30k)
Cycle time	$6\ s < t \leq 7\ s$
Temperature	80 °C
Anode/cathode RH	100%/100%
Anode gas/cathode gas	H_2/air
Pressure	100 kPa total partial pressure of $N_2 + O_2$

Metric	Frequency	Target
Catalytic activity[a)]	At beginning and end of test minimum	≤40% loss of initial catalytic activity
H_2-air performance at 0.80 V[b),c)]	After 0, 100, 1k, 5k, 10k, and 30k cycles	0.300 A cm^{-2} at beginning of test (BOT)
H_2-air performance at high current density[c)–e)]	After 0, 100, 1k, 5k, 10k, and 30k cycles	≤30 mV loss at 0.80 A cm^{-2}

a) Current density (A cm^2) and mass activity (A $mg_{catalyst}{}^{-1}$) at 0.90 V (*iR-corrected*) in H_2–O_2 fuel cell, 100 kPa partial pressure of O_2, 100% RH (anode and cathode), 80 °C; report average and individual values for three consecutive curves (value for each curve should be an average of negative- and positive-going plots); acquire H_2–O_2 polarization curves from OCV to 0.60 V in 25 mV steps (smaller if needed), hold time of 45 seconds/point; use constant anode/cathode gas flows of 0.7/0.35 SLPM for a 5 cm^2 differential cell, corresponding to *ca.* 10/10 stoic at 2 A cm^{-2} for H_2/O_2 (use proportionately larger flow rates for larger cells); report average current during the last 15 seconds. Multiple points required at high voltage; 0.90 V *iR-free* voltage should be determined based on these points.

b) Current density (A cm^{-2}) at 0.80 V in H_2-air fuel cell, 100 kPa total pressure of $N_2 + O_2$, 100% RH, 80 °C. Average values, hold time and average current should be reported as in (*a*). Acquire polarization curve from 0.20 to 0.70 V in 50 mV steps and from 0.70 V to OCV in 25 mV (smaller if needed) steps. Record both H_2-air polarization curves at constant anode/cathode gas flows of 0.7/1.7 SLPM for a 5 cm^2 cell corresponding to *ca.* 10/10 stoic at 2 A cm^{-2} for H_2/air (use proportionately larger flow rates for larger cells).

c) MEA of the same type as used in activity testing (membrane, electrode, catalyst loading, etc.).

d) H_2-air fuel cell, 100 kPa total partial pressure of $N_2 + O_2$, 100% RH, 80 °C. Average values total hold time, and average current should be reported as in (*a*).

e) Before durability testing and after 30k cycles: polarization curve at 250 kPa, 75% RH, 95 °C (these conditions are subject to updating based on the results of initial testing). Average values, total hold time, and average current should be reported as in (*a*).

Source: Reproduced from Ref. [19] with the permission of Springer Nature.

up to 30k cycles. They should be performed at 80 °C with a 100% RH for both the anode and cathode with 100 kPa total partial pressure of N_2.

Experimental details are tailored for use with a single differential cell ≥5 cm² operating with a non-stabilized membrane. Fuel cell polarization curves should be acquired in two regions: (i) from 0.20 to 0.70 V using 50 mV steps and (ii) from 0.70 V to OCV in 25 mV steps or smaller, as necessary. Polarization ought to be recorded under both H_2/air and H_2/O_2 at constant anode/cathode gas flows of 0.7/1.7 SLPM in a 5 cm² cell. Gas flows should be adjusted proportionately for larger cells.

Changes in catalyst performance from initial to the end of cycling can be evaluated using polarization curves to compare current density ($A\ cm^{-2}$) and mass activity ($A\ mg_{catalyst}^{-1}$) at 0.90 V vs. RHE (*iR*-corrected) in H_2/O_2 fuel cell. For this protocol, polarization curves are conducted at 80 °C with 100 kPa partial pressure of O_2 at cathode and 100% RH for anode and cathode. Average and individual measurements corresponding to three consecutive curves should be reported. Reported curves are recommended to include the average of negative- and positive-going plots. Hold times at each voltage should be 45 seconds/point, where the reported current is the average current measured during the last 15 seconds of the hold. A catalyst is considered as meeting durability standards if it shows an initial activity loss of no more than 40%.

PGM-free electrocatalyst performance in MEA should be assessed at 0.8 V vs. RHE and at high current densities. Performance tests are to be performed after cycling at 0, 100, 1k, 5k, 10k, and 30k cycles. Current density ($A\ cm^{-2}$) measurements need to be collected at 0.80 V vs. RHE in a H_2/air fuel cell at 80 °C with 100 kPa total partial pressure of N_2 and O_2 and at 100% RH using the same type of MEA for all testing. Reported values should follow the same voltage average, total hold time, and current average as used for activity assessments. The target H_2/air performance at 0.80 V vs. RHE is 0.300 $A\ cm^{-2}$ at BOT. For H_2/air performance at high current densities, polarization curve should be collected before durability testing and after 30k cycles. The polarization curve should be acquired at 250 kPa, 75% RH at 95 °C. Average values, hold times, and current densities need to be reported. Target H_2/air performance loss should be no more than 30 mV at 0.80 $A\ cm^{-2}$ after 30k cycles. Cyclic voltammetry under N_2 (or other inert gas) should be collected after 0, 100, 1k, 5k, 10k, and 30k cycles with a scan rate of 20 $mV\ s^{-1}$ at 80 °C at 100% RH.

13.4.4 Recommendations for ORR Electrocatalyst Evaluation in RRDE in O_2 and in an Inert Gas

While not obligatory, electrocatalyst testing via RRDE is advantageous because performance and durability of the catalyst can be assessed on a smaller scale, without using large amounts of catalyst required for MEA testing in fuel cells. The tests should be in a three-electrode setup, including a PGM-free counter electrode (e.g. glassy carbon rod) and a reference electrode. The recommended catalyst loading on the glassy carbon disk (the working electrode) is 0.6 $mg\ cm^{-2}$ with reported ionomer content. The ideal electrolyte for this test is 0.5 M H_2SO_4 or another non-adsorbing electrolyte saturated with O_2 or inert gas.

ORR polarization curves should be collected using square-wave voltammetry in O_2-saturated solution with steps between 0.060 V vs. RHE and OCP, with a rise time of 0.5 seconds or less and at 900 rpm. Average and individual values from three consecutive polarization curves should be reported. Cyclic voltammetry should also be recorded under N_2 (or another inert gas) in the same potential range as used for recording ORR polarization curves at a scan rate of 20 mV s^{-1}. Steady-state potential step experiments are recommended to be corrected for double-layer charging.

Durability cycling experiments in RRDE should be conducted at the same intervals as fuel cell testing, with performance checks after 0, 100, 1k, 3k, 10k, and 30k cycles. Total number of cycles should continue until OCP drops below 0.80 V vs. RHE or up to 30k cycles. After each cycling set, changes in $E_{1/2}$ (V vs. RHE), OCP (V vs. RHE), and double-layer capacitance should be recorded. From RRDE ring data, H_2O_2 yield should be collected across entire potential range of the RDE experiment from 0 to 1.0 V vs. RHE after each cycling set. Detailed conditions and metrics for durability testing are outlined in supplementary information of the work by Zhang et al. [19].

13.4.5 Electrocatalyst Corrosion

Accelerated carbon corrosion occurs at high potentials, >0.90 V vs. RHE, and can lead to loss of electrochemically active surface area, detachment of catalyst, and mass transport issues via porosity loss. RRDE tests can be performed using a three-electrode set-up, including a PGM-free counter electrode (e.g. glassy carbon rod) and reference electrode. A catalyst loading of 0.6 mg cm^{-2} on the glassy carbon disk of the RRDE is recommended. Ionomer content of the working electrode should also be reported. Electrochemistry experiments should be collected in 0.5 M H_2SO_4 or another electrolyte saturated with atmospheric pressure O_2.

The corrosion polarization curves collected using triangle sweep voltammetry in N_2-saturated solution (or another inert gas) at 500 mV s^{-1} with steps between 1.0 V vs. RHE and 1.5 V vs. RHE and at a disk rotation of 900 rpm is recommended. Total number of cycles should continue until OCP drops below 0.80 V vs. RHE or up to 5000 cycles. Average and individual values from three consecutive polarization curves should be reported. Slow cyclic voltammetry scans at 20 mV s^{-1} and steady-state potential step experiments should be corrected for double-layer charging.

Durability cycling experiments in RRDE ought to be conducted at the same intervals as fuel cell testing with performance checks after 0, 100, 1k, and 5k cycles. After each cycling set, changes in $E_{1/2}$ (V vs. RHE), OCP (V vs. RHE) and double-layer capacitance should be recorded. From RRDE data, H_2O_2 yield should be assessed across entire potential range of the RDE experiment, from 1.0 to 1.5 V vs. RHE, after each cycling set. Details of conditions and metrics for durability testing are outlined in the text by Zhang et al. [19].

13.5 Low-PGM Catalyst Degradation

Large-scale application of fuel cells for automotive transportation has been hindered by the high cost and limited durability of PGM catalysts required at loadings

>0.3 mg_{Pt} cm_{elec}^{-2}. Understanding the pathways of PGM catalyst performance degradation is necessary to overcome catalyst lifetime durability and cost constraints, as well as to pave the way for further catalyst development and their broad implementation in fuel cells. Strategies for improving performance of PGM catalysts have focused on: (i) material development for greater catalyst mass activity [58, 59], (ii) electrode design and fabrication [60, 61], (iii) reduction of oxygen transport resistance [62–64], and (iv) dispersion and stability of small Pt particle sizes [65]. However, the challenge of improved performance without hampering durability remains at the forefront [66–69].

PGM catalyst degradation in PEFCs is linked to the loss of ECSA resulting from primary degradation processes such as: (i) Pt dissolution and (ii) carbon support corrosion, which can initiate secondary processes, including (iii) detachment of Pt particles, (iv) size-driven dissolution of small Pt particles and re-deposition onto larger Pt particles, known as Ostwald ripening, and (v) migration and coalescence of Pt particles on a carbon support [66, 67, 69–76]. These processes are illustrated in Figure 13.6. Although all these degradation paths are supported in the literature by experimental evidence, the dominating drivers for PEFC losses at voltage values >0.8 V are Pt dissolution and carbon support corrosion. The pathways leading to low-PGM catalyst performance losses are discussed below.

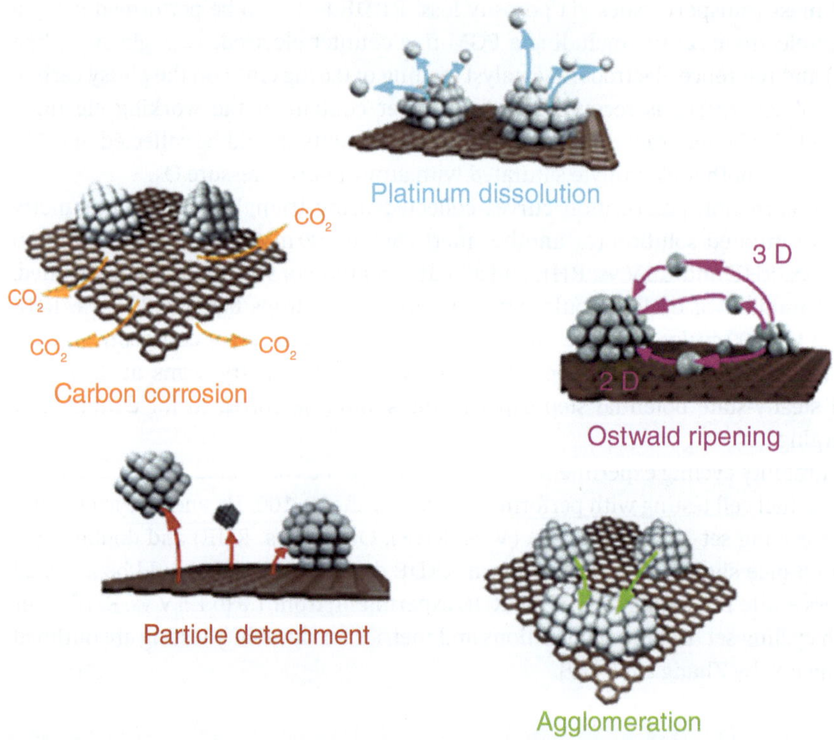

Figure 13.6 Illustration depicting the proposed degradation mechanisms for carbon-supported Pt catalysts in fuel cells. Source: Reproduced from Ref. [74]. Beilstein-Institut/CC BY 2.0.

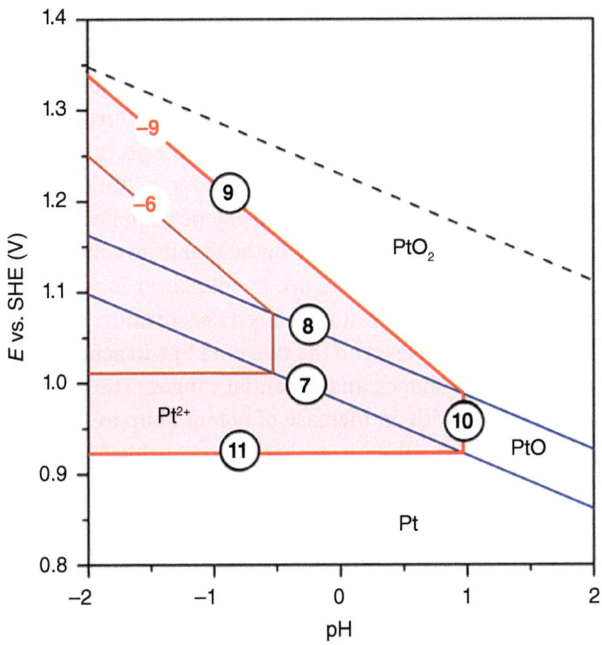

Figure 13.7 Potential-pH Pourbaix diagram describing Pt dissolution at 298 K, where the highlighted area (red) denotes the regions of Pt^{2+} stability at concentrations on the order of 10^{-6} and 10^{-9} mol dm^{-3} (red numbers). The lines delineate the reactions (13.7–13.11) in the text at equilibrium. Source: Reproduced from Ref. [67] with the permission of Elsevier.

13.5.1 Pt Dissolution

Conditions that favor Pt dissolution can be predicted from Pourbaix diagrams that describe the potential-pH equilibria (Figure 13.7). Relevant reactions involve Pt oxidation to PtO_3 (13.7–13.9), as well as Pt electrochemical pathways (13.10–13.12) [67].

$$Pt + H_2O \rightarrow PtO + 2H^+ + 2e^- \qquad E^\circ(Pt^\circ/PtO) = 0.98–0.059\,pH \tag{13.7}$$

$$PtO + H_2O \rightarrow PtO_2 + 2H^+ + 2e^- \qquad E^\circ(PtO/PtO_2) = 1.045–0.059\,pH \tag{13.8}$$

$$PtO_2 + H_2O \rightarrow PtO_3 + 2H^+ + 2e^- \qquad E^\circ(PtO_2/PtO_3) = 2.0–0.059\,pH \tag{13.9}$$

$$PtO + 2H^+ \rightarrow Pt^{2+} + H_2O \qquad \log(Pt^{2+}) = -7.06–2\,pH \tag{13.10}$$

$$Pt \rightarrow Pt^{2+} + 2e^- \qquad E^\circ(Pt^\circ/Pt^{2+}) = 1.2 + 0.29\,\log(Pt^{2+}) \tag{13.11}$$

$$PtO_2 + 4H^+ + 2e^- \rightarrow Pt^{2+} + 2H_2O \qquad E^\circ(PtO_2/Pt^{2+}) = 0.84–0.12\,pH–0.029\,\log(Pt^{2+}) \tag{13.12}$$

In the Pourbaix diagram, the circled numbers correspond to different redox or chemical reactions participating in Pt speciation and the red numbers describe the concentration of dissolved Pt. From this diagram, one can see that an increase of electrode potential at micromolar concentrations of dissolved Pt^{2+} at pH 0 results in the formation of PtO_2 without further loss of Pt into solution. However, at nanomolar concentrations of dissolved Pt^{2+}, an increase in potential at pH 0 leads to further Pt^{2+} dissolution. While these diagrams represent oversimplifications of the processes that result in Pt dissolution and subsequent catalyst performance loss in PEFCs, they are useful for investigating possible phenomena influencing practical applications of Pt-based catalysts. For example, there has been a debate in the literature concerning the mechanisms of Pt dissolution, including the nature of soluble Pt ions and Pt oxide intermediates that are hypothesized to result in catalyst deactivation [75–78].

Early studies by Mitsushima et al. [78] focused on the rate of Pt loss in acidic environments using various potential-cycle shapes and potential ranges. The authors found that the rate of Pt loss increased with an increase of potential up to 1.8 V vs. RHE. They also report a dependence of Pt solution speciation to Pt oxide formation via symmetric or asymmetric cycling conditions and scan rate. A symmetric cycle was described by the same scan times between the high potential (E_H) and low potential (E_L), and an asymmetric scan includes a changing scan rate for the E_H and E_L potential sweeps, resulting in variable scan time between potentials. These cases are illustrated in Figure 13.8. For slow scan rate of cathodic triangular wave ($0.5\,V\,s^{-1}$), the electron transfer number (n) was found to be two, which suggests that Pt oxidation occurs before the reduction of PtO_2 to Pt^{2+} described in Eq. 13.12.

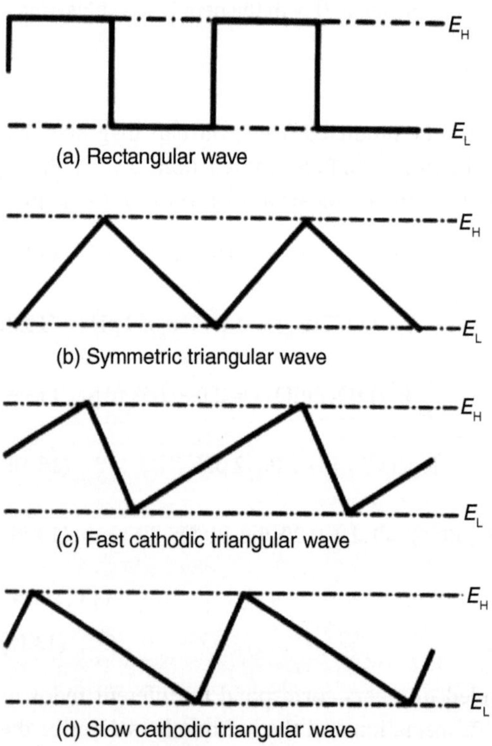

Figure 13.8 Illustration of potential wave forms from high potential (E_H) to low potential (E_L) as a function of time for (a) rectangular wave, (b) symmetric triangular wave, (c) fast cathodic triangular wave, and (d) slow cathodic triangular wave. Source: Reproduced from Ref. [78] with the permission of IOP Publishing.

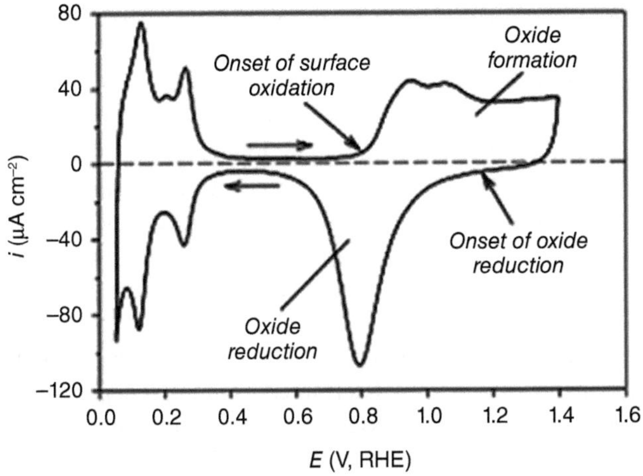

Figure 13.9 Cyclic voltammogram of polycrystalline Pt electrode in 0.5 M H_2SO_4 at a 50 mV s^{-1} scan rate and 298 K from surface oxide growth of Pt electrodes. Source: Reproduced from Ref. [89] with the permission of Elsevier.

Whereas, for symmetric waves or fast scan (20 V s^{-1}) cathodic triangular waves n was found to be four. The overall dissolution reaction with a four-electron transfer was proposed to occur via the oxidation reaction described in Eq. (13.3), which is independent of Pt oxide formation.

$$Pt \rightarrow Pt^{4+} + 4e^- \tag{13.13}$$

These findings point toward a relationship between Pt oxide formation and Pt dissolution. However, there is disagreement on the role of Pt oxides in Pt dissolution. Some investigators speculate that Pt dissolution takes place during positive-going potential sweeps that result in Pt oxides, with Pt loss occurring via either Pt oxide reduction [79–81] or chemical Pt dissolution [82–84] of Pt oxides into solution. A competing opinion suggests that Pt oxide forms during Pt dissolution and acts as a protective passivating layer [79, 80, 85, 86].

Platinum oxide formation is thought to contribute to Pt dissolution by either detachment of Pt atoms or formation of soluble Pt species [87, 88]. Cyclic voltammograms of a polycrystalline platinum electrode in 0.5 M H_2SO_4 solution show the onset of surface oxidation at 0.85 V vs. RHE and reduction of the platinum oxide starting at 1.2 V vs. RHE (Figure 13.9). Platinum oxide formation on the electrode is proposed to result from: (i) chemisorbed OH radicals from H_2O deprotonation [90–92], (ii) absorbed O species [89, 93], or (iii) point defects [94, 95].

PEFC end-of-life (EOL) tests show platinum particles enrichment at the anode–membrane interface. The mechanism and potential for Pt dissolution and subsequent particle formation are currently under debate. Pt dissolution mechanisms that can lead to particle formation are presumed to result from three major pathways, including (i) Ostwald ripening, where Pt species in solution or in ionomer deposit onto already-formed particles in the membrane, (ii) competing particle migration mechanism resulting in an agglomeration of Pt nanoparticles, and (iii) agglomeration of Pt particles resulting from carbon support corrosion [96–100].

Recent studies utilizing *in situ* ICP-MS have provided better insight into PGM catalyst durability and dissolution. In a study coupling ICP-MS with RRDE experiments, Lopes et al. [101] measured the Pt dissolution rates to find that the degree of platinum stability depends on several factors, including the degree and strength of Pt coordination environment, as well as covalent and non-covalent interactions of species in solution. In that study, it was determined that the low coordinated Pt(110) structure was subject to more Pt dissolution, whereas the Pt(111) structure was found to be most stable to dissolution.

Additional studies have revealed a correlation of dissolution rate with potential [83] and particle size [84]. Cherevko et al. [83] determined a potential dependence of Pt dissolution mechanisms, where anodic dissolution is the dominant process, with traces of dissolved platinum detected at 0.85 V vs. RHE via ICP-MS. In a secondary process, platinum dissolution was also detected at low potentials during the reduction of accumulated platinum oxide. In a temperature-dependence study utilizing electrochemical quartz crystal nanobalance (EQCN) technique, by Inzelt et al. [87], Pt dissolution was observed during platinum oxide reduction. The extent of Pt dissolution varied with potential, and scan rate, but was less affected by changes in temperature.

Sandbeck et al. [65] have reported the effect of Pt particle size on platinum dissolution. In that study, Pt particles from 2 to 10 nm were deposited on glassy carbon substrate and subjected to an accelerated stability testing protocol to initiate Pt dissolution. The accelerated stability testing protocol utilized CVs, from 0.6 V to 1.5 V vs. RHE at 500 mV s^{-1} for 1000 cycles, to cause Pt dissolution with minimal carbon corrosion [102]. The authors found that decreasing particle size resulted in increasing dissolution rates. Decreasing particle size resulted in an increase in metal oxophilicity and the formation of a Pt oxide protective layer. Increasing particle size led to a decrease in electrochemically active surface area and Pt dissolution.

Understanding the factors that initiate Pt dissolution in PEFCs and the degradation pathways leading to performance losses represent important challenge in the field. Advancements of *in situ* spectroscopic techniques have provided important insights in that regard. Future progress in the field of PGM electrocatalysis must focus on addressing catalyst durability issues before widespread implantation of PEFCs in transportation applications.

13.5.2 Carbon Support Corrosion

Porous carbons are widely used as catalyst supports due to their chemical and physical stability, high surface area, and conductivity. However, electrochemical testing of carbon-supported PGM catalysts has shown that these supports are susceptible to carbon oxidation reaction (COR, reaction 13.14), resulting in CO_2 formation [103].

$$C + 2\,H_2O \rightarrow CO_2 + 4\,H^+ + 4e^- \qquad E = 0.207\ \text{V vs.RHE} \qquad (13.14)$$

COR is thermodynamically favorable above a low potential of 0.207 V vs. RHE, but is kinetically handicapped, allowing carbon-based materials to operate in PEFCs at

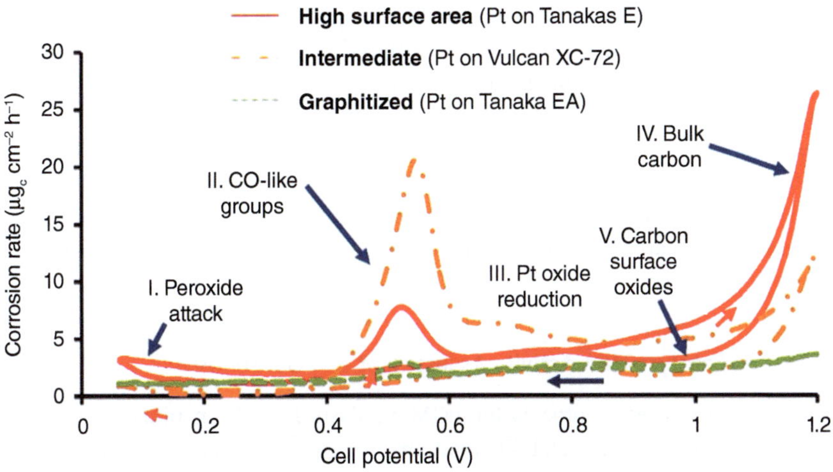

Figure 13.10 Various corrosion steps on three different types of carbon used cathode catalyst layer including Pt on high surface area (red), Pt on Vulcan (orange, dotted, and Pt on graphitized carbon support (LSAC) (green, dotted). Source: Reproduced from Ref. [109] with the permission of IOP Publishing.

much higher potentials (0.6–1.2 V vs. RHE). The rate of COR becomes challenging at high potentials 1.0–1.1 V vs. RHE in acidic conditions [66, 104, 105]. These challenging conditions, which occur especially during the PEFC startup and shutdown, can initiate COR, ultimately resulting in collapse of the electrode structure [106, 107]. Roen et al. [108] examined the influence of varying platinum concentration, catalyst type, and temperature on carbon corrosion, utilizing an in-line mass spectroscopy during cyclic voltammetry scans. They observed an increase in CO_2 generation rates with increasing Pt mass fraction, suggesting that Pt can catalyze COR.

In a study by Macauley et al. [109], the effects of carbon corrosion were quantified by monitoring CO_2 generation by non-dispersive infrared spectroscopy (NDIR) in various potential ranges. These studies were conducted on three different carbon supports with different levels of graphitization (Figure 13.10). In these experiments, peak V has been attributed to oxidation of carbon surface oxides, peak IV has been associated with the electrochemical oxidation of carbon by water to generate CO_2, and peak III has been related to the oxidation of carbon via reduction of platinum oxides. Peaks I and II have been assigned to carbon oxidation by hydrogen peroxide [104]. Based on these experiments, the authors have concluded corrosion rates are reduced with highly graphitized carbons, making such materials suitable for durable carbon supports for PGM materials.

Carbon support corrosion has been also suggested to contribute to performance losses through platinum particle detachment and decreased electrical conductivity. Ioroi et al. [110] utilized accelerated degradation testing to demonstrate how cathode corrosion can dramatically decrease the catalyst layer thickness. CVs recorded following degradation tests show losses of ECSA of platinum, suggesting corrosion of the carbon support leads to decreased conductivity. The loss of Pt particles via detachment also contributes to secondary degradation pathways that result in losses of ECSA. An investigation by Zhang et al. [111] found that the effects of carbon corrosion also influence mass transfer losses by damaging electrode microstructure and enhancing hydrophilicity. The increased hydrophilicity leads in turn to micropore flooding and increased mass-transport resistance.

Carbon support corrosion is of major importance for PGM catalysts durability. Oxidation of the carbon support during start-up and shutdown is implicated in the losses of ECSA from Pt particle loss and increasing material resistance. Understanding how different types of carbon resist corrosion is important for minimizing corrosion rates under operating conditions.

13.5.3 Pt Catalyst MEA Activity Assessment and Durability

Accelerated stress test protocols for PGM catalysts have been updated as better understanding of structure–stability relationships of Pt catalysts has emerged. While spectacular advancement in activity of Pt electrocatalysts has been demonstrated using rotating disk electrode (RDE), these catalysts do not often retain their high performance once incorporated into fuel cell MEAs. Conditions used for MEA processing such as ink formulation, ionomer content, or carbon support significantly impact the catalyst performance and durability during PEFC operation. Herein, we describe the most recent suggested protocols for conditioning and testing of PGM catalyst durability, utilizing ASTs in PEFCs [64, 112].

13.5.4 PGM Electrocatalyst MEA Conditioning in H_2/Air

A typical MEA conditioning protocol includes a voltage hold, voltage/load cycling, and performance measurements to check effectiveness of the approach used. MEA conditioning and its impact on performance were recently investigated by Kabir et al. [64] in an effort to develop standardized test protocol and achieve optimal performance. The authors investigated conditioning by voltage cycling, followed by a low-voltage hold (0.1 V) under over-saturated conditions (150% RH) and at low cell temperature (40 °C). These conditioning steps reduced both local oxygen transport and ionomer resistance, resulting in improved accessibility of Pt sites. The voltage hold was found to aid in the removal of surface-blocking species, sulfates (bisulfates), and oxides, to improve mass activities by a factor of between two and three. The suggested MEA conditioning protocol reported by Kabir et al. [64] is provided in Table 13.4.

The MEA break-in and catalyst activation is to be performed in an H_2/air fuel cell test station at 80 °C and 150 kPa absolute pressure. The cell voltage changes between 0.9 and 0.725 to 0.625 V. A detailed break-in protocol and voltage sequence are summarized in Table 13.4 [64]. Next, the MEA is held at OCV for 30 seconds at 100% RH, followed by holds at 0.9 and 0.6 V. The RH is changed during a 0.7 V hold to 32%, followed by a voltage hold cycle at OCV, 0.9, and 0.6 V. The RH is increased to 100% after a 0.7 V hold, for a final cycle of voltage holds at OCV, 0.9, and 0.6 V.

The BOL cyclic voltammograms (CVs) are conducted to measure Pt ECSA, determined by integrating the hydrogen underpotential deposition region of the

Table 13.4 Break-in protocol for PGM-based catalysts conducted in H_2/air, 80 °C, 150 kPa.

Step	Procedure	Time	Voltage (V)	T_{cell} (°C)	RH (%)	BP (kPa_{abs})	Flow rate (slpm) A	C
	Stabilization	2 min	0.600	80	100	150	1.0	2.5
	Cycle cell voltage between 0.900 V and 0.725 to 0.625 V	4 min	0.725	80	100	150	1.4	3.3
		3 min	0.900					
		4 min	0.700					
		3 min	0.900					
		4 min	0.675					
		3 min	0.900					
		4 min	0.650					
		3 min	0.900					
		4 min	0.625					
	Open circuit voltage (OCV)	30 s		80	100	150	1.5	3.7
×10 100% RH	0.9 V hold	3 min	0.900					
	0.6 V hold	4 min	0.600					
	0.7 V hold for RH change	2 min	0.700	80	32	150	1.5	3.7
	Open circuit voltage (OCV)	30 s		80	32	150	1.5	3.7
×5 32% RH	0.9 V hold	3 min	0.900					
	0.6 V hold	4 min	0.600					
	0.7 V hold for RH change	2 min	0.700	80	100	150	1.5	3.7
	Open circuit voltage (OCV)	30 s		80	100	150	1.5	3.7
×10 100% RH	0.9 V hold	3 min	0.900					
	0.6 V hold	4 min	0.600					

Source: Reproduced from Ref. [64] with the permission of the American Chemical Society.

voltammogram. CVs should be collected at 30 °C, in H_2/N_2 cell at 150 kPa and 100% RH, between 0.06 and 1.2 V, at 50 or 20 mV s^{-1} scan rates [64].

A voltage recovery (VR) step is important for the removal of surface contaminants such as sulfates. The first VR includes a potential hold at 0.85, 0.6, and 0.1 V at various times under H_2/air (0.9/0.5 SLPM) at 40 °C and 150% RH. The protocol details

Table 13.5 Voltage recovery protocol for PGM-based catalysts for catalyst conditioning monitoring.

Step	Procedure	Time	Voltage (V)	T_{cell} (°C)	RH (%)	BP (kPa)	Flow rate (slpm) H_2/Air A	C
×2 Voltage recovery	0.85 V hold	5 min	0.850	<u>40</u>	<u>150</u>	150	2.3	2.5
	0.60 V hold	10 min	0.600				0.9	0.5
	0.10 V hold	1 h	0.100					

Source: Reproduced from Ref. [64] with the permission of the American Chemical Society.

are provided in Table 13.5 [64]. Cell performance is assessed and monitored utilizing H_2/O_2 and H_2/air polarization measurements at 80 °C, 100 kPa O_2 partial pressure (150 kPa cell pressure) and 100% RH for four minutes per point. *I–V* curves should be collected from 0.75 V to OCV (positive-going voltage scan). As at the BOL point of testing, the EOL Pt ECSA should be determined from CVs recorded at 30 °C and 100% RH (H_2/N_2 cell). The scans should be collected between 0.05 and 1.2 V at a scan rate of 20 or 50 mV s^{-1}.

13.5.5 Accelerated Stress Test of PGM Electrocatalyst Durability

AST protocols are recommended to study degradation of PGM catalysts via dissolution, catalyst particle agglomeration, carbon corrosion, and by other mechanisms in operating PEFC MEAs. Various test protocols have been reported in the literature, typically recommending the voltage range between 0.6 and 1.0 V (or 0.95 V) for determining the effect of Pt dissolution and the range from 1.0 to 1.5 V to assess the effect of carbon corrosion. The variations in AST protocols found in the literature make it difficult to compare catalyst performance and durability. Herein, we summarize an AST that has been adopted by the US Department of Energy (DOE), which is believed to accurately simulate material degradation, including loss of ECSA, Pt metal dissolution, and carbon support corrosion in an MEA under actual fuel cell cathode operating conditions (see Table 13.6 for details) [112, 113].

The DOE-adopted AST is designed to evaluate electrocatalyst durability by applying square-wave voltage cycles. This AST uses the voltage range proposed for automotive fuel cell operation, in which the catalyst surface undergoes oxidation and reduction, leading to catalyst degradation. The voltage range is designed to minimize carbon support corrosion and isolate performance losses specific to the PGM catalyst. The AST should be conducted with carbon-supported PGM catalyst in an MEA using square-wave cycles between 0.6 and 0.95 V, hold time of 3 seconds at both voltages, and a rise/drop time of no more than 0.5 seconds. The AST is to be conducted for 30k cycles in H_2/N_2 cell at 80 °C, 100% RH (H_2 at 200 standard cubic centimeter per minute, sccm; and N_2 at 75 sccm; 50 cm^2 cell).

Catalyst mass activity should be determined at the beginning and end of testing, with a target loss of initial catalytic activity of no more than 40%. To monitor activity

Table 13.6 Accelerated stress test protocol to assess PGM-based catalyst durability in PEFCs.

	Electrocatalyst cycle and metrics
Cycle	Square wave cycle: steps between 0.6 V (3 s) and 0.95 V (3 s) with rise time of ~0.5 s or less; run polarization curve and ECSA at specified intervals. Single cell 25–50 cm^2
Number	30,000 cycles
Cycle time	6 s
Temperature	80 °C
Relative humidity	Anode/cathode 100/100%
Fuel/oxidant	H_2/N_2 (H_2 at 200 sccm and N_2 at 75 sccm for a 50-cm^2 cell)
Pressure	Atmospheric pressure

Metric[a)]	Frequency	Target
Catalytic mass activity[b)]	At beginning and end of test, minimum	<40% loss of Initial catalytic activity
Polarization curve from 0 to >1.5 $A\,cm^{-2}$[c)]	After 0, 1k, 5k, 10k, and 30k cycles	<30 mV loss at 0.8 $A\,cm^{-2}$
ECSA/cyclic voltammetry	After 10, 100, 1k, 3k, 10k, 20k, and 30k cycles	<40% loss of initial area

a) A protocol such as the one shown in Table 13.7 should be used to recover reversible losses prior to measuring each metric.
b) Mass activity in $A\,mg^{-1}$ @150 kPa abs back pressure at 900 mV *iR*-corrected on H_2/O_2, 100% RH, 80 °C, anode stoichiometry 2; cathode stoichiometry 9.5. A minimum hold time of 15 minutes is recommended, with the mass activity calculated based on the average current during the last one minute. Multiple points should be measured at low current, and the 0.9 V *iR*-free potential should be determined based on these measurements. Measured ORR current may be corrected for H_2 crossover. Based on the protocol published by Gasteiger et al. [9].
c) Polarization curve per protocol in this table.
Source: Reproduced from Ref. [113]. US Department of Energy Report.

losses, polarization curves at current densities between 0 and 1.5 $A\,cm^{-2}$ should be conducted after 0, 1k, 5k, 10k, and 30k cycles, with a target voltage loss of no more than 30 mV at a current density of 0.8 $A\,cm^{-2}$. The ECSA should be determined by cyclic voltammetry after 10, 100, 1k, 3k, 10k, 20k, and 30k cycles, with the target area loss of no more than 40%.

13.6 Conclusion

Despite the remarkable progress in the past three decades, improvements in catalyst durability and activity are still necessary to reduce fuel cell stack cost and make fuel cell power systems competitive for automotive applications. Thanks to

Table 13.7 MEA recovery protocol[a)].

Step	Step name	Anode comp.	Anode flow (SLPM)	Cathode comp.	Cathode flow (SLPM)	Duration (s)
1	N_2 soak	100% N_2	2	100% N_2	4	120
2	Air soak	N/A	0	Air	4	900
3	N_2 soak	100% N_2	2	100% N_2	4	120
4	H_2 soak	100% H_2	2	N/A	0	600
5	H_2-air back on	100% H_2	2	Air	4	5

a) The following parameters are constant throughout the test anode and cathode inlet RH = 100%, anode and cathode outlet pressure = 150 kPa abs. Cell temperature should be set to the temperature of the upcoming diagnostic test,

recent advancements in activity, PGM-free electrocatalysts have largely reduced the performance gap between PGM-based materials. However, both classes of catalysts continue to suffer from performance losses during operation, calling for improvements to durability.

Ongoing research into catalyst degradation pathways for both PGM and PGM-free catalysts is crucial for gaining a better understanding of the structure–activity relationships and develop durable catalysts for transportation applications. The major durability challenges of both PGM and PGM-free catalysts include the impact from both carbon corrosion and metal-loss pathways. For PGM catalysts, Pt dissolution and carbon corrosion can spur secondary effects such as detachment of Pt particles from the support or Pt agglomeration. Similarly, PGM-free catalysts are also susceptible to metal loss via active site demetallation and carbon corrosion. Methods to minimize and understand degradation pathways for both PGM and PGM-free catalysts are at the center of current research.

Performance and durability test protocols for fuel cell ORR catalysts have been designed to address degradation pathways that are specific to each class of catalysts. PGM-free catalyst degradation has been reported to proceed mainly through (i) Fe loss or demetallation, and (ii) carbon oxidation via hydrogen peroxide attack, while performance losses from other mechanisms, like (iii) competing protonation events at the active site or neighboring N-atoms or (iv) micropore flooding, continue to be debated in the literature [15, 17, 18, 29–32]. In contrast, PGM catalyst degradation in PEFCs is linked to the loss of electrochemically active surface area, resulting from primary degradation processes such as (i) Pt dissolution and (ii) carbon support corrosion, which can initiate secondary processes, including (iii) detachment of Pt particles, (iv) size-driven dissolution of small Pt particles and re-deposition onto larger Pt particles, known as Ostwald ripening, and (v) migration and coalescence of Pt particles on a carbon support [66, 67, 69–76].

While the degradation mechanisms for PGM and PGM-free catalysts are distinct, they share performance degradation due to loss of metal sites and carbon oxidation/corrosion. The activity drop due to carbon corrosion has been correlated with a decrease in metal content, thus, ORR active sites in PGM catalysts and presumed active sites in PGM-free catalysts. While in the case of Fe-based PGM-free catalysts, it is hypothesized that carbon corrosion is a result of hydrogen peroxide production via Fenton reactions at Fe-surface sites, the corrosion of carbon support in PGM catalysts is thought to proceed via a Pt-catalyzed carbon oxidation rather than soluble Pt species. Despite the differences in mechanism, carbon corrosion has a similarly negative effect on the durability of both PGM and PGM-free catalysts. Another similarity between these two classes of catalysts is the loss of transition metal species via demetallation or dissolution pathways. While differences in specific thermodynamic relationships between metal species, oxidation state, and conditions of the cathode operation exist between low-PGM and PGM-free catalysts, the conditions that promote metal oxide formation or carbon oxidation can also accelerate the loss of metal species into solution in both catalyst classes.

There remains a need for broad adoption of standardized protocols for assessing the stability and durability of PGM-free and PGM catalysts. When developing such protocols, it has been important to consider the differences in operational requirements, degradation mechanisms, metrics for performance assessment for each class of catalysts. To address the electrocatalyst performance evaluation challenge, in this chapter we have summarized testing protocols developed in the fuel cell research program supported by the US Department of Energy. The protocols include recommendations on testing catalyst durability and assessing carbon corrosion effects under fuel cell operating conditions, as well as in an aqueous electrolytes using the rotating ring-disk electrode. Large-scale implementation of protocols would facilitate standardization of PGM-free and PGM-based catalyst testing that would in turn allow for easier comparison of catalysts in the literature and help to identify materials or application in fuel cell stacks for automotive transportation.

Acronyms

AST	accelerated stress tests
BOT	beginning of testing
CHE	computational hydrogen electrode
COR	carbon oxidation reaction
CVs	cyclic voltammograms
DEMS	differential mass spectroscopy
DFT	density functional theory
ECSA	electrochemically active surface area
E_H	high potential
E_L	low potential
EQCN	electrochemical quartz crystal nanobalance
EXAFS	extended X-ray absorption fine structure
HOR	hydrogen oxidation reaction
ICP-MS	inductively coupled plasma mass spectrometry
MEA	membrane electrode assembly
MOF	metal organic framework

NDIR	non-dispersive infrared spectroscopy
OCV	open circuit voltage
ORR	oxygen reduction reaction
PEFC	polymer electrolyte fuel cells
PGM	platinum group metal
PGM-free	platinum group metal-free
POP	porous organic polymers
RH	relative humidity
RHE	reversible hydrogen electrode
ROS	reactive oxygen species
RRDE	rotating ring disk electrode
SFC	scanning flow cell
SHE	standard hydrogen electrode
SLPM	standard liters per minute
TOF	turnover frequency
VR	voltage recovery
XAS	X-ray absorption

References

1 Mazloomi, K. and Gomes, C. (2012). Hydrogen as an energy carrier: prospects and challenges. *Renewable and Sustainable Energy Reviews* 16 (5): 3024–3033. https://doi.org/10.1016/j.rser.2012.02.028.

2 da Silva Veras, T., Mozer, T.S., da, dos Costa Rubim Messeder, Santos, D., and da Silva César, A. (2017). Hydrogen: trends, production and characterization of the main process worldwide. *International Journal of Hydrogen Energy* 42 (4): 2018–2033. https://doi.org/10.1016/j.ijhydene.2016.08.219.

3 Sharaf, O.Z. and Orhan, M.F. (2014). An overview of fuel cell technology: fundamentals and applications. *Renewable and Sustainable Energy Reviews* 32: 810–853. https://doi.org/10.1016/j.rser.2014.01.012.

4 Carmo, M., Fritz, D.L., Mergel, J., and Stolten, D. (2013). A comprehensive review on PEM water electrolysis. *International Journal of Hydrogen Energy* 38 (12): 4901–4934. https://doi.org/10.1016/j.ijhydene.2013.01.151.

5 Shao, M., Chang, Q., Dodelet, J.-P., and Chenitz, R. (2016). Recent advances in electrocatalysts for oxygen reduction reaction. *Chemical Reviews* 116 (6): 3594–3657. https://doi.org/10.1021/acs.chemrev.5b00462.

6 Martinez, U., Komini Babu, S., Holby, E.F. et al. (2019). Progress in the development of Fe-based PGM-free electrocatalysts for the oxygen reduction reaction. *Advanced Materials* 31 (31): 1806545. https://doi.org/10.1002/adma.201806545.

7 Wu, G., More, K.L., Johnston, C.M., and Zelenay, P. (2011). High-performance electrocatalysts for oxygen reduction derived from polyaniline, iron, and cobalt. *Science* 332 (6028): 443. https://doi.org/10.1126/science.1200832.

8 Chen, Z., Higgins, D., Yu, A. et al. (2011). A review on non-precious metal electrocatalysts for PEM fuel cells. *Energy & Environmental Science* 4 (9): 3167–3192. https://doi.org/10.1039/C0EE00558D.

9 Gasteiger, H.A., Kocha, S.S., Sompalli, B., and Wagner, F.T. (2005). Activity benchmarks and requirements for Pt, Pt-alloy, and non-Pt oxygen reduction catalysts for PEMFCs. *Applied Catalysis B: Environmental* 56 (1): 9–35. https://doi.org/10.1016/j.apcatb.2004.06.021.

10 Gasteiger, H.A., Panels, J.E., and Yan, S.G. (2004). Dependence of PEM fuel cell performance on catalyst loading. *Journal of Power Sources* 127 (1): 162–171. https://doi.org/10.1016/j.jpowsour.2003.09.013.

11 Vielstich, W., Lamm, A., and Gasteiger, H.A. (ed.) (2003). *Handbook of Fuel Cells: Fundamentals, Technology, and Applications*. Chichester, England: Wiley.

12 Jaouen, F., Proietti, E., Lefèvre, M. et al. (2011). Recent advances in non-precious metal catalysis for oxygen-reduction reaction in polymer electrolyte fuel cells. *Energy & Environmental Science* 4 (1): 114–130. https://doi.org/10.1039/C0EE00011F.

13 Thompson, S.T. and Papageorgopoulos, D. (2019). Platinum group metal-free catalysts boost cost competitiveness of fuel cell vehicles. *Nature Catalysis* 2 (7): 558–561. https://doi.org/10.1038/s41929-019-0291-x.

14 Thompson, S.T., Wilson, A.R., Zelenay, P. et al. (2018). ElectroCat: DOE's approach to PGM-free catalyst and electrode R&D. *Solid State Ionics* 319: 68–76. https://doi.org/10.1016/j.ssi.2018.01.030.

15 Choi, C.H., Lim, H.-K., Chung, M.W. et al. (2018). The Achilles' heel of iron-based catalysts during oxygen reduction in an acidic medium. *Energy & Environmental Science* 11 (11): 3176–3182. https://doi.org/10.1039/C8EE01855C.

16 Osmieri, L., Park, J., Cullen, D.A. et al. (2021). Status and challenges for the application of platinum group metal-free catalysts in proton-exchange membrane fuel cells. *Current Opinion in Electrochemistry* 25: 100627. https://doi.org/10.1016/j.coelec.2020.08.009.

17 Shao, Y., Dodelet, J.-P., Wu, G., and Zelenay, P. (2019). PGM free cathode catalysts for PEM fuel cells: a mini-review on stability challenges. *Advanced Materials* 31 (31): 1807615. https://doi.org/10.1002/adma.201807615.

18 Singh, H., Zhuang, S., Ingis, B. et al. (2019). Carbon-based catalysts for oxygen reduction reaction: a review on degradation mechanisms. *Carbon* 151: 160–174. https://doi.org/10.1016/j.carbon.2019.05.075.

19 Zhang, H., Osmieri, L., Park, J.H. et al. (2022). Standardized protocols for evaluating platinum group metal-free oxygen reduction reaction electrocatalysts in polymer electrolyte fuel cells. *Nature Catalysis* 5 (5): 455–462. https://doi.org/10.1038/s41929-022-00778-3.

20 Jasinski, R. (1964). A new fuel cell cathode catalyst. *Nature* 201 (4925): 1212–1213. https://doi.org/10.1038/2011212a0.

21 Sa, Y.J., Seo, D.-J., Woo, J. et al. (2016). A general approach to preferential formation of active Fe–N_x sites in Fe–N/C electrocatalysts for efficient oxygen reduction reaction. *Journal of the American Chemical Society* 138 (45): 15046–15056. https://doi.org/10.1021/jacs.6b09470.

22 Koslowski, U.I., Abs-Wurmbach, I., Fiechter, S., and Bogdanoff, P. (2008). Nature of the catalytic centers of porphyrin-based electrocatalysts for the ORR: a correlation of kinetic current density with the site density of Fe−N_4 centers.

The Journal of Physical Chemistry C 112 (39): 15356–15366. https://doi.org/10.1021/jp802456e.

23 Chen, M., He, Y., Spendelow, J.S., and Wu, G. (2019). Atomically dispersed metal catalysts for oxygen reduction. *ACS Energy Letters* 4 (7): 1619–1633. https://doi.org/10.1021/acsenergylett.9b00804.

24 Barkholtz, H.M. and Liu, D.-J. (2017). Advancements in rationally designed PGM-free fuel cell catalysts derived from metal–organic frameworks. *Materials Horizons* 4 (1): 20–37. https://doi.org/10.1039/C6MH00344C.

25 K. Strickland, E. Miner, Q. Jia, U. Tylus, N. Ramaswamy, W. Liang, M.-T. Sougrati, F. Jaouen, and S. Mukerjee, "Highly active oxygen reduction non-platinum group metal electrocatalyst without direct metal–nitrogen coordination," *Nature Communications*, vol. 6, no. 1, p. 7343, 2015. https://doi.org/10.1038/ncomms8343.

26 Singh, D., Tian, J., Mamtani, K. et al. (2014). A comparison of N-containing carbon nanostructures (CN_x) and N-coordinated iron–carbon catalysts (FeNC) for the oxygen reduction reaction in acidic media. *Journal of Catalysis* 317: 30–43. https://doi.org/10.1016/j.jcat.2014.05.025.

27 Chen, Y., Ji, S., Chen, C. et al. (2018). Single-atom catalysts: synthetic strategies and electrochemical applications. *Joule* 2 (7): 1242–1264. https://doi.org/10.1016/j.joule.2018.06.019.

28 C. H. Choi, H.-K. Lim, M. W. Chung, J. C. Park, H. Shin, H. Kim, and S. I. Woo, "Long-range electron transfer over graphene-based catalyst for high-performing oxygen reduction reactions: importance of size, N-doping, and metallic impurities," *Journal of the American Chemical Society*, vol. 136, no. 25, pp. 9070–9077, 2014. https://doi.org/10.1021/ja5033474.

29 Yin, X. and Zelenay, P. (2018). (Invited) Kinetic models for the degradation mechanisms of PGM-free ORR catalysts. *ECS Transactions* 85 (13): 1239–1250. https://doi.org/10.1149/08513.1239ecst.

30 Wang, W., Jia, Q., Mukerjee, S., and Chen, S. (2019). Recent insights into the oxygen-reduction electrocatalysis of Fe/N/C materials. *ACS Catalysis* 9 (11, 11): 10126–10141. https://doi.org/10.1021/acscatal.9b02583.

31 Weiss, J., Zhang, H., and Zelenay, P. (2020). Recent progress in the durability of Fe-N-C oxygen reduction electrocatalysts for polymer electrolyte fuel cells. *Journal of Electroanalytical Chemistry* 875: 114696. https://doi.org/10.1016/j.jelechem.2020.114696.

32 N. Yang, L. Peng, L. Li, J. Li, Q. Liao, M. Shao, and Z. Wei, "Theoretically probing the possible degradation mechanisms of an FeNC catalyst during the oxygen reduction reaction," *Chemical Science*, 12, 37, 12476–12484. https://doi.org/10.1039/D1SC02901K.

33 Proietti, E., Jaouen, F., Lefèvre, M. et al. (2011). Iron-based cathode catalyst with enhanced power density in polymer electrolyte membrane fuel cells. *Nature Communications* 2 (1): 416. https://doi.org/10.1038/ncomms1427.

34 Mineva, T., Matanovic, I., Atanassov, P. et al. (2019). Understanding active sites in Pyrolyzed Fe–N–C catalysts for fuel cell cathodes by bridging density

functional theory calculations and ^{57}Fe Mössbauer spectroscopy. *ACS Catalysis* 9 (10): 9359–9371. https://doi.org/10.1021/acscatal.9b02586.

35 X. Ao, W. Zhang, Z. Li, L. Lv, Y. Ruan, H.-H. Wu, W.-H. Chiang, C. Wang, M. Liu, and X. C. Zeng, "Unraveling the high-activity nature of Fe–N–C electrocatalysts for the oxygen reduction reaction: the extraordinary synergy between Fe–N_4 and Fe_4N," *Journal of Materials Chemistry A*, vol. 7, no. 19, pp. 11792–11801, 2019. https://doi.org/10.1039/C9TA02338K.

36 H. T. Chung, D. A. Cullen, D. Higgins, B. T. Sneed, E. F. Holby, K. L. More, and P. Zelenay, "Direct atomic-level insight into the active sites of a high-performance PGM-free ORR catalyst," *Science*, vol. 357, no. 6350, p. 479, 2017. https://doi.org/10.1126/science.aan2255.

37 U. I. Kramm, J. Herranz, N. Larouche, T. M. Arruda, M. Lefèvre, F. Jaouen, P. Bogdanoff, S. Fiechter, I. Abs-Wurmbach, S. Mukerjee, and J.-P. Dodelet, "Structure of the catalytic sites in Fe/N/C-catalysts for O_2-reduction in PEM fuel cells," *Physical Chemistry Chemical Physics*, vol. 14, no. 33, pp. 11673–11688, 2012. https://doi.org/10.1039/C2CP41957B.

38 Biddinger, E.J. and Ozkan, U.S. (2010). Role of graphitic edge plane exposure in carbon nanostructures for oxygen reduction reaction. *The Journal of Physical Chemistry C* 114 (36): 15306–15314. https://doi.org/10.1021/jp104074t.

39 Liu, G., Li, X., Ganesan, P., and Popov, B.N. (2009). Development of non-precious metal oxygen-reduction catalysts for PEM fuel cells based on N-doped ordered porous carbon. *Applied Catalysis B: Environmental* 93 (1): 156–165. https://doi.org/10.1016/j.apcatb.2009.09.025.

40 Franke, R., Ohms, D., and Wiesener, K. (1989). Investigation of the influence of thermal treatment on the properties of carbon materials modified by N_4-chelates for the reduction of oxygen in acidic media. *Journal of Electroanalytical Chemistry and Interfacial Electrochemistry* 260 (1): 63–73. https://doi.org/10.1016/0022-0728(89)87099-8.

41 Wiesener, K. (1986). N_4-chelates as electrocatalyst for cathodic oxygen reduction. *Electrochimica Acta* 31 (8): 1073–1078. https://doi.org/10.1016/0013-4686(86)80022-6.

42 A. Zitolo, V. Goellner, V. Armel, M.-T. Sougrati, T. Mineva, L. Stievano, E. Fonda, and F. Jaouen, "Identification of catalytic sites for oxygen reduction in iron- and nitrogen-doped graphene materials," *Nature Materials*, vol. 14, no. 9, pp. 937–942, 2015. https://doi.org/10.1038/nmat4367.

43 K. Artyushkova, S. Rojas-Carbonell, C. Santoro, E. Weiler, A. Serov, R. Awais, R. R. Gokhale, and P. Atanassov, "Correlations between synthesis and performance of Fe-based PGM-free catalysts in acidic and alkaline media: evolution of surface chemistry and morphology," *ACS Applied Energy Materials*, vol. 2, no. 8, pp. 5406–5418, 2019. https://doi.org/10.1021/acsaem.9b00331.

44 Choi, C.H., Baldizzone, C., Polymeros, G. et al. (2016). Minimizing operando demetallation of Fe-N-C electrocatalysts in acidic medium. *ACS Catalysis* 6 (5): 3136–3146. https://doi.org/10.1021/acscatal.6b00643.

45 Holby, E.F., Wang, G., and Zelenay, P. (2020). Acid stability and demetalation of PGM-free ORR electrocatalyst structures from density functional theory: a model for "single-atom catalyst" dissolution. *ACS Catalysis* 10 (24): 14527–14539. https://doi.org/10.1021/acscatal.0c02856.

46 Chenitz, R., Kramm, U.I., Lefèvre, M. et al. (2018). A specific demetalation of Fe–N_4 catalytic sites in the micropores of NC_Ar + NH_3 is at the origin of the initial activity loss of the highly active Fe/N/C catalyst used for the reduction of oxygen in PEM fuel cells. *Energy & Environmental Science* 11 (2): 365–382. https://doi.org/10.1039/C7EE02302B.

47 Choi, C.H., Baldizzone, C., Grote, J.-P. et al. (2015). Stability of Fe-N-C catalysts in acidic medium studied by operando spectroscopy. *Angewandte Chemie International Edition* 54 (43): 12753–12757. https://doi.org/10.1002/anie.201504903.

48 V. Goellner, V. Armel, A. Zitolo, E. Fonda, and F. Jaouen, "Degradation by hydrogen peroxide of metal-nitrogen-carbon catalysts for oxygen reduction," *Journal of the Electrochemical Society*, vol. 162, no. 6, pp. H403-H414, 2015. https://doi.org/10.1149/2.1091506jes.

49 Naumov, G.B., Ryzhenko, B.N., and Khodakovskii, I.L. (1974). *Handbook of Thermodynamic Data* (ed. I. Barnes and V. Speltz). Menlo Park, CA; Springfield, VA: U.S. Geological Survey, Water Resources Division; [Distributed by] N.T.I.S. (in English.).

50 Xia, D., Yu, C., Zhao, Y. et al. (2021). Degradation and regeneration of Fe–N_x active sites for the oxygen reduction reaction: the role of surface oxidation, Fe demetallation and local carbon microporosity. *Chemical Science* 12 (34): 11576–11584. https://doi.org/10.1039/D1SC03754D.

51 Lefèvre, M. and Dodelet, J.-P. (2003). Fe-based catalysts for the reduction of oxygen in polymer electrolyte membrane fuel cell conditions: determination of the amount of peroxide released during electroreduction and its influence on the stability of the catalysts. *Electrochimica Acta* 48 (19): 2749–2760. https://doi.org/10.1016/S0013-4686(03)00393-1.

52 L. Gubler, S. M. Dockheer, and W. H. Koppenol, "Radical (HO•, H• and HOO•) formation and ionomer degradation in polymer electrolyte fuel cells," *Journal of the Electrochemical Society*, vol. 158, no. 7, p. B755, 2011. https://doi.org/10.1149/1.3581040.

53 Zhang, G., Chenitz, R., Lefèvre, M. et al. (2016). Is iron involved in the lack of stability of Fe/N/C electrocatalysts used to reduce oxygen at the cathode of PEM fuel cells? *Nano Energy* 29: 111–125. https://doi.org/10.1016/j.nanoen.2016.02.038.

54 Choi, J.-Y., Yang, L., Kishimoto, T. et al. (2017). Is the rapid initial performance loss of Fe/N/C non precious metal catalysts due to micropore flooding? *Energy & Environmental Science* 10 (1): 296–305. https://doi.org/10.1039/C6EE03005J.

55 Herranz, J., Jaouen, F., Lefèvre, M. et al. (2011). Unveiling N-protonation and anion-binding effects on Fe/N/C catalysts for O_2 reduction in proton-exchange-membrane fuel cells. *The Journal of Physical Chemistry C* 115 (32): 16087–16097. https://doi.org/10.1021/jp2042526.

56 Liang, Z., Luo, M., Chen, M. et al. (2020). Evaluating the catalytic activity of transition metal dimers for the oxygen reduction reaction. *Journal of Colloid and Interface Science* 568: 54–62. https://doi.org/10.1016/j.jcis.2020.02.034.

57 Osmieri, L., Cullen, D.A., Chung, H.T. et al. (2020). Durability evaluation of a Fe–N–C catalyst in polymer electrolyte fuel cell environment via accelerated stress tests. *Nano Energy* 78: 105209. https://doi.org/10.1016/j.nanoen.2020.105209.

58 Huang, X., Zhao, Z., Cao, L. et al. (2015). High-performance transition metal–doped Pt_3Ni octahedra for oxygen reduction reaction. *Science* 348 (6240): 1230. https://doi.org/10.1126/science.aaa8765.

59 Han, B., Carlton, C.E., Kongkanand, A. et al. (2015). Record activity and stability of dealloyed bimetallic catalysts for proton exchange membrane fuel cells. *Energy & Environmental Science* 8 (1, 266): 258. https://doi.org/10.1039/C4EE02144D.

60 Sassin, M.B., Garsany, Y., Gould, B.D., and Swider-Lyons, K.E. (2017). Fabrication method for laboratory-scale high-performance membrane electrode assemblies for fuel cells. *Analytical Chemistry* 89 (1): 511–518. https://doi.org/10.1021/acs.analchem.6b03005.

61 S. Litster and G. McLean, "PEM fuel cell electrodes," *Journal of Power Sources*, vol. 130, no. 1, pp. 61–76, 2004. https://doi.org/10.1016/j.jpowsour.2003.12.055.

62 Kongkanand, A. and Mathias, M.F. (2016). The priority and challenge of high-power performance of low-platinum proton-exchange membrane fuel cells. *The Journal of Physical Chemistry Letters* 7 (7): 1127–1137. https://doi.org/10.1021/acs.jpclett.6b00216.

63 Nonoyama, N., Okazaki, S., Weber, A.Z. et al. (2011). Analysis of oxygen-transport diffusion resistance in proton-exchange-membrane fuel cells. *Journal of the Electrochemical Society* 158 (4): B416. https://doi.org/10.1149/1.3546038.

64 Kabir, S., Myers, D.J., Kariuki, N. et al. (2019). Elucidating the dynamic nature of fuel cell electrodes as a function of conditioning: an ex situ material characterization and in situ electrochemical diagnostic study. *ACS Applied Materials & Interfaces* 11 (48): 45016–45030. https://doi.org/10.1021/acsami.9b11365.

65 Sandbeck, D.J.S., Secher, N.M., Speck, F.D. et al. (2020). Particle size effect on platinum dissolution: considerations for accelerated stability testing of fuel cell catalysts. *ACS Catalysis* 10 (11): 6281–6290. https://doi.org/10.1021/acscatal.0c00779.

66 Borup, R., Meyers, J., Pivovar, B. et al. (2007). Scientific aspects of polymer electrolyte fuel cell durability and degradation. *Chemical Reviews* 107 (10): 3904–3951. https://doi.org/10.1021/cr050182l.

67 S. Cherevko, N. Kulyk, and K. J. J. Mayrhofer, "Durability of platinum-based fuel cell electrocatalysts: dissolution of bulk and nanoscale platinum," *Nano Energy*, vol. 29, pp. 275–298, 2016. https://doi.org/10.1016/j.nanoen.2016.03.005.

68 Papadias, D.D., Ahluwalia, R.K., Kariuki, N. et al. (2018). Durability of Pt-Co alloy polymer electrolyte fuel cell cathode catalysts under accelerated stress tests. *Journal of the Electrochemical Society* 165 (6): F3166–F3177. https://doi.org/10.1149/2.0171806jes.

69 Lopes, P.P., Li, D., Lv, H. et al. (2020). Eliminating dissolution of platinum-based electrocatalysts at the atomic scale. *Nature Materials* 19 (11): 1207–1214. https://doi.org/10.1038/s41563-020-0735-3.

70 Meier, J.C., Galeano, C., Katsounaros, I. et al. (2012). Degradation mechanisms of Pt/C fuel cell catalysts under simulated start–stop conditions. *ACS Catalysis* 2 (5): 832–843. https://doi.org/10.1021/cs300024h.

71 I. Takahashi and S. S. Kocha, "Examination of the activity and durability of PEMFC catalysts in liquid electrolytes," *Journal of Power Sources*, vol. 195, no. 19, pp. 6312–6322, 2010. https://doi.org/10.1016/j.jpowsour.2010.04.052.

72 T. Ioroi, Z. Siroma, S.-I. Yamazaki, and K. Yasuda, "Electrocatalysts for PEM fuel cells," *Advanced Energy Materials*, vol. 9, no. 23, p. 1801284, 2019. https://doi.org/10.1002/aenm.201801284.

73 Myers, D.J., Wang, X., Smith, M.C., and More, K.L. (2018). Potentiostatic and potential cycling dissolution of polycrystalline platinum and platinum nano-particle fuel cell catalysts. *Journal of the Electrochemical Society* 165 (6): F3178–F3190. https://doi.org/10.1149/2.0211806jes.

74 Meier, J.C., Galeano, C., Katsounaros, I. et al. (2014). Design criteria for stable Pt/C fuel cell catalysts, (in eng). *Beilstein Journal of Nanotechnology* 5: 44–67. https://doi.org/10.3762/bjnano.5.5.

75 J. Zhang, Y. Yuan, L. Gao, G. Zeng, M. Li, and H. Huang, "Stabilizing Pt-based electrocatalysts for oxygen reduction reaction: fundamental understanding and design strategies," *Advanced Materials*, vol. 33, no. 20, p. 2006494, 2021. https://doi.org/10.1002/adma.202006494.

76 P. Ren, P. Pei, Y. Li, Z. Wu, D. Chen, and S. Huang, "Degradation mechanisms of proton exchange membrane fuel cell under typical automotive operating conditions," *Progress in Energy and Combustion Science*, vol. 80, p. 100859, 2020. https://doi.org/10.1016/j.pecs.2020.100859.

77 Guilminot, E., Corcella, A., Charlot, F. et al. (2007). Detection of Pt^{z+} ions and Pt nanoparticles inside the membrane of a used PEMFC. *Journal of the Electrochemical Society* 154 (1): B96. https://doi.org/10.1149/1.2388863.

78 Mitsushima, S., Kawahara, S., Ota, K.-I., and Kamiya, N. (2007). Consumption rate of Pt under potential cycling. *Journal of the Electrochemical Society* 154 (2): B153. https://doi.org/10.1149/1.2400596.

79 Sugawara, Y., Okayasu, T., Yadav, A.P. et al. (2012). Dissolution mechanism of platinum in sulfuric acid solution. *Journal of the Electrochemical Society* 159 (11): F779–F786. https://doi.org/10.1149/2.017212jes.

80 Uchimura, M., Sugawara, S., Suzuki, Y. et al. (2008). Electrocatalyst durability under simulated automotive drive cycles. *ECS Transactions* 16 (2): 225–234. https://doi.org/10.1149/1.2981858.

81 Gilbert, J.A., Kariuki, N.N., Subbaraman, R. et al. (2012). In situ anomalous small-angle X-ray scattering studies of platinum nanoparticle fuel cell electrocatalyst degradation. *Journal of the American Chemical Society* 134 (36): 14823–14833. https://doi.org/10.1021/ja3038257.

82 Umeda, M., Kuwahara, Y., Nakazawa, A., and Inoue, M. (2009). Pt degradation mechanism in concentrated sulfuric acid studied using rotating ring–disk

electrode and electrochemical quartz crystal microbalance. *The Journal of Physical Chemistry C* 113 (35): 15707–15713. https://doi.org/10.1021/jp903956b.

83 Cherevko, S., Keeley, G.P., Geiger, S. et al. (2015). Dissolution of platinum in the operational range of fuel cells. *ChemElectroChem* 2 (10): 1471–1478. https://doi.org/10.1002/celc.201500098.

84 Rinaldo, S.G., Stumper, J., and Eikerling, M. (2010). Physical theory of platinum nanoparticle dissolution in polymer electrolyte fuel cells. *The Journal of Physical Chemistry C* 114 (13): 5773–5785. https://doi.org/10.1021/jp9101509.

85 X. Wang, R. Kumar, and D. J. Myers, "Effect of voltage on platinum dissolution," *Electrochemical and Solid-State Letters*, vol. 9, no. 5, p. A225, 2006. https://doi.org/10.1149/1.2180536.

86 S. Cherevko, A. R. Zeradjanin, G. P. Keeley, and K. J. J. Mayrhofer, "A comparative study on gold and platinum dissolution in acidic and alkaline media," *Journal of the Electrochemical Society*, vol. 161, no. 12, pp. H822-H830, 2014. https://doi.org/10.1149/2.0881412jes.

87 G. Inzelt, B. Berkes, and A. Kriston, "Temperature dependence of two types of dissolution of platinum in acid media. An electrochemical nanogravimetric study," (in English), *Electrochimica Acta*, vol. 55, no. 16, pp. 4742–4749, 2010. https://doi.org/10.1016/j.electacta.2010.03.074.

88 Nagy, Z. and You, H. (2002). Applications of surface X-ray scattering to electrochemistry problems. *Electrochimica Acta* 47 (19): 3037–3055. https://doi.org/10.1016/S0013-4686(02)00223-2.

89 Jerkiewicz, G., Vatankhah, G., Lessard, J. et al. (2004). Surface-oxide growth at platinum electrodes in aqueous H_2SO_4: reexamination of its mechanism through combined cyclic-voltammetry, electrochemical quartz-crystal nanobalance, and Auger electron spectroscopy measurements. *Electrochimica Acta* 49 (9): 1451–1459. https://doi.org/10.1016/j.electacta.2003.11.008.

90 Reddy, A.K.N., Genshaw, M.A., and Bockris, J.O.M. (1968). Ellipsometric study of oxygen-containing films on platinum anodes. *The Journal of Chemical Physics* 48 (2): 671–675. https://doi.org/10.1063/1.1668699.

91 Vetter, K.J. and Schultze, J.W. (1972). The kinetics of the electrochemical formation and reduction of monomolecular oxide layers on platinum in 0.5 *M* H_2SO_4: Part II. Galvanostatic pulse measurements and the model of oxide growth. *Journal of Electroanalytical Chemistry and Interfacial Electrochemistry* 34 (1): 141–158. https://doi.org/10.1016/S0022-0728(72)80510-2.

92 Angerstein-Kozlowska, H., Conway, B.E., and Sharp, W.B.A. (1973). The real condition of electrochemically oxidized platinum surfaces: Part I. Resolution of component processes. *Journal of Electroanalytical Chemistry and Interfacial Electrochemistry* 43 (1): 9–36. https://doi.org/10.1016/S0022-0728(73)80307-9.

93 G. Jerkiewicz, M. Alsabet, M. Grden, H. Varela, and G. Tremiliosi-Filho, "Erratum to: "comprehensive study of the growth of thin oxide layers on Pt electrodes under well-defined temperature, potential, and time conditions" [Journal of Electroanalytical Chemistry, 589 (2006) 120–127]," *Journal of Electroanalytical Chemistry*, vol. 625, no. 2, pp. 172–174, 2009. https://doi.org/10.1016/j.jelechem.2008.10.015.

94 Mao, F., Sharifi-Asl, S., Yu, J., and Macdonald, D.D. (2014). Diagnosis of the mechanism of anodic oxide film growth on platinum in H_2SO_4. *Journal of the Electrochemical Society* 161 (5): C254–C260. https://doi.org/10.1149/2.062405jes.

95 Sun, A., Franc, J., and Macdonald, D.D. (2006). Growth and properties of oxide films on platinum. *Journal of the Electrochemical Society* 153 (7): B260. https://doi.org/10.1149/1.2200156.

96 Ahluwalia, R.K., Arisetty, S., Peng, J.-K. et al. (2014). Dynamics of particle growth and electrochemical surface area loss due to platinum dissolution. *Journal of the Electrochemical Society* 161 (3): F291–F304. https://doi.org/10.1149/2.051403jes.

97 Ahluwalia, R.K., Arisetty, S., Wang, X. et al. (2013). Thermodynamics and kinetics of platinum dissolution from carbon-supported electrocatalysts in aqueous media under potentiostatic and potentiodynamic conditions. *Journal of the Electrochemical Society* 160 (4): F447–F455. https://doi.org/10.1149/2.018306jes.

98 Holby, E.F., Shao-Horn, Y., Sheng, W., and Morgan, D. (2019). Oxide formation and removal on Pt surfaces under potential cycled conditions. *ECS Transactions* 25 (1): 583–592. https://doi.org/10.1149/1.3210609.

99 Shao-Horn, Y., Sheng, W.C., Chen, S. et al. (2007). Instability of supported platinum nanoparticles in low-temperature fuel cells. *Topics in Catalysis* 46 (3): 285–305. https://doi.org/10.1007/s11244-007-9000-0.

100 Liu, Z.Y., Zhang, J.L., Yu, P.T. et al. (2010). Transmission electron microscopy observation of corrosion behaviors of platinized carbon blacks under thermal and electrochemical conditions. *Journal of the Electrochemical Society* 157 (6): B906. https://doi.org/10.1149/1.3391737.

101 Lopes, P.P., Strmcnik, D., Tripkovic, D. et al. (2016). Relationships between atomic level surface structure and stability/activity of platinum surface atoms in aqueous environments. *ACS Catalysis* 6 (4): 2536–2544. https://doi.org/10.1021/acscatal.5b02920.

102 Pizzutilo, E., Geiger, S., Grote, J.P. et al. (2016). On the need of improved accelerated degradation protocols (ADPs): examination of platinum dissolution and carbon corrosion in half-cell tests. *Journal of the Electrochemical Society* 163 (14): F1510–F1514. https://doi.org/10.1149/2.0731614jes.

103 Pourbaix, M. (1966). *Atlas of Electrochemical Equilibria in Aqueous Solutions*, First English edition. Oxford: Pergamon Press.

104 Maass, S., Finsterwalder, F., Frank, G. et al. (2008). Carbon support oxidation in PEM fuel cell cathodes. *Journal of Power Sources* 176 (2): 444–451. https://doi.org/10.1016/j.jpowsour.2007.08.053.

105 Castanheira, L., Silva, W.O., Lima, F.H.B. et al. (2015). Carbon corrosion in proton-exchange membrane fuel cells: effect of the carbon structure, the degradation protocol, and the gas atmosphere. *ACS Catalysis* 5 (4): 2184–2194. https://doi.org/10.1021/cs501973j.

106 Dillet, J., Spernjak, D., Lamibrac, A. et al. (2014). Impact of flow rates and electrode specifications on degradations during repeated startups and shutdowns in polymer-electrolyte membrane fuel cells. *Journal of Power Sources* 250: 68–79. https://doi.org/10.1016/j.jpowsour.2013.10.141.

107 Seo, D., Park, S., Jeon, Y. et al. (2010). Physical degradation of MEA in PEM fuel cell by on/off operation under nitrogen atmosphere. *Korean Journal of Chemical Engineering* 27 (1): 104–109. https://doi.org/10.1007/s11814-009-0312-6.

108 Roen, L.M., Paik, C.H., and Jarvi, T.D. (2004). Electrocatalytic corrosion of carbon support in PEMFC cathodes. *Electrochemical and Solid-State Letters* 7 (1): A19. https://doi.org/10.1149/1.1630412.

109 Macauley, N., Papadias, D.D., Fairweather, J. et al. (2018). Carbon corrosion in PEM fuel cells and the development of accelerated stress tests. *Journal of the Electrochemical Society* 165 (6): F3148–F3160. https://doi.org/10.1149/2.0061806jes.

110 Ioroi, T., Senoh, H., Yamazaki, S.-I. et al. (2008). Stability of corrosion-resistant Magnéli-phase Ti_4O_7-supported PEMFC catalysts at high potentials. *Journal of the Electrochemical Society* 155 (4): B321. https://doi.org/10.1149/1.2833310.

111 Zhang, X., Yang, Y., Guo, L., and Liu, H. (2017). Effects of carbon corrosion on mass transfer losses in proton exchange membrane fuel cells. *International Journal of Hydrogen Energy* 42 (7): 4699–4705. https://doi.org/10.1016/j.ijhydene.2016.08.223.

112 Stariha, S., Macauley, N., Sneed, B.T. et al. (2018). Recent advances in catalyst accelerated stress tests for polymer electrolyte membrane fuel cells. *Journal of the Electrochemical Society* 165 (7): F492–F501. https://doi.org/10.1149/2.0881807jes.

113 DOE Fuel Cell Technologies Office. (2016). Fuel cell technologies office multi-year research, development, and demonstration plan. *Section 3.4 Fuel Cells.* https://energy.gov/sites/prod/files/2016/10/f33/fcto_myrdd_fuel_cells.pdf (accessed 30 August 2021)

Part VII

Systems

14

Modeling of Polymer Electrolyte Membrane Fuel Cells

Andrea Baricci[1], Andrea Casalegno[1], Dario Maggiolo[2], Federico Moro[3], Matteo Zago[1], and Massimo Guarnieri[3]

[1] *Politecnico di Milano, Dipartimento di Energia, via Lambruschini 4a, 20156 Milano, Italy*
[2] *Chalmers University of Technology, Department of Applied Mechanics, Hörsalsvägen 7A, SE-412 96 Gothenburg, Sweden*
[3] *University of Padova, Department of Industrial Engineering, via Giovanni Gradenigo 6/A, 35131 Padova, Italy*

14.1 Introduction

Proton exchange membrane fuel cells (PEMFCs) are considered nowadays as very promising power sources for battery electric vehicles (BEV) and other services due to high energy conversion efficiency, high power density, and zero emissions [1, 2]. PEMFCs convert the chemical energy of supplied reactants (hydrogen and oxygen) into electrical energy and generate water and heat as a byproduct. The typical PEMFC power range is from a few milliwatts to a few hundred kilowatts, with a stack efficiency of the order of 50% and power density up to $1\,kW\,kg^{-1}$. These features, combined to a relatively low operating temperature (below 100 °C), make PEMFCs particularly suited for transport applications. The main drawbacks are a complex heat and water management (e.g. membrane hydration must be kept at RH > 80% at 80 °C), relatively short lifetime (~6000 hours), and the use of precious-metal-based catalysts and perfluorinated membranes, which dramatically affects the PEMFC cost.

A PEMFC stack consists of a series of elementary cells. Each cell is made of more layers of different materials, Figure 14.1. The membrane electrode assembly (MEA) is sandwiched between the anode current collector (ACC) and the cathode current collector (CCC), which connect the cells in series and to the external electric circuit. The MEA consists of five elementary layers: anode and cathode diffusion layers (ADL, CDL), which are the porous section of the electrode that facilitate the gas distribution to the active sites; anode and cathode catalyst layers (ACL, CCL), which include the catalysts particles constituting the active reaction zones and are devoted to accelerate the oxidation and reduction reactions, at the interface between electrodes and membrane; the proton exchange membrane (PEM), which is made typically of a perfluorinated polymer capable of carrying protons and is a very good electronic insulator ensuring the separation between half-reactions.

Electrocatalysis for Membrane Fuel Cells: Methods, Modeling, and Applications, First Edition.
Edited by Nicolas Alonso-Vante and Vito Di Noto.

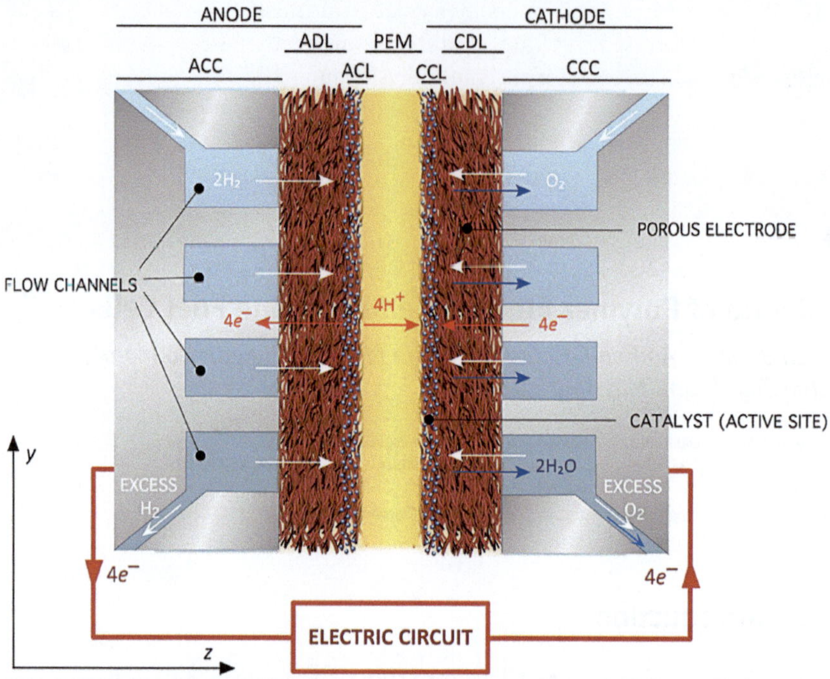

Figure 14.1 Scheme of PEMFC with anode and cathode current collector (ACC, CCC), diffusion layers (ADL, CDL), catalyst layers (ACL, CCL), and proton exchange membrane (PEM). Anode and cathode current collectors connect the fuel cell to an external electric circuit.

Reactants (hydrogen H_2 at the anode and oxygen O_2 at the cathode) are supplied through flow channels carved in the ACC and the CCC. Oxidation at the anode (negative electrode) releases four electrons e^-, which flow through the external circuit to the cathode (positive electrode), whereas protons H^+ flow through the PEM. The four e^- coming from the ACC through the electrical circuit, four H^+ migrating from the ACC through PEM and oxygen molecules combine at the CCL to form two water molecules (H_2O). The hydrogen oxidation reaction (HOR) at the ACL and the oxygen reduction reaction (ORR) at the CCL are:

$$\begin{aligned} \text{anode (HOR)}: &\quad 2H_2 \rightarrow 4H^+ + 4e^- \\ \text{cathode (ORR)}: &\quad O_2 + 4H^+ + 4e^- \rightarrow 2H_2O \end{aligned} \tag{14.1}$$

In the process, heat is generated as well, in a considerable amount, since the PEMFC efficiency is around 50%, i.e. half of the chemical energy generated by oxidation is dissipated.

14.2 General Equations for PEMFC Models

14.2.1 Analytical and Numerical Modeling

Mathematical modeling and simulation are major tools for the design and optimization of cells, cell stacks, and the whole power systems. They can also provide insight into physical phenomena occurring inside the fuel cell, which cannot be easily assessed by direct *in situ* measurements. The performance of a PEMFC depends on several complex multiphysics and multiscale phenomena such as multiphase flow, heat and mass transport in porous media, electrical conduction, and electrochemical kinetics. Partial differential equations (PDEs), which describe physical phenomena inside the device, are typically fully coupled and highly non-linear. Moreover, the layers of different material forming a PEMFC cell are thin and present a high aspect ratio. All these features combined together make the numerical modeling of PEMFCs very challenging. For instance, typical thicknesses of MEA layers are: ~0.1 mm for diffusion layers (DLs, also named gas diffusion layers – GDLs), 10 μm for catalyst layer (CLs), and a few tens of μm for the PEM. On the other hand, the MEA area can be 20 cm × 20 cm in a fuel cell with 500 A rated current (1 A cm^{-2} current density) and even more in large devices.

Nowadays, after more than 30 years of research, a number of mathematical models have been presented in scientific and technical literature for addressing this complexity. Based on the modeling approach, these models can be subdivided into two main categories: *numerical models*, which require the numerical solution of PDEs through discretization techniques such as the finite element method (FEM) and the finite volume method (FVM); *analytical models*, which make use of closed-form solutions obtained under suitable physical and geometrical simplifying assumptions. Fully 3D numerical models account for the real PEMFC geometry and, by using computational fluid dynamics (CFD), are capable to solve the complete set of transport equations governing mass, momentum, species, energy, and charge conservation [3]. The resulting simulations are, however, highly sensitive to input variables and boundary conditions. Convergence of non-linear solvers of discretized PDEs is typically very difficult to obtain and is highly problem dependent. Due to the high complexity of the geometry and the multiphysics nature of PDEs, three-dimensional (3D) models discretized by FEM or FVM often consist of several millions of degrees of freedom (DOFs) and need large computational resources for their solution. Conversely, analytical or semi-analytical one-dimensional (1D) or two-dimensional (2D) models can provide a simplest although less accurate information about electrical and chemical physical quantities at the local scale; nevertheless, all major PEMFC transport and kinetics process are still taken into account. A 1D model enables the description of the physical phenomena on a single spatial axis, which is often chosen as the direction of gas diffusion (z-axis, in Figure 14.1). With this type of model, electrical, fluid, and thermal phenomena can be resolved along the z-axis. For instance, the 1D model developed by Springer [4] also includes a detailed description of the water transport in the PEM. The spatial 1D assumption limits the analysis to mass flow directed along the z-axis, neglecting the gas flow channels at current collectors (CCs) which are directed along the y-axis in Figure 14.1. Moreover, current density hot spots which may occur across the PEM due to uneven membrane, catalyst, pores or contact cannot be considered. All these aspects are of central importance in the design of performing fuel cells. A 2D model can be developed in two different ways: along the channel (AT) or through the membrane (TM) to the gas flow direction in the gas channel, while the other modeling dimension is across the membrane in both cases. Models based on the first geometry (parallel to the main gas flow direction in the channels, i.e. y–z plane in Figure 14.1) consider the influence of fluid behavior in the channel; while in the second mode (perpendicular to the gas flow direction, i.e. x–z plane) the interdigitated flow pattern can be easily investigated. In semi-analytical models, closed-form solutions adopted in the second mode can be combined with CFD for an accurate reconstruction of the flow field. For instance, Wang et al. [5] presented a 2D AT model capable of investigating two-phase flow and transport of reactants and products in the cathode. Overpotential variation on the CL was not accounted for by this model and the influence of the flow-field geometry on fuel cell electrochemical performance could not be assessed. To analyze such an influence of the bipolar plate rib or land, a 2D TM model was developed by West and Fuller [6].

In order to provide the fundamentals of fuel cell simulation, a 1D analytical (steady-state isothermal) model for reconstructing the polarization curve is presented. It can be used to investigate how main geometrical and physical parameters affect the PEMFC performance. The main equations governing mass and charge transfer, and electrochemical kinetics are here directly introduced for the sake of conciseness. More details can be found in the literature, e.g. in [1] and [4]. Note that dynamic effects such as the charge accumulation at electrodes (double-layer capacitance) and variations in the load current and reactant supply are not accounted for by steady-state analysis. In such a case the steady-state model can be extended by introducing first-order lumped circuit elements [7, 8].

14.2.2 Reversible Electromotive Force

The *cell reversible emf* (electromotive force) is a potential difference that is generated between the electrode of a cell in absence of dissipative effects, namely in no-load conditions. The reversible emf (E) of a PEMFC depends on temperature T and on gas partial pressures at the anode and cathode p_{H_2} and p_{O_2} according to the following expression derived from the Nernst equation:

$$E = E^0 + \Delta E_T(T) + \Delta E_p(T, p_{H_2}, p_{O_2}) \tag{14.2}$$

where $E^0 = 1.229\ V$ is the value at standard temperature $T^0 = 298.15$ K and pressure $p^0 = 100$ kPa, and ΔE_T and ΔE_p are the entropic variation from standard conditions due to temperature and gas activities (hence pressures), respectively. The entropic emf variation due to temperature allows for an accurate modeling at different temperatures and is calculated as:

$$\Delta E_T(T) = \frac{\Delta \hat{s}}{nF}(T - T^0) \tag{14.3}$$

where $\Delta \hat{s}$ is the molar reaction entropy, n is the number of charge carriers per reaction, and $F = 96\,485\ \mathrm{C\,mol^{-1}}$ is the Faraday constant. In the case of reactions (14.1), $n = 2$ and $\Delta \hat{s} = -44.34\ \mathrm{J\ mol^{-1}\ K^{-1}}$. If the fuel cell is operated below 100 °C (e.g. low-temperature PEMFCs), liquid water is produced at the cathode and the water activity in Nernst equation can be set to unit, while other gas activities can be replaced by their partial pressures. Therefore, the entropic emf variation due to pressure change with respect to standard conditions is obtained:

$$\Delta E_p(T, p_{H_2}, p_{O_2}) = \frac{R\,T}{2\,F} \log\left(\frac{p_{H_2}}{p^0}\sqrt{\frac{p_{O_2}}{p^0}}\right) \tag{14.4}$$

where $R = 8.314\,\mathrm{J\,mol^{-1}\,K^{-1}}$ is the ideal gas constant.

14.2.3 Fuel Cell Voltage

In steady-state load conditions, the cell voltage V differs from the cell reversible emf due to overpotentials related to thermodynamic losses which depend on the cell current density j, as

$$V(j) = E - \eta_{act}(j) - \eta_{ohm}(j) - \eta_{con}(j) \tag{14.5}$$

where η_{act} is the activation overpotential due to electrochemical kinetics of reactions, η_{ohm} is the ohmic overpotential due to joule losses mainly occurring inside the PEM, and η_{con} is the concentration overpotential due to mass transport losses in flow channels and DLs. Equation (14.5) governs the polarization curve of a PEMFC and can be used to simulate its electrical behavior in system circuit models, which are used, e.g. in designing the power conditioning unit needed that interface the fuel cell stack to an electric load or in sizing the PEMFC from a specified power demand.

Figure 14.2 shows a typical polarization curve obtained from (14.5). The first part of the characteristic, at low current density, is dominated by activation losses. In particular, low-temperature fuel cells such as PEMFCs exhibit a sharp drop of the voltage at near zero current densities. The constant slope in the middle part of the characteristic is due to the constant resistance value of the PEM. This means that the membrane hydration and, thus, its proton conductivity can be considered to be

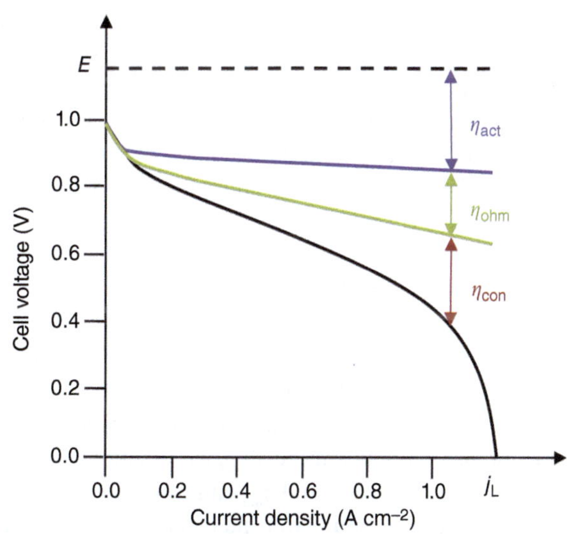

Figure 14.2 Typical polarization curve of a PEMFC according to (14.5): E is the cell reversible emf, η_{act} is the activation overpotential, η_{ohm} is the ohmic overpotential, and η_{con} is the concentration overpotential.

approximately constant. With a higher current density, concentration losses produce a sudden voltage drop due to reactant depletion. The limiting current density j_L at which voltage vanishes is an important figure of merit for design purposes because it provides the maximum operating range for a FC. In order to account for gas crossover (notably hydrogen) through the PEM, internal current losses, and unwanted side reactions, (14.5) can be corrected by replacing current density entry j with $j + j_{leak}$, where j_{leak} is the so-called *leakage current density* due to parasitic effects, see, e.g. [1].

14.2.4 Activation Overpotential

The activation overpotential η_{act} is the voltage drop associated with activation of electrochemical reactions in the CLs. This type of polarization dominates other losses at low current density and the amount of voltage drop depends on the amount of catalyst and on its performance (in order to minimize it, Pt-based catalysts are typically adopted in PEMFCs). The Butler–Volmer equation expresses the exponential relation between current density produced by the reaction and the overpotential in steady-state conditions (see Chapter 3 for details):

$$j = j_{0i}^0 \left[\frac{\overline{c}_{Ri}}{\overline{c}_{Ri}^0} e^{\alpha_i n \frac{F}{RT} \eta_{act,i}} - \frac{\overline{c}_{Pi}}{\overline{c}_{Pi}^0} e^{-(1-\alpha_i) n \frac{F}{RT} \eta_{act,i}} \right] \tag{14.6}$$

where α_i is the transfer coefficient of each half-reaction (anode i = A or cathode i = C), $\overline{c}_{Ri}$ and $\overline{c}_{Pi}$ are the actual concentrations of reactant and product at the active sites, and $\overline{c}_{Ri}^0$ and $\overline{c}_{Pi}^0$ are the corresponding reference concentrations used to measure the exchange current density j_{0i}^0. The latter is a function of temperature, catalyst loading, and catalyst-specific surface area. For instance, a temperature increment and an active surface area increment both lead to a higher exchange current density, i.e. to a better fuel cell performance because of a smaller activation energy (see Chapter 3). Due to the structure of active sites at CLs (the so-called *triple-phase boundary*, TPB), the total active area A_{TPB}, where the exchange current is produced, is much larger than the fuel cell cross-section A. For instance, $A_{TPB}/A \approx 10^3$ for catalysts of common use. For PEMFCs, which typically operate at low temperature and high current density, the forward reaction dynamics dominates so that its activation overpotential is large and the second term in (14.6) can be neglected (see Chapter 3 for details). By using this assumption, both activation overpotentials are obtained from the Tafel equation, as:

$$\eta_{act,i} = \frac{RT}{\alpha_i n_i F} \log\left(\frac{j}{j_{0i}}\right) \tag{14.7}$$

where $j_{0i} = j_{0i}^0\, \overline{c}_{Ri}/\overline{c}_{Ri}^0$ is the actual exchange current density at anode or cathode TPBs. It should be noted that, however, in PEMFCs the anode activation losses are usually much smaller than the cathode activation losses since the oxygen reduction reaction at CCL has a more sluggish kinetics. Therefore, the anode overpotential $\eta_{act,A}$ can be neglected, and η_{act} can be computed by considering the cathode only, i.e. by setting $n_C = 4$ in (14.7) and by taking the oxygen concentration to get j_{0i}.

14.2.5 Ohmic Overpotential – PEM Model

The most used material for PEMs is Nafion®, which consists of polytetrafluoroethylene (PTFE) backbone chains bearing perfluoroethereal side chains terminated with sulfonic acid groups [2]. This material exhibits an excellent behavior as a protonic conductor when properly hydrated. The ohmic overpotential is the voltage drop associated with the ohmic losses occurring in the cell, which are mainly due to the PEM, $\eta_{ohm} = \text{ASR}_\text{M} j$, when an ionic current density j is present. ASR_M is the *area-specific resistance* of the membrane, which is defined as:

$$\text{ASR}_\text{M} = \int_0^{\delta_\text{M}} \frac{1}{\sigma(\lambda(z), T)} \text{d}z \tag{14.8}$$

where δ_M is the membrane thickness. The proton conductivity of the PEM σ increases with temperature T and water content λ. In the case of Nafion, λ can be expressed locally as:

$$\lambda = c_\text{w}/c_\text{s} \tag{14.9}$$

where c_w is the water concentration and $c_\text{s} = 1970\ \text{mol m}^{-3}$ is the sulfonic acid group concentration. Depending on c_w, λ varies in the range 0–22 and, in practice, ion conduction occurs for $\lambda > 2$. To achieve an optimal fuel cell performance, with reduced ohmic losses, it is crucial to maintain the membrane fully hydrated, $\lambda \approx 22$. On the other hand, at high current density, cathode flooding and anode dehydration may occur, which are side effects to be avoided. Consequently, a proper water management must be ensured and a proper numerical modeling is required.

An experimental correlation of the proton conductivity of Nafion, originally proposed by Springer [4], is adopted in a number of works in literature. It can be used for $\lambda > 2$ and proposes a linear dependence on λ (obtained after experimental data interpolation) and an exponential dependence on T (according to an Arrhenius-like behavior):

$$\sigma(\lambda, T) = \sigma_{303}(\lambda)\ \exp\left[1268\left(\frac{1}{303} - \frac{1}{T}\right)\right] \tag{14.10}$$

where $\sigma_{303}(\lambda)$ is the ionic conductivity at the reference temperature $T = 303$ K:

$$\sigma_{303}(\lambda) = (0.005139\,\lambda - 0.00326)\ \ (\text{S cm}^{-1}) \tag{14.11}$$

Since λ varies inside the membrane, $\sigma(\lambda,T)$ is a function of the spatial coordinates. This distribution of λ inside the PEM depends on two transport mechanisms: *electroosmotic drag* of water from the anode driven by proton current density j and *back diffusion* from the cathode of the water produced at the cathode by the ORR of (14.1).

The former consists in a water flux induced by a movement of solvated ions (producing current density j) that drag water molecules in their solvation sheet and can be considered proportional to both λ and j. The second effect can be modeled with Fick's first law. By assuming a 1D mass flow along the membrane thickness (along the z-axis in Figure 14.1), i.e. assuming a $\lambda(z)$ dependence with no variation of λ along the y-axis, the net water flux in the membrane N_w^M is obtained as:

$$N_\text{w}^\text{M} = n_\text{drag} \frac{\lambda}{22} \frac{j}{F} - D_\text{w}^\text{M} \frac{\rho_\text{dry}}{\text{EW}_\text{M}} \frac{\text{d}\lambda}{\text{d}z} \tag{14.12}$$

where $n_{drag} \approx 2.5$ is the number of dragged water molecules per proton ($\xi_{drag} \approx \lambda \cdot 2.5/22$ is the drag coefficient at 303 K), ρ_{dry} is the density of dry Nafion, EW_M is the Nafion equivalent weight, and D_w^M is the water diffusivity in the membrane. This parameter depends on the water content and temperature, according to the following empirical expression, which holds for $\lambda > 4$:

$$D_w^M(\lambda, T) = D_{w303}^M(\lambda) \exp\left[2416\left(\frac{1}{353} - \frac{1}{T}\right)\right] \tag{14.13}$$

where $D_{w353}^M(\lambda)$ is the diffusivity at the reference temperature $T = 353$ K:

$$D_{w353}^M(\lambda) = 0.41\frac{\lambda}{25}\left(1 + \tanh\frac{\lambda - 2.5}{14}\right) \ (\text{cm}^2\ \text{s}^{-1}) \tag{14.14}$$

The water content profile across the PEM can be reconstructed by integrating (14.12) under proper boundary conditions λ_A and λ_C at the two CLs–PEM interfaces. At this aim, λ can be related to the water vapor activity $a_w = p_w/p_{sat}$, which is the ratio between the actual water vapor pressure p_w and the saturation water vapor pressure p_{sat}. The latter varies with temperature and can be computed with the empirical equation:

$$\log_{10} p_{sat}(T) = -2.1794 + 0.029\,53\,T - 9.1837 \cdot 10^{-5}\,T^2 + 1.445 \cdot 10^{-7}\,T^3 \tag{14.15}$$

where p_{sat} is in bars and T in degrees Celsius. An empirical correlation between λ and a_w is:

$$\lambda = \begin{cases} 0.043 + 17.81\,a_w - 39.85\,a_w^2 + 36\,a_w^3 & 0 \le a_w \le 1 \\ 14 + 1.4\,(a_w - 1) & 1 < a_w \le 3 \end{cases} \tag{14.16}$$

The water partial pressure in the PEMFC can be obtained by solving a mass transfer model in the whole fuel cell. The water flux in (14.12) can be quantified by using the net transport water coefficient $\nu_w = N_w^M/N_{H_2}^A$, where$N_{H_2}^A = j/(2F)$ is the stoichiometric hydrogen flow at the anode. The net transport coefficient and hydration boundary values have to be determined by an iterative procedure such as that one proposed in [9]. Other strategies, e.g. based on CFD, are also possible. A more complete model to determine the water distribution inside the cell is presented in Section 14.3.

14.2.6 Concentration Overpotential

In load conditions, the electric current generation at the CLs is sustained by the inflow of reagents and outflow of products, which in turn are provided by convective mass flow in the flow channels within the current collectors and diffusive mass flow in the DLs. These mass flows produce gradients in the reactant and product concentrations, making bulk values at the DLs different from those at CLs. For any species k and medium m, the diffusion molar flux at DLs is computed by Fick's first law $\boldsymbol{N}_k^m = -D_k^m \nabla c_k$, where D_k^m is a diffusion coefficient (diffusivity) of the species k in the medium m, and c_k is the bulk species molar concentration. Assuming a 1D mass flow and noting that in steady-state conditions the concentrations vary linearly with the distance across the DL, Fick's first law for anode and cathode reactants at the respective DLs becomes:

$$N_{Ri}^{DL} = D_{Ri}^{DL} \frac{c_{Ri} - \bar{c}_{Ri}}{\delta_{DL}} \tag{14.17}$$

where δ_{DL} is the DL thickness and D_{Ri}^{DL} is its effective diffusivity in the DL. In simplified models, the latter is usually computed by means of the Bruggeman correction $D_{Ri}^{DL} = \varepsilon^{1.5} D_{Ri}$, where ε is the porosity of the DL medium. From (14.12) and by posing $N_{Ri}^{DL} = j/(n_i F)$, concentrations of reacting gases (hydrogen at the anode, oxygen at the cathode) at the CLs are related to the current density as:

$$\bar{c}_{Ri} = c_{Ri} - \frac{j\delta_{DL}}{n_i F D_{Ri}^{DL}} \tag{14.18}$$

The maximum theoretical value of the current density compatible with (14.18) occurs when the reactant concentration drops to zero at the CL, i.e. $\bar{c}_{Ri} = 0$. This maximum value is called *limiting current density* $j_{L,i} = n_i F D_{Ri}^{DL} c_{Ri}/\delta_{DL}$ and represents a design constraint. By using $j_{L,i}$, (14.18) can be rewritten as:

$$\bar{c}_{Ri} = c_{Ri} \left(1 - \frac{j}{j_{L,i}}\right) \tag{14.19}$$

which indicates that the reactant is entirely depleted at the CL when the current density reaches the limiting value. In order to account for depletion of reactant at CLs, the Nernst potential variation with respect to reference conditions, used for computing the Butler–Volmer equation, has to be considered. By using (14.18), this variation can be rewritten as:

$$\eta_{i,con} = \left(E^0 - \frac{RT}{n_i F} \log \frac{1}{c_{Ri}}\right) - \left(E^0 - \frac{RT}{n_i F} \log \frac{1}{\bar{c}_{Ri}}\right) = \frac{RT}{n_i F} \log \left(\frac{j_{L,i}}{j_{L,i} - j}\right) \tag{14.20}$$

Again, it can be noted that anode kinetics can be neglected so that in computing the total concentration overvoltage η_{con} of (14.5) only the cathode contribution is accounted for with $i = C$ and $n_C = 4$.

A more comprehensive computation of the mass flow can be obtained by considering also the convective mass transfer which occurs along flow channels. To this aim, Fick's first law can be modified as $N_{Ri}^{DL} = \kappa_{Ri}(c_{Ri} - \bar{c}_{Ri})$, where the mass transfer coefficient κ_{Ri} takes into account also transport losses related to diffusion in the flow channels. It is defined as a series of equivalent hydraulic conductance values, as:

$$\kappa_{Ri} = \left(\frac{L}{\mathrm{Sh}\, D_{Ri}^{GC}} + \frac{\delta_{DL}}{D_{Ri}^{DL}}\right)^{-1} \tag{14.21}$$

where L is the flow channel length, Sh is the (dimensionless) Sherwood number, and D_{Ri}^{GC} is the diffusivity of the reactant i within the flow channel. The corresponding limiting current density is $j_{L,i} = n_i F \kappa_{Ri} c_{Ri}$.

A more detailed modeling of mass transport in porous media accounting for non-linear variation of species concentration inside the DLs is provided in Section 14.4.

14.2.7 Examples of Fuel Cell Modeling

The analytical model synthesized in (14.5) depends only on a few physical parameters, namely:

- cell operative temperature T
- cell reversible emf E
- cathode exchange current density j_{OC}
- cathode activation coefficient α_C
- area-specific resistance ASR_M
- limiting current density j_L
- leakage current density j_{leak}

As an example of application, physical parameters reported in Table 14.1 are used to compute the polarization curve of a low-temperature PEMFC. For the sake of comparison, a solid oxide fuel cell (SOFC) with the same limiting current density value (2 A cm^{-2}) is considered. Both polarization curves are plotted in Figure 14.3. In the case of a PEMFC, Figure 14.3a, given its low-operating temperature (350 K), the Tafel approximation for the cathode kinetics is acceptable. In the case of a SOFC, Figure 14.3b, the polarization curve has a constant slope below the limiting current density value, because the activation overpotential is negligible due to a much higher operating temperature (1000 K) compared to PEMFC and ohmic losses prevail at low current density as well.

The model presented in this section provides a simple description of the fuel cell performance, since only electrical quantities are involved. The dependence on the reactant concentration is considered in fact through the definition of the limiting

Table 14.1 Physical parameters of the 1-D analytical model for PEMFC and SOFC.

	T (K)	E (V)	n_a	n_c	j_{OC} (A cm^{-2})	α_C	ASR_M (Ω cm^2)	j_L (A cm^{-2})	j_{leak} (A cm^{-2})
PEMFC	350	1.22	2	4	10^{-4}	0.3	0.01	2	10^{-2}
SOFC	1000	1.06	2	4	0.4	0.3	0.04	2	10^{-2}

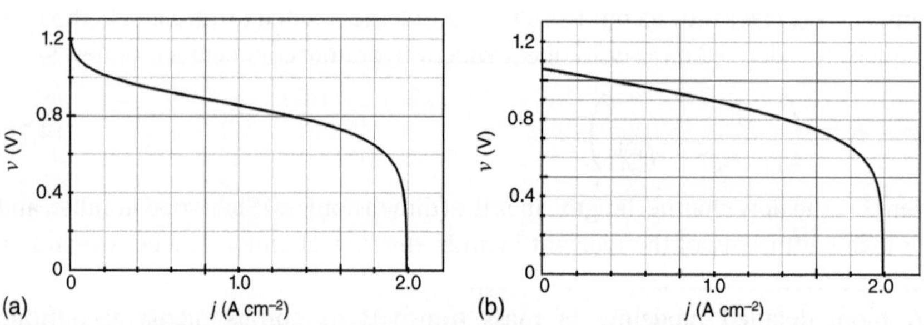

Figure 14.3 Polarization curves for (a) PEMFC and (b) SOFC computed with the 1-D analytical model.

current density. The main advantage is that electrical parameters can be easily identified from experimental polarization curves. For a more detailed simulation of the fuel cells behavior, multiphysics 1D models accounting also for mass transport parameters and anode kinetics must be adopted. In doing so, accurate stochastic techniques for non-linear fitting are required for the parameter identification, such as that in [10].

14.3 Multiphase Water Transport Model for PEMFCs

In polymer electrolyte membrane fuel cells, performance and degradation are strongly linked to the concentration of water in the core components, where it is found in three phases. In the gas phase water vapor dilutes both hydrogen at anode side and air at the cathode side. When water vapor is present at a partial pressure equal to or higher than saturation pressure, a local condensation occurs with formation of a liquid phase. The liquid phase is initially formed in the porous media (DLs and CLs) where it may prevent the reactants from reaching the catalyst active sites. This phenomenon, known as liquid water flooding [11], limits the performance of electrocatalyst specifically under low oxygen concentration or under high current density, which requires a large flux of oxygen to feed the electrochemical reaction. Flooding is a result of several possible events: at the anode side, a local temperature lower than the gas dew point as well as a fast increase in pressure may occur which may result in a local condensation. More frequently, flooding occurs at the cathode due to a high local production of water from oxygen reduction reaction, causing a performance loss associated with a local depletion of oxygen on the cathode catalyst. When liquid water impedes hydrogen access to the anode catalyst active sites, instead, a local fuel starvation occurs, which may potentially lead to an unstable performance and a severe damage of the catalyst. Because of the severity of this phenomenon at the anode side, relative humidity and temperature are generally kept under control during transients, by managing the hydrogen pressure or flow rate. Liquid water from the porous media eventually also invades the flow channels of the bipolar plates, causing slug flow that is detected from the cell voltage measurement as oscillating and unstable values. The third phase in which water may be present in fuel cell systems is adsorbed phase into the membrane. The perfluorosulfonic acid ionomers, that allow ion conduction in state-of-the-art PEMFCs, show a significant dependence of its conductivity on the water content in the membrane. Even though several theories coexist, and no overall consensus has been achieved in the scientific community [12], it is agreed that ionic groups tend to form clusters in the membrane phase, which are distant and poorly connected when the polymer is equilibrated with a low relative humidity gas. At increasing relative humidity, these clusters tend to form interconnections that become effective paths for ion and water transport. All above considerations clarify that water management is a crucial issue in PEMFC system control and mathematical models are necessary to disentangle its complexity, in particular at high current density, where the transport phenomena play a major role. In this section the presence of water in the solid phase (ice) is not considered, which also could be found in operations at subzero temperature. Concerning this last scenario, a relevant literature is available for the interested readers [13].

From a mathematical point of view, multiphase water transfer in PEM fuel cells is modeled by coupling the conservation equations of water in each phase with a phenomenological equation describing water flux. This method has the advantage of integrating fundamental physics-based equations with a large amount of *ex situ* experimental measurements on transport properties of materials. The conservation equation for water transport in the membrane phase is based on the water content λ of (14.9) that expresses the water uptake in the polymer [14]. In early investigations, λ was considered in integral form, so that it was expressed as the moles of water molecules N_{H_2O} adsorbed in the polymer per mole of sulfonic acid group $N_{SO_3^-}$. This definition can be related to the mass ratio of water/ionomer in the membrane M_{H_2O}/M_M, which can be measured by means of gravimetric techniques:

$$\lambda = \frac{N_{H_2O}}{N_{SO_3^-}} = \frac{EW_M}{M_w^{H_2O}} \frac{M_{H_2O}}{M_M} \tag{14.22}$$

where EW_M is the ionomer equivalent weight and $M_w^{H_2O}$ is the water molecular weight. Consistently with the initial integral definition and along with the development of more sophisticated fuel cell models, it was recognized that water content had to be evaluated locally in the membrane to take into account non-uniform conditions. Consequently, a molar conservation equation of water in the adsorbed phase was introduced [15]:

$$\frac{\partial N_w^M}{\partial z} + \frac{\rho_{dry}}{EW_M} \frac{\partial \lambda}{\partial t} = S_{ORR} + S_{v \to \lambda} \tag{14.23}$$

where S_{ORR} is the source terms that represents the water generated in the ionomer phase, due to ORR in (14.1), and $S_{v \to \lambda}$ is the exchanged volumetric rate of water between the adsorbed phase and the vapor phase. The water flux in the adsorbed phase N_w^M was derived from experimental evidence indicating that water transfer occurred across the membrane phase because of two major mechanisms: the first one is the electroosmotic drag and the diffusion in the membrane due to concentration gradients, as already stated, which are formulated in (14.12). Additional terms are sometimes included. One is the *hydraulic permeation*, a water flux associated with capillary pressure gradients from anode to cathode side (as described below); the other phenomenon is the *thermo-osmotic flux*, a transfer of water induced by temperature gradient across the membrane phase. These phenomena require further investigation to be fully understood, so they are generally neglected in present models. However, the temperature dependence is considered, as highlighted in the equations presented in Section 14.2, although they are not expressed here for the sake of simplicity.

The values of the material parameter used in the equation depend on the polymer composition (specifically, molecular weight of proprietary ionomer formulation) and on the operating conditions (specifically, temperature and water content). In the case of Nafion, experimental characterization results can be found in the literature [15], notably yielding the expressions already been presented in (14.10), (14.11), (14.13), and (14.14). It is worth noticing that the dependence on λ and T involves a strong non-linearity that generates a numerical stiffness in the solution of fuel cell models.

Estimated activation energies account for the effect of temperature on these properties [15]. In order to correctly simulate the water content in PEM fuel cells, a proper adsorption and desorption rate expression for water from gas/liquid phase must be introduced in the CLs, where the water is present in three phases. This effect is accounted for by the source term $S_{v\to\lambda}$ of (14.23) that represents the water exchanged with the vapor phase. Non-equilibrium between phases is generally verified in PEMFC multiphase models adopting kinetic laws for water adsorption/desorption. These laws rely generally on kinetic constants which represent the interfacial resistance for adsorption. The nature of this term remains elusive since few experimental works tried to measure it in fuel cell systems [16]. A simple example of such kinetic laws is reported here:

$$S_{v\to\lambda} = k_{v\to\lambda}\frac{\rho_{\mathrm{dry}}}{\mathrm{EW_M}}(\lambda_{\mathrm{eq}} - \lambda) \tag{14.24}$$

where the coefficient $k_{v\to\lambda}$ was evaluated by Ge et al. [16], who reported a non-negligible dependence on water content and temperature, to confirm that this feature is not accurately captured by the simple law reported in (14.24). In [15] the water content in equilibrium conditions λ_{eq} is obtained from experimental data of adsorption isotherm and given as a function of relative humidity and temperature in a form of a polynomial fitting. This allows to reproduce the non-linear shape with a clear increase in slope in the region close to the fully saturated condition, where the formation of interconnected domains in the polymer structure facilitates the adsorption of water in the bulk portion of the membrane. In operation, a deviation from this equilibrium condition is found under high current density, as a consequence of the interfacial resistance between phases. Numerical solution requires boundary conditions for water content and a no-flux condition can be set at the interface between the CLs and the DLs, where the ionomer phase is no more present.

For the sake of example, steady-state simulations performed with a 1D model that implements the equations described above [17, 18] are reported in Figure 14.4a that allows to investigate how water transport in the adsorbed phase impacts on fuel cell performance. The effect of the membrane thickness was evaluated under different operating conditions: a fully humidified case was compared with a dry one to get insight on the impact of water management. A clear change at intermediate current density was found that is connected to a higher ohmic loss in thicker membranes. It was also found that the voltage drop in low relative humidity increases as the membrane thickness increases, indicating a lower tolerance toward dry conditions. To better understand this aspect, the water content profile across membrane and CLs was reported for the dry case, Figure 14.4b. Non-uniform water content was observed across the membrane thickness, due to the water transfer mechanisms: the maximum water content was found in the CCL where water is produced by ORR while the minimum is in the ACL. A linear profile was also simulated considering

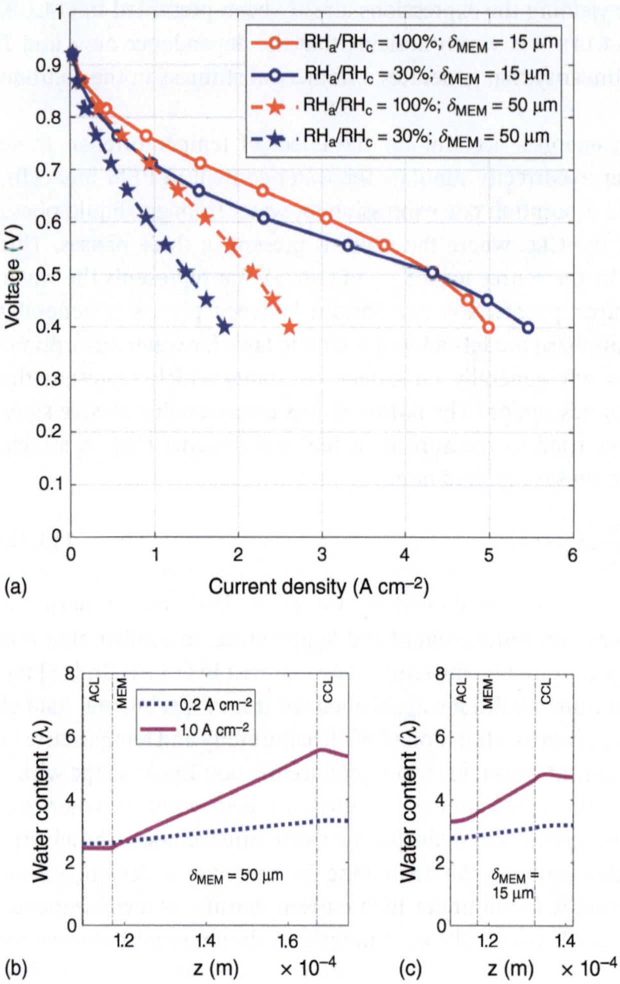

Figure 14.4 Simulations performed with a 1D model implementing multiphase water transport equations. Effect of membrane thickness changed from 15 to 50 μm under dry operative conditions (relative humidity RH 30% at both anode and cathode) and fully humid (RH 100%) at 2.3 bar absolute pressure and 80 °C temperature; (a) polarization curve; (b) and (c) water content profiles under dry conditions for the thick and thin membrane, respectively.

the balance between water drag flux and back diffusion of water from cathode to anode. Under low current density (dotted curve), the average water content in the membrane reduced in consequence of a low water production rate at the cathode. Comparing the thin and the thick membrane cases, Figure 14.4b,c, a significant difference is predicted by the model, specifically at high current density, where water back diffusion is expected to be enhanced, thus decreasing the water content in the CCL but leading to more uniform values across the membrane. For this reason, at higher current density the impact of the dry condition is mitigated in the case of a thin membrane.

Liquid water transport in PEMFC porous media is approximated with a continuum approach based on the definition of liquid water saturation, which is the fraction of porous medium volume occupied by liquid water. The mass conservation equation for liquid water is:

$$\frac{\partial N_{\text{liq}}}{\partial z} + \varepsilon \frac{\rho_{\text{liq}}}{\text{EW}_{\text{liq}}} \frac{\partial s}{\partial t} = S_{\text{liq}} \tag{14.25}$$

where ε is the medium porosity, ρ_{liq} is the liquid water density, EW_{liq} is the liquid water equivalent weight, and s is the liquid water saturation. The source term S_{liq} represents the condensation/evaporation rate computed in all the porous domains. The liquid water flux N_{liq} is assumed to be driven by capillarity as generally observed in fuel cell porous media, where the capillary pressure is defined as:

$$P_{\text{cap}} = P_{\text{liq}} - P_{\text{gas}} \tag{14.26}$$

between the liquid and gas phase pressures P_{liq} and P_{gas} which coexist in the porous medium, specifically the non-wetting phase (liquid in PEMFC) and wetting one (gas in PEMFC). Liquid water flux in the porous medium is thus computed according to the phenomenological expression:

$$N_{\text{liq}} = -\frac{\rho_{\text{liq}}}{\text{EW}_{\text{liq}}} \frac{K_{\text{r}} K_{\text{abs}}}{\mu_{\text{liq}}} \frac{\text{d}P_{\text{liq}}}{\text{d}z} \tag{14.27}$$

where K_{abs} and K_{r} are the absolute and relative medium permeability to liquid water and μ_{liq} is its dynamic viscosity. Several models are available in the literature to simulate the relative permeability K_{r} of the liquid phase as a function of liquid water saturation s, as well as specific functions that link the capillary pressure to the liquid water saturation, among which the Leverett function is the most adopted in the literature, but not the only one available [15]. To reproduce the effect of liquid water flooding on the gas phase diffusivity, the latter is corrected by a factor that depends on the saturation s, in the form $(1-s)^{\psi}$ where ψ is a coefficient in the range 1–4 depending on the properties of the porous medium [15].

Steady-state simulations performed with a 1D model are reported in Figure 14.5a to investigate the impact of liquid water transport on fuel cell performance. The impact of CL thickness is simulated under dry and humid conditions as done in the previous case shown for membrane thickness. It is observed that large CL thickness determines an increase in performance in the ORR activation region because of the higher catalyst loading that is achieved. On the other hand, in the high current density region a significant lower limiting current is found, which is attributed to oxygen transport limitations consequent to the formation of liquid water saturation. The saturation profile simulated under fully humidified condition is reported in Figure 14.5b,c at 2.0 A cm^{-2}. The saturation shows a clear peak in the CCL, where water is produced from ORR and a lower contact angle is found compared to GDL and microporous layer (MPL) that are more hydrophobic. Under low relative humidity, instead, a higher potential drop of the polarization curve occurs in the case of a thick CL. This loss is associated to the ionic potential, whose profile is also reported in Figure 14.5b,c. A large potential drop is found for a thick CL, in consequence of

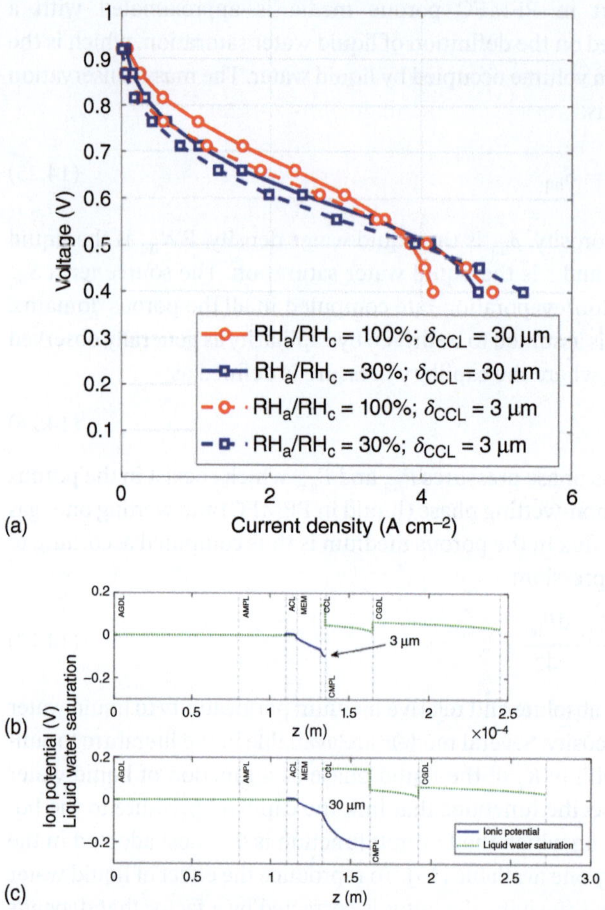

Figure 14.5 Simulations performed with a 1D model implementing multiphase water transport equations. Effect of cathode catalyst layer thickness changed from 3 to 30 μm under dry operative conditions (RH 30% both anode and cathode) and fully humid (RH 100%) at 2.3 bar absolute pressure and 80 °C temperature; (a) polarization curve; (b) and (c) ionic potential and liquid water saturation profiles under fully humidified conditions for the thick and thin catalyst layers, respectively.

ion transport limitations that are found in this case. As a conclusion, the advantage of a thick CL is evident only in the kinetic region, where transport phenomena play a minor role.

14.4 Fluid Mechanics in PEMFC Porous Media: From 3D Simulations to 1D Models

Models of mass transport in fuel cells aim at describing the physical phenomena which determine the spatial and temporal distribution of reactants and products in the two cell compartments. Given the complexity of these phenomena, which involve multiphase flows and a wide range of length scales (from macro- till micro- and nano-scale), mass transport description in such models unavoidably relies on some simplifications that are based on fundamental fluid-mechanic principles.

Thanks to the recent increase in computational power, numerical simulations represent nowadays a solution of choice to identify such principles and to draw up guidelines for the conceptualization of simple and effective transport models. Recent innovative numerical methodologies allow to solve the fundamental fluid equations at the microscale, and thus, to establish and assess the link between three-dimensional microscopic transport and the fuel cell macroscopic 1D models applied at the system level. Among the methodologies, the lattice Boltzmann method has shown great potentiality to model the complex physics involved in multiphase flows (an example of simulation results and a metric scale is presented in Figure 14.6). This "bottom-up" approach is particularly relevant for the characterization and prediction of mass transport and phase change phenomena in the fuel cell porous materials [20–22].

The DLs are key components for regulating mass transport inside the cell. Their porous microstructure is fundamental in determining the dynamics and distribution of fluids, two physical phenomena which have a major effect on the management of the produced water, and, in turn, on the cell performance [23].

In the DLs, the fluid is often present in more than one phase, i.e. gas and liquid, and interacts with the porous intricate microstructure in a complex manner. Consequently, a rigorous fluid-mechanic description would require the formulation of several intermittent, heterogeneous and non-linear phenomena characterizing multiphase flows in porous media. Suffice it to say that the motion at the contact line at

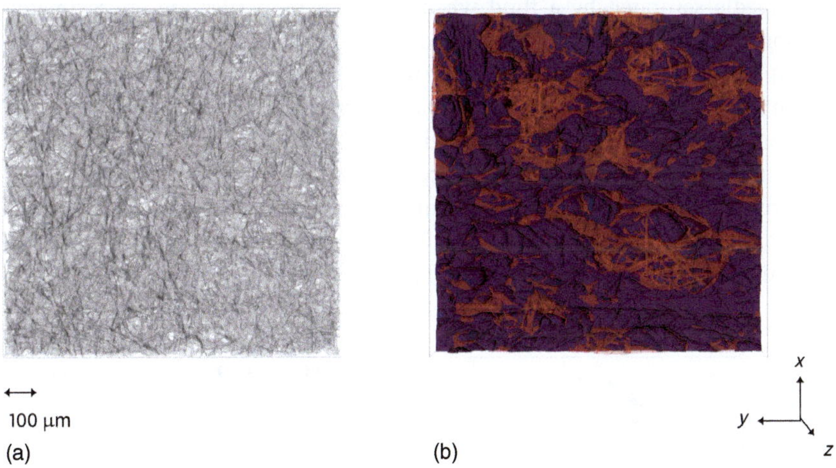

Figure 14.6 An example of lattice Boltzmann simulation of water infiltration into a 1 mm^2 fuel cell DL reconstructed via X-ray computed tomography, panel (a). Such simulations can solve liquid intrusion into the DL for different wetting conditions, such as hydrophilic (marked with red color) and hydrophobic (blue), panel (b). The approach allows a great resolution of the order of few micrometers. Source: Adapted from Ref. [19].

the liquid–gas–solid interface is a physical mode of transport whose exact mathematical description depends on the scale of observation and still stimulates a great debate [24]. It follows that in order to construct efficient models of mass transport in fuel cells, one has to get rid of such complexity and adopt some simplifications that allow to predict the relevant mass transport aspects without the need of solving the full 3D fluid-mechanic problem. In the present section, some important fluid mechanic principles and mechanisms of transport in fuel cell porous media are presented, and their application in 1D models is discussed.

14.4.1 From Micro- to Macroscopic Models

One of the most popular approaches to model flows in porous media is the Darcy equation. It linearly relates the macroscopic fluid velocity to the pressure gradient that drives the fluid through the medium:

$$u = -\frac{K}{\mu}\nabla P \tag{14.28}$$

where, as already stated, μ is the fluid dynamic viscosity (not to be confused with the electromagnetic permeability) and K is the medium hydraulic permeability that expresses the ability of the porous material to accommodate a certain flow and mass transport. The main benefits of the Darcy model are its simplicity and generality: provided that the characteristic permeability of the porous medium is known, the rate of mass transport can be predicted on the basis of the driving force, for instance of the pressure difference imposed between bipolar plate gas flow channels (GCs) and catalysts layer (CL), Figure 14.1, which delimit the DL [25].

The Darcy equation is originally conceived for single-phase flows, that is, for porous media saturated by a fluid whose physical properties are approximated as uniform in space, such as density and viscosity. When the flow is present in different phases, i.e. liquid and gas, as happens in fuel cells, the relationship between flow velocity and driving force is more complex and the Darcy equation is no longer enough accurate for its description, Figure 14.7. One of the main reasons behind such inadequacy is the increasing importance that the microscopic and microstructural variables have in determining the global, macroscopic transport behavior in the presence of multiphase flows. It is possible to obtain alternative formulations, which however also encompass the microscopic description of the

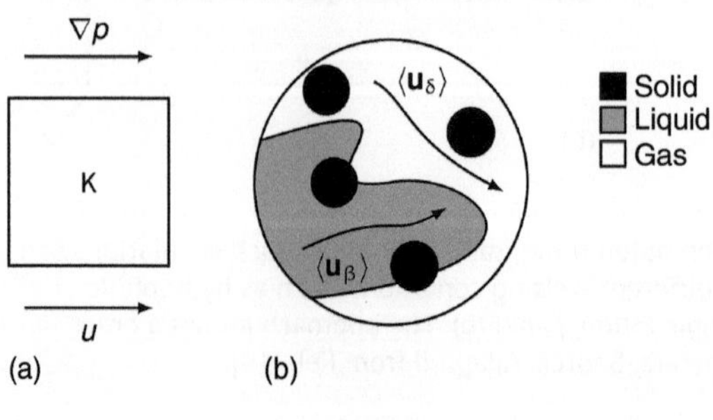

Figure 14.7 (a) Darcy's model for mass transport in porous media, which links the variables at the macroscopic scale acting like a "black-box" and (b) the volume-averaging perspective, where the transport is described for each phase at the microscale.

flows at the microscale. For the sake of example, a good step in this direction can be achieved through the application of the volume-averaging technique, a mathematical upscaling procedure that aims at describing the microscopic fluid phenomena in terms of macroscopic averaged variables [26].

The volume averaging is performed in a representative elementary volume where the different phases, i.e. liquid, gas, and solid, are present, leading to the so-called phase average for a generic variable φ_β belonging to phase β:

$$\langle \varphi_\beta \rangle = \frac{1}{\tau} \int_{\tau_\beta} \varphi_\beta \mathrm{d}\tau \tag{14.29}$$

where τ is the representative elementary volume and τ_β is the volume occupied by the phase β. The volume fraction for each phase is thus indicated as:

$$\varepsilon_\beta = \tau_\beta / \tau \tag{14.30}$$

By applying this averaging procedure to the equations of fluid motion of each phase, together with the proper boundary conditions at the phase interfaces, one obtains:

$$\langle \boldsymbol{u}_\beta \rangle = \frac{\boldsymbol{K}_\beta}{\mu_\beta} \cdot \nabla \langle P_\beta \rangle \varepsilon_\beta^{-1} + \boldsymbol{K}_{\beta\delta} \cdot \langle \boldsymbol{u}_\delta \rangle \tag{14.31}$$

where the brackets $\langle\rangle$ indicate the volume-averaging operator which defines the fluid macroscopic variables, such as the fluid velocity vectors $\boldsymbol{u}_\beta$, $\boldsymbol{u}_\delta$, and the pressure P_β, relative to the phase β or δ. (bold symbols indicate 3D vectors). For the mathematical details about the volume averaging procedure, the reader is encouraged to explore the relevant scientific literature [26].

14.4.2 Porous Medium Anisotropy

This first level of complexity is introduced by the intrinsic 3D nature of the volume-averaged equation. The traditional Darcy's formulation of flows in porous media defines the medium permeability as a scalar quantity. While this approximation can hold for isotropic materials, it fails to represent the anisotropy of the porous microstructures that exhibit a certain kind of orientation. The case of DLs for fuel cells certainly falls in the latter category since the fibers orientation in these materials shows often anisotropic features.

A DL presents one dimension (the thickness) much smaller than the others and the fibers are usually preferentially aligned along the larger dimensions, inducing the so-called in-plane orientation. The compression of the DL can also affect the final fibers orientation. Therefore, an important step toward the formulation of an effective mass transport model in fuel cells is the characterization of the permeability vector $\boldsymbol{K}_\beta$, and, consequently, of the effect of anisotropy on mass transport.

For the understanding of the phenomena involved, it is preferable to use mathematical methods that consider the full morphology of the porous microstructure to solve the fluid equations at the relevant scales. For instance, by combining the generation of porous morphologies with differently oriented fibers and lattice Boltzmann simulations, it was shown how the anisotropy of the medium can affect flow and dispersion of reactants, and the effective diffusive coefficients for different anisotropic configurations were estimated [27]. By means of a similar approach, a water droplet effective permeability was extracted for differently oriented fibrous media [28].

14.4.3 Fluid–Fluid Viscous Drag

The physical picture of multiphase flows in porous media is not only complicated by their intrinsic 3D patterns. Additionally, mechanisms acting on a pore-scale level can contribute to the macroscopic mass transport phenomena. By looking again at (14.31), we can reason about the meaning of the fluid-dynamic parameters involved. The mathematical details concerning $\boldsymbol{K}_\beta$ and $\boldsymbol{K}_{\beta\delta}$ are derivable through the solution of complex, coupled boundary-value problems for tensor variables. It is however possible to interpret them from an intuitive point of view. While the permeability vector $\boldsymbol{K}_\beta$ can be interpreted as an effective permeability of the phase β as a function of the saturation of the medium, the two-phase permeability $\boldsymbol{K}_{\beta\delta}$ – as the name suggests – is instead also dependent on the flow dynamics of the other phase δ. This quantity, when multiplied by the velocity of the other phase, yields an additional transport term to the equation. The classical interpretation is that this term conveys information about the importance of the viscous forces of one fluid upon the other, as regards the viscous forces exerted by the solid and the surface tension forces [26].

Following this reasoning, one would expect that for very large differences in the values of viscosities of the two phases, such a term could be considered negligible. For large values of μ_β compared to μ_δ the phase β would act as a solid with respect to δ and the effect of the flow of one phase on the other can be disregarded, provided that the reduction in porosity induced by the presence of the other phase is taken into account. By looking at (14.31) one can indeed assume:

$$\langle \boldsymbol{u}_\beta \rangle = -\frac{\boldsymbol{K}_\beta}{\mu_\beta} \cdot \nabla \langle P_\beta \rangle \epsilon_\beta^{-1}, \quad \text{for } \frac{\mu_\beta}{\mu_\delta} \gg 1 \tag{14.32}$$

where the last term of (14.31) has been disregarded. This is the fundamental reason that motivates the choice of mathematical formulations where the permeabilities and diffusivities of each phase are dependent on the relative phase volume fraction, as it does the Bruggeman equation for dispersion of the gas phase [29] or the Leverett model for the liquid-phase transport [30]. While such simplification surely holds in fuel cells in the case of low-humidity conditions, where the liquid gas viscosities ratio is of the order of 100, it can be debatable in high-humidity conditions, where the value of such a ratio approaches the unity.

14.4.4 Surface Tension and Capillary Pressure

There is a further complication in the formulation of Darcy-like models: the exact quantification of the volume-average pressure. Knowing and predicting such a quantity is necessary for completing the mathematical formulation of the mass transport model. The traditional way to close the system of equations for two-phase transport makes use of the principle of capillary pressure, i.e. the difference between the pressures of the two phases:

$$P_{\text{cap}} = -\langle P_\beta \rangle \varepsilon_\beta^{-1} + \langle P_\delta \rangle \varepsilon_\delta^{-1} = \pm\gamma \langle c \rangle_{\beta\delta} + O\left(\frac{\mu_\beta \langle \boldsymbol{u}_\beta \rangle}{l_\beta}\right) \tag{14.33}$$

where γ is the surface tension between the two phases and $\langle c \rangle_{\beta\delta}$ is the average curvature of the two-phase interface. The last term of (14.33) represents the order of magnitude of the contribution of viscous forces (with l_β the characteristic size of the more viscous phase β). If one disregards the latter, in the presence of a flat interface the curvature is null and the pressures are equal; when the curvature is concave with respect to the phase, the sign on the right-hand side is positive, whereas it is negative when it is convex. The quantification of the average curvature is one of the main historical difficulties when dealing with a multiphase mass transport model. Such a geometrical parameter is profoundly affected by the affinity that the solid material exhibits with the different phases (quantified by the contact angle) and by a great number of parameters that define the porous microstructure. Not only the morphology of the two-phase interface exhibits also a dynamic behavior, it has also been shown to depend on the fluid velocities [31]. The contact angle is related to the two-phase curvature. By applying the Young and Young–Laplace equations to a pore throat of radius r_{p} delimited by two fibers when the two-phase are at thermodynamic equilibrium, the pressure difference (14.33) can be written as:

$$P_{\text{cap}} = \pm\gamma \langle c \rangle_{\beta\delta} = \frac{\gamma \cos(\theta)}{r_{\text{p}}} \tag{14.34}$$

where values of θ lower than 90° indicate a hydrophilic medium, whereas higher values indicate a hydrophobic one. A sketch of the possible configurations of the two-phase flow as function of the material affinity with the fluid phases is shown in Figure 14.8.

In fuel cells the surface tension and the capillary pressure forces are relevant for the determination of the mass transport and its spatial distribution. Several studies have suggested that the liquid movement by means of capillary forces is the dominating mechanism in DLs [32]. The exact measurement of the contact angle

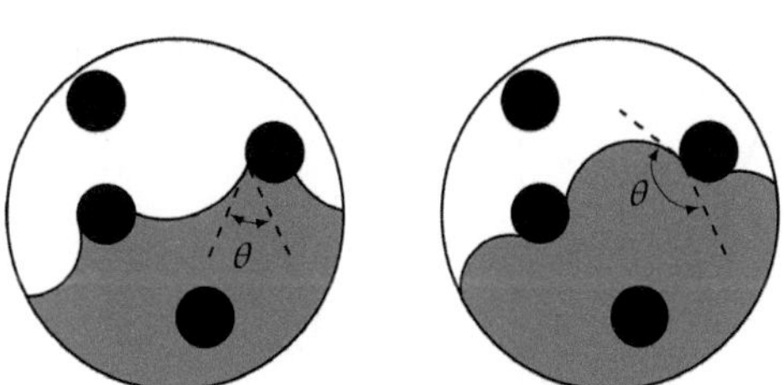

(a) (b)

Figure 14.8 An example of two-phase flow configuration within a hydrophilic (a) and hydrophobic (b) medium. The gray-shaded areas indicate the denser phase. The fluid–fluid-two-phase interface curvature is determined by the capillary pore throat radius and the contact angle.

characterizing the DL fibrous material is difficult to achieve in practice. A crude estimation of the capillary pressure in a fuel cell DL can be based on the mean pore radius, by approximating the average curvature with the inverse of r_p, resulting of the order of 10–100 μm. Since most of the fibrous layers used as DL is from slightly to highly hydrophobic [33], we can intuitively consider that when the two-phase interface crosses a pore throat, its meniscus forms a convex semi-circle whose radius is r_p (Figure 14.8 shows a figurative example). Thus, we can also assume an effective contact angle of roughly 180°, which represents the effect of pore constriction.

By inserting such values in (14.34), we can estimate a negative capillary pressure in the order of 0.7–7 kPa (with liquid–gas surface tension $\gamma = 0.07\,\mathrm{N\,m^{-1}}$). This is not a negligible value and consequently it shows that it is crucial to incorporate the effect of capillary pressure in the formulation of a mass transport model in fuel cells.

The porous microstructure plays again a pivotal role, by determining the capillary pressure forces. This observation further motivates the use of pore-scale models to extract quantities that effectively conveys information of the microscopic multiphase dynamics. Recent examples of such an approach make use of X-ray computed tomography to accurately reconstruct the full morphology of the DL and combine such a technique with lattice-Boltzmann-based multiphase flow simulations [19] or pore network models [34]. An example of the obtained flow simulations is presented in Figure 14.6, and the extracted macroscopic effective quantities are also reported in Figure 14.9.

These simulations show the key role that the capillary pressure has on the flow dynamics at the pore-scale. The liquid infiltration rate, that is, the variation of porous medium saturation in time is shown to follow a time-dependent power law, with the mean liquid front infiltration depth along the direction of the flow increasing with the square root of time. This relationship is the mathematical representation of a fluid intruding into a porous medium at a rate that is inversely proportional to its average infiltration depth. The power law scaling can be derived from (14.32). Let us assume a viscous liquid infiltrating into a medium initially filled with a much less viscous fluid (a gas). Under these circumstances, we can neglect the viscous pressure drop in the gas phase and the fluid–fluid viscous drag since $\mu_\beta \gg \mu_\delta$. We can then

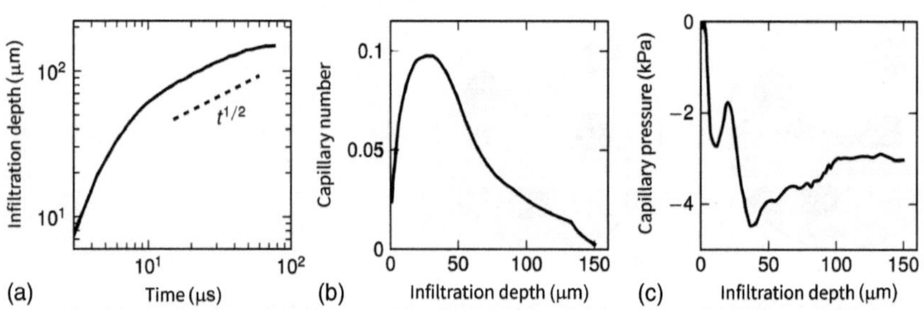

Figure 14.9 Macroscopic effective quantities extracted from X-ray reconstruction and lattice Boltzmann simulations. (a) Infiltration depth. (b) Capillarity number. (c) Capillarity pressure. Source: Adapted from Ref. [19].

also assume the flow to be one-dimensional and write the average velocity along the main flow direction z as:

$$\langle u_{\beta z} \rangle = \frac{\mathrm{d}\langle h_{\beta z}\rangle}{\mathrm{d}t} \tag{14.35}$$

with $h_{\beta z}$ the average liquid front position within the medium (Figure 14.9). In the simplest assumption that the flow is driven by a pressure difference ΔP (only viscous contribution) between inlet and outlet, it is possible to rewrite (14.32) as:

$$\frac{\mathrm{d}\langle h_{\beta z}\rangle}{\mathrm{d}t} = \frac{K_{\beta z}}{\mu_\beta}\frac{\Delta P}{\langle h_{\beta z}\rangle} \tag{14.36}$$

By separating the variables and integrating we obtain:

$$\langle h_{\beta z}\rangle = \left(K_{\beta z}\Delta P \mu_\beta^{-1} t\right)^{1/2} \tag{14.37}$$

which is the aforementioned power-law scaling. An average velocity and a phase capillary number that are inversely proportional to the liquid infiltration depth correspond to this scaling.

However, in the model presented in (14.36) and (14.37), we should also consider the effects induced by the capillary pressure. This pressure exhibits a dynamic behavior and approaches a quasi-constant value after an initial fast stage. This value, around −3 kPa, is indeed close to the approximation based on the mean pore radius, which, in the case of the specific geometry under consideration (a DL AvCarb MGL370), results in a size of 40 μm. The capillary pressure acts adversely to the flow, as expected, thus indicating a two-phase interface concave with respect to the liquid phase, induced by the motion of the interface passing the pore constrictions. Following this observation and, neglecting the pressure drop in the gamma phase, since $\mu_\beta \gg \mu_\delta$, the two-phase system of equations can be closed by rewriting the pressure gradient via (14.33) as:

$$-\nabla_z \langle P_\beta\rangle \varepsilon_\beta^{-1} \approx \left(-\frac{\gamma}{r_\mathrm{p}} + \Delta P\right)\frac{1}{\langle h_{\beta z}\rangle} \tag{14.38}$$

where the two terms under brackets represent the resistance to the flow induced by the presence of pore constrictions $P_\mathrm{cap} = -\gamma/r_\mathrm{p}$ and by purely viscous forces (then proportional to the applied pressure difference). We can rewrite the equation of motion by adding the further contribution of the microstructure-induced capillary pressure:

$$\langle h_{\beta z}\rangle = \left[\frac{r_\mathrm{p}^2}{4\mu_\beta}\left(\Delta P - \frac{\gamma}{r_\mathrm{p}}\right)\right]^{1/2} t^{1/2} \tag{14.39}$$

In writing (14.39) we have also assumed the phase permeability as $K_{\beta z} = \frac{r_\mathrm{p}^2}{4}$, a value valid for infiltration in a capillary tube of radius r_p. Equation (14.39) can be conceptualized as a system of parallel circular pores, of average radius r_p, and contact angle 180°, within which the liquid phase flows in an equivalent way as it does within the complex DL microstructure (Figure 14.10).

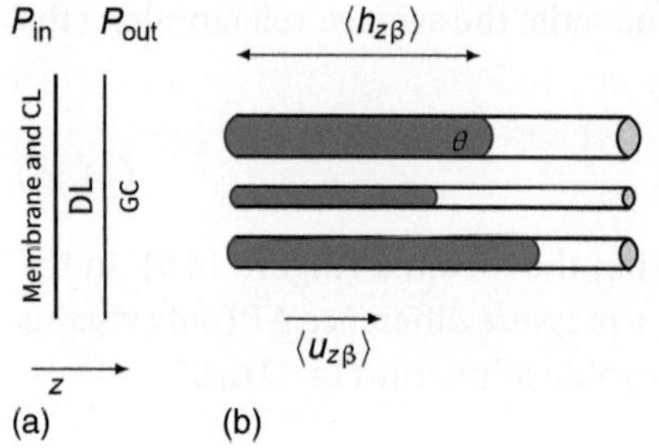

Figure 14.10 (a) Diffusion layer (DL) placed between membrane and catalyst layer (CL) and the gas channels (GC), where the boundary conditions determining Eq. (14.9) are the pressures at the layers' interfaces. (b) The bundle of capillary tubes model which mimics the liquid transport through the DL.

It is worth noticing that (14.39) has been obtained without considering the hydrophobicity of the porous material surface, that is, by setting the contact angle in the numerical simulation to 90°, in order to mimic a non-treated material. The measured capillary pressure is then purely induced by the microstructure, in contrast to the simplified view that predicts a flat interface for neutrally wet solid surfaces and a null contribution of the capillary pressure.

14.5 Physical-based Modeling for Electrochemical Impedance Spectroscopy

14.5.1 Experimental Measurement and Modeling Approaches

The electrochemical impedance spectroscopy (EIS) is a powerful *in situ* measurement technique that permits to evaluate and distinguish between kinetic and mass transport effects occurring in an electrochemical system and allows validating dynamic numerical models [35]. It consists in perturbing a steady-state bias operation condition (I_0, V_0) of the polarization curve with a small sinusoidal current signal $i_{AC}(\omega t) = I\sin(\omega t)$ at frequencies $f = \omega/2\pi$ which are varied over a wide range (typically 10^{-3} to 10^3 Hz) and in measuring the corresponding voltage response $v_{AC}(\omega t) = V\sin(\omega t + \varphi)$. Assumed the linearity of the system, the electrical impedance $Z(f)$ at a given frequency is the complex operator defined as the ratio between the complex representation of the voltage sinusoid $v_{AC}(\omega t)$ and the complex representation of the current sinusoid $i_{AC}(\omega t)$. Consequently, the impedance module is given by the amplitude ratio V/I and the impedance phase is given by the phase shift φ.[1]

At different frequencies of the AC perturbation, different physical phenomena are excited depending on their typical characteristic time and this results in changes in the impedance module and phase, giving useful information on the entity of internal fuel cell effects [36]. The membrane produces the dominant loss effects at very high frequencies ($f = 10^3$ Hz), electrochemical reactions respond around 10–100 Hz, and mass transport phenomena appear at lower frequencies

1 The underlying theory consists of the symbolic method of circuit analysis based on the Steinmetz transform.

($f = 10^{-2}$ Hz). The impedance values are usually reported in Nyquist and Bode plots. The former reports the opposite of the imaginary part of the impedance $-Z_I(f)$ vs. the real part $Z_R(f)$, while the latter usually illustrates $Z_R(f)$ and $-Z_I(f)$ as functions of the frequency f. Typical Nyquist and Bode plots of a PEMFC are reported in Figure 14.11 and described in Section 14.5.3.

Despite EIS providing a lot of useful information, the interpretation of measurements is not trivial and modeling plays a fundamental role in the analysis of experimental data. In the literature, the interpretation of impedance measurements usually resorts to the equivalent circuit method (ECM) [37]. ECM postulates a representation of the fuel cell in terms of electric elements, independently of linking them to specific physical phenomena. Even though simple and fast, this method provides only qualitative information since the choice of the equivalent circuit is not unique.

A different approach consists in correlating the equivalent circuit elements with the physical parameters of the system [38]. This methodology is fast and reliable and permits a quantitative interpretation of EIS measurements. However, it can be applied only in a few limiting cases when it is possible to derive an analytical solution of the constitutive equations of the system [39].

For this reason, physical-based modeling in time domain is fundamental to assess the origin of the different element features for complex systems, such as PEMFC. This method aims at replicating the experimental measurements, simulating several harmonic periods $T = 1/f$ for each frequency of the applied current perturbation. However, the numerical resolution of the resulting system usually requires huge computational costs, so that two techniques have been developed to tackle this issue: the current relaxation approach and the Laplace transform in frequency domain.

14.5.2 Physical-based Modeling

14.5.2.1 Current Relaxation

The current relaxation approach consists in studying the transient response in very fast voltage time-step (typically shorter than 10^{-6} seconds). This method is based on the principle that current relaxation in the time domain provides the same information as the EIS [40]. It requires less computational effort compared to the simulation of several harmonic periods for each frequency because just the time evolution of current and voltage needs to be simulated. The duration of the time steps has to be smaller than the characteristic time of the fastest process of the system, while the duration of the simulation needs to be longer than the characteristic time of the slowest process.

The method requires to put the system in a steady-state (bias) working point (I_0,V_0) of the polarization curve, to apply a voltage time-step $v(t) = V\delta_{-1}(t)$ of small amplitude V in order to ensure the linear behavior of the system and to acquire the corresponding current $i(t)$ that constitutes the system response. The Fourier transform is then applied to the time functions of the voltage $v(t)$ and current $i(t)$:

$$V(f) = \int_{-\infty}^{+\infty} v(t)\exp(-j2\pi ft)\mathrm{d}t$$
$$I(f) = \int_{-\infty}^{+\infty} i(t)\exp(-j2\pi ft)\mathrm{d}t \tag{14.40}$$

where j is the imaginary unit. From these Fourier transform $V(f)$ and $I(f)$, the impedance as a function of f is simply computed as:

$$Z(f) = \frac{V(f)}{I(f)} \tag{14.41}$$

Only the stimulus $v(t)$ and its response $i(t)$ must be considered in (14.40) and to this aim the steady-state terms V_0 and I_0 have to be subtracted from the total voltage and current. In order to reduce the computational time, for the time integration in (14.40) advance numerical methods with adaptive time steps can be adopted. Further details on this modeling approach can be found in [40].

14.5.2.2 Laplace Transform

This method consists in linearizing the system about a steady-state point (I_0, V_0) and in perturbing each variable $a_k(t)$ with a sufficiently low sinusoidal disturbance:

$$a_k(t) = A_{0k} + \Re e\{\tilde{A}_k \exp(j2\pi ft)\} = A_{0k} + A\cos(2\pi ft + \alpha_k) \tag{14.42}$$

where A_{0k} is the steady-state value (i.e. I_0 or V_0) and $\tilde{A}_k = A_k \exp(j\alpha_k)$ is a complex number representing the amplitude and the phase shift of the sinusoidal perturbation. Neglecting the terms given by products of disturbances and subtracting the steady-state terms, it is possible to obtain a system of linear equations in the complex perturbation $\tilde{A}_k$ that needs to be solved at each frequency. The Laplace transform allows replacing time derivatives d/dt with the multiplication by the complex variable s, $\mathcal{L}[\mathrm{d}a_k/\mathrm{d}t] = s\tilde{A}_k$, thus algebrizing the differential terms appearing in the equations of dynamic systems. In this application, the complex variable is reduced to a pure imaginary one, $s = j2\pi f = j\omega$, since sinusoidal perturbations are considered. The resulting system consists of ordinary differential equations instead of partial differential equations that can be solved with conventional numerical methods. The cell impedance at each frequency is then evaluated as the ratio between the complex representations of the oscillating voltage and current, in agreement with (14.41). The main drawback of the Laplace transform consists in the strong manipulation induced by the linearization of the system. An application of this methodology for EIS modeling in PEMFC can be found in [41].

14.5.3 Typical Impedance Features of PEMFC

Since the results of physical-based models do not depend on the adopted equivalent electric circuit, the presented approach allows to directly assess the origin of the different impedance features. As described in [42], a physical-based model can be reduced by neglecting the contribution to the impedance of some phenomena, permitting to evaluate the effect of one specific physical process on the impedance

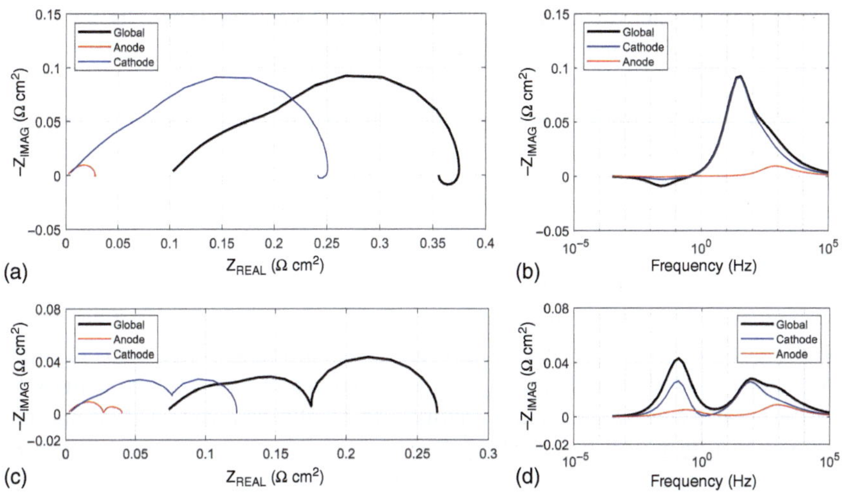

Figure 14.11 Typical impedance response of a PEM fuel cell simulated at 80 °C, $RH_A = 50\%$, $RH_C = 80\%$, $\lambda_A = 2$, $\lambda_C = 4$: (a) Nyquist plot at 0.2 A cm^{-2}; (b) Bode plot at 0.2 A cm^{-2}; (c) Nyquist plot at 2 A cm^{-2}; (d) Bode plot at 2 A cm^{-2}.

behavior and to quantify the corresponding loss. Figure 14.11 illustrates typical Nyquist and Bode plot of a PEMFC biased at 0.2 A cm^{-2} (Figure 14.11a,b) and 2 A cm^{-2} (Figure 14.11c,d) with a 1D model, highlighting the separated contribution of anode and cathode [18].

The impedance spectrum at 0.2 A cm^{-2} (Figure 14.11a,b) presents a dominant capacitive behavior that is mainly due to the contribution of ORR at CCL [41]. When the performance at the CCL is not limited by any mass transport issues, the cathode impedance is represented by an ideal semicircle inversely proportional to the operating bias current density, i.e. directly proportional to the Tafel slope, defined as the slope of electrode potential vs. the logarithm of current density [43]. This trend is confirmed analyzing the cathode contribution shown in Figure 14.11c,d, which show that the capacitive effect at high frequency is smaller at 2 A cm^{-2} than at 0.2 A cm^{-2}. Due to mass transport within the CLs and DLs, the capacitive feature appears elongated and at high operating current density (Figure 14.11c,d) a second feature appears at lower frequencies, in particular when liquid water is present.

The contribution of ACL and CCL is minor at both 0.2 and 2 A cm^{-2}: this is due to the faster kinetic of HOR and the greater diffusivity of hydrogen, also confirmed by the higher characteristic frequencies (Figure 14.11b,d) compared to the cathodic contribution [44].

The real axis intercept at high frequency is named high-frequency resistance (HFR) and is related to the ohmic protonic resistance of the membrane and the electronic resistance of the DLs. The latter is usually negligible at low operating current. Thus, the HFR mainly depends on the membrane proton conductivity, which is a non-linear function of membrane water content as shown in (14.9) and its distribution across the membrane thickness. As current increases, a higher hydration of the membrane occurs due to the increased water production that leads to a lower HFR (Figure 14.11a,c).

The effect of proton conductivity in the CL, described by Ohm's law $\eta_{CL} = ASR_{CL}j$, manifests itself in the impedance spectrum as a 45° linear branch at high frequency and it is strongly affected by the hydration state of the ionomer in the CL. Low operating current or low RH values usually imply an elongation of this impedance feature, as evident in Figurc 14.11a.

At low operating current dcnsities and low frequency (0.05–0.001 Hz), another typical impedance feature is the inductive behavior (Figure 14.11a,b), which is due to both membrane and cathode contributions. The former is related to the change in membrane conductivity induced by oscillations in membrane water content [45], while the latter is due to a variation in conductivity resulting from the oscillating current density (i.e. water production) [44]. This interpretation is in agreement with the characteristic time of water transport through ionomer, usually included between 100 and 1000 seconds [46].

14.5.4 Application of EIS Modeling to PEMFC Diagnostic

The application of physical-based models is useful in the analysis of PEMFC losses, especially when there is a strong interplay between mass transport issues, reaction kinetics, and cell degradation. All these phenomena cannot be clearly separated analyzing voltage loss of polarization curves, and therefore EIS physical-based modeling plays a crucial role.

In the following, the physical-based approach is used to understand the effect of cathode electrochemical surface area (ECSA) loss in PEMFCs with low catalyst loading. Lowering the catalyst loading is fundamental in order to tackle the technology cost target, but the reduction of Pt loading limits the performance due to the low CCL roughness factor. As reported in [17], the effect of low roughness factor can be considered introducing a constant mass transport resistance that represents sluggish diffusion of oxygen in the ionomer/carbon microstructure, further reducing oxygen concentration at the active sites, according to diffusion effects as expressed by (14.12). This oxygen transport limitation is a very fast phenomenon involving diffusion in a very small volume and it determines an apparent increase in the charge transfer resistance, appearing as an elongated semicircle, which is visible in the spectrum at beginning of life (BoL) in Figure 14.12.

Moreover, PEMFCs with low Pt loading undergo severe degradation. In particular, cathode ECSA loss is not homogeneous through the electrode thickness due to presence of a platinum-depleted region next to the membrane and the corresponding voltage losses are not fully understood and need to be further investigated. Introducing in the model a non-homogeneous distribution of active area resembling the one measured at the end of life (EoL) after an accelerated stress test (AST) for catalyst degradation [17], it is possible to provide an insight into the origin of voltage loss. As illustrated in Figure 14.12, impedance spectrum measured at the EoL shows a significant increase of the capacitive effects, but at 1.4 A cm^{-2} ECSA loss and mass

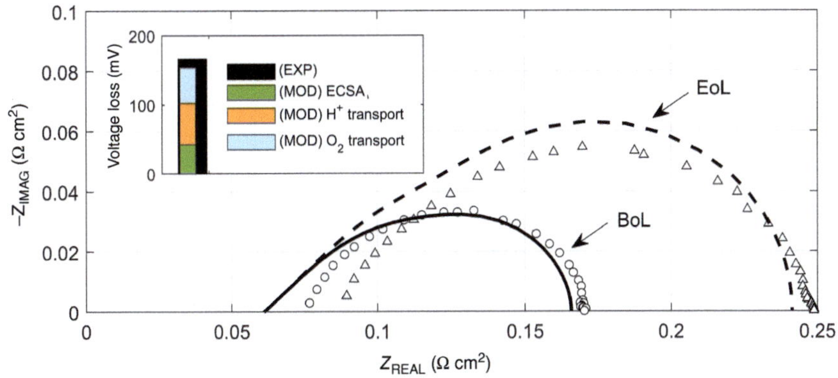

Figure 14.12 Comparison between simulated (line) and experimental (markers) EIS at $0.8\,A\,cm^{-2}$ before (BoL) and after (EoL) accelerated stress test. The inset highlights the breakdown of performance loss at $1.4\,A\,cm^{-2}$. Source: Reproduced from Ref. [17] with the permission of Elsevier.

transport issues due to the low roughness factor can explain roughly 60% of the measured performance decrease (inset in Figure 14.12). The localized depletion of platinum next to the membrane implies a shift of the reaction rate toward the DL interface, resulting in additional proton transport limitations, which are responsible for nearly 40% of the total voltage loss. As explained in Section 14.5.3, proton transport in the ACL does not affect the HFR but manifests itself as a 45° linear branch at high frequency: in agreement with experiments this feature increases in the spectrum at the EoL. The analysis of voltage loss in polarization curves does not permit to clearly distinguish its origin, which could be therefore misattributed to a wrong physical process.

The presented application of the EIS physical-based model highlights the potentiality and flexibility of the approach. Other relevant examples can be found in [47, 48].

14.5.5 Approximations of 1D Approach

The adoption of a 1D model domain neglects two main physical phenomena that are relevant in PEMFC operation, especially under low stoichiometry condition: the distribution of oxygen concentration under the rib and the oxygen depletion along the distribution channel.

As reported in [49], the region of the MEA under the rib of the distribution channel is characterized by lower oxygen concentration due to the increased reaction rate promoted by the electric contact between the CC and the DL. The non-uniform oxygen distribution determines two major changes in the spectrum, illustrated in Figure 14.13a. Firstly, in the low-frequency region a distortion of the spectrum appears due to oxygen transport limitations under the rib. Secondly, the HFR increases due to the additional electrical resistance in the DL, related to

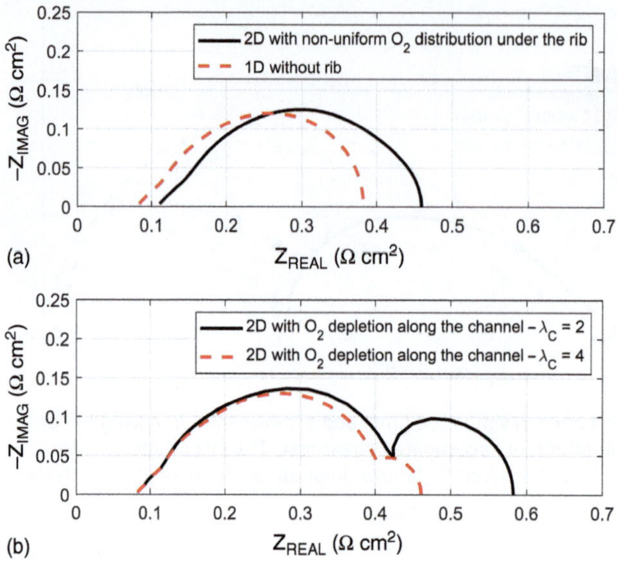

Figure 14.13 2D effects on impedance spectra not considered with the 1D modeling approach: (a) non-uniform oxygen distribution under the rib; (b) oxygen depletion along the distribution channel. Source: Reproduced from Ref. [49] with the permission of Elsevier.

the longer electrical path required to connect the regions under the channel. If this 2D effect is not considered, model calibration may imply an overestimation of the membrane conductivity to simulate the experimental HFR.

Instead, the oxygen depletion along the distribution channel leads to an additional impedance contribution appearing at frequencies between 10 and 0.1 Hz [11, 50], reported in Figure 14.13b. In sinusoidal regime, oxygen transport due to advection along the distribution channel implies the oscillation of oxygen concentration along the channel, and this leads to the additional contribution of the channel to the impedance. The characteristic frequency of this impedance feature is inversely proportional to the airflow velocity and increases when oxygen concentration diminishes significantly between the inlet and the outlet of the channel. Nevertheless, even under high stoichiometry (i.e. negligible consumption of oxygen along the channel), a distortion in the low-frequency region of the spectrum is observed (Figure 14.13b).

14.6 Conclusions and Perspectives

PEMFC modeling conjugates principles of electrochemistry with heat and mass transfer, electric conduction, multiphase flow, and interfacial vapor adsorption in a multidisciplinary approach. Complexity is further increased by the non-linearity and dependence of material properties on the microstructure of the porous materials. Microscopic simulations are required to estimate the values of real macroscopic transport properties at the device scale. This chapter dedicated particular attention to water management and specifically on water transport in the adsorbed and liquid phases, which significantly impacts the cathode catalyst layer performance. Adsorbed water in the ionomer phase, which is simulated by means of the water content (λ), is crucial in determining the ion potential in the membrane and catalyst layers and consequently the ohmic loss. The presence of liquid water also affects ion potential by improving the hydration and the ion conductivity, but also hinders oxygen transfer across porous media because of liquid water flooding. Analyses of electrochemical impedance spectroscopy, which allows to distinguish physical phenomena occurring at different time scales, is a powerful tool in such a context. Ion and electron transport losses are identified by specific events that occur in the high-frequency range, therefore being easily distinguished from gas/vapor diffusion which affects the low-frequency part of the spectrum. Gas diffusion determines either a low-frequency capacitive contribution, when oxygen transport is a limiting factor, or a low-frequency inductive feature, if related to water transport in the adsorbed phase. 1D model of performance and impedance permits to improve the understanding of PEMFC operation and the estimation of material properties of the catalyst layers thanks to the fast computational times. Application of 1D models in real-world data is anyway hindered by the geometrical features of the flow field that induce a 3D distribution of gas, liquid, temperature, and potentials in the MEA. However, the validity of the 1D approach can be verified with EIS, because additional features can be detected when boundary conditions are not uniform over the sample area. The development of innovative testing hardware, i.e. zero-gradient or differential cells, is also a crucial point for model validation that will further support the development of 1D models and contribute to new material characterization. Zero-gradient cell hardware combined with the 1D models is expected to provide additional insights into the interpretation of experimental data, allowing a simplification and improvement in data interpretation. Simulation of EIS and development of innovative testing hardware share the potential to provide an insight in PEMFC technology for material development.

Acronyms

1D	one-dimensional
2D	two-dimensional
3D	three-dimensional
AC	alternating current
ACC	anode current collector
ACL	anode catalyst layers
ADL	anode diffusion layers
ASR	area-specific resistance
AST	accelerated stress test
AT	along the channel
BEV	battery electric vehicles

BoL	beginning of life
CC	current collector
CCC	cathode current collector
CCL	cathode catalyst layer
CDL	cathode diffusion layer
CFD	computational fluid dynamics
CL	catalyst layer
DL	diffusion layer
DOF	degrees of freedom
ECM	equivalent circuit method
ECSA	electrochemical surface area
EIS	electrochemical impedance spectroscopy
EoL	end of life
FC	fuel cell
FEM	finite element method
FVM	finite volume method
GC	gas flow channel
GDL	gas diffusion layer
HFR	high-frequency resistance
HOR	hydrogen oxidation reaction
MEA	membrane electrode assembly
MPL	microporous layer
ORR	oxygen reduction reaction
PDE	partial differential equation
PEM	proton exchange membrane
PEMFC	proton exchange membrane fuel cell
PTFE	polytetrafluoroethylene
RH	relative humidity
SOFC	oxide fuel cell
TM	through the membrane
TPB	three-phase boundary

Symbols

A	fuel cell cross-section
A_{TPB}	area of *triple-phase boundary*
ASR	area-specific resistance
a_{w}	water vapor activity
$c_{\mathrm{R}}, c_{\mathrm{P}}$	actual concentrations of bulk reactant and product
$\bar{c}_{\mathrm{R}}, \bar{c}_{\mathrm{P}}$	actual concentrations of reactant and product at the CL active sites
$\bar{c}_{\mathrm{R}}^{0},\ \bar{c}_{\mathrm{P}}^{0}$	reference concentrations of reactant and product at the CL active sites
c_{s}	sulfonic acid concentration
c_{w}	water concentration
$D_{\mathrm{R}}^{\mathrm{DL}}$	effective reactant diffusivity in the DL

D_R^{GC}	reactant diffusivity in the gas flow channel
D^m	diffusion coefficient (diffusivity) in medium m
D_w^M	water diffusivity in the membrane
E	reversible electromotive force
E^0	standard potential
EW_M	Nafion equivalent weight
f	frequency
F	Faraday constant
$h_{\beta z}$	1D intrusion of phase β along the z direction in the porous medium
$I(f)$	Fourier transform of current
i_{AC}	AC current density signal
j	cell current density
j_L	limiting current density
j_{leak}	leakage current density
j_0	actual exchange current density
j_0^0	exchange current density at reference reactant concentration
$k_{v\rightarrow\lambda}$	adsorption rate coefficient for phase change with vapor
K_r	relative permeability of porous medium
K_{abs}	absolute permeability of porous medium
$\boldsymbol{K}_\beta, \boldsymbol{K}_\delta$	3D permeability vectors of phase β and δ (e.g. liquid and gas) in the porous medium
$\boldsymbol{K}_{\beta\delta}$	3D two-phase permeability vector in the porous medium
$K_{\beta z}$	1D permeability of phase β along the z direction in the porous medium
M_{H_2O}	mass of adsorbed water in the ionomer
M_M	mass of ionomer
n	number of charge carriers per reaction
n_{drag}	number of dragged water molecules per proton
$N_{H_2}^A$	anodic stoichiometric hydrogen flow
N_R^{DL}	DL stoichiometric reactant flow
N_{H_2O}	moles of adsorbed water in the ionomer
N_{liq}	molar flux of liquid water in porous media
$\mathbf{N}^m$	gas diffusion molar flux
N_w^M	water flux in the membrane
$N_{SO_3^-}$	moles of sulfonic acid groups in the ionomer
p^0	standard pressure (100 kPa)
p_{H_2}	hydrogen partial pressure
p_{O_2}	oxygen partial pressure
p_{sat}	saturation water vapor pressure
P_{cap}	capillary pressure
P_{gas}	pressure of gas phase
P_{liq}	pressure of liquid phase
P_β, P_δ	pressures of phases β and δ
r_p	mean pore radius
R	ideal gas constant

s	saturation
Sh	Sherwood number
S_{liq}	volumetric phase change rate between vapor and liquid phases
S_{ORR}	volumetric water generation due to ORR
$S_{v\rightarrow\lambda}$	volumetric phase change rate between vapor and adsorbed phases
t	time
T	temperature
T^0	standard temperature (298.15 K)
$\boldsymbol{u}_\beta$, $\boldsymbol{u}_\delta$	3D velocity vector of the β and δ phases
$u_{\beta z}$	1D velocity of the phase β along the z direction in the porous medium
V	cell voltage
$V(f)$	Fourier transform of cell voltage
v_{AC}	AC voltage signal
z	through-plane spatial coordinate in the MEA
Z_I	imaginary part of electrochemical impedance
Z_R	real part of electrochemical impedance
$Z(f)$	electrochemical impedance
α_i	transfer coefficient of each half-reaction
β, δ	phases in the porous medium
γ	surface tension
δ_{DL}	DL thickness
δ_{M}	membrane thickness
ΔE_T	entropic variation from standard conditions due to temperature
ΔE_p	entropic variation from standard conditions due to pressure
$\Delta\widehat{s}$	molar reaction entropy
ΔP	pressure difference at the inlet and outlet of the porous medium
ε	porosity of the porous medium
ε_β, ε_δ	volume fractions of the β and δ phases in the porous medium
η_{act}	activation overpotential due to electrochemical kinetics of reactions
η_{con}	concentration overpotential due to mass transport losses
η_{ohm}	ohmic overpotential due to joule losses
κ	mass transfer coefficient
λ	water content
λ_{eq}	equilibrium water content from adsorption isotherm
μ_{liq}	liquid water dynamic viscosity
μ_β, μ_δ	dynamic viscosities of the β and δ phases
ν_{w}	net transport water coefficient
ρ_{dry}	density of dry Nafion
ρ_{liq}	density of liquid water
σ	proton conductivity
τ	volume
φ	phase shift between of AC voltage and current
ω	angular frequency
$\langle\,\rangle$	volume averaging operator

References

1 O'Hayre, R., Cha, S.W., Coltella, W., and Prinz, F.B. (2009). *Fuel Cell Fundamentals*. NY: Wiley.

2 Dicks, A.L. and Rand, A.J. (2018). *Fuel Cell Systems Explained*, 3e. NY: Wiley.

3 Demuren, A. and Edwards, R.L. (2020). Modeling proton exchange membrane fuel cells—A review. In: 50 Years of CFD in Engineering Sciences (ed. A. Runchal), 513–547. Singapore: Springer https://doi.org/10.1007/978-981-15-2670-1_15.

4 Springer, T.E., Zawodzinski, T.A., and Gottesfeld, S. (1991). Polymer electrolyte fuel cell models. *Journal of Electrochemical Society* 138 (8): 2334–2341.

5 Wang, Z.H., Wang, C.Y., and Chen, K.S. (2001). Two-phase flow and transport in the air cathode of proton exchange membrane fuel cells. *Journal of Power Sources* 94 (1): 40–50.

6 West, A.C. and Fuller, T.F. (1996). Influence of rib spacing in proton-exchange membrane electrode assemblies. *Journal of Applied Electrochemistry* 26 (6): 557–565.

7 Caisheng, W., Nehrir, M.H., and Shaw, S.R. (2005). Dynamic models and model validation for PEM fuel cells using electrical circuits. *IEEE Transactions on Energy Conversion* 20 (2): 442–451.

8 Guarnieri, M. and Moro, F. (2010). A novel circuit model of a proton exchange membrane fuel cell. *COMPEL: The International Journal for Computation and Mathematics in Electrical and Electronic Engineering* 29 (6): 1562–1572.

9 Falcão, D.S., Oliveira, V.B., Rangel, C.M. et al. (2009). Water transport through a PEM fuel cell: a one-dimensional model with heat transfer effects. *Chemical Engineering Science* 64 (9): 2216–2225.

10 Guarnieri, M., Negro, E., Di Noto, V., and Alotto, P. (2016). A selective hybrid stochastic strategy for fuel-cell multi-parameter identification. *Journal of Power Sources* 332 (1): 249–264.

11 Weber, A.Z., Borup, R.L., Darling, R.M. et al. (2014). A critical review of modeling transport phenomena in polymer-electrolyte fuel cells. *Journal of the Electrochemical Society* 161 (12): F1254–F1299.

12 Kusoglu, A. and Weber, A.Z. (2017). New insights into perfluorinated sulfonic-acid ionomers. *Chemical Reviews* 117 (3): 987–1104.

13 Wan, Z., Chang, H., Shu, S. et al. (2014). A review on cold start of proton exchange membrane fuel cells. *Energies* 7 (5): 3179–3203.

14 Springer, T.E., Zawodzinski, T.A., and Gottesfeld, S. (1991). Polymer electrolyte fuel cell model. *Journal of the Electrochemical Society* 138 (8): 2334–2342.

15 Vetter, R. and Schumacher, J.O. (2019). Experimental parameter uncertainty in proton exchange membrane fuel cell modeling. Part I: scatter in material parameterization. *Journal of Power Sources* 438: 227018.

16 Ge, S., Li, X., Yi, B., and Hsing, I.-M. (2005). Absorption, desorption, and transport of water in polymer electrolyte membranes for fuel cells. *Journal of the Electrochemical Society* 152 (6): A1149.

17 Baricci, A., Bonanomi, M., Yu, H. et al. (2018). Modelling analysis of low platinum polymer fuel cell degradation under voltage cycling: gradient catalyst layers with improved durability. *Journal of Power Sources* 405: 89–100.

18 Grimaldi, A. (2017). Modelling and analysis in 1D and 3D of dehydration and flooding phenomena in PEMFC with simulation of impedance spectra. Master thesis. Politecnico di Milano.

19 Farzaneh, M., Ström, H., Zanini, F. et al. (2020). Pore-scale transport and two-phase fluid structures in fibrous porous layers: application to fuel cells and beyond. *Transport in Porous Media* 136: 1–26.

20 Succi, S. (2001). *The Lattice Boltzmann Equation: For Fluid Dynamics and Beyond*. Oxford University Press.

21 Kim, K.N., Kang, J.H., Lee, S.G. et al. (2015). Lattice Boltzmann simulation of liquid water transport in microporous and gas diffusion layers of polymer electrolyte membrane fuel cells. *Journal of Power Sources* 278: 703–717.

22 Maggiolo, D., Marion, A., and Guarnieri, M. (2014). Lattice Boltzmann modeling of water cumulation at the gas channel-gas diffusion layer interface in polymer electrolyte membrane fuel cells. *Journal of Fuel Cell Science and Technology* 11 (6).

23 Majlan, E.H., Rohendi, D., Daud, W.R.W. et al. (2018). Electrode for proton exchange membrane fuel cells: a review. *Renewable and Sustainable Energy Reviews* 89: 117–134.

24 Snoeijer, J.H. and Andreotti, B. (2008). A microscopic view on contact angle selection. *Physics of Fluids* 20 (5): 057101.

25 Pasaogullari, U. and Wang, C.Y. (2004). Liquid water transport in gas diffusion layer of polymer electrolyte fuel cells. *Journal of the Electrochemical Society* 151 (3): A399.

26 Whitaker, S. (1986). Flow in porous media II: the governing equations for immiscible, two-phase flow. *Transport in Porous Media* 1 (2): 105–125.

27 Maggiolo, D., Picano, F., Zanini, F. et al. (2020). Solute transport and reaction in porous electrodes at high Schmidt numbers. *Journal of Fluid Mechanics* 896.

28 Molaeimanesh, G.R. and Akbari, M.H. (2016). Role of wettability and water droplet size during water removal from a PEMFC GDL by lattice Boltzmann method. *International Journal of Hydrogen Energy* 41 (33): 14872–14884.

29 Tjaden, B., Cooper, S.J., Brett, D.J. et al. (2016). On the origin and application of the Bruggeman correlation for analysing transport phenomena in electrochemical systems. *Current Opinion in Chemical Engineering* 12: 44–51.

30 Pasquier, S., Quintard, M., and Davit, Y. (2017). Modeling two-phase flow of immiscible fluids in porous media: Buckley-Leverett theory with explicit coupling terms. *Physical Review Fluids* 2 (10): 104101.

31 Weitz, D.A., Stokes, J.P., Ball, R.C., and Kushnick, A.P. (1987). Dynamic capillary pressure in porous media: origin of the viscous-fingering length scale. *Physical Review Letters* 59 (26): 2967.

32 Gostick, J.T., Ioannidis, M.A., Fowler, M.W., and Pritzker, M.D. (2008). Direct measurement of the capillary pressure characteristics of water–air–gas diffusion layer systems for PEM fuel cells. *Electrochemistry Communications* 10 (10): 1520–1523.

33 Lim, C. and Wang, C.Y. (2004). Effects of hydrophobic polymer content in GDL on power performance of a PEM fuel cell. *Electrochimica Acta* 49 (24): 4149–4156.

34 Tranter, T.G., Boillat, P., Mularczyk, A. et al. (2020). Pore network modelling of capillary transport and relative diffusivity in gas diffusion layers with patterned wettability. *Journal of the Electrochemical Society* 167 (11): 114512.

35 Barsoukov, E. and Macdonald, J.R. (2005). *Impedance Spectroscopy: Theory, Experiment, and Applications*. Hoboken, NJ: Wiley.

36 Yuan, X.Z., Song, C., Wang, H. et al. (2010). *Electrochemical Impedance Spectroscopy in PEM Fuel Cells*. London: Springer.

37 Rezaei Niya, S.M. and Hoorfar, M. (2013). Study of proton exchange membrane fuel cells using electrochemical impedance spectroscopy technique – a review. *Journal of Power Sources* 240: 281–293.

38 Orazem, M.E. and Tribollet, B. (2008). *Electrochemical Impedance Spectroscopy*. Hoboken, NJ: Wiley.

39 Baricci, A. and Casalegno, A. (2015). A simple analytical approach to simulate the electrochemical impedance response of flooded agglomerates in polymer fuel cells. *Electrochimica Acta* 157: 324–332.

40 Bessler, W.G. (2007). Rapid impedance modeling via potential step and current relaxation simulations. *Journal of the Electrochemical Society* 154 (11): B1186–B1191.

41 Springer, T.E., Zawodzinski, T.A., Wilson, M.S., and Golfesfeld, S. (1996). Characterization of polymer electrolyte fuel cells using AC impedance spectroscopy. *Journal of the Electrochemical Society* 143 (2): 587–599.

42 Gewies, S. and Bessler, W.G. (2008). Physically based impedance modeling of Ni/YSZ cermet anodes. *Journal of the Electrochemical Society* 155 (9): B937–B952.

43 Kulikovsky, A.A. (2010). *Analytical Modelling of Fuel Cells*. Amsterdam: Elsevier.

44 Wiezell, K., Holmström, N., and Lindbergh, G. (2012). Studying low-humidity effects in PEFCs using EIS II. Modelling. *Journal of the Electrochemical Society* 159 (8): F379–F392.

45 Schneider, I.A., Bayer, M.H., Wokaun, A. et al. (2008). Impedance response of the proton exchange membrane in polymer electrolyte fuel cells. *Journal of the Electrochemical Society* 155 (8): B783–B792.

46 Wu, H., Li, X., and Berg, P. (2009). On the modeling of water transport in polymer electrolyte membrane fuel cells. *Electrochimica Acta* 54 (27): 6913–6927.

47 Baricci, A., Zago, M., and Casalegno, A. (2016). Modelling analysis of heterogeneity of ageing in high temperature polymer electrolyte fuel cells: insight into the evolution of electrochemical impedance spectra. *Electrochimica Acta* 222: 596–607.

48 Baricci, A., Bisello, A., Serov, A. et al. (2019). Analysis of the effect of catalyst layer thickness on the performance and durability of platinum group metal-free catalysts for polymer electrolyte membrane fuel cells. *Sustainable Energy & Fuels* 3 (12): 3375–3386.

49 Baricci, A., Mereu, R., Messaggi, M. et al. (2017). Application of computational fluid dynamics to the analysis of geometrical features in PEM fuel cells flow fields with the aid of impedance spectroscopy. *Applied Energy* 205: 670–682.

50 Baricci, A., Zago, M., and Casalegno, A. (2014). A quasi 2D model of a high temperature polymer fuel cell for the interpretation of impedance spectra. *Fuel Cells* 14 (6): 926–937.

15

Physics-based Modeling of Polymer Electrolyte Membrane Fuel Cells: From Cell to Automotive Systems

Andrea Baricci[1], Matteo Zago[1], Simone Buso[2], Marco Sorrentino[3], and Andrea Casalegno[1]

[1] *Politecnico di Milano, Dipartimento di Energia, via Lambruschini, 4a, 20156 Milano, Italy*
[2] *Università di Padova, Dipartimento di Ingegneria dell'informazione, Via G. Gradenigo, 6b, 35131 Padova, Italy*
[3] *Università degli studi di Salerno, Dipartimento di Ingegneria Industriale, Via Giovanni Paolo II, 132, 84084 Fisciano, Salerno, Italy*

15.1 Polymer Fuel Cell Model for Stack Simulation

15.1.1 General Characteristics of a Fuel Cell System for Automotive Applications

Polymer electrolyte membrane fuel cells (PEMFCs) are facing an increasing global interest for applications in the mobility sector. At the time the authors are writing this book, several carmakers already introduced on the market hydrogen fuel cell vehicles in the European car segments D and E (Toyota, Hyundai, Honda, and Mercedes), while others are ending the validation process (BMW and Audi). These initial demonstrations are the visible tip of a large movement that aims at the electrification in the transport sector by means of hydrogen power systems, specifically for trucks, rail, maritime, and aeronautical applications. Even though there are some features that are common in all the fuel cell system configurations, several differences are found in the design among carmakers [1]; this occurs because of the early stage of development and the variety of applications. A typical PEMFC system for automotive applications is reported in Figure 15.1. Three auxiliary subsystems are integrated in the power train and require proprietary control algorithms that allow to safely operate the stack within a specific range of operating conditions, e.g. temperature and pressure. These control strategies have a significant impact on both performance and durability of the PEMFC stack, as already highlighted in the initial chapters of this book, thus requiring some theoretical analysis that will be carried out in this chapter.

In the diagram of Figure 15.1, three subsystems, that were mentioned before, are highlighted. The air management system includes all the components that supply the stack with the oxidant, which is ambient air that is filtered, compressed, and finally humidified to the required level. The hydrogen supply subsystem includes

Electrocatalysis for Membrane Fuel Cells: Methods, Modeling, and Applications, First Edition.
Edited by Nicolas Alonso-Vante and Vito Di Noto.

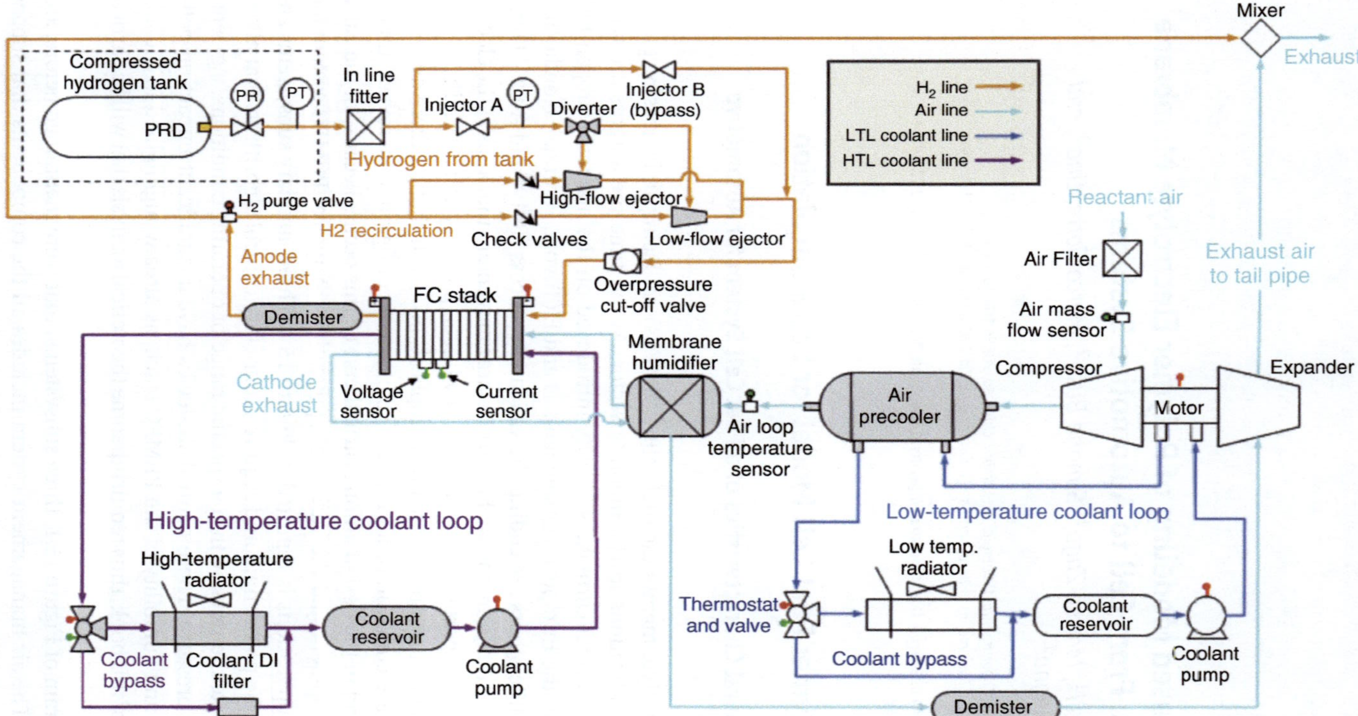

Figure 15.1 Schematic of a hydrogen PEMFC power system for automotive applications. Air subsystem, hydrogen subsystem, and thermal managements subsystem are identified. Source: Reproduced from Ref. [1] with the permission of Elsevier.

the high-pressure storage tank, injectors for pressure and flow control, and also a device that allows recirculation of the anode outlet stream to improve hydrogen utilization. The thermal management subsystem, finally, is necessary to properly control the temperature of the stack and is composed of a tank to store the cooling fluid, a heat exchanger (radiator), and a recirculating pump. The complete system allows to control the fuel cell stack operating conditions, which are the temperature of the stack, the pressure, the flow rate, and the relative humidity of air and hydrogen feed. The balance of plant (BoP) is then composed of all these parts, plus the electric machines and the battery systems that are also present in hybridized systems. The current section discusses the modeling of the fuel cell stack, while Section 15.2 deals with the modeling of the auxiliary subsystems, focusing on the PEMFC power system only, while the electric machines are analyzed in Section 15.3. The discussion is then generalized for other transport sectors in the last part (Section 15.4) by introducing the effect of hybridization.

From a system point of view, the modeling of the polymer fuel cell stack consists of a code that computes the output power and flow rate of fluids that are exchanged with the other components of the balance of plant: hydrogen, air, and the cooling fluid flows. The inlet conditions of the fluids are controlled and provided by the auxiliary subsystems model, while the outlet values are a result of the stack model. However, the stack model has three levels of complexity that justify the large effort that was dedicated to the development and numerical implementation of this tool. First, the model must consider a large set of phenomena spanning over electrochemistry, heat transfer, mass transfer and water phase change kinetics between vapor, liquid and adsorbed phases, as discussed in Chapter 14. These phenomena can hardly be neglected if a realistic result is expected by the simulation; consequently, model validation must be performed on a large set of experimental data. Secondly, automotive operation is strongly dynamic, thus requiring a transient model to consider the unsteady conditions in the stack. Thanks to the fact that the fuel cell systems are always hybridized, fast dynamics are in general not necessary, and only phenomena that have a slow response time must be included (>1 seconds). Lastly, computational requirements are always stringent: system simulations are performed during the design phase, aiming at the optimal configuration of the system and the control algorithms. Multiple runs of the code are therefore necessary within a working day. Some of these requirements conflict with each other, so a fast and realistic validated stack model is still an unresolved challenge to the authors' knowledge. One approach for stack modeling, which prioritizes the computational cost, is based on empirical or semiempirical models that are parametrized over a large set of experimental data [2, 3]. These models can be based on some analytical solutions [4] or equivalent circuits that linearize the stack response to load changes. These models are generally very fast but limited in their prediction capabilities; in particular, they are not reliable for the transient simulations and not accurate to predict heat and flow rates exchanged by the stack with the balance of plant: thermal power, water transfer from anode to cathode, or hydrogen consumption. The alternative tool is represented by physics-based models [5–9] that are highly representative of the real stack operation and do not require experimental data from the system, but from the components. On the other hand, they are quite demanding from a computational point of view. Some hybrid models also exist that combine both approaches but were never applied to system simulations [10].

In this section, the attention will be devoted mostly to physics-based modeling of stack, consistently with the physical approach proposed in Chapter 14. As shown there, the transport equations are also implemented in the stack model, but here an additional problem raises: since the stack geometry is three-dimensional, severe limitations would result by the use of a one-dimensional model that would not consider the effect of the real stack geometry [11]; on the other hand, it would be impossible to run a full dynamic simulation of the whole stack with a 3D code within acceptable time. For this reason, some simplifications are exploited to successfully improve the computational time but maintain representativeness of the simulations.

The first simplification consists in simulating a single cell in the stack and consider it as representative of the whole, thus neglecting possible differences among cells that may result from nonoptimal design of the stack or nonuniform distribution of reactants among cells. These aspects become relevant after aging of the stack, otherwise are generally negligible. Focusing on a single cell, the geometry of the flow field is typically made of parallel channels that distribute hydrogen, air, and the coolant over the whole surface. The influence of the features of the flow field geometry on the simulation of the stack performance is then carried out in Sections 15.1.2 and 15.1.3.

15.1.2 Analysis of the Channel Geometry for Stack Performance Modeling

A geometrical feature that determines a deviation between 1D and 3D PEMFC models is associated with the region where the electrical contact between the gas diffusion layer (GDL) and the bipolar plate takes place. This region is called *land* or *rib* and is distinguished by the *channel* region, which is the projected area of the membrane electrode assembly (MEA) under the channel. A bidimensional schematic of this geometrical feature is represented in Figure 15.2a, where the position of the MEA is also indicated. The geometrical ratio between the length of the rib and the channel (R/C) is generally taken as a parameter to analyze this feature in modeling. Some simulations performed with a 3D model [12] are reported in Figure 15.2 to discuss the effect of the rib size on PEMFC stack performance: five cases were simulated with R/C parameter ranging from 1 : 7 (narrow rib) to 7 : 1 (large rib), by keeping constant PEMFC area. The simulation results are summarized by looking at the oxygen mole fraction contour plot in Figure 15.2c, where it is observed that the oxygen diffusion under the rib is significantly hindered in the case with a large rib geometry. The reason for this limitation is found in the longer diffusion path imposed to oxygen to reach the active catalyst site under the rib region. Oxygen diffusion is also impeded by the liquid water condensation that occurs here: water is produced from the oxygen reduction reaction (ORR); thus, water flux incurs a higher transport resistance from under the rib than under the channel, because it is more distant, but also because high water vapor concentration in this region may lead to a

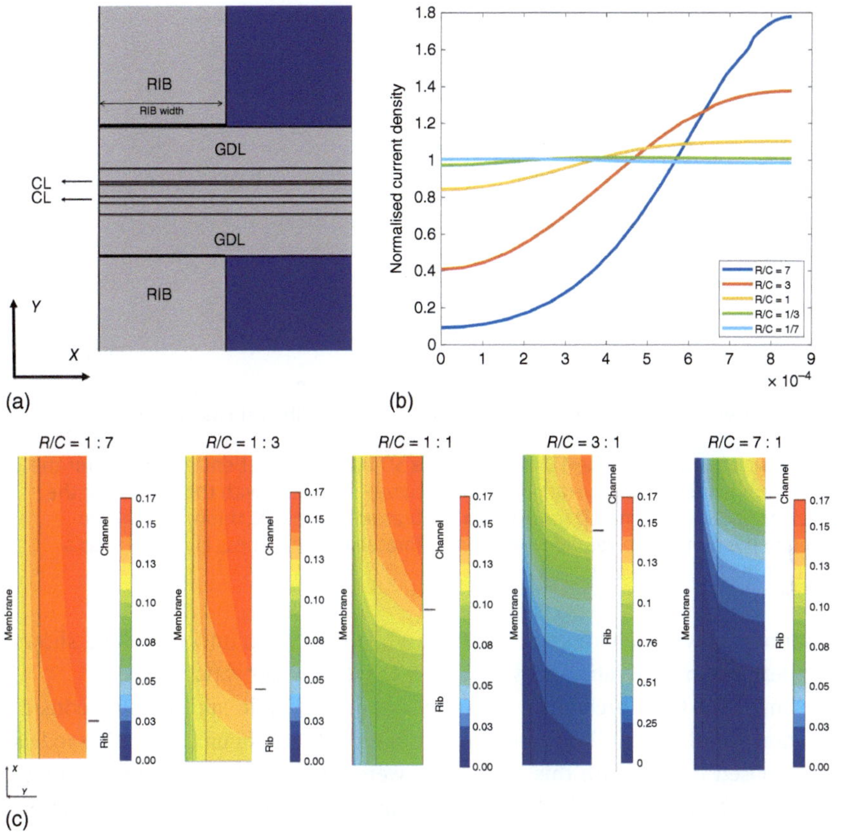

Figure 15.2 Effect of channel/rib geometry on PEMFC performance: (a) schematic of the geometry for the 3D simulation; (b) normalized current density distribution between the rib and channel regions; (c) contour of oxygen mole fraction for R/C values ranging between 1 : 7 and 7 : 1.

possible condensation, which further discriminates the performance of this region. The impact of this nonuniform oxygen distribution inside the GDL is reported in Figure 15.2b, where normalized current density distribution is plotted as a function of the local coordinate. The contribution of the region under the land relatively to the one under the rib is shown. In the case of a thin rib, the current density distribution becomes uniform to the value that is obtained under the channel. On the other side, when R/C parameter becomes higher than 1, a clear difference appears in the current density value between the rib and the channel. In the simulation reported, a higher current density is found under the channel regions, where oxygen is effectively supplied. Further minor effects are also found in the simulation that are not reported here for sake of brevity, which include electric effects, since the region under the rib is closer to the electrical contact with the bipolar plate [13]; thermal effects that are analogous to the electric ones, since the heat generated in the catalyst layer is mainly removed through the contact with the bipolar plate; and additional effects are also found under extremely low relative humidity in the gas feed, since the region under

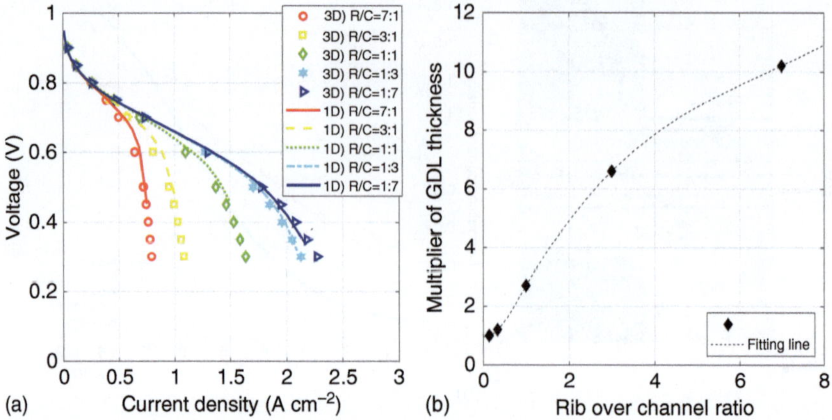

Figure 15.3 Calibration of the 1D model on the 3D model simulations to account for the effect of rib/channel geometry: (a) polarization curves simulated with the 3D model; the simulations obtained by the 1D model after calibration of the GDL thickness are also included in the plot; (b) correlation between calibrated GDL thickness and rib size as a function of the R/C parameter.

the channel shows a dehydrated condition, while the region under the land shows higher performance, because of its capability to retain water vapor.

The impact of the rib/channel geometry is not negligible and must be considered when simulating PEMFC stack performance. Even though 2D + 1D models were proposed to deal with this issue, they were never applied to system simulations, mostly because of the high computational cost required. Thus, an alternative method was proposed to have a realistic performance of PEMFC stack geometry. This method is derived by looking at the polarization curves simulated for the geometries discussed in Figure 15.2. The polarization curves are reported in Figure 15.3a; it is evident that the rib/channel geometry has a relevant impact on the limiting current, thus on the high-power density region. This observation, combined with the considerations derived from Figure 15.2, suggests a modification to the 1D model. This modification consists in increasing the GDL thickness in order to reproduce the longer "average" diffusion length highlighted by the 3D model results. After calibration of the 1D model, a good consistency is found with the polarization curves simulated by the 3D model. Calibrated GDL thickness is also reported in Figure 15.3b, where a nonlinear trend is found for extreme values of R/C, but a linear approximation would be possible in the R/C = 0–3 range.

15.1.3 Analysis of the Air and Hydrogen Utilization for Stack Performance Modeling

The air utilization is another parameter that determines a relevant deviation between 1D and 3D PEMFC models. It is typical to adopt the inverse of the utilization, a parameter that is known as the stoichiometry and is defined as the ratio between the air flow rate supplied by the BoP and the actual consumption

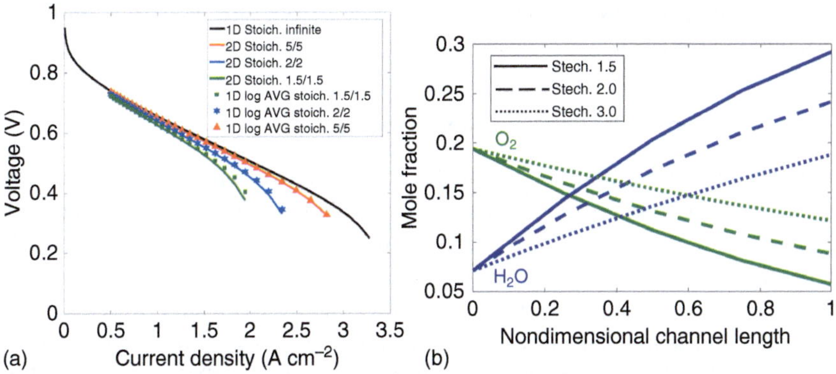

Figure 15.4 Effect of air and hydrogen stoichiometry on the polarization curve, simulations performed with a 2D model: (a) polarization curves; (b) evolution of water vapor and oxygen mole fraction along the channel coordinate.

of oxygen due to the ORR, as measured by the stack current (crossover and gas leakage are not considered because they are practically negligible). In laboratory testing of single cells, the stoichiometry is generally set to a very high value in order to avoid nonuniform distribution of reactants on the PEMFC active area, which would complicate the interpretation of the results of the test. In systems, however, high stoichiometry would increase the power consumption from the auxiliary components, in particular the air compressor, affecting dramatically the system efficiency. Because of this, the stoichiometry is generally set in the range 1.5–2. In Figure 15.4a, it is shown how the stoichiometry affects PEMFC performance in the range 1.5–5, as obtained by a 2D model. It is observed that the major effects are found in the high current density range, where a lower air stoichiometry implies a lower current density. Even though both hydrogen and oxygen stoichiometries have been changed in the simulation, the air flow is always affecting performance in a more visible way. In Figure 15.4b, the evolution of oxygen and water vapor mole fractions along the channel direction are reported. In the air inlet region (left) the oxygen mole fraction is much higher, but a low value of water partial pressure is also observed, since air is fed from the BoP with a relative humidity around 30%. For this reason, this zone suffers from dehydration in the intermediate current density region but shows high performance in the high-power-density region, where good hydration is guaranteed by the water that is produced by ORR. The opposite is found in the air outlet region (right) where both low oxygen partial pressure and high relative humidity determine the local oxygen depletion on the catalyst active site.

A possible approach to account for the effect of stoichiometry in a 1D model is to modify the boundary condition at the channel side in a way that considers the average value of mole fraction between the inlet and outlet conditions. A simulation with a 1D model applying the logarithmic average between the inlet and outlet mole fractions is also reported in Figure 15.4a, and an acceptable consistency with the 2D model is reported. This method, however, is not indicated when transient simulations are performed, because it neglects the delay time associated with gas distribution in the stack.

15.1.4 Introduction to Transient Stack Models

The importance of a dynamic stack model for automotive applications is critically described in this section. For this reason, simulations analyzing the transient response of a PEMFC stack after a current density step are reported in Figure 15.5, where Figure 15.5a reports the voltage response. A comparison is performed between 1D and 2D models, and it is concluded that the 2D model is necessary to capture the transient voltage response in the range above 1 second, which is generally associated with the dynamics of water transport in the PEMFC. In particular, the dynamics of the water content in the membrane is reported in Figure 15.5b for the 1D model in three MEA positions in the anode and cathode catalyst layers and the membrane. The transient response of water content indicates that the

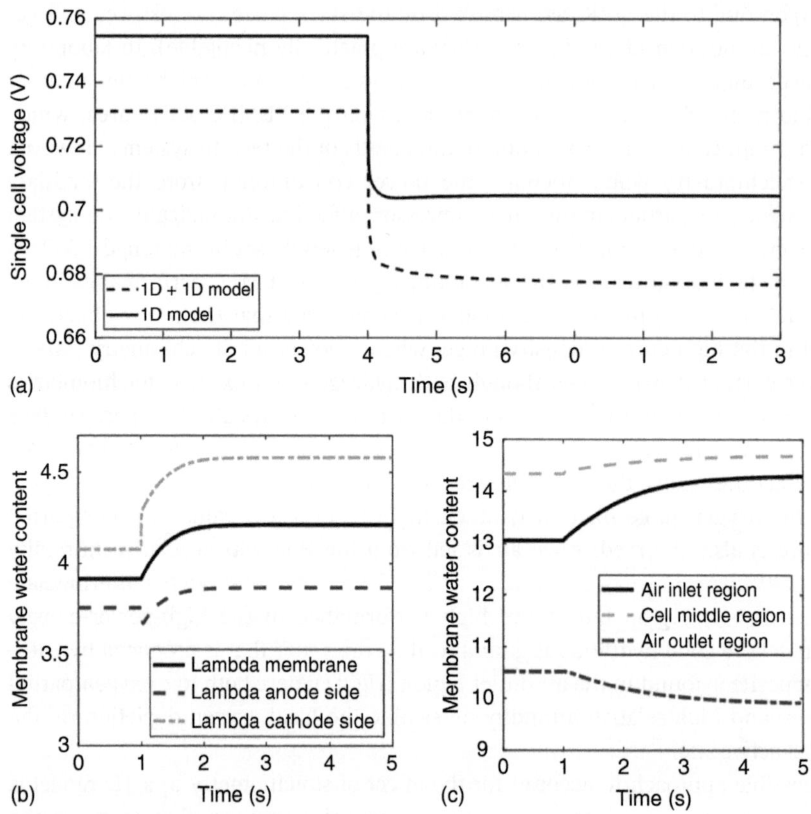

Figure 15.5 Transient simulations of a PEMFC stack after a current step was performed from 0.4 to 0.7 A cm^{-2}: (a) voltage response and comparison between a 1D and a 2D model; (b) transient water content in three positions covering the anode and cathode catalyst layers and the membrane; (c) transient membrane water content at several positions along the channel.

electroosmotic drag dominates at the anode side, where an initial dehydration process is observed after the step. A different trend is found at the cathode side, where also water production, due to ORR, plays a key role in the water balance. If this simulation is compared with the same one performed with a 2D model, reported in Figure 15.5c, it is inferred that the further delay in the transient response of the 2D case is due to the slow evolution of the gas composition along the channel, which increases the time required to reach the steady state. A significant difference in the absolute value of water content estimated by the 1D and 2D models is also observed comparing Figure 15.5a,b: in the 1D model a dehydrated condition is obtained by the simulation, because RH values in air and hydrogen channels are set to their inlet values, disregarding thus the evolution of water vapor mole fraction that occurs along the channel. Such a condition does not occur in 1D + 1D model that integrates the evolution of RH along the channel. Under finite (or low) stoichiometry, 1D model leads to a nonrealistic result, because of the counterflow configuration that is actually implemented in real systems, which allows to avoid this issue.

In conclusion, a physics-based stack model for automotive applications was presented. The challenges at the current state of the art are connected to the correct representation of the geometrical features of the flow field and the requirement for a dynamics model, specifically in the relevant range for automotive application (>1 second). The transient evolution of water among phases was identified as a critical aspect that can be captured with physical models.

15.2 Auxiliary Subsystems Modeling

Modeling of the auxiliary subsystems that were introduced in Section 15.1 is described qualitatively in this section. A complete description of the implementation and the theory is not reported for sake of brevity, but it can be found in the literature [14, 15]. The analysis is focused on the air loop, the hydrogen loop, and the thermal management subsystems. The other components of the balance of plant, including the electric motor, the generator, the electric machines, and all the auxiliaries for hybridization, will be described in Section 15.3.

15.2.1 Air Management Subsystem

The key components that are present in the air loop are the air compressor and the humidifier, which are relevant because they affect the operating conditions of the stack (pressure and relative humidity of oxidant) and impact the system efficiency. Several configurations for the air compressor were considered in automotive applications, and the choice depends on a combination of efficiency and cost. Several types of compressors were investigated, roots, scroll, radial, or screw-type compressors, each one having some weaknesses [16]. In particular, the requirements are high efficiency over the entire flow range, particularly in the low flow range where the fuel cell system is frequently run, i.e. under idle load. Considering that a unique standard has not emerged yet from automotive industry [17], in this work a centrifugal compressor is adopted. The compressor model consists in a static map that is generated depending on the geometrical characteristics of the compressor: these maps are reported for the specific case of this application in Figure 15.6a,b. The dynamics of the air discharge flow is associated with the inertia of the impeller that is driven by an electric motor, while all the transient effects, due to fluid dynamics in the compressor, are neglected.

The air humidifier is a component that permits to increase the air dew point in order to supply humid air to the stack. Not all the PEMFC hydrogen power trains include this component; e.g., the Toyota Mirai 2015 supplies dry air flow directly into the stack, but this component is included in this work, consistently with the scheme reported in Figure 15.1. Several alternatives exist in general for air humidification, some of which require liquid water and are not adopted in the automotive field because it is difficult to condensate and recover water from the air exhaust. A planar membrane humidifier that allows to recover water vapor from the cathodic outflow is considered, with reference from the literature [18]. The membrane is the

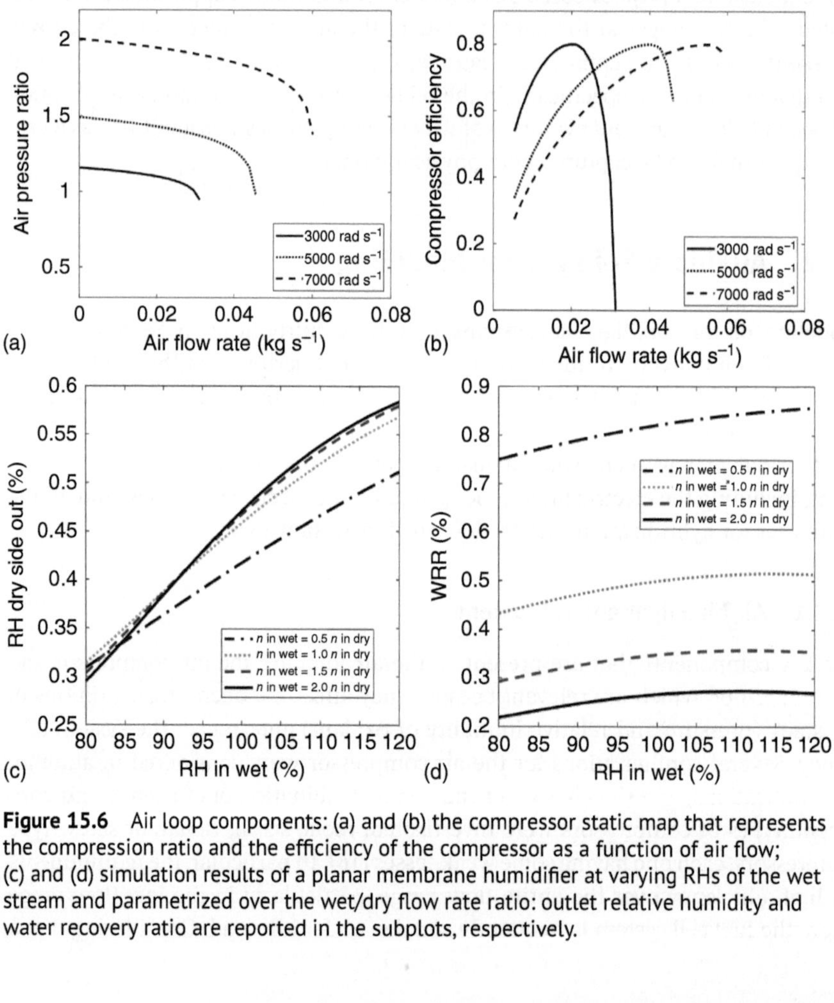

Figure 15.6 Air loop components: (a) and (b) the compressor static map that represents the compression ratio and the efficiency of the compressor as a function of air flow; (c) and (d) simulation results of a planar membrane humidifier at varying RHs of the wet stream and parametrized over the wet/dry flow rate ratio: outlet relative humidity and water recovery ratio are reported in the subplots, respectively.

active element that selectively adsorbs water from the wet stream to humidify the dry one, and the physics-based model considers heat and mass transfer between the wet and dry streams. Polymer materials that are adopted for the membrane are perfluorosulfonated polymers, derived from fuel cell membrane materials, but with some modifications in their properties in order to facilitate the water transfer. In Figure 15.6, some simulations performed on the humidifier are reported, in which the relative humidity of the wet flow was varied in the range of 80% to oversaturated conditions: in (a) the relative humidity value achieved for the air entering the stack is reported (all RH values are referred to the stack nominal temperature of 80 °C); in (b) the water recovery ratio (*WRR*) is also reported, defined as the fraction of water vapor recovered at the dry side over the water supplied at the wet side, according to the following equation:

$$\mathrm{WRR} = \frac{\dot{m}_{\mathrm{H_2O,dry\text{-}out}} - \dot{m}_{\mathrm{H_2O,dry\text{-}in}}}{\dot{m}_{\mathrm{H_2O,wet\text{-}in}}} \tag{15.1}$$

The dynamic response of the humidifier depends largely on the time that takes the air to cross the humidifier volume but also the delay due to the water adsorption/desorption kinetics and the inertia of the membrane. This last aspect becomes less relevant when thin membranes are implemented.

Other components are present in the air loop that are not discussed in detail. The intercooler is a heat exchanger between the air compressor and the humidifier that lowers the air temperature that may be too high after compression and could damage the stack or the humidifier. The backpressure valve is needed to pressurize the airflow to the set-point pressure. The stack manifolds and the pipe volumes simulate the delay that occurs when the gas must fill the unactive volumes in the air loop. A bypass valve on the stack is also present, and it prevents the air from entering the stack during stops. Air filtering is also not included in the simulation.

15.2.2 Hydrogen Management Subsystem

The hydrogen loop consists in the high-pressure tank from which hydrogen is expanded across injectors in order to fill the manifolds and the anodic volume. A recirculation system is always present to improve hydrogen utilization. In this simulation a hydrogen pump is simulated, even though the use of ejectors is becoming more common, since they do not require electric power and thus reduce the auxiliary power consumption and improve the system efficiency. The main problem resulting from the use of ejectors is the limitations in the hydrogen recirculation when no flow is taken from the tank, for example during stops. Because of this, some hydrogen blowers are also installed for specific cases of systems shut down. Another important element in the hydrogen loop is the purge valve that is a valve that periodically opens to purge the anodic flow whenever excess water or nitrogen permeating from the air side accumulates. The control logic of the purge valve relies on the sensors that are available in the system, but the duration of the time interval between purges is known to depend on both the current density and the desired level of dilutants in the hydrogen feed [19]. A demister is also present to separate the liquid phase, if present.

15.2.3 Thermal Management Subsystem

The thermal management subsystem is the last one presented in this section, but one of the most relevant for automotive applications, as already highlighted in the *incipit* of this book. Stack cooling is crucial when a high-power load is required. Differently from internal combustion engines, most of the thermal power is removed from the coolant and not from the exhaust gas, whose contribution is almost negligible, being at low temperature (<100 °C). For this reason, the design of the cooling system is strategic for both performance optimization and also durability. Temperature is in fact one of the major stressors for all degradation mechanisms that affect membrane, GDL, and electrocatalyst. The thermal management subsystem consists of a pump that circulates the liquid coolant and a radiator that exchanges heat between the coolant and ambient air, which is forced into the heat exchanger proportionally to the vehicle speed and also by fans that are activated to boost the heat transfer when needed. Both fans and pump require an electric power, whose contributions must be computed in the auxiliary power consumption. Heat exchange is simulated according to a literature model [20]. Temperature dynamics in the balance of plant are very relevant and are due to the inertia of the cooling liquid. Additional modifications of this layout are introduced to manage the specific case of the subzero temperature start, where ice formation can occur, but this case is not considered in this chapter.

15.2.4 PEMFC System Simulation

The system that is described in this section is combined with the stack model that was presented in Section 15.1 to simulate a *new European driving cycle* (NEDC). For this simulation, the hybridization is not considered, and it will be discussed in Section 15.4 of this chapter. The scope here is to show how a PEMFC stack operates in connection with the balance of plant. A stack of 370 cells of 237 cm^2 of area is simulated, applying a counterflow configuration between air and hydrogen streams, while the coolant flow is co-flow with air. In Figure 15.7, the gross power and the net power are reported; the first considers even the power consumption due to auxiliaries, including compressor, fan, and the hydrogen pump power. The power peaks observed are due to the acceleration of the vehicle that requires a higher power than in the cruise mode.

In Figure 15.8, the stack voltage and current are reported for the NEDC. Looking at the stack voltage, a transient behavior is observed during the acceleration periods, while an almost steady condition is achieved during idling. If the local membrane water content is plotted, Figure 15.9, unsteady operation is actually found in the air and hydrogen inlet zones. In these extreme regions a lower water content is reported, if compared to the central area of the PEMFC, where the membrane is well humidified. At the air inlet, in particular, the idle periods tend to dehydrate the membrane, while the same occurs at the anode side at high power, because of the electroosmotic drag effect that was observed in Section 15.1.4, which is typical for counter-flow configuration when the reactants are not fed under fully humidified condition. Membrane dehydration also occurs at anode inlet because this region

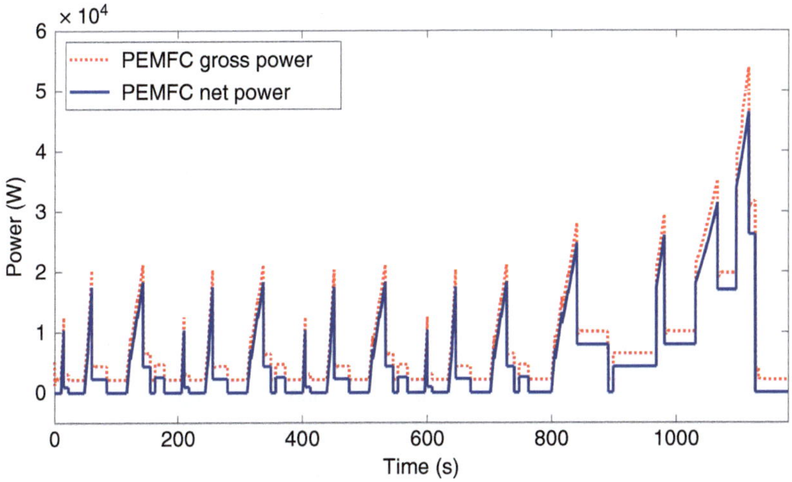

Figure 15.7 Stack gross and net power generated by the PEMFC stack in a NEDC cycle performed without considering hybridization.

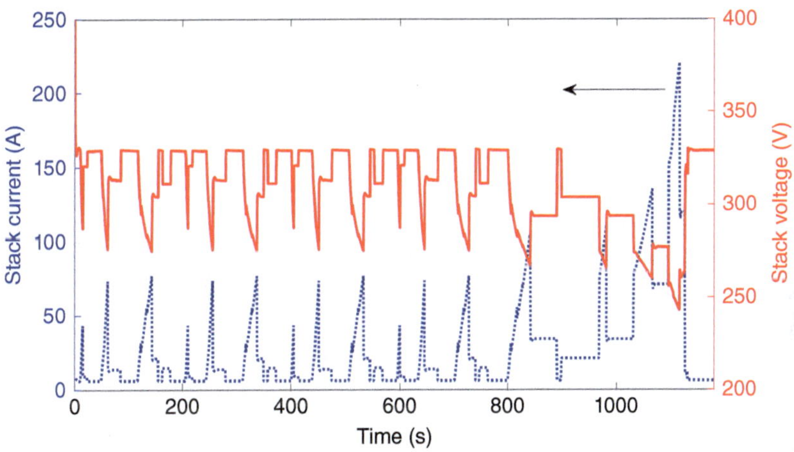

Figure 15.8 Stack voltage and current simulated during the simulation of the NEDC.

experiences the highest temperature values, as reported in Figure 15.10. As expected, the temperature profile during time follows the power peaks, having a maximum while operated at high power. The thermal management subsystem demonstrates its deficiency in stack cooling when the coolant inlet temperature deviates from the set-point value of 343.15 K, which is also observed under high-power load. Further information on the transient operation of the humidifier is also reported in Figure 15.11, where the relative humidity at the outlet of the humidifier, thus at the air inlet side of the stack, is reported. Relative humidity in the range between 40% and 50% is obtained by the simulation. It is concluded that the balance of plant contributes to the dynamic response of the stack, because of the time required by the control system to keep the controlled variables within the set-point values.

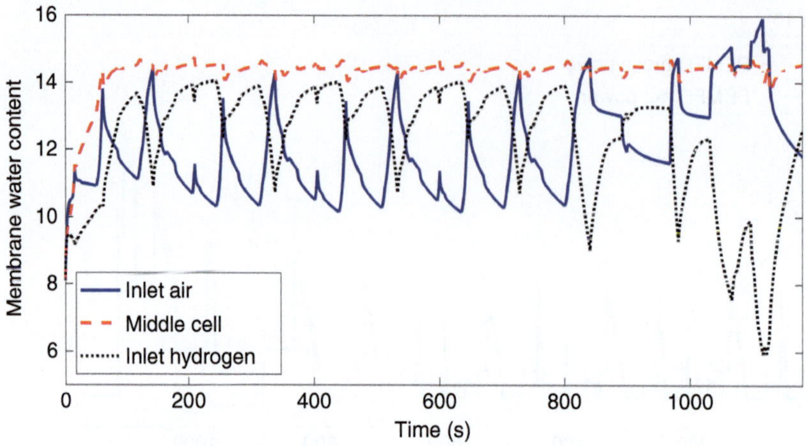

Figure 15.9 Local membrane water content simulated during the simulation of the NEDC.

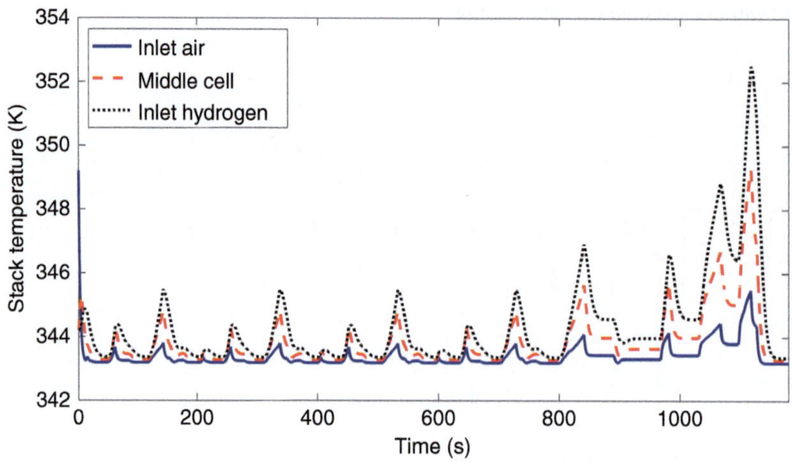

Figure 15.10 Coolant local temperature at several positions in the stack simulated during the NEDC.

Finally, the system and stack efficiency are computed according to the following definitions, which make use of the hydrogen utilization factor (UF_{H_2}, see Chapter 3 for details):

$$\eta_{stack} = \frac{\Delta V}{E_{id}} \tag{15.2}$$

$$\eta_{aux} = \frac{P_{gross} - P_{aux}}{P_{gross}} \tag{15.3}$$

$$\eta_{system} = \eta_{stack} \cdot UF_{H_2} \cdot \eta_{aux} \tag{15.4}$$

The system efficiency, which is reported in Figure 15.12, is always lower than the stack efficiency and becomes unacceptably low at low power, because of the effect of

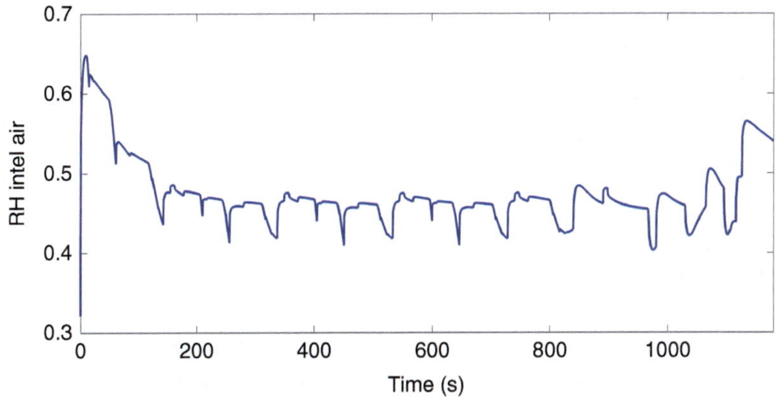

Figure 15.11 Relative humidity in the air stream at the stack inlet (after the humidifier) as simulated during the NEDC.

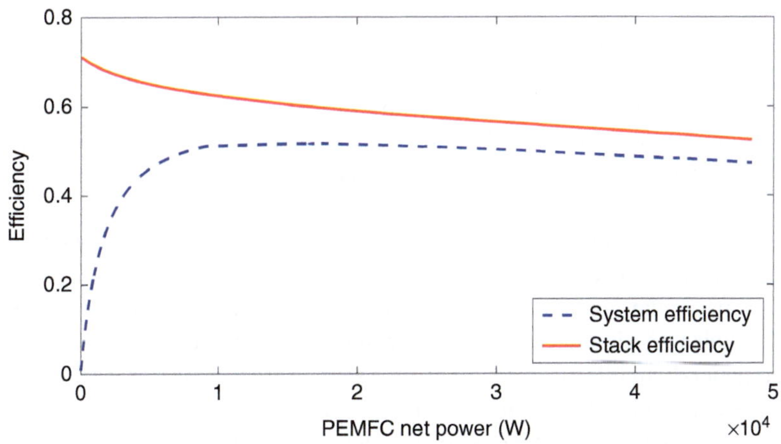

Figure 15.12 Stack and system efficiency computed by the NEDC simulation at different power levels.

the auxiliaries. The system efficiency shows a broad peak that is achieved at approximately from 10% to 40% of the maximum stack power.

15.3 Electronic Power Converters for Fuel Cell-powered Vehicles

As widely discussed in Section 15.1, the use of PEMFCs in electrical vehicles is a potentially interesting alternative to the use of large batteries, i.e. having nominal capacities in the several tens of kilowatt hour range. However, as compared to the latter, the PEMFCs present a limited adaptability to variable electrical loads. The slow dynamics of reactant gases within the stack, together with the intrinsically limited dynamic performance of pumps and the other mechanical components of the generator, impose a severe limitation to the output power rate of change. On the other hand, a vehicle power consumption is constantly changing, often exhibiting relatively fast variation rates.

Such an intrinsic mismatch between the *source* and the *load characteristics* is a problem also common to other electrical systems, encountered whenever the generator has limited (or, at times, zero) load *adaptability*. In vehicle applications of PEMFC, this condition may lead to severe and potentially harmful disequilibrium in the cell electrochemical processes, mostly resulting in stack flooding, when the electrical load is reduced, or starving, when the load is increased too rapidly for the PEMFC control system to adapt.

Another equally important problem is represented by the *electrical* mismatch between the input and output sides of the power conversion architecture. This problem is actually threefold: on the one hand, the PEMFC is, typically, a low-voltage, high-current source, while the load is, typically, a voltage source inverter (VSI)-based motor drive, which can be generally categorized as a high-voltage, low-current load.[1]

On the other hand, the PEMFC is intrinsically a dc power generator, as opposed to the load, i.e. the motor propelling the vehicle, that is intrinsically an ac-powered electrical machine. Finally, the PEMFC is intrinsically unilateral and cannot drain power, whereas the motor can operate bilaterally, i.e. not only draining electrical power to accelerate the vehicle but also regenerating it from braking.

In this chapter, we analyze the implications of both the limited load adaptability and the electrical mismatches that characterize the vehicle application of PEMFC, focusing, in particular, on the options offered by power electronics technologies to mitigate their undesirable consequences without penalizing the vehicle performance. In general, architectural solutions at system and/or converter level allow to improve safety and reliability by including high-power-density energy storage devices, like, e.g. an ultracapacitor or a small battery, allow it to perform more safely and reliably. Control solutions are also applicable at converter level to solve the electrical mismatch problems with minimum hardware usage and, consequently, reduced cost.

In this application field of power electronics, many other challenges are encountered that, however, are not specific of the PEMFC but rather common to any automotive application. Among these, it is worth mentioning the need for minimum weight and volume occupation, often expressed by very demanding power density constraints, and the complexity of power circuit thermal management that derives from this. The electromagnetic compatibility of the power converter architecture, both in terms of interference generation and susceptibility to external disturbances, is another very critical area of system design. These topics, however, represent the main object of power electronic design textbooks and, therefore, will not be covered here.

---

1 These characteristics are not intrinsic properties of PEMFCs and VSIs but rather the result of cost minimization at design level.

15.3.1 Power Converter Architecture

A large number of studies have been made to identify the optimal system architecture for an electric vehicle where a fuel cell stack is used to generate the requested electrical power. As mentioned above, the ideal architecture aims at mitigating, as much as economically and technically possible, the limited load adaptability of the source and its inherent electrical mismatch with the load. Let us go through each of them briefly.

15.3.2 Load Adaptability

The solution of this problem calls for the integration in the system of one or more additional energy tanks, not large in capacity but capable of delivering and absorbing considerable power, i.e. characterized by high power density. The objective is to implement a concept known as peak-power shaving, schematically illustrated in Figure 15.13. As can be seen, the additional energy tank is used to supply, or to drain, the load power variations that are too fast for the PEMFC stack to track safely.

We can elaborate on this concept defining a characteristic power *slew rate* for the PEMFC stack, i.e. the maximum output power variation that can take place over a given time interval t_{FC} without excessively stressing it or inducing dangerous disequilibrium conditions, e.g.

$$\mathrm{SR_{FC}} = \frac{[\Delta P_{FC}]_{MAX}}{t_{FC}} \tag{15.5}$$

From Eq. (15.5), it is possible to define the worst case stack response time t_S, that is the quantity:

$$t_S = \left[k \cdot \frac{\Delta P}{\mathrm{SR_{FC}}}\right]_{MAX} \tag{15.6}$$

In Eq. (15.6), the function to be maximized indicates the amount of time needed by the PEMFC stack to vary its output by the quantity ΔP without exceeding its

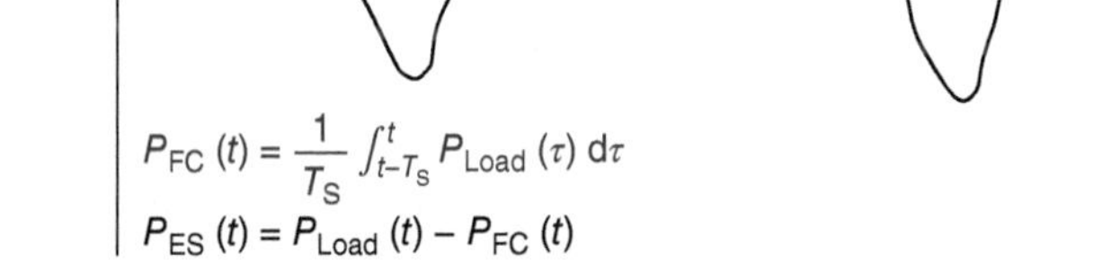

Figure 15.13 Peak shaving concept: the PEMFC is only delivering the long-term, *average* load power demand, while the energy storage device delivers and absorbs the rapid load power variations.

characteristic slew rate, SR_{FC}. The k coefficient appearing in Eq. (15.6) indicates the design safety margin between the allowed maximum slew rate and the PEMFC structural one, as defined by Eq. (15.5). Obviously, the worst case, i.e. the longest, response time results from the largest expected power variation, represented by a full throttle acceleration from zero to maximum speed. Less-conservative approaches are actually based on more realistic load power profiles, estimated from vehicle dynamic simulations.

According to the peak shaving approach, over time spans shorter than t_S, the PEMFC should always be operating in *quasi-static* conditions, which can only happen if any faster transient is made, by some means, imperceptible to the stack. This is the purpose of the high-power-density energy tank. The ideal, instantaneous fuel cell power output can be calculated as the average of the load power over a time window of duration t_S, i.e.

$$P_{FC}(t) = \frac{1}{t_S} \int_{t-t_S}^{t} P_{load}(\tau)\mathrm{d}\tau \tag{15.7}$$

which, in turn, allows to define the instantaneous power requested to the high-power-density energy storage device, namely

$$P_{ES}(t) = P_{FC}(t) - P_{load}(t) \tag{15.8}$$

The latter quantity $P_{ES}(t)$, by integration over time, yields the energy sizing of the tank, i.e. its requested capacity in Joule. Despite their apparent simplicity, the system of Eqs. (15.5)–(15.8) calls for a sophisticated system-level analysis to be properly set. The PEMFC design and the vehicle design teams need to strictly cooperate to achieve a sensible sizing of the generator and the associated energy storage system, taking into account the expected operating conditions of the vehicle with adequate safety margins.

Having briefly outlined the energy processing architecture we are dealing with, let us now examine the power electronic components involved in its management and control.

15.3.3 Power Electronic System Components

From the power electronics engineering perspective, the above outlined system architecture can only be managed by a multiport energy conversion system. Indeed, three energy exchange flows (i.e. powers) need to be properly regulated, namely (i) the energy flow supplied by the PEMFC stack; (ii) the energy flow requested by the load; and (iii) the energy flow exchanged by the system with the storage device. The ideal electronic power converter is, therefore, a three-port one, capable of mixing the energy flows as necessary to feed the load, absorb from the PEMFC stack only the fraction (15.3) of the load demand, and compensating the difference between load demand and PEMFC output by modulating the power absorbed, or delivered, by the energy storage device, as per (15.4).

A sketch of the power conversion system is shown in Figure 15.14. The power electronics designer aims not just at strictly achieving the above functionality but

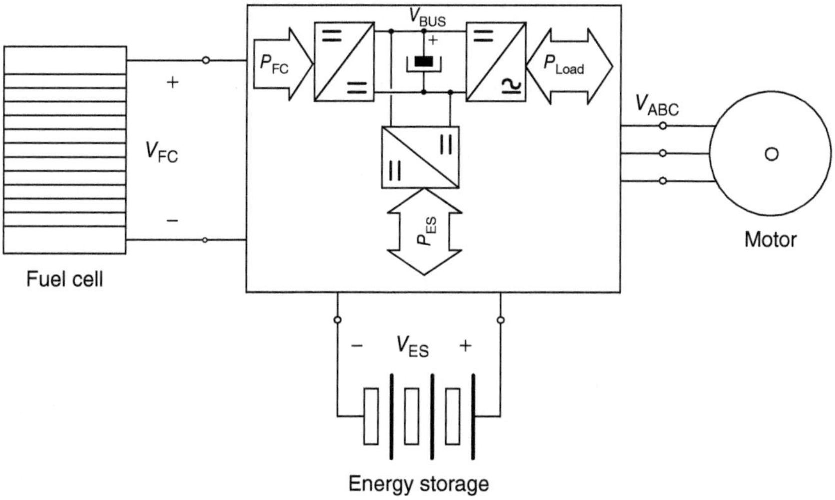

Figure 15.14 The three-port converter architecture. The unidirectional energy flow from the PEMFC stack and the two bidirectional flows of the storage device and the motor are balanced instantaneously by the integration of three power converters, coupled at a high-voltage dc bus. When more than one motor is needed, the motor port is served by more VSIs, one per motor, parallel connected at the input.

also to maximize the conversion efficiency, cutting to the minimum the power losses intrinsic to any energy transformation based on solid-state electronic devices, and, furthermore, to design a highly reliable, standard compliant, and minimum cost conversion system.

Clearly, the problem involves a long optimization process, where all the mentioned design constraints are weighted against one another in an attempt to balance cost and performance. It is outside the scope of this discussion to delve into the details of power converter engineering; nevertheless, it is important to underline the complexity of the considered application case, which explains the large number of proposed solutions any interested reader can find in the technical literature.

To simplify the exposition, the power electronic architecture can be considered as made up of different energy conversion circuits, physically interconnected and adequately controlled, so as to appear from the outside as a single unit with the above-outlined capabilities. The level up to which the integration of the different unit components can be pushed depends on several, application-specific, factors including reliability, cost, weight, and volume constraints, not to mention electromagnetic compatibility. For the sake of clarity, in the following sections we just analyze each of them as a separate entity, sharing either the input or the output port with one of the other units, but completely functional as stand-alone circuits. In reality, a partial or full integration of two or more units can be achieved, where each of them ceases to be independently functional, while the resulting circuit simultaneously accomplishes, as a whole, the functions of each component. The merging is done sharing some of the active and/or passive components of the independent units, typically at the cost of a carefully considered loss of flexibility.

15.3.3.1 Port Interface Converters

As mentioned above, the internal structure of the power converter architecture in Figure 15.14 is functionally equivalent to the interconnection of three independent port converters, namely the fuel cell interface, the energy storage interface, and the motor drive converter. Let us consider the main characteristics of each of these units.

15.3.3.2 The PEMFC Interface Converter

Most of the times, the PEMFC design is such that the stack can be considered a low-voltage (between 100 and 300 V), high-current (sometimes up to 1 kA) dc source. The PEMFC interface converter is primarily designed to smooth the electrical mismatch with the vehicle motor, which, by construction, is a high-voltage, low-current ac load. To this purpose, a dc–dc converter of boost type, i.e. capable of delivering power to a high-voltage sink from a low-voltage source, is typically used at the fuel cell (FC) port. In order to manage the high input current, the circuit exploits an interleaved multiphase architecture, offering current-sharing and ripple-cancelation effects. Besides boosting the voltage, this stage is capable of tightly regulating the PEMFC output current, with minimal sensitivity to the very significant and time-variant output impedance of the source. Thanks to suitable closed-loop current regulators, the PEMFC output current can be adjusted with virtually zero tracking error within a suitably limited bandwidth, automatically filtering any transient faster than t_S. At the same time, dangerous overcurrent conditions can be prevented by the limitation of the current loop set point. It is worth noting that the interconnection between the PEMFC port and motor port converters inherently defines an inner dc bus, whose voltage V_{BUS} is typically set around either 400 or 800 V. A relatively small capacitor is used as the high-voltage energy tank, which is requested to decouple the three port converters from one another.

15.3.3.3 The Motor Interface Converter

The motor converter, generally indicated as the inverter, is connected between the high-voltage dc bus and the motor port[2] to complete the source–load adaptation. It is typically structured as a three-phase, voltage source, and full bridge circuit. Its local controller is able not only to operate the dc–ac conversion required to drive the load but also to deliver power to the motor, once again, in a tightly controlled way. Motor currents are indeed actively regulated, turning the converter into a large bandwidth torque regulator, as is essential to optimize the vehicle performance (e.g. by torque steering algorithms). At the same time, the active torque control prevents overcurrent conditions in the motor, by limiting the inverter output current whenever necessary. Above the torque regulation loop, a conventional electrical drive system has speed, and, in some cases, even position regulators. The arrangement is a typical nested, multiloop configuration, where the position loop, if any, drives the speed loop which, finally, drives the torque loop. A variety of algorithms can be

2 The use of more than one motor, as requested by some vehicle designs, does not change the general sense of the discussion. To manage more motors, additional VSI drives are parallelly connected to the high-voltage dc bus.

used to map the desired vehicle dynamic performance into suitable position, speed, and torque set points for the motor(s).

15.3.3.4 The Energy Storage Interface

As explained above, the instantaneous compensation of the mismatch between input and output power flows calls for a third port, connected to high-power-density energy storage.[3] The energy storage port is served by a dedicated dc–dc converter, whose second port is again the internal high-voltage dc bus of the system. The main converter task is to keep the dc bus voltage V_{BUS} tightly regulated at its set point. In order to do that, the control system typically implements a double loop regulator, where the dc voltage controller drives an inner, and faster, input current regulator. By doing so, any time the load demand is higher than the fuel cell power output, which would make the dc bus voltage decrease, the converter increases the current flowing out of the energy storage, supplying the additional power. Similarly, when energy flows backward from the motor, i.e. during regenerative braking, the converter, in response to the natural dc bus voltage increase, commands a negative (i.e. recharging) current in the energy storage, draining the power in excess. The storage current controller prevents automatically the circulation of excessive currents in the storage device, thus limiting the stress to the minimum. Topology-wise, the storage interface converter is generally of boost or buck-boost type, the latter being employed when the nominal voltage of the storage device is close to the dc bus voltage set point. In some cases, galvanic isolation is introduced to allow safe access to the storage device even when the power system is energized.

15.3.3.5 Supervisory Control

The activity of the three power interface converters needs to be automatically coordinated, in order to prevent undesirable and potentially fatal transients, overvoltage or overcurrent conditions. In order to do that, the local controllers of the interface converters usually communicate with a master controller, which is itself coupled to a higher level vehicle dynamic controller. Starting from the vehicle driver's inputs, the dynamic controller generates position, speed, and torque set points for the motor that the master controller delivers to the inverter controller. At the same time, the master controller sets the PEMFC output current set point of the PEMFC interface converter, so as to match the load power demand. Ideally, the power from the PEMFC, in the absence of further perturbations, will slowly become equal to the power delivered to the motors. During the relatively long PEMFC transient, however, the requested power is automatically supplied by the energy storage, keeping a perfect balance in the power flows. Another important task of the master controller is to regulate the amount of energy available in the storage, charging or discharging it whenever necessary. The goal is to maintain a sufficient amount of energy to compensate sudden load demands, but without fully

3 In Figure 15.14, the energy storage device is shown as a battery just for simplicity. Actually, different storage technologies are compatible with PEMFC-powered vehicle applications, and are often combined, ranging from ultracapacitors to flywheel mechanical energy reservoirs.

charging the device. Indeed, enough capacity must be maintained at any time to recover the energy coming from the motor during deceleration phases. This is a very delicate part of the system management, often dealt with by means of sophisticated, adaptive, and predictive controllers. The quality of these storage management strategies has a direct impact of the fuel efficiency of the vehicle on its performance and reliability.

The use of PEMFC in vehicles is a challenging application field of power electronics, due to the slow fuel cell response time and the source–load characteristic mismatch. The typical power system architecture, which includes the PEMFC, an energy storage device, and the electrical motor(s), calls for a three-port power converter organization. A possible, basic configuration is the one that comprises a unidirectional boost converter between the PEMFC and the high-voltage dc bus, a bidirectional boost or buck-boost converter between the energy storage device and the same dc bus, and, finally, one or more three-phase voltage source inverters between the bus and the motors. Sophisticated controllers are needed at all system levels, to regulate each converter's electrical variable; to control the PEMFC state, the energy storage state, and the motor state; and, finally, to rapidly match the electrical state of the power system with the dynamic state of the vehicle.

15.4 Fuel Cell Powertrains for Mobility Use

This section focuses on the high potential offered by fuel cell-based propulsion systems to address both energy savings and environmental constraints in all major transport sectors. The main motivations that justify the adoption of such a technology, in conjunction with an expected increased recourse to hydrogen as the cleanest energy vector and fuel of the future, are firstly recalled. Afterward, the main pros and cons associated with the deployment of fuel cell hybrid architectures in each specific sector are briefly presented and discussed. The section ends with a selected case study, through which it will be discussed how model-based approach can significantly contribute to solving the most relevant open issues associated with fuel cell mobility, particularly those related to optimal sizing and energy management of powertrains, in view of cost-effectiveness and market acceptance targets achievement.

15.4.1 Transport Application Scenarios

At the time the authors are writing this book, the European Commission has in 2020 [21] recently defined an ambitious, yet realistic strategy, to contextualize the application areas of hydrogen-based energy conversion systems as well as point out the major expected benefits. The recent 2022 historical events clearly further highlighted the relevance of adopting such a strategy and turn it into effective actions as soon as possible. More in detail, particularly, the envisioned switch from mostly linear and wasteful energy flows to a more circular and integrated energy system was identified as a key breakthrough toward the pursued carbon neutral European Union

(EU) by 2050 [21]. Such a circular approach will surely promote a stronger synergy between stationary generation and mobility, with the aim of making both as clean and less consuming as possible, mainly relying on hydrogen as the energy vector. In fact, the green hydrogen generated from more renewable-based energy mixes will not only be used in industrial processes but also in intersectorial hubs aimed at supporting and promoting the usage of fuel cell vehicles (FCVs). Fuel cell technology is considered as the most promising one to successfully achieve more hydrogen-based circular energy system and economy at the EU level, mainly due to the following most relevant positive features:

- High efficiency, which in turn positively impacts greenhouse gas (GHG) emission levels from a more general fuel feeding perspective.
- Continuous operation: same as internal combustion engines and different than batteries, whose charging exhibits several infrastructural, commercial, and technological barriers.
- High modularity in meeting different applications in terms of average and/or peak power demand.
- Combustion is avoided and consequently no emission of NO_x, soot and unburnt hydrocarbons occurs.
- Silent operation, as air blower/compressor is the only and main moving component.

Among the several transport means, light-duty vehicles (LDVs) surely play the most relevant role in terms of GHG emissions [22]. This aspect reflected over the recent past years into an increasing attention paid by relevant automakers [23] toward the development and early commercialization of hydrogen-fed fuel cell-based LDVs. Such an early market scenario appears as nowadays technologically viable, since high technology readiness levels (TRLs) were achieved already [24]. Nevertheless, against the selection of LDV EU fleets, as the first transport category to be partially converted into FCVs, are two main drawbacks: high costs, due to the difficulty of reaching the needed production volumes in due time, and the difficulty to achieve massive spreading of the refueling infrastructure [25]. Therefore, the EU commission, together with specific research groups and partnerships (e.g. Clean Hydrogen for Europe [CH2E] and Clean Aviation), decided to orientate research, manufacturing, and financing efforts toward heavy-duty and long-distance transport applications over a short-term timeframe. In the following paragraphs, the main technological, infrastructural, and commercial aspects, to be accounted for when investigating the potential offered by the spreading of fuel cell (especially PEMFC) systems in all the main transport applications but the one (i.e. LDV) already commented above, are reviewed. Fuel cell hybrid buses and/or coaches exhibit a very interesting TRL (up to 8, as for LDV [24]), whereas current cost of hydrogen fuel still does not match the targeted value of 5–7 €/kg [26]. One of the most relevant strengths in fuel cell technology deployment for buses hybridization relates to the refueling phase, which can be centralized by equipping every bus fleet depot with a dedicated hydrogen refueling station (HRS) [27]. Complete tank fill can be accomplished in less than 10 minutes nowadays, thus confirming the high potential offered by this application area. It is indeed worth remarking here that in many cities (e.g. London) fuel cell buses have already reached a considerable number of working hours, as reported in [28].

Moving to heavy-duty tracks/vehicles (HDVs), HRS- and refueling time-related considerations similar to those made above for buses can be done. Moreover, the high energy density ensured by hydrogen storage adds significant advantages, as compared to batteries, when aiming at full electrification of long-haul HDVs [28]. The above aspects justify the inclusion of fuel cell-based hybridization of long-haul and logistic HDVs as the most significant contributor toward transport-related GHG emissions reduction, especially when considering the fact that road freight transport amounts up to 75% of the entire EU freight transport [29]. Therefore, current forecasts indicate that the request for fuel cell HDVs will triple by 2050, thus encouraging developers and global automotive original equipment manufacturers (OEMs) worldwide to go beyond the current TRL (i.e. 6) within a short timeframe [30].

The maritime sector also shows a high interest in spreading fuel cell technology (in this case also high temperature, such as solid oxide fuel cells [31]) in a variety of applications, ranging from small vessels to large cruise ships [30]. This is due to the very relevant potential offered by ships fuel cell-based electrification to cope with the increasingly stringent emission and noise reduction requirements, especially in harbor areas worldwide [32]. The above motivations will encourage developers and manufacturers and researchers worldwide to provide key contributions to quickly improve the current low TRL of maritime fuel cell systems [30].

As for trains, hydrogen-powered propulsion is currently considered the best solution for nonelectrified regional connections [33]. Once again, such a choice is justified by the increased range and reduced downtime for recharging ensured by fuel cell trains, as compared to battery-based ones [34]. The currently achieved TRL (i.e. around 6) justifies the positive projection of 1–5 new trains being fuel-cell powered by 2030 [30]. Current configurations feature 200 kW highly integrated fuel cell power modules, capable of meeting load requirements of 300 passenger capacity trains, with a maximum range setting around 1000 km [34].

Finally, the aviation sector could tremendously benefit from the very high energy density of hydrogen, thus pushing the most relevant OEMs [30] in drafting an ambitious, yet solid roadmap toward massive usage of hydrogen fuel cells. Particularly, they will be used in conjunction with the incumbent airplane propulsion technology (i.e. gas turbine) to develop more efficient aviation engines. Current TRLs still set to moderate values [30], whereas assumed 2030 objectives (i.e. development of first demonstrations of liquid hydrogen-fueled FC-turbine propulsion systems, increase the efforts devoted to FC-based auxiliary power units R&D, and meeting specific tanks gravimetric index requirements [30]) will surely speed up the technological growing process.

15.4.2 Tools for the Codesign of Transport Fuel Cell Systems and Energy Management Strategies

The previously discussed motivations and limitations, characterizing every possible deployment of PEMFC technology in the heterogeneous transport sector, lead to the determination of relevant research and development gaps. Particularly, the EU commission, through the Horizon 2020 (H2020) work program [35], underlined the relevance of exploring and developing new vehicle integration concepts, mainly aiming at accounting for the fact that there is no single electrified powertrain technology [36], entitled to becoming the panacea solution to environmental and energetic issues of transportation worldwide. Moreover, once the technology is identified, manufacturers will develop the mobility product by carefully accounting for the specific destination of that powertrain; e.g. an LDV is destined to be mostly driven in urban or suburban areas. To cope with these requirements, recent H2020 funding calls [35] identified, among several relevant targets, the need of providing developers and manufacturers with smart tools, also relying on model-based approach, to automatically identify key design parameters, such as the degree of hybridization (DH) [37], of electrified powertrains. In the specific case of fuel cell hybrid vehicle (FCHVs), this means correctly individuating the relationship between battery (i.e. $P_{\mathrm{B,n}}$) and fuel cell system (i.e. $P_{\mathrm{FCS,n}}$) nominal power (see Eq. (15.9)). DH and optimal sizing in general were also shown to be strictly correlated with adopted control strategies [38].

$$\mathrm{DH} = \frac{P_{\mathrm{B,n}}}{P_{\mathrm{B,n}} + P_{\mathrm{FCS,n}}} \tag{15.9}$$

The following Section 15.4.2.1 are thus devoted to describing a case study, whose results demonstrate the appropriateness and the physical meaningfulness of applying codesign methodologies to address the afore-discussed research needs. The genericity and flexibility of the developed tools for subsequent extension to other transport application areas, as well as to accounting for other effects (e.g. components degradation, fault tolerability, and stringent safety constraints) that significantly impact design cost-effectiveness, will also be highlighted.

15.4.2.1 Automotive Case Study: Optimal Codesign of an LDV FCHV Powertrain

The current case study, described in detail in previous contributions [39, 40], deals with the simultaneous optimization of powertrain components and associated control strategies for an LDV FCHV. Figure 15.15 provides a synoptic representation of the main procedure workflow. The initial conditions of the design parameters, see Table 15.1, along with the selected driving cycle (i.e. the NEDC), are the external inputs. The developed optimization procedure iteratively modifies the values of the variables to be optimized, which are also used to cause specification-independent control rules (developed offline, as described in [41]) to adapt to currently investigated solution. On the basis of the cost function value and constraints satisfaction (i.e. yes or no), the process continues or stops. In the latter case, provided that convergence was achieved, the procedure will yield on output both final FCHV powertrain design and the specification-dependent control strategy, to be exploited for further model-based analyses (e.g. to assess LDV fuel economy in different driving cycles). More details on the aforementioned specification-independent rule-based control rule, specifically applied to appropriately load demand between battery and PEMFC system, along with its adaptation to studied configurations, can be retrieved from [39].

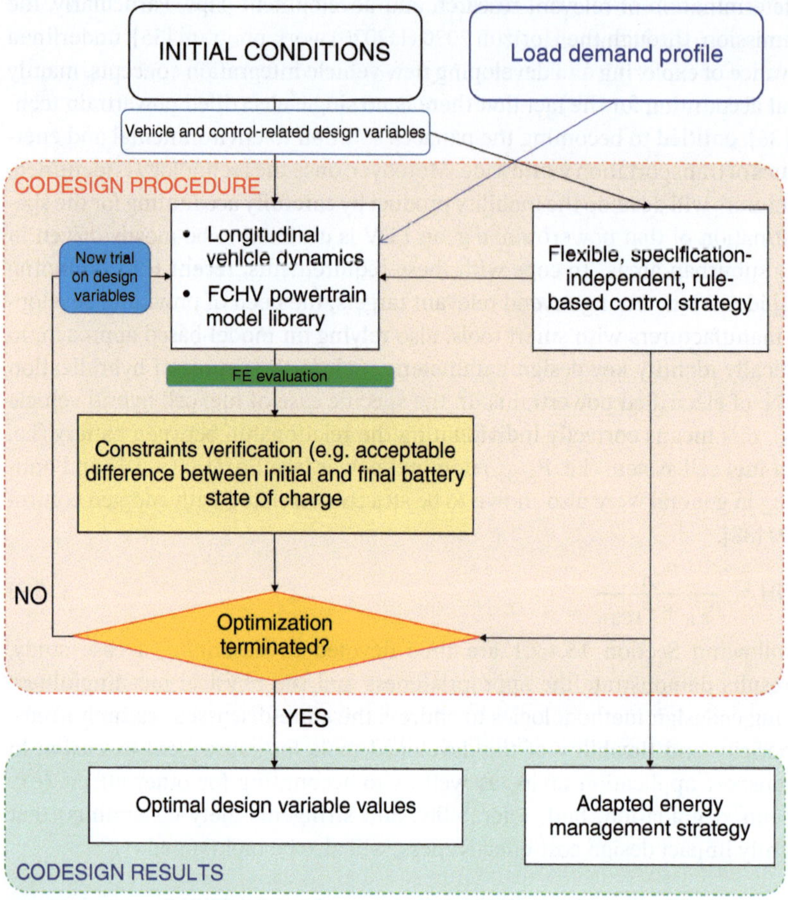

Figure 15.15 The synoptic diagram illustrating the entire FCHV codesign procedure described in Section 15.4.2.

Table 15.1 Resume of the main design outcomes yielded on output by the proposed FCHV codesign procedure.

	Optimal design output				Constraints-related outputs		
Case	FE (km kg^{-1})	$P_{FCS,n}$	E_{batt}	$t_{horizon}$ (s)	t_{max} (s)	$t_{PD,charge}$	SOC_f
1	136.7	100	100	438	100	100	100, i.e. 85% of initial SOC
2	138	67	104	800	300	356	77

The variables t_{max} and $t_{PD,charge}$ correspond to the maximum post-driving recharging time (to increase SOC_f up to initial SOC value) and solution-specific post-driving recharging time, respectively. E_{batt} and $t_{horizon}$, instead, refer to the optimal battery size and average load prediction time horizon. It is finally worth remarking that some (i.e. those related to $P_{FCS,n}$ [kW], E_{batt} [kWh], $t_{PD,charge}$ [s], and SOC_f) of the units specified in the "Case" row are dropped off in the values reported in the corresponding column, where they are expressed as a percentage of Case 1.
Source: Adapted from Refs. [39, 40].

The evaluation of the abovementioned cost function entails appropriately assessing the vehicle fuel economy (i.e. FE), computed, see Eq. (15.10), as the ratio between driven distance (i.e. S [km]) and consumed hydrogen (i.e. m_{H_2} [kg]). To this aim the model library presented in the following Section "Modeling Approach" was used.

$$FE = \frac{S}{m_{H_2}} \quad (15.10)$$

Modeling Approach

The fuel consumption is estimated as a function of load demand, which in the specific case of LDV FCHVs is related to the vehicle longitudinal dynamics. Figure 15.16 synthetically links the longitudinal dynamics equation, used to compute current traction power demand, to the vehicle specifications. The latter data are provided in Figure 15.16 for the reference conventional vehicle, which was assumed to derive, through the vehicle mass model also detailed in Figure 15.16, the mass and electric motor size of the FCHV counterpart. The latter process assumes that the nominal power of the PEMFC system (i.e. $P_{FCS,n}$ in Eq. (15.9)) is known. This explains why the overall model associated with Figures 15.16 and 15.17 is completely run in the inner part of Figure 15.15.

Once the traction and regenerative braking power are known, the m_{H_2} variable, see Eq. (15.10), will be computed, this time by referring to the powertrain schematic of Figure 15.17. This time, efficiency maps describing PEMFC system, battery pack (B), and electric motor (EM) performance are adopted, so as to convert the wheel load request into hydrogen consumption. Efficiency maps were either inferred from experimental data (i.e. fuel cell system efficiency variation, estimated as a function of current net power provided) or derived from literature data (i.e. electric motor efficiency dependence on operating torque and speed). The battery pack state of charge is instead simulated by applying Kirchhoff's law to a generator connected in series with a resistance, whose associated losses are estimated considering literature-derived internal resistance data for high-performance Li-ion technology [39]. Battery mass (transferred to the mass model of Figure 15.16) is estimated through the shown Ragone plot and current battery energy density considered in the iterative optimization of Figure 15.15.

Optimization Outcomes

The procedure described in Section 15.4.2 was run on the vehicle model of Figure 15.16 for a driving mission composed of 10 NEDC in series. The objective was to maximize the fuel economy, see Eq. (15.10), by strengthening (case 1 in Table 15.1) and relaxing (i.e. case 2 in Table 15.1) the charge sustaining battery management constraint. This way, as extensively discussed in [39], both pure hybrid and more plug-in (or also more range–extender like) solutions could be explored. The latter consideration reflects in the different PEMFC system and battery sizes shown in Table 15.1. In the latter table, it also emerges how the third design variable (i.e. the average load prediction time horizon), which indeed is more related to the adopted control strategy, well represents the effect of changing FCHV design objective. Particularly, if more hybrid solutions are sought, the time horizon sets

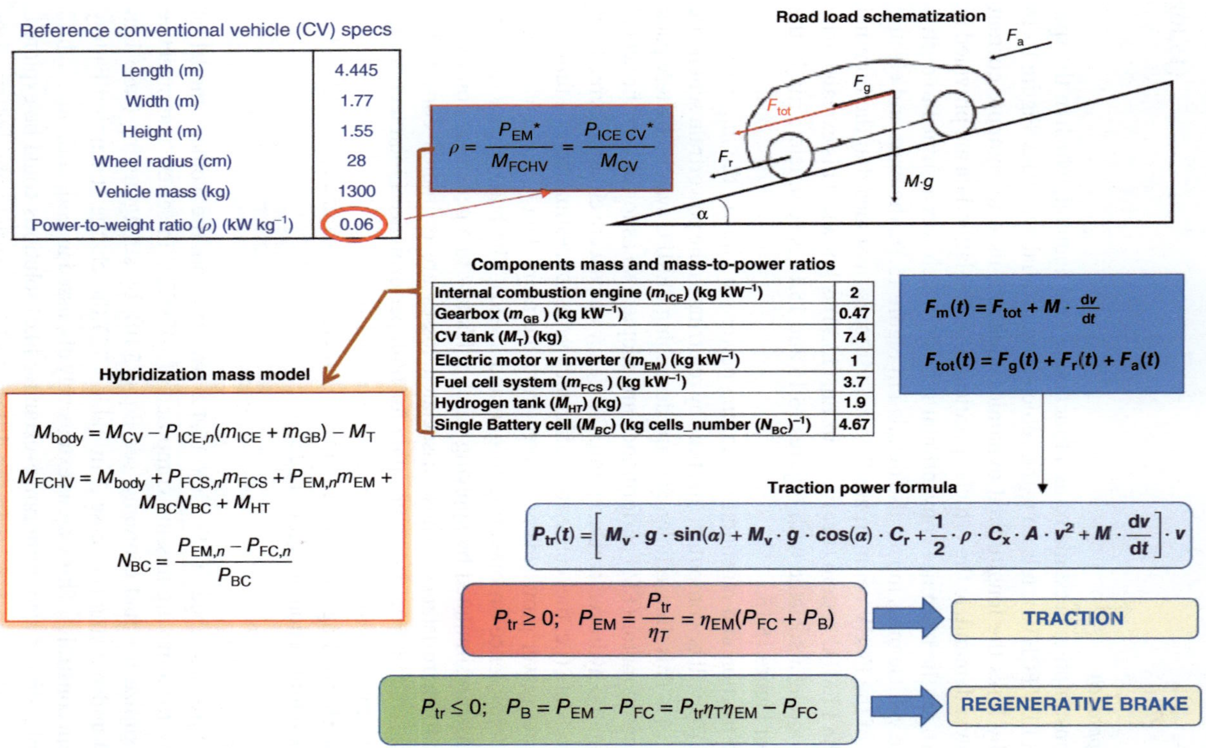

Figure 15.16 The longitudinal dynamic model used in the FCHV codesign procedure to simulate power demand at wheels as a function of selected driving cycle (i.e. NEDC) and currently investigated (through the iterative procedure shown in Figure 15.15) configuration solution. The actual configuration is determined as a function of selected design parameters (i.e. fuel cell system and battery energy density), through suitable vehicle mass adaptation. The latter process is carried out by means of the mass model shown in the current figure, provided the reference vehicle and targeted performance (expressed in terms of power-to-weight ratio) are defined (such data [39] also are provided in the figure). Source: Reproduced from Ref. [39] with the permission of Elsevier.

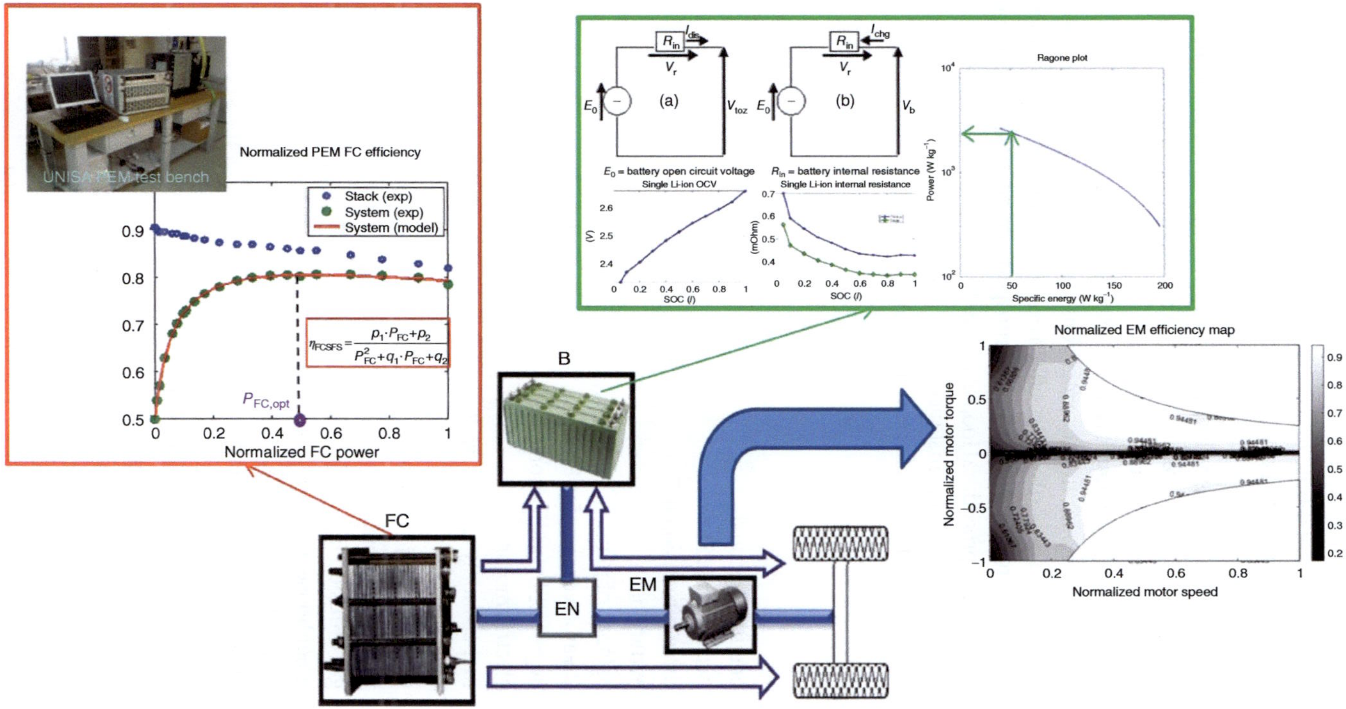

Figure 15.17 Model library utilized to estimate the fuel consumption and battery state of charge for the fuel cell hybrid vehicle configuration that is currently investigated by the codesign procedure of Section 15.4.2. Source: Reproduced from Ref. [39] with the permission of Elsevier.

to a fraction of the driving cycle (about one-third) that allows well distinguishing between urban and suburban driving routes of the overall NEDC. On the other hand, if the constraint on final battery state of charge (i.e. SOC_f in Table 15.1) is relaxed (i.e. case 2), the optimal time horizon sets to about two-thirds of the entire NEDC. This is due to the fact that in case 2 the optimization procedure does not aim at reaching an SOC_f value closer to the initial one. Therefore, the load prediction timeframe can be extended up to a value that allows avoiding overdischarging the battery, which might result in increased energy losses and eventual difficulty in meeting the load demand with the assumed DH.

The described procedure was shown versatile enough in each section it consists of. Particularly, among the inputs, further exogenous variables can be included, such as eventual incentives for supporting cost-effectiveness or different driving schedules/load demand, so as to adapt to different transport sectors (e.g. aviation and maritime). Inside the main procedural loop of Figure 15.15, both constraint and control strategies can be modified, for instance to include safety requirements, space constraints, or to account for aging effects in the control strategy.

Acronyms

1D	one-dimensional
2D	two-dimensional
3D	three-dimensional
B	battery pack
BoP	balance of plant
CH2E	Clean Hydrogen for Europe
DH	degree of hybridization
EM	electric motor
FE	fuel economy
FC	fuel cell
FCHV	fuel cell hybrid vehicle
FCV	fuel cell vehicles
GDL	gas diffusion layer
GHG	greenhouse gas
HDV	heavy-duty vehicle
HRS	hydrogen refueling station
LDV	light-duty vehicle
MEA	membrane electrode assembly
NEDC	new European driving cycle
OEM	global automotive original equipment manufacturer
ORR	oxygen reduction reaction
PEMFC	polymer electrolyte membrane fuel cells
R/C	ratio between the length of the rib and the channel
RH	relative humidity
SOC_f	final battery state of charge

TRL	technology readiness level
VSI	voltage source inverter
WRR	water recovery ratio

Symbols

E_{id}	ideal thermodynamic potential
m_{H_2}	consumed hydrogen mass in the driving cycle
$\dot{m}_{H_2O,dry\text{-}in}$	inlet water mass flow rate on the dry side of air humidifier
$\dot{m}_{H_2O,dry\text{-}out}$	outlet water mass flow rate on the dry side of air humidifier
$\dot{m}_{H_2O,wet\text{-}in}$	inlet water mass flow rate on the wet side of air humidifier
P_{aux}	auxiliary power consumption
$P_{B,n}$	nominal (i.e. installed) battery pack power
$P_{FC}(t)$	instantaneous fuel cell output power
$P_{FCS,n}$	nominal (i.e. installed) fuel cell system power
$P_{ES}(t)$	instantaneous energy storage output power
P_{gross}	gross stack power
S	driving distance
SR_{FC}	slew rate
UF_{H_2}	hydrogen utilization factor
t_{FC}	arbitrary time interval where PEMFC output power variation is measured
t_S	worst case stack response time
ΔV	cell voltage
η_{aux}	auxiliary efficiency
η_{stack}	stack efficiency

References

1 Thompson, S.T., James, B.D., Huya-Kouadio, J.M. et al. (2018). Direct hydrogen fuel cell electric vehicle cost analysis: system and high-volume manufacturing description, validation, and outlook. *Journal of Power Sources* 399: 304–313. https://doi.org/10.1016/j.jpowsour.2018.07.100.

2 Najafi, B., Bonomi, P., Casalegno, A. et al. (2020). Rapid fault diagnosis of PEM fuel cells through optimal electrochemical impedance spectroscopy tests. *Energies* 13: 3643. https://www.mdpi.com/1996-1073/13/14/3643.

3 Polverino, P., Frisk, E., Jung, D. et al. (2017). Model-based diagnosis through structural analysis and causal computation for automotive polymer electrolyte membrane fuel cell systems. *Journal of Power Sources* 357: 26–40. https://www.sciencedirect.com/science/article/pii/S037877531730592X.

4 Vivona, D., Casalegno, A., and Baricci, A. (2019). Validation of a pseudo 2D analytical model for high temperature PEM fuel cell impedance valid at typical operative conditions. *Electrochimica Acta* 310: 122–135.

5 Najafi, B., Haghighat Mamaghani, A., Baricci, A. et al. (2015). Mathematical modelling and parametric study on a 30 kW_{el} high temperature PEM fuel cell based residential micro cogeneration plant. *International Journal of Hydrogen Energy* 40: 1569–1583.

6 Baricci, A., Zago, M., and Casalegno, A. (2016). Modelling analysis of heterogeneity of ageing in high temperature polymer electrolyte fuel cells: insight into the evolution of electrochemical impedance spectra. *Electrochimica Acta* 222: 596–607. http://www.sciencedirect.com/science/article/pii/S0013468616323465.

7 Baricci, A., Zago, M., and Casalegno, A. (2014). A quasi 2D model of a high temperature polymer fuel cell for the interpretation of impedance spectra. *Fuel Cells* 14: 926–937. http://www.scopus.com/inward/record.url?eid=2-s2.0-84918532327&partnerID=MN8TOARS.

8 Baricci, A., Bonanomi, M., Yu, H. et al. (2018). Modelling analysis of low platinum polymer fuel cell degradation under voltage cycling: gradient catalyst layers with improved durability. *Journal of Power Sources* 405: 89–100. https://doi.org/10.1016/j.jpowsour.2018.09.092.

9 Futter, G.A., Gazdzicki, P., Friedrich, K.A. et al. (2018). Physical modeling of polymer-electrolyte membrane fuel cells: understanding water management and impedance spectra. *Journal of Power Sources* 391: 148–161. http://linkinghub.elsevier.com/retrieve/pii/S0378775318304130.

10 Nandjou, F., Poirot-Crouvezier, J.-P., Chandesris, M. et al. (2016). A pseudo-3D model to investigate heat and water transport in large area PEM fuel cells – part 2: application on an automotive driving cycle. *International Journal of Hydrogen Energy* 41: 15573–15584. http://www.sciencedirect.com/science/article/pii/S0360319916303901.

11 Baricci, A., Mereu, R., Messaggi, M. et al. (2017). Application of computational fluid dynamics to the analysis of geometrical features in PEM fuel cells flow fields with the aid of impedance spectroscopy. *Applied Energy* 205: 670–682. https://doi.org/10.1016/j.apenergy.2017.08.017.

12 Grimaldi, A. (2017). Modelling and analysis in 1D and 3D of dehydration and flooding phenomena in PEMFC with simulation of impedance spectra. Politecnico di Milano.

13 Giovanazzi, F. (2020). Numerical modeling of PEM fuel cell through 3D and 1D approaches: an analysis of geometrical features and relative humidity effects. Politecnico di Milano.

14 Montauti, M. and Garancini, T. (2019). Preliminary modelling of hydrogen PEMFC system for automotive applications focusing on air. Politecnico di Milano.

15 Pukrushpan, J.T., Stefanopoulou, A.G., and Peng, H. (2004). *Control of Fuel Cell Power Systems*. Springer-Verlag.

16 Hasegawa, T., Imanishi, H., Nada, M., and Ikogi, Y. (2016). Development of the fuel cell system in the Mirai FCV. *SAE Technical Papers*. https://www.sae.org/content/2016-01-1185 (accessed 13 January 2021).

17 Wu, D., Peng, C., Yin, C., and Tang, H. (2020). Review of system integration and control of proton exchange membrane fuel cells. *Electrochemical Energy Reviews* 3 (3): 466–505. http://link.springer.com/10.1007/s41918-020-00068-1.

18 Ahluwalia, R.K., Wang, X., Johnson, W.B. et al. (2015). Performance of a cross-flow humidifier with a high flux water vapor transport membrane. *Journal of Power Sources* 17 (291): 225–238.

19 Steinberger, M., Geiling, J., Oechsner, R., and Frey, L. (2018). Anode recirculation and purge strategies for PEM fuel cell operation with diluted hydrogen feed gas. *Applied Energy* 232 (September): 572–582. https://doi.org/10.1016/j.apenergy.2018.10.004.

20 Jung, D. and Assanis, D.N. (2006). Numerical modeling of cross flow compact heat exchanger with louvered fins Using thermal resistance concept. *SAE Technical Papers.* https://www.sae.org/content/2006-01-0726

21 EU Commission. (2020) EU Energy System Integration Strategy. https://ec.europa.eu/commission/presscorner/detail/en/FS_20_1295 (accessed 13 January 2021).

22 Usai, L., Hung, C.R., Vásquez, F. et al. (2021). Life cycle assessment of fuel cell systems for light duty vehicles, current state-of-the-art and future impacts. *Journal of Cleaner Production* 280: 125086.

23 Kast, J., Vijayagopal, R., Gangloff, J.J. Jr., and Marcinkoski, J. (2017). Clean commercial transportation: medium and heavy duty fuel cell electric trucks. *International Journal of Hydrogen Energy* 42: 4508–4517.

24 Kampker, A., Ayvaz, P., Schön, C. et al. (2020). Challenges towards large-scale fuel cell production: results of an expert assessment study. *International Journal of Hydrogen Energy* 45: 29288–29296.

25 Fuel Cell Hydrogen Joint Undertaking (FCH) (2015). Fuel cell electric buses – potential for sustainable public transport in Europe. https://www.fch.europa.eu/sites/default/files/150909_FINAL_Bus_Study_Report_OUT_0.PDF (accessed 13 January 2021).

26 Hydrogen Council (2020). Path to hydrogen competitiveness. A cost perspective. https://hydrogencouncil.com/wp-content/uploads/2020/01/Path-to-Hydrogen-Competitiveness_Full-Study-1.pdf. pp. 17 and 29 (accessed 13 January 2021).

27 Fuel Cell Electric Buses (2020). About fuel cell electric buses. https://www.fuelcellbuses.eu/wiki/fuel-cell-electric-buses-fuel-cell-electric-buses/about-fuel-cell-electric-buses (accessed 13 January 2021).

28 Ballard (2020). Fuel cell trucks: solution to heavy duty transport emissions?. https://blog.ballard.com/fuel-cell-truck (accessed 13 January 2021).

29 Eurostat (2020). Freight transport statistics. https://ec.europa.eu/eurostat/statistics-explained/index.php/Freight_transport_statistics (accessed 13 January 2021).

30 Clean Hydrogen Joint Undertaking (2022). Strategic Research and Innovation Agenda. https://www.clean-hydrogen.europa.eu/about-us/key-documents/strategic-research-and-innovation-agenda_en (accessed 8 September 2022).

31 Samsung Heavy Industries (SHI) (2019). SHI to become the first shipbuilder to develop COTs powered by fuel cells. http://www.samsungshi.com/Eng/pr/news_

view.aspx?Seq=1127&mac=3b071cb26f3a5cbd0b63a5ac0857622e (accessed 13 January 2021).

32 SAFETY4SEA (2018). New study focuses on hydrogen fuel cell research vessel. https://safety4sea.com/new-study-focuses-on-hydrogen-fuel-cell-research-vessel (accessed 13 January 2021).

33 FCH (2017). Development of business cases for fuel cells and hydrogen applications for regions and cities. https://www.fch.europa.eu/sites/default/files/171121_FCH2JU_Application-Package_WG1_Trains%20%28ID%202910561%29%20%28ID%202911647%29.pdf (accessed 13 January 2021).

34 Cummins (2020). Trains: on the right track. https://www.cummins.com/new-power/applications/trains (accessed 13 January 2021).

35 European Commission (2018). Smart, green and integrated transport – Work Programme 2018–2020. https://ec.europa.eu/research/participants/data/ref/h2020/wp/2018-2020/main/h2020-wp1820-transport_en.pdf (accessed 13 January 2021).

36 Shirres, D. (2018). Could hydrogen trains be the future of rail?. https://www.imeche.org/news/news-article/could-hydrogen-trains-be-the-future-of-rail (accessed 13 January 2021).

37 Arsie, I., Di Domenico, A., Pianese, C., and Sorrentino, M. (2007). Modeling and analysis of transient behavior of polymer electrolyte membrane fuel cell hybrid vehicles. *ASME Journal of Fuel Cell Science and Technology* 4: 261–271.

38 Feng, Y. and Dong, Z. (2020). Integrated design and control optimization of fuel cell hybrid mining truck with minimized lifecycle cost. *Applied Energy* 270: 115164.

39 Sorrentino, M., Cirillo, V., and Nappi, L. (2019). Development of flexible procedures for co-optimizing design and control of fuel cell hybrid vehicles. *Energy Conversion and Management* 185: 537–551.

40 Sorrentino, M. and Capaldo, D. (2019). Simultaneous optimization of real-time control strategies and powertrain design for fuel cell hybrid vehicles. *SAE Technical Paper 2019-24-0199*. https://doi.org/10.4271/2019-24-0199.

41 Sorrentino, M., Pianese, C., and Cilento, M. (2016). A specification independent control strategy for simultaneous optimization of fuel cell hybrid vehicles design and energy management. *IFAC-PapersOnLine* 49: 369–376.

Index

Electrocatalysis for Membrane Fuel Cells: Methods, Modeling, and Applications, First Edition.
Edited by Nicolas Alonso-Vante and Vito Di Noto.

n

o

p